Relations for Discrete Cash Flows with End-of-Period Compounding

Type	Find/Given	Factor Notation and Formula	Relation	Sample Cash Flow Diagram
Single Amount	F/P Compound amount	$(F/P,i,n) = (1 + i)^n$	$F = P(F/P,i,n)$	
	P/F Present worth	$(P/F,i,n) = \dfrac{1}{(1 + i)^n}$	$P = F(P/F,i,n)$ (Sec. 2.1)	
Uniform Series	P/A Present worth	$(P/A,i,n) = \dfrac{(1 + i)^n - 1}{i(1 + i)^n}$	$P = A(P/A,i,n)$	
	A/P Capital recovery	$(A/P,i,n) = \dfrac{i(1 + i)^n}{(1 + i)^n - 1}$	$A = P(A/P,i,n)$ (Sec. 2.2)	
	F/A Compound amount	$(F/A,i,n) = \dfrac{(1 + i)^n - 1}{i}$	$F = A(F/A,i,n)$	
	A/F Sinking fund	$(A/F,i,n) = \dfrac{i}{(1 + i)^n - 1}$	$A = F(A/F,i,n)$ (Sec. 2.3)	
Arithmetic Gradient	P_G/G Present worth	$(P/G,i,n) = \dfrac{(1 + i)^n - in - 1}{i^2(1 + i)^n}$	$P_G = G(P/G,i,n)$	
	A_G/G Uniform series	$(A/G,i,n) = \dfrac{1}{i} - \dfrac{n}{(1 + i)^n - 1}$ (Gradient only)	$A_G = G(A/G,i,n)$ (Sec. 2.5)	
Geometric Gradient	P_g/A_1 and g Present worth	$P_g = \begin{cases} \dfrac{A_1\left[1 - \left(\dfrac{1 + g}{1 + i}\right)^n\right]}{i - g} & g \neq i \\ A_1\dfrac{n}{1 + i} & g = i \end{cases}$ (Gradient and base A_1)	$g \neq i$ $g = i$ (Sec. 2.6)	

Seventh Edition

ENGINEERING ECONOMY

Seventh Edition

ENGINEERING ECONOMY

Leland Blank, P. E.
Texas A & M University
American University of Sharjah, United Arab Emirates

Anthony Tarquin, P. E.
University of Texas at El Paso

Mc Graw Hill

Connect
Learn
Succeed™

ENGINEERING ECONOMY: SEVENTH EDITION

Published by McGraw-Hill, a business unit of The McGraw-Hill Companies, Inc., 1221 Avenue of the Americas, New York, NY 10020. Copyright © 2012 by The McGraw-Hill Companies, Inc. All rights reserved. Previous editions © 2005, 2002, and 1998. No part of this publication may be reproduced or distributed in any form or by any means, or stored in a database or retrieval system, without the prior written consent of The McGraw-Hill Companies, Inc., including, but not limited to, in any network or other electronic storage or transmission, or broadcast for distance learning.

Some ancillaries, including electronic and print components, may not be available to customers outside the United States.

� This book is printed on recycled, acid-free paper containing 10% postconsumer waste.

1 2 3 4 5 6 7 8 9 0 QDB/QDB 1 0 9 8 7 6 5 4 3 2 1

ISBN 978-0-07-337630-1
MHID 0-07-337630-2

Vice President & Editor-in-Chief: *Marty Lange*
Vice President EDP/Central Publishing Services: *Kimberly Meriwether David*
Global Publisher: *Raghothaman Srinivasan*
Sponsoring Editor: *Peter E. Massar*
Senior Marketing Manager: *Curt Reynolds*
Development Editor: *Lorraine K. Buczek*
Senior Project Manager: *Jane Mohr*
Design Coordinator: *Brenda A. Rolwes*
Cover Designer: *Studio Montage, St. Louis, Missouri*
Cover Image: © *Brand X Pictures/PunchStock RF*
Buyer: *Kara Kudronowicz*
Media Project Manager: *Balaji Sundararaman*
Compositor: *MPS Limited, a Macmillan Company*
Typeface: *10/12 Times*
Printer: *Quad/Graphics-Dubuque*

All credits appearing on page or at the end of the book are considered to be an extension of the copyright page.

Library of Congress Cataloging-in-Publication Data

Blank, Leland T.
  Engineering economy / Leland Blank, Anthony Tarquin. — 7th ed.
    p. cm.
  Includes bibliographical references and index.
  ISBN-13: 978-0-07-337630-1 (alk. paper)
  ISBN-10: 0-07-337630-2
  1. Engineering economy. I. Tarquin, Anthony J. II. Title.
  TA177.4.B58 2012
  658.15—dc22
                                    2010052297

This book is dedicated to Dr. Frank W. Sheppard, Jr. His lifelong commitment to education, fair financial practices, international outreach, and family values has been an inspiration to many—one person at a time.

McGraw-Hill Create™ ● ● ●

Craft your teaching resources to match the way you teach! With McGraw-Hill Create™, www .mcgrawhillcreate.com, you can easily rearrange chapters, combine material from other content sources, and quickly upload content you have written like your course syllabus or teaching notes. Find the content you need in Create by searching through thousands of leading McGraw-Hill textbooks. Arrange your book to fit your teaching style. Create even allows you to personalize your book's appearance by selecting the cover and adding your name, school, and course information. Order a Create book and you'll receive a complimentary print review copy in 3–5 business days or a complimentary electronic review copy (eComp) via email in minutes. Go to www. mcgrawhillcreate.com today and register to experience how McGraw-Hill Create™ empowers you to teach *your* students *your* way.

McGraw-Hill Higher Education and Blackboard Have Teamed Up ● ● ●

Blackboard, the Web-based course-management system, has partnered with McGraw-Hill to better allow students and faculty to use online materials and activities to complement face-to-face teaching. Blackboard features exciting social learning and teaching tools that foster more logical, visually impactful and active learning opportunities for students. You'll transform your closed-door classrooms into communities where students remain connected to their educational experience 24 hours a day.

This partnership allows you and your students access to McGraw-Hill's Create™ right from within your Blackboard course – all with one single sign-on. McGraw-Hill and Blackboard can now offer you easy access to industry leading technology and content, whether your campus hosts it, or we do. Be sure to ask your local McGraw-Hill representative for details.

Electronic Textbook Options ● ● ●

This text is offered through CourseSmart for both instructors and students. CourseSmart is an online resource where students can purchase the complete text online at almost half the cost of a traditional text. Purchasing the eTextbook allows students to take advantage of CourseSmart's web tools for learning, which include full text search, notes and highlighting, and email tools for sharing notes between classmates. To learn more about CourseSmart options, contact your sales representative or visit www.CourseSmart.com.

CONTENTS

LEARNING STAGE 2   **BASIC ANALYSIS TOOLS**

LEARNING STAGE 2   EPILOGUE: SELECTING THE BASIC ANALYSIS TOOL

LEARNING STAGE 3   MAKING BETTER DECISIONS

LEARNING STAGE 4	ROUNDING OUT THE STUDY

This edition includes the time-tested approach and topics of previous editions and introduces significantly new print and electronic features useful to learning about and successfully applying the exciting field of engineering economics. Money makes a huge difference in the life of a corporation, an individual, and a government. Learning to understand, analyze, and manage the money side of any project is vital to its success. To be professionally successful, every engineer must be able to deal with the time value of money, economic facts, inflation, cost estimation, tax considerations, as well as spreadsheet and calculator use. This book is a great help to the learner and the instructor in accomplishing these goals by using easy-to-understand language, simple graphics, and online features.

What's New and What's Best ● ● ●

This seventh edition is full of new information and features. Plus the supporting online materials are new and updated to enhance the teaching and learning experience.

New topics:
- **Ethics** and the economics of engineering
- **Service sector projects** and their evaluation
- **Real options** development and analysis
- **Value-added taxes** and how they work
- **Multiple rates of return** and ways to eliminate them using spreadsheets
- **No tabulated factors** needed for equivalence computations (Appendix D)

New features in print and online:
- **Totally new design** to highlight important terms, concepts, and decision guidelines
- **Progressive examples** that continue throughout a chapter
- **Downloadable online presentations** featuring voice-over slides and animation
- **Vital concepts and guidelines** identified in margins; brief descriptions available (Appendix E)
- **Fresh spreadsheet displays** with on-image comments and function details
- **Case studies** (21 of them) ranging in topics from ethics to energy to simulation

Retained features:
- **Many end-of-chapter problems** (over 90% are new or revised)
- **Easy-to-read language** to enhance understanding in a variety of course environments
- **Fundamentals of Engineering (FE) Exam** review questions that double as additional or review problems for quizzes and tests
- **Hand and spreadsheet solutions** presented for many examples
- **Flexible chapter ordering** after fundamental topics are understood
- **Complete solutions manual** available online (with access approval for instructors)

How to Use This Text ● ● ●

This textbook is best suited for a one-semester or one-quarter undergraduate course. Students should be at the sophomore level or above with a basic understanding of engineering concepts and terminology. A course in calculus is not necessary; however, knowledge of the concepts in advanced mathematics and elementary probability will make the topics more meaningful.

Practitioners and professional engineers who need a refresher in economic analysis and cost estimation will find this book very useful as a reference document as well as a learning medium.

Chapter Organization and Coverage Options ● ● ●

The textbook contains 19 chapters arranged into four learning stages, *as indicated in the flowchart* on the next page, and five appendices. Each chapter starts with a statement of purpose and a specific learning outcome (ABET style) for each section. Chapters include a summary, numerous

CHAPTERS IN EACH LEARNING STAGE

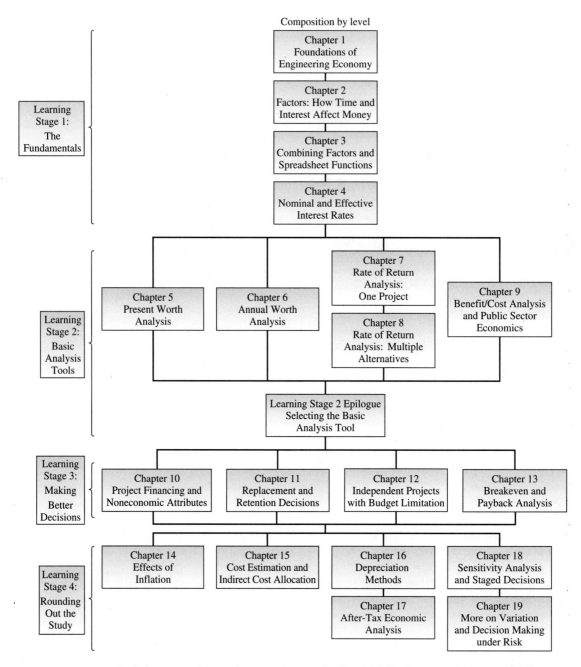

end-of-chapter problems (essay and numerical), multiple-choice problems useful for course review and FE Exam preparation, and a case study.

The appendices are important elements of learning for this text:

Appendix A Spreadsheet layout and functions (Excel is featured)

Appendix B Accounting reports and business ratios

Appendix C Code of Ethics for Engineers (from NSPE)

Appendix D Equivalence computations using calculators and geometric series; no tables

Appendix E Concepts, guidelines, terms, and symbols for engineering economics

There is considerable flexibility in the sequencing of topics and chapters once the first six chapters are covered, *as shown in the progression graphic* on the next page. If the course is designed to emphasize sensitivity and risk analysis, Chapters 18 and 19 can be covered immediately

CHAPTER AND TOPIC PROGRESSION OPTIONS

Topics may be introduced at the point indicated or any point thereafter
(Alternative entry points are indicated by ◄ – – –)

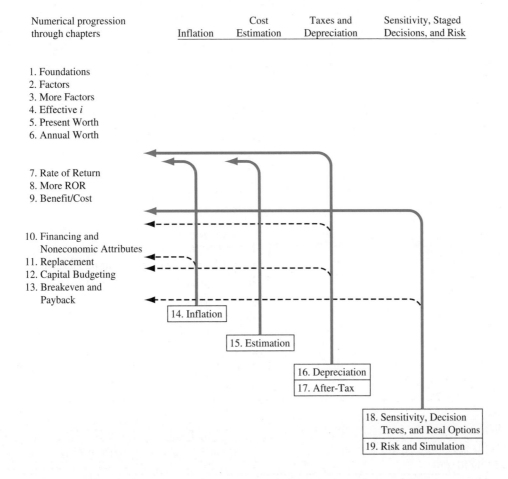

after Learning Stage 2 (Chapter 9) is completed. If depreciation and tax emphasis are vitally important to the goals of the course, Chapters 16 and 17 can be covered once Chapter 6 (annual worth) is completed. The progression graphic can help in the design of the course content and topic ordering.

SAMPLE OF RESOURCES FOR

LEARNING OUTCOMES

Each chapter begins with a purpose, list of topics, and learning outcomes (ABET style) for each corresponding section. This behavioral-based approach sensitizes the reader to what is ahead, leading to improved understanding and learning.

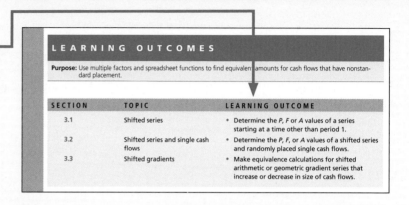

LEARNING OUTCOMES

Purpose: Use multiple factors and spreadsheet functions to find equivalent amounts for cash flows that have nonstandard placement.

SECTION	TOPIC	LEARNING OUTCOME
3.1	Shifted series	• Determine the P, F or A values of a series starting at a time other than period 1.
3.2	Shifted series and single cash flows	• Determine the P, F, or A values of a shifted series and randomly placed single cash flows.
3.3	Shifted gradients	• Make equivalence calculations for shifted arithmetic or geometric gradient series that increase or decrease in size of cash flows.

CONCEPTS AND GUIDELINES

To highlight the fundamental building blocks of the course, a checkmark and title in the margin call attention to particularly important concepts and decision-making guidelines. Appendix E includes a brief description of each fundamental concept.

☑ Time value of money	It is a well-known fact that money **makes** money. The time value of money explains the change in the amount of money **over time** for funds that are owned (invested) or owed (borrowed). This is the most important concept in engineering economy.

IN-CHAPTER EXAMPLES

Numerous in-chapter examples throughout the book reinforce the basic concepts and make understanding easier. Where appropriate, the example is solved using separately marked hand and spreadsheet solutions.

EXAMPLE 4.6

A dot-com company plans to place money in a new venture capital fund that currently returns 18% per year, compounded daily. What effective rate is this (*a*) yearly and (*b*) semiannually?

Solution

(*a*) Use Equation [4.7], with $r = 0.18$ and $m = 365$.

$$\text{Effective } i\% \text{ per year} = \left(1 + \frac{0.18}{365}\right)^{365} - 1 = 19.716\%$$

(*b*) Here $r = 0.09$ per 6 months and $m = 182$ days.

$$\text{Effective } i\% \text{ per 6 months} = \left(1 + \frac{0.09}{182}\right)^{182} - 1 = 9.415\%$$

PROGRESSIVE EXAMPLES

Several chapters include a progressive example—a more detailed problem statement introduced at the beginning of the chapter and expanded upon throughout the chapter in specially marked examples. This approach illustrates different techniques and some increasingly complex aspects of a real-world problem.

PE

Water for Semiconductor Manufacturing Case: The worldwide contribution of semiconductor sales is about $250 billion per year, or about 10% of the world's GDP (gross domestic product). This industry produces the microchips used in many of the communication, entertainment, transportation, and computing devices we use every day. Depending upon the type and size of fabrication plant (fab), the need for ultrapure water (UPW) to manufacture these tiny integrated circuits is high, ranging from 500 to 2000 gpm (gallons per minute). Ultrapure water is obtained by special processes that commonly include reverse osmosis/deionizing resin bed technologies. Potable water obtained from purifying seawater or brackish groundwater may cost from $2 to $3 per 1000 gallons, but to obtain UPW on-site for semiconductor manufacturing may cost an additional $1 to $3 per 1000 gallons.

A fab costs upward of $2.5 billion to construct, with approximately 1% of this total, or $25 million, required to provide the ultrapure water needed, including the necessary wastewater and recycling equipment.

A newcomer to the industry, Angular Enterprises, has estimated the cost profiles for two options to supply its anticipated fab with water. It is fortunate to

have the option of desalinated seawater or purified groundwater sources in the location chosen for its new fab. The initial cost estimates for the UPW system are given below.

Source	Seawater (S)	Groundwater (G)
Equipment first cost, $M	−20	−22
AOC, $M per year	−0.5	−0.3
Salvage value, % of first cost	5	10
Cost of UPW, $ per 1000 gallons	4	5

Angular has made some initial estimates for the UPW system.

Life of UPW equipment	10 years
UPW needs	1500 gpm
Operating time	16 hours per day for 250 days per year

This case is used in the following topics (Sections) and problems of this chapter:

PW analysis of equal-life alternatives (Section 5.2)
PW analysis of different-life alternatives (Section 5.3)
Capitalized cost analysis (Section 5.5)
Problems 5.20 and 5.34

INSTRUCTORS AND STUDENTS

ONLINE PRESENTATIONS

An icon in the margin indicates the availability of an animated voice-over slide presentation that summarizes the material in the section and provides a brief example for learners who need a review or prefer video-based materials. Presentations are keyed to the sections of the text.

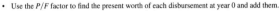

3.1 Calculations for Uniform Series That Are Shifted ● ● ●

When a uniform series begins at a time other than at the end of period 1, it is called a **shifted series**. In this case several methods can be used to find the equivalent present worth P. For example, P of the uniform series shown in Figure 3–1 could be determined by any of the following methods:

- Use the P/F factor to find the present worth of each disbursement at year 0 and add them.
- Use the F/P factor to find the future worth of each disbursement in year 13, add them, and then find the present worth of the total, using $P = F(P/F,i,13)$.
- Use the F/A factor to find the future amount $F = A(F/A,i,10)$, and then compute the present worth, using $P = F(P/F,i,13)$.
- Use the P/A factor to compute the "present worth" $P_3 = A(P/A,i,10)$ (which will be located in year 3, not year 0), and then find the present worth in year 0 by using the $(P/F,i,3)$ factor.

Figure 7–12
Spreadsheet application of ROIC method using Goal Seek, Example 7.6.

Figure 13–11
Payback period analysis, Example 13.8

SPREADSHEETS

The text integrates spreadsheets to show how easy they are to use in solving virtually any type of engineering economic analysis problem. Cell tags or full cells detail built-in functions and relations developed to solve a specific problem.

FE EXAM AND COURSE REVIEWS

Each chapter concludes with several multiple-choice, FE Exam–style problems that provide a simplified review of chapter material. Additionally, these problems cover topics for test reviews and homework assignments.

ADDITIONAL PROBLEMS AND FE EXAM REVIEW QUESTIONS

8.38 When conducting a rate of return (ROR) analysis involving multiple mutually exclusive alternatives, the first step is to:
- (a) Rank the alternatives according to decreasing initial investment cost
- (b) Rank the alternatives according to increasing initial investment cost
- (c) Calculate the present worth of each alternative using the MARR
- (d) Find the LCM between all of the alternatives

8.39 In comparing mutually exclusive alternatives by the ROR method, you should:
- (a) Find the ROR of each alternative and pick the one with the *highest* ROR

8.43 For these alternatives, the sum of the incremental cash flows is:

Year	A	B
0	−10,000	−14,000
1	+2,500	+4,000
2	+2,500	+4,000
3	+2,500	+4,000
4	+2,500	+4,000
5	+2,500	+4,000

- (a) $2500
- (b) $3500
- (c) $6000
- (d) $8000

RENEWABLE ENERGY SOURCES FOR ELECTRICITY GENERATION

Background

Pedernales Electric Cooperative (PEC) is the largest member-owned electric co-op in the United States with over 232,000 meters in 12 Central Texas counties. PEC has a capacity of approximately 1300 MW (megawatts) of power, of which 277 MW, or about 21%, is from renewable sources. The latest addition is 60 MW of power from a wind farm in south Texas close to the city of Corpus Christi. A constant question is how much of PEC's generation capacity should be from renewable sources, especially given the environmental issues with coal-generated electricity and the rising costs of hydrocarbon fuels.

Wind and nuclear sources are the current consideration for the PEC leadership as Texas is increasing its generation by nuclear power and the state is the national leader in wind farm–produced electricity.

Consider yourself a member of the board of directors of PEC. You are an engineer who has been newly elected by the PEC membership to serve a 3-year term as a director-at-large. As such, you do not represent a specific district within the entire service area; all other directors do represent a specific district. You have many questions about the operations of PEC, plus you are interested in the economic and societal benefits of pursuing more renewable source generation capacity.

Information

Here are some data that you have obtained. The information is sketchy, as this point, and the numbers are very approximate. Electricity generation cost estimates are national in scope, not PEC-specific, and are provided in cents per kilowatt-hour (¢/kWh).

	Generation Cost, ¢/kWh	
Fuel Source	**Likely Range**	**Reasonable Average**
Coal	4 to 9	7.4
Natural gas	4 to 10.5	8.6
Wind	4.8 to 9.1	8.2
Solar	4.5 to 15.5	8.8

National average cost of electricity to residential customers: 11¢/kWh

PEC average cost to residential customers: 10.27 ¢/kWh (from primary sources) and 10.92 ¢/kWh (renewable sources)

Expected life of a generation facility: 20 to 40 years (it is likely closer to 20 than 40)

Time to construct a facility: 2 to 5 years

Capital cost to build a generation facility: $900 to $1500 per kW

You have also learned that the PEC staff uses the well-recognized *levelized energy cost* (LEC) method to determine the price of electricity that must be charged to customers to break even. The formula takes into account the capital cost of the generation facilities, the cost of capital of borrowed money, annual maintenance and operation (M&O) costs, and the expected life of the facility. The LEC formula, expressed in dollars per kWh for $(t = 1, 2, \ldots, n)$, is

$$LEC = \frac{\sum_{t=1}^{t=n} \dfrac{P_t + A_t + C_t}{(1 + i)^t}}{\sum_{t=1}^{t=n} \dfrac{E_t}{(1 + i)^t}}$$

where P_t = capital investments made in year t
$\quad A_t$ = annual maintenance and operating (M&O) costs for year t
$\quad C_t$ = fuel costs for year t
$\quad E_t$ = amount of electricity generated in year t
$\quad n$ = expected life of facility
$\quad i$ = discount rate (cost of capital)

Case Study Exercises

1. If you wanted to know more about the new arrangement with the wind farm in south Texas for the additional 60 MW per year, what types of questions would you ask of a staff member in your first meeting with him or her?

2. Much of the current generation capacity of PEC facilities utilizes coal and natural gas as the primary fuel source. What about the ethical aspects of the government's allowance for these plants to continue polluting the atmosphere with the emissions that may cause health problems for citizens and further the effects of global warming? What types of regulations, if any, should be developed for PEC (and other generators) to follow in the future?

CASE STUDIES

New and updated case studies at the end of most chapters present real-world, in-depth treatments and exercises in the engineering profession. Each case includes a background, relevant information, and an exercise section.

ACKNOWLEDGMENT OF CONTRIBUTORS

It takes the input and efforts of many individuals to make significant improvements in a textbook. We wish to give special thanks to the following persons for their contributions to this edition.

Paul Askenasy, Texas Commission on Environmental Quality
Jack Beltran, Bristol-Myers Squibb
Robert Lundquist, Ohio State University
William Peet, Infrastructure Coordination, Government of Niue
Sallie Sheppard, Texas A&M University

We thank the following individuals for their comments, feedback, and review of material to assist in making this edition a real success.

Ahmed Alim, University of Houston
Alan Atalah, Bowling Green State University
Fola Michael Ayokanmbi, Alabama A&M University
William Brown, West Virginia University at Parkersburg
Hector Carrasco, Colorado State University–Pueblo
Robert Chiang, California State University, Pomona
Ronald Cutwright, Florida State University
John F. Dacquisto, Gonzaga University
Houshang Darabi, University of Illinois at Chicago
Freddie Davis, West Texas A&M University
Edward Lester Dollar, Southern Polytechnic State University
Ted Eschenbach, University of Alaska
Clara Fang, University of Hartford
Abel Fernandez, University of the Pacific
Daniel A. Franchi, California Polytechnic State University, San Luis Obispo
Mark Frascatore, Clarkson University
Benjamin M. Fries, University of Central Florida
Nathan Gartner, University of Massachusetts–Lowell
Johnny R. Graham, University of North Carolina–Charlotte
Liling Huang, Northern Virginia Community College
David Jacobs, University of Hartford
Adam Jannik, Northwestern State University
Peter E. Johnson, Valparaiso University
Justin W. Kile, University of Wisconsin–Platteville
John Kushner, Lawrence Technological University
Clifford D. Madrid, New Mexico State University
Saeed Manafzadeh, University of Illinois at Chicago
Quamrul Mazumder, University of Michigan–Flint
Deb McAvoy, Ohio University
Gene McGinnis, Christian Brothers University
Bruce V. Mutter, Bluefield State College
Hong Sioe Oey, University of Texas at El Paso
Richard Palmer, University of Massachusetts
Michael J. Rider, Ohio Northern University
John Ristroph, University of Louisiana at Lafayette
Saeid L. Sadri, Georgia Institute of Technology
Scott Schultz, Mercer University
Kyo D. Song, Norfolk State University
James Stevens, University of Colorado at Colorado Springs
John A. Stratton, Rochester Institute of Technology
Mathias J. Sutton, Purdue University
Pete Weiss, Valparaiso University

Greg Wiles, Southern Polytechnic State University
Richard Youchak, University of Pittsburgh at Johnstown
William A. Young, II, Ohio University

If you discover errors that require correction in the next printing of the textbook or in updates of the online resources, please contact us. We hope you find the contents of this edition helpful in your academic and professional activities.

Leland Blank   lelandblank@yahoo.com
Anthony Tarquin   atarquin@utep.edu

LEARNING STAGE 1

The Fundamentals

The fundamentals of engineering economy are introduced in these chapters. When you have completed stage 1, you will be able to understand and work problems that account for the **time value of money, cash flows** occurring at different times with different amounts, and **equivalence** at different interest rates. The techniques you master here form the basis of how an engineer in any discipline can take **economic value** into account in virtually any project environment.

The factors commonly used in all engineering economy computations are introduced and applied here. Combinations of these factors assist in moving monetary values forward and backward through time and at different interest rates. Also, after these chapters, you should be comfortable using many of the spreadsheet functions.

Many of the terms common to economic decision making are introduced in learning stage 1 and used in later chapters. A checkmark icon in the margin indicates that a new **concept or guideline** is introduced at this point.

CHAPTER 1

Foundations of Engineering Economy

LEARNING OUTCOMES

Purpose: Understand and apply fundamental concepts and use the terminology of engineering economics.

SECTION	TOPIC	LEARNING OUTCOME
1.1	Description and role	• Define engineering economics and describe its role in decision making.
1.2	Engineering economy study approach	• Understand and identify the steps in an engineering economy study.
1.3	Ethics and economics	• Identify areas in which economic decisions can present questionable ethics.
1.4	Interest rate	• Perform calculations for interest rates and rates of return.
1.5	Terms and symbols	• Identify and use engineering economic terminology and symbols.
1.6	Cash flows	• Understand cash flows and how to graphically represent them.
1.7	Economic equivalence	• Describe and calculate economic equivalence.
1.8	Simple and compound interest	• Calculate simple and compound interest amounts for one or more time periods.
1.9	MARR and opportunity cost	• State the meaning and role of Minimum Attractive Rate of Return (MARR) and opportunity costs.
1.10	Spreadsheet functions	• Identify and use some Excel functions commonly applied in engineering economics.

T he need for engineering economy is primarily motivated by the work that engineers do in performing analyses, synthesizing, and coming to a conclusion as they work on projects of all sizes. In other words, engineering economy is at the heart of **making decisions**. These decisions involve the fundamental elements of **cash flows of money, time, and interest rates**. This chapter introduces the basic concepts and terminology necessary for an engineer to combine these three essential elements in organized, mathematically correct ways to solve problems that will lead to better decisions.

1.1 Engineering Economics: Description and Role in Decision Making ● ● ●

Decisions are made routinely to choose one alternative over another by individuals in everyday life; by engineers on the job; by managers who supervise the activities of others; by corporate presidents who operate a business; and by government officials who work for the public good. Most decisions involve money, called **capital** or **capital funds**, which is usually limited in amount. The decision of where and how to invest this limited capital is motivated by a primary goal of **adding value** as future, anticipated results of the selected alternative are realized. Engineers play a vital role in capital investment decisions based upon their ability and experience to design, analyze, and synthesize. The factors upon which a decision is based are commonly a combination of economic and noneconomic elements. Engineering economy deals with the economic factors. By definition,

Engineering economy involves formulating, estimating, and evaluating the expected economic outcomes of alternatives designed to accomplish a defined purpose. Mathematical techniques simplify the economic evaluation of alternatives.

Because the formulas and techniques used in engineering economics are applicable to all types of money matters, they are equally useful in business and government, as well as for individuals. Therefore, besides applications to projects in your future jobs, what you learn from this book and in this course may well offer you an economic analysis tool for making personal decisions such as car purchases, house purchases, major purchases on credit, e.g., furniture, appliances, and electronics.

Other terms that mean the same as *engineering economy* are *engineering economic analysis, capital allocation study, economic analysis,* and similar descriptors.

People make decisions; computers, mathematics, concepts, and guidelines assist people in their decision-making process. Since most decisions affect what will be done, the time frame of engineering economy is primarily the **future**. Therefore, the numbers used in engineering economy are **best estimates of what is expected to occur**. The estimates and the decision usually involve four essential elements:

Cash flows

Times of occurrence of cash flows

Interest rates for time value of money

Measure of economic worth for selecting an alternative

Since the estimates of cash flow amounts and timing are about the future, they will be somewhat different than what is actually observed, due to changing circumstances and unplanned events. In short, the variation between an amount or time estimated now and that observed in the future is caused by the stochastic (random) nature of all economic events. **Sensitivity analysis** is utilized to determine how a decision might change according to varying estimates, especially those expected to vary widely. Example 1.1 illustrates the fundamental nature of variation in estimates and how this variation may be included in the analysis at a very basic level.

EXAMPLE 1.1

An engineer is performing an analysis of warranty costs for drive train repairs within the first year of ownership of luxury cars purchased in the United States. He found the average cost (to the nearest dollar) to be $570 per repair from data taken over a 5-year period.

Year	2006	2007	2008	2009	2010
Average Cost, $/repair	525	430	619	650	625

What range of repair costs should the engineer use to ensure that the analysis is sensitive to changing warranty costs?

Solution

At first glance the range should be approximately –25% to +15% of the $570 average cost to include the low of $430 and high of $650. However, the last 3 years of costs are higher and more consistent with an average of $631. The observed values are approximately ±3% of this more recent average.

If the analysis is to use the most recent data and trends, a range of, say, ±5% of $630 is recommended. If, however, the analysis is to be more inclusive of historical data and trends, a range of, say, ±20% or ±25% of $570 is recommended.

The criterion used to select an alternative in engineering economy for a specific set of estimates is called a **measure of worth**. The measures developed and used in this text are

Present worth (PW)	Future worth (FW)	Annual worth (AW)
Rate of return (ROR)	Benefit/cost (B/C)	Capitalized cost (CC)
Payback period	Economic value added (EVA)	Cost Effectiveness

All these measures of worth account for the fact that money makes money over time. This is the concept of the **time value of money.**

Time value of money

> It is a well-known fact that money **makes** money. The time value of money explains the change in the amount of money **over time** for funds that are owned (invested) or owed (borrowed). This is the most important concept in engineering economy.

The time value of money is very obvious in the world of economics. If we decide to invest capital (money) in a project today, we inherently expect to have more money in the future than we invested. If we borrow money today, in one form or another, we expect to return the original amount plus some additional amount of money.

Engineering economics is equally well suited for the future and for the **analysis of past cash flows** in order to determine if a specific criterion (measure of worth) was attained. For example, assume you invested $4975 exactly 3 years ago in 53 shares of IBM stock as traded on the New York Stock Exchange (NYSE) at $93.86 per share. You expect to make 8% per year appreciation, not considering any dividends that IBM may declare. A quick check of the share value shows it is currently worth $127.25 per share for a total of $6744.25. This increase in value represents a rate of return of 10.67% per year. (These type of calculations are explained later.) This past investment has well exceeded the 8% per year criterion over the last 3 years.

1.2 Performing an Engineering Economy Study ● ● ●

An engineering economy study involves many elements: problem identification, definition of the objective, cash flow estimation, financial analysis, and decision making. Implementing a structured procedure is the best approach to select the best solution to the problem.

The steps in an engineering economy study are as follows:

1. Identify and understand the problem; identify the objective of the project.
2. Collect relevant, available data and define viable solution alternatives.
3. Make realistic cash flow estimates.
4. Identify an economic measure of worth criterion for decision making.

5. Evaluate each alternative; consider noneconomic factors; use sensitivity analysis as needed.
6. Select the best alternative.
7. Implement the solution and monitor the results.

Technically, the last step is not part of the economy study, but it is, of course, a step needed to meet the project objective. There may be occasions when the best economic alternative requires more capital funds than are available, or significant noneconomic factors preclude the most economic alternative from being chosen. Accordingly, steps 5 and 6 may result in selection of an alternative different from the economically best one. Also, sometimes more than one project may be selected and implemented. This occurs when projects are independent of one another. In this case, steps 5 through 7 vary from those above. Figure 1–1 illustrates the steps above for one alternative. Descriptions of several of the elements in the steps are important to understand.

Problem Description and Objective Statement   A succinct statement of the problem and primary objective(s) is very important to the formation of an alternative solution. As an illustration, assume the problem is that a coal-fueled power plant must be shut down by 2015 due to the production of excessive sulfur dioxide. The objectives may be to generate the forecasted electricity

Figure 1–1
Steps in an engineering economy study.

needed for 2015 and beyond, plus to not exceed all the projected emission allowances in these future years.

Alternatives   These are stand-alone descriptions of viable solutions to problems that can meet the objectives. Words, pictures, graphs, equipment and service descriptions, simulations, etc. define each alternative. The best estimates for parameters are also part of the alternative. Some parameters include equipment first cost, expected life, salvage value (estimated trade-in, resale, or market value), and annual operating cost (AOC), which can also be termed *maintenance and operating* (M&O) *cost,* and subcontract cost for specific services. If changes in income (revenue) may occur, this parameter must be estimated.

Detailing all viable alternatives at this stage is crucial. For example, if two alternatives are described and analyzed, one will likely be selected and implementation initiated. If a third, more attractive method that was available is later recognized, a wrong decision was made.

Cash Flows   All cash flows are estimated for each alternative. Since these are future expenditures and revenues, the results of step 3 usually prove to be inaccurate when an alternative is actually in place and operating. When cash flow estimates for specific parameters are expected to vary significantly from a *point estimate* made now, risk and sensitivity analyses (step 5) are needed to improve the chances of selecting the best alternative. Sizable variation is usually expected in estimates of revenues, AOC, salvage values, and subcontractor costs. Estimation of costs is discussed in Chapter 15, and the elements of variation (risk) and sensitivity analysis are included throughout the text.

Engineering Economy Analysis   The techniques and computations that you will learn and use throughout this text utilize the cash flow estimates, time value of money, and a selected measure of worth. The result of the analysis will be one or more numerical values; this can be in one of several terms, such as money, an interest rate, number of years, or a probability. In the end, a selected measure of worth mentioned in the previous section will be used to select the best alternative.

Before an economic analysis technique is applied to the cash flows, some decisions about what to include in the analysis must be made. Two important possibilities are taxes and inflation. Federal, state or provincial, county, and city taxes will impact the costs of every alternative. An after-tax analysis includes some additional estimates and methods compared to a before-tax analysis. If taxes and inflation are expected to impact all alternatives equally, they may be disregarded in the analysis. However, if the size of these projected costs is important, taxes and inflation should be considered. Also, if the impact of inflation over time is important to the decision, an additional set of computations must be added to the analysis; Chapter 14 covers the details.

Selection of the Best Alternative   The measure of worth is a primary basis for selecting the best economic alternative. For example, if alternative A has a rate of return (ROR) of 15.2% per year and alternative B will result in an ROR of 16.9% per year, B is better economically. However, there can always be **noneconomic** or **intangible factors** that must be considered and that may alter the decision. There are many possible noneconomic factors; some typical ones are

- Market pressures, such as need for an increased international presence
- Availability of certain resources, e.g., skilled labor force, water, power, tax incentives
- Government laws that dictate safety, environmental, legal, or other aspects
- Corporate management's or the board of director's interest in a particular alternative
- Goodwill offered by an alternative toward a group: employees, union, county, etc.

As indicated in Figure 1–1, once all the economic, noneconomic, and risk factors have been evaluated, a final decision of the "best" alternative is made.

At times, only one viable alternative is identified. In this case, the **do-nothing (DN) alternative** may be chosen provided the measure of worth and other factors result in the alternative being a poor choice. The do-nothing alternative maintains the status quo.

Whether we are aware of it or not, we use criteria every day to choose between alternatives. For example, when you drive to campus, you decide to take the "best" route. But how did you define *best*? Was the best route the safest, shortest, fastest, cheapest, most scenic, or what? Obviously, depending upon which criterion or combination of criteria is used to identify the best, a different route might be selected each time. In economic analysis, **financial units (dollars or other currency)** are generally used as the tangible basis for evaluation. Thus, when there are several ways of accomplishing a stated objective, the alternative with the lowest overall cost or highest overall net income is selected.

1.3 Professional Ethics and Economic Decisions ● ● ●

Many of the fundamentals of engineering ethics are intertwined with the roles of money and economics-based decisions in the making of professionally ethical judgments. Some of these integral connections are discussed here, plus sections in later chapters discuss additional aspects of ethics and economics. For example, Chapter 9, Benefit/Cost Analysis and Public Sector Economics, includes material on the ethics of public project contracts and public policy. Although it is very limited in scope and space, it is anticipated that this coverage of the important role of economics in engineering ethics will prompt further interest on the part of students and instructors of engineering economy.

The terms **morals** and **ethics** are commonly used interchangeably, yet they have slightly different interpretations. Morals usually relate to the underlying tenets that form the character and conduct of a person in judging right and wrong. Ethical practices can be evaluated by using a code of morals or **code of ethics** that forms the standards to guide decisions and actions of individuals and organizations in a profession, for example, electrical, chemical, mechanical, industrial, or civil engineering. There are several different levels and types of morals and ethics.

Universal or common morals   These are fundamental moral beliefs held by virtually all people. Most people agree that to steal, murder, lie, or physically harm someone is wrong.

It is possible for **actions** and **intentions** to come into conflict concerning a common moral. Consider the World Trade Center buildings in New York City. After their collapse on September 11, 2001, it was apparent that the design was not sufficient to withstand the heat generated by the firestorm caused by the impact of an aircraft. The structural engineers who worked on the design surely did not have the intent to harm or kill occupants in the buildings. However, their design actions did not foresee this outcome as a measurable possibility. Did they violate the common moral belief of not doing harm to others or murdering?

Individual or personal morals   These are the moral beliefs that a person has and maintains over time. These usually parallel the common morals in that stealing, lying, murdering, etc. are immoral acts.

It is quite possible that an individual strongly supports the common morals and has excellent personal morals, but these may conflict from time to time when decisions must be made. Consider the engineering student who genuinely believes that cheating is wrong. If he or she does not know how to work some test problems, but must make a certain minimum grade on the final exam to graduate, the decision to cheat or not on the final exam is an exercise in following or violating a personal moral.

Professional or engineering ethics   Professionals in a specific discipline are guided in their decision making and performance of work activities by a formal standard or code. The code states the commonly accepted standards of honesty and integrity that each individual is expected to demonstrate in her or his practice. There are codes of ethics for medical doctors, attorneys, and, of course, engineers.

Although each engineering profession has its own code of ethics, the **Code of Ethics for Engineers** published by the National Society of Professional Engineers (NSPE) is very commonly used and quoted. This code, reprinted in its entirety in Appendix C, includes numerous sections that have direct or indirect economic and financial impact upon the designs, actions, and decisions that engineers make in their professional dealings. Here are three examples from the Code:

> "Engineers, in the fulfillment of their duties, shall hold paramount the *safety, health, and welfare of the public*." (section I.1)

> "Engineers shall *not accept financial or other considerations*, including free engineering designs, from material or equipment suppliers for specifying their product." (section III.5.a)

> "Engineers using designs supplied by a client recognize that the *designs remain the property of the client* and may not be duplicated by the engineer for others without express permission." (section III.9.b)

As with common and personal morals, conflicts can easily rise in the mind of an engineer between his or her own ethics and that of the employing corporation. Consider a manufacturing engineer who has recently come to firmly disagree morally with war and its negative effects on human beings. Suppose the engineer has worked for years in a military defense contractor's facility and does the detailed cost estimations and economic evaluations of producing fighter jets for the Air Force. The Code of Ethics for Engineers is silent on the ethics of producing and using war materiel. Although the employer and the engineer are not violating any ethics code, the engineer, as an individual, is stressed in this position. Like many people during a declining national economy, retention of this job is of paramount importance to the family and the engineer. Conflicts such as this can place individuals in real dilemmas with no or mostly unsatisfactory alternatives.

At first thought, it may not be apparent how activities related to engineering economics may present an ethical challenge to an individual, a company, or a public servant in government service. Many money-related situations, such as those that follow, can have ethical dimensions.

In the design stage:

- Safety factors are compromised to ensure that a price bid comes in as low as possible.
- Family or personal connections with individuals in a company offer unfair or insider information that allows costs to be cut in strategic areas of a project.
- A potential vendor offers specifications for company-specific equipment, and the design engineer does not have sufficient time to determine if this equipment will meet the needs of the project being designed and costed.

While the system is operating:

- Delayed or below-standard maintenance can be performed to save money when cost overruns exist in other segments of a project.
- Opportunities to purchase cheaper repair parts can save money for a subcontractor working on a fixed-price contract.
- Safety margins are compromised because of cost, personal inconvenience to workers, tight time schedules, etc.

A good example of the last item—safety is compromised while operating the system—is the situation that arose in 1984 in Bhopal, India (Martin and Schinzinger 2005, pp. 245–8). A Union Carbide plant manufacturing the highly toxic pesticide chemical methyl isocyanate (MIC) experienced a large gas leak from high-pressure tanks. Some 500,000 persons were exposed to inhalation of this deadly gas that burns moist parts of the body. There were 2500 to 3000 deaths within days, and over the following 10-year period, some 12,000 death claims and 870,000 personal injury claims were recorded. Although Union Carbide owned the facility, the Indian government had only Indian workers in the plant. Safety practices clearly eroded due to cost-cutting measures, insufficient repair parts, and reduction in personnel to save salary money. However, one of the surprising practices that caused unnecessary harm to workers was the fact that masks, gloves, and other protective gear were not worn by workers in close proximity to the tanks containing MIC. Why? Unlike in plants in the United States and other countries, there was no air conditioning in the Indian plant, resulting in high ambient temperatures in the facility.

Many ethical questions arise when corporations operate in international settings where the corporate rules, worker incentives, cultural practices, and costs in the home country differ from those in the host country. Often these ethical dilemmas are fundamentally based in the economics that provide cheaper labor, reduced raw material costs, less government oversight, and a host of other cost-reducing factors. When an engineering economy study is performed, it is important for the engineer performing the study to consider all ethically related matters to ensure that the cost and revenue estimates reflect what is likely to happen once the project or system is operating.

It is important to understand that the translation from universal morals to personal morals and professional ethics does vary from one culture and country to another. As an example, consider the common belief (universal moral) that the awarding of contracts and financial arrangements for services to be performed (for government or business) should be accomplished in a fair and transparent fashion. In some societies and cultures, corruption in the process of contract making is common and often "overlooked" by the local authorities, who may also be involved in the affairs. Are these immoral or unethical practices? Most would say, "Yes, this should not be allowed. Find and punish the individuals involved." Yet, such practices do continue, thus indicating the differences in interpretation of common morals as they are translated into the ethics of individuals and professionals.

EXAMPLE 1.2

Jamie is an engineer employed by Burris, a United States–based company that develops subway and surface transportation systems for medium-sized municipalities in the United States and Canada. He has been a registered professional engineer (PE) for the last 15 years. Last year, Carol, an engineer friend from university days who works as an individual consultant, asked Jamie to help her with some cost estimates on a metro train job. Carol offered to pay for his time and talent, but Jamie saw no reason to take money for helping with data commonly used by him in performing his job at Burris. The estimates took one weekend to complete, and once Jamie delivered them to Carol, he did not hear from her again; nor did he learn the identity of the company for which Carol was preparing the estimates.

Yesterday, Jamie was called into his supervisor's office and told that Burris had not received the contract award in Sharpstown, where a metro system is to be installed. The project estimates were prepared by Jamie and others at Burris over the past several months. This job was greatly needed by Burris, as the country and most municipalities were in a real economic slump, so much so that Burris was considering furloughing several engineers if the Sharpstown bid was not accepted. Jamie was told he was to be laid off immediately, not because the bid was rejected, but because he had been secretly working without management approval for a prime consultant of Burris' main competitor. Jamie was astounded and angry. He knew he had done nothing to warrant firing, but the evidence was clearly there. The numbers used by the competitor to win the Sharpstown award were the same numbers that Jamie had prepared for Burris on this bid, and they closely matched the values that he gave Carol when he helped her.

Jamie was told he was fortunate, because Burris' president had decided to not legally charge Jamie with unethical behavior and to not request that his PE license be rescinded. As a result, Jamie was escorted out of his office and the building within one hour and told to not ask anyone at Burris for a reference letter if he attempted to get another engineering job.

Discuss the ethical dimensions of this situation for Jamie, Carol, and Burris' management. Refer to the NSPE Code of Ethics for Engineers (Appendix C) for specific points of concern.

Solution

There are several obvious errors and omissions present in the actions of Jamie, Carol, and Burris' management in this situation. Some of these mistakes, oversights, and possible code violations are summarized here.

Jamie

- Did not learn identity of company Carol was working for and whether the company was to be a bidder on the Sharpstown project
- Helped a friend with confidential data, probably innocently, without the knowledge or approval of his employer
- Assisted a competitor, probably unknowingly, without the knowledge or approval of his employer
- Likely violated, at least, Code of Ethics for Engineers section II.1.c, which reads, "Engineers shall not reveal facts, data, or information without the prior consent of the client or employer except as authorized or required by law or this Code."

Carol

- Did not share the intended use of Jamie's work
- Did not seek information from Jamie concerning his employer's intention to bid on the same project as her client
- Misled Jamie in that she did not seek approval from Jamie to use and quote his information and assistance
- Did not inform her client that portions of her work originated from a source employed by a possible bid competitor
- Likely violated, at least, Code of Ethics for Engineers section III.9.a, which reads, "Engineers shall, whenever possible, name the person or persons who may be individually responsible for designs, inventions, writings, or other accomplishments."

Burris' management

- Acted too fast in dismissing Jamie; they should have listened to Jamie and conducted an investigation
- Did not put him on administrative leave during a review
- Possibly did not take Jamie's previous good work record into account

These are not all ethical considerations; some are just plain good business practices for Jamie, Carol, and Burris.

1.4 Interest Rate and Rate of Return ● ● ●

Interest is the manifestation of the time value of money. Computationally, interest is the difference between an ending amount of money and the beginning amount. If the difference is zero or negative, there is no interest. There are always two perspectives to an amount of interest—interest paid and interest earned. These are illustrated in Figure 1–2. Interest is **paid** when a person or organization borrowed money (obtained a loan) and repays a larger amount over time. Interest is **earned** when a person or organization saved, invested, or lent money and obtains a return of a larger amount over time. The numerical values and formulas used are the same for both perspectives, but the interpretations are different.

Interest paid on borrowed funds (a loan) is determined using the original amount, also called the *principal,*

$$\text{Interest} = \text{amount owed now} - \text{principal} \qquad [1.1]$$

When interest paid over a *specific time unit* is expressed as a percentage of the principal, the result is called the **interest rate.**

$$\text{Interest rate } (\%) = \frac{\text{interest accrued per time unit}}{\text{principal}} \times 100\% \qquad [1.2]$$

The time unit of the rate is called the **interest period.** By far the most common interest period used to state an interest rate is 1 year. Shorter time periods can be used, such as 1% per month. Thus, the interest period of the interest rate should always be included. If only the rate is stated, for example, 8.5%, a 1-year interest period is assumed.

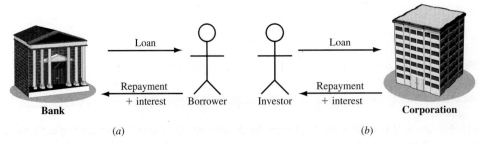

Figure 1–2
(*a*) Interest paid over time to lender. (*b*) Interest earned over time by investor.

EXAMPLE 1.3

An employee at LaserKinetics.com borrows $10,000 on May 1 and must repay a total of $10,700 exactly 1 year later. Determine the interest amount and the interest rate paid.

Solution

The perspective here is that of the borrower since $10,700 repays a loan. Apply Equation [1.1] to determine the interest paid.

$$\text{Interest paid} = \$10,700 - 10,000 = \$700$$

Equation [1.2] determines the interest rate paid for 1 year.

$$\text{Percent interest rate} = \frac{\$700}{\$10,000} \times 100\% = 7\% \text{ per year}$$

EXAMPLE 1.4

Stereophonics, Inc., plans to borrow $20,000 from a bank for 1 year at 9% interest for new recording equipment. (*a*) Compute the interest and the total amount due after 1 year. (*b*) Construct a column graph that shows the original loan amount and total amount due after 1 year used to compute the loan interest rate of 9% per year.

Solution

(*a*) Compute the total interest accrued by solving Equation [1.2] for interest accrued.

$$\text{Interest} = \$20,000(0.09) = \$1800$$

The total amount due is the sum of principal and interest.

$$\text{Total due} = \$20,000 + 1800 = \$21,800$$

(*b*) Figure 1–3 shows the values used in Equation [1.2]: $1800 interest, $20,000 original loan principal, 1-year interest period.

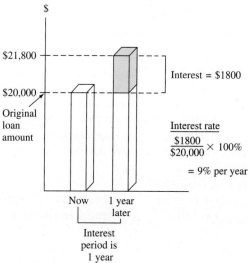

Figure 1–3
Values used to compute an interest rate of 9% per year. Example 1.4.

Comment

Note that in part (*a*), the total amount due may also be computed as

$$\text{Total due} = \text{principal}(1 + \text{interest rate}) = \$20,000(1.09) = \$21,800$$

Later we will use this method to determine future amounts for times longer than one interest period.

From the perspective of a saver, a lender, or an investor, **interest earned** (Figure 1–2b) is the final amount minus the initial amount, or principal.

$$\text{Interest earned} = \text{total amount now} - \text{principal} \qquad [1.3]$$

Interest earned over a specific period of time is expressed as a percentage of the original amount and is called **rate of return (ROR).**

$$\text{Rate of return (\%)} = \frac{\text{interest accrued per time unit}}{\text{principal}} \times 100\% \qquad [1.4]$$

The time unit for rate of return is called the **interest period,** just as for the borrower's perspective. Again, the most common period is 1 year.

The term **return on investment (ROI)** is used equivalently with ROR in different industries and settings, especially where large capital funds are committed to engineering-oriented programs.

The numerical values in Equations [1.2] and [1.4] are the same, but the term *interest rate paid* is more appropriate for the borrower's perspective, while the *rate of return earned* is better for the investor's perspective.

EXAMPLE 1.5

(a) Calculate the amount deposited 1 year ago to have $1000 now at an interest rate of 5% per year.
(b) Calculate the amount of interest earned during this time period.

Solution

(a) The total amount accrued ($1000) is the sum of the original deposit and the earned interest. If X is the original deposit,

$$\text{Total accrued} = \text{deposit} + \text{deposit(interest rate)}$$

$$\$1000 = X + X(0.05) = X(1 + 0.05) = 1.05X$$

The original deposit is

$$X = \frac{1000}{1.05} = \$952.38$$

(b) Apply Equation [1.3] to determine the interest earned.

$$\text{Interest} = \$1000 - 952.38 = \$47.62$$

In Examples 1.3 to 1.5 the interest period was 1 year, and the interest amount was calculated at the end of one period. When more than one interest period is involved, e.g., the amount of interest after 3 years, it is necessary to state whether the interest is accrued on a *simple* or *compound* basis from one period to the next. This topic is covered later in this chapter.

Since **inflation** can significantly increase an interest rate, some comments about the fundamentals of inflation are warranted at this early stage. By definition, inflation represents a decrease in the value of a given currency. That is, $10 now will not purchase the same amount of gasoline for your car (or most other things) as $10 did 10 years ago. The changing value of the currency affects market interest rates.

Inflation

> In simple terms, interest rates reflect two things: a so-called real rate of return *plus* the expected inflation rate. The real rate of return allows the investor to purchase more than he or she could have purchased before the investment, while inflation raises the real rate to the market rate that we use on a daily basis.

The safest investments (such as government bonds) typically have a 3% to 4% real rate of return built into their overall interest rates. Thus, a market interest rate of, say, 8% per year on a bond means that investors expect the inflation rate to be in the range of 4% to 5% per year. Clearly, inflation causes interest rates to rise.

From the borrower's perspective, the rate of inflation is another interest rate *tacked on to the real interest rate.* And from the vantage point of the saver or investor in a fixed-interest account, inflation *reduces the real rate of return* on the investment. Inflation means that cost and revenue cash flow estimates increase over time. This increase is due to the changing value of money that is forced upon a country's currency by inflation, thus making a unit of currency (such as the dollar) worth less relative to its value at a previous time. We see the effect of inflation in that money purchases less now than it did at a previous time. Inflation contributes to

- A reduction in purchasing power of the currency
- An increase in the CPI (consumer price index)
- An increase in the cost of equipment and its maintenance
- An increase in the cost of salaried professionals and hourly employees
- A reduction in the real rate of return on personal savings and certain corporate investments

In other words, inflation can materially contribute to changes in corporate and personal economic analysis.

Commonly, engineering economy studies assume that inflation affects all estimated values equally. Accordingly, an interest rate or rate of return, such as 8% per year, is applied throughout the analysis without accounting for an additional inflation rate. However, if inflation were explicitly taken into account, and it was reducing the value of money at, say, an average of 4% per year, then it would be necessary to perform the economic analysis using an inflated interest rate. (The rate is 12.32% per year using the relations derived in Chapter 14.)

1.5 Terminology and Symbols ● ● ○

The equations and procedures of engineering economy utilize the following terms and symbols. Sample units are indicated.

$P =$ value or amount of money at a time designated as the present or time 0. Also P is referred to as present worth (PW), present value (PV), net present value (NPV), discounted cash flow (DCF), and capitalized cost (CC); monetary units, such as dollars

$F =$ value or amount of money at some future time. Also F is called future worth (FW) and future value (FV); dollars

$A =$ series of consecutive, equal, end-of-period amounts of money. Also A is called the annual worth (AW) and equivalent uniform annual worth (EUAW); dollars per year, euros per month

$n =$ number of interest periods; years, months, days

$i =$ interest rate per time period; percent per year, percent per month

$t =$ time, stated in periods; years, months, days

The symbols P and F represent one-time occurrences: A occurs with the same value in each interest period for a specified number of periods. It should be clear that a present value P represents a single sum of money at some time prior to a future value F or prior to the first occurrence of an equivalent series amount A.

It is important to note that the symbol A always represents a uniform amount (i.e., the same amount each period) that extends through *consecutive* interest periods. Both conditions must exist before the series can be represented by A.

The interest rate i is expressed in percent per interest period, for example, 12% per year. Unless stated otherwise, assume that the rate applies throughout the entire n years or interest periods. The decimal equivalent for i is always used in formulas and equations in engineering economy computations.

All engineering economy problems involve the element of time expressed as n and interest rate i. In general, every problem will involve at least four of the symbols P, F, A, n, and i, with at least three of them estimated or known.

Additional symbols used in engineering economy are defined in Appendix E.

EXAMPLE 1.6

Today, Julie borrowed $5000 to purchase furniture for her new house. She can repay the loan in either of the two ways described below. Determine the engineering economy symbols and their value for each option.

(*a*) Five equal annual installments with interest based on 5% per year.
(*b*) One payment 3 years from now with interest based on 7% per year.

Solution

(*a*) The repayment schedule requires an equivalent annual amount *A*, which is unknown.

$$P = \$5000 \qquad i = 5\% \text{ per year} \qquad n = 5 \text{ years} \qquad A = ?$$

(*b*) Repayment requires a single future amount *F*, which is unknown.

$$P = \$5000 \qquad i = 7\% \text{ per year} \qquad n = 3 \text{ years} \qquad F = ?$$

EXAMPLE 1.7

You plan to make a lump-sum deposit of $5000 now into an investment account that pays 6% per year, and you plan to withdraw an equal end-of-year amount of $1000 for 5 years, starting next year. At the end of the sixth year, you plan to close your account by withdrawing the remaining money. Define the engineering economy symbols involved.

Solution

All five symbols are present, but the future value in year 6 is the unknown.

$$P = \$5000$$

$$A = \$1000 \text{ per year for 5 years}$$

$$F = ? \text{ at end of year 6}$$

$$i = 6\% \text{ per year}$$

$$n = 5 \text{ years for the } A \text{ series and 6 for the } F \text{ value}$$

EXAMPLE 1.8

Last year Jane's grandmother offered to put enough money into a savings account to generate $5000 in interest this year to help pay Jane's expenses at college. (*a*) Identify the symbols, and (*b*) calculate the amount that had to be deposited exactly 1 year ago to earn $5000 in interest now, if the rate of return is 6% per year.

Solution

(*a*) Symbols *P* (last year is −1) and *F* (this year) are needed.

$$P = ?$$

$$i = 6\% \text{ per year}$$

$$n = 1 \text{ year}$$

$$F = P + \text{interest} = ? + \$5000$$

(*b*) Let *F* = total amount now and *P* = original amount. We know that *F* − *P* = $5000 is accrued interest. Now we can determine *P*. Refer to Equations [1.1] through [1.4].

$$F = P + Pi$$

The $5000 interest can be expressed as

$$\text{Interest} = F - P = (P + Pi) - P$$

$$= Pi$$

$$\$5000 = P(0.06)$$

$$P = \frac{\$5000}{0.06} = \$83,333.33$$

1.6 Cash Flows: Estimation and Diagramming ● ● ●

As mentioned in earlier sections, cash flows are the amounts of money estimated for future projects or observed for project events that have taken place. All cash flows occur during specific time periods, such as 1 month, every 6 months, or 1 year. Annual is the most common time period. For example, a payment of $10,000 once every year in December for 5 years is a series of 5 outgoing cash flows. And an estimated receipt of $500 every month for 2 years is a series of 24 incoming cash flows. Engineering economy bases its computations on the timing, size, and direction of cash flows.

> **Cash inflows** are the receipts, revenues, incomes, and savings generated by project and business activity. A **plus sign** indicates a cash inflow.

Cash flow

> **Cash outflows** are costs, disbursements, expenses, and taxes caused by projects and business activity. A **negative or minus sign** indicates a cash outflow. When a project involves only costs, the minus sign may be omitted for some techniques, such as benefit/cost analysis.

Of all the steps in Figure 1–1 that outline the engineering economy study, estimating cash flows (step 3) is the most difficult, primarily because it is an attempt to predict the future. Some examples of cash flow estimates are shown here. As you scan these, consider how the cash inflow or outflow may be estimated most accurately.

Cash Inflow Estimates

Income: +$150,000 per year from sales of solar-powered watches

Savings: +$24,500 tax savings from capital loss on equipment salvage

Receipt: +$750,000 received on large business loan plus accrued interest

Savings: +$150,000 per year saved by installing more efficient air conditioning

Revenue: +$50,000 to +$75,000 per month in sales for extended battery life iPhones

Cash Outflow Estimates

Operating costs: −$230,000 per year annual operating costs for software services

First cost: −$800,000 next year to purchase replacement earthmoving equipment

Expense: −$20,000 per year for loan interest payment to bank

Initial cost: −$1 to −$1.2 million in capital expenditures for a water recycling unit

All of these are **point estimates,** that is, *single-value estimates* for cash flow elements of an alternative, except for the last revenue and cost estimates listed above. They provide a **range estimate,** because the persons estimating the revenue and cost do not have enough knowledge or experience with the systems to be more accurate. For the initial chapters, we will utilize point estimates. The use of risk and sensitivity analysis for range estimates is covered in the later chapters of this book.

Once all cash inflows and outflows are estimated (or determined for a completed project), the **net cash flow** for each time period is calculated.

> Net cash flow = cash inflows − cash outflows   [1.5]
> $$\text{NCF} = R - D$$ [1.6]

where NCF is net cash flow, R is receipts, and D is disbursements.

At the beginning of this section, the *timing, size, and direction of cash flows* were mentioned as important. Because cash flows may take place at any time during an interest period, as a matter of convention, all cash flows are assumed to occur at the end of an interest period.

> The end-of-period convention means that all cash inflows and all cash outflows are assumed to take place at the **end of the interest period** in which they actually occur. When several inflows and outflows occur within the same period, the *net* cash flow is assumed to occur at the *end* of the period.

End-of-period convention

Figure 1–4
A typical cash flow time scale for 5 years.

Figure 1–5
Example of positive and negative cash flows.

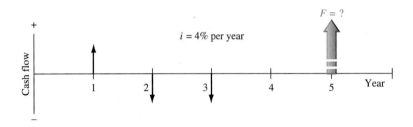

In assuming end-of-period cash flows, it is important to understand that future (F) and uniform annual (A) amounts are located at the end of the interest period, which is not necessarily December 31. If in Example 1.7 the lump-sum deposit took place on July 1, 2011, the withdrawals will take place on July 1 of each succeeding year for 6 years. Remember, end of the period means end of interest period, not end of calendar year.

The **cash flow diagram** is a very important tool in an economic analysis, especially when the cash flow series is complex. It is a graphical representation of cash flows drawn on the y axis with a time scale on the x axis. The diagram includes what is known, what is estimated, and what is needed. That is, once the cash flow diagram is complete, another person should be able to work the problem by looking at the diagram.

Cash flow diagram time $t = 0$ is the present, and $t = 1$ is the end of time period 1. We assume that the periods are in years for now. The time scale of Figure 1–4 is set up for 5 years. Since the end-of-year convention places cash flows at the ends of years, the "1" marks the end of year 1.

While it is not necessary to use an exact scale on the cash flow diagram, you will probably avoid errors if you make a neat diagram to approximate scale for both time and relative cash flow magnitudes.

The direction of the arrows on the diagram is important to differentiate income from outgo. A vertical arrow pointing up indicates a positive cash flow. Conversely, a down-pointing arrow indicates a negative cash flow. **We will use a bold, colored arrow to indicate what is unknown and to be determined.** For example, if a future value F is to be determined in year 5, a wide, colored arrow with $F = ?$ is shown in year 5. The interest rate is also indicated on the diagram. Figure 1–5 illustrates a cash inflow at the end of year 1, equal cash outflows at the end of years 2 and 3, an interest rate of 4% per year, and the unknown future value F after 5 years. The arrow for the unknown value is generally drawn in the opposite direction from the other cash flows; however, the engineering economy computations will determine the actual sign on the F value.

Before the diagramming of cash flows, a perspective or vantage point must be determined so that + or − signs can be assigned and the economic analysis performed correctly. Assume you borrow $8500 from a bank today to purchase an $8000 used car for cash next week, and you plan to spend the remaining $500 on a new paint job for the car two weeks from now. There are several perspectives possible when developing the cash flow diagram—those of the borrower (that's you), the banker, the car dealer, or the paint shop owner. The cash flow signs and amounts for these perspectives are as follows.

Perspective	Activity	Cash flow with Sign, $	Time, week
You	Borrow	+8500	0
	Buy car	−8000	1
	Paint job	−500	2
Banker	Lender	−8500	0
Car dealer	Car sale	+8000	1
Painter	Paint job	+500	2

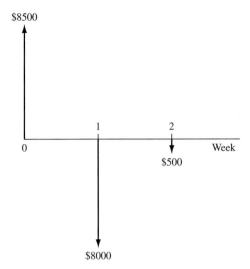

Figure 1–6
Cash flows from perspective of borrower for loan and purchases.

One, and only one, of the perspectives is selected to develop the diagram. For your perspective, all three cash flows are involved and the diagram appears as shown in Figure 1–6 with a time scale of weeks. Applying the end-of-period convention, you have a receipt of +$8500 now (time 0) and cash outflows of −$8000 at the end of week 1, followed by −$500 at the end of week 2.

EXAMPLE 1.9

Each year Exxon-Mobil expends large amounts of funds for mechanical safety features throughout its worldwide operations. Carla Ramos, a lead engineer for Mexico and Central American operations, plans expenditures of $1 million *now* and each of the next 4 years just for the improvement of field-based pressure-release valves. Construct the cash flow diagram to find the equivalent value of these expenditures at the end of year 4, using a cost of capital estimate for safety-related funds of 12% per year.

Solution

Figure 1–7 indicates the uniform and negative cash flow series (expenditures) for five periods, and the unknown F value (positive cash flow equivalent) at exactly the same time as the fifth expenditure. Since the expenditures start immediately, the first $1 million is shown at time 0, not time 1. Therefore, the last negative cash flow occurs at the end of the fourth year, when F also occurs. To make this diagram have a full 5 years on the time scale, the addition of the year −1 completes the diagram. This addition demonstrates that year 0 is the end-of-period point for the year −1.

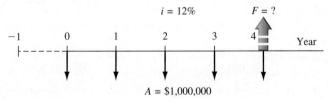

Figure 1–7
Cash flow diagram, Example 1.9.

EXAMPLE 1.10

An electrical engineer wants to deposit an amount P now such that she can withdraw an equal annual amount of $A_1 = \$2000$ per year for the first 5 years, starting 1 year after the deposit, and a different annual withdrawal of $A_2 = \$3000$ per year for the following 3 years. How would the cash flow diagram appear if $i = 8.5\%$ per year?

Solution

The cash flows are shown in Figure 1–8. The negative cash outflow P occurs now. The withdrawals (positive cash inflow) for the A_1 series occur at the end of years 1 through 5, and A_2 occurs in years 6 through 8.

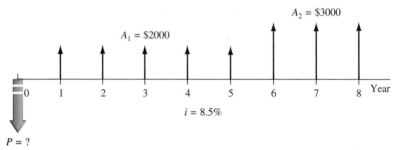

Figure 1–8
Cash flow diagram with two different A series, Example 1.10.

EXAMPLE 1.11

A rental company spent $2500 on a new air compressor 7 years ago. The annual rental income from the compressor has been $750. The $100 spent on maintenance the first year has increased each year by $25. The company plans to sell the compressor at the end of next year for $150. Construct the cash flow diagram from the company's perspective and indicate where the present worth now is located.

Solution

Let now be time $t = 0$. The incomes and costs for years -7 through 1 (next year) are tabulated below with net cash flow computed using Equation [1.5]. The net cash flows (one negative, eight positive) are diagrammed in Figure 1–9. Present worth P is located at year 0.

End of Year	Income	Cost	Net Cash Flow
−7	$ 0	$2500	$−2500
−6	750	100	650
−5	750	125	625
−4	750	150	600
−3	750	175	575
−2	750	200	550
−1	750	225	525
0	750	250	500
1	750 + 150	275	625

Figure 1–9
Cash flow diagram, Example 1.11.

1.7 Economic Equivalence ● ● ●

Economic equivalence is a fundamental concept upon which engineering economy computations are based. Before we delve into the economic aspects, think of the many types of equivalency we may utilize daily by transferring from one scale to another. Some example transfers between scales are as follows:

Length:
12 inches = 1 foot   3 feet = 1 yard   39.370 inches = 1 meter
100 centimeters = 1 meter   1000 meters = 1 kilometer   1 kilometer = 0.621 mile

Pressure:
1 atmosphere = 1 newton/meter^2 = 10^3 pascal = 1 kilopascal

Often equivalency involves two or more scales. Consider the equivalency of a *speed* of 110 kilometers per hour (kph) into miles per minute using conversions between distance and time scales with three-decimal accuracy.

Speed:
1 mile = 1.609 kilometers   1 hour = 60 minutes
110 kph = 68.365 miles per hour (mph)   68.365 mph = 1.139 miles per minute

Four scales—time in minutes, time in hours, length in miles, and length in kilometers—are combined to develop these equivalent statements on speed. Note that throughout these statements, the fundamental relations of 1 mile = 1.609 kilometers and 1 hour = 60 minutes are applied. If a fundamental relation changes, the entire equivalency is in error.

Now we consider economic equivalency.

Economic equivalence is a combination of **interest rate** and **time value of money** to determine the different amounts of money at different points in time that are equal in economic value.

Economic equivalence

As an illustration, if the interest rate is 6% per year, $100 today (present time) is equivalent to $106 one year from today.

$$\text{Amount accrued} = 100 + 100(0.06) = 100(1 + 0.06) = \$106$$

If someone offered you a gift of $100 today or $106 one year from today, it would make no difference which offer you accepted from an economic perspective. In either case you have $106 one year from today. However, the two sums of money are equivalent to each other *only* when the interest rate is 6% per year. At a higher or lower interest rate, $100 today is not equivalent to $106 one year from today.

In addition to future equivalence, we can apply the same logic to determine equivalence for previous years. A total of $100 now is equivalent to $100/1.06 = $94.34 one year ago at an interest rate of 6% per year. From these illustrations, we can state the following: $94.34 last year, $100 now, and $106 one year from now are equivalent at an interest rate of 6% per year. The fact that these sums are equivalent can be verified by computing the two interest rates for 1-year interest periods.

$$\frac{\$6}{\$100} \times 100\% = 6\% \text{ per year}$$

and

$$\frac{\$5.66}{\$94.34} \times 100\% = 6\% \text{ per year}$$

The cash flow diagram in Figure 1–10 indicates the amount of interest needed each year to make these three different amounts equivalent at 6% per year.

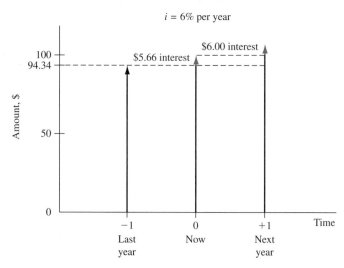

Figure 1–10
Equivalence of money at 6% per year interest.

EXAMPLE 1.12

Manufacturers make backup batteries for computer systems available to Batteries+ dealers through privately owned distributorships. In general, batteries are stored throughout the year, and a 5% cost increase is added each year to cover the inventory carrying charge for the distributorship owner. Assume you own the City Center Batteries+ outlet. Make the calculations necessary to show which of the following statements are true and which are false about battery costs.

(*a*) The amount of $98 now is equivalent to a cost of $105.60 one year from now.
(*b*) A truck battery cost of $200 one year ago is equivalent to $205 now.
(*c*) A $38 cost now is equivalent to $39.90 one year from now.
(*d*) A $3000 cost now is equivalent to $2887.14 one year earlier.
(*e*) The carrying charge accumulated in 1 year on an investment of $20,000 worth of batteries is $1000.

Solution

(*a*) Total amount accrued = 98(1.05) = $102.90 ≠ $105.60; therefore, it is false. Another way to solve this is as follows: Required original cost is 105.60/1.05 = $100.57 ≠ $98.
(*b*) Equivalent cost 1 year ago is 205.00/1.05 = $195.24 ≠ $200; therefore, it is false.
(*c*) The cost 1 year from now is $38(1.05) = $39.90; true.
(*d*) Cost now is 2887.14(1.05) = $3031.50 ≠ $3000; false.
(*e*) The charge is 5% per year interest, or $20,000(0.05) = $1000; true.

Comparison of alternative cash flow series requires the use of equivalence to determine when the series are economically equal or if one is economically preferable to another. The keys to the analysis are the interest rate and the timing of the cash flows. Example 1.13 demonstrates how easy it is to be misled by the size and timing of cash flows.

EXAMPLE 1.13

Howard owns a small electronics repair shop. He wants to borrow $10,000 now and repay it over the next 1 or 2 years. He believes that new diagnostic test equipment will allow him to work on a wider variety of electronic items and increase his annual revenue. Howard received 2-year repayment options from banks A and B.

	Amount to pay, $ per year	
Year	Bank A	Bank B
1	−5,378.05	−5,000.00
2	−5,378.05	−5,775.00
Total paid	−10,756.10	−10,775.00

After reviewing these plans, Howard decided that he wants to repay the $10,000 after only 1 year based on the expected increased revenue. During a family conversation, Howard's brother-in-law offered to lend him the $10,000 now and take $10,600 after exactly 1 year. Now Howard has three options and wonders which one to take. Which one is economically the best?

Solution

The repayment plans for both banks are economically equivalent at the interest rate of 5% per year. (This is determined by using computations that you will learn in Chapter 2.) Therefore, Howard can choose either plan even though the bank B plan requires a slightly larger sum of money over the 2 years.

The brother-in-law repayment plan requires a total of $600 in interest 1 year later plus the principal of $10,000, which makes the interest rate 6% per year. Given the two 5% per year options from the banks, this 6% plan should not be chosen as it is not economically better than the other two. Even though the sum of money repaid is smaller, the timing of the cash flows and the interest rate make it less desirable. The point here is that cash flows themselves, or their sums, cannot be relied upon as the primary basis for an economic decision. The interest rate, timing, and economic equivalence must be considered.

1.8 Simple and Compound Interest ● ● ●

The terms *interest, interest period,* and *interest rate* (introduced in Section 1.4) are useful in calculating equivalent sums of money for one interest period in the past and one period in the future. However, for more than one interest period, the terms *simple interest* and *compound interest* become important.

Simple interest is calculated using the principal only, ignoring any interest accrued in preceding interest periods. The total simple interest over several periods is computed as

$$\text{Simple interest} = (\text{principal})(\text{number of periods})(\text{interest rate}) \qquad [1.7]$$
$$I = Pni$$

where I is the amount of interest earned or paid and the interest rate i is expressed in decimal form.

EXAMPLE 1.14

GreenTree Financing lent an engineering company $100,000 to retrofit an environmentally unfriendly building. The loan is for 3 years at 10% per year simple interest. How much money will the firm repay at the end of 3 years?

Solution

The interest for each of the 3 years is

$$\text{Interest per year} = \$100,000(0.10) = \$10,000$$

Total interest for 3 years from Equation [1.7] is

$$\text{Total interest} = \$100,000(3)(0.10) = \$30,000$$

The amount due after 3 years is

Total due = $100,000 + 30,000 = $130,000

The interest accrued in the first year and in the second year does not earn interest. The interest due each year is $10,000 calculated only on the $100,000 loan principal.

In most financial and economic analyses, we use **compound interest** calculations.

For *compound interest,* the interest accrued for each interest period is calculated on the **principal plus the total amount of interest accumulated in all previous periods.** Thus, compound interest means interest on top of interest.

Compound interest reflects the effect of the time value of money on the interest also. Now the interest for one period is calculated as

Compound interest = (principal + all accrued interest)(interest rate)   **[1.8]**

In mathematical terms, the interest I_t for time period t may be calculated using the relation.

$$I_t = \left(P + \sum_{j=1}^{j=t-1} I_j \right)(i)$$ [1.9]

EXAMPLE 1.15

Assume an engineering company borrows $100,000 at 10% per year compound interest and will pay the principal and all the interest after 3 years. Compute the annual interest and total amount due after 3 years. Graph the interest and total owed for each year, and compare with the previous example that involved simple interest.

Solution
To include compounding of interest, the annual interest and total owed each year are calculated by Equation [1.8].

Interest, year 1:   100,000(0.10) = $10,000
Total due, year 1:   100,000 + 10,000 = $110,000
Interest, year 2:   110,000(0.10) = $11,000
Total due, year 2:   110,000 + 11,000 = $121,000
Interest, year 3:   121,000(0.10) = $12,100
Total due, year 3:   121,000 + 12,100 = $133,100

The repayment plan requires no payment until year 3 when all interest and the principal, a total of $133,100, are due. Figure 1–11 uses a cash flow diagram format to compare end-of-year (*a*) simple and (*b*) compound interest and total amounts owed. The differences due to compounding are clear. An extra $133,100 – 130,000 = $3100 in interest is due for the compounded interest loan.

Note that while simple interest due each year is constant, the compounded interest due grows geometrically. Due to this geometric growth of compound interest, the difference between simple and compound interest accumulation increases rapidly as the time frame increases. For example, if the loan is for 10 years, not 3, the extra paid for compounding interest may be calculated to be $59,374.

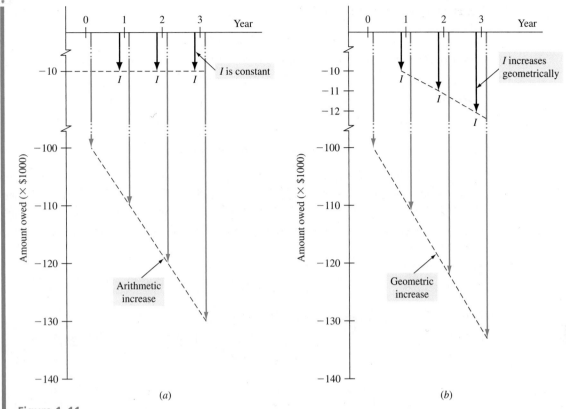

Figure 1–11

Interest I owed and total amount owed for (*a*) simple interest (Example 1.14) and (*b*) compound interest (Example 1.15).

A more efficient way to calculate the total amount due after a number of years in Example 1.15 is to utilize the fact that compound interest increases geometrically. This allows us to skip the year-by-year computation of interest. In this case, the **total amount due at the end of each year** is

Year 1: $\$100,000(1.10)^1 = \$110,000$

Year 2: $\$100,000(1.10)^2 = \$121,000$

Year 3: $\$100,000(1.10)^3 = \$133,100$

This allows future totals owed to be calculated directly without intermediate steps. The general form of the equation is

$$\text{Total due after } n \text{ years} = \text{principal}(1 + \text{interest rate})^{n\ \text{years}} \qquad [1.10]$$
$$= P(1 + i)^n$$

where i is expressed in decimal form. Equation [1.10] was applied above to obtain the $\$133,100$ due after 3 years. This fundamental relation will be used many times in the upcoming chapters.

We can combine the concepts of interest rate, compound interest, and equivalence to demonstrate that different loan repayment plans may be equivalent, but differ substantially in amounts paid from one year to another and in the total repayment amount. This also shows that there are many ways to take into account the time value of money.

EXAMPLE 1.16

Table 1–1 details four different loan repayment plans described below. Each plan repays a $5000 loan in 5 years at 8% per year compound interest.

- **Plan 1: Pay all at end.** No interest or principal is paid until the end of year 5. Interest accumulates each year on the total of principal and all accrued interest.
- **Plan 2: Pay interest annually, principal repaid at end.** The accrued interest is paid each year, and the entire principal is repaid at the end of year 5.
- **Plan 3: Pay interest and portion of principal annually.** The accrued interest and one-fifth of the principal (or $1000) are repaid each year. The outstanding loan balance decreases each year, so the interest (column 2) for each year decreases.
- **Plan 4: Pay equal amount of interest and principal.** Equal payments are made each year with a portion going toward principal repayment and the remainder covering the accrued interest. Since the loan balance decreases at a rate slower than that in plan 3 due to the equal end-of-year payments, the interest decreases, but at a slower rate.

TABLE 1–1	Different Repayment Schedules Over 5 Years for $5000 at 8% Per Year Compound Interest			
(1) End of Year	**(2)** Interest Owed for Year	**(3)** Total Owed at End of Year	**(4)** End-of-Year Payment	**(5)** Total Owed After Payment
Plan 1: Pay All at End				
0				$5000.00
1	$400.00	$5400.00	—	5400.00
2	432.00	5832.00	—	5832.00
3	466.56	6298.56	—	6298.56
4	503.88	6802.44	—	6802.44
5	544.20	7346.64	$−7346.64	
Total			$−7346.64	
Plan 2: Pay Interest Annually; Principal Repaid at End				
0				$5000.00
1	$400.00	$5400.00	$−400.00	5000.00
2	400.00	5400.00	−400.00	5000.00
3	400.00	5400.00	−400.00	5000.00
4	400.00	5400.00	−400.00	5000.00
5	400.00	5400.00	−5400.00	
Total			$−7000.00	
Plan 3: Pay Interest and Portion of Principal Annually				
0				$5000.00
1	$400.00	$5400.00	$−1400.00	4000.00
2	320.00	4320.00	−1320.00	3000.00
3	240.00	3240.00	−1240.00	2000.00
4	160.00	2160.00	−1160.00	1000.00
5	80.00	1080.00	−1080.00	
Total			$−6200.00	
Plan 4: Pay Equal Annual Amount of Interest and Principal				
0				$5000.00
1	$400.00	$5400.00	$−1252.28	4147.72
2	331.82	4479.54	−1252.28	3227.25
3	258.18	3485.43	−1252.28	2233.15
4	178.65	2411.80	−1252.28	1159.52
5	92.76	1252.28	−1252.28	
Total			$−6261.40	

(*a*) Make a statement about the *equivalence* of each plan at 8% compound interest.

(*b*) Develop an 8% per year *simple* interest repayment plan for this loan using the same approach as plan 2. Comment on the total amounts repaid for the two plans.

Solution

(*a*) The amounts of the annual payments are different for each repayment schedule, and the total amounts repaid for most plans are different, even though each repayment plan requires exactly 5 years. The difference in the total amounts repaid can be explained by the time value of money and by the partial repayment of principal prior to year 5.

A loan of $5000 at time 0 made at 8% per year compound interest is equivalent to each of the following:

Plan 1 $7346.64 at the end of year 5

Plan 2 $400 per year for 4 years and $5400 at the end of year 5

Plan 3 Decreasing payments of interest and partial principal in years 1 ($1400) through 5 ($1080)

Plan 4 $1252.28 per year for 5 years

An engineering economy study typically uses plan 4; interest is compounded, and a constant amount is paid each period. This amount covers accrued interest and a partial amount of principal repayment.

(*b*) The repayment schedule for 8% per year simple interest is detailed in Table 1–2. Since the annual accrued interest of $400 is paid each year and the principal of $5000 is repaid in year 5, the schedule is exactly the same as that for 8% per year compound interest, and the total amount repaid is the same at $7000. In this unusual case, simple and compound interest result in the same total repayment amount. Any deviation from this schedule will cause the two plans and amounts to differ.

TABLE 1–2	A 5-Year Repayment Schedule of $5000 at 8% per Year Simple Interest			
End of Year	Interest Owed for Year	Total Owed at End of Year	End-of-Year Payment	Total Owed After Payment
0				$5000
1	$400	$5400	$−400	5000
2	400	5400	−400	5000
3	400	5400	−400	5000
4	400	5400	−400	5000
5	400	5400	−5400	0
Total			$−7000	

1.9 Minimum Attractive Rate of Return ● ● ●

For any investment to be profitable, the investor (corporate or individual) expects to receive more money than the amount of capital invested. In other words, a fair *rate of return,* or *return on investment,* must be realizable. The definition of ROR in Equation [1.4] is used in this discussion, that is, amount earned divided by the principal.

Engineering alternatives are evaluated upon the prognosis that a reasonable ROR can be expected. Therefore, some reasonable rate must be established for the selection criteria (step 4) of the engineering economy study (Figure 1–1).

Minimum Attractive Rate
of Return (MARR)

> The Minimum Attractive Rate of Return (MARR) is a reasonable rate of return established for the evaluation and selection of alternatives. A project is not economically viable unless it is **expected to return at least the MARR.** MARR is also referred to as the *hurdle rate, cutoff rate, benchmark rate,* and *minimum acceptable rate of return.*

Figure 1–12 indicates the relations between different rate of return values. In the United States, the current U.S. Treasury Bill return is sometimes used as the benchmark safe rate. The MARR will always be higher than this, or a similar, safe rate. The MARR is not a rate that is calculated as a ROR. The MARR is established by (financial) managers and is used as a criterion against which an alternative's ROR is measured, when making the accept/reject investment decision.

To develop a foundation-level understanding of how a MARR value is established and used to make investment decisions, we return to the term **capital** introduced in Section 1.1. Although the MARR is used as a criterion to decide on investing in a project, the size of MARR is fundamentally connected to how much it costs to obtain the needed capital funds. It always costs money in the form of interest to raise capital. The interest, expressed as a percentage rate per year, is called the **cost of capital.** As an example on a personal level, if you want to purchase a new widescreen HDTV, but do not have sufficient money (capital), you could obtain a bank loan for, say, a cost of capital of 9% per year and pay for the TV in cash now. Alternatively, you might choose to use your credit card and pay off the balance on a monthly basis. This approach will probably cost you at least 15% per year. Or, you could use funds from your savings account that earns 5% per year and pay cash. This approach means that you also forgo future returns from these funds. The 9%, 15%, and 5% rates are your cost of capital estimates to raise the capital for the system by different methods of capital financing. In analogous ways, corporations estimate the **cost of capital** from different sources to raise funds for engineering projects and other types of projects.

Cost of capital

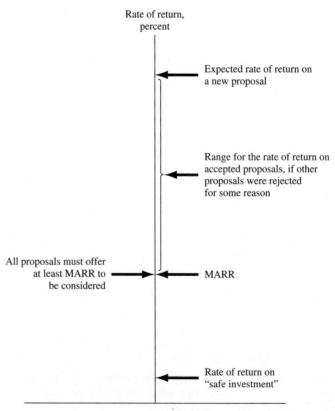

Figure 1–12
Size of MAAR relative to other rate of return values.

In general, capital is developed in two ways—equity financing and debt financing. A combination of these two is very common for most projects. Chapter 10 covers these in greater detail, but a snapshot description follows.

Equity financing   The corporation uses its own funds from cash on hand, stock sales, or retained earnings. Individuals can use their own cash, savings, or investments. In the example above, using money from the 5% savings account is equity financing.

Debt financing   The corporation borrows from outside sources and repays the principal and interest according to some schedule, much like the plans in Table 1–1. Sources of debt capital may be bonds, loans, mortgages, venture capital pools, and many others. Individuals, too, can utilize debt sources, such as the credit card (15% rate) and bank options (9% rate) described above.

Combinations of debt-equity financing mean that a **weighted average cost of capital (WACC)** results. If the HDTV is purchased with 40% credit card money at 15% per year and 60% savings account funds earning 5% per year, the weighted average cost of capital is $0.4(15) + 0.6(5) = 9\%$ per year.

For a corporation, the *established MARR* used as a criterion to accept or reject an investment alternative will usually be *equal to or higher than the WACC* that the corporation must bear to obtain the necessary capital funds. So the inequality

$$\text{ROR} \geq \text{MARR} > \text{WACC} \qquad\qquad [1.11]$$

must be correct for an accepted project. Exceptions may be government-regulated requirements (safety, security, environmental, legal, etc.), economically lucrative ventures expected to lead to other opportunities, etc.

Often there are many alternatives that are expected to yield a ROR that exceeds the MARR as indicated in Figure 1–12, but there may not be sufficient capital available for all, or the project's risk may be estimated as too high to take the investment chance. Therefore, new projects that are undertaken usually have an expected return at least as great as the return on another alternative that is not funded. The expected rate of return on the unfunded project is called the **opportunity cost.**

The opportunity cost is the rate of return of a forgone opportunity caused by the inability to pursue a project. Numerically, it is the **largest rate of return of all the projects not accepted (forgone) due to the lack of capital funds or other resources.** When no specific MARR is established, the de facto MARR is the opportunity cost, i.e., the ROR of the first project not undertaken due to unavailability of capital funds.

Opportunity cost

As an illustration of opportunity cost, refer to Figure 1–12 and assume a MARR of 12% per year. Further, assume that a proposal, call it A, with an expected ROR = 13% is not funded due to a lack of capital. Meanwhile, proposal B has a ROR = 14.5% and is funded from available capital. Since proposal A is not undertaken due to the lack of capital, its estimated ROR of 13% is the *opportunity cost;* that is, the opportunity to make an additional 13% return is forgone.

1.10 Introduction to Spreadsheet Use ● ● ●

The functions on a computer spreadsheet can greatly reduce the amount of hand work for equivalency computations involving *compound interest* and the terms P, F, A, i, and n. The use of a calculator to solve most simple problems is preferred by many students and professors as described in Appendix D. However, as cash flow series become more complex, the spreadsheet offers a good alternative. Microsoft Excel is used throughout this book because it is readily available and easy to use. Appendix A is a primer on using spreadsheets and Excel. The functions used in engineering economy are described there in detail, with explanations of all the parameters. Appendix A also includes a section on spreadsheet layout that is useful when the economic analysis is presented to someone else—a coworker, a boss, or a professor.

A total of seven Excel functions can perform most of the fundamental engineering economy calculations. The functions are great supplemental tools, but they do not replace the understanding of engineering economy relations, assumptions, and techniques. Using the symbols P, F, A, i, and n defined in the previous section, the functions most used in engineering economic analysis are formulated as follows.

> To find the present value P: = **PV**($i\%,n,A,F$)
>
> To find the future value F: = **FV**($i\%,n,A,P$)
>
> To find the equal, periodic value A: = **PMT**($i\%,n,P,F$)
>
> To find the number of periods n: = **NPER**($i\%,A,P,F$)
>
> To find the compound interest rate i: = **RATE**(n,A,P,F)
>
> To find the compound interest rate i: = **IRR**(**first_cell:last_cell**)
>
> To find the present value P of any series: = **NPV**($i\%$, **second_cell:last_cell**) + **first_cell**

If some of the parameters don't apply to a particular problem, they can be omitted and zero is assumed. For readability, spaces can be inserted between parameters within parentheses. If the parameter omitted is an interior one, the comma must be entered. The last two functions require that a series of numbers be entered into contiguous spreadsheet cells, but the first five can be used with no supporting data. In all cases, the function must be preceded by an equals sign (=) in the cell where the answer is to be displayed.

To understand how the spreadsheet functions work, look back at Example 1.6a, where the equivalent annual amount A is unknown, as indicated by $A = ?$. (In Chapter 2, we learn how engineering economy factors calculate A, given P, i, and n.) To find A using a spreadsheet function, simply enter the PMT function = PMT(5%,5,5000). Figure 1–13 is a screen image of a spreadsheet with the PMT function entered into cell B4. The answer ($1154.87) is displayed. The answer may appear in red and in parentheses, or with a minus sign on your screen to indicate a negative amount from the perspective of a reduction in the account balance. The right side of Figure 1–13 presents the solution to Example 1.6b. The future value F is determined by using the FV function. The FV function appears in the formula bar; and many examples throughout this text will include cell tags, as shown here, to indicate the format of important entries.

The following example demonstrates the use of a spreadsheet to develop relations (not built-in functions) to calculate interest and cash flows. Once set up, the spreadsheet can be used to perform sensitivity analysis for estimates that are subject to change. We will illustrate the use of spreadsheets throughout the chapters. (*Note:* The spreadsheet examples may be omitted, if spreadsheets are not used in the course. A solution by hand is included in virtually all examples.)

Figure 1–13
Use of spreadsheet functions PMT and FV, Example 1.6.

EXAMPLE 1.17

A Japan-based architectural firm has asked a United States–based software engineering group to infuse GPS sensing capability via satellite into monitoring software for high-rise structures in order to detect greater than expected horizontal movements. This software could be very beneficial as an advance warning of serious tremors in earthquake-prone areas in Japan and the United States. The inclusion of accurate GPS data is estimated to increase annual revenue over that for the current software system by $200,000 for each of the next 2 years, and by $300,000 for each of years 3 and 4. The planning horizon is only 4 years due to the rapid advances made internationally in building-monitoring software. Develop spreadsheets to answer the questions below.

(a) Determine the total interest and total revenue after 4 years, using a compound rate of return of 8% per year.
(b) Repeat part (a) if estimated revenue increases from $300,000 to $600,000 in years 3 and 4.
(c) Repeat part (a) if inflation is estimated to be 4% per year. This will decrease the *real rate of return* from 8% to 3.85% per year (Chapter 14 shows why).

Solution by Spreadsheet

Refer to Figure 1–14a to d for the solutions. All the spreadsheets contain the same information, but some cell values are altered as required by the question. (Actually, all the questions can be answered on one spreadsheet by changing the numbers. Separate spreadsheets are shown here for explanation purposes only.)

The Excel functions are constructed with reference to the cells, not the values themselves, so that sensitivity analysis can be performed without function changes. This approach treats the value in a cell as a *global variable* for the spreadsheet. For example, the 8% rate in cell B2 will be referenced in all functions as B2, not 8%. Thus, a change in the rate requires only one alteration in the cell B2 entry, not in every relation where 8% is used. See Appendix A for additional information about using cell referencing and building spreadsheet relations.

(a) Figure 1–14a shows the results, and Figure 1–14b presents all spreadsheet relations for estimated interest and revenue (yearly in columns C and E, cumulative in columns D and F). As an illustration, for year 3 the interest I_3 and revenue plus interest R_3 are

$$I_3 = \text{(cumulative revenue through year 2)(rate of return)}$$
$$= \$416{,}000(0.08)$$
$$= \$33{,}280$$

$$R_3 = \text{revenue in year 3} + I_3$$
$$= \$300{,}000 + 33{,}280$$
$$= \$333{,}280$$

The detailed relations shown in Figure 1–14b calculate these values in cells C8 and E8.

Cell C8 relation for I_3: = F7*B2
Cell E8 relation for CF_3: = B8 + C8

The equivalent amount after 4 years is $1,109,022, which is comprised of $1,000,000 in total revenue and $109,022 in interest compounded at 8% per year. The shaded cells in Figure 1–14a and b indicate that the sum of the annual values and the last entry in the cumulative columns must be equal.

(b) To determine the effect of increasing estimated revenue for years 3 and 4 to $600,000, use the same spreadsheet and change the entries in cells B8 and B9 as shown in Figure 1–14c. Total interest increases 22%, or $24,000, from $109,222 to $133,222.

(c) Figure 1–14d shows the effect of changing the original i value from 8% to an inflation-adjusted rate of 3.85% in cell B2 on the first spreadsheet. [Remember to return to the $300,000 revenue estimates for years 3 and 4 after working part (b).] Inflation has now reduced total interest by 53% from $109,222 to $51,247, as shown in cell C10.

	A	B	C	D	E	F
1			Part (a) - Find totals in year 4			
2	i =	8.0%				
3 4	End of Year	Revenue at end of year, $	Interest earned during year, $	Cumulative interest, $	Revenue during year with interest, $	Cumulative revenue with interest, $
5	0					
6	1	200,000	0	0	200,000	200,000
7	2	200,000	16,000	16,000	216,000	416,000
8	3	300,000	33,280	49,280	333,280	749,280
9	4	300,000	59,942	109,222	359,942	1,109,222
10			109,222		1,109,222	

(a) Total interest and revenue for base case, year 4

	A	B	C	D	E	F
1			Part (a) - Find totals in year 4			
2	i =	0.08				
3 4	End of Year	Revenue at end of year, $	Interest earned during year, $	Cumulative interest, $	Revenue during year with interest, $	Cumulative revenue with interest, $
5	0					
6	1	200000	0	=C6	=B6 + C6	=E6
7	2	200000	=F6*$B$1	=C7 + D6	=B7 + C7	=E7 + F6
8	3	300000	=F7*$B$1	=C8 + D7	=B8 + C8	=E8 + F7
9	4	300000	=F8*$B$1	=C9 + D8	=B9 + C9	=E9 + F8
10			=SUM(C6:C9)		=SUM(E6:E9)	

(b) Spreadsheet relations for base case

	A	B	C	D	E	F
1			Part (b) - Find totals in year 4 with increased revenues			
2	i =	8.0%				
3 4	End of Year	Revenue at end of year, $	Interest earned during year, $	Cumulative interest, $	Revenue during year with interest, $	Cumulative revenue with interest, $
5	0					
6	1	200,000	0	0	200,000	200,000
7	2	200,000	16,000	16,000	216,000	416,000
8	3	600,000	33,280	49,280	633,280	1,049,280
9	4	600,000	83,942	133,222	683,942	1,733,222
10			133,222		1,733,222	
11						
12		Revenue changed				
13						

(c) Totals with increased revenue in years 3 and 4

	A	B	C	D	E	F	
1			Part (c) - Find totals in year 4 considering 4% inflation				Rate of return changed
2	i =	3.85%					
3 4	End of Year	Revenue at end of year, $	Interest earned during year, $	Cumulative interest, $	Revenue during year with interest, $	Cumulative revenue with interest, $	
5	0						
6	1	200,000	0	0	200,000	200,000	
7	2	200,000	7,700	7,700	207,700	407,700	
8	3	300,000	15,696	23,396	315,696	723,396	
9	4	300,000	27,851	51,247	327,851	1,051,247	
10			51,247		1,051,247		

(d) Totals with inflation of 4% per year considered

Figure 1–14
Spreadsheet solutions with sensitivity analysis, Example 1.17a to c.

Comment

Later we will learn how to utilize the NPV and FV Excel financial functions to obtain the same answers determined in Figure 1–14, where we developed each basic relation.

When you are working with an Excel spreadsheet, it is possible to display all of the entries and functions on the screen as shown in Figure 1–14b by simultaneously touching the <Ctrl> and <`> keys, which may be in the upper left of the keyboard on the key with <~>.

CHAPTER SUMMARY

Engineering economy is the application of economic factors and criteria to evaluate alternatives, considering the time value of money. The engineering economy study involves computing a specific economic measure of worth for estimated cash flows over a specific period of time.

The concept of *equivalence* helps in understanding how different sums of money at different times are equal in economic terms. The differences between simple interest (based on principal only) and compound interest (based on principal and interest upon interest) have been described in formulas, tables, and graphs. This power of compounding is very noticeable, especially over extended periods of time, and for larger sums of money.

The MARR is a reasonable rate of return established as a hurdle rate to determine if an alternative is economically viable. The MARR is always higher than the return from a safe investment and the cost to acquire needed capital.

Also, we learned a lot about cash flows:

End-of-year convention for cash flow location

Net cash flow computation

Different perspectives in determining the cash flow sign

Construction of a cash flow diagram

Difficulties with estimating future cash flows accurately

PROBLEMS

Basic Concepts

1.1   List the four essential elements involved in decision making in engineering economic analysis.

1.2   What is meant by (*a*) limited capital funds and (*b*) sensitivity analysis?

1.3   List three measures of worth that are used in engineering economic analysis.

1.4   Identify the following factors as either economic (tangible) or noneconomic (intangible): first cost, leadership, taxes, salvage value, morale, dependability, inflation, profit, acceptance, ethics, interest rate.

Ethics

1.5   Stefanie is a design engineer with an international railroad locomotive manufacturing company in Illinois. Management wants to return some of the engineering design work to the United States rather than export all of it to India, where the primary design work has been accomplished for the last decade. This transfer will employ more people locally and could improve the economic conditions for families in and around Illinois.

Stefanie and her design team were selected as a test case to determine the quality and speed of the design work they could demonstrate on a more fuel-efficient diesel locomotive. Neither she nor any of her team members have done such a significant design job, because their jobs had previously entailed only the interface with the subcontracted engineers in India. One of her team members had a great design idea on a key element that will improve fuel efficiency by approximately 15%. She told Stefanie it came from one of the Indian-generated documents, but that it would probably be okay for the team to use it and remain silent as to its origin, since it was quite clear the U.S. management was about to cancel the foreign contract. Although reluctant at first, Stefanie did go forward with a design that included the efficiency improvement, and no mention of the origin of the idea was made at the time of the oral presentation or documentation delivery. As a result, the Indian contract was canceled and full design responsibility was transferred to Stefanie's group.

Consult the NSPE Code of Ethics for Engineers (Appendix C) and identify sections that are points of concern about Stefanie's decisions and actions.

1.6   Consider the common moral precept that stealing is wrong. Hector is with a group of friends in a local supermarket. One of Hector's buddies takes a high-energy drink from a six-pack on the shelf, opens it, drinks it, and returns the empty can to the package, with no intention of paying for it. He then invites the others to do the same, saying, "It's only one drink. Others do it all the time." All the others, except Hector, have now consumed a drink of their choice. Personally, Hector believes this is a form of stealing. State three actions that Hector can take, and evaluate them from the personal moral perspective.

1.7 While going to work this morning off site from his office, an engineer accidently ran a stop sign and was in a car accident that resulted in the death of a 5-year-old child. He has a strong belief in the universal moral that it is wrong to do serious harm to another person. Explain the conflict that can arise for him between the universal moral and his personal moral about doing serious harm, given the accident was deemed his fault.

1.8 Claude is a fourth-year engineering university student who has just been informed by his instructor that he made a very low grade on his Spanish language final test for the year. Although he had a passing score prior to the final, his final grade was so low that he has now flunked the entire year and will likely have to extend his graduation another semester or two.

Throughout the year, Claude, who hated the course and his instructor, has copied homework, cheated on tests, and never seriously studied for anything in the course. He did realize during the semester that he was doing something that even he considered wrong morally and ethically. He knew he had done badly on the final. The classroom was reconfigured for the final exam in a way that he could not get any answers from classmates, and cell phones were collected prior to the exam, thus removing texting possibilities to friends outside the classroom who might help him on the final exam. Claude is now face to face with the instructor in her office. The question to Claude is, "What have you been doing throughout this year to make passing scores repeatedly, but demonstrate such a poor command of Spanish on the final exam?"

From an ethical viewpoint, what options does Claude have in his answer to this question? Also, discuss some of the possible effects that this experience may have upon Claude's future actions and moral dilemmas.

Interest Rate and Rate of Return

1.9 RKI Instruments borrowed $3,500,000 from a private equity firm for expansion of its manufacturing facility for making carbon monoxide monitors/controllers. The company repaid the loan after 1 year with a single payment of $3,885,000. What was the interest rate on the loan?

1.10 Emerson Processing borrowed $900,000 for installing energy-efficient lighting and safety equipment in its La Grange manufacturing facility. The terms of the loan were such that the company could pay *interest only* at the end of each year for up to 5 years, after which the company would have to pay the entire amount due. If the interest rate on the loan was 12% per year and the company paid only the interest for 4 years, determine the following:

(a) The amount of each of the four interest payments
(b) The amount of the final payment at the end of year 5

1.11 Which of the following 1-year investments has the highest rate of return?

(a) $12,500 that yields $1125 in interest,
(b) $56,000 that yields $6160 in interest, or
(c) $95,000 that yields $7600 in interest.

1.12 A new engineering graduate who started a consulting business borrowed money for 1 year to furnish the office. The amount of the loan was $23,800, and it had an interest rate of 10% per year. However, because the new graduate had not built up a credit history, the bank made him buy loan-default insurance that cost 5% of the loan amount. In addition, the bank charged a loan setup fee of $300. What was the effective interest rate the engineer paid for the loan?

1.13 When the inflation rate is expected to be 8% per year, what is the market interest rate likely to be?

Terms and Symbols

1.14 The symbol P represents an amount of money at a time designated as present. The following symbols also represent a present amount of money and require similar calculations. Explain what each symbol stands for: PW, PV, NPV, DCF, and CC.

1.15 Identify the four engineering economy symbols and their values from the following problem statement. Use a question mark with the symbol whose value is to be determined.

Thompson Mechanical Products is planning to set aside $150,000 now for possibly replacing its large synchronous refiner motors whenever it becomes necessary. If the replacement is not needed for 7 years, how much will the company have in its investment set-aside account, provided it achieves a rate of return of 11% per year?

1.16 Identify the four engineering economy symbols and their values from the following problem statement.

Use a question mark with the symbol whose value is to be determined.

Atlas Long-Haul Transportation is considering installing Valutemp temperature loggers in all of its refrigerated trucks for monitoring temperatures during transit. If the systems will reduce insurance claims by $100,000 two years from now, how much should the company be willing to spend now, if it uses an interest rate of 12% per year?

1.17 Identify the four engineering economy symbols and their values from the following problem statement. Use a question mark with the symbol whose value is to be determined.

A green algae, *Chlamydomonas reinhardtii*, can produce hydrogen when temporarily deprived of sulfur for up to 2 days at a time. A small company needs to purchase equipment costing $3.4 million to commercialize the process. If the company wants to earn a rate of return of 10% per year and recover its investment in 8 years, what must be the net value of the hydrogen produced each year?

1.18 Identify the four engineering economy symbols and their values from the following problem statement. Use a question mark with the symbol whose value is to be determined.

Vision Technologies, Inc., is a small company that uses ultra-wideband technology to develop devices that can detect objects (including people) inside of buildings, behind walls, or below ground. The company expects to spend $100,000 per year for labor and $125,000 per year for supplies before a product can be marketed. At an interest rate of 15% per year, what is the total equivalent future amount of the company's expenses at the end of 3 years?

Cash Flows

1.19 What is meant by end-of-period convention?

1.20 Identify the following as cash inflows or outflows to commercial air carriers: fuel cost, pension plan contributions, fares, maintenance, freight revenue, cargo revenue, extra-bag charges, water and sodas, advertising, landing fees, seat preference fees.

1.21 Many credit unions use semiannual interest periods to pay interest on customer savings accounts. For a credit union that uses June 30 and December 31 as its semiannual interest periods, determine the end-of-period amounts that will be recorded for the deposits shown in the table.

Month	Deposit, $
Jan	50
Feb	70
Mar	—
Apr	120
May	20
June	—
July	150
Aug	90
Sept	—
Oct	—
Nov	40
Dec	110

1.22 For a company that uses a year as its interest period, determine the *net cash flow* that will be recorded at the *end of the year* from the cash flows shown.

Month	Receipts, $1000	Disbursements, $1000
Jan	500	300
Feb	800	500
Mar	200	400
Apr	120	400
May	600	500
June	900	600
July	800	300
Aug	700	300
Sept	900	500
Oct	500	400
Nov	400	400
Dec	1800	700

1.23 Construct a cash flow diagram for the following cash flows: $25,000 outflow at time 0, $9000 per year inflow in years 1 through 5 at an interest rate of 10% per year, and an unknown future amount in year 5.

1.24 Construct a cash flow diagram to find the present worth in year 0 at an interest rate of 15% per year for the following situation.

Year	Cash Flow, $
0	−19,000
1–4	+8,100

1.25 Construct a cash flow diagram that represents the amount of money that will be accumulated in 15 years from an investment of $40,000 now at an interest rate of 8% per year.

Equivalence

1.26 At an interest rate of 15% per year, an investment of $100,000 one year ago is equivalent to how much now?

1.27   During a recession, the price of goods and services goes down because of low demand. A company that makes Ethernet adapters is planning to expand its production facility at a cost of $1,000,000 one year from now. However, a contractor who needs work has offered to do the job for $790,000 if the company will do the expansion now instead of 1 year from now. If the interest rate is 15% per year, how much of a discount is the company getting?

1.28   As a principal in the consulting firm where you have worked for 20 years, you have accumulated 5000 shares of company stock. One year ago, each share of stock was worth $40. The company has offered to buy back your shares for $225,000. At what interest rate would the firm's offer be equivalent to the worth of the stock last year?

1.29   A design/build engineering company that usually gives year-end bonuses in the amount of $8000 to each of its engineers is having cash flow problems. The company said that although it could not give bonuses this year, it would give each engineer two bonuses next year, the regular one of $8000 plus an amount equivalent to the $8000 that each engineer should have gotten this year. If the interest rate is 8% per year, what will be the total amount of bonus money the engineers should get next year?

1.30   University tuition and fees can be paid by using one of two plans.

       Early-bird: Pay total amount due *1 year in advance* and get a 10% discount.

       On-time: Pay total amount due when classes start.

       The cost of tuition and fees is $10,000 per year.

       (*a*) How much is paid in the early-bird plan?
       (*b*) What is the equivalent amount of the savings compared to the on-time payment at the time that the on-time payment is made?

Simple and Compound Interest

1.31   If a company sets aside $1,000,000 now into a contingency fund, how much will the company have in 2 years, if it does not use any of the money and the account grows at a rate of 10% per year?

1.32   Iselt Welding has extra funds to invest for future capital expansion. If the selected investment pays simple interest, what interest rate would be required for the amount to grow from $60,000 to $90,000 in 5 years?

1.33   To finance a new product line, a company that makes high-temperature ball bearings borrowed $1.8 million at 10% per year interest. If the com-

pany repaid the loan in a lump sum amount after 2 years, what was (*a*) the amount of the payment and (*b*) the amount of interest?

1.34   Because market interest rates were near all-time lows at 4% per year, a hand tool company decided to call (i.e., pay off) the high-interest bonds that it issued 3 years ago. If the interest rate on the bonds was 9% per year, how much does the company have to pay the bond holders? The face value (principal) of the bonds is $6,000,000.

1.35   A solid waste disposal company borrowed money at 10% per year interest to purchase new haulers and other equipment needed at the company-owned landfill site. If the company got the loan 2 years ago and paid it off with a single payment of $4,600,000, what was the principal amount *P* of the loan?

1.36   If interest is compounded at 20% per year, how long will it take for $50,000 to accumulate to $86,400?

1.37   To make CDs look more attractive than they really are, some banks advertise that their rates are higher than their competitors' rates; however, the fine print says that the rate is a simple interest rate. If a person deposits $10,000 at 10% per year simple interest, what compound interest rate would yield the same amount of money in 3 years?

MARR and Opportunity Cost

1.38   Give three other names for minimum attractive rate of return.

1.39   Identify the following as either equity or debt financing: bonds, stock sales, retained earnings, venture capital, short-term loan, capital advance from friend, cash on hand, credit card, home equity loan.

1.40   What is the weighted average cost of capital for a corporation that finances an expansion project using 30% retained earnings and 70% venture capital? Assume the interest rates are 8% for the equity financing and 13% for the debt financing.

1.41   Managers from different departments in Zenith Trading, a large multinational corporation, have offered six projects for consideration by the corporate office. A staff member for the chief financial officer used key words to identify the projects and then listed them in order of projected rate of return as shown below. If the company wants to grow rapidly through high leverage and uses only 10% equity financing that has a cost of equity capital of 9% and 90% debt financing with a cost of debt capital of 16%, which projects should the company undertake?

Project ID	Projected ROR, % per year
Inventory	30
Technology	28.4
Warehouse	19
Products	13.1
Energy	9.6
Shipping	8.2

Spreadsheet Functions

1.42 State the purpose for each of the following built-in spreadsheet functions.
   (a) PV($i\%,n,A,F$)
   (b) FV($i\%,n,A,P$)
   (c) RATE(n,A,P,F)
   (d) IRR(first_cell:last_cell)
   (e) PMT($i\%,n,P,F$)
   (f) NPER($i\%,A,P,F$)

1.43 What are the values of the engineering economy symbols P, F, A, i, and n in the following functions? Use a question mark for the symbol that is to be determined.
   (a) NPER(8%,−1500,8000,2000)
   (b) FV(7%,102000,−9000)
   (c) RATE(10,1000,−12000,2000)
   (d) PMT(11%,20,,14000)
   (e) PV(8%,15,−1000,800)

1.44 Write the engineering economy symbol that corresponds to each of the following spreadsheet functions.
   (a) PMT   (b) FV   (c) NPER   (d) PV   (e) IRR

1.45 In a built-in spreadsheet function, if a certain parameter is not present, (a) under what circumstances can it be left blank and (b) when must a comma be entered in its place?

1.46 Sheryl and Marcelly both invest $1000 at 10% per year for 4 years. Sheryl receives simple interest and Marcelly gets compound interest. Use a spreadsheet and cell reference formats to develop relations that show a total of $64 more interest for Marcelly at the end of the 4 years. Assume no withdrawals or further deposits are made during the 4 years.

ADDITIONAL PROBLEMS AND FE EXAM REVIEW QUESTIONS

1.47 The concept that different sums of money at different points in time can be said to be equal to each other is known as:
   (a) Evaluation criterion
   (b) Equivalence
   (c) Cash flow
   (d) Intangible factors

1.48 The evaluation criterion that is usually used in an economic analysis is:
   (a) Time to completion
   (b) Technical feasibility
   (c) Sustainability
   (d) Financial units (dollars or other currency)

1.49 All of the following are examples of cash outflows, *except*:
   (a) Asset salvage value
   (b) Income taxes
   (c) Operating cost of asset
   (d) First cost of asset

1.50 In most engineering economy studies, the best alternative is the one that:
   (a) Will last the longest time
   (b) Is most politically correct
   (c) Is easiest to implement
   (d) Has the lowest cost

1.51 The following annual maintenance and operation (M&O) costs for a piece of equipment were collected over a 5-year period: $12,300, $8900, $9200, $11,000, and $12,100. The average is $10,700. In conducting a sensitivity analysis, the most reasonable range of costs to use (i.e., percent from the average) is:
   (a) ±5%   (b) ±11%   (c) ±17%   (d) ±25%

1.52 At an interest rate of 10% per year, the equivalent amount of $10,000 *one year ago* is closest to:
   (a) $8264   (b) $9091   (c) $11,000   (d) $12,000

1.53 Assume that you and your best friend each have $1000 to invest. You invest your money in a fund that pays 10% per year *compound* interest. Your friend invests her money at a bank that pays 10% per year *simple* interest. At the end of 1 year, the difference in the total amount for each of you is:
   (a) You have $10 more than she does
   (b) You have $100 more than she does
   (c) You both have the same amount of money
   (d) She has $10 more than you do

1.54 The time it would take for a given sum of money to double at 4% per year *simple* interest is closest to:
   (a) 30 years   (b) 25 years
   (c) 20 years   (d) 10 years

1.55 All of the following are examples of equity financing, *except*:
  (*a*) Mortgage
  (*b*) Money from savings
  (*c*) Cash on hand
  (*d*) Retained earnings

1.56 To finance a new project costing $30 million, a company borrowed $21 million at 16% per year interest and used retained earnings valued at 12% per year for the remainder of the investment. The company's weighted average cost of capital for the project was closest to:
  (*a*) 12.5% (*b*) 13.6% (*c*) 14.8% (*d*) 15.6%

CASE STUDY

RENEWABLE ENERGY SOURCES FOR ELECTRICITY GENERATION

Background

Pedernales Electric Cooperative (PEC) is the largest member-owned electric co-op in the United States with over 232,000 meters in 12 Central Texas counties. PEC has a capacity of approximately 1300 MW (megawatts) of power, of which 277 MW, or about 21%, is from renewable sources. The latest addition is 60 MW of power from a wind farm in south Texas close to the city of Corpus Christi. A constant question is how much of PEC's generation capacity should be from renewable sources, especially given the environmental issues with coal-generated electricity and the rising costs of hydrocarbon fuels.

Wind and nuclear sources are the current consideration for the PEC leadership as Texas is increasing its generation by nuclear power and the state is the national leader in wind farm–produced electricity.

Consider yourself a member of the board of directors of PEC. You are an engineer who has been newly elected by the PEC membership to serve a 3-year term as a director-at-large. As such, you do not represent a specific district within the entire service area; all other directors do represent a specific district. You have many questions about the operations of PEC, plus you are interested in the economic and societal benefits of pursuing more renewable source generation capacity.

Information

Here are some data that you have obtained. The information is sketchy, as this point, and the numbers are very approximate. Electricity generation cost estimates are national in scope, not PEC-specific, and are provided in cents per kilowatt-hour (¢/kWh).

Fuel Source	Generation Cost, ¢/kWh	
	Likely Range	Reasonable Average
Coal	4 to 9	7.4
Natural gas	4 to 10.5	8.6
Wind	4.8 to 9.1	8.2
Solar	4.5 to 15.5	8.8

National average cost of electricity to residential customers: 11¢/kWh

PEC average cost to residential customers: 10.27 ¢/kWh (from primary sources) and 10.92 ¢/kWh (renewable sources)

Expected life of a generation facility: 20 to 40 years (it is likely closer to 20 than 40)

Time to construct a facility: 2 to 5 years

Capital cost to build a generation facility: $900 to $1500 per kW

You have also learned that the PEC staff uses the well-recognized *levelized energy cost* (LEC) method to determine the price of electricity that must be charged to customers to break even. The formula takes into account the capital cost of the generation facilities, the cost of capital of borrowed money, annual maintenance and operation (M&O) costs, and the expected life of the facility. The LEC formula, expressed in dollars per kWh for ($t = 1, 2, \ldots, n$), is

$$\text{LEC} = \frac{\sum\limits_{t=1}^{t=n} \dfrac{P_t + A_t + C_t}{(1 + i)^t}}{\sum\limits_{t=1}^{t=n} \dfrac{E_t}{(1 + i)^t}}$$

where P_t = capital investments made in year t
  A_t = annual maintenance and operating (M&O) costs for year t
  C_t = fuel costs for year t
  E_t = amount of electricity generated in year t
  n = expected life of facility
  i = discount rate (cost of capital)

Case Study Exercises

1. If you wanted to know more about the new arrangement with the wind farm in south Texas for the additional 60 MW per year, what types of questions would you ask of a staff member in your first meeting with him or her?

2. Much of the current generation capacity of PEC facilities utilizes coal and natural gas as the primary fuel source. What about the ethical aspects of the government's allowance for these plants to continue polluting the atmosphere with the emissions that may cause health problems for citizens and further the effects of global warming? What types of regulations, if any, should be developed for PEC (and other generators) to follow in the future?

3. You developed an interest in the LEC relation and the publicized cost of electricity of 10.27¢/kWh for this year. You wonder if the addition of 60 MW of wind-sourced electricity will make any difference in the LEC value for this next year. You did learn the following:

This is year $t = 11$ for LEC computation purposes

$n = 25$ years

$i = 5\%$ per year

$E_{11} = 5.052$ billion kWh

LEC last year was 10.22 ¢/kWh (last year's breakeven cost to customers)

From these sketchy data, can you determine the value of unknowns in the LEC relation for this year? Is it possible to determine if the wind farm addition of 60 MW makes any difference in the electricity rate charged to customers? If not, what additional information is necessary to determine the LEC with the wind source included?

CASE STUDY

REFRIGERATOR SHELLS

Background

Large refrigerator manufacturers such as Whirlpool, General Electric, Frigidaire, and others may subcontract the molding of their plastic liners and door panels. One prime national subcontractor is Innovations Plastics. Because of improvements in mechanical properties, the molded plastic can sustain increased vertical and horizontal loading, thus significantly reducing the need for attached metal anchors for some shelving. However, improved molding equipment is needed to enter this market now. The company president wants a recommendation on whether Innovations should offer the new technology to the major manufacturers and an estimate of the necessary capital investment to enter this market.

You work as an engineer for Innovations. At this stage, you are not expected to perform a complete engineering economic analysis, for not enough information is available. You are asked to formulate reasonable alternatives, determine what data and estimates are needed for each one, and ascertain what criteria (economic and noneconomic) should be utilized to make the final decision.

Information

Some information useful at this time is as follows:

- The technology and equipment are expected to last about 10 years before new methods are developed.
- Inflation and income taxes will not be considered in the analysis.
- The expected returns on capital investment used for the last three new technology projects were compound rates of 15%, 5%, and 18%. The 5% rate was the criterion for enhancing an employee-safety system on an existing chemical-mixing process.
- Equity capital financing beyond $5 million is not possible. The amount of debt financing and its cost are unknown.
- Annual operating costs have been averaging 8% of first cost for major equipment.
- Increased annual training costs and salary requirements for handling the new plastics and operating new equipment can range from $800,000 to $1.2 million.

There are two manufacturers working on the new-generation equipment. You label these options as alternatives A and B.

Case Study Exercises

1. Use the first four steps of the decision-making process to generally describe the alternatives and identify what economic-related estimates you will need to complete an engineering economy analysis for the president.
2. Identify any noneconomic factors and criteria to be considered in making the alternative selection.
3. During your inquiries about alternative B from its manufacturer, you learn that this company has already produced a prototype molding machine and has sold it to a company in Germany for $3 million (U.S. dollars). Upon inquiry, you further discover that the German company already has unused capacity on the equipment for manufacturing plastic shells. The company is willing to sell time on the equipment to Innovations immediately to produce its own shells for U.S. delivery. This could allow immediate market entry into the United States. Consider this as alternative C, and develop the estimates necessary to evaluate C at the same time as alternatives A and B.

CHAPTER 2

Factors: How Time and Interest Affect Money

LEARNING OUTCOMES

Purpose: Derive and use the engineering economy factors to account for the time value of money.

SECTION	TOPIC	LEARNING OUTCOME
2.1	*F/P* and *P/F* factors	• Derive and use factors for single amounts—compound amount (*F/P*) and present worth (*P/F*) factor.
2.2	*P/A* and *A/P* factors	• Derive and use factors for uniform series—present worth (*P/A*) and capital recovery (*A/P*) factors.
2.3	*F/A* and *A/F* factors	• Derive and use factors for uniform series—compound amount (*F/A*) and sinking fund (*A/F*) factors.
2.4	Factor values	• Use linear interpolation in factor tables or spreadsheet functions to determine factor values.
2.5	Arithmetic gradient	• Use the present worth (*P/G*) and uniform annual series (*A/G*) factors for arithmetic gradients.
2.6	Geometric gradient	• Use the geometric gradient series factor (*P/A,g*) to find present worth.
2.7	Find *i* or *n*	• Use equivalence relations to determine *i* (interest rate or rate of return) or *n* for a cash flow series.

T he cash flow is fundamental to every economic study. Cash flows occur in many configurations and amounts—isolated single values, series that are uniform, and series that increase or decrease by constant amounts or constant percentages. This chapter develops derivations for all the commonly used **engineering economy factors** that take the time value of money into account.

The application of factors is illustrated using their mathematical forms and a standard notation format. Spreadsheet functions are used in order to rapidly work with cash flow series and to perform sensitivity analysis.

If the derivation and use of factors are not covered in the course, alternate ways to perform time value of money calculations are summarized in Appendix D.

(PE)

The Cement Factory Case: Votorantim Cimentos North America, Inc., is a subsidiary of a Brazil-based company that recently announced plans to develop a new cement factory in Houston County in the state of Georgia. The plant will be called Houston American Cement, or HAC. The location is ideal for cement making because of the large deposit of limestone in the area.

The plant investment, expected to amount to $200 million, has been planned for 2012; however, it is currently delayed due to the economic downturn in construction. When the plant is completed and operating at full capacity,

based upon the projected needs and cost per metric ton, it is possible that the plant could generate as much as $50,000,000 annually in revenue. All analysis will use a planning horizon of 5 years commencing when the plant begins operation.

This case is used in the following topics (and sections) of this chapter:

Single-amount factors (2.1)

Uniform series factors (2.2 and 2.3)

Arithmetic gradient factors (2.5)

Geometric gradient factors (2.6)

Determining unknown n values (2.7)

2.1 Single-Amount Factors (F/P and P/F) ● ● ●

The **most fundamental factor in engineering economy** is the one that determines the amount of money F accumulated after n years (or periods) from a *single* present worth P, with interest compounded one time per year (or period). Recall that compound interest refers to interest paid on top of interest. Therefore, if an amount P is invested at time $t = 0$, the amount F_1 accumulated 1 year hence at an interest rate of i percent per year will be

$$F_1 = P + Pi$$

$$= P(1 + i)$$

where the interest rate is expressed in decimal form. At the end of the second year, the amount accumulated F_2 is the amount after year 1 plus the interest from the end of year 1 to the end of year 2 on the entire F_1.

$$F_2 = F_1 + F_1 i$$

$$= P(1 + i) + P(1 + i)i \qquad [2.1]$$

The amount F_2 can be expressed as

$$F_2 = P(1 + i + i + i^2)$$

$$= P(1 + 2i + i^2)$$

$$= P(1 + i)^2$$

Similarly, the amount of money accumulated at the end of year 3, using Equation [2.1], will be

$$F_3 = F_2 + F_2 i$$

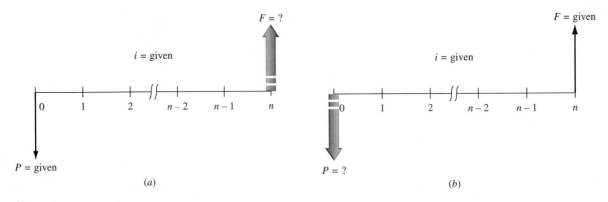

Figure 2–1
Cash flow diagrams for single-payment factors: (*a*) find *F*, given *P*, and (*b*) find *P*, given *F*.

Substituting $P(1 + i)^2$ for F_2 and simplifying, we get

$$F_3 = P(1 + i)^3$$

From the preceding values, it is evident by mathematical induction that the formula can be generalized for *n* years. To find *F*, given *P*,

$$F = P(1 + i)^n \qquad\qquad [2.2]$$

The factor $(1 + i)^n$ is called the *single-payment compound amount factor* (SPCAF), but it is usually referred to as the **F/P factor**. This is the conversion factor that, when multiplied by *P*, yields the future amount *F* of an initial amount *P* after *n* years at interest rate *i*. The cash flow diagram is seen in Figure 2–1*a*.

Reverse the situation to **determine the P value for a stated amount F** that occurs *n* periods in the future. Simply solve Equation [2.2] for *P*.

$$P = F\left[\frac{1}{(1 + i)^n}\right] = F(1 + i)^{-n} \qquad\qquad [2.3]$$

The expression $(1 + i)^{-n}$ is known as the *single-payment present worth factor* (SPPWF), or the **P/F factor**. This expression determines the present worth *P* of a given future amount *F* after *n* years at interest rate *i*. The cash flow diagram is shown in Figure 2–1*b*.

Note that the two factors derived here are for *single payments;* that is, they are used to find the present or future amount when only one payment or receipt is involved.

> A standard notation has been adopted for all factors. The notation includes two cash flow symbols, the interest rate, and the number of periods. It is always in the general form (*X/Y,i,n*). The letter *X* represents what is sought, while the letter *Y* represents what is given. For example, *F/P* means *find F when given P*. The *i* is the interest rate in percent, and *n* represents the number of periods involved.

Using this notation, (*F/P*,6%,20) represents the factor that is used to calculate the future amount *F* accumulated in 20 periods if the interest rate is 6% per period. The *P* is given. The standard notation, simpler to use than formulas and factor names, will be used hereafter.

Table 2–1 summarizes the standard notation and equations for the *F/P* and *P/F* factors. This information is also included inside the front cover.

TABLE 2–1	*F/P* and *P/F* Factors: Notation and Equations				
Factor			**Standard Notation**	**Equation**	**Excel**
Notation	**Name**	**Find/Given**	**Equation**	**with Factor Formula**	**Function**
(*F/P,i,n*)	Single-payment compound amount	*F/P*	$F = P(F/P,i,n)$	$F = P(1 + i)^n$	$= FV(i\%,n,,P)$
(*P/F,i,n*)	Single-payment present worth	*P/F*	$P = F(P/F,i,n)$	$P = F(1 + i)^{-n}$	$= PV(i\%,n,,F)$

To simplify routine engineering economy calculations, tables of factor values have been prepared for interest rates from 0.25% to 50% and time periods from 1 to large n values, depending on the i value. These tables, found at the rear of the book, have a colored edge for easy identification. They are arranged with factors across the top and the number of periods n down the left side. The word *discrete* in the title of each table emphasizes that these tables utilize the end-of-period convention and that interest is compounded once each interest period. For a given factor, interest rate, and time, the correct factor value is found at the intersection of the factor name and n. For example, the value of the factor $(P/F,5\%,10)$ is found in the P/F column of Table 10 at period 10 as 0.6139. This value is determined by using Equation [2.3].

$$(P/F,5\%,10) = \frac{1}{(1+i)^n}$$

$$= \frac{1}{(1.05)^{10}}$$

$$= \frac{1}{1.6289} = 0.6139$$

$$P = P_0 + A\,(P/A, i, n)$$

For **spreadsheets**, a future value F is calculated by the FV function using the format

$$= FV(i\%,n,,P) \qquad [2.4]$$

A present amount P is determined using the PV function with the format

$$= PV(i\%,n,,F) \qquad [2.5]$$

These functions are included in Table 2–1. Refer to Appendix A or Excel online help for more information on the use of FV and PV functions.

EXAMPLE 2.1

Sandy, a manufacturing engineer, just received a year-end bonus of $10,000 that will be invested immediately. With the expectation of earning at the rate of 8% per year, Sandy hopes to take the entire amount out in exactly 20 years to pay for a family vacation when the oldest daughter is due to graduate from college. Find the amount of funds that will be available in 20 years by using (*a*) hand solution by applying the factor formula and tabulated value and (*b*) a spreadsheet function.

Solution

The cash flow diagram is the same as Figure 2–1*a*. The symbols and values are

$$P = \$10,000 \qquad F = ? \qquad i = 8\% \text{ per year} \qquad n = 20 \text{ years}$$

(*a*) *Factor formula:* Apply Equation [2.2] to find the future value F. Rounding to four decimals, we have

$$F = P(1+i)^n = 10,000(1.08)^{20} = 10,000(4.6610)$$
$$= \$46,610$$

Standard notation and tabulated value: Notation for the F/P factor is $(F/P,i\%,n)$.

$$F = P(F/P,8\%,20) = 10,000(4.6610)$$
$$= \$46,610$$

Table 13 provides the tabulated value. Round-off errors can cause a slight difference in the final answer between these two methods.

(*b*) *Spreadsheet:* Use the FV function to find the amount 20 years in the future. The format is that shown in Equation [2.4]; the numerical entry is $= \text{FV}(8\%,20,,10000)$. The spreadsheet will appear similar to that in the right side of Figure 1–13, with the answer ($46,609.57) displayed. (You should try it on your own computer now.) The FV function has performed the computation in part (*a*) and displayed the result.

The equivalency statement is: If Sandy invests $10,000 now and earns 8% per year every year for 20 years, $46,610 will be available for the family vacation.

EXAMPLE 2.2 The Cement Factory Case

As discussed in the introduction to this chapter, the Houston American Cement factory will require an investment of $200 million to construct. Delays beyond the anticipated implementation year of 2012 will require additional money to construct the factory. Assuming that the cost of money is 10% per year, compound interest, use both **tabulated factor values** and **spreadsheet functions** to determine the following for the board of directors of the Brazilian company that plans to develop the plant.

(*a*) The equivalent investment needed if the plant is built in 2015.
(*b*) The equivalent investment needed had the plant been constructed in the year 2008.

Solution
Figure 2–2 is a cash flow diagram showing the expected investment of $200 million ($200 M) in 2012, which we will identify as time $t = 0$. The required investments 3 years in the future and 4 years in the past are indicated by $F_3 = ?$ and $P_{-4} = ?$, respectively.

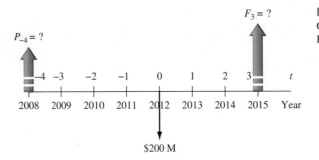

Figure 2–2
Cash flow diagram for Example 2.2*a* and *b*.

(*a*) To find the equivalent investment required in 3 years, apply the F/P factor. Use $1 million units and the tabulated value for 10% interest (Table 15).

$$F_3 = P(F/P,i,n) = 200(F/P,10\%,3) = 200(1.3310)$$
$$= \$266.2 \quad (\$266,200,000)$$

Now, use the FV function on a spreadsheet to find the same answer, $F_3 = \$266.20$ million. (Refer to Figure 2–3, left side.)

Figure 2–3
Spreadsheet functions for Example 2.2.

(*b*) The year 2008 is 4 years prior to the planned construction date of 2012. To determine the equivalent cost 4 years earlier, consider the $200 M in 2012 ($t = 0$) as the future value F and apply the P/F factor for $n = 4$ to find P_{-4}. (Refer to Figure 2–2.) Table 15 supplies the tabulated value.

$$P_{-4} = F(P/F,i,n) = 200(P/F,10\%,4) = 200(0.6830)$$
$$= \$136.6 \quad (\$136,600,000)$$

The PV function $= PV(10\%,4,,200)$ will display the same amount as shown in Figure 2–3, right side.

This equivalence analysis indicates that at $136.6 M in 2008, the plant would have cost about 68% as much as in 2012, and that waiting until 2015 will cause the price tag to increase about 33% to $266 M.

2.2 Uniform Series Present Worth Factor and Capital Recovery Factor (P/A and A/P) ●●●

The equivalent present worth P of a uniform series A of end-of-period cash flows (investments) is shown in Figure 2–4a. An expression for the present worth can be determined by considering each A value as a future worth F, calculating its present worth with the P/F factor, Equation [2.3], and summing the results.

$$P = A\left[\frac{1}{(1+i)^1}\right] + A\left[\frac{1}{(1+i)^2}\right] + A\left[\frac{1}{(1+i)^3}\right] + \cdots$$
$$+ A\left[\frac{1}{(1+i)^{n-1}}\right] + A\left[\frac{1}{(1+i)^n}\right]$$

The terms in brackets are the P/F factors for years 1 through n, respectively. Factor out A.

$$P = A\left[\frac{1}{(1+i)^1} + \frac{1}{(1+i)^2} + \frac{1}{(1+i)^3} + \cdots + \frac{1}{(1+i)^{n-1}} + \frac{1}{(1+i)^n}\right] \qquad [2.6]$$

To simplify Equation [2.6] and obtain the P/A factor, multiply the n-term geometric progression in brackets by the $(P/F,i\%,1)$ factor, which is $1/(1+i)$. This results in Equation [2.7]. Now subtract the two equations, [2.6] from [2.7], and simplify to obtain the expression for P when $i \neq 0$ (Equation [2.8]).

$$\frac{P}{1+i} = A\left[\frac{1}{(1+i)^2} + \frac{1}{(1+i)^3} + \frac{1}{(1+i)^4} + \cdots + \frac{1}{(1+i)^n} + \frac{1}{(1+i)^{n+1}}\right] \qquad [2.7]$$

$$\frac{1}{1+i}P = A\left[\frac{1}{(1+i)^2} + \frac{1}{(1+i)^3} + \cdots + \frac{1}{(1+i)^n} + \frac{1}{(1+i)^{n+1}}\right]$$

$$- \qquad P = A\left[\frac{1}{(1+i)^1} + \frac{1}{(1+i)^2} + \cdots + \frac{1}{(1+i)^{n-1}} + \frac{1}{(1+i)^n}\right]$$

$$\frac{-i}{1+i}P = A\left[\frac{1}{(1+i)^{n+1}} - \frac{1}{(1+i)^1}\right]$$

$$P = \frac{A}{-i}\left[\frac{1}{(1+i)^n} - 1\right]$$

$$P = A\left[\frac{(1+i)^n - 1}{i(1+i)^n}\right] \qquad i \neq 0 \qquad\qquad [2.8]$$

The term in brackets in Equation [2.8] is the conversion factor referred to as the *uniform series present worth factor* (USPWF). It is the **P/A factor** used to calculate the **equivalent P value in year 0** for a uniform end-of-period series of A values beginning at the end of period 1 and extending for n periods. The cash flow diagram is Figure 2–4a.

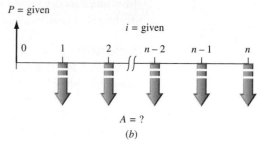

Figure 2–4
Cash flow diagrams used to determine (a) P, given a uniform series A, and (b) A, given a present worth P.

TABLE 2–2 P/A and A/P Factors: Notation and Equations

Factor Notation	Factor Name	Find/Given	Factor Formula	Standard Notation Equation	Excel Function
$(P/A,i,n)$	Uniform series present worth	P/A	$\dfrac{(1+i)^n - 1}{i(1+i)^n}$	$P = A(P/A,i,n)$	$= \mathrm{PV}(i\%,n,A)$
$(A/P,i,n)$	Capital recovery	A/P	$\dfrac{i(1+i)^n}{(1+i)^n - 1}$	$A = P(A/P,i,n)$	$= \mathrm{PMT}(i\%,n,P)$

To reverse the situation, the present worth P is known and the equivalent uniform series amount A is sought (Figure 2–4b). The first A value occurs at the end of period 1, that is, one period after P occurs. Solve Equation [2.8] for A to obtain

$$A = P\left[\frac{i(1+i)^n}{(1+i)^n - 1}\right] \qquad\qquad [2.9]$$

The term in brackets is called the *capital recovery factor* (CRF), or **A/P factor**. It calculates the **equivalent uniform annual worth A** over n years for a given P in year 0, when the interest rate is i.

Placement of P

> The P/A and A/P factors are derived with the present worth P and the first uniform annual amount A **one year (period) apart.** That is, the present worth P *must always* be located **one period prior to the first** A.

The factors and their use to find P and A are summarized in Table 2–2 and inside the front cover. The standard notations for these two factors are $(P/A,i\%,n)$ and $(A/P,i\%,n)$. Tables at the end of the text include the factor values. As an example, if $i = 15\%$ and $n = 25$ years, the P/A factor value from Table 19 is $(P/A,15\%,25) = 6.4641$. This will find the equivalent present worth at 15% per year for any amount A that occurs uniformly from years 1 through 25.

Spreadsheet functions can determine both P and A values in lieu of applying the P/A and A/P factors. The PV function calculates the P value for a given A over n years and a separate F value in year n, if it is given. The format, is

$$= \mathrm{PV}(i\%,n,A,F) \qquad\qquad [2.10]$$

Similarly, the A value is determined by using the PMT function for a given P value in year 0 and a separate F, if given. The format is

$$= \mathrm{PMT}(i\%,n,P,F) \qquad\qquad [2.11]$$

Table 2–2 includes the PV and PMT functions.

EXAMPLE 2.3

How much money should you be willing to pay now for a guaranteed $600 per year for 9 years starting next year, at a rate of return of 16% per year?

Solution

The cash flows follow the pattern of Figure 2–4a, with $A = \$600$, $i = 16\%$, and $n = 9$. The present worth is

$$P = 600(P/A,16\%,9) = 600(4.6065) = \$2763.90$$

The PV function $= \mathrm{PV}(16\%,9,600)$ entered into a single spreadsheet cell will display the answer $P = (\$2763.93)$.

EXAMPLE 2.4 The Cement Factory Case

As mentioned in the chapter introduction of this case, the Houston American Cement plant may generate a revenue base of $50 million per year. The president of the Brazilian parent company Votorantim Cimentos may have reason to be quite pleased with this projection for the simple reason that over the 5-year planning horizon, the expected revenue would total $250 million, which is $50 million more than the initial investment. With money worth 10% per year, address the following question from the president: Will the initial investment be recovered over the 5-year horizon with the time value of money considered? If so, by how much extra in present worth funds? If not, what is the equivalent annual revenue base required for the recovery plus the 10% return on money? Use both tabulated factor values and spreadsheet functions.

Solution

Tabulated value: Use the P/A factor to determine whether $A = \$50$ million per year for $n = 5$ years starting 1 year after the plant's completion ($t = 0$) at $i = 10\%$ per year is equivalently less or greater than $200 M. The cash flow diagram is similar to Figure 2–4a, where the first A value occurs 1 year after P. Using $1 million units and Table 15 values,

$$P = 50(P/A,10\%,5) = 50(3.7908)$$

$$= \$189.54 \quad (\$189,540,000)$$

The present worth value is less than the investment plus a 10% per year return, so the president should not be satisfied with the projected annual revenue.

To determine the minimum required to realize a 10% per year return, use the A/P factor. The cash flow diagram is the same as Figure 2–4b, where A starts 1 year after P at $t = 0$ and $n = 5$.

$$A = 200(A/P,10\%,5) = 200(0.26380)$$

$$= \$52.76 \text{ per year}$$

The plant needs to generate $52,760,000 per year to realize a 10% per year return over 5 years.

Spreadsheet: Apply the PV and PMT functions to answer the question. Figure 2–5 shows the use of $= \text{PV}(i\%,n,A,F)$ on the left side to find the present worth and the use of $= \text{PMT}(i\%,n,P,F)$ on the right side to determine the minimum A of $52,760,000 per year. Because there are no F values, it is omitted from the functions. The minus sign placed before each function name forces the answer to be positive, since these two functions always display the answer with the opposite sign entered on the estimated cash flows.

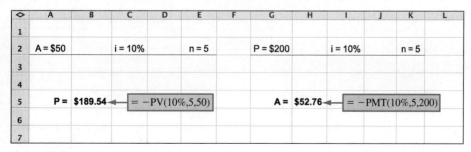

Figure 2–5
Spreadsheet functions to find P and A for the cement factory case, Example 2.4.

2.3 Sinking Fund Factor and Uniform Series Compound Amount Factor (A/F and F/A) ●●●

The simplest way to derive the A/F factor is to substitute into factors already developed. If P from Equation [2.3] is substituted into Equation [2.9], the following formula results.

$$A = F\left[\frac{1}{(1+i)^n}\right]\left[\frac{i(1+i)^n}{(1+i)^n - 1}\right]$$

$$A = F\left[\frac{i}{(1+i)^n - 1}\right] \tag{2.12}$$

The expression in brackets in Equation [2.12] is the **A/F or sinking fund factor**. It determines the **uniform annual series** A that is equivalent to a given future amount F. This is shown graphically in Figure 2–6a, where A is a uniform annual investment.

Placement of F

> The uniform series A begins at the **end of year (period) 1** and continues **through the year of the given F.** The last A value and F occur at the same time.

Equation [2.12] can be rearranged to find F for a stated A series in periods 1 through n (Figure 2–6b).

$$F = A\left[\frac{(1+i)^n - 1}{i}\right] \tag{2.13}$$

The term in brackets is called the *uniform series compound amount factor* (USCAF), or **F/A factor**. When multiplied by the given uniform annual amount A, it yields the **future worth of the uniform series**. It is important to remember that the future amount F occurs in the same period as the last A.

Standard notation follows the same form as that of other factors. They are $(F/A,i,n)$ and $(A/F,i,n)$. Table 2–3 summarizes the notations and equations, as does the inside front cover.

As a matter of interest, the uniform series factors can be symbolically determined by using an abbreviated factor form. For example, $F/A = (F/P)(P/A)$, where cancellation of the P is correct. Using the factor formulas, we have

$$(F/A,i,n) = [(1+i)^n]\left[\frac{(1+i)^n - 1}{i(1+i)^n}\right] = \frac{(1+i)^n - 1}{i}$$

For solution by spreadsheet, the FV function calculates F for a stated A series over n years. The format is

$$= \text{FV}(i\%,n,A,P) \tag{2.14}$$

The P may be omitted when no separate present worth value is given. The PMT function determines the A value for n years, given F in year n and possibly a separate P value in year 0. The format is

$$= \text{PMT}(i\%,n,P,F) \tag{2.15}$$

If P is omitted, the comma must be entered so the function knows the last entry is an F value.

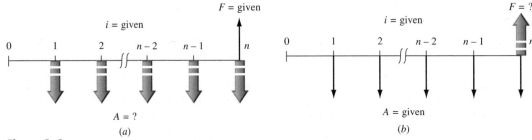

Figure 2–6
Cash flow diagrams to (a) find A, given F, and (b) find F, given A.

TABLE 2–3	*F/A* and *A/F* Factors: Notation and Equations				
Factor			**Factor**	**Standard Notation**	**Excel**
Notation	**Name**	**Find/Given**	**Formula**	**Equation**	**Functions**
$(F/A,i,n)$	Uniform series compound amount	F/A	$\dfrac{(1 + i)^n - 1}{i}$	$F = A(F/A,i,n)$	$= FV(i\%,n,A)$
$(A/F,i,n)$	Sinking fund	A/F	$\dfrac{i}{(1 + i)^n - 1}$	$A = F(A/F,i,n)$	$= PMT(i\%,n,F)$

EXAMPLE 2.5

The president of Ford Motor Company wants to know the equivalent future worth of a $1 million capital investment each year for 8 years, starting 1 year from now. Ford capital earns at a rate of 14% per year.

Solution

The cash flow diagram (Figure 2–7) shows the annual investments starting at the end of year 1 and ending in the year the future worth is desired. In $1000 units, the *F* value in year 8 is found by using the *F/A* factor.

$$F = 1000(F/A,14\%,8) = 1000(13.2328) = \$13,232.80$$

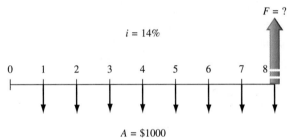

Figure 2–7
Diagram to find *F* for a uniform series, Example 2.5.

EXAMPLE 2.6 The Cement Factory Case   PE

Once again, consider the HAC case presented at the outset of this chapter, in which a projected $200 million investment can generate $50 million per year in revenue for 5 years starting 1 year after start-up. A 10% per year time value of money has been used previously to determine *P*, *F*, and *A* values. Now the president would like the answers to a couple of new questions about the estimated annual revenues. Use tabulated values, factor formulas, or spreadsheet functions to provide the answers.

(*a*) What is the equivalent future worth of the estimated revenues after 5 years at 10% per year?

(*b*) Assume that, due to the economic downturn, the president predicts that the corporation will earn only 4.5% per year on its money, not the previously anticipated 10% per year. What is the required amount of the annual revenue series over the 5-year period to be economically equivalent to the amount calculated in (*a*)?

Solution

(*a*) Figure 2–6*b* is the cash flow diagram with $A = \$50$ million. Note that the last *A* value and $F = ?$ both occur at the end of year $n = 5$. We use tabulated values and the spreadsheet function to find *F* in year 5.

Tabulated value: Use the *F/A* factor and 10% interest factor table. In $1 million units, the future worth of the revenue series is

$$F = 50(F/A,10\%,5) = 50(6.1051)$$
$$= \$305.255 \quad (\$305,255,000)$$

Figure 2–8
Spreadsheet functions to find F and A at $i = 4.5\%$ for the cement factory case, Example 2.6.

If the rate of return on the annual revenues were 0%, the total amount after 5 years would be $250,000,000. The 10% per year return is projected to grow this value by 22%.

Spreadsheet: Apply the FV factor in the format $= -FV(10\%,5,,50)$ to determine $F = \$305.255$ million. Because there is no present amount in this computation, P is omitted from the factor. See Figure 2–8, left side. (As before, the minus sign forces the FV function to result in a positive value.)

(b) The president of the Brazilian company planning to develop the cement plant in Georgia is getting worried about the international economy. He wants the revenue stream to generate the equivalent that it would at a 10% per year return, that is, $305.255 million, but thinks that only a 4.5% per year return is achievable.

Factor formula: The A/F factor will determine the required A for 5 years. Since the factor tables do not include 4.5%, use the formula to answer the question. In $1 million units,

$$A = 305.255(A/F,4.5\%,5) = 305.255\left[\frac{0.045}{(1.045)^5 - 1}\right] = 305.255(0.18279)$$
$$= \$55.798$$

The annual revenue requirement grows from $50 million to nearly $55,800,000. This is a significant increase of 11.6% each year.

Spreadsheet: It is easy to answer this question by using the $= PMT(i\%,n,,F)$ function with $i = 4.5\%$ and $F = \$305.255$ found in part (a). We can use the cell reference method (described in Appendix A) for the future amount F. Figure 2–8, right side, displays the required A of $55.798 per year (in $1 million units).

2.4 Factor Values for Untabulated *i* or *n* Values ●●○

Often it is necessary to know the correct numerical value of a factor with an i or n value that is not listed in the compound interest tables in the rear of the book. Given specific values of i and n, there are several ways to obtain any factor value.

- Use the formula listed in this chapter or the front cover of the book,
- Use an Excel function with the corresponding P, F, or A value set to 1.
- Use linear interpolation in the interest tables.

When the **formula** is applied, the factor value is accurate since the specific i and n values are input. However, it is possible to make mistakes since the formulas are similar to each other, especially when uniform series are involved. Additionally, the formulas become more complex when gradients are introduced, as you will see in the following sections.

A **spreadsheet function** determines the factor value if the corresponding P, A, or F argument in the function is set to 1 and the other parameters are omitted or set to zero. For example, the P/F factor is determined using the PV function with A omitted (or set to 0) and $F = 1$, that is, $PV(i\%,n,,1)$ or $PV(i\%,n,0,1)$. A minus sign preceding the function identifier causes the factor to have a positive value. Functions to determine the six common factors are as follows.

Factor	To Do This	Excel Function
P/F	Find *P*, given *F*.	$= -PV(i\%,n,,1)$
F/P	Find *F*, given *P*.	$= -FV(i\%,n,,1)$
P/A	Find *P*, given *A*.	$= -PV(i\%,n,1)$
A/P	Find *A*, given *P*.	$= -PMT(i\%,n,1)$
F/A	Find *F*, given *A*.	$= -FV(i\%,n,1)$
A/F	Find *A*, given *F*.	$= -PMT(i\%,n,,1)$

Figure 2–9 shows a spreadsheet developed explicitly to determine these factor values. When it is made live in Excel, entering any combination of *i* and *n* displays the exact value for all six factors. The values for *i* = 3.25% and *n* = 25 years are shown here. As we already know, these same functions will determine a final *P*, *A*, or *F* value when actual or estimated cash flow amounts are entered.

Linear interpolation for an untabulated interest rate *i* or number of years *n* takes more time to complete than using the formula or spreadsheet function. Also interpolation introduces some level of inaccuracy, depending upon the distance between the two boundary values selected for *i* or *n*, as the formulas themselves are nonlinear functions. Interpolation is included here for individuals who wish to utilize it in solving problems. Refer to Figure 2–10 for a graphical description of the following explanation. First, select two tabulated values (x_1 and x_2) of the parameter for which the factor is requested, that is, *i* or *n*, ensuring that the two values surround and are not too distant from the required value *x*. Second, find the corresponding tabulated factor values (f_1 and f_2). Third, solve for the unknown, linearly interpolated value *f* using the formulas below, where the differences in parentheses are indicated in Figure 2–10 as *a* through *c*.

	A	B	C	D
1	i =	n =		
2	**3.25%**	**25.00**	Enter requested i and n	
3			**Value obtained**	
4	**Factor**	**Value**	**with this function**	
5	P/F	0.44952	`= -PV($A$2,$B$2,,1)`	
6	P/A	16.93786	`= -PV($A$2,$B$2,1)`	
7				
8	F/P	2.22460	`= -FV($A$2,$B$2,,1)`	
9	F/A	37.67993	`= -FV($A$2,$B$2,1)`	
10				
11	A/F	0.02654	`= -PMT($A$2,$B$2,,1)`	
12	A/P	0.05904	`= -PMT($A$2,$B$2,1)`	
13				

Figure 2–9
Use of Excel functions to display factor values for any *i* and *n* values.

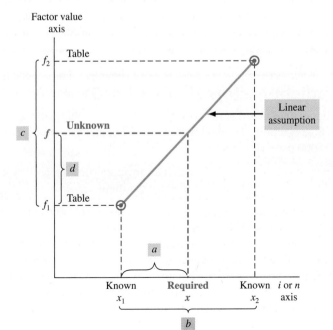

Figure 2–10
Linear interpolation in factor value tables.

$$f = f_1 + \frac{(x - x_1)}{(x_2 - x_1)}(f_2 - f_1) \qquad\qquad [2.16]$$

$$f = f_1 + \frac{a}{b}c = f_1 + d \qquad\qquad [2.17]$$

The value of d will be positive or negative if the factor is increasing or decreasing, respectively, in value between x_1 and x_2.

EXAMPLE 2.7

Determine the P/A factor value for $i = 7.75\%$ and $n = 10$ years, using the three methods described previously.

Solution

Factor formula: Apply the formula from inside the front cover of the book for the P/A factor. Showing 5-decimal accuracy,

$$(P/A,7.75\%,10) = \frac{(1 + i)^n - 1}{i(1 + i)^n} = \frac{(1.0775)^{10} - 1}{0.0775(1.0775)^{10}} = \frac{1.10947}{0.16348}$$

$$= 6.78641$$

Spreadsheet: Utilize the spreadsheet function in Figure 2–9, that is, $= -PV(7.75\%,10,1)$, to display 6.78641.

Linear interpolation: Use Figure 2–10 as a reference for this solution. Apply the Equation [2.16] and [2.17] sequence, where x is the interest rate i, the bounding interest rates are $i_1 = 7\%$ and $i_2 = 8\%$, and the corresponding P/A factor values are $f_1 = (P/A,7\%,10) = 7.0236$ and $f_2 = (P/A,8\%,10) = 6.7101$. With 4-place accuracy,

$$f = f_1 + \frac{(i - i_1)}{(i_2 - i_1)}(f_2 - f_1) = 7.0236 + \frac{(7.75 - 7)}{(8 - 7)}(6.7101 - 7.0236)$$

$$= 7.0236 + (0.75)(-0.3135) = 7.0236 - 0.2351$$

$$= 6.7885$$

Comment

Note that since the P/A factor value decreases as i increases, the linear adjustment is negative at -0.2351. As is apparent, linear interpolation provides an approximation to the correct factor value for 7.75% and 10 years, plus it takes more calculations than using the formula or spreadsheet function. It is possible to perform two-way linear interpolation for untabulated i and n values; however, the use of a spreadsheet or factor formula is recommended.

Online
Presentation

2.5 Arithmetic Gradient Factors (*P/G* and *A/G*) ●●●

Assume a manufacturing engineer predicts that the cost of maintaining a robot will increase by $5000 per year until the machine is retired. The cash flow series of maintenance costs involves a constant gradient, which is $5000 per year.

An **arithmetic gradient** series is a cash flow series that either increases or decreases by a **constant amount** each period. The amount of change is called the **gradient**.

Formulas previously developed for an *A* series have year-end amounts of equal value. In the case of a gradient, each year-end cash flow is different, so new formulas must be derived. First, assume that the cash flow at the end of year 1 is the **base amount** of the cash flow series and, therefore, not part of the gradient series. This is convenient because in actual applications, the base amount is usually significantly different in size compared to the gradient. For example, if you purchase a used car with a 1-year warranty, you might expect to pay the gasoline and insurance costs during the first year of operation. Assume these cost $2500; that is, $2500 is the base amount. After the first year, you absorb the cost of repairs, which can be expected to increase

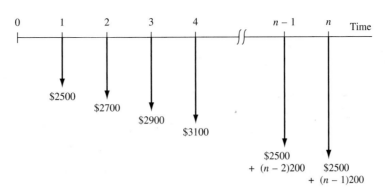

Figure 2–11
Cash flow diagram of an arithmetic gradient series.

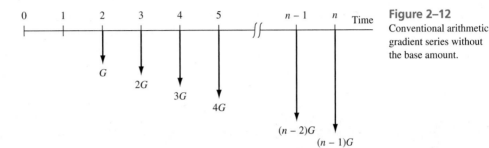

Figure 2–12
Conventional arithmetic gradient series without the base amount.

each year. If you estimate that total costs will increase by $200 each year, the amount the second year is $2700, the third $2900, and so on to year n, when the total cost is $2500 + (n - 1)200$. The cash flow diagram is shown in Figure 2–11. Note that the gradient ($200) is first observed between year 1 and year 2, and the base amount ($2500 in year 1) is not equal to the gradient.

Define the symbols G for gradient and CF_n for cash flow in year n as follows.

> G = constant arithmetic change in cash flows from one time period to the next; G may be positive or negative.

$$CF_n = \text{base amount} + (n - 1)G \qquad [2.18]$$

It is important to realize that the base amount defines a uniform cash flow series of the size A that occurs eash time period. We will use this fact when calculating equivalent amounts that involve arithmetic gradients. If the base amount is ignored, a generalized arithmetic (increasing) gradient cash flow diagram is as shown in Figure 2–12. Note that the gradient begins between years 1 and 2. This is called a **conventional gradient**.

EXAMPLE 2.8

A local university has initiated a logo-licensing program with the clothier Holister, Inc. Estimated fees (revenues) are $80,000 for the first year with uniform increases to a total of $200,000 by the end of year 9. Determine the gradient and construct a cash flow diagram that identifies the base amount and the gradient series.

Solution

The year 1 base amount is $CF_1 = \$80,000$, and the total increase over 9 years is

$$CF_9 - CF_1 = 200,000 - 80,000 = \$120,000$$

Equation [2.18], solved for G, determines the arithmetic gradient.

$$G = \frac{(CF_9 - CF_1)}{n - 1} = \frac{120,000}{9 - 1}$$
$$= \$15,000 \text{ per year}$$

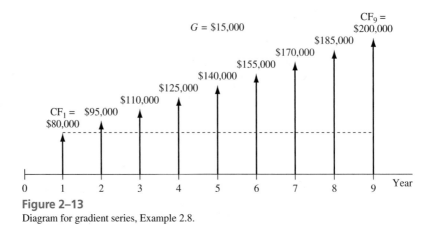

Figure 2–13
Diagram for gradient series, Example 2.8.

The cash flow diagram (Figure 2–13) shows the base amount of $80,000 in years 1 through 9 and the $15,000 gradient starting in year 2 and continuing through year 9.

The **total present worth P_T** for a series that includes a base amount A and conventional arithmetic gradient must consider the present worth of both the uniform series defined by A and the arithmetic gradient series. The addition of the two results in P_T.

$$P_T = P_A \pm P_G \qquad\qquad [2.19]$$

where P_A is the present worth of the uniform series only, P_G is the present worth of the gradient series only, and the $+$ or $-$ sign is used for an increasing $(+G)$ or decreasing $(-G)$ gradient, respectively.

The corresponding equivalent annual worth A_T is the sum of the base amount series annual worth A_A and gradient series annual worth A_G, that is,

$$A_T = A_A \pm A_G \qquad\qquad [2.20]$$

Three factors are derived for arithmetic gradients: the P/G factor for present worth, the A/G factor for annual series, and the F/G factor for future worth. There are several ways to derive them. We use the single-payment present worth factor $(P/F,i,n)$, but the same result can be obtained by using the F/P, F/A, or P/A factor.

In Figure 2–12, the present worth at year 0 of only the gradient is equal to the sum of the present worths of the individual cash flows, where each value is considered a future amount.

$$P = G(P/F,i,2) + 2G(P/F,i,3) + 3G(P/F,i,4) + \cdots$$
$$+ [(n - 2)G](P/F,i,n - 1) + [(n - 1)G](P/F,i,n)$$

Factor out G and use the P/F formula.

$$P = G\left[\frac{1}{(1 + i)^2} + \frac{2}{(1 + i)^3} + \frac{3}{(1 + i)^4} + \cdots + \frac{n - 2}{(1 + i)^{n-1}} + \frac{n - 1}{(1 + i)^n}\right] \qquad [2.21]$$

Multiplying both sides of Equation [2.21] by $(1 + i)^1$ yields

$$P(1 + i)^1 = G\left[\frac{1}{(1 + i)^1} + \frac{2}{(1 + i)^2} + \frac{3}{(1 + i)^3} + \cdots + \frac{n - 2}{(1 + i)^{n-2}} + \frac{n - 1}{(1 + i)^{n-1}}\right] \quad [2.22]$$

Subtract Equation [2.21] from Equation [2.22] and simplify.

$$iP = G\left[\frac{1}{(1 + i)^1} + \frac{1}{(1 + i)^2} + \cdots + \frac{1}{(1 + i)^{n-1}} + \frac{1}{(1 + i)^n}\right] - G\left[\frac{n}{(1 + i)^n}\right] \qquad [2.23]$$

The left bracketed expression is the same as that contained in Equation [2.6], where the P/A factor was derived. Substitute the closed-end form of the P/A factor from Equation [2.8]

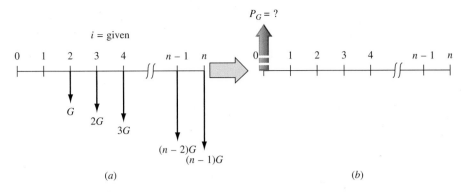

Figure 2–14
Conversion diagram from an arithmetic gradient to a present worth.

into Equation [2.23] and simplify to solve for P_G, the present worth of the gradient series only.

$$P_G = \frac{G}{i}\left[\frac{(1+i)^n - 1}{i(1+i)^n} - \frac{n}{(1+i)^n}\right] \qquad [2.24]$$

Equation [2.24] is the general relation to **convert an arithmetic gradient G (not including the base amount) for n years into a present worth at year 0**. Figure 2–14a is converted into the equivalent cash flow in Figure 2–14b. The *arithmetic gradient present worth factor*, or **P/G factor,** may be expressed in two forms:

$$(P/G,i,n) = \frac{1}{i}\left[\frac{(1+i)^n - 1}{i(1+i)^n} - \frac{n}{(1+i)^n}\right]$$

or

$$(P/G,i,n) = \frac{(1+i)^n - in - 1}{i^2(1+i)^n} \qquad [2.25]$$

Remember: The conventional arithmetic gradient starts in year 2, and P is located in year 0.

Placement of gradient P_G

Equation [2.24] expressed as an engineering economy relation is

$$P_G = G(P/G,i,n) \qquad [2.26]$$

which is the rightmost term in Equation [2.19] to calculate total present worth. The G carries a minus sign for decreasing gradients.

The equivalent uniform annual series A_G for an arithmetic gradient G is found by multiplying the present worth in Equation [2.26] by the $(A/P,i,n)$ formula. In standard notation form, the equivalent of algebraic cancellation of P can be used.

$$A_G = G(P/G,i,n)(A/P,i,n)$$
$$= G(A/G,i,n)$$

In equation form,

$$A_G = \frac{G}{i}\left[\frac{(1+i)^n - 1}{i(1+i)^n} - \frac{n}{(1+i)^n}\right]\left[\frac{i(1+i)^n}{(1+i)^n - 1}\right]$$

$$A_G = G\left[\frac{1}{i} - \frac{n}{(1+i)^n - 1}\right] \qquad [2.27]$$

which is the rightmost term in Equation [2.20]. The expression in brackets in Equation [2.27] is called the *arithmetic gradient uniform series factor* and is identified by **(A/G,i,n)**. This factor converts Figure 2–15a into Figure 2–15b.

The P/G and A/G factors and relations are summarized inside the front cover. Factor values are tabulated in the two rightmost columns of factor values at the rear of this text.

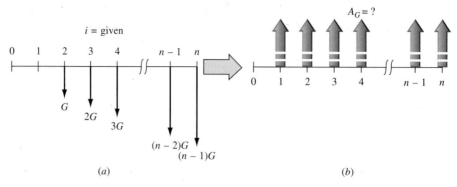

Figure 2–15
Conversion diagram of an arithmetic gradient series to an equivalent uniform annual series.

There is no direct, single-cell spreadsheet function to calculate P_G or A_G for an arithmetic gradient. Use the NPV function to display P_G and the PMT function to display A_G after entering all cash flows (base and gradient amounts) into contiguous cells. General formats for these functions are

$$= \text{NPV}(i\%, second\_cell:last\_cell) + first\_cell \qquad [2.28]$$
$$= \text{PMT}(i\%, n, cell\_with\_P_G) \qquad [2.29]$$

The word entries in italic are cell references, not the actual numerical values. (See Appendix A, Section A.2, for a description of cell reference formatting.) These functions are demonstrated in Example 2.10.

An **F/G factor** (*arithmetic gradient future worth factor*) to calculate the future worth F_G of a gradient series can be derived by multiplying the *P/G* and *F/P* factors. The resulting factor, (*F/G,i,n*), in brackets, and engineering economy relation is

$$F_G = G\left[\left(\frac{1}{i}\right)\left(\frac{(1 + i)^n - 1}{i}\right) - n\right]$$

EXAMPLE 2.9

Neighboring parishes in Louisiana have agreed to pool road tax resources already designated for bridge refurbishment. At a recent meeting, the engineers estimated that a total of $500,000 will be deposited at the end of next year into an account for the repair of old and safety-questionable bridges throughout the area. Further, they estimate that the deposits will increase by $100,000 per year for only 9 years thereafter, then cease. Determine the equivalent (*a*) present worth and (*b*) annual series amounts, if public funds earn at a rate of 5% per year.

Solution

(*a*) The cash flow diagram of this conventional arithmetic gradient series from the perspective of the parishes is shown in Figure 2–16. According to Equation [2.19], two computations must be made and added: the first for the present worth of the base amount P_A and the second for the present worth of the gradient P_G. The total present worth P_T occurs in year 0. This is illustrated by the partitioned cash flow diagram in Figure 2–17. In $1000 units, the total present worth is

$$P_T = 500(P/A,5\%,10) + 100(P/G,5\%,10)$$

$$= 500(7.7217) + 100(31.6520)$$

$$= \$7026.05 \quad (\$7,026,050)$$

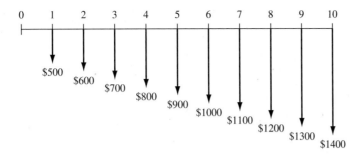

Figure 2–16
Cash flow series with a conventional arithmetic gradient (in $1000 units),
Example 2.9.

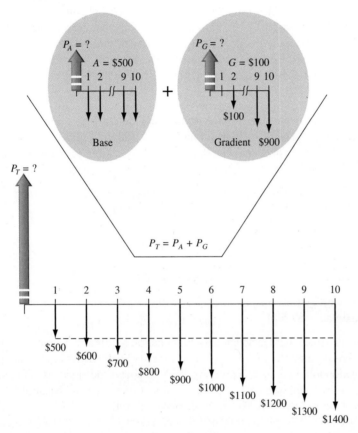

Figure 2–17
Partitioned cash flow diagram (in $1000 units), Example 2.9.

(b) Here, too, it is necessary to consider the gradient and the base amount separately. The total annual series A_T is found by Equation [2.20] and occurs in years 1 through 10.

$$A_T = 500 + 100(A/G,5\%,10) = 500 + 100(4.0991)$$

$$= \$909.91 \text{ per year} \quad (\$909,910)$$

Comment

Remember: The P/G and A/G factors determine the present worth and annual series of the *gradient only.* Any other cash flows must be considered separately.

If the present worth is already calculated [as in part (*a*)], P_T can be multiplied by an A/P factor to get A_T. In this case, considering round-off error,

$$A_T = P_T(A/P,5\%,10) = 7026.05(0.12950)$$

$$= \$909.873 \quad (\$909,873)$$

EXAMPLE 2.10 The Cement Factory Case

The announcement of the HAC cement factory states that the $200 million (M) investment is planned for 2012. Most large investment commitments are actually spread out over several years as the plant is constructed and production is initiated. Further investigation may determine, for example, that the $200 M is a present worth in the year 2012 of anticipated investments during the next 4 years (2013 through 2016). Assume the amount planned for 2013 is $100 M with constant decreases of $25 M each year thereafter. As before, assume the time value of money for investment capital is 10% per year to answer the following questions using tabulated factors and spreadsheet functions, as requested below.

(a) In equivalent present worth values, does the planned decreasing investment series equal the announced $200 M in 2012? Use both tabulated factors and spreadsheet functions.
(b) Given the planned investment series, what is the equivalent annual amount that will be invested from 2013 to 2016? Use both tabulated factors and spreadsheet functions.
(c) (This optional question introduces **Excel's Goal Seek tool**.) What must be the amount of yearly constant decrease through 2016 to have a present worth of exactly $200 M in 2012, provided $100 M is expended in 2013? Use a spreadsheet.

Solution

(a) The investment series is a decreasing arithmetic gradient with a base amount of $100 M in year 1 (2013) and $G = \$-25$ M through year 4 (2016). Figure 2–18 diagrams the cash flows with the shaded area showing the constantly declining investment each year. The P_T value at time 0 at 10% per year is determined by using tables and a spreadsheet.

Tabulated factors: Equation [2.19] with the minus sign for negative gradients determines the total present worth P_T. Money is expressed in $1 million units.

$$P_T = P_A - P_G = 100(P/A,10\%,4) - 25(P/G,10\%,4) \qquad [2.30]$$

$$= 100(3.1699) - 25(4.3781)$$

$$= \$207.537 \quad (\$207,537,000)$$

In present worth terms, the planned series will exceed the equivalent of $200 M in 2012 by approximately $7.5 M.

Spreadsheet: Since there is no spreadsheet function to directly display present worth for a gradient series, enter the cash flows in a sequence of cells (rows or columns) and use the NPV function to find present worth. Figure 2–19 shows the entries and function NPV(i%,second_cell:last_cell). There is no first_cell entry here, because there is no investment per se in year 0. The result displayed in cell C9, $207.534, is the total P_T for the planned series. (Note that the NPV function does not consider two separate series of cash flows as is necessary when using tabulated factors.)

The interpretation is the same as in part (a); the planned investment series exceeds the $200 M in present worth terms by approximately $7.5 M.

(b) *Tabulated factors:* There are two equally correct ways to find A_T. First, apply Equation [2.20] that utilizes the A/G factor, and second, use the P_T value obtained above and the A/P factor. Both relations are illustrated here, in $1 million units,

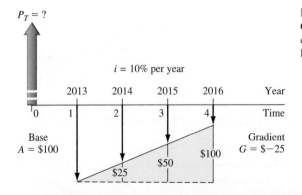

Figure 2–18
Cash flow diagram for decreasing gradient in $1 million units, Example 2.10.

	A	B	C	D	E	F	G
1	in $1 M units	G =	-25.000				
2							
3	Year	Time, t	Investment, $1 M				
4	2012	0					
5	2013	1	100.000				
6	2014	2	75.000				
7	2015	3	50.000				
8	2016	4	25.000				
9	P @ 10%		207.534				
10							
11	A of inv @ 10%		65.471				
12							
13							

Present worth of investments
= NPV(10%,C5:C8)

Annual worth of investments
= −PMT(10%,4,C9)

Figure 2–19
Spreadsheet solution for Example 2.10a and b.

Use Equation [2.20]:

$$A_T = 100 - 25(A/G,10\%,4) = 100 - 25(1.3812)$$

$$= \$65.471 \quad (\$65,471,000 \text{ per year})$$

Use P_T:

$$A_T = 207.537(A/P,10\%,4) = 207.537(0.31547)$$

$$= \$65.471 \text{ per year}$$

Spreadsheet: Apply the PMT function in Equation [2.29] to obtain the same $A_T = \$65.471$ per year (Figure 2–19).

(c) (Optional) The **Goal Seek tool** is described in Appendix A. It is an excellent tool to apply when one cell entry must equal a specific value and only one other cell can change. This is the case here; the NPV function (cell C9 in Figure 2–19) must equal $200, and the gradient G (cell C1) is unknown. This is the same as stating $P_T = 200$ in Equation [2.30] and solving for G. All other parameters retain their current value.

Figure 2–20 (top) pictures the same spreadsheet used previously with the Goal Seek template added and loaded. When OK is clicked, the solution is displayed; $G = \$-26.721$. Refer to Figure 2–20 again. This means that if the investment is decreased by a constant annual amount of $26.721 M, the equivalent total present worth invested over the 4 years will be exactly $200 M.

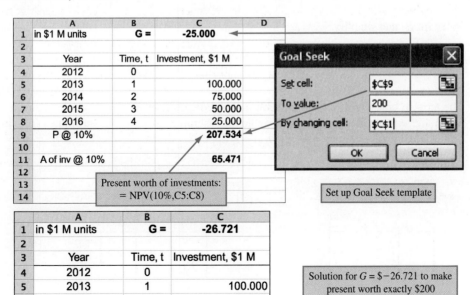

Figure 2–20
Solution for arithmetic gradient using Goal Seek, Example 2.10c.

2.6 Geometric Gradient Series Factors ● ● ○

It is common for annual revenues and annual costs such as maintenance, operations, and labor to go up or down by a constant percentage, for example, $+5\%$ or -3% per year. This change occurs every year on top of a starting amount in the first year of the project. A definition and description of new terms follow.

> A **geometric gradient** series is a cash flow series that either increases or decreases by a **constant percentage** each period. The uniform change is called the **rate of change**.
>
> g = **constant rate of change**, in decimal form, by which cash flow values increase or decrease from one period to the next. The gradient g can be $+$ or $-$.
>
> A_1 = **initial cash flow in year 1** of the geometric series
>
> P_g = **present worth** of the entire geometric gradient series, including the initial amount A_1

Note that the initial cash flow A_1 **is not considered separately** when working with geometric gradients.

Figure 2–21 shows increasing and decreasing geometric gradients starting at an amount A_1 in time period 1 with present worth P_g located at time 0. The relation to determine the total present worth P_g **for the entire cash flow series** may be derived by multiplying each cash flow in Figure 2–21a by the P/F factor $1/(1 + i)^n$.

$$P_g = \frac{A_1}{(1 + i)^1} + \frac{A_1(1 + g)}{(1 + i)^2} + \frac{A_1(1 + g)^2}{(1 + i)^3} + \cdots + \frac{A_1(1 + g)^{n-1}}{(1 + i)^n}$$

$$= A_1\left[\frac{1}{1 + i} + \frac{1 + g}{(1 + i)^2} + \frac{(1 + g)^2}{(1 + i)^3} + \cdots + \frac{(1 + g)^{n-1}}{(1 + i)^n}\right] \qquad [2.31]$$

Multiply both sides by $(1 + g)/(1 + i)$, subtract Equation [2.31] from the result, factor out P_g, and obtain

$$P_g\left(\frac{1 + g}{1 + i} - 1\right) = A_1\left[\frac{(1 + g)^n}{(1 + i)^{n+1}} - \frac{1}{1 + i}\right]$$

Solve for P_g and simplify.

$$P_g = A_1\left[\frac{1 - \left(\frac{1 + g}{1 + i}\right)^n}{i - g}\right] \qquad g \neq i \qquad [2.32]$$

The term in brackets in Equation [2.32] is the $(P/A,g,i,n)$ or *geometric gradient series present worth factor* for values of g not equal to the interest rate i. When $g = i$, substitute i for g in Equation [2.31] and observe that the term $1/(1 + i)$ appears n times.

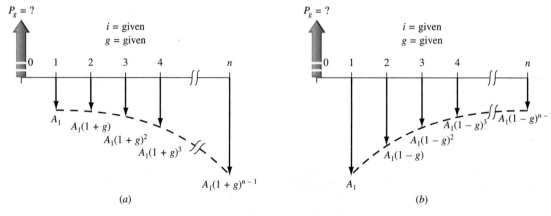

Figure 2–21
Cash flow diagram of (a) increasing and (b) decreasing geometric gradient series and present worth P_g.

$$P_g = A_1\left(\frac{1}{(1+i)} + \frac{1}{(1+i)} + \frac{1}{(1+i)} + \cdots + \frac{1}{(1+i)}\right)$$

$$P_g = \frac{nA_1}{(1+i)} \qquad\qquad\qquad\qquad\qquad [2.33]$$

The $(P/A,g,i,n)$ factor calculates P_g in period $t = 0$ for a geometric gradient series **starting in period 1** in the amount A_1 and increasing by a constant rate of g each period.

Placement of
Gradient P_g

The equation for P_g and the $(P/A,g,i,n)$ factor formula are

$$P_g = A_1(P/A,g,i,n) \qquad\qquad\qquad [2.34]$$

$$(P/A,g,i,n) = \begin{cases} \dfrac{1 - \left(\dfrac{1+g}{1+i}\right)^n}{i-g} & g \neq i \\[3ex] \dfrac{n}{1+i} & g = i \end{cases} \qquad [2.35]$$

It is possible to derive factors for the equivalent A and F values; however, it is easier to determine the P_g amount and then multiply by the A/P or F/P factor.

As with the arithmetic gradient series, there are no direct spreadsheet functions for geometric gradient series. Once the cash flows are entered, P and A are determined using the NPV and PMT functions, respectively.

EXAMPLE 2.11

A coal-fired power plant has upgraded an emission control valve. The modification costs only $8000 and is expected to last 6 years with a $200 salvage value. The maintenance cost is expected to be high at $1700 the first year, increasing by 11% per year thereafter. Determine the equivalent present worth of the modification and maintenance cost by hand and by spreadsheet at 8% per year.

Solution by Hand

The cash flow diagram (Figure 2–22) shows the salvage value as a positive cash flow and all costs as negative. Use Equation [2.35] for $g \neq i$ to calculate P_g. Total P_T is the sum of three present worth components.

$$\begin{aligned} P_T &= -8000 - P_g + 200(P/F,8\%,6) \\ &= -8000 - 1700\left[\frac{1 - (1.11/1.08)^6}{0.08 - 0.11}\right] + 200(P/F,8\%,6) \\ &= -8000 - 1700(5.9559) + 126 = \$-17{,}999 \end{aligned}$$

Figure 2–22
Cash flow diagram of a geometric gradient, Example 2.11.

Solution by Spreadsheet

Figure 2–23 details the spreadsheet operations to find the geometric gradient present worth P_g and total present worth P_T. To obtain $P_T = \$-17,999$, three components are summed—first cost, present worth of estimated salvage in year 6, and P_g. Cell tags detail the relations for the second and third components; the first cost occurs at time 0.

Comment

The relation that calculates the $(P/A,g,i\%,n)$ factor is rather complex, as shown in the cell tag and formula bar for C9. If this factor is used repeatedly, it is worthwhile using cell reference formatting so that A_1, i, g, and n values can be changed and the correct value is always obtained. Try to write the relation for cell C9 in this format.

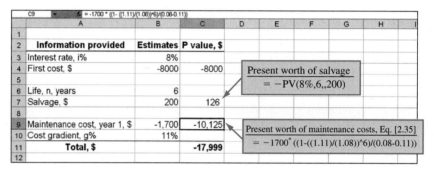

Figure 2–23
Geometric gradient and total present worth calculated via spreadsheet, Example 2.11.

EXAMPLE 2.12 The Cement Factory Case
PE

Now let's go back to the proposed Houston American Cement plant in Georgia. The revenue series estimate of $50 million annually is quite optimistic, especially since there are many other cement product plants operating in Florida and Georgia on the same limestone deposit. (The website for the HAC plant shows where they are located currently; it is clear that keen competition will be present.) Therefore, it is important to be sensitive in our analysis to possibly declining and increasing revenue series, depending upon the longer-term success of the plant's marketing, quality, and reputation. Assume that revenue may start at $50 million by the end of the first year, but then decreases geometrically by 12% per year through year 5. Determine the **present worth** and **future worth** equivalents of all revenues during this 5-year time frame at the same rate used previously, that is, 10% per year.

Solution

The cash flow diagram appears much like Figure 2–21b, except that the arrows go up for revenues. In year 1, $A_1 = \$50$ M and revenues decrease in year 5 to

$$A_1(1 - g)^{n-1} = 50\,M(1 - 0.12)^{5-1} = 50\,M(0.88)^4 = \$29.98\,M$$

First, we determine P_g in year 0 using Eq. [2.35] with $i = 0.10$ and $g = -0.12$, then we calculate F in year 5. In $1 million units,

$$P_g = 50\left[\frac{1 - \left(\frac{0.88}{1.10}\right)^5}{0.10 - (-0.12)}\right] = 50[3.0560]$$

$$= \$152.80$$

$$F = 152.80(F/P,10\%,5) = 152.80(1.6105)$$

$$= \$246.08$$

This means that the decreasing revenue stream has a 5-year future equivalent worth of $246.080 M. If you look back to Example 2.6, we determined that the F in year 5 for the uniform revenue series of $50 M annually is $305.255 M. In conclusion, the 12% declining geometric gradient has lowered the future worth of revenue by $59.175 M, which is a sizable amount from the perspective of the owners of Votorantim Cimentos North America, Inc.

2.7 Determining *i* or *n* for Known Cash Flow Values ● ● ●

When all the cash flow values are known or have been estimated, the *i* value (interest rate or rate of return) or *n* value (number of years) is often the unknown. An example for which *i* is sought may be stated as follows: A company invested money to develop a new product. After the net annual income series is known following several years on the market, determine the rate of return *i* on the investment. There are several ways to find an unknown *i* or *n* value, depending upon the nature of the cash flow series and the method chosen to find the unknown. The simplest case involves only single amounts (*P* and *F*) and solution utilizing a spreadsheet function. The most difficult and complex involves finding *i* or *n* for irregular cash flows mixed with uniform and gradient series utilizing solution by hand and calculator. The solution approaches are summarized below, followed by examples.

Single Amounts—*P* and *F* Only

Hand or Calculator Solution   Set up the equivalence relation and (1) solve for the variable using the factor formula, or (2) find the factor value and interpolate in the tables.

Spreadsheet Solution   Use the IRR or RATE function to find *i* or the NPER function to find *n*. (See below and Appendix A for details.)

Uniform Series—*A* Series

Hand or Calculator Solution   Set up the equivalence relation using the appropriate factor (*P/A, A/P, F/A,* or *A/F*), and use the second method mentioned above.

Spreadsheet Solution   Use the IRR or RATE function to find *i* or the NPER function to find *n*.

Mixed *A* Series, Gradients, and/or Isolated Values

Hand or Calculator Solution   Set up the equivalence relation and use (1) trial and error or (2) the calculator functions.

Spreadsheet Solution   Use the IRR or RATE function to find *i* or the NPER function to find *n*. (This is the recommended approach.)

Besides the PV, FV, and NPV functions, other spreadsheet functions useful in determining *i* are IRR (internal rate of return) and RATE, and NPER (number of periods) to find *n*. The formats are shown here and the inside front cover with a detailed explanation in Appendix A. In all three of these functions, at least one cash flow entry must have a sign opposite that of others in order to find a solution.

$$= \text{IRR(first\_cell:last\_cell)} \qquad\qquad [2.36]$$

To use IRR to find *i*, enter all cash flows into contiguous cells, including zero values.

$$= \text{RATE}(n,A,P,F) \qquad\qquad [2.37]$$

The single-cell RATE function finds *i* when an *A* series and single *P* and/or *F* values are involved.

$$= \text{NPER}(i\%,A,P,F) \qquad\qquad [2.38]$$

NPER is a single-cell function to find *n* for single *P* and *F* values, or with an *A* series.

EXAMPLE 2.13

If Laurel made a $30,000 investment in a friend's business and received $50,000 5 years later, determine the rate of return.

Solution

Since only single amounts are involved, i can be determined directly from the P/F factor.

$$P = F(P/F,i,n) = F\frac{1}{(1 + i)^n}$$

$$30{,}000 = 50{,}000\,\frac{1}{(1 + i)^5}$$

$$0.600 = \frac{1}{(1 + i)^5}$$

$$i = \left(\frac{1}{0.6}\right)^{0.2} - 1 = 0.1076 \quad (10.76\%)$$

Alternatively, the interest rate can be found by setting up the standard P/F relation, solving for the factor value, and interpolating in the tables.

$$P = F(P/F,i,n)$$

$$30{,}000 = 50{,}000(P/F,i,5)$$

$$(P/F,i,5) = 0.60$$

From the interest tables, a P/F factor of 0.6000 for $n = 5$ lies between 10% and 11%. Interpolate between these two values to obtain $i = 10.76\%$.

EXAMPLE 2.14

Pyramid Energy requires that for each of its offshore wind power generators $5000 per year be placed into a capital reserve fund to cover unexpected major rework on field equipment. In one case, $5000 was deposited for 15 years and covered a rework costing $100,000 in year 15. What rate of return did this practice provide to the company? Solve by hand and spreadsheet.

Solution by Hand

The cash flow diagram is shown in Figure 2–24. Either the A/F or F/A factor can be used. Using A/F,

$$A = F(A/F,i,n)$$

$$5000 = 100{,}000(A/F,i,15)$$

$$(A/F,i,15) = 0.0500$$

From the A/F interest tables for 15 years, the value 0.0500 lies between 3% and 4%. By interpolation, $i = 3.98\%$.

$F = \$100,000$ **Figure 2–24**
Diagram to determine the rate of return, Example 2.14.

Solution by Spreadsheet

Refer to the cash flow diagram (Figure 2–24) while completing the spreadsheet (Figure 2–25). A single-cell solution using the RATE function can be applied since $A = \$-5000$ occurs each year and $F = \$100{,}000$ takes place in the last year of the series. The function $= \text{RATE}(15,-5000,,100000)$ displays the value $i = 3.98\%$. This function is fast, but it allows only limited sensitivity analysis because all the A values have to change by the same amount. The IRR function is much better for answering "what if" questions.

To apply the IRR function, enter the value 0 in a cell (for year 0), followed by –5000 for 14 years and in year 15 enter +95,000 (Figure 2–25). In any cell enter the IRR function. The answer $i = 3.98\%$ is displayed. It is advisable to enter the year numbers 0 through n (15 in this example) in the column immediately to the left of the cash flow entries. The IRR function does not need these numbers, but it makes the cash flow entry activity easier and more accurate. Now any cash flow can be changed, and a new rate will be displayed immediately via IRR.

	A	B	C	D	E
1				Year	Cash flow, $
2	**Using RATE**			0	0
3	3.98%			1	-5000
4				2	-5000
5	*i* using RATE function			3	-5000
6	= RATE(15,-5000,,100000)			4	-5000
7				5	-5000
8				6	-5000
9	**Using IRR**			7	-5000
10	3.98%			8	-5000
11				9	-5000
12	*i* using IRR function			10	-5000
13	= IRR(E2:E17))			11	-5000
14				12	-5000
15				13	-5000
16				14	-5000
17				15	95000
18					

Figure 2–25

Use of RATE and IRR functions to determine *i* value for a uniform series, Example 2.14.

EXAMPLE 2.15 The Cement Factory Case   PE

From the introductory comments about the HAC plant, the annual revenue is planned to be $50 million. All analysis thus far has taken place at 10% per year; however, the parent company has made it clear that its other international plants are able to show a 20% per year return on the initial investment. Determine the number of years required to generate 10%, 15%, and 20% per year returns on the $200 million investment at the Georgia site.

Solution

If hand solution is utilized, the present worth relation can be established and the n values interpolated in the tables for each of the three rate of return values. In $1 million units, the relation is

$$P = -200 + 50(P/A,i\%,n) \qquad (i = 10\%, 15\%, 20\%)$$

$$(P/A,i\%,n) = 4.00$$

This is a good opportunity to utilize a spreadsheet and repeated NPER functions from Equation [2.38], since several i values are involved. Figure 2–26 shows the single-cell $= \text{NPER}(i\%,50,-200)$ function for each rate of return. The number of years (rounded up) to produce at least the required returns are

Return, *i*%	Years
10	6
15	7
20	9

◇	A	B	C	D
1				
2	i value	10%	15%	20%
3				
4	NPER function	`= NPER(10%,50,-200)	`= NPER(15%,50,-200)	`= NPER(20%,50,-200)
5				
6	**Result, years**	5.36	6.56	8.83
7				

Figure 2–26
Use of NPER function to find *n* values for various rate of return requirements,
Example 2.15.

CHAPTER SUMMARY

Formulas and factors derived and applied in this chapter perform equivalence calculations for present, future, annual, and gradient cash flows. Capability in using these formulas and their standard notation manually and with spreadsheets is critical to complete an engineering economy study. Using these formulas and spreadsheet functions, you can convert single cash flows into uniform cash flows, gradients into present worths, and much more. Additionally, you can solve for rate of return *i* or time *n*.

PROBLEMS

Use of Interest Tables

2.1 Look up the numerical value for the following factors from the interest tables.
1. $(P/F,6\%,8)$
2. $(A/P,10\%,10)$
3. $(A/G,15\%,20)$
4. $(A/F,2\%,30)$
5. $(P/G,35\%,15)$

Determination of *F*, *P*, and *A*

2.2 How much can Haydon Rheosystems, Inc., afford to spend now on an energy management system if the software will save the company $21,300 per year for the next 5 years? Use an interest rate of 10% per year.

2.3 A manufacturer of off-road vehicles is considering the purchase of dual-axis inclinometers for installation in a new line of tractors. The distributor of the inclinometers is temporarily overstocked and is offering them at a 40% discount from the regular cost of $142. If the purchaser gets them now instead of 2 years from now, which is when they will be needed, what is the present worth of the savings per unit? The company would pay the regular price, if purchased in 2 years. Assume the interest rate is 10% per year.

2.4 The Moller Skycar M400 is a flying car known as a personal air vehicle (PAV) that is expected to be FAA-certified by December 31, 2011. The cost is $985,000, and a $100,000 deposit will hold one of the first 100 "cars." Assume a buyer pays the $885,000 balance 3 years after making the $100,000 deposit. At an interest rate of 10% per year, what is the effective total cost of the PAV in year 3?

2.5 A family that won a $100,000 prize on America's Funniest Home Videos decided to put one-half of the money in a college fund for their child who was responsible for the prize. If the fund earned interest at 6% per year, how much was in the account 14 years after it was started?

2.6 One of the biggest vulnerabilities in a control system is network devices, such as Ethernet-based network switches that are located in unsecured locations and accessible to everyone. DeltaX switches, manufactured by Dahne Security, allow the user to automatically lock and unlock the port access to all switches in the network. The company is considering expanding its manufacturing lines now or doing it in 3 years. If the cost now would be $1.9 million, what equivalent amount could the company afford to spend in 3 years? The interest rate is 15% per year.

2.7 A company that sells high-purity laboratory chemicals is considering investing in new equipment that will reduce cardboard costs by better matching the size of the products to be shipped to the size of the shipping container. If the new equipment will cost $220,000 to purchase and install, how much must the company save each year for 3 years in order to justify the investment, if the interest rate is 10% per year?

2.8 Red Valve Co. of Carnegie, Pennsylvania, makes a control pinch valve that provides accurate, repeatable control of abrasive and corrosive slurries, outlasting gate, plug, ball, and even satellite coated valves. How much can the company afford to spend now on new equipment in lieu of spending $75,000 four years from now? The company's rate of return is 12% per year.

2.9 If GHD Plastics purchases a new building now for $1.3 million for its corporate headquarters, what must the building be worth in 10 years? The company expects all expenditures to earn a rate of return of at least 18% per year.

2.10 CGK Rheosystems makes high-performance rotational viscometers capable of steady shear and yield stress testing in a rugged, compact footprint. How much could the company afford to spend now on new equipment in lieu of spending $200,000 one year from now and $300,000 three years from now, if the company uses an interest rate of 15% per year?

2.11 Five years ago a consulting engineer purchased a building for company offices constructed of bricks that were not properly fired. As a result, some of the bricks were deteriorated from their exposure to rain and snow. Because of the problem with the bricks, the selling price of the building was 25% below the price of comparable, structurally sound buildings. The engineer repaired the damaged bricks and arrested further deterioration by applying an extra-strength solvent-based RTV elastomeric sealant. This resulted in restoring the building to its fair market value. If the depressed purchase price of the building was $600,000 and the cost of getting it repaired was $25,000, what is the equivalent value of the "forced appreciation" today, if the interest rate is 8% per year?

2.12 Metso Automation, which manufactures addressable quarter-turn electric actuators, is planning to set aside $100,000 now and $150,000 one year from now for possible replacement of the heating and cooling systems in three of its larger manufacturing plants. If the replacement won't be needed for 4 years, how much will the company have in the account, if it earns interest at a rate of 8% per year?

2.13 Syringe pumps often fail because reagents adhere to the ceramic piston and deteriorate the seal. Trident Chemical developed an integrated polymer dynamic seal that provides a higher sealing force on the sealing lip, resulting in extended seal life. One of Trident's customers expects to reduce downtime by 30% as a result of the new seal design. If lost production would have cost the company $110,000 per year for the next 4 years, how much could the company afford to spend now on the new seals, if it uses an interest rate of 12% per year?

2.14 China spends an estimated $100,000 per year on cloud seeding efforts, which includes using anti-aircraft guns and rocket launchers to fill the sky with silver iodide. In the United States, utilities that run hydroelectric dams are among the most active cloud seeders, because they believe it is a cost-effective way to increase limited water supplies by 10% or more. If the yields of cash crops will increase by 4% each year for the next 3 years because of extra irrigation water captured behind dams during cloud seeding, what is the maximum amount the farmers should spend now on the cloud seeding activity? The value of the cash crops without the extra irrigation water would be $600,000 per year. Use an interest rate of 10% per year.

2.15 The Public Service Board (PSB) awarded two contracts worth a combined $1.07 million to improve (i.e., deepen) a retention basin and reconstruct the spillway that was severely damaged in a flood 2 years ago. The PSB said that, because of the weak economy, the bids came in $950,000 lower than engineers expected. If the projects are assumed to have a 20-year life, what is the annual worth of the savings at an interest rate of 6% per year?

2.16 The National Highway Traffic Safety Administration raised the average fuel efficiency standard to 35.5 miles per gallon for cars and light trucks by the year 2016. The rules will cost consumers an average of $434 extra per vehicle in the 2012 model year. If a person purchases a new car in 2012 and keeps it for 5 years, how much must be saved in fuel costs each year to justify the extra cost? Use an interest rate of 8% per year.

2.17 In an effort to reduce childhood obesity by reducing the consumption of sugared beverages, some states have imposed taxes on soda and other soft drinks. A survey by Roland Sturm of 7300 fifth-graders revealed that if taxes averaged 4 cents on each dollar's worth of soda, no real difference in overall consumption was noticed. However, if taxes were increased to 18 cents on the dollar, Sturm calculated they would make a significant difference. For a student who consumes 100 sodas per year, what is the future worth of the extra cost from 4 cents to 18 cents per soda? Assume the student consumes sodas from grade 5 through graduation in grade 12. Use an interest rate of 6% per year.

2.18 The Texas Tomorrow Fund (TTF) is a program started in 1996 in Texas wherein parents could prepay their child's college tuition when the child was young. Actuaries set the price based on costs and investment earnings at that time. Later, the Texas legislature allowed universities to set their own tuition rates; tuition costs jumped dramatically. The cost for entering a newborn in 1996 was $10,500. If the TTF fund grew at a rate of 4% per year, while tuition costs increased at 7% per year, determine the state's shortfall when a newborn enters college 18 years later.

2.19 Henry Mueller Supply Co. sells tamperproof, normally open thermostats (i.e., thermostat closes as temperature rises). Annual cash flows are shown in the table below. Determine the future worth of the net cash flows at an interest rate of 10% per year.

Year	1	2	3	4	5	6	7	8
Income, $1000	200	200	200	200	200	200	200	200
Cost, $1000	90	90	90	90	90	90	90	90

2.20 A company that makes self-clinching fasteners expects to purchase new production-line equipment in 3 years. If the new units will cost $350,000, how much should the company set aside each year, if the account earns 10% per year?

Factor Values

2.21 Find the numerical value of the following factors using (a) interpolation and (b) the formula.
   1. $(A/P,13\%,15)$
   2. $(P/G,27\%,10)$

2.22 Find the numerical value of the following factors using (a) interpolation, (b) the formula, and (c) a spreadsheet function.
   1. $(F/P,14\%,62)$
   2. $(A/F,1\%,45)$

2.23 For the factor $(F/P,10\%,43)$, find the percent difference between the interpolated and formula-calculated values, assuming the formula-calculated value is the correct one.

2.24 For the factor $(F/A,15\%,52)$, find the percent difference between the interpolated and formula-

calculated values, assuming the formula-calculated value is the correct one.

Arithmetic Gradient

2.25 Profits from recycling paper, cardboard, aluminum, and glass at a liberal arts college have increased at a constant rate of $1100 in each of the last 3 years. If this year's profit (end of year 1) is expected to be $6000 and the profit trend continues through year 5, (a) what will the profit be at the end of year 5 and (b) what is the present worth of the profit at an interest rate of 8% per year?

2.26 A report by the Government Accountability Office (GAO) shows that the GAO expects the U.S. Postal Service to lose a record $7 billion at the end of this year, and if the business model is not changed, the losses will total $241 billion by the end of year 10. If the losses increase *uniformly* over the 10-year period, determine the following:
   (a) The expected increase in losses each year
   (b) The loss 5 years from now
   (c) The equivalent uniform worth of the losses at an interest rate of 8% per year

2.27 Rolled ball screws are suitable for high-precision applications such as water jet cutting. Their total manufacturing cost is expected to decrease because of increased productivity, as shown in the table. Determine the equivalent annual cost at an interest rate of 8% per year.

Year	1	2	3	4	5	6	7	8
Cost, $1000	200	195	190	185	180	175	170	165

2.28 Western Hydra Systems makes a panel milling machine with a 2.7-m-diameter milling head that emits low vibration and processes stress-relieved aluminum panels measuring up to 6000 mm long. The company wants to borrow money for a new production/warehouse facility. If the company offers to repay the loan with $60,000 in year 1 and amounts increasing by $10,000 each year through year 5, how much can the company borrow at an interest rate of 10% per year?

2.29 GKX Industries expects sales of its hydraulic seals (in inch and metric sizes) to increase according to the cash flow sequence $70 + 4k, where k is in years and cash flow is in $1000.
   (a) What is the amount of the cash flow in year 3?
   (b) What is the future worth of the entire cash flow series in year 10? Let $i = 10\%$ per year.

2.30   For the cash flows below, determine the amount in year 1, if the annual worth in years 1 through 9 is $601.17 and the interest rate is 10% per year.

Year	1	2	3	4	5	6	7	8	9
Cost, $1000	A	$A + 30$	$A + 60$	$A + 90$	$A + 120$	$A + 150$	$A + 180$	$A + 210$	$A + 240$

2.31   Apple Computer wants to have $2.1 billion available 5 years from now to finance production of a handheld "electronic brain" that, based on your behavior, will learn how to control nearly all the electronic devices in your home, such as the thermostat, coffee pot, TV, and sprinkler system. The company expects to set aside uniformly increasing amounts of money each year to meet its goal. If the amount set aside at the end of year 1 is $50 million, how much will the constant increase G have to be each year? Assume the investment account grows at a rate of 18% per year.

2.32   Tacozza Electric, which manufactures brush dc servomotors, budgeted $75,000 per year to pay for certain components over the next 5 years. If the company expects to spend $15,000 in year 1, how much of a uniform (arithmetic) increase each year is the company expecting in the cost of this part? Assume the company uses an interest rate of 10% per year.

Geometric Gradient

2.33   There are no tables in the back of your book for the geometric gradient series factors. Calculate the first two annual worth factor values, that is, A values for $n = 1$ and 2, that would be in a 10% interest table for a growth rate of 4% per year.

2.34   Determine the present worth of a geometric gradient series with a cash flow of $50,000 in year 1 and increases of 6% each year through year 8. The interest rate is 10% per year.

2.35   Determine the difference in the present worth values of the following two commodity contracts at an interest rate of 8% per year.

Contract 1 has a cost of $10,000 in year 1; costs will escalate at a rate of 4% per year for 10 years. Contract 2 has the same cost in year 1, but costs will escalate at 6% per year for 11 years.

2.36   El Paso Water Utilities (EPWU) purchases surface water for treatment and distribution to EPWU customers from El Paso County Water Improvement District during the irrigation season. A new contract between the two entities resulted in a reduction in future price increases in the cost of the water from 8% per year to 4% per year for the next 20 years. If the cost of water next year (which is year 1 of the new contract) will be $260 per acre-foot, what is the present worth of the savings (in $/acre-ft) to the utility between the old and the new contracts? Let the interest rate equal 6% per year.

2.37   Determine the present worth of a maintenance contract that has a cost of $30,000 in year 1 and annual increases of 6% per year for 10 years. Use an interest rate of 6% per year.

Interest Rate and Rate of Return

2.38   Gesky Industrial Products manufactures brushless blowers for boilers, food service equipment, kilns, and fuel cells. The company borrowed $18,000,000 for a plant expansion and repaid the loan in seven annual payments of $3,576,420, with the first payment made 1 year after the company received the money. What was the interest rate on the loan? Use hand and spreadsheet solutions.

2.39   If the value of Jane's retirement portfolio increased from $170,000 to $813,000 over a 15-year period, with no deposits made to the account over that period, what annual rate of return did she make?

2.40   A person's credit score is important in determining the interest rate on a home mortgage. According to Consumer Credit Counseling Service, a homeowner with a $100,000 mortgage and a 520 credit score will pay $110,325 more in interest charges over the life of a 30-year loan than a homeowner with the same mortgage and a credit score of 720. How much higher would the interest rate per year have to be in order to account for this much difference in interest charges, if the $100,000 loan is repaid in a single lump-sum payment at the end of 30 years?

2.41   During a period when the real estate market in Phoenix, Arizona, was undergoing a significant downturn, CSM Consulting Engineers made an agreement with a distressed seller to purchase an office building under the following terms: total price of $1.2 million with a down payment of $200,000 now and no payments for 4 years, after which the remaining balance of $1 million would be paid. CSM was able to make this deal because of poor market conditions at the time of purchase, and, at the same time, planning to sell the building in 4 years (when market conditions would probably be better) and move to a larger office building in Scottsdale, Arizona. If CSM was able to sell the building in exactly 4 years for $1.9 million, what rate of return per year did the company make on the investment?

2.42 A start-up company that makes hydraulic seals borrowed $800,000 to expand its packaging and shipping facility. The contract required the company to repay the investors through an innovative mechanism called *faux dividends,* a series of uniform annual payments over a fixed period of time. If the company paid $250,000 per year for 5 years, what was the interest rate on the loan?

2.43 Bessimer Electronics manufactures addressable actuators in one of its Maquiladora plants in Mexico. The company believes that by investing $24,000 each year in years 1, 2, and 3, it will avoid spending $87,360 in year 3. If the company does make the annual investments, what rate of return will it realize?

2.44 UV curable epoxy resins are used in sealing, in gap filling, and as a clear coating. Your boss saw a report submitted by the chief financial officer (CFO) that said the equivalent annual worth of maintaining the equipment used in producing the resins was $48,436 over the last 5 years. The report showed that the cost in year 1 was $42,000, and it increased arithmetically by $4000 each year. Your boss thought $48,436 was too high, so she asked you to determine what interest rate the CFO used in making the calculations. What was the interest rate?

Number of Years

2.45 Acme Bricks, a masonry products company, wants to have $600,000 on hand before it invests in new conveyors, trucks, and other equipment. If the company sets aside $80,000 per year in an account that increases in value at a rate of 15% per year, how many years will it be before Acme can purchase the equipment?

2.46 An engineer who was contemplating retirement had $1.6 million in his investment portfolio. However, a severe recession caused his portfolio to decrease to only 55% of the original amount, so he kept working. If he was able to invest his money at a rate of return of 9% per year after the recession ended, how many years did it take for his account to get back to the $1.6 million value?

2.47 You own a small engineering consulting company. If you invest $200,000 of the company's money in a natural gas well that is expected to provide income of $29,000 per year, how long must the well produce at that rate in order to get the money back plus a rate of return of 10% per year?

2.48 A perceptive engineer started saving for her retirement 15 years ago by diligently saving $18,000 each year through the present time. She invested in a stock fund that averaged a 12% rate of return over that period. If she makes the same annual investment and gets the same rate of return in the future, how long will it be from now (time zero) before she has $1,500,000 in her retirement fund?

2.49 A mechanical engineering graduate who wanted to have his own business borrowed $350,000 from his father as start-up money. Because he was family, his father charged interest at only 4% per year. If the engineer was able to pay his father $15,000 in year 1, $36,700 in year 2, and amounts *increasing* by $21,700 each year, how many years did it take for the engineer to repay the loan?

2.50 The energy costs of a company involved in powder coating of outdoor furniture are expected to increase at a rate of $400 per year. The cost at the end of the next year (year 1) is expected to be $13,000. How many years will it be from now before the equivalent annual cost is $16,000 per year, if interest is 8% per year?

2.51 In cleaning out some files that were left behind by the engineer who preceded you in your current job, you found an old report that had a calculation for the present worth of certain maintenance costs for state highways. The report contained the following equation (with cost in $1 million):

$$12\{1 - [(1 + 0.03)/(1 + 0.06)]^x\}/(0.06 - 0.03) = 140$$

The value of x that was used in the calculation was illegible. What is its value?

2.52 The equivalent annual worth of an increasing arithmetic gradient is $135,300. If the cash flow in year 1 is $35,000 and the gradient amount is $19,000, what is the value of n at an interest rate of 10% per year?

2.53 You are told that the present worth of an increasing geometric gradient is $88,146. If the cash flow in year 1 is $25,000 and the gradient increase is 18% per year, what is the value of n? The interest rate is 10% per year.

ADDITIONAL PROBLEMS AND FE EXAM REVIEW QUESTIONS

2.54 The amount of money that Diamond Systems can spend now for improving productivity in lieu of spending $30,000 three years from now at an interest rate of 12% per year is closest to:
- (*a*) $15,700
- (*b*) $17,800
- (*c*) $19,300
- (*d*) $21,350

2.55 A manufacturing company spent $30,000 on a new conveyor belt. If the conveyor belt resulted in cost savings of $4200 per year, the length of time it would take for the company to recover its investment at 8% per year is closest to:
- (*a*) Less than 9 years
- (*b*) 9 to 10 years
- (*c*) 11 to 12 years
- (*d*) Over 12 years

2.56 Levi Strauss has some of its jeans stone-washed under a contract with independent U.S. Garment Corp. If U.S. Garment's operating cost per machine is $22,000 for year 1 and increases by a constant $1000 per year through year 5, what is the equivalent uniform annual cost per machine for the 5 years at an interest rate of 8% per year?
- (*a*) $23,850
- (*b*) $24,650
- (*c*) $25,930
- (*d*) Over $26,000

2.57 The *F/G* factor values can be derived by multiplying:
- (*a*) (*P/F*) and (*A/G*) factor values
- (*b*) (*F/P*) and (*A/G*) factor values
- (*c*) (*P/F*) and (*P/G*) factor values
- (*d*) (*F/P*) and (*P/G*) factor values

2.58 At $i = 4\%$ per year, *A* for years 1 through 6 of the cash flows shown below is closest to:
- (*a*) $300
- (*b*) $560
- (*c*) $800
- (*d*) $1040

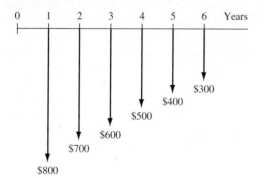

2.59 The value of the factor (*P/F*,*i*,10) can be found by getting the factor values for (*P/F*,*i*,4) and (*P/F*,*i*,6) and:
- (*a*) Adding the values for (*P/F*,*i*,4) and (*P/F*,*i*,6)
- (*b*) Multiplying the values for (*P/F*,*i*,4) and (*P/F*,*i*,6)
- (*c*) Dividing the value for (*P/F*,*i*,6) by the value for (*P/F*,*i*,4)
- (*d*) None of the above

2.60 A small construction company is considering the purchase of a used bulldozer for $61,000. If the company purchases the dozer now, the equivalent future amount in year 4 that the company is paying for the dozer at 4% per year interest is closest to:
- (*a*) $52,143
- (*b*) $65,461
- (*c*) $71,365
- (*d*) Over $72,000

2.61 The cost of lighting and maintaining the tallest smokestack in the United States (at a shuttered ASARCO refinery) is $90,000 per year. At an interest rate of 10% per year, the present worth of maintaining the smokestack for 10 years is closest to:
- (*a*) $1,015,000
- (*b*) $894,000
- (*c*) $712,000
- (*d*) $553,000

2.62 An enthusiastic new engineering graduate plans to start a consulting firm by borrowing $100,000 at 10% per year interest. The loan payment each year to pay off the loan in 7 years is closest to:
- (*a*) $18,745
- (*b*) $20,540
- (*c*) $22,960
- (*d*) $23,450

2.63 An engineer who believed in "save now and play later" wanted to retire in 20 years with $1.5 million. At 10% per year interest, to reach the $1.5 million goal, starting 1 year from now, the engineer must annually invest:
- (*a*) $26,190
- (*b*) $28,190
- (*c*) $49,350
- (*d*) $89,680

2.64 The cost of a border fence is $3 million per mile. If the life of such a fence is assumed to be 10 years, the equivalent annual cost of a

10-mile-long fence at an interest rate of 10% per year is closest to:
- (a) $3.6 million
- (b) $4.2 million
- (c) $4.9 million
- (d) Over $5.0 million

2.65 An investment of $75,000 in equipment that will reduce the time for machining self-locking fasteners will save $20,000 per year. At an interest rate of 10% per year, the number of years required to recover the initial investment is closest to:
- (a) 6 years
- (b) 5 years
- (c) 4 years
- (d) 3 years

2.66 The number of years required for an account to accumulate $650,000 if Ralph deposits $50,000 each year and the account earns interest at a rate of 6% per year is closest to:
- (a) 13 years
- (b) 12 years
- (c) 11 years
- (d) 10 years

2.67 Aero Serve, Inc., manufactures cleaning nozzles for reverse-pulse jet dust collectors. The company spent $40,000 on a production control system that will increase profits by $13,400 per year for 5 years. The rate of return per year on the investment is closest to:
- (a) 20%
- (b) 18%
- (c) 16%
- (d) Less than 15%

2.68 Energy costs for a green chemical treatment have been increasing uniformly for 5 years. If the cost in year 1 was $26,000 and it increased by $2000 per year through year 5, the present worth of the costs at an interest rate of 10% per year is closest to:
- (a) $102,900
- (b) $112,300
- (c) $122,100
- (d) $195,800

2.69 In planning for your retirement, you expect to save $5000 in year 1, $6000 in year 2, and amounts increasing by $1000 each year through year 20. If your investments earn 10% per year, the amount you will have at the end of year 20 is closest to:
- (a) $242,568
- (b) $355,407
- (c) $597,975
- (d) $659,125

2.70 Income from a precious metals mining operation has been decreasing uniformly for 5 years. If income in year 1 was $300,000 and it decreased by $30,000 per year through year 4, the annual worth of the income at 10% per year is closest to:
- (a) $310,500
- (b) $258,600
- (c) $203,900
- (d) $164,800

2.71 If you are able to save $5000 in year 1, $5150 in year 2, and amounts increasing by 3% each year through year 20, the amount you will have at the end of year 20 at 10% per year interest is closest to:
- (a) $60,810
- (b) $102,250
- (c) $351,500
- (d) Over $410,000

CASE STUDY

TIME MARCHES ON; SO DOES THE INTEREST RATE

Background

During the last week, Sundara has read about different situations that involve money, interest rate, and different amounts of time. She has gotten interested in the major effects that time and interest rates have on the amount of money necessary to do things and the significant growth in the amount of money when a large number of years are considered. In all cases, the interest focuses on the amount of money at the end of the time period.

Information

The four situations are described here.

A. Manhattan Island was purchased in 1626 for $24. After 385 years in 2011, at 6% per year compounded interest, the current value must be very large.

B. At the age of 22, if she saved only $2000 per year for the next 10 years (starting next year) and made a return of 6% per year, by today's standards, she would have accumulated a nice sum at the age of 70.

C. A corporation invested $2 million in developing and marketing a new product in 1945 (just after World War II, this was a lot of money) and has made a steady net cash flow of $300,000 per year for some 65 years. Sundara estimated the annual rate of return must be quite good, especially given that she is lucky to earn 4% per year on her own investments these days.

D. A friend who is not good with money, went to a pawn shop and borrowed $200 for one week and paid $30 in interest. Sundara thought this might be a pretty good deal, in case she ever ran low on cash. However, she did not know whether the interest was simple or compounded monthly, and how much may be owed were this loan not paid off for 1 year.

Case Study Exercises

1. What is the annual interest rate for each situation? Include both the annual simple and the compound rates for situation D.

2. Calculate and observe the total amount of money involved in each situation at the end of the time periods compared to the starting amount. Is the ending amount larger or smaller than you would expect it to be prior to making any computations?

3. Think of a situation for yourself that may be similar to any of those above. Determine the interest rate, the time period, and the starting and ending amounts of money.

Combining Factors and Spreadsheet Functions

ost estimated cash flow series do not fit exactly the series for which the factors, equations, and spreadsheet functions in Chapter 2 were developed. For a given sequence of cash flows, there are usually several correct ways to determine the equivalent present worth P, future worth F, or annual worth A. This chapter explains how to combine engineering economy factors and spreadsheet functions to address more complex situations involving shifted uniform series, gradient series, and single cash flows.

3.1 Calculations for Uniform Series That Are Shifted ● ● ●

When a uniform series begins at a time other than at the end of period 1, it is called a **shifted series**. In this case several methods can be used to find the equivalent present worth P. For example, P of the uniform series shown in Figure 3–1 could be determined by any of the following methods:

- Use the P/F factor to find the present worth of each disbursement at year 0 and add them.
- Use the F/P factor to find the future worth of each disbursement in year 13, add them, and then find the present worth of the total, using $P = F(P/F,i,13)$.
- Use the F/A factor to find the future amount $F = A(F/A,i,10)$, and then compute the present worth, using $P = F(P/F,i,13)$.
- Use the P/A factor to compute the "present worth" $P_3 = A(P/A,i,10)$ (which will be located in year 3, not year 0), and then find the present worth in year 0 by using the $(P/F,i,3)$ factor.

Typically the last method is used for calculating the present worth of a uniform series that does not begin at the end of period 1. For Figure 3–1, the "present worth" obtained using the P/A factor is located in year 3. This is shown as P_3 in Figure 3–2. Note that a P value is always located *1 year or period prior* to the beginning of the first series amount. Why? Because the P/A factor was derived with P in time period 0 and A beginning at the end of period 1. The most common mistake made in working problems of this type is improper placement of P. Therefore, it is extremely important to remember:

> The present worth is always located **one period prior** to the first uniform series amount when using the P/A factor.

Placement of P

To determine a future worth or F value, recall that the F/A factor derived in Section 2.3 had the F located in the *same* period as the last uniform series amount. Figure 3–3 shows the location of the future worth when F/A is used for Figure 3–1 cash flows.

> The future worth is always located in the **same period as the last** uniform series amount when using the F/A factor.

Placement of F

It is also important to remember that the number of periods n in the P/A or F/A factor is equal to the number of uniform series values. It may be helpful to *renumber* the cash flow diagram to avoid errors in counting. Figures 3–2 and 3–3 show Figure 3–1 renumbered to determine $n = 10$.

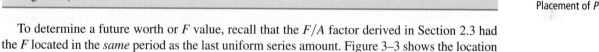

Figure 3–1
A uniform series that is shifted.

Figure 3–2
Location of present worth and renumbering for n for the shifted uniform series in Figure 3–1.

Figure 3–3
Placement of *F* and renumbering for *n* for the shifted uniform series of Figure 3–1.

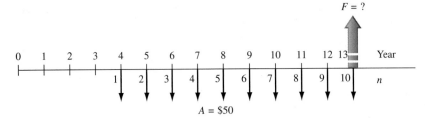

As stated above, several methods can be used to solve problems containing a uniform series that is shifted. However, it is generally more convenient to use the uniform series factors than the single-amount factors. Specific steps should be followed to avoid errors:

1. Draw a diagram of the positive and negative cash flows.
2. Locate the present worth or future worth of each series on the cash flow diagram.
3. Determine *n* for each series by renumbering the cash flow diagram.
4. Draw another cash flow diagram representing the desired equivalent cash flow.
5. Set up and solve the equations.

These steps are illustrated below.

EXAMPLE 3.1

The offshore design group at Bechtel just purchased upgraded CAD software for $5000 now and annual payments of $500 per year for 6 years starting 3 years from now for annual upgrades. What is the present worth in year 0 of the payments if the interest rate is 8% per year?

Solution

The cash flow diagram is shown in Figure 3–4. The symbol P_A is used throughout this chapter to represent the present worth of a uniform annual series *A*, and P_A' represents the present worth at a time other than period 0. Similarly, P_T represents the total present worth at time 0. The correct placement of P_A' and the diagram renumbering to obtain *n* are also indicated. *Note that P_A' is located in actual year 2, not year 3. Also, n = 6, not 8, for the P/A factor.* First find the value of P_A' of the shifted series.

$$P_A' = \$500(P/A,8\%,6)$$

Since P_A' is located in year 2, now find P_A in year 0.

$$P_A = P_A'(P/F,8\%,2)$$

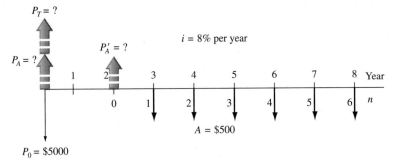

Figure 3–4
Cash flow diagram with placement of *P* values, Example 3.1.

The total present worth is determined by adding P_A and the initial payment P_0 in year 0.

$$P_T = P_0 + P_A$$
$$= 5000 + 500(P/A,8\%,6)(P/F,8\%,2)$$
$$= 5000 + 500(4.6229)(0.8573)$$
$$= \$6981.60$$

The more complex that cash flow series become, the more useful are the spreadsheet functions. When the uniform series A is shifted, the NPV function is used to determine P, and the PMT function finds the equivalent A value. The NPV function, like the PV function, determines the P values, but NPV can handle any combination of cash flows directly from the cells. As we learned in Chapter 2, enter the net cash flows in contiguous cells (column or row), making sure to enter "0" for all zero cash flows. Use the format

$$\text{NPV}(i\%, \text{second\_cell:last\_cell}) + \text{first\_cell}$$

First_cell contains the cash flow for year 0 and must be listed separately for NPV to correctly account for the time value of money. The cash flow in year 0 may be 0.

The easiest way to find an equivalent A over n years for a shifted series is with the PMT function, where the P value is from the NPV function above. The format is the same as we learned earlier; the entry for P is a cell reference, not a number.

$$\text{PMT}(i\%, n, \text{cell\_with\_}P, F)$$

Alternatively, the same technique can be used when an F value was obtained using the FV function. Now the last entry in PMT is "cell_with_F."

It is very fortunate that any parameter in a spreadsheet function can itself be a function. Thus, it is possible to write the PMT function in a single cell by embedding the NPV function (and FV function, if needed). The format is

$$\text{PMT}(i\%, n, \text{NPV}(i\%, \text{second\_cell:last\_cell}) + \text{first\_cell}, F) \qquad [3.1]$$

Of course, the answer for A is the same for the two-cell operation or a single-cell, embedded function. All three of these functions are illustrated in Example 3.2.

EXAMPLE 3.2

Recalibration of sensitive measuring devices costs $8000 per year. If the machine will be recalibrated for each of 6 years starting 3 years after purchase, calculate the 8-year equivalent uniform series at 16% per year. Show hand and spreadsheet solutions.

Solution by Hand

Figure 3–5a and b shows the original cash flows and the desired equivalent diagram. To convert the $8000 shifted series to an equivalent uniform series over all periods, first convert the uniform series into a present worth or future worth amount. Then either the A/P factor or the A/F factor can be used. Both methods are illustrated here.

Present worth method. (Refer to Figure 3–5a.) Calculate P'_A for the shifted series in year 2, followed by P_T in year 0. There are 6 years in the A series.

$$P'_A = 8000(P/A, 16\%, 6)$$

$$P_T = P'_A(P/F, 16\%, 2) = 8000(P/A, 16\%, 6)(P/F, 16\%, 2)$$

$$= 8000(3.6847)(0.7432) = \$21{,}907.75$$

The equivalent series A' *for 8 years* can now be determined via the A/P factor.

$$A' = P_T(A/P, 16\%, 8) = \$5043.60$$

Future worth method. (Refer to Figure 3–5a.) First calculate the future worth F in year 8.

$$F = 8000(F/A, 16\%, 6) = \$71{,}820$$

The A/F factor is now used to obtain A' over all 8 years.

$$A' = F(A/F, 16\%, 8) = \$5043.20$$

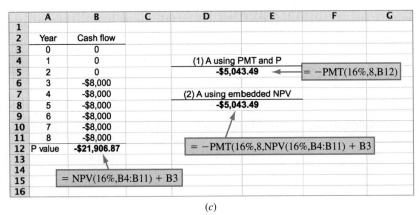

(c)

Figure 3–5
(a) Original and (b) equivalent cash flow diagrams; and (c) spreadsheet functions to determine
P and A, Example 3.2.

Solution by Spreadsheet
(Refer to Figure 3–5c.) Enter the cash flows in B3 through B11 with entries of "0" in the first
three cells. Use the NPV function to display P = $21,906.87.

There are two ways to obtain the equivalent A over 8 years. Of course, only one of these
PMT functions needs to be entered. (1) Enter the PMT function making direct reference to
the P value (see cell tag for D/E5), or (2) use Equation [3.1] to embed the NPV function into
the PMT function (see cell tag for D/E8).

Online
Presentation

3.2 Calculations Involving Uniform Series and Randomly Placed Single Amounts ● ● ●

When a cash flow includes both a uniform series and randomly placed single amounts, the procedures of Section 3.1 are applied to the uniform series and the single-amount formulas are applied to the one-time cash flows. This approach, illustrated in Examples 3.3 and 3.4, is merely a
combination of previous ones. For spreadsheet solutions, it is necessary to enter the net cash
flows before using the NPV and other functions.

EXAMPLE 3.3

An engineering company in Wyoming that owns 50 hectares of valuable land has decided to
lease the mineral rights to a mining company. The primary objective is to obtain long-term
income to finance ongoing projects 6 and 16 years from the present time. The engineering
company makes a proposal to the mining company that it pay $20,000 per year for 20 years
beginning 1 year from now, plus $10,000 six years from now and $15,000 sixteen years from
now. If the mining company wants to pay off its lease immediately, how much should it pay
now if the investment is to make 16% per year?

Solution

The cash flow diagram is shown in Figure 3–6 from the owner's perspective. Find the present worth of the 20-year uniform series and add it to the present worth of the two one-time amounts to determine P_T.

$$P_T = 20{,}000(P/A{,}16\%{,}20) + 10{,}000(P/F{,}16\%{,}6) + 15{,}000(P/F{,}16\%{,}16)$$

$$= \$124{,}075$$

Note that the $20,000 uniform series starts at the end of year 1, so the P/A factor determines the present worth at year 0.

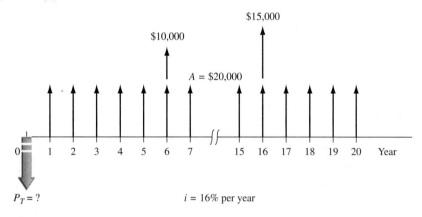

Figure 3–6
Diagram including a uniform series and single amounts, Example 3.3.

When you calculate the A value for a cash flow series that includes randomly placed single amounts and uniform series, **first convert everything to a present worth or a future worth.** Then you obtain the A value by multiplying P or F by the appropriate A/P or A/F factor. Example 3.4 illustrates this procedure.

EXAMPLE 3.4

A design-build-operate engineering company in Texas that owns a sizable amount of land plans to lease the drilling rights (oil and gas only) to a mining and exploration company. The contract calls for the mining company to pay $20,000 per year for 20 years beginning 3 years from now (i.e., beginning at the end of year 3 and continuing through year 22) plus $10,000 six years from now and $15,000 sixteen years from now. Utilize engineering economy relations by hand and by spreadsheet to determine the five *equivalent values* listed below at 16% per year.

1. Total present worth P_T in year 0
2. Future worth F in year 22
3. Annual series over all 22 years
4. Annual series over the first 10 years
5. Annual series over the last 12 years

Solution by Hand

Figure 3–7 presents the cash flows with equivalent P and F values indicated in the correct years for the P/A, P/F, and F/A factors.

1. **P_T in year 0:** First determine P'_A of the series in year 2. Then P_T is the sum of three P values: the series present worth value moved back to $t = 0$ with the P/F factor, and the two P values at $t = 0$ for the two single amounts in years 6 and 16.

$$P'_A = 20,000(P/A,16\%,20)$$

$$P_T = P'_A(P/F,16\%,2) + 10,000(P/F,16\%,6) + 15,000(P/F,16\%,16)$$

$$= 20,000(P/A,16\%,20)(P/F,16\%,2) + 10,000(P/F,16\%,6)$$
$$+ 15,000(P/F,16\%,16)$$

$$= \$93,625 \tag{3.2}$$

2. **F in year 22:** To determine F in year 22 from the original cash flows (Figure 3–7), find F for the 20-year series and add the two F values for the two single amounts. Be sure to carefully determine the n values for the single amounts: $n = 22 - 6 = 16$ for the $10,000 amount and $n = 22 - 16 = 6$ for the $15,000 amount.

$$F = 20,000(F/A,16\%,20) + 10,000(F/P,16\%,16) + 15,000(F/P,16\%,6) \tag{3.3}$$

$$= \$2,451,626$$

3. **A over 22 years:** Multiply $P_T = \$93,625$ from (1) above by the A/P factor to determine an equivalent 22-year A series, referred to as A_{1-22} here.

$$A_{1-22} = P_T(A/P,16\%,22) = 93,625(0.16635) = \$15,575 \tag{3.4}$$

An alternate way to determine the 22-year series uses the F value from (2) above. In this case, the computation is $A_{1-22} = F(A/F,16\%,22) = \$15,575$.

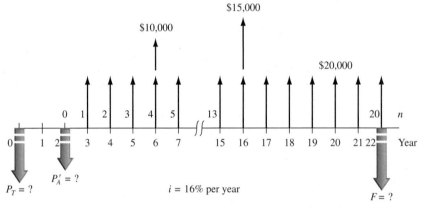

Figure 3–7
Diagram for Example 3.4.

4. **A over years 1 to 10:** This and (5), which follows, are special cases that often occur in engineering economy studies. The equivalent A series is calculated for a number of years different from that covered by the original cash flows. This occurs when a defined *study period* or *planning horizon* is preset for the analysis. (More is mentioned about study periods later.) To determine the equivalent A series for years 1 through 10 only (call it A_{1-10}), the P_T *value must be used* with the A/P factor for $n = 10$. This computation transforms Figure 3–7 into the equivalent series A_{1-10} in Figure 3–8a.

$$A_{1-10} = P_T(A/P,16\%,10) = 93,625(0.20690) = \$19,371 \tag{3.5}$$

5. **A over years 11 to 22:** For the equivalent 12-year series for years 11 through 22 (call it A_{11-22}), the F *value must be used* with the A/F factor for 12 years. This transforms Figure 3–7 into the 12-year series A_{11-22} in Figure 3–8b.

$$A_{11-22} = F(A/F,16\%,12) = 2,451,626(0.03241) = \$79,457 \tag{3.6}$$

Notice the huge difference of more than $60,000 in equivalent annual amounts that occurs when the present worth of $93,625 is allowed to compound at 16% per year for the first 10 years. This is another demonstration of the time value of money.

Solution by Spreadsheet

Figure 3–9 is a spreadsheet image with answers for all five questions. The $20,000 series and the two single amounts have been entered into separate columns, B and C. The zero cash flow

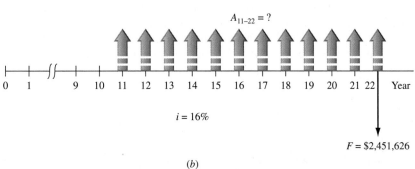

Figure 3–8
Cash flows of Figure 3–7 converted to equivalent uniform series for (*a*) years 1 to 10 and (*b*) years 11 to 22.

	A	B	C	D	E	F
1		Interest rate		16.00%		
2						
3		Cash flows		Value	Spreadsheet function used	
4	Year	Series	Single	calculated	to obtain result in column F	Result
5	0	0	0	Present worth		
6	1	0	0	of series	`= NPV(D1,B6:B27) + B5	$88,122
7	2	0	0			
8	3	$ 20,000	0			
9	4	$ 20,000	0	Present worth		
10	5	$ 20,000	0	of singles	`= NPV(D1,C6:C27) + C5	$5,500
11	6	$ 20,000	$ 10,000			
12	7	$ 20,000	0			
13	8	$ 20,000	0			
14	9	$ 20,000	0	1. Present worth	`= F6 + F10 or	$93,622
15	10	$ 20,000	0	total	`= NPV(D1,B6:B27) + NPV(D1,C6:C27)	$93,622
16	11	$ 20,000	0			
17	12	$ 20,000	0	2. Future worth		
18	13	$ 20,000	0	total	`= FV(D1,22,0,-F14)	$2,451,621
19	14	$ 20,000	0			
20	15	$ 20,000	0	3. Annual series		
21	16	$ 20,000	$ 15,000	over 22 years	`= PMT(D1,22,-F14)	$15,574
22	17	$ 20,000	0			
23	18	$ 20,000	0	4. Annual series		
24	19	$ 20,000	0	for first 10 years	`= PMT(D1,10,-F14)	$19,370
25	20	$ 20,000	0			
26	21	$ 20,000	0	5. Annual series		
27	22	$ 20,000	0	for last 12 years	`= PMT(D1,12,0,-F18)	$79,469

Figure 3–9
Spreadsheet using cell reference format. Example 3.4.

values are all entered so that the functions will work correctly. This is an excellent example demonstrating the versatility of the NPV, FV, and PMT functions. To prepare for sensitivity analysis, the functions are developed using cell reference format or global variables, as indicated in the column *E* function. This means that virtually any number—the interest rate, any cash flow estimate in the series or the single amounts, and the timing within the 22-year time frame—can be changed and the new answers will be immediately displayed.

1. Present worth values for the series and single amounts are determined in cells F6 and F10, respectively, using the NPV function. The sum of these in F14 is $P_T = \$93,622$, which corresponds to the value in Equation [3.2]. Alternatively, P_T can be determined directly via the sum of two NPV functions, shown in row 15.

2. The FV function in row 18 uses the *P* value in F14 (preceded by a minus sign) to determine *F* twenty-two years later. This is significantly easier than Equation [3.3].

3. To find the 22-year *A* series starting in year 1, the PMT function in row 21 references the *P* value in cell F14. This is effectively the same procedure used in Equation [3.4] to obtain A_{1-22}.

For the spreadsheet enthusiast, it is possible to find the 22-year A series value directly by using the PMT function with embedded NPV functions. The cell reference format is $= \text{PMT(D1,22,−(NPV(D1,B6:B27)+B5 + NPV(D1,C6:C27)+C5))}$.

4. and 5.   It is quite simple to determine an equivalent uniform series over any number of periods using a spreadsheet, provided the series starts one period after the P value is located or ends in the same period that the F value is located. These are both true for the series requested here. The results in F24 and F27 are the same as A_{1-10} and A_{11-22} in Equations [3.5] and [3.6], respectively.

Comment

Remember that **round-off error** will always be present when comparing hand and spreadsheet results. The spreadsheet functions carry more decimal places than the tables during calculations. Also, be very careful when constructing spreadsheet functions. It is easy to miss a value, such as the P or F in PMT and FV functions, or a minus sign between entries. Always check your function entries carefully before touching <Enter>.

Online Presentation

Placement of gradient P

3.3 Calculations for Shifted Gradients ●●○

In Section 2.5, we derived the relation $P = G(P/G,i,n)$ to determine the present worth of the **arithmetic gradient** series. The P/G factor, Equation [2.25], was derived for a present worth in year 0 with the gradient first appearing in year 2.

> The present worth of an **arithmetic gradient** will always be located **two periods before the gradient starts**.

Refer to Figure 2–14 as a refresher for the cash flow diagrams.

The relation $A = G(A/G,i,n)$ was also derived in Section 2.5. The A/G factor in Equation [2.27] performs the equivalence transformation of a **gradient only** into an A series from years 1 through n (Figure 2–15). Recall that the base amount must be treated separately. Then the equivalent P or A values can be summed to obtain the equivalent total present worth P_T and total annual series A_T.

A conventional gradient series starts between periods 1 and 2 of the cash flow sequence. A gradient starting at any other time is called a **shifted gradient**. The n value in the P/G and A/G factors for a shifted gradient is determined by renumbering the time scale. The period in which the *gradient first appears is labeled period 2*. The n value for the gradient factor is determined by the renumbered period where the last gradient increase occurs.

Partitioning a cash flow series into the arithmetic gradient series and the remainder of the cash flows can make very clear what the gradient n value should be. Example 3.5 illustrates this partitioning.

EXAMPLE 3.5

Fujitsu, Inc. has tracked the average inspection cost on a robotics manufacturing line for 8 years. Cost averages were steady at $100 per completed unit for the first 4 years, but have increased consistently by $50 per unit for each of the last 4 years. Analyze the gradient increase, using the P/G factor. Where is the present worth located for the gradient? What is the general relation used to calculate total present worth in year 0?

Solution

The cash flow diagram in Figure 3–10a shows the base amount $A = \$100$ and the arithmetic gradient $G = \$50$ starting between periods 4 and 5. Figure 3–10b and c partitions these two series. Gradient year 2 is placed in actual year 5 of the entire sequence in Figure 3–10c. It is clear that $n = 5$ for the P/G factor. The $P_G = ?$ arrow is correctly placed in gradient year 0, which is year 3 in the cash flow series.

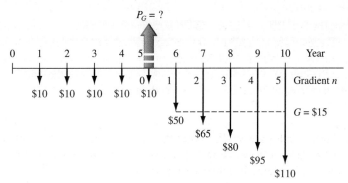

Figure 3–10
Partitioned cash flow, (*a*) = (*b*) + (*c*), Example 3.5.

The general relation for P_T is taken from Equation [2.19]. The uniform series $A = \$100$ occurs for all 8 years, and the $G = \$50$ gradient present worth P_G appears in year 3.

$$P_T = P_A + P_G = 100(P/A,i,8) + 50(P/G,i,5)(P/F,i,3)$$

It is important to note that the A/G factor *cannot* be used to find an equivalent A value in periods 1 through n for cash flows involving a shifted gradient. Consider the cash flow diagram of Figure 3–11. To find the equivalent annual series in years 1 through 10 for the gradient series only, first find the present worth P_G of the gradient in actual year 5, take this present worth back to year 0, and annualize the present worth for 10 years with the A/P factor. If you apply the annual series gradient factor $(A/G,i,5)$ directly, the gradient is converted into an equivalent annual series over years 6 through 10 only, not years 1 through 10, as requested.

> **To find the equivalent A series of a shifted gradient through all the n periods, first find the present worth of the gradient at actual time 0, then apply the $(A/P,i,n)$ factor.**

Figure 3–11
Determination of G and n values used in factors for a shifted gradient.

EXAMPLE 3.6

Set up the engineering economy relations to compute the equivalent annual series in years 1 through 7 for the cash flow estimates in Figure 3–12.

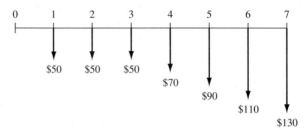

Figure 3–12
Diagram of a shifted gradient, Example 3.6.

Solution

The base amount annual series is $A_B = \$50$ for all 7 years (Figure 3–13). Find the present worth P_G in actual year 2 of the $20 gradient that starts in actual year 4. The gradient n is 5.

$$P_G = 20(P/G,i,5)$$

Bring the gradient present worth back to actual year 0.

$$P_0 = P_G(P/F,i,2) = 20(P/G,i,5)(P/F,i,2)$$

Annualize the gradient present worth from year 1 through year 7 to obtain A_G.

$$A_G = P_0(A/P,i,7)$$

Finally, add the base amount to the gradient annual series.

$$A_T = 20(P/G,i,5)(P/F,i,2)(A/P,i,7) + 50$$

For a spreadsheet solution, enter the original cash flows into adjacent cells, say, B3 through B10, and use an embedded NPV function in the PMT function. The single-cell function is $= \text{PMT}(i\%,7,-\text{NPV}(i\%, \text{B3:B10}))$.

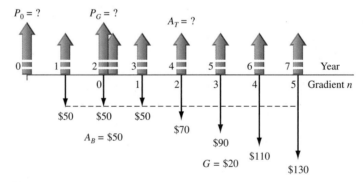

Figure 3–13
Diagram used to determine A for a shifted gradient, Example 3.6.

In Section 2.6, we derived the relation $P_g = A_1(P/A,g,i,n)$ to determine the present worth of a **geometric gradient** series, including the initial amount A_1. The factor was derived to find the present worth in year 0, with A_1 in year 1 and the first gradient appearing in year 2.

Placement of gradient P

> The present worth of a **geometric gradient series** will always be located **two periods before the gradient stars**, and the **initial amount is included** in the resulting present worth. Refer to Figure 2–21 as a refresher for the cash flows.

Equation [2.35] is the formula used for the factor. It is not tabulated.

EXAMPLE 3.7

Chemical engineers at a Coleman Industries plant in the Midwest have determined that a small amount of a newly available chemical additive will increase the water repellency of Coleman's tent fabric by 20%. The plant superintendent has arranged to purchase the additive through a 5-year contract at $7000 per year, starting 1 year from now. He expects the annual price to increase by 12% per year thereafter for the next 8 years. Additionally, an initial investment of $35,000 was made now to prepare a site suitable for the contractor to deliver the additive. Use $i = 15\%$ per year to determine the equivalent total present worth for all these cash flows.

Solution

Figure 3–14 presents the cash flows. The total present worth P_T is found using $g = 0.12$ and $i = 0.15$. Equations [2.34] and [2.35] are used to determine the present worth P_g for the entire geometric series at actual year 4, which is moved to year 0 using $(P/F,15\%,4)$.

$$P_T = 35,000 + A(P/A,15\%,4) + A_1(P/A,12\%,15\%,9)(P/F,15\%,4)$$

$$= 35,000 + 7000(2.8550) + \left[7000 \frac{1 - (1.12/1.15)^9}{0.15 - 0.12}\right](0.5718)$$

$$= 35,000 + 19,985 + 28,247$$

$$= \$83,232$$

Note that $n = 4$ in the $(P/A,15\%,4)$ factor because the $7000 in year 5 is the cash flow of the initial amount A_1.

For solution by spreadsheet, enter the cash flows of Figure 3–14. If cells B1 through B14 are used, the function to find $P = \$83,230$ is

$$NPV(15\%,B2:B14)+B1$$

The fastest way to enter the geometric series is to enter $7000 for year 5 (into cell B6) and set up each succeeding cell multiplied by 1.12 for the 12% increase.

Figure 3–14
Cash flow diagram including a geometric gradient with $g = 12\%$, Example 3.7.

Decreasing arithmetic and geometric gradients are common, and they are often **shifted gradient series**. That is, the constant gradient is $-G$ or the percentage change is $-g$ from one period to the next, and the first appearance of the gradient is at some time period (year) other than year 2 of the series. Equivalence computations for present worth P and annual worth A are basically the same as discussed in Chapter 2, except for the following.

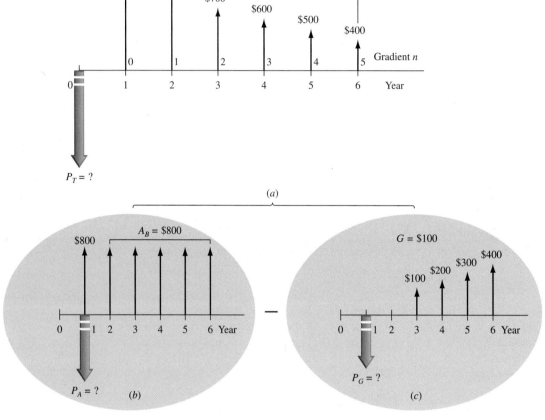

Figure 3–15
Partitioned cash flow of a shifted arithmetic gradient, $(a) = (b) - (c)$.

For shifted, decreasing gradients:

- The base amount A (arithmetic) or initial amount A_1 (geometric) is the *largest amount* in the first year of the series.
- The gradient amount is *subtracted from* the previous year's amount, not added to it.
- The amount used in the factors is $-G$ for arithmetic and $-g$ for geometric gradient series.
- The present worth P_G or P_g is located 2 years prior to the appearance of the first gradient; however, a P/F factor is necessary to find the present worth in year 0.

Figure 3–15 partitions a decreasing arithmetic gradient series with $G = \$-100$ that is shifted 1 year forward in time. P_G occurs in actual year 1, and P_T is the sum of three components.

$$P_T = \$800(P/F,i,1) + 800(P/A,i,5)(P/F,i,1) - 100(P/G,i,5)(P/F,i,1)$$

EXAMPLE 3.8

Morris Glass Company has decided to invest funds for the next 5 years so that development of "smart" glass is well funded in the future. This type of new-technology glass uses electro-chrome coating to allow rapid adjustment to sun and dark in building glass, as well as assisting with internal heating and cooling cost reduction. The financial plan is to invest first, allow appreciation to occur, and then use the available funds in the future. All cash flow estimates are in $1000 units, and the interest rate expectation is 8% per year.

Years 1 through 5: Invest $7000 in year 1, decreasing by $1000 per year through year 5.

Years 6 through 10: No new investment and no withdrawals.

Years 11 through 15: Withdraw $20,000 in year 11, decreasing 20% per year through year 15.

Determine if the anticipated withdrawals will be covered by the investment and appreciation plans. If the withdrawal series is over- or underfunded, what is the exact amount available in year 11, provided all other estimates remain the same?

Solution by Hand
Figure 3–16 presents the cash flow diagram and the placement of the equivalent P values used in the solution. Calculate the present worth of both series in actual year 0 and add them to determine if the investment series is adequate to fund the anticipated withdrawals.

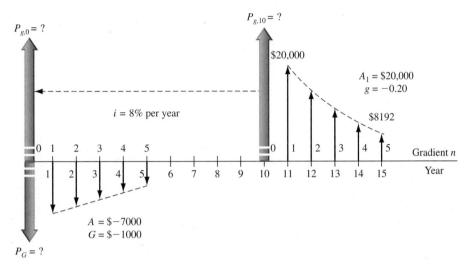

Figure 3–16
Investment and withdrawal series, Example 3.8.

Investment series: Decreasing, conventional arithmetic series starting in year 2 with $A = \$-7000$, $G = \$-1000$, and gradient $n = 5$ years. The present worth in year 0 is

$$P_G = -[7000(P/A,8\%,5) - 1000(P/G,8\%,5)]$$
$$= \$-20,577$$

Withdrawal series: Decreasing, shifted geometric series starting in year 12 with $A_1 = \$20,000$, $g = -0.20$, and gradient $n = 5$ years. If the present worth in year 10 is identified as $P_{g,10}$, the present worth in year 0 is $P_{g,0}$. Use Equation [2.35] for the $(P/A,-20\%,8\%,5)$ factor.

$$P_{g,0} = P_{g,10}(P/F,i,n) = A_1(P/A,g,i,n)(P/F,i,n) \qquad [3.7]$$

$$= 20,000 \left\{ \frac{1 - \left[\frac{1 + (-0.20)}{1 + 0.08} \right]^5}{0.08 - (-0.20)} \right\} (0.4632)$$

$$= 20,000(2.7750)(0.4632)$$

$$= \$25,707$$

The net total present worth is

$$P_T = -20,577 + 25,707 = \$+5130$$

This means that more is withdrawn than the investment series earns. Either additional funds must be invested or less must be withdrawn to make the series equivalent at 8% per year.

To find the exact amount of the initial withdrawal series to result in $P_T = 0$, let A_1 be an unknown in Equation [3.7] and set $P_{g,0} = -P_G = 20,577$.

$$20,577 = A_1(2.7750)(0.4632)$$
$$A_1 = \$16,009 \text{ in year 11}$$

The geometric series withdrawal would be 20% less each year.

(a)   (b)

Figure 3–17
Spreadsheet solution, Example 3.8. (*a*) Cash flows and NPV function and (*b*) Goal Seek to determine initial withdrawal amount in year 11.

Solution by Spreadsheet

See Figure 3–17*a*. To determine if the investment series will cover the withdrawal series, enter the cash flows (in column B using the functions shown in column C) and apply the NPV function shown in the cell tag to display $P_T = \$+5130$ directly. As above, the $+$ sign indicates that, from a time value of money perspective, there is more withdrawn than invested and earned.

The Goal Seek tool is very handy in determining the initial withdrawal amount that results in $P_T = 0$ (cell B19). Figure 3–17*b* shows the template and result $A_1 = \$16,009$ in year 11. Each succeeding withdrawal is 80% of the previous one.

Comment

If the withdrawal series is fixed as estimated initially and the investment series base amount A can be increased, the Goal Seek tool can be used to, again, set $P_T = 0$ (cell B19). However, now establish the entry in B4 as the changing cell. The response is $A = \$-8285$, and, as before, succeeding investments are $1000 less.

CHAPTER SUMMARY

In Chapter 2, we derived the equations to calculate the present, future, or annual worth of specific cash flow series. In this chapter, we learned how to use these equations on cash flow series that are shifted in time from those for which the basic relations are derived. For example, when a uniform series does not begin in period 1, we still use the P/A factor to find the "present worth" of the series, except the P value is located one period ahead of the first A value, not at time 0. For arithmetic and geometric gradients, the P value is two periods ahead of where the gradient starts. With this information, it is possible to solve for P, A, or F for any conceivable cash flow series.

We have used the power of spreadsheet functions in determining P, F, and A values by single-cell entries and using cash flow estimates entered into a series of spreadsheet cells. Though spreadsheet solutions are fast, they do remove some of the understanding of how the time value of money and the factors change the equivalent value of money.

PROBLEMS

Present Worth Calculations

3.1   Industrial Electric Services has a contract with the U.S. Embassy in Mexico to provide maintenance for scanners and other devices in the building. If the first payment of $12,000 is received *now*, what is the present worth of the contract, provided the company will receive a total of 10 payments (i.e., years 0 through 9) and the interest rate is 10% per year?

3.2   Civil engineering consulting firms that provide services to outlying communities are vulnerable to a number of factors that affect the financial condition of the communities, such as bond issues and real estate developments. A small consulting firm entered into a fixed-price contract with a large developer, resulting in a stable income of $260,000 per year in years 1 through 3. At the end of that time, a mild recession slowed the development, so the parties signed another contract for $190,000 per year for 2 more years. Determine the present worth of the two contracts at an interest rate of 10% per year.

3.3   The net cash flow associated with development and sale of a new product is shown. Determine the present worth at an interest rate of 12% per year. The cash flow is in $1000 units. Show (a) hand and (b) spreadsheet solutions.

Year	1	2	3	4	5	6	7	8	9
Cash Flow, $	−120	−100	−40	+50	+50	+80	+80	+80	+80

3.4   Standby power for water utility pumps and other electrical devices is provided by diesel-powered generators. As an alternative, the utility can use natural gas to power the generators, but it will be a few years before the gas is available at remote sites. The utility estimates that by switching to gas, it will save $22,000 per year, starting 3 years from now. At an interest rate of 8% per year, determine the present worth in year 0 of the projected savings that will occur in years 3 through 10.

3.5   Find the present worth at $i = 10\%$ per year for the cash flow series shown below.

3.6   The subsidized cost of producing water at the El Paso Water Utilities (EPWU) Kay Bailey Hutchison (KBH) desalting plant is $1.56 per 1000 gallons. Under a contract that EPWU has with Fort Bliss, the utility sells water to the Army post at a discounted price of $1.28 per 1000 gallons. (The KBH plant was built on Army property.) If Fort Bliss uses 2 billion gallons of water each year, what is the present worth of the discount for a 20-year period at an interest rate of 6% per year?

3.7   The rising cost of athletic programs at major universities has induced college presidents and athletic directors to devise innovative ways to finance cash-strapped sports programs. One of the latest schemes for big-time athletics is the "sports mortgage." At the University of Kansas, Jayhawk fans can sign up to pay $105,000 now, or over a 10-year period, for the right to buy top seats for football games during the next 30 years. In return, the seats themselves will stay locked in at current-year prices. Season tickets in tier 1 are currently selling for $350 each. A fan plans to purchase the sports mortgage along with a current-season ticket and pay for both *now*, then buy a ticket each year for the next 30 years. What is the total present worth of the pricing plan at an interest rate of 10% per year?

3.8   The cash flow associated with making self-locking fasteners is shown below. Determine the net present worth (year 0) at an interest rate of 10% per year.

Year	0	1	2	3	4	5	6	7	8	9
Income, $1000	20	20	20	20	30	30	30	30	30	30
Cost, $1000	8	8	8	8	12	12	12	12	12	25

3.9   Beckman Technologies, a relatively small manufacturer of precision laboratory equipment, borrowed $2 million to renovate one of its testing labs. In an effort to pay off the loan quickly, the company made four payments in years 1 through 4, with each payment being twice as large as the preceding one. At an interest rate of 10% per year, what was the size of the first payment?

Annual Worth Calculations

3.10   Revenue from the sale of ergonomic hand tools was $300,000 in years 1 through 4 and $465,000 in years 5 through 9. Determine the equivalent annual revenue in years 1 through 9 at an interest rate of 10% per year.

3.11   Two engineering graduates who recently got married are planning for their early retirement 20 years from now. They believe that they will need $2,000,000 in year 20. Their plan is to live on one of their salaries and invest the other. They already have $25,000 in their investment account. (a) How much will they have to invest each year if the account grows at a rate of 10% per year? (b) If the maximum they have available to invest each year is $40,000, will they reach their goal of $2 million by year 20?

3.12   Costs associated with the manufacture of miniature high-sensitivity piezoresistive pressure transducers are $73,000 per year. A clever industrial engineer found that by spending $16,000 now to reconfigure the production line and reprogram two of the robotic arms, the cost will go down to $58,000 next year and $52,000 in years 2 through 5. Using an interest rate of 10% per year, determine (a) the equivalent annual *cost* of the manufacturing operations and (b) the equivalent annual *savings* in years 1 through 5.

3.13 Calculate the equivalent annual cost in years 1 through 9 of the following series of disbursements. Use an interest rate of 10% per year. Show (a) hand and (b) spreadsheet solutions.

Year	Disbursement, $	Year	Disbursement, $
0	8000	5	4000
1	4000	6	5000
2	4000	7	5000
3	4000	8	5000
4	4000	9	5000

3.14 For the cash flows below, find the value of x that makes the equivalent annual worth in *years 1 through 7* equal to $300 per year. Use an interest rate of 10% per year. Show solutions (a) by hand and (b) using the Goal Seek tool.

Year	Cash Flow, $	Year	Cash Flow, $
0	200	4	x
1	200	5	200
2	200	6	200
3	200	7	200

3.15 Precision Instruments, Inc. manufactures high-sensitivity mini accelerometers designed for modal analysis testing. The company borrowed $10,000,000 with the understanding that it would make a $2,000,000 payment at the end of year 1 and then make equal annual payments in years 2 through 5 to pay off the loan. If the interest rate on the loan was 9% per year, how much was each payment in years 2 through 5?

3.16 A construction management company is examining its cash flow requirements for the next 7 years. The company expects to replace office machines and computer equipment at various times over the 7-year planning period. Specifically, the company expects to spend $6000 one year from now, $9000 three years from now, and $10,000 six years from now. What is the annual worth (in years 1 through 7) of the planned expenditures, at an interest rate of 10% per year?

3.17 Find the equivalent annual worth for the cash flows shown, using an interest rate of 12% per year. Monetary units are $1000.

Year	1	2	3	4	5	6	7	8	9
Cash Flow, $	20	20	20	20	60	60	60	60	60

3.18 Calculate the net annual worth in years 1 through 10 of the following series of incomes and expenses, if the interest rate is 10% per year.

Year	Income, $/Year	Expense, $/Year
0	0	−2500
1–4	700	−200
5–10	2000	−300

3.19 The city of El Paso gave the El Paso Tennis and Swim Club a 20-year lease to continue using a 10-acre arroyo park for its facilities. The club will pay $1000 per year plus make $350,000 in improvements at the park. In addition, the club will open its tennis courts to the public from 1 to 5 P.M. Monday through Thursday. If the club makes $100,000 worth of improvements now and then $50,000 worth each year for the next 5 years, what is the equivalent annual cost of the lease to the club at an interest rate of 10% per year?

3.20 Stadium Capital Financing Group is a Chicago company that conceived the so-called sports mortgage wherein fans agree to pay a relatively large amount of money over a 10- to 30-year period for the right to buy top seats for football games for up to the next 50 years. In return, season ticket prices stay locked in at current-year prices, and the package can be sold in the secondary market, while taking a tax write-off for donating to a school. Assume a fan buys a sports mortgage at West Virginia University for $150,000 that is to be paid over a 10-year period with the right to buy two season tickets for $300 each for the next 30 years. The first payment is made now (i.e., beginning-of-year payment), and an additional nine payments are to be made at the end of each year for the next 9 years. Assume the fan also purchases the two season tickets (also beginning-of-year payments). What is the total amount of the payment each year in years *0 through 9*? Use an interest rate of 10% per year.

Future Worth Calculations

3.21 The expansion plans of Acme Granite, Stone & Brick call for the company to add capacity for a new product in 5 years. The company wants to have $360,000 available before it announces the product. If the company sets aside $55,000 now and $90,000 in year 2, what uniform annual amount will it have to put in an account in years 3 through 5 to have the $360,000? Assume the account earns interest at 8% per year.

3.22 For the cash flows shown, calculate the future worth in *year 8* using $i = 10\%$ per year.

Year	0	1	2	3	4	5	6
Cash Flow, $	100	100	100	100	100	300	300

3.23 Assume you plan to start an annuity plan by making your first deposit *now*. If you make annual deposits of a uniform amount A into the account that earns interest at a rate of 7% per year, how many years from now will it be until the value of the account is equal to 10 times the value of a single deposit?

3.24 New actuator element technology enables engineers to simulate complex computer-controlled movements in any direction. If the technology results in cost savings in the design of amusement park rides, what is the future worth in year 5 of savings of $70,000 now and $20,000 per year in years 1 through 3 at an interest rate of 10% per year?

3.25 Austin Utilities is planning to install solar panels to provide some of the electricity for its groundwater desalting plant. The project would be done in two phases. The first phase will cost $4 million in year 1 and $5 million in year 2. This investment will result in energy savings (phase 2) of $540,000 in year 3, $546,000 in year 4, and amounts increasing by $6000 each year through year 10. Let $i = 10\%$ per year.

(*a*) What is the future worth of the *savings*?
(*b*) Is the cost of the solar project justified by the savings? (Hint: Calculate the difference between savings and cost).

3.26 For the cash flow diagram shown, determine the value of W that will render the equivalent future worth in year 8 equal to $\$-500$ at an interest rate of 10% per year.

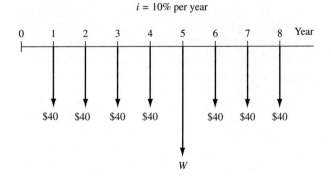

3.27 For the cash flows shown in the diagram, determine the value of x that will make the future worth in year 8 equal to $\$-70,000$.

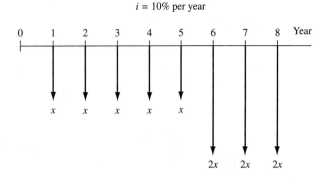

Random Single Amounts and Uniform Series

3.28 A small oil company is planning to replace its Coriolis flowmeters with Emerson Hastelloy flowmeters. The replacement process will cost the company $50,000 three years from now. How much money must the company set aside each year beginning *now* (year 0) in order to have the total amount available immediately after making the last deposit at the end of year 3? Assume the company can invest its funds at 15% per year.

3.29 A company that manufactures air-operated drain valve assemblies budgeted $74,000 per year to pay for plastic components over a 5-year period. If the company spent only $42,000 in year 1, what uniform annual amount should the company expect to spend in each of the next *4 years* to expend the entire budget? Assume the company uses an interest rate of 10% per year.

3.30 A recently hired chief executive officer wants to reduce future production costs to improve the company's earnings, thereby increasing the value of the company's stock. The plan is to invest $40,000 now and $40,000 in each of the next 2 years to improve productivity. By how much must annual costs decrease in years 3 through 7 to recover the investment plus a return of 12% per year?

3.31 Use the cash flow diagram to determine the single amount of money Q_4 in year 4 that is equivalent to all of the cash flows shown. Use $i = 10\%$ per year.

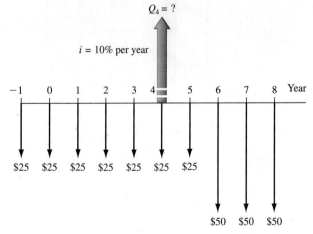

3.32 For the following series of income and expenses, find the equivalent value in year 9 at an interest rate of 12% per year. Show (*a*) hand and (*b*) spreadsheet solutions.

Years	Income, $	Expense, $
0	0	−70,000
1–6	9,000	−13,000
7–9	28,000	−14,000
10–16	38,000	−19,000

3.33 An investor just purchased property under a unique financing agreement with the seller. The contract price is $1.6 million. The payment plan is *Z* dollars now, 2*Z* dollars in year 2, and 3*Z* dollars in years 3 through 5. If the interest rate on the transaction is 10% per year, how much is the payment in year 2?

3.34 A foursome of entrepreneurial electrical engineering graduates has a plan to start a new solar power equipment company based on STE (solar thermal electric) technology. They have recently approached a group of investors with their idea and asked for a loan of $5 million. Within the agreement, the loan is to be repaid by allocating 80% of the company's profits each year for the first 4 years to the investors. In the fifth year, the company will pay the balance remaining on the loan in cash. The company's business plan indicates that they expect to make no profit for the first year, but in years 2 through 4, they anticipate profits to be $1.5 million per year. If the investors accept the deal at an interest rate of 15% per year, and the business plan works to perfection, what is the expected amount of the last loan payment (in *year 5*)?

Shifted Gradients

3.35 Find the present worth in year 0 for the cash flows shown. Let *i* = 10% per year.

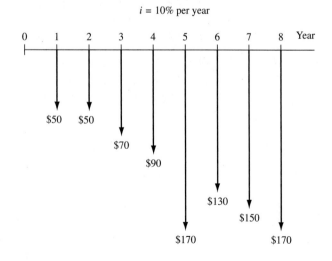

i = 10% per year

3.36 For the cash flows shown, determine the present worth in year 0, if the interest rate is 12% per year.

Year	1	2	3	4	5	6	7	8	9	10
Cash Flow, $	13	13	13	13	16	19	22	25	28	31

3.37 A low-cost noncontact temperature measuring tool may be able to identify railroad car wheels that are in need of repair long before a costly structural failure occurs. If BNF Railroad saves $100,000 in years 1 through 5, $110,000 in year 6, and constant amounts increasing by $10,000 each year through year 20, what is the equivalent annual worth over the 20 years of the savings? The interest rate is 10% per year.

3.38 Determine the present worth in year 0 of the cash flows shown at an interest rate of 15% per year.

Year	1	2	3	4	5	6	7	8	9	10
Cash Flow, $	90	90	90	85	80	75	70	65	60	55

3.39 The Pedernales Electric Cooperative estimates that the present worth *now* of income from an investment in renewable energy sources is $12,475,000. There will be no income in years 1 and 2, but in year 3 income will be $250,000, and thereafter it will increase according to an arithmetic gradient through year 15. What is the required gradient, if the interest rate is 15% per year? Solve (*a*) by hand and (*b*) using a spreadsheet.

3.40 Calculate the annual cost in years 1 through 9 of the following series of disbursements. Use an interest rate of 10% per year.

Year	Disbursement, $	Year	Disbursement, $
0	5000	5	7500
1	5500	6	8000
2	6000	7	8500
3	6500	8	9000
4	7000	9	9500

3.41 The cost associated with manufacturing high-performance lubricants closely follows the cost of crude oil. For the last 10 years, a small, independent refiner had a cost of $3.4 million in years 1 through 3, after which the cost increased by 3% per year through this year. Determine the current equivalent worth (i.e., *now*) of the manufacturing cost, using an interest rate of 10% per year. Show both (*a*) hand and (*b*) spreadsheet solutions.

3.42 Find the future worth in year 10 of $50,000 in year 0 and amounts increasing by 15% per year through year 10 at an interest rate of 10% per year.

3.43 The cost of tuition at public universities has been steadily increasing for many years. One Midwestern university pledged to keep the tuition constant for

4 years for all students who finished in the top 3% of their class. One such student who liked research planned to enroll at the university and continue there until earning a PhD degree (a total time of 9 years). If the tuition for the first 4 years will be $7200 per year and it increases by 5% per year for the next 5 years, what is the present worth of the tuition cost at an interest rate of 8% per year?

3.44 A private equity firm purchased a cable company and assumed the company's debt as part of the transaction. The deal was structured such that the private equity firm received $3 million immediately after the deal was closed (in year 0) through the sale of some assets. This year (year 1) income was $3.36 million, and it is projected to increase by 12% each year through expansion of the customer base. What was the present worth in the year of purchase of the income stream over a 10-year period? The firm's expected rate of return for any purchase is 15% per year.

3.45 Calculate the present worth in year 0 of a series of cash flows that starts at $150,000 in year 0 and increases by 10% per year through year 5. Assume $i = 10\%$ per year.

Shifted Decreasing Gradients

3.46 For the cash flows shown, determine the value of G such that the present worth in year 0 equals $16,000 at an interest rate of 10% per year.

Year	0	1	2	3	4	5
Cash Flow, $	0	8000	8000	8000−G	8000−2G	8000−3G

3.47 For these cash flows, find the equivalent annual worth in *years 1 through 7* at an interest rate of 10% per year.

Year	Cash Flow, $	Year	Cash Flow, $
0	850	4	650
1	800	5	600
2	750	6	550
3	700	7	500

3.48 The pumping cost for delivering water from the Ohio River to Wheeling Steel for cooling hot-

rolled steel was $1.8 million for the first 3 years. An effective energy conservation program resulted in a reduced cost to $1.77 million in year 4, $1.74 million in year 5, and amounts decreasing by $30,000 each year through year 9. What is the equivalent annual worth of the pumping costs over the 9 years at an interest rate of 12% per year?

3.49 Income from the mining of mineral deposits usually decreases as the resource becomes more difficult to extract. Determine the future worth in year 10 of a mineral lease that yielded income of $14,000 in years 1 through 4 and then amounts that decreased by 5% per year through year 10. Use an interest rate of 18% per year. Show (*a*) hand and (*b*) spreadsheet solutions.

3.50 For the cash flows shown in the diagram, determine the future worth in year 8 at an interest rate of 10% per year.

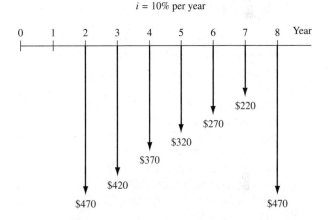

$i = 10\%$ per year

3.51 Income from the sale of application software (apps) is usually constant for several years and then decreases quite rapidly as the market gets close to saturation. Income from one smart phone app was $38,000 in years 1 through 3 and then decreased geometrically by 15% per year through year 7. Determine the equivalent annual income in years 1 through 7, using an interest rate of 10% per year.

3.52 Determine the future worth in year 10 of a cash flow series that starts in year 0 at $100,000 and decreases by 12% per year. Use an interest rate of 12% per year.

ADDITIONAL PROBLEMS AND FE EXAM REVIEW QUESTIONS

3.53 A manufacturer of toilet flush valves wants to have $1,900,000 available 3 years from now so that a new product line can be initiated. If the company plans to deposit money each year, starting *now*, the equation that represents the deposit each year at 8% per year interest is:
(a) $1,900,000(A/F,8\%,3)$
(b) $1,900,000(A/F,8\%,4)$
(c) $1,900,000 + 1,900,000(A/F,8\%,3)$
(d) $1,900,000 + 1,900,000(A/F,8\%,2)$

3.54 The present worth in year 0 of a lease that requires a payment of $9000 *now* and amounts increasing by 5% per year through year 10 at 8% per year interest is closest to:
(a) $73,652
(b) $79,939
(c) $86,335
(d) Over $87,000

3.55 For the diagram shown, the respective values of *n* for the following equation are:

$$P_0 = 100(P/A,10\%,n)(P/F,10\%,n)$$

(a) 6 and 1
(b) 6 and 2
(c) 7 and 1
(d) 7 and 2

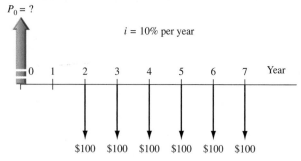

3.56 Summit Metals is planning to expand its Wichita, Kansas, manufacturing operation 5 years from now at a cost of $10,000,000. If the company plans to deposit money into an account each year for 4 years beginning 2 years from now (first deposit is in year 2) to pay for the expansion, the equation that represents the amount of the deposit at 9% per year interest is:
(a) $A = 10,000,000(A/F,9\%,5)$
(b) $A = 10,000,000(A/F,9\%,4)$
(c) $A = 10,000,000(A/P,9\%,4)$
(d) $A = 10,000,000(A/F,9\%,4)(P/F,9\%,1)$

3.57 The amount of money a person must deposit 3 years from now in order to withdraw $10,000 per year for 10 years beginning 15 years from now at an interest rate of 10% per year is closest to:
(a) $15,500   (b) $17,200
(c) $19,300   (d) $21,500

3.58 Cindy wants to deposit money for 4 consecutive years starting 3 years from now so she can withdraw $50,000 twelve years from now. Assume the interest rate is 8% per year. The annual deposit is closest to:
(a) $6990
(b) $7670
(c) $8530
(d) $10,490

3.59 The net present worth in year 0 of the following series of incomes and expenses at 8% per year is closest to:

Years	Income, $	Expenses, $
0	12,000	−1000
1–6	700	−100
7–11	900	−200

(a) $14,300   (b) $15,500
(c) $16,100   (d) $16,500

3.60 For the cash flows shown, the equivalent annual worth in periods 1 through 5 at an interest rate of 10% per year is closest to:

Annual Period	Amount, $
0	1000
1	1000
2	1000
3	1000
4	1000
5	1500

(a) $1120   (b) $1240
(c) $1350   (d) $1490

3.61 For the cash flows shown, the value of *X* that will make the present worth in year 0 equal to $5000 at an interest rate of 10% per year is closest to:

Year	0	1	2	3	4	5	6	7	8	9
Cash Flow, $	200	300	400	500	600	700	800	900	1000	X

(a) $2895   (b) $3125
(c) $3305   (d) $3765

3.62 In order to have cash available for unforeseen emergencies, Baring Systems, a military contractor, wants to have $2,000,000 in a contingency fund 4 years from now. The amount the company must deposit each year in years 0 through 4 at an interest rate of 10% per year is closest to:
(a) $420,100   (b) $327,600
(c) $284,600   (d) $206,900

CASE STUDY

PRESERVING LAND FOR PUBLIC USE

Background and Information

The Trust for Public Land (TPL) is a national organization that purchases and oversees the improvement of large land sites for government agencies at all levels. Its mission is to ensure the preservation of the natural resources, while providing necessary, but minimal, development for recreational use by the public. All TPL projects are evaluated at 7% per year, and TPL reserve funds earn 7% per year.

A southern U.S. state, which has long-term groundwater problems, has asked the TPL to manage the purchase of 10,000 acres of aquifer recharge land and the development of three parks of different use types on the land. The 10,000 acres will be acquired in increments over the next 5 years with $4 million expended immediately on purchases. Total annual purchase amounts are expected to decrease 25% each year through year 5 and then cease for this particular project.

A city with 1.5 million citizens immediately to the southeast of this acreage relies heavily on the aquifer's water. Its citizens passed a bond issue last year, and the city government now has available $3 million for the purchase of land. The bond interest rate is an effective 7% per year.

The engineers working on the park plan intend to complete all the development over a 3-year period starting in year 4, when the amount budgeted is $550,000. Increases in construction costs are expected to be $100,000 each year through year 6.

At a recent meeting, the following agreements were made:

- Purchase the initial land increment now. Use the bond issue funds to assist with this purchase. Take the remaining amount from TPL reserves.
- Raise the remaining project funds over the next 2 years in equal annual amounts.
- Evaluate a financing alternative (suggested informally by one individual at the meeting) in which the TPL provides all funds, except the $3 million available now, until the parks development is initiated in year 4.

Case Study Exercises

1. For each of the 2 years, what is the equivalent annual amount necessary to supply the remaining project funds?

2. If the TPL did agree to fund all costs except the $3 million bond proceeds now available, determine the equivalent annual amount that must be raised in years 4 through 6 to supply all remaining project funds. Assume the TPL will not charge any extra interest over the 7% to the state or city on the borrowed funds.

3. Review the TPL website (www.tpl.org). Identify some economic and noneconomic factors that you believe must be considered when the TPL is deciding to purchase land to protect it from real estate development.

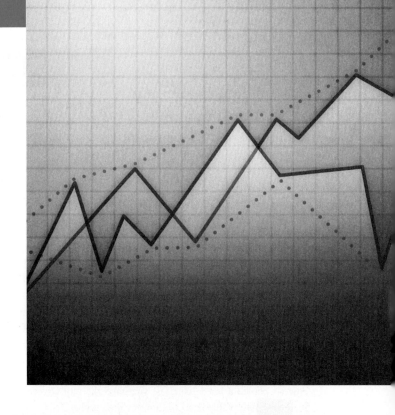

CHAPTER 4

Nominal and Effective Interest Rates

LEARNING OUTCOMES

Purpose: Make computations for interest rates and cash flows that are on a time basis other than a year.

SECTION	TOPIC	LEARNING OUTCOME
4.1	Statements	• Understand interest rate statements that include nominal and effective rates.
4.2	Effective annual rate	• Derive and use the formula for an effective annual interest rate.
4.3	Effective rate	• Determine the effective interest rate for any stated time period.
4.4	Payment period and compounding period	• Determine the payment period (PP) and compounding period (CP) for equivalence computations.
4.5	Single cash flows with PP ≥ CP	• Perform equivalence calculations for single-amount cash flows and PP ≥ CP.
4.6	Series cash flows with PP ≥ CP	• Perform equivalence calculations for series and gradient cash flows and PP ≥ CP.
4.7	Single amounts and series with PP < CP	• Perform equivalence calculations for cash flows with PP < CP.
4.8	Continuous compounding	• Derive and use the effective interest rate formula for interest rates that are compounded continuously.
4.9	Varying rates	• Perform equivalency calculations for interest rates that vary from one time period to another.

I n all engineering economy relations developed thus far, the interest rate has been a constant, annual value. For a substantial percentage of the projects evaluated by professional engineers in practice, the interest rate is compounded more frequently than once a year; frequencies such as semiannually, quarterly, and monthly are common. In fact, weekly, daily, and even continuous compounding may be experienced in some project evaluations. Also, in our own personal lives, many of our financial considerations—loans of all types (home mortgages, credit cards, automobiles, boats), checking and savings accounts, investments, stock option plans, etc.—have interest rates compounded for a time period shorter than 1 year. This requires the introduction of two new terms—**nominal and effective interest rates.**

This chapter explains how to understand and use nominal and effective interest rates in engineering practice and in daily life situations. Equivalence calculations for any compounding frequency in combination with any cash flow frequency are presented.

PE

The Credit Card Offer Case: Today, Dave received a special offer of a new credit card from Chase Bank linked with the major airline that he flies frequently. It offers a generous bonus package for signing up by a specific date about 60 days from now. The bonus package includes extra airline points (once the first purchase is made), priority airport check-in services (for 1 year), several free checked-bag allowances (for up to 10 check-ins), extra frequent-flyer points on the airline, access to airline lounges (provided he uses the card on a set time basis), plus several other rewards (rental car discounts, cruise trip amenities, and floral order discounts).

The annual fee of $85 for membership does not start until the second year, and balance transfers from other credit cards have a low transfer fee, provided they are made at the time of initial membership.

Dave has a credit card currently with a bank that he is planning to leave due to its poor customer service and high monthly fees. If he enrolls, he will transfer the $1000 balance on the current card to the new Chase Bank card.

In the page that accompanies the offer letter, "pricing information" is included. This includes interest rates, interest charges, and fees. A summary of several of these rates and fees follows.

APR (annual percentage rate) for purchases and balance transfers*	**14.24% per year** (sum of the current U.S. Government prime rate of 3.25% and 10.99%, which is the APR added to determine the balance transfer APR for Chase Bank)
APR for cash and overdraft advances*	**19.24% per year**
Penalty APR for late minimum payment, exceeding credit limit, and returned unpaid payments*†	**29.99% per year (maximum penalty APR)**

Fees are listed as follows:

Annual membership	$85; free the first year
Balance transfers	$5 or 3% of each transfer, whichever is greater
Cash advances	$10 or 3% of each advance, whichever is greater
Late payment	$39 each occurrence, if balance exceeds $250
Over the credit limit	$39 each occurrence
Returned check or payment	$39 each occurrence

*All APR rates are variable, based on a current 3.25% prime rate with 10.99% added to determine purchase/balance transfer APR; with 15.99% added to determine cash/overdraft APR; and with 26.99% added to determine penalty APR.
†The penalty APR applies indefinitely to future transactions. If no minimum payment is received within 60 days, the penalty APR applies to all outstanding balances and all future transactions on the account.

(Continued)

This case is used in the following topics (and sections) of this chapter:

Nominal and effective interest rate statements (4.1)

Effective annual interest rates (4.2)

Equivalence relations: Series with PP ≥ CP (4.6)

Online Presentation

4.1 Nominal and Effective Interest Rate Statements ●●●

In Chapter 1, we learned that the primary difference between simple interest and compound interest is that compound interest includes interest on the interest earned in the previous period, while simple interest does not. Here we discuss *nominal and effective interest rates*, which have the same basic relationship. The difference here is that the concepts of nominal and effective must be used when interest is compounded more than once each year. For example, if an interest rate is expressed as 1% per month, the terms *nominal* and *effective* interest rates must be considered.

To understand and correctly handle effective interest rates is important in engineering practice as well as for individual finances. The interest amounts for loans, mortgages, bonds, and stocks are commonly based upon interest rates compounded more frequently than annually. The engineering economy study must account for these effects. In our own personal finances, we manage most cash disbursements and receipts on a nonannual time basis. Again, the effect of compounding more frequently than once per year is present. First, consider a **nominal interest rate**.

Nominal interest rate *r*

A nominal interest rate *r* is an interest rate that **does not account** for compounding. By definition,

$$r = \text{interest rate per time period} \times \text{number of periods} \qquad [4.1]$$

A nominal rate may be calculated for *any time period longer than the time period stated* by using Equation [4.1]. For example, the interest rate of 1.5% per month is the same as each of the following nominal rates.

Time Period	Nominal Rate by Equation [4.1]	What This Is
24 months	1.5 × 24 = 36%	Nominal rate per 2 years
12 months	1.5 × 12 = 18%	Nominal rate per 1 year
6 months	1.5 × 6 = 9%	Nominal rate per 6 months
3 months	1.5 × 3 = 4.5%	Nominal rate per 3 months

Note that none of these rates mention anything about compounding of interest; they are all of the form "*r*% per time period." These nominal rates are calculated in the same way that simple rates are calculated using Equation [1.7], that is, interest rate *times* number of periods.

After the nominal rate has been calculated, the **compounding period (CP)** must be included in the interest rate statement. As an illustration, again consider the nominal rate of 1.5% per month. If we define the CP as 1 month, the nominal rate statement is 18% per year, *compounded monthly*, or 4.5% per quarter, *compounded monthly*. Now we can consider an **effective interest rate**.

Effective interest rate *i*

An effective interest rate *i* is a rate wherein the **compounding of interest is taken into account**. Effective rates are commonly expressed on an annual basis as an effective annual rate; however, any time basis may be used.

The most common form of interest rate statement when compounding occurs over time periods shorter than 1 year is "% per time period, compounded CP-ly," for example, 10% per year, compounded monthly, or 12% per year, compounded weekly. An effective rate may not always include the compounding period in the statement. If the CP is not mentioned, it is understood to be the same as the time period mentioned with the interest rate. For example, an interest rate of "1.5% per month" means that interest is compounded each month; that is, CP is 1 month. An equivalent effective rate statement, therefore, is 1.5% per month, compounded monthly.

All of the following are effective interest rate statements because either **they state they are effective** or the **compounding period is not mentioned**. In the latter case, the CP is the same as the time period of the interest rate.

Statement	CP	What This Is
$i = 10\%$ per year	CP not stated; CP = year	Effective rate per year
$i =$ effective 10% per year, compounded monthly	CP stated; CP = month	Effective rate per year
$i = 1\frac{1}{2}\%$ per month	CP not stated; CP = month	Effective rate per month
$i =$ effective $1\frac{1}{2}\%$ per month, compounded monthly	CP stated; CP = month	Effective rate per month; terms *effective* and *compounded monthly* are redundant
$i =$ effective 3% per quarter, compounded daily	CP stated; CP = day	Effective rate per quarter

All nominal interest rates can be converted to effective rates. The formula to do this is discussed in the next section.

> All interest formulas, factors, tabulated values, and spreadsheet functions must use an effective interest rate to properly account for the time value of money.

The term **APR (Annual Percentage Rate)** is often stated as the annual interest rate for credit cards, loans, and house mortgages. This is the same as the **nominal rate**. An APR of 15% is the same as a nominal 15% per year or a nominal 1.25% on a monthly basis. Also the term **APY (Annual Percentage Yield)** is a commonly stated annual rate of return for investments, certificates of deposit, and saving accounts. This is the same as an **effective rate**. The names are different, but the interpretations are identical. As we will learn in the following sections, the effective rate is always greater than or equal to the nominal rate, and similarly APY ≥ APR.

Based on these descriptions, there are always three time-based units associated with an interest rate statement.

Interest period (t)—The period of time over which the interest is expressed. This is the *t* in the statement of *r%* per time period *t*, for example, 1% *per month*. The time unit of 1 year is by far the most common. It is assumed when not stated otherwise.

Compounding period (CP)—The shortest time unit over which interest is charged or earned. This is defined by the compounding term in the interest rate statement, for example, 8% per year, *compounded monthly*. If CP is not stated, it is assumed to be the same as the interest period.

Compounding frequency (m)—The number of times that compounding occurs within the interest period *t*. If the compounding period CP and the time period *t* are the same, the compounding frequency is 1, for example, 1% *per month, compounded monthly*.

Consider the (nominal) rate of 8% per year, compounded monthly. It has an interest period *t* of 1 year, a compounding period CP of 1 month, and a compounding frequency *m* of 12 times per year. A rate of 6% per year, compounded weekly, has *t* = 1 year, CP = 1 week, and *m* = 52, based on the standard of 52 weeks per year.

In previous chapters, all interest rates had *t* and CP values of 1 year, so the compounding frequency was always *m* = 1. This made them all effective rates, because the interest period and compounding period were the same. Now, it will be necessary to express a nominal rate as an effective rate on the same time base as the compounding period.

An effective rate can be determined from a nominal rate by using the relation

$$\text{Effective rate per CP} = \frac{r\% \text{ per time period } t}{m \text{ compounding periods per } t} = \frac{r}{m} \qquad [4.2]$$

As an illustration, assume $r = 9\%$ per year, compounded monthly; then $m = 12$. Equation [4.2] is used to obtain the effective rate of $9\%/12 = 0.75\%$ per month, compounded monthly.

Note that changing the interest period t does not alter the compounding period, which is 1 month in this illustration. Therefore, $r = 9\%$ per year, compounded monthly, and $r = 4.5\%$ per 6 months, compounded monthly, are two expression of the same interest rate.

EXAMPLE 4.1

Three different bank loan rates for electric generation equipment are listed below. Determine the effective rate on the basis of the compounding period for each rate.

(a) 9% per year, compounded quarterly.
(b) 9% per year, compounded monthly.
(c) 4.5% per 6 months, compounded weekly.

Solution

Apply Equation [4.2] to determine the effective rate per CP for different compounding periods. The graphic in Figure 4–1 indicates the effective rate per CP and how the interest rate is distributed over time.

Figure 4–1
Relations between interest period t, compounding period CP, and effective interest rate per CP.

Sometimes it is not obvious whether a stated rate is a nominal or an effective rate. Basically there are three ways to express interest rates, as detailed in Table 4–1. The right column includes a statement about the effective rate. For the first format, a nominal interest rate is given and the compounding period is stated. The effective rate must be calculated (discussed in the next sections). In the second format, the stated rate is identified as effective (or APY could also be used), so the rate is used directly in computations.

In the third format, no compounding period is identified, for example, 8% per year. This rate is effective over a compounding period equal to the stated interest period of 1 year in this case. The effective rate for any other time period must be calculated.

TABLE 4–1	Various Ways to Express Nominal and Effective Interest Rates	
Format of Rate Statement	**Example of Statement**	**What about the Effective Rate?**
Nominal rate stated, compounding period stated	8% per year, compounded quarterly	Find effective rate for any time period (next two sections)
Effective rate stated	Effective 8.243% per year, compounded quarterly	Use effective rate of 8.243% per year directly for annual cash flows
Interest rate stated, no compounding period stated	8% per year	Rate is effective for CP equal to stated interest period of 1 year; find effective rate for all other time periods

EXAMPLE 4.2   The Credit Card Offer Case   **PE**

As described in the introduction to this case, Dave has been offered what is described as a credit card deal that should not be refused—at least that is what the Chase Bank offer letter implies. The balance transfer APR interest rate of 14.24% is an annual rate, with no compounding period mentioned. Therefore, it follows the format of the third entry in Table 4–1, that is, interest rate stated, no CP stated. Therefore, we should conclude that the CP is 1 year, the same as the annual interest period of the APR. However, as Dave and we all know, credit card payments are required monthly.

(a) First, determine the effective interest rates for *compounding periods of 1 year and 1 month* so Dave knows some effective rates he might be paying when he transfers the $1000 balance from his current card.

(b) Second, assume that immediately after he accepts the card and completes the $1000 transfer, Dave gets a bill that is due 1 month later. What is the amount of the total balance he owes?

Now, Dave looks a little closer at the fine print of the "pricing information" sheet and discovers a small-print statement that Chase Bank uses the daily balance method (including new transactions) to determine the balance used to calculate the interest due at payment time.

(c) We will reserve the implication of this new finding until later, but for now help Dave by determining the *effective daily interest rate* that may be used to calculate interest due at the end of 1 month, provided the CP is 1 day.

Solution

(a) The interest period is 1 year. Apply Equation [4.2] for both CP values of 1 year ($m = 1$ compounding period per year) and 1 month ($m = 12$ compounding periods per year).

    CP of year: Effective rate per year = 14.24/1 = 14.24%

    CP of month: Effective rate per month = 14.24/12 = 1.187%

(b) The interest will be at the monthly effective rate, plus the balance transfer fee of 3%.

$$\text{Amount owed after 1 month} = 1000 + 1000(0.01187) + 0.03(1000)$$
$$= 1000 + 11.87 + 30$$
$$= \$1041.87$$

Including the $30 fee, this represents an interest rate of (41.87/1000)(100%) = 4.187% for only the 1-month period.

(c) Again apply Equation [4.2], now with $m = 365$ compounding periods per year.

    CP of day: Effective rate per day = 14.24/365 = 0.039%

4.2 Effective Annual Interest Rates ● ● ●

In this section, effective **annual interest rates** are calculated. Therefore, the year is used as the interest period t, and the compounding period CP can be any time unit less than 1 year. For example, we will learn that a *nominal* 18% per year, compounded quarterly is the same as an *effective* rate of 19.252% per year.

Online Presentation

The symbols used for nominal and effective interest rates are

    r = nominal interest rate per year

  CP = time period for each compounding

    m = number of compounding periods per year

     i = effective interest rate per compounding period = r/m

  i_a = effective interest rate per year

The relation $i = r/m$ is exactly the same as Equation [4.2].

Figure 4–2
Future worth calculation
at a rate i, compounded m
times in a year.

As mentioned earlier, treatment for nominal and effective interest rates parallels that of simple and compound interest. Like compound interest, an effective interest rate at any point during the year includes (compounds) the interest rate for all previous compounding periods during the year. Therefore, the derivation of an effective interest rate formula directly parallels the logic used to develop the future worth relation $F = P(1 + i)^n$. We set $P = \$1$ for simplification.

The future worth F **at the end of 1 year** is the principal P plus the interest $P(i)$ through the year. Since interest may be compounded several times during the year, use the effective annual rate symbol i_a to write the relation for F with $P = \$1$.

$$F = P + Pi_a = 1(1 + i_a)$$

Now consider Figure 4–2. The effective rate i per CP must be compounded through all m periods to obtain the total effect of compounding by the end of the year. This means that F can also be written as

$$F = 1(1 + i)^m$$

Equate the two expressions for F and solve for i_a. The **effective annual interest rate formula** for i_a is

$$i_a = (1 + i)^m - 1 \qquad \text{[4.3]}$$

Equation [4.3] calculates the effective annual interest rate i_a for any number of compounding periods per year when i is the rate for one compounding period.

If the effective annual rate i_a and compounding frequency m are known, Equation [4.3] can be solved for i to determine the *effective interest rate per compounding period.*

$$i = (1 + i_a)^{1/m} - 1 \qquad \text{[4.4]}$$

As an illustration, Table 4–2 utilizes the nominal rate of 18% per year for different compounding periods (year to week) to determine the effective annual interest rate. In each case, the effective rate i per CP is applied m times during the year. Table 4–3 summarizes the effective annual rate for frequently quoted nominal rates using Equation [4.3]. A standard of 52 weeks and 365 days per year is used throughout. The values in the continuous-compounding column are discussed in Section 4.8.

TABLE 4–2 Effective Annual Interest Rates Using Equation [4.3]

$r = 18\%$ per year, compounded CP-ly

Compounding Period, CP	Times Compounded per Year, m	Rate per Compound Period, $i\%$	Distribution of i over the Year of Compounding Periods	Effective Annual Rate, $i_a = (1 + i)^m - 1$
Year	1	18	18% 1	$(1.18)^1 - 1 = 18\%$
6 months	2	9	9% 9% 1 2	$(1.09)^2 - 1 = 18.81\%$
Quarter	4	4.5	4.5% 4.5% 4.5% 4.5% 1 2 3 4	$(1.045)^4 - 1 = 19.252\%$
Month	12	1.5	1.5% in each 1 2 3 4 5 6 7 8 9 10 11 12	$(1.015)^{12} - 1 = 19.562\%$
Week	52	0.34615	0.34615% in each 1 2 3 24 26 28 50 52	$(1.0034615)^{52} - 1 = 19.684\%$

TABLE 4–3 Effective Annual Interest Rates for Selected Nominal Rates

Nominal Rate r%	Semiannually ($m = 2$)	Quarterly ($m = 4$)	Monthly ($m = 12$)	Weekly ($m = 52$)	Daily ($m = 365$)	Continuously ($m = \infty$; $e^r - 1$)
0.25	0.250	0.250	0.250	0.250	0.250	0.250
0.50	0.501	0.501	0.501	0.501	0.501	0.501
1.00	1.003	1.004	1.005	1.005	1.005	1.005
1.50	1.506	1.508	1.510	1.511	1.511	1.511
2	2.010	2.015	2.018	2.020	2.020	2.020
3	3.023	3.034	3.042	3.044	3.045	3.046
4	4.040	4.060	4.074	4.079	4.081	4.081
5	5.063	5.095	5.116	5.124	5.126	5.127
6	6.090	6.136	6.168	6.180	6.180	6.184
7	7.123	7.186	7.229	7.246	7.247	7.251
8	8.160	8.243	8.300	8.322	8.328	8.329
9	9.203	9.308	9.381	9.409	9.417	9.417
10	10.250	10.381	10.471	10.506	10.516	10.517
12	12.360	12.551	12.683	12.734	12.745	12.750
15	15.563	15.865	16.076	16.158	16.177	16.183
18	18.810	19.252	19.562	19.684	19.714	19.722
20	21.000	21.551	21.939	22.093	22.132	22.140
25	26.563	27.443	28.073	28.325	28.390	28.403
30	32.250	33.547	34.489	34.869	34.968	34.986
40	44.000	46.410	48.213	48.954	49.150	49.182
50	56.250	60.181	63.209	64.479	64.816	64.872

EXAMPLE 4.3

Janice is an engineer with Southwest Airlines. She purchased Southwest stock for $6.90 per share and sold it exactly 1 year later for $13.14 per share. She was very pleased with her investment earnings. Help Janice understand exactly what she earned in terms of (a) effective annual rate and (b) effective rate for quarterly compounding, and for monthly compounding. Neglect any commission fees for purchase and selling of stock and any quarterly dividends paid to stockholders.

Solution

(a) The effective annual rate of return i_a has a compounding period of 1 year, since the stock purchase and sales dates are exactly 1 year apart. Based on the purchase price of $6.90 per share and using the definition of interest rate in Equation [1.2],

$$i_a = \frac{\text{amount of increase per 1 year}}{\text{original price}} \times 100\% = \frac{6.24}{6.90} \times 100\% = 90.43\% \text{ per year}$$

(b) For the effective annual rates of 90.43% per year, compounded quarterly, and 90.43%, compounded monthly, apply Equation [4.4] to find corresponding effective rates on the basis of each compounding period.

Quarter: $m = 4$ times per year $i = (1.9043)^{1/4} - 1 = 1.17472 - 1 = 0.17472$

This is 17.472% per quarter, compounded quarterly.

Month: $m = 12$ times per year $i = (1.9043)^{1/12} - 1 = 1.05514 - 1 = 0.05514$

This is 5.514% per month, compounded monthly.

Comment

Note that these quarterly and monthly rates are less than the effective annual rate divided by the number of quarters or months per year. In the case of months, this would be 90.43%/12 = 7.54% per month. This computation is incorrect because it neglects the fact that compounding takes place 12 times during the year to result in the effective annual rate of 90.43%.

The spreadsheet function that displays the result of Equation [4.3], that is, the **effective annual rate** i_a, is the EFFECT function. The format is

$$= \text{EFFECT(nominal\_rate\_per\_year, compounding\_frequency)}$$

$$= \text{EFFECT}(r\%, m) \hspace{4cm} [4.5]$$

Note that the rate entered in the EFFECT function is the **nominal annual rate $r\%$ per year**, not the effective rate $i\%$ per compounding period. The function automatically finds i for use in Equation [4.3]. As an example, assume the nominal annual rate is $r = 5.25\%$ per year, compounded quarterly, and you want to find the effective annual rate i_a. The correct input is $= \text{EFFECT}(5.25\%, 4)$ to display $i_a = 5.354\%$ per year. This is the spreadsheet equivalent of Equation [4.3] with $i = 5.25/4 = 1.3125\%$ per quarter with $m = 4$.

$$i_a = (1 + 0.013125)^4 - 1 = 0.05354 \quad (5.354\%)$$

The thing to remember about using the EFFECT function is that the nominal rate r entered must be expressed over the same period of time as that of the effective rate required, which is 1 year here.

The NOMINAL spreadsheet function finds the **nominal annual rate r.** The format is

$$= \text{NOMINAL(effective\_rate, compounding\_frequency\_per\_year)}$$

$$= \text{NOMINAL}(i_a\%, m) \hspace{4cm} [4.6]$$

This function is designed to display **only** nominal **annual** rates. Accordingly, the m entered must be the number of times interest is compounded per year. For example, if the effective annual rate is 10.381% per year, compounded quarterly, and the nominal annual rate is requested, the function is $= \text{NOMINAL}(10.381\%, 4)$ to display $r = 10\%$ per year, compounded quarterly. The nominal rates for shorter time periods than 1 year are determined by using Equation [4.1]. For example, the quarterly rate is $10\%/4 = 2.5\%$.

The things to remember when using the NOMINAL function are that the answer is always a nominal annual rate, the rate entered must be an effective annual rate, and the m must equal the number of times interest is compounded annually.

EXAMPLE 4.4   The Credit Card Offer Case   (PE)

In our Progressive Example, Dave is planning to accept the offer for a Chase Bank credit card that carries an APR (nominal rate) of 14.24% per year, or 1.187% per month. He will transfer a balance of $1000 and plans to pay it and the transfer fee of $30, due at the end of the first month. Let's assume that Dave makes the transfer, and only days later his employer has a 1-year assignment for him in the country of the Cameroon in northwestern Africa. Dave accepts the employment offer, and in his hurried, excited departure, he forgets to send the credit card service company a change of address. Since he is now out of mail touch, he does not pay his monthly balance due, which we calculated in Example 4.2 to be $1041.87.

(a) If this situation continues for a total of 12 months, determine the total due after 12 months and the effective annual rate of interest Dave has accumulated. Remember, the fine print on the card's interest and fee information states a penalty APR of 29.99% per year after one late payment of the minimum payment amount, plus a late payment fee of $39 per occurrence.

(b) If there were no penalty APR and no late-payment fee, what effective annual interest rate would be charged for this year? Compare this rate with the answer in part (a).

Solution

(a) Because Dave did not pay the first month's amount, the new balance of $1041.87 and all future monthly balances will accumulate interest at the higher monthly rate of

$$29.99\%/12 = 2.499\% \text{ per month}$$

Additionally, the $39 late-payment fee will be added each month, starting with the second month, and interest will be charged on these fees also each month thereafter. The first 3 months and last 2 months are detailed below. Figure 4–3 details the interest and fees for all 12 months.

	A	B	C	D	E	F	G	H	I	J	K
1		Amount	Monthly		Amount						
2	Month	of credit	interest	Fees	due					Interest rates	
3	0	1000.00		0	1000.00					Regular	Penalty
4	1	1000.00	11.87	30	1041.87				Per year	0.1424	0.2999
5	2	1041.87	26.04	39	1106.90				Per month	0.01187	0.02499
6	3	1106.90	27.66	39	1173.57	= SUM(B6:D6)					
7	4	1173.57	29.33	39	1241.90						
8	5	1241.90	31.04	39	1311.93					= J4/12	= K4/12
9	6	1311.93	32.79	39	1383.72						
10	7	1383.72	34.58	39	1457.30						
11	8	1457.30	36.42	39	1532.72						
12	9	1532.72	38.31	39	1610.03						
13	10	1610.03	40.24	39	1689.27						
14	11	1689.27	42.22	39	1770.48						
15	12	1770.48	44.25	39	1853.73						
16											
17			Interest rate per month				Month 1: transfer fee				
18			Month 1: 1.187%				Months 2-12: late-payment fee				
19			Months 2-12: penalty 2.499%								
20											

Figure 4–3
Monthly amounts due for a credit card, Example 4.4.

Month 1:	$1000 + 1000(0.01187) + 30 = \$1041.87$
Month 2:	$1041.87 + 1041.87(0.02499) + 39 = \$1106.91$
Month 3:	$1106.91 + 1106.91(0.02499) + 39 = \$1173.57$

$$\vdots$$

Month 11:	$1689.25 + 1689.25(0.02499) + 39 = \$1770.46$
Month 12:	$1770.46 + 1770.46(0.02499) + 39 = \$1853.71$

The *effective monthly rate* is determined by using the F/P factor to find the i value at which $1000 now is equivalent to $1853.71 after 12 periods.

$$1853.71 = 1000(F/P,i,12) = 1000(1+i)^{12}$$

$$1+i = (1.85371)^{1/12} = 1.05278$$

$$i = 5.278\% \text{ per month}$$

Since the compounding period is 1 month, use Equation [4.3] to determine the *effective annual rate* of 85.375% per year, compounded monthly.

$$i_a = (1+i)^m - 1 = (1.05278)^{12} - 1$$

$$= 0.85375 \quad (85.375\%)$$

(*b*) If there were no penalty fees for late payments and the nominal annual rate of 14.24% (or 1.187% per month) were applied throughout the 12 months, the effective annual rate would be 15.207% per year, compounded monthly. By Equation [4.3], with a small rounding error included,

$$i_a = (1+i)^m - 1 = (1.01187)^{12} - 1 = 0.15207$$

First, Dave will not pay at the stated rate of 14.24%, because this is the APR (nominal rate), not the APY (effective rate) of 15.207%. Second, and much more important, is the huge difference made by (1) the increase in rate to an APR of 29.99% and (2) the monthly fees of $39 for not making a payment. These large fees become part of the credit balance and accumulate interest at the penalty rate of 29.99% per year. The result is an effective annual rate jump from 15.207% to 85.375% per year, compounded monthly.

Comment

This is but one illustration of why the best advice to an individual or company in debt is to spend down the debt. The quoted APR by credit card, loan, and mortgage institutions can be quite deceiving; plus, the addition of penalty fees increases the effective rate very rapidly.

When Equation [4.3] is applied to find i_a the result is usually not an integer. Therefore, the engineering economy factor cannot be obtained directly from the interest factor tables. There are alternative ways to find the factor value.

- Use the factor formula with the i_a rate substituted for i.
- Use the spreadsheet function with i_a (as discussed in Section 2.4).
- Linearly interpolate between two tabulated rates (as discussed in Section 2.4).

4.3 Effective Interest Rates for Any Time Period ●●●

Online Presentation

Equation [4.3] in Section 4.2 calculates an effective interest rate per year from any effective rate over a shorter time period. We can generalize this equation to determine the **effective interest rate for any time period** (shorter or longer than 1 year).

$$\text{Effective } i \text{ per time period} = \left(1 + \frac{r}{m}\right)^m - 1 \qquad [4.7]$$

where i = effective rate for specified time period (say, semiannual)

r = nominal interest rate for same time period (semiannual)

m = number of times interest is compounded per stated time period (times per 6 months)

The term r/m is always the effective interest rate over a compounding period CP, and m is always the number of times that interest is compounded per the time period on the left of the equals sign in Equation [4.7]. Instead of i_a, this general expression uses i as the symbol for the effective interest rate, which conforms to the use of i throughout the remainder of this text. Examples 4.5 and 4.6 illustrate the use of this equation.

EXAMPLE 4.5

Tesla Motors manufactures high-performance battery electric vehicles. An engineer is on a Tesla committee to evaluate bids for new-generation coordinate-measuring machinery to be directly linked to the automated manufacturing of high-precision vehicle components. Three bids include the interest rates that vendors will charge on unpaid balances. To get a clear understanding of finance costs, Tesla management asked the engineer to determine the effective semiannual and annual interest rates for each bid. The bids are as follows:

    Bid 1:   9% per year, compounded quarterly

    Bid 2:   3% per quarter, compounded quarterly

    Bid 3:   8.8% per year, compounded monthly

(*a*) Determine the effective rate for each bid on the basis of semiannual periods.

(*b*) What are the effective annual rates? These are to be a part of the final bid selection.

(*c*) Which bid has the lowest effective annual rate?

Solution

(*a*) Convert the nominal rates to a semiannual basis, determine m, then use Equation [4.7] to calculate the effective semiannual interest rate i. For bid 1,

$$r = 9\% \text{ per year} = 4.5\% \text{ per 6 months}$$

$$m = 2 \text{ quarters per 6 months}$$

$$\text{Effective } i\% \text{ per 6 months} = \left(1 + \frac{0.045}{2}\right)^2 - 1 = 1.0455 - 1 = 4.55\%$$

Table 4–4 (left section) summarizes the effective semiannual rates for all three bids.

TABLE 4–4	Effective Semiannual and Annual Interest Rates for Three Bid Rates, Example 4.5					
	Semiannual Rates			Annual Rates		
Bid	Nominal r per 6 Months, %	CP per 6 Months, m	Equation [4.7], Effective i, %	Nominal r per Year, %	CP per Year, m	Equation [4.7], Effective i, %
1	4.5	2	4.55	9	4	9.31
2	6.0	2	6.09	12	4	12.55
3	4.4	6	4.48	8.8	12	9.16

(b) For the effective annual rate, the time basis in Equation [4.7] is 1 year. For bid 1,

$$r = 9\% \text{ per year} \qquad m = 4 \text{ quarters per year}$$

$$\text{Effective } i\% \text{ per year} = \left(1 + \frac{0.09}{4}\right)^4 - 1 = 1.0931 - 1 = 9.31\%$$

The right section of Table 4–4 includes a summary of the effective annual rates.

(c) Bid 3 includes the lowest effective annual rate of 9.16%, which is equivalent to an effective semiannual rate of 4.48% when interest is compounded monthly.

EXAMPLE 4.6

A dot-com company plans to place money in a new venture capital fund that currently returns 18% per year, compounded daily. What effective rate is this (a) yearly and (b) semiannually?

Solution

(a) Use Equation [4.7], with $r = 0.18$ and $m = 365$.

$$\text{Effective } i\% \text{ per year} = \left(1 + \frac{0.18}{365}\right)^{365} - 1 = 19.716\%$$

(b) Here $r = 0.09$ per 6 months and $m = 182$ days.

$$\text{Effective } i\% \text{ per 6 months} = \left(1 + \frac{0.09}{182}\right)^{182} - 1 = 9.415\%$$

Online Presentation

4.4 Equivalence Relations: Payment Period and Compounding Period ● ● ◐

Now that the procedures and formulas for determining effective interest rates with consideration of the compounding period are developed, it is necessary to consider the **payment period**.

The payment period (PP) is the length of time between cash flows (inflows or outflows). It is common that the lengths of the payment period and the compounding period (CP) do not coincide. It is important to determine if PP = CP, PP > CP, or PP < CP.

If a company deposits money each month into an account that earns at the nominal rate of 8% per year, compounded semiannually, the cash flow deposits define a payment period of 1 month and the nominal interest rate defines a compounding period of 6 months. These time periods are shown in Figure 4–4. Similarly, if a person deposits a bonus check once a year into an account that compounds interest quarterly, PP = 1 year and CP = 3 months.

Figure 4–4
One-year cash flow diagram for a monthly payment period (PP) and semiannual compounding period (CP).

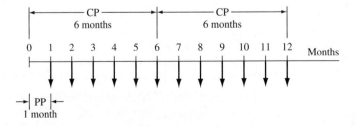

TABLE 4–5	Section References for Equivalence Calculations Based on Payment Period and Compounding Period Comparison	
Length of Time	**Involves Single Amounts (P and F Only)**	**Involves Uniform Series or Gradient Series (A, G, or g)**
PP = CP	Section 4.5	Section 4.6
PP > CP	Section 4.5	Section 4.6
PP < CP	Section 4.7	Section 4.7

As we learned earlier, to correctly perform equivalence calculations, an effective interest rate is needed in the factors and spreadsheet functions. Therefore, it is essential that the time periods of the interest rate and the payment period be on the same time basis. The next three sections (4.5 to 4.7) describe procedures to determine correct i and n values for engineering economy factors and spreadsheet functions. First, compare the length of PP and CP, then identify the cash flows as only single amounts (P and F) or as a series (A, G, or g). Table 4–5 provides the section reference. The section references are the same when PP = CP and PP > CP, because the procedures to determine i and n are the same, as discussed in Sections 4.5 and 4.6.

A general principle to remember throughout these equivalence computations is that when cash actually flows, it is necessary to account for the time value of money. For example, assume that cash flows occur every 6 months and that interest is compounded quarterly. After 3 months there is no cash flow and no need to determine the effect of quarterly compounding. However, at the 6-month time point, it is necessary to consider the interest accrued during the previous two quarters.

4.5 Equivalence Relations: Single Amounts with PP ≥ CP ●●●

Online Presentation

With only P and F estimates defined, the payment period is not specifically identified. In virtually all situations, PP will be equal to or greater than CP. The length of the PP is defined by the interest period in the stated interest rate. If the rate is 8% per year, for example, PP = CP = 1 year. However, if the rate is 10% per year, compounded quarterly, then PP is 1 year, CP is 1 quarter or 3 months, and PP > CP. The procedures to perform equivalence computations are the same for both situations, as explained here.

When only single-amount cash flows are involved, there are two equally correct ways to determine i and n for P/F and F/P factors. Method 1 is easier to apply, because the interest tables in the back of the text can usually provide the factor value. Method 2 likely requires a factor formula calculation, because the resulting effective interest rate is not an integer. For spreadsheets, either method is acceptable; however, method 1 is usually easier.

Method 1: Determine the effective interest rate over the ***compounding period CP***, and set n equal to the number of compounding periods between P and F. The relations to calculate P and F are

$$P = F(P/F, \text{effective } i\% \text{ per CP, total number of periods } n) \qquad [4.8]$$

$$F = P(F/P, \text{effective } i\% \text{ per CP, total number of periods } n) \qquad [4.9]$$

For example, assume that the stated credit card rate is nominal 15% per year, compounded monthly. Here CP is 1 month. To find P or F over a 2-year span, calculate the effective monthly rate of 15%/12 = 1.25% and the total months of 2(12) = 24. Then 1.25% and 24 are used in the P/F and F/P factors.

Any time period can be used to determine the effective interest rate; however, the interest rate that is associated with the CP is typically the best because it is usually a whole number. Therefore, the factor tables in the back of the text can be used.

Method 2: Determine the **effective interest rate for the *time period t of the nominal rate,*** and set n equal to the total number of periods, using this same time period.

The P and F relations are the same as in Equations [4.8] and [4.9] with the term *effective i% per t* substituted for the interest rate. For a credit card rate of 15% per year, compounded monthly, the time period t is 1 year. The effective rate over 1 year and the n values are

$$\text{Effective } i\% \text{ per year} = \left(1 + \frac{0.15}{12}\right)^{12} - 1 = 16.076\%$$

$$n = 2 \text{ years}$$

The P/F factor is the same by both methods: $(P/F,1.25\%,24) = 0.7422$ using Table 5 in the rear of the text; and $(P/F,16.076\%,2) = 0.7422$ using the P/F factor formula.

EXAMPLE 4.7

Over the past 10 years, Gentrack has placed varying sums of money into a special capital accumulation fund. The company sells compost produced by garbage-to-compost plants in the United States and Vietnam. Figure 4–5 is the cash flow diagram in $1000 units. Find the amount in the account now (after 10 years) at an interest rate of 12% per year, compounded semiannually.

Solution
Only P and F values are involved. Both methods are illustrated to find F in year 10.

Method 1:   Use the semiannual CP to express the effective semiannual rate of 6% per 6-month period. There are $n = (2)$(number of years) semiannual periods for each cash flow. Using tabulated factor values, the future worth by Equation [4.9] is

$$F = 1000(F/P,6\%,20) + 3000(F/P,6\%,12) + 1500(F/P,6\%,8)$$

$$= 1000(3.2071) + 3000(2.0122) + 1500(1.5938)$$

$$= \$11,634 \quad (\$11.634 \text{ million})$$

Method 2:   Express the effective annual rate, based on semiannual compounding.

$$\text{Effective } i\% \text{ per year} = \left(1 + \frac{0.12}{2}\right)^2 - 1 = 12.36\%$$

The n value is the actual number of years. Use the factor formula $(F/P,i,n) = (1.1236)^n$ and Equation [4.9] to obtain the same answer as above.

$$F = 1000(F/P,12.36\%,10) + 3000(F/P,12.36\%,6) + 1500(F/P,12.36\%,4)$$

$$= 1000(3.2071) + 3000(2.0122) + 1500(1.5938)$$

$$= \$11,634 \quad (\$11.634 \text{ million})$$

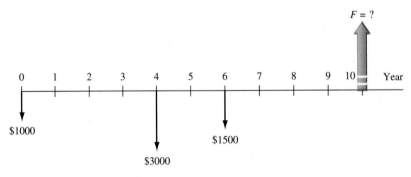

Figure 4–5
Cash flow diagram, Example 4.7.

TABLE 4–6	Examples of *n* and *i* Values Where PP = CP or PP > CP		
Cash Flow Series	Interest Rate	What to Find; What Is Given	Standard Notation
$500 semiannually for 5 years	16% per year, compounded semiannually	Find *P*; given *A*	$P = 500(P/A,8\%,10)$
$75 monthly for 3 years	24% per year, compounded monthly	Find *F*; given *A*	$F = 75(F/A,2\%,36)$
$180 quarterly for 15 years	5% per quarter	Find *F*; given *A*	$F = 180(F/A,5\%,60)$
$25 per month increase for 4 years	1% per month	Find *P*; given *G*	$P = 25(P/G,1\%,48)$
$5000 per quarter for 6 years	1% per month	Find *A*; given *P*	$A = 5000(A/P,3.03\%,24)$

4.6 Equivalence Relations: Series with PP ≥ CP ●●●

When uniform or gradient series are included in the cash flow sequence, the procedure is basically the same as method 2 above, except that PP is now defined by the length of time between cash flows. This also establishes the time unit of the effective interest rate. For example, if cash flows occur on a *quarterly* basis, PP is 1 *quarter* and the effective *quarterly* rate is necessary. The *n* value is the total number of *quarters*. If PP is a quarter, 5 years translates to an *n* value of 20 quarters. This is a direct application of the following general guideline:

Online Presentation

When cash flows involve a series (i.e., *A*, *G*, *g*) and the payment period equals or exceeds the compounding period in length:

- Find the effective *i* per payment period.
- Determine *n* as the total number of payment periods.

In performing equivalence computations for series, *only* these values of *i* and *n* can be used in interest tables, factor formulas, and spreadsheet functions. In other words, there are no other combinations that give the correct answers, as there are for *single-amount cash flows*.

   Table 4–6 shows the correct formulation for several cash flow series and interest rates. Note that *n* is always equal to the total number of payment periods and *i* is an effective rate expressed over the same time period as *n*.

EXAMPLE 4.8

For the past 7 years, Excelon Energy has paid $500 every 6 months for a software maintenance contract. What is the equivalent total amount after the last payment, if these funds are taken from a pool that has been returning 8% per year, compounded quarterly?

Solution

The cash flow diagram is shown in Figure 4–6. The payment period (6 months) is longer than the compounding period (quarter); that is, PP > CP. Applying the guideline, we need to determine an **effective semiannual interest rate.** Use Equation [4.7] with *r* = 4% per 6-month period and *m* = 2 quarters per semiannual period.

$$\text{Effective } i\% \text{ per 6 months} = \left(1 + \frac{0.04}{2}\right)^2 - 1 = 4.04\%$$

The effective semiannual interest rate can also be obtained from Table 4–3 by using the *r* value of 4% and *m* = 2 to get *i* = 4.04%.

The value $i = 4.04\%$ seems reasonable, since we expect the effective rate to be slightly higher than the nominal rate of 4% per 6-month period. The total number of semiannual payment periods is $n = 2(7) = 14$. The relation for F is

$$F = A(F/A,4.04\%,14)$$

$$= 500(18.3422)$$

$$= \$9171.09$$

To determine the F/A factor value 18.3422 using a spreadsheet, enter the FV function from Figure 2–9, that is, $= -\text{FV}(4.04\%,14,1)$. Alternatively, the final answer of $9171.09 can be displayed directly using the function $= -\text{FV}(4.04\%,14,500)$.

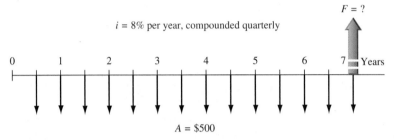

Figure 4–6
Diagram of semiannual deposits used to determine F, Example 4.8.

EXAMPLE 4.9 Credit Card Offer Case   PE

In our continuing credit card saga of Dave and his job transfer to Africa, let's assume he did remember that the total balance is $1030, including the $30 balance transfer fee, and he wants to set up a monthly automatic checking account transfer to pay off the entire amount in 2 years. Once he learned that the minimum payment is $25 per month, Dave decided to be sure the monthly transfer exceeds this amount to avoid any further penalty fees and the penalty APR of 29.99% per year. What amount should he ask to be transferred by the due date each month? What is the APY he will pay, if this plan is followed exactly and Chase Bank does not change the APR during the 2-year period? Also, assume he left the credit card at home and will charge no more to it.

Solution
The monthly A series is needed for a total of $n = 2(12) = 24$ payments. In this case, $PP = CP = 1$ month, and the effective monthly rate is $i = 14.24\%/12 = 1.187\%$ per month.

Solution by hand: Use a calculator or hand computation to determine the A/P factor value.

$$A = P(A/P,i,n) = 1030(A/P,1.187\%,24) = 1030(0.04813)$$

$$= \$49.57 \text{ per month for 24 months}$$

Solution by Spreadsheet: Use the function $= -\text{PMT}(1.187\%,24,1)$ to determine the factor value 0.04813 to determine A for $n = 24$ payments. Alternatively use the function $= \text{PMT}(1.187\%,24,1030)$ to directly display the required monthly payment of $A = \$-49.57$.

The *effective annual interest rate or APY* is computed using Equation [4.7] with $r = 14.24\%$ per year, compounded monthly, and $m = 12$ times per year.

$$\text{Effective } i \text{ per year} = \left(1 + \frac{0.1424}{12}\right)^{12} - 1 = 1.15207 - 1$$

$$= 15.207\% \text{ per year}$$

This is the same effective annual rate i_a determined in Example 4.4b.

EXAMPLE 4.10

The Scott and White Health Plan (SWHP) has purchased a robotized prescription fulfillment system for faster and more accurate delivery to patients with stable, pill-form medication for chronic health problems, such as diabetes, thyroid, and high blood pressure. Assume this high-volume system costs $3 million to install and an estimated $200,000 per year for all materials, operating, personnel, and maintenance costs. The expected life is 10 years. An SWHP bio-medical engineer wants to estimate the total revenue requirement for each 6-month period that is necessary to recover the investment, interest, and annual costs. Find this semiannual A value both by hand and by spreadsheet, if capital funds are evaluated at 8% per year, using two different compounding periods:

Rate 1. 8% per year, compounded *semiannually*.
Rate 2. 8% per year, compounded *monthly*.

Solution

Figure 4–7 shows the cash flow diagram. Throughout the 20 semiannual periods, the annual cost occurs every other period, and the capital recovery series is sought for every 6-month period. This pattern makes the solution by hand quite involved if the P/F factor, not the P/A factor, is used to find P for the 10 annual $200,000 costs. The spreadsheet solution is recommended in cases such as this.

Solution by hand—rate 1: Steps to find the semiannual A value are summarized below:

PP = CP at 6 months; find the effective rate per semiannual period.

Effective semiannual $i = 8\%/2 = 4\%$ per 6 months, compounded semiannually.

Number of semiannual periods $n = 2(10) = 20$.

Calculate P, using the P/F factor for $n = 2, 4, \ldots, 20$ periods because the costs are annual, not semiannual. Then use the A/P factor over 20 periods to find the semiannual A.

$$P = 3{,}000{,}000 + 200{,}000 \left[\sum_{k=2,4}^{20} (P/F,4\%,k) \right]$$

$$= 3{,}000{,}000 + 200{,}000(6.6620) = \$4{,}332{,}400$$

$$A = \$4{,}332{,}400(A/P,4\%,20) = \$318{,}778$$

Conclusion: Revenue of $318,778 is necessary every 6 months to cover all costs and interest at 8% per year, compounded semiannually.

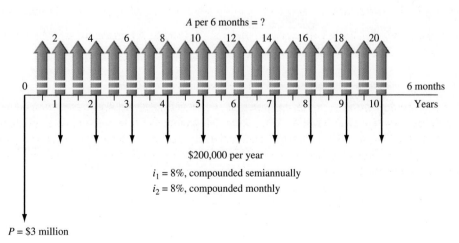

Figure 4–7
Cash flow diagram with two different compounding periods, Example 4.10.

Solution by hand—rate 2: The PP is 6 months, but the CP is now monthly; therefore, PP > CP. To find the effective semiannual rate, the effective interest rate Equation [4.7] is applied with $r = 4\%$ and $m = 6$ months per semiannual period.

$$\text{Effective semiannual } i = \left(1 + \frac{0.04}{6}\right)^6 - 1 = 4.067\%$$

$$P = 3,000,000 + 200,000 \left[\sum_{k=2,4}^{20} (P/F,4.067\%,k)\right]$$

$$= 3,000,000 + 200,000(6.6204) = \$4,324,080$$

$$A = \$4,324,080(A/P,4.067\%,20) = \$320,064$$

Now, \$320,064, or \$1286 more semiannually, is required to cover the more frequent compounding of the 8% per year interest. Note that all P/F and A/P factors must be calculated with factor formulas at 4.067%. This method is usually more calculation-intensive and error-prone than the spreadsheet solution.

Solution by spreadsheet—rates 1 and 2: Figure 4–8 presents a general solution for the problem at both rates. (Several rows at the bottom of the spreadsheet are not printed. They continue the cash flow pattern of \$200,000 every other 6 months through cell B32.) The functions in C8 and E8 are general expressions for the effective rate per PP, expressed in months. This allows some sensitivity analysis to be performed for different PP and CP values. Note the functions in C7 and E7 to determine m for the effective rate relations. This technique works well for spreadsheets once PP and CP are entered in the time unit of the CP.

Each 6-month period is included in the cash flows, including the \$0 entries, so the NPV and PMT functions work correctly. The final A values in D14 (\$318,784) and F14 (\$320,069) are the same (except for rounding) as those above.

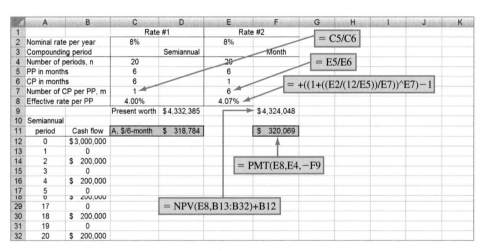

Figure 4–8
Spreadsheet solution for semiannual A series for different compounding periods, Example 4.10.

4.7 Equivalence Relations: Single Amounts and Series with PP < CP ● ● ●

If a person deposits money each month into a savings account where interest is compounded quarterly, do all the monthly deposits earn interest before the next quarterly compounding time? If a person's credit card payment is due with interest on the 15th of the month, and if the full payment is made on the 1st, does the financial institution reduce the interest owed, based on early payment? The usual answers are no. However, if a monthly payment on a \$10 million, quarterly compounded, bank loan were made early by a large corporation, the corporate financial officer would likely insist that the bank reduce the amount of interest due, based on early payment. These are examples of PP < CP. The timing of cash flow transactions between compounding points introduces the question of how **interperiod compounding** is handled. Fundamentally, there are two policies: interperiod cash flows earn *no interest,* or they earn *compound interest.*

> For a no-interperiod-interest policy, negative cash flows (deposits or payments, depending on the perspective used for cash flows) are all regarded as *made at the end of the compounding period,* and positive cash flows (receipts or withdrawals) are all regarded as *made at the beginning.*

As an illustration, when interest is compounded quarterly, all monthly deposits are moved to the end of the quarter (no interperiod interest is earned), and all withdrawals are moved to the beginning (no interest is paid for the entire quarter). This procedure can significantly alter the distribution of cash flows before the effective quarterly rate is applied to find P, F, or A. This effectively forces the cash flows into a PP = CP situation, as discussed in Sections 4.5 and 4.6. Example 4.11 illustrates this procedure and the economic fact that, within a one-compounding-period time frame, there is no interest advantage to making payments early. Of course, noneconomic factors may be present.

EXAMPLE 4.11

Last year AllStar Venture Capital agreed to invest funds in Clean Air Now (CAN), a start-up company in Las Vegas that is an outgrowth of research conducted in mechanical engineering at the University of Nevada–Las Vegas. The product is a new filtration system used in the process of carbon capture and sequestration (CCS) for coal-fired power plants. The venture fund manager generated the cash flow diagram in Figure 4–9a in $1000 units from AllStar's perspective. Included are payments (outflows) to CAN made over the first year and receipts (inflows) from CAN to AllStar. The receipts were unexpected this first year; however, the product has great promise, and advance orders have come from eastern U.S. plants anxious to become zero-emission coal-fueled plants. The interest rate is 12% per year, compounded quarterly, and AllStar uses the no-interperiod-interest policy. How much is AllStar in the "red" at the end of the year?

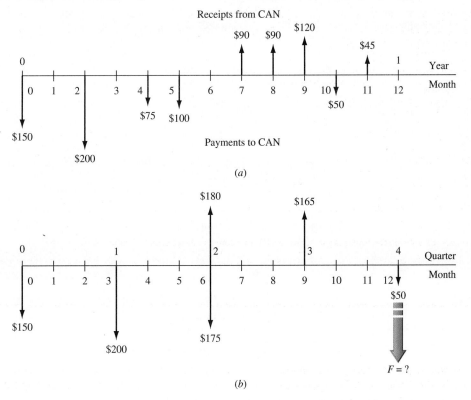

Figure 4–9
(*a*) Actual and (*b*) moved cash flows (in $1000) for quarterly compounding periods using no interperiod interest, Example 4.11.

Solution

With no interperiod interest considered, Figure 4–9b reflects the moved cash flows. All nega-tive cash flows (payments to CAN) are moved to the end of the respective quarter, and all positive cash flows (receipts) are moved to the beginning of the respective quarter. Calculate the F value at $12\%/4 = 3\%$ per quarter.

$$F = 1000[-150(F/P,3\%,4) - 200(F/P,3\%,3) + (-175 + 180)(F/P,3\%,2)$$

$$+ 165(F/P,3\%,1) -50]$$

$$= \$-262,111$$

AllStar has a net investment of $262,111 in CAN at the end of the year.

If PP $<$ CP and interperiod compounding is earned, then the cash flows are not moved, and the equivalent P, F, or A values are determined using the effective interest rate per payment period. The engineering economy relations are determined in the same way as in the previous two sections for PP $\geq$ CP. The effective interest rate formula will have an m value less than 1, because there is only a fractional part of the CP within one PP. For example, weekly cash flows and quarterly com-pounding require that $m = 1/13$ of a quarter. When the nominal rate is 12% per year, compounded quarterly (the same as 3% per quarter, compounded quarterly), the effective rate per PP is

$$\text{Effective weekly } i\% = (1.03)^{1/13} - 1 = 0.228\% \text{ per week}$$

4.8 Effective Interest Rate for Continuous Compounding ● ● ●

If we allow compounding to occur more and more frequently, the compounding period becomes shorter and shorter and m, the number of compounding periods per payment period, increases.

Continuous compounding is present when the duration of CP, the compounding period, becomes infinitely small and m, the number of times interest is compounded per period, becomes infinite. Businesses with large numbers of cash flows each day consider the interest to be continuously compounded for all transactions.

As m approaches infinity, the effective interest rate Equation [4.7] must be written in a new form. First, recall the definition of the natural logarithm base.

$$\lim_{h \to \infty}\left(1 + \frac{1}{h}\right)^h = e = 2.71828+ \qquad\qquad [4.10]$$

The limit of Equation [4.7] as m approaches infinity is found by using $r/m = 1/h$, which makes $m = hr$.

$$\lim_{m \to \infty} i = \lim_{m \to \infty}\left(1 + \frac{r}{m}\right)^m - 1$$

$$= \lim_{h \to \infty}\left(1 + \frac{1}{h}\right)^{hr} - 1 = \lim_{h \to \infty}\left[\left(1 + \frac{1}{h}\right)^h\right]^r - 1$$

$$i = e^r - 1 \qquad\qquad [4.11]$$

Equation [4.11] is used to compute the **effective continuous interest rate,** when the time periods on i and r are the same. As an illustration, if the nominal annual $r = 15\%$ per year, the effective continuous rate per year is

$$i\% = e^{0.15} - 1 = 16.183\%$$

For convenience, Table 4–3 includes effective continuous rates for the nominal rates listed.

To find an effective or nominal interest rate for continuous compounding using the spread-sheet functions **EFFECT** or **NOMINAL,** enter a very large value for the compounding fre-quency m in Equation [4.5] or [4.6], respectively. A value of 10,000 or higher provides sufficient accuracy. Both functions are illustrated in Example 4.12.

EXAMPLE 4.12

(*a*) For an interest rate of 18% per year, compounded continuously, calculate the effective monthly and annual interest rates.

(*b*) An investor requires an effective return of at least 15%. What is the minimum annual nominal rate that is acceptable for continuous compounding?

Solution by Hand

(*a*) The nominal monthly rate is $r = 18\%/12 = 1.5\%$, or 0.015 per month. By Equation [4.11], the effective monthly rate is

$$i\% \text{ per month} = e^r - 1 = e^{0.015} - 1 = 1.511\%$$

Similarly, the effective annual rate using $r = 0.18$ per year is

$$i\% \text{ per year} = e^r - 1 = e^{0.18} - 1 = 19.722\%$$

(*b*) Solve Equation [4.11] for r by taking the natural logarithm.

$$e^r - 1 = 0.15$$

$$e^r = 1.15$$

$$\ln e^r = \ln 1.15$$

$$r = 0.13976$$

Therefore, a rate of 13.976% per year, compounded continuously, will generate an effective 15% per year return. The general formula to find the nominal rate, given the effective continuous rate i, is $r = \ln(1 + i)$.

Solution by Spreadsheet

(*a*) Use the EFFECT function with the nominal monthly rate $r = 1.5\%$ and annual rate $r = 18\%$ with a large m to display effective i values. The functions to enter on a spreadsheet and the responses are as follows:

Monthly:	= EFFECT(1.5%,10000)	effective i = 1.511% per month
Annual:	= EFFECT(18%,10000)	effective i = 19.722% per year

(*b*) Use the function in Equation [4.6] in the format = NOMINAL(15%,10000) to display the nominal rate of 13.976% per year, compounded continuously.

EXAMPLE 4.13

Engineers Marci and Suzanne both invest $5000 for 10 years at 10% per year. Compute the future worth for both individuals if Marci receives annual compounding and Suzanne receives continuous compounding.

Solution

Marci: For annual compounding the future worth is

$$F = P(F/P,10\%,10) = 5000(2.5937) = \$12,969$$

Suzanne: Using Equation [4.11], first find the effective i per year for use in the F/P factor.

$$\text{Effective } i\% = e^{0.10} - 1 = 10.517\%$$

$$F = P(F/P,10.517\%,10) = 5000(2.7183) = \$13,591$$

Continuous compounding causes a $622 increase in earnings. For comparison, daily compounding yields an effective rate of 10.516% ($F = \$13,590$), only slightly less than the 10.517% for continuous compounding.

For some business activities, cash flows occur throughout the day. Examples of costs are energy and water costs, inventory costs, and labor costs. A realistic model for these activities is to increase the frequency of the cash flows to become continuous. In these cases, the economic analysis can be performed for **continuous cash flow** (also called continuous funds flow) and the continuous compounding of interest as discussed above. Different expressions must be derived for the factors for these cases. In fact, the monetary differences for continuous cash flows relative to the discrete cash flow and discrete compounding assumptions are usually not large. Accordingly, most engineering economy studies do not require the analyst to utilize these mathematical forms to make a sound economic decision.

4.9 Interest Rates That Vary over Time ● ● ●

Real-world interest rates for a corporation vary from year to year, depending upon the financial health of the corporation, its market sector, the national and international economies, forces of inflation, and many other elements. Loan rates may increase from one year to another. Home mortgages financed using ARM (adjustable-rate mortgage) interest is a good example. The mortgage rate is slightly adjusted annually to reflect the age of the loan, the current cost of mortgage money, etc.

When P, F, and A values are calculated using a constant or average interest rate over the life of a project, rises and falls in i are neglected. If the variation in i is large, the equivalent values will vary considerably from those calculated using the constant rate. Although an engineering economy study can accommodate varying i values mathematically, it is more involved computationally to do so.

To determine the P value for future cash flow values (F_t) at different i values (i_t) for each year t, we will assume *annual compounding*. Define

$$i_t = \text{effective annual interest rate for year } t \qquad (t = \text{years 1 to } n)$$

To determine the present worth, calculate the P of each F_t value, using the applicable i_t, and sum the results. Using standard notation and the P/F factor,

$$P = F_1(P/F,i_1,1) + F_2(P/F,i_1,1)(P/F,i_2,1) + \cdots$$
$$+ F_n(P/F,i_1,1)(P/F,i_2,1) \cdots (P/F,i_n,1) \qquad [4.12]$$

When only single amounts are involved, that is, one P and one F in the final year n, the last term in Equation [4.12] is the expression for the present worth of the future cash flow.

$$P = F_n(P/F,i_1,1)(P/F,i_2,1) \cdots (P/F,i_n,1) \qquad [4.13]$$

If the equivalent uniform series A over all n years is needed, first find P, using either of the last two equations; then substitute the symbol A for each F_t symbol. Since the equivalent P has been determined numerically using the varying rates, this new equation will have only one unknown, namely, A. Example 4.14 illustrates this procedure.

EXAMPLE 4.14

CE, Inc., leases large earth tunneling equipment. The net profit from the equipment for each of the last 4 years has been decreasing, as shown below. Also shown are the annual rates of return on invested capital. The return has been increasing. Determine the present worth P and equivalent uniform series A of the net profit series. Take the annual variation of rates of return into account.

Year	1	2	3	4
Net Profit	$70,000	$70,000	$35,000	$25,000
Annual Rate	7%	7%	9%	10%

Solution

Figure 4–10 shows the cash flows, rates for each year, and the equivalent P and A. Equation [4.12] is used to calculate P. Since for both years 1 and 2 the net profit is $70,000 and the annual rate is 7%, the P/A factor can be used for these 2 years only.

$$P = [70(P/A,7\%,2) + 35(P/F,7\%,2)(P/F,9\%,1)$$
$$+ 25(P/F,7\%,2)(P/F,9\%,1)(P/F,10\%,1)](1000)$$

$$= [70(1.8080) + 35(0.8013) + 25(0.7284)](1000)$$

$$= \$172{,}816 \qquad\qquad [4.14]$$

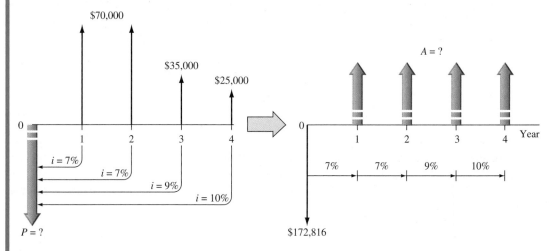

Figure 4–10
Equivalent P and A values for varying interest rates, Example 4.14.

To determine an equivalent annual series, substitute the symbol A for all net profit values on the right side of Equation [4.14], set it equal to $P = \$172{,}816$, and solve for A. This equation accounts for the varying i values each year. See Figure 4–10 for the cash flow diagram transformation.

$$\$172{,}816 = A[(1.8080) + (0.8013) + (0.7284)] = A[3.3377]$$

$$A = \$51{,}777 \text{ per year}$$

Comment

If the average of the four annual rates, that is, 8.25% is used, the result is $A = \$52{,}467$. This is a $690 per year overestimate of the equivalent annual net profit.

When there is a cash flow in year 0 and interest rates vary annually, this cash flow must be included to determine P. In the computation for the equivalent uniform series A over all years, including year 0, it is important to include this initial cash flow at $t = 0$. This is accomplished by inserting the factor value for $(P/F,i_0,0)$ into the relation for A. This factor value is always 1.00. It is equally correct to find the A value using a future worth relation for F in year n. In this case, the A value is determined using the F/P factor, and the cash flow in year n is accounted for by including the factor $(F/P,i_n,0) = 1.00$.

CHAPTER SUMMARY

Since many real-world situations involve cash flow frequencies and compounding periods other than 1 year, it is necessary to use nominal and effective interest rates. When a nominal rate r is stated, the effective interest rate i per payment period is determined by using the effective interest rate equation.

$$\text{Effective } i = \left(1 + \frac{r}{m}\right)^m - 1$$

The m is the number of compounding periods (CP) per interest period. If interest compounding becomes more and more frequent, then the length of a CP approaches zero, continuous compounding results, and the effective i is $e^r - 1$.

All engineering economy factors require the use of an effective interest rate. The i and n values placed in a factor depend upon the type of cash flow series. If only single amounts (P and F) are present, there are several ways to perform equivalence calculations using the factors. However, when series cash flows (A, G, and g) are present, only one combination of the effective rate i and number of periods n is correct for the factors. This requires that the relative lengths of PP and CP be considered as i and n are determined. The interest rate and payment periods must have the same time unit for the factors to correctly account for the time value of money.

From one year (or interest period) to the next, interest rates will vary. To accurately perform equivalence calculations for P and A when rates vary significantly, the applicable interest rate should be used, not an average or constant rate.

PROBLEMS

Nominal and Effective Rates

4.1 From the interest statement 18% per year, compounded monthly, determine the values for interest period, compounding period, and compounding frequency.

4.2 From the interest statement 1% per month, determine the values for interest period, compounding period, and compounding frequency.

4.3 Determine the number of times interest would be compounded in 6 months from the interest statements (*a*) 18% per year, compounded monthly, (*b*) 1% per month, and (*c*) 2% per quarter.

4.4 For an interest rate of 1% per 2 months, determine the number of times interest would be compounded in (*a*) 2 months, (*b*) two semiannual periods, and (*c*) 3 years.

4.5 Identify the compounding period for the following interest statements: (*a*) 3% per quarter; (*b*) 10% per year, compounded semiannually; (*c*) nominal 7.2% per year, compounded monthly; (*d*) effective 3.4% per quarter, compounded weekly; and (*e*) 2% per month, compounded continuously.

4.6 Identify the following interest rate statements as either nominal or effective: (*a*) 1.5% per month, compounded daily; (*b*) 17% per year, compounded quarterly; (*c*) effective 15% per year, compounded monthly; (*d*) nominal 0.6% per month, compounded weekly; (*e*) 0.3% per week, compounded weekly; and (*f*) 8% per year.

4.7 Convert the given interest rates in the left-hand column into the nominal rates listed in the right-hand column. (Assume 4 weeks/month.)

Given Interest Rate	Desired Interest Rate
1% per month	Nominal rate per year
3% per quarter	Nominal rate per 6 months
2% per quarter	Nominal rate per year
0.28% per week	Nominal rate per quarter
6.1% per 6 months	Nominal rate per 2 years

4.8 What nominal interest rate per year is equivalent to 11.5% per year, compounded monthly?

4.9 For a Federal Credit Union that offers an interest rate of 8% per year, compounded quarterly, determine the nominal rate per 6 months.

4.10 The Second National Bank of Fullertum advertises an APR of 14% compounded monthly for student loans. Determine the APY. Show hand and spreadsheet solutions.

4.11 For an effective annual rate i_a of 15.87% compounded quarterly, determine (*a*) the effective quarterly rate and (*b*) the nominal annual rate. (*c*) What is the spreadsheet function to find the nominal annual rate above?

4.12 High-tech companies such as IBM, AMD, and Intel have been using nanotechnology for several years to make microchips with faster speeds while using less power. A less well-known company in the chip business has been growing fast enough that the company uses a minimum attractive rate of return of 60% per year. If this MARR is an effective annual rate compounded monthly, determine the effective monthly rate.

4.13 An interest rate of 21% per year, compounded every 4 months, is equivalent to what effective rate per year? Show hand and spreadsheet solutions.

4.14   An interest rate of 8% per 6 months, compounded monthly, is equivalent to what effective rate per quarter?

4.15   A small company that makes modular bevel gear drives with a tight swing ratio for optimizing pallet truck design was told that the interest rate on a mortgage loan would be an *effective* 4% per quarter, compounded monthly. The owner was confused by the terminology and asked you to help. What are (*a*) the APR and (*b*) the APY?

4.16   In 'N Out Payday Loans advertises that for a fee of only $10, you can immediately borrow up to $200 for one month. If a person accepts the offer, what are (*a*) the nominal interest rate per year and (*b*) the effective rate per year?

4.17   A government-required truth-in-lending document showed that the APR was 21% and the APY was 22.71%. Determine the compounding frequency at which the two rates are equivalent.

4.18   Julie has a low credit rating, plus she was furloughed from her job 2 months ago. She has a new job starting next week and expects a salary to start again in a couple of weeks. Since she is a little short on money to pay her rent, she decided to borrow $100 from a loan company, which will charge her only $10 interest if the $110 is paid no more than 1 week after the loan is made. What are the (*a*) nominal annual and (*b*) effective annual interest rates that she will pay on this loan?

Equivalence When PP ≥ CP

4.19   Assume you deposit 25% of your monthly check of $5500 into a savings account at a credit union that compounds interest semiannually. (*a*) What are the payment and compounding periods? (*b*) Is the payment period greater than or less than the compounding period?

4.20   Interest is compounded quarterly, and single-payment cash flows (that is, *F* and *P*) are separated by 5 years. What must the *compounding period* be on the interest rate, if the value of *n* in the *P/F* or *F/P* equation is (*a*) $n = 5$, (*b*) $n = 20$, or (*c*) $n = 60$?

4.21   When interest is compounded quarterly and a uniform series cash flow of $4000 occurs every 6 months, what time periods on *i* and *n* must be used?

4.22   Pinpoint Laser Systems is planning to set aside $260,000 now for possibly replacing its 4-pole, 38-MW synchronous motors when it becomes necessary. If the replacement is expected to take place in 3 years, how much will the company have in its investment set-aside account? Assume the company can achieve a rate of return of 12% per year, compounded quarterly.

4.23   Wheeling-Pittsburgh Steel is investigating whether it should replace some of its basic oxygen furnace equipment now or wait to do it later. The cost later (i.e., 3 years from now) is estimated to be $1,700,000. How much can the company afford to spend now, if its minimum attractive rate of return is 1.5% per month?

4.24   How much can Wells Fargo lend to a developer who will repay the loan by selling 6 view lots at $190,000 each 2 years from now? Assume the bank will lend at a nominal 14% per year, compounded semiannually.

4.25   How much will be in a high-yield account at the National Bank of Arizona 12 years from now if you deposit $5000 now and $7000 five years from now? The account earns interest at a rate of 8% per year, compounded quarterly.

4.26   Loadstar Sensors is a company that makes load/force sensors based on capacitive sensing technology. The company wants to have $28 million for a plant expansion 4 years from now. If the company has already set aside $12 million in an investment account for the expansion, how much more must the company add to the account next year (i.e., 1 year from now) so that it will have the $28 million 4 years from now? The account earns interest at 12% per year, compounded quarterly.

4.27   A structural engineering consulting company is examining its cash flow requirements for the next 6 years. The company expects to replace office machines and computer equipment at various times over the 6-year planning period. Specifically, the company expects to spend $21,000 two years from now, $24,000 three years from now, and $10,000 five years from now. What is the present worth of the planned expenditures at an interest rate of 10% per year, compounded semiannually?

4.28   Irvin Aerospace of Santa Ana, California, was awarded a 5-year contract to develop an advanced space capsule airbag landing attenuation system for NASA's Langley Research Center. The company's computer system uses fluid structure interaction modeling to test and analyze each airbag design concept. What is the present worth of the contract at 16% per year, compounded quarterly, if the quarterly cost in years 1 through 5 is $2 million per quarter?

4.29   Heyden Motion Solutions ordered $7 million worth of seamless tubes for its drill collars from the Timken Company of Canton, Ohio. (A drill collar is the heavy tubular connection between a drill pipe and a drill bit.) At 12% per year, compounded semiannually, what is the equivalent uniform cost per semiannual period over a 5-year amortization period?

4.30   In 2010, the National Highway Traffic Safety Administration raised the average fuel efficiency standard to 35.5 miles per gallon (mpg) for cars and light trucks by the year 2016. The rules will cost consumers an average of $926 extra per vehicle in the 2016 model year. Assume Yolanda will purchase a new car in 2016 and plans to keep it for 5 years. How much will the monthly savings in the cost of gasoline have to be to recover Yolanda's extra cost? Use an interest rate of 0.75% per month.

4.31   Fort Bliss, a U.S. Army military base, contributed $3.3 million of the $87 million capital cost for a desalting plant constructed and operated by El Paso Water Utilities (EPWU). In return, EPWU agreed to sell water to Fort Bliss at $0.85 per 1000 gallons for 20 years. If the army base uses 200 million gallons of water per month, what is the Army's cost per month for water? The $3.3 million capital cost is amortized at an interest rate of 6% per year, compounded monthly.

4.32   Beginning in 2011, city hall, administrative offices, and municipal courts in the city of El Paso, Texas, will go on a 10 hour/day, 4-day workweek from the beginning of May through the end of September. The shortened workweek will affect 25% of the city's 6100 employees and will save $42,600 per month for those 5 months, because of reduced fuel, utility, and janitorial expenses. If this work schedule continues for the next 10 years, what is the future worth of the savings at the end of that time (i.e., end of year 2020)? Use an interest rate of 0.5% per month.

4.33   In October 2009, Wal-Mart started selling caskets on its website that undercut many funeral homes. Prices ranged from $999 for steel models such as *Dad Remembered* to $3199 for the *Sienna Bronze* casket. Part of the business model is to get people to plan ahead, so the company is allowing people to pay for the caskets over a 12-month period with no interest. An individual purchased a *Sienna Bronze* casket and made 12 equal monthly payments (in months 1 through 12) at no interest. How much did this person save each month compared to another person who paid an interest rate of 6% per year, compounded monthly?

4.34   NRG Energy plans to construct a giant solar plant in Santa Teresa, New Mexico, to supply electricity to West Texas Electric. The plant will have 390,000 heliostats to concentrate sunlight onto 32 water towers to generate steam. It will provide enough power for 30,000 homes. If the company spends $28 million per month for 2 years in constructing the plant, how much will the company have to make each month in years 3 through 22 (that is, 20 years) to recover its investment plus 18% per year, compounded monthly?

4.35   Many college students have Visa credit cards that carry an interest rate of "simple 24% per year" (that is, 2% per month). When the balance on such a card is $5000, the minimum payment is $110.25.
  (*a*)   What is the amount of interest in the first payment?
  (*b*)   How long will it take, in months, to pay off the balance, if the cardholder continues to make payments of $110.25 per month and adds no other charges to the card?

4.36   Bart is an engineering graduate who did not take the engineering economy elective during his B.S. degree course work. After working for a year or so, he found himself in financial trouble, and he borrowed $500 from a friend in the finance department at his office. Bart agreed to repay the loan principal plus $75 interest 1 month later. The two got separated doing different jobs, and 1 year went by. The friend e-mailed Bart after exactly 1 year and asked for the loan repayment plus the interest with monthly compounding, since Bart had made no effort to repay the loan during the year. (*a*) What does Bart now owe his friend? (*b*) What effective annual interest rate did Bart pay on this $500 loan?

4.37   AT&T announced that the early termination fee for smart phones will jump from $175 to $375. The fee will be reduced by $10 each month of the 2-year contract. Wireless carriers justify the fees by pointing out that the cost of a new phone is heavily discounted from what the carrier pays the manufacturer. Assume AT&T pays $499 for an iPhone that it sells for $199. How much profit would the company have to make each month (i.e., prior to the termination), if it wanted to make a rate of return of 1.5% per month on its $300 investment in a customer who terminates the contract after 12 months?

4.38   What is the future worth of a present cost of $285,000 to Monsanto, Inc. 5 years from now at an interest rate of 2% per month?

4.39 Environmental recovery company RexChem Partners plans to finance a site reclamation project that will require a 4-year cleanup period. The company plans to borrow $3.6 million now. How much will the company have to get in a lump-sum payment when the project is over in order to earn 24% per year, compounded quarterly, on its investment?

4.40 For the cash flows shown below, determine the equivalent uniform worth in years 1 through 5 at an interest rate of 18% per year, compounded monthly.

Year	1	2	3	4	5
Cash Flow, $	200,000	0	350,000	0	400,000

4.41 For the cash flow diagram shown, solve for F, using an interest rate of 1% per month.

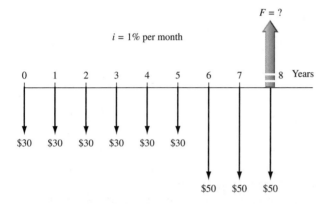

4.42 According to the Government Accountability Office (GAO), if the U.S. Postal Service does not change its business model, it will lose $480 million next month and $500 million the month after that, and the losses will increase by $20 million per month for the next 10 years. At an interest rate of 0.25% per month, what is the equivalent uniform amount per month of the losses through year 10?

4.43 Equipment maintenance costs for manufacturing explosion-proof pressure switches are projected to be $100,000 in year 1 and increase by 4% each year through year 5. What is the present worth of the maintenance costs at an interest rate of 10% per year, compounded quarterly?

Equivalence When PP < CP

4.44 If you deposit $1000 per month into an investment account that pays interest at a rate of 6% per year, compounded quarterly, how much will be in the account at the end of 5 years? There is no interperiod compounding.

4.45 Freeport McMoran purchased two model MTVS peristaltic pumps (to inject sulfuric acid and King Lee antiscalant) for use at its nanofiltration water conditioning plant. The cost of the pumps was $950 each. If the chemical cost is $11 per day, determine the equivalent cost *per month* at an interest rate of 12% per year, compounded monthly. Assume 30 days per month and a 3-year pump life.

4.46 Income from recycling paper and cardboard at the U.S. Army's Fort Benning Maneuver Center has averaged $3000 per month for 2½ years. What is the future worth of the income (after the 2½ years) at an interest rate of 6% per year, compounded quarterly? Assume there is no interperiod compounding.

4.47 The Autocar E3 refuse truck has an energy recovery system developed by Parker Hannifin LLC that is expected to reduce fuel consumption by 50%. Pressurized fluid flows from carbon fiber-reinforced accumulator tanks to two hydrostatic motors that propel the vehicle forward. (The truck recharges the accumulators when it brakes.) The fuel cost for a regular refuse truck is $900 per month. How much can a private waste-hauling company afford to spend now on the recovery system, if it wants to recover its investment in 3 years plus a return of 14% per year, compounded semiannually? Assume no interperiod compounding.

Continuous Compounding

4.48 What effective interest rate per year is equal to 1.2% per month, compounded continuously? Show hand and spreadsheet solutions.

4.49 What effective interest rate per quarter is equal to a nominal 1.6% per month, compounded continuously?

4.50 What nominal rate per month is equivalent to an effective 1.3% per month, compounded continuously?

4.51 Companies such as GE that have huge amounts of cash flow every day base their financial calculations on continuous compounding. If the company wants to make an effective 25% per year, compounded continuously, what nominal daily rate of return has to be realized? Assume 365 days per year.

4.52 U.S. Steel is planning a plant expansion that is expected to cost $13 million. How much money must the company set aside now in a lump-sum investment to have the money in 2 years? Capital funds earn interest at a rate of 12% per year, compounded continuously.

4.53 Periodic outlays for inventory control software at Baron Chemicals are expected to be $150,000 immediately, $200,000 in 1 year, and $350,000 in 2 years. What is the present worth of the costs at an interest rate of 10% per year, compounded continuously?

Varying Interest Rates

4.54 Many small companies use accounts receivable as collateral to borrow money for continuing operations and meeting payrolls. If a company borrows $300,000 now at an interest rate of 1% per month, but the rate changes to 1.25% per month after 4 months, how much will the company owe at the end of 1 year?

4.55 The maintenance cost for furnaces at a copper smelting plant have been constant at $140,000 per year for the past 5 years. If the interest rate was 8% per year for the first 3 years and then it increased to 10% in years 4 and 5, what is the equivalent future worth (in year 5) of the maintenance cost? Show hand and spreadsheet solutions.

4.56 By filling carbon nanotubes with miniscule wires made of iron and iron carbide, incredibly thin nanowires can be extruded by blasting the carbon nanotubes with an electron beam. If Gentech Technologies plans to spend $1.7 million in year 1, $2.1 million in year 2, and $3.4 million in year 3 to develop the technology, determine the present worth of the investments in year 0, if the interest rate in year 1 is 10% and in years 2 and 3 it is 12% per year.

4.57 Find (*a*) the present worth *P*, and (*b*) the equivalent uniform annual worth *A* for the cash flows shown below.

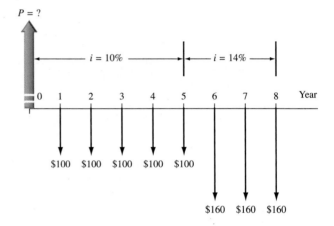

ADDITIONAL PROBLEMS AND FE EXAM REVIEW QUESTIONS

4.58 The term *annual percentage rate* is the same as:
  (*a*) Effective rate
  (*b*) Nominal rate
  (*c*) Annual percentage yield
  (*d*) All of the above

4.59 An interest rate is an effective rate under all of the following conditions, *except when:*
  (*a*) The compounding period is not stated
  (*b*) The interest period and compounding period are the same
  (*c*) The interest statement says that the interest rate is effective
  (*d*) The interest period is shorter than the compounding period

4.60 An interest rate of nominal 12% per year, compounded weekly, is:
  (*a*) An effective rate per year
  (*b*) An effective rate per week
  (*c*) A nominal rate per year
  (*d*) A nominal rate per week

4.61 An interest rate of 1.5% per month, compounded continuously, is the same as:
  (*a*) An effective 1.5% per month
  (*b*) 4.5% per quarter, compounded continuously

  (*c*) 6.0% per quarter, compounded continuously
  (*d*) 9% per 6 months

4.62 An effective 12.68% per year, compounded monthly, is the closest to:
  (*a*) 12% per year
  (*b*) 12% per year, compounded annually
  (*c*) 1% per month
  (*d*) 1% per month, compounded annually

4.63 For an interest rate of 2% per month, the effective *semiannual* interest rate is:
  (*a*) 2.02%
  (*b*) 12.005%
  (*c*) 12.31%
  (*d*) 12.62%

4.64 If you make quarterly deposits for 3 years into an account that compounds interest at 1% per month, the value of *n* in the *F/A* factor that will determine *F* at the end of the 3-year period is:
  (*a*) 3   (*b*) 12
  (*c*) 36   (*d*) None of these

4.65 Identify the following interest rates as nominal or effective.

Rate 1:   1.5% per quarter

Rate 2:   1.5% per quarter, compounded monthly

(a) Both are nominal rates.

(b) Rate 1 is nominal and rate 2 is effective.

(c) Rate 1 is effective and rate 2 is nominal.

(d) Both are effective.

4.66 In solving uniform series problems, n in the standard factor notation equation is equal to the number of arrows (i.e., cash flows) in the original cash flow diagram when:

(a) The payment period (PP) is longer than the compounding period (CP).

(b) The compounding period is equal to the payment period.

(c) Both (a) and (b) are correct.

(d) The compounding period is longer than the payment period.

4.67 An engineer who is saving for her retirement plans to deposit $500 every *quarter*, starting one quarter from now, into an investment account. If the account pays interest at 6% per year, compounded semiannually, the total she will have at the end of 25 years is closest to:

(a) $50,000   (b) $56,400

(c) $79,700   (d) $112,800

4.68 A company that makes flange-mount, motorized rotary potentiometers expects to spend $50,000 for a certain machine 4 years from now. At an interest rate of 12% per year, compounded quarterly, the present worth of the machine's cost is represented by the following equation:

(a) $P = 50{,}000(P/F, 3\%, 16)$

(b) $P = 50{,}000(P/F, \text{effective } i/6 \text{ months}, 8)$

(c) $P = 50{,}000(P/F, \text{effective } i/\text{year}, 4)$

(d) Any of the above

4.69 For the cash flow diagram shown, the unit of the payment period (PP) is:

(a) Months   (b) Quarters

(c) Semiannual   (d) Years

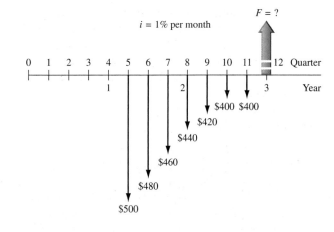

4.70 A small company plans to spend $10,000 in year 2 and $10,000 in year 5. At an interest rate of *effective* 10% per year, compounded semiannually, the equation that represents the equivalent annual worth A in years 1 through 5 is:

(a) $A = 10{,}000(P/F,10\%,2)(A/P,10\%,5)$
$+ 10{,}000(A/F,10\%,5)$

(b) $A = 10{,}000(A/P,10\%,4) + 10{,}000(A/F,10\%,5)$

(c) $A = 10{,}000(P/F,5\%,2)(A/P,5\%,10)$
$+ 10{,}000(A/F,5\%,10)$

(d) $A = [10{,}000(F/P,10\%,5)$
$+ 10{,}000](A/F,10\%,5)$

4.71 Assume you make *monthly* deposits of $200 starting one month from now into an account that pays 6% per year, compounded semiannually. If you want to know the total after *4 years*, the value of n you should use in the F/A factor is:

(a) 2   (b) 4   (c) 8   (d) 12

4.72 The cost of replacing part of a cell phone video-chip production line in 6 years is estimated to be $500,000. At an interest rate of 14% per year, compounded semiannually, the uniform amount that must be deposited into a sinking fund every 6 months is closest to:

(a) $21,335   (b) $24,825

(c) $27,950   (d) $97,995

CASE STUDY

IS OWNING A HOME A NET GAIN OR NET LOSS OVER TIME?

Background

The Carroltons are deliberating whether to purchase a house or continue to rent for the next 10 years. They are assured by both of their employers that no transfers to new locations will occur for at least this number of years. Plus, the high school that their children attend is very good for their college prep education, and they all like the neighborhood where they live now.

They have a total of $40,000 available now and estimate that they can afford up to $2850 per month for the total house payment.

If the Carroltons do not buy a house, they will continue to rent the house they currently occupy for $2700 per month. They will also place the $40,000 into an investment instrument that is expected to earn at the rate of 6% per year. Additionally, they will add to this investment *at the end of each year* the same amount as the monthly 15-year mortgage payments. This alternative is called the rent–don't buy plan.

Information

Two financing plans using fixed-rate mortgages are currently available. The details are as follows.

Plan	Description
A	30-year fixed rate of 5.25% per year interest; 10% down payment
B	15-year fixed rate of 5.0% per year interest; 10% down payment

Other information:

- Price of the house is $330,000.
- Taxes and insurance (T&I) are $500 per month.
- Up-front fees (origination fee, survey fee, attorney's fee, etc.) are $3000.

Any money not spent on the down payment or monthly payment will be invested and return at a rate of 6% per year (0.5% per month).

The Carroltons anticipate selling the house after 10 years and plan for a 10% increase in price, that is, $363,000 (after all selling expenses are paid)

Case Study Exercises

1. The 30-year fixed-rate mortgage (plan A) is analyzed below. No taxes are considered on proceeds from the savings or investments.

    Perform a similar analysis for the 15-year loan (plan B) and the rent–don't buy plan. The Carroltons decided to use the *largest future worth after 10 years* to select the best of the plans. Do the analysis for them and select the best plan.

Plan A analysis: 30-year fixed-rate loan

Amount of money required for closing costs:	
Down payment (10% of $330,000)	$33,000
Up-front fees (origination fee, attorney's fee, survey, filing fee, etc.)	3,000
Total	$36,000

The amount of the loan is $297,000, and equivalent monthly principal and interest (P&I) is determined at $5.25\%/12 = 0.4375\%$ per month for $30(12) = 360$ months.

$$A = 297,000(A/P,0.4375\%,360) = 297,000(0.005522)$$
$$= \$1640$$

Add the T&I of $500 for a total monthly payment of

$$\text{Payment}_A = \$2140 \text{ per month}$$

The future worth of plan A is the sum of three future worth components: remainder of the $40,000 available for the closing costs (F_{1A}); left-over money from that available for monthly payments (F_{2A}); and increase in the house value when it is sold after 10 years (F_{3A}). These are calculated here.

$$F_{1A} = (40,000 - 36,000)(F/P,0.5\%,120)$$
$$= \$7278$$

Money available each month to invest after the mortgage payment, and the future worth after 10 years is

$$2850 - 2140 = \$710$$
$$F_{2A} = 710(F/A,0.5\%,120)$$
$$= \$116,354$$

Net money from the sale in 10 years (F_{3A}) is the difference between the net selling price ($363,000) and the remaining balance on the loan.

Loan balance

$$= 297,000(F/P,0.4375\%,120)$$
$$- 1640(F/A,0.4375\%,120)$$
$$= 297,000(1.6885) - 1640(157.3770)$$
$$= \$243,386$$

$F_{3A} = 363,000 - 243,386 = \$119,614$

Total future worth of plan A is

$$F_A = F_{1A} + F_{2A} + F_{3A}$$
$$= 7278 + 116,354 + 119,614$$
$$= \$243,246$$

2. Perform this analysis if all estimates remain the same, except that when the house sells 10 years after purchase, the bottom has fallen out of the housing market and the net selling price is only 70% of the purchase price, that is, $231,000.

LEARNING STAGE 2

Basic Analysis Tools

An engineering project or alternative is formulated to make or purchase a **product,** to develop a **process,** or to provide a **service** with specified results. An engineering economic analysis evaluates cash flow estimates for parameters such as initial cost, annual costs and revenues, nonrecurring costs, and possible salvage value over an estimated useful life of the product; process, or service. The chapters in this Learning Stage develop and demonstrate the basic tools and techniques to evaluate one or more alternatives using the factors, formulas, and spreadsheet functions learned in Stage 1.

After completing these chapters, you will be able to evaluate most engineering project proposals using a well-accepted economic analysis technique, such as present worth, future worth, capitalized cost, life-cycle costing, annual worth, rate of return, or benefit/cost analysis.

The epilogue to this stage provides an approach useful in selecting the engineering economic method that will provide the best analysis for the estimates and conditions present once the mutually exclusive alternatives are defined.

Important note: If depreciation and/or after-tax analysis is to be considered along with the evaluation methods in Chapters 5 through 9, Chapter 16 and/or Chapter 17 should be covered, preferably after Chapter 6.

CHAPTER 5

Present Worth Analysis

LEARNING OUTCOMES

Purpose: Utilize different present worth techniques to evaluate and select alternatives.

SECTION	TOPIC	LEARNING OUTCOME
5.1	Formulate alternatives	• Identify mutually exclusive and independent projects; define revenue and cost alternatives.
5.2	PW of equal-life alternatives	• Select the best of equal-life alternatives using present worth analysis.
5.3	PW of different-life alternatives	• Select the best of different-life alternatives using present worth analysis.
5.4	FW analysis	• Select the best alternative using future worth analysis.
5.5	CC analysis	• Select the best alternative using capitalized cost (CC) analysis.

A future amount of money converted to its equivalent value now has a present worth (PW) that is always less than that of the future cash flow, because all *P/F* factors have a value less than 1.0 for any interest rate greater than zero. For this reason, present worth values are often referred to as *discounted cash flows (DCF)*, and the interest rate is referred to as the *discount rate*. Besides PW, two other terms frequently used are *present value (PV)* and *net present value (NPV)*. Up to this point, present worth computations have been made for one project or alternative. In this chapter, techniques for comparing two or more mutually exclusive alternatives by the present worth method are treated. Two additional applications are covered here—**future worth and capitalized cost.** Capitalized costs are used for projects with very long expected lives or long planning horizons.

To understand how to organize an economic analysis, this chapter begins with a description of independent and mutually exclusive projects as well as revenue and cost alternatives.

PE

Water for Semiconductor Manufacturing Case: The worldwide contribution of semiconductor sales is about $250 billion per year, or about 10% of the world's GDP (gross domestic product). This industry produces the microchips used in many of the communication, entertainment, transportation, and computing devices we use every day. Depending upon the type and size of fabrication plant (fab), the need for ultrapure water (UPW) to manufacture these tiny integrated circuits is high, ranging from 500 to 2000 gpm (gallons per minute). Ultrapure water is obtained by special processes that commonly include reverse osmosis/deionizing resin bed technologies. Potable water obtained from purifying seawater or brackish groundwater may cost from $2 to $3 per 1000 gallons, but to obtain UPW on-site for semiconductor manufacturing may cost an additional $1 to $3 per 1000 gallons.

A fab costs upward of $2.5 billion to construct, with approximately 1% of this total, or $25 million, required to provide the ultrapure water needed, including the necessary wastewater and recycling equipment.

A newcomer to the industry, Angular Enterprises, has estimated the cost profiles for two options to supply its anticipated fab with water. It is fortunate to have the option of desalinated seawater or purified groundwater sources in the location chosen for its new fab. The initial cost estimates for the UPW system are given below.

Source	Seawater (S)	Groundwater (G)
Equipment first cost, $M	−20	−22
AOC, $M per year	−0.5	−0.3
Salvage value, % of first cost	5	10
Cost of UPW, $ per 1000 gallons	4	5

Angular has made some initial estimates for the UPW system.

Life of UPW equipment	10 years
UPW needs	1500 gpm
Operating time	16 hours per day for 250 days per year

This case is used in the following topics (Sections) and problems of this chapter:

PW analysis of equal-life alternatives (Section 5.2)

PW analysis of different-life alternatives (Section 5.3)

Capitalized cost analysis (Section 5.5)

Problems 5.20 and 5.34

5.1 Formulating Alternatives ● ● ●

The evaluation and selection of economic proposals require **cash flow estimates** over a stated period of time, mathematical techniques to calculate the **measure of worth** (review Example 1.2 for possible measures), and a **guideline for selecting the best proposal.** From all the

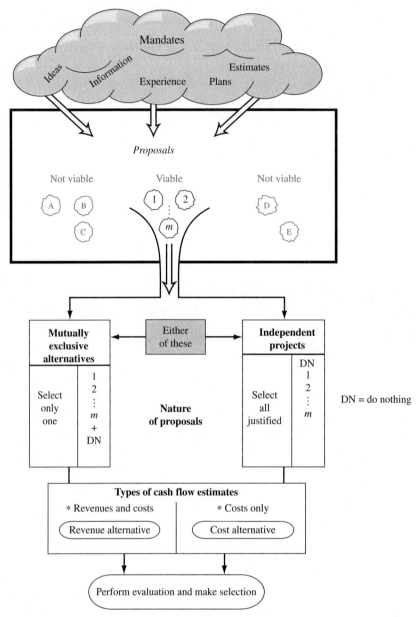

Figure 5–1
Progression from proposals to economic evaluation to selection.

proposals that may accomplish a stated purpose, the alternatives are formulated. This progression is detailed in Figure 5–1. Up front, some proposals are viable from technological, economic, and/or legal perspectives; others are not viable. Once the obviously nonviable ideas are eliminated, the remaining viable proposals are fleshed out to form the alternatives to be evaluated. Economic evaluation is one of the primary means used to select the best alternative(s) for implementation.

The nature of the economic proposals is always one of two types:

Mutually exclusive alternatives: Only one of the proposals can be selected. For terminology purposes, each viable proposal is called an *alternative*.

Independent projects: More than one proposal can be selected. Each viable proposal is called a *project*.

The **do-nothing (DN)** proposal is usually understood to be an option when the evaluation is performed.

Do nothing

The DN alternative or project means that the **current approach is maintained;** nothing new is initiated. No new costs, revenues, or savings are generated.

If it is absolutely required that one or more of the defined alternatives be selected, do nothing is not considered. This may occur when a mandated function must be installed for safety, legal, government, or other purposes.

Mutually exclusive alternatives and independent projects are selected in completely different ways. A mutually exclusive selection takes place, for example, when an engineer must select *the* best diesel-powered engine from several available models. Only one is chosen, and the rest are rejected. If none of the alternatives are economically justified, then all can be rejected and, by default, the DN alternative is selected. For independent projects one, two or more, in fact, all of the projects that are economically justified can be accepted, provided capital funds are available. This leads to the two following fundamentally different evaluation bases:

Mutually exclusive alternatives **compete with one another** and are compared pairwise.

Independent projects are evaluated one at a time and **compete only with the DN project.**

Any of the techniques in Chapters 5 through 9 can be used to evaluate either type of proposal—mutually exclusive or independent. When performed correctly as described in each chapter, any of the techniques will reach the same conclusion of which alternative or alternatives to select. This chapter covers the present worth method.

A parallel can be developed between independent and mutually exclusive evaluation. Assume there are *m* independent projects. Zero, one, two, or more may be selected. Since each project may be in or out of the selected group of projects, there are a total of 2^m **mutually exclusive alternatives.** This number includes the DN alternative, as shown in Figure 5–1. For example, if the engineer has three diesel engine models (A, B, and C) and may select any number of them, there are $2^3 = 8$ alternatives: DN, A, B, C, AB, AC, BC, ABC. Commonly, in real-world applications, there are restrictions, such as an upper budgetary limit, that eliminate many of the 2^m alternatives. Independent project analysis without budget limits is discussed in this chapter and through Chapter 9. Chapter 12 treats independent projects with a budget limitation; this is called capital budgeting.

Finally, it is important to recognize the nature of the cash flow estimates before starting the computation of a measure of worth that leads to the final selection. Cash flow estimates determine whether the alternatives are revenue- or cost-based. All the alternatives or projects must be of the same type when the economic study is performed. Definitions for these types follow:

Revenue: Each alternative generates cost (cash outflow) and revenue (cash inflow) estimates, and possibly savings, also considered cash inflows. Revenues can vary for each alternative.

Cost: Each alternative has only cost cash flow estimates. Revenues or savings are assumed equal for all alternatives; thus they are not dependent upon the alternative selected. These are also referred to as **service alternatives.**

Revenue or cost alternative

Although the exact procedures vary slightly for revenue and cost cash flows, all techniques and guidelines covered through Chapter 9 apply to both. Differences in evaluation methodology are detailed in each chapter.

5.2 Present Worth Analysis of Equal-Life Alternatives ● ● ●

Online Presentation

The PW comparison of alternatives with equal lives is straightforward. The present worth *P* is renamed PW of the alternative. The present worth method is quite popular in industry because all future costs and revenues are transformed to **equivalent monetary units NOW;** that is, all future cash flows are converted (discounted) to present amounts (e.g., dollars) at a specific rate of return, which is the MARR. This makes it very simple to determine which alternative has the best economic advantage. The required conditions and evaluation procedure are as follows:

If the alternatives have the same capacities for the same time period (life), the **equal-service requirement** is met. Calculate the PW value at the stated MARR for each alternative.

Equal-service requirement

For **mutually exclusive (ME)** alternatives, whether they are revenue or cost alternatives, the following guidelines are applied to justify a single project or to select one from several alternatives.

Project evaluation
ME alternative
selection

> **One alternative:** If PW ≥ 0, the requested MARR is met or exceeded and the alternative is economically justified.
>
> **Two or more alternatives:** Select the alternative with the PW that is **numerically largest,** that is, less negative or more positive. This indicates a lower PW of cost for cost alternatives or a larger PW of net cash flows for revenue alternatives.

Note that the guideline to select one alternative with the lowest cost or highest revenue uses the criterion of **numerically largest.** This is not the absolute value of the PW amount, because the sign matters. The selections below correctly apply the guideline for two alternatives A and B.

PW_A	PW_B	Selected Alternative
$-2300	$-1500	B
-500	+1000	B
+2500	+2000	A
+4800	-400	A

For **independent** projects, each PW is considered separately, that is, compared with the DN project, which always has PW = 0. The selection guideline is as follows:

Independent project
selection

> **One or more independent projects:** Select all projects with PW ≥ 0 at the MARR.

The independent projects must have positive and negative cash flows to obtain a PW value that can exceed zero; that is, they must be revenue projects.

All PW analyses require a MARR for use as the i value in the PW relations. The bases used to establish a realistic MARR were summarized in Chapter 1 and are discussed in detail in Chapter 10.

EXAMPLE 5.1

A university lab is a research contractor to NASA for in-space fuel cell systems that are hydrogen- and methanol-based. During lab research, three equal-service machines need to be evaluated economically. Perform the present worth analysis with the costs shown below. The MARR is 10% per year.

	Electric-Powered	Gas-Powered	Solar-Powered
First cost, $	-4500	-3500	-6000
Annual operating cost (AOC), $/year	-900	-700	-50
Salvage value S, $	200	350	100
Life, years	8	8	8

Solution

These are cost alternatives. The salvage values are considered a "negative" cost, so a + sign precedes them. (If it costs money to dispose of an asset, the estimated disposal cost has a − sign.) The PW of each machine is calculated at $i = 10\%$ for $n = 8$ years. Use subscripts E, G, and S.

$$PW_E = -4500 - 900(P/A,10\%,8) + 200(P/F,10\%,8) = \$-9208$$
$$PW_G = -3500 - 700(P/A,10\%,8) + 350(P/F,10\%,8) = \$-7071$$
$$PW_S = -6000 - 50(P/A,10\%,8) + 100(P/F,10\%,8) = \$-6220$$

The solar-powered machine is selected since the PW of its costs is the lowest; it has the numerically largest PW value.

EXAMPLE 5.2 Water for Semiconductor Manufacturing Case

As discussed in the introduction to this chapter, ultrapure water (UPW) is an expensive commodity for the semiconductor industry. With the options of seawater or groundwater sources, it is a good idea to determine if one system is more economical than the other. Use a MARR of 12% per year and the present worth method to select one of the systems.

Solution

An important first calculation is the cost of UPW per year. The general relation and estimated costs for the two options are as follows:

$$\textbf{UPW cost relation: } \frac{\$}{\text{year}} = \left(\frac{\text{cost in } \$}{1000 \text{ gallons}}\right)\left(\frac{\text{gallons}}{\text{minute}}\right)\left(\frac{\text{minutes}}{\text{hour}}\right)\left(\frac{\text{hours}}{\text{day}}\right)\left(\frac{\text{days}}{\text{year}}\right)$$

$$\text{Seawater: } (4/1000)(1500)(60)(16)(250) = \$1.44 \text{ M per year}$$
$$\text{Groundwater: } (5/1000)(1500)(60)(16)(250) = \$1.80 \text{ M per year}$$

Calculate the PW at $i = 12\%$ per year and select the option with the lower cost (larger PW value). In $1 million units:

PW relation: PW = first cost − PW of AOC − PW of UPW + PW of salvage value

$$\begin{aligned} PW_S &= -20 - 0.5(P/A,12\%,10) - 1.44(P/A,12\%,10) + 0.05(20)(P/F,12\%,10) \\ &= -20 - 0.5(5.6502) - 1.44(5.6502) + 1(0.3220) \\ &= \$-30.64 \end{aligned}$$

$$\begin{aligned} PW_G &= -22 - 0.3(P/A,12\%,10) - 1.80(P/A,12\%,10) + 0.10(22)(P/F,12\%,10) \\ &= -22 - 0.3(5.6502) - 1.80(5.6502) + 2.2(0.3220) \\ &= \$-33.16 \end{aligned}$$

Based on this present worth analysis, the seawater option is cheaper by $2.52 M.

5.3 Present Worth Analysis of Different-Life Alternatives ● ● ●

Online Presentation

When the present worth method is used to compare mutually exclusive alternatives that have different lives, the equal-service requirement must be met. The procedure of Section 5.2 is followed, with one exception:

> The PW of the alternatives must be compared over the **same number of years** and must end at the same time to satisfy the equal-service requirement.

Equal-service requirement

This is necessary, since the present worth comparison involves calculating the equivalent PW of all future cash flows for each alternative. A fair comparison requires that PW values represent cash flows associated with equal service. For cost alternatives, failure to compare equal service will always favor the shorter-lived mutually exclusive alternative, even if it is not the more economical choice, because fewer periods of costs are involved. The equal-service requirement is satisfied by using either of two approaches:

> **LCM:** Compare the PW of alternatives over a period of time equal to the **least common multiple (LCM)** of their estimated lives.
>
> **Study period:** Compare the PW of alternatives using a **specified study period of n years.** This approach does not necessarily consider the useful life of an alternative. The study period is also called the *planning horizon*.

LCM or study period

For either approach, calculate the PW at the MARR and use the same selection guideline as that for equal-life alternatives. The LCM approach makes the cash flow estimates extend to the same period, as required. For example, lives of 3 and 4 years are compared over a 12-year period.

The first cost of an alternative is reinvested at the beginning of each life cycle, and the estimated salvage value is accounted for at the end of each life cycle when calculating the PW values over the LCM period. Additionally, the LCM approach requires that some assumptions be made about subsequent life cycles.

The assumptions when using the LCM approach are that

1. The service provided will be needed over the entire LCM years or more.
2. The selected alternative can be repeated over each life cycle of the LCM in exactly the same manner.
3. Cash flow estimates are the same for each life cycle.

As will be shown in Chapter 14, the third assumption is valid only when the cash flows are expected to change by exactly the inflation (or deflation) rate that is applicable through the LCM time period. If the cash flows are expected to change by any other rate, then the PW analysis must be conducted using constant-value dollars, which considers inflation (Chapter 14).

A study period analysis is necessary if the first assumption about the length of time the alternatives are needed cannot be made. For the study period approach, a time horizon is chosen over which the economic analysis is conducted, and only those cash flows which occur during that time period are considered relevant to the analysis. All cash flows occurring beyond the study period are ignored. An estimated market value at the end of the study period must be made. The time horizon chosen might be relatively short, especially when short-term business goals are very important. The study period approach is often used in replacement analysis (Chapter 11). It is also useful when the LCM of alternatives yields an unrealistic evaluation period, for example, 5 and 9 years.

EXAMPLE 5.3

National Homebuilders, Inc., plans to purchase new cut-and-finish equipment. Two manufacturers offered the estimates below.

	Vendor A	Vendor B
First cost, $	−15,000	−18,000
Annual M&O cost, $ per year	−3,500	−3,100
Salvage value, $	1,000	2,000
Life, years	6	9

(a) Determine which vendor should be selected on the basis of a present worth comparison, if the MARR is 15% per year.
(b) National Homebuilders has a standard practice of evaluating all options over a 5-year period. If a study period of 5 years is used and the salvage values are not expected to change, which vendor should be selected?

Solution

(a) Since the equipment has different lives, compare them over the LCM of 18 years. For life cycles after the first, the first cost is repeated in year 0 of each new cycle, which is the last year of the previous cycle. These are years 6 and 12 for vendor A and year 9 for B. The cash flow diagram is shown in Figure 5–2. Calculate PW at 15% over 18 years.

$$PW_A = -15,000 - 15,000(P/F,15\%,6) + 1000(P/F,15\%,6)$$
$$-15,000(P/F,15\%,12) + 1000(P/F,15\%,12) + 1000(P/F,15\%,18)$$
$$-3,500(P/A,15\%,18)$$
$$= \$-45,036$$

$$PW_B = -18,000 - 18,000(P/F,15\%,9) + 2000(P/F,15\%,9)$$
$$+ 2000(P/F,15\%,18) - 3100(P/A,15\%,18)$$
$$= \$-41,384$$

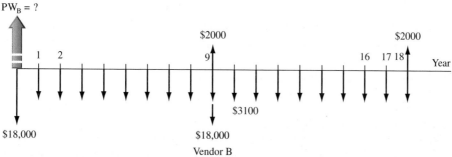

Figure 5–2
Cash flow diagram for different-life alternatives, Example 5.3a.

Vendor B is selected, since it costs less in PW terms; that is, the PW_B value is numerically larger than PW_A.

(b) For a 5-year study period, no cycle repeats are necessary. The PW analysis is

$$PW_A = -15,000 - 3500(P/A,15\%,5) + 1000(P/F,15\%,5)$$
$$= \$-26,236$$

$$PW_B = -18,000 - 3100(P/A,15\%,5) + 2000(P/F,15\%,5)$$
$$= \$-27,397$$

Vendor A is now selected based on its smaller PW value. This means that the shortened study period of 5 years has caused a switch in the economic decision. In situations such as this, the standard practice of using a fixed study period should be carefully examined to ensure that the appropriate approach, that is, LCM or fixed study period, is used to satisfy the equal-service requirement.

EXAMPLE 5.4   Water for Semiconductor Manufacturing Case   PE

When we discussed this case in the introduction, we learned that the initial estimates of equipment life were 10 years for both options of UPW (ultrapure water)—seawater and groundwater. As you might guess, a little research indicates that seawater is more corrosive and the equipment life is shorter—5 years rather than 10. However, it is expected that, instead of complete replacement, a total refurbishment of the equipment for $10 M after 5 years will extend the life through the anticipated 10th year of service.

With all other estimates remaining the same, it is important to determine if this 50% reduction in expected usable life and the refurbishment expense may alter the decision to go with the seawater option, as determined in Example 5.2. For a complete analysis, consider both a 10-year and a 5-year option for the expected use of the equipment, regardless of the source of UPW.

	A	B	C	D	E	F	G
1		PW analysis over LCM of 10 years					
2							
3	Year	Seawater	Groundwater		Seawater option		
4	0	-20.00	-22.00		= AOC + UPW cost/year		
5	1	-1.94	-2.10		= −0.5 − 1.44		
6	2	-1.94	-2.10				
7	3	-1.94	-2.10				
8	4	-1.94	-2.10		Groundwater option		
9	5	-11.94	-2.10		= AOC + UPW cost/year		
10	6	-1.94	-2.10		= −0.3 − 1.80		
11	7	-1.94	-2.10				
12	8	-1.94	-2.10		Both options		
13	9	-1.94	-2.10		Salvage values		
14	10	-0.94	0.10		included here		
15							
16	NPV function	`=NPV(12%,B5:B14)+B4	`=NPV(12%,C5:C14)+C4				
17							
18	NPV ($1 M)	(36.31)	(33.16)				
19							
20							
21		PW analysis over 5-year study period					
22							
23	Year	Seawater	Groundwater				
24	0	-20.00	-22.00				
25	1	-1.94	-2.10				
26	2	-1.94	-2.10				
27	3	-1.94	-2.10		Salvage values are the		
28	4	-1.94	-2.10		same after 5 and 10 years		
29	5	-0.94	0.10				
30							
31	NPV function	`=NPV(12%,B25:B29)+B24	`=NPV(12%,C25:C29)+C24				
32							
33	NPV ($1 M)	(26.43)	(28.32)				

Figure 5–3
PW analyses using LCM and study period approaches for water for semiconductor manufacturing case, Example 5.4.

Solution

A spreadsheet and the NPV function are a quick and easy way to perform this dual analysis. The details are presented in Figure 5–3.

LCM of 10 years: In the top part of the spreadsheet, the LCM of 10 years is necessary to satisfy the equal-service requirement; however, the first cost in year 5 is the refurbishment cost of $\$-10$ M, not the $\$-20$ M expended in year 0. Each year's cash flow is entered in consecutive cells; the $\$-11.94$ M in year 5 accounts for the continuing AOC and annual UPW cost of $\$-1.94$ M, plus the $\$-10$ M refurbishment cost. The NPV functions shown on the spreadsheet determine the 12% per year PW values in $1 million units.

$$PW_S = \$-36.31 \qquad PW_G = \$-33.16$$

Now, the *groundwater option is cheaper*; the economic decision is reversed with this new estimate of life and year 5 refurbishment expense.

Study period of 5 years: The lower portion of Figure 5–3 details a PW analysis using the second approach to evaluating different-life alternatives, that is, a specific study period, which is 5 years in this case study. Therefore, all cash flows after 5 years are neglected.

Again the economic decision is reversed as the 12% per year PW values *favor the seawater option*.

$$PW_S = \$-26.43 \qquad PW_G = \$-28.32$$

Comments

The decision switched between the LCM and study period approaches. Both are correct answers given the decision of how the equal-service requirement is met. This analysis demonstrates how important it is to compare mutually exclusive alternatives over time periods that are believable and to take the time necessary to make the most accurate cost, life, and MARR estimates when the evaluation is performed.

> If the PW evaluation is incorrectly performed using the respective lives of the two options, the equal-service requirement is violated, and PW values favor the shorter-lived option, that is, seawater. The PW values are
>
> Option S: $n = 5$ years, $PW_S = \$-26.43$ M, from the bottom left calculation in Figure 5–3.
> Option G: $n = 10$ years, $PW_G = \$-33.16$ M, from the top right calculation in Figure 5–3.

For **independent projects**, use of the LCM approach is unnecessary since each project is compared to the do-nothing alternative, not to each other, and satisfying the equal-service requirement is not a problem. Simply use the MARR to determine the PW over the respective life of each project, and **select all projects with a PW ≥ 0.**

5.4 Future Worth Analysis ● ● ○

The future worth (FW) of an alternative may be determined directly from the cash flows, or by multiplying the PW value by the F/P factor, at the established MARR. The n value in the F/P factor is either the LCM value or a specified study period. Analysis of alternatives using FW values is especially applicable to large capital investment decisions when a prime goal is to maximize the *future wealth* of a corporation's stockholders.

Future worth analysis over a specified study period is often utilized if the asset (equipment, a building, etc.) might be sold or traded at some time before the expected life is reached. Suppose an entrepreneur is planning to buy a company and expects to trade it within 3 years. FW analysis is the best method to help with the decision to sell or keep it 3 years hence. Example 5.5 illustrates this use of FW analysis. Another excellent application of FW analysis is for projects that will come online at the end of a multiyear investment period, such as electric generation facilities, toll roads, airports, and the like. They are analyzed using the FW value of investment commitments made during construction.

> The selection guidelines for FW analysis are the same as for PW analysis; FW ≥ 0 means the MARR is met or exceeded. For two or more mutually exclusive alternatives, select the one with the numerically largest FW value.

ME alternative
selection

EXAMPLE 5.5

A British food distribution conglomerate purchased a Canadian food store chain for £75 million 3 years ago. There was a net loss of £10 million at the end of year 1 of ownership. Net cash flow is increasing with an arithmetic gradient of £+5 million per year starting the second year, and this pattern is expected to continue for the foreseeable future. This means that breakeven net cash flow was achieved this year. Because of the heavy debt financing used to purchase the Canadian chain, the international board of directors expects a MARR of 25% per year from any sale.

(a) The British conglomerate has just been offered £159.5 million by a French company wishing to get a foothold in Canada. Use FW analysis to determine if the MARR will be realized at this selling price.

(b) If the British conglomerate continues to own the chain, what selling price must be obtained at the end of 5 years of ownership to just make the MARR?

Solution

(a) Set up the future worth relation in year 3 (FW_3) at $i = 25\%$ per year and an offer price of £159.5 million. Figure 5–4a presents the cash flow diagram in million £ units.

$$FW_3 = -75(F/P,25\%,3) - 10(F/P,25\%,2) - 5(F/P,25\%,1) + 159.5$$

$$= -168.36 + 159.5 = \pounds-8.86 \text{ million}$$

No, the MARR of 25% will not be realized if the £159.5 million offer is accepted.

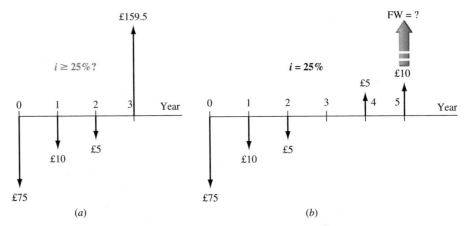

Figure 5–4
Cash flow diagrams for Example 5.5. (a) i = ?; (b) FW = ?.

(b) Determine the future worth 5 years from now at 25% per year. Figure 5–4b presents the cash flow diagram. The A/G and F/A factors are applied to the arithmetic gradient.

$$\text{FW}_5 = -75(F/P,25\%,5) - 10(F/A,25\%,5) + 5(A/G,25\%,5)(F/A,25\%,5)$$
$$= £-246.81 \text{ million}$$

The offer must be for at least £246.81 million to make the MARR. This is approximately 3.3 times the purchase price only 5 years earlier, in large part based on the required MARR of 25%.

5.5 Capitalized Cost Analysis ● ● ●

Many public sector projects such as bridges, dams, highways and toll roads, railroads, and hydro-electric and other power generation facilities have very long expected useful lives. A **perpetual or infinite life** is the effective planning horizon. Permanent endowments for charitable organizations and universities also have perpetual lives. The economic worth of these types of projects or endowments is evaluated using the present worth of the cash flows.

Capitalized Cost (CC) is the present worth of a project that has a very long life (more than, say, 35 or 40 years) or when the planning horizon is considered very long or infinite.

The formula to calculate CC is derived from the PW relation $P = A(P/A,i\%,n)$, where $n = \infty$ time periods. Take the equation for P using the P/A factor and divide the numerator and denominator by $(1 + i)^n$ to obtain

$$P = A\left[\frac{1 - \dfrac{1}{(1 + i)^n}}{i}\right]$$

As n approaches ∞, the bracketed term becomes $1/i$. We replace the symbols P and PW with CC as a reminder that this is a capitalized cost equivalence. Since the A value can also be termed AW for annual worth, the capitalized cost formula is simply

$$\text{CC} = \frac{A}{i} \quad \text{or} \quad \text{CC} = \frac{\text{AW}}{i} \qquad \text{[5.1]}$$

Solving for A or AW, the amount of new money that is generated each year by a capitalization of an amount CC is

$$\text{AW} = \text{CC}(i) \qquad \text{[5.2]}$$

This is the same as the calculation $A = P(i)$ for an infinite number of time periods. Equation [5.2] can be explained by considering the time value of money. If $20,000 is invested now (this is the capitalization) at 10% per year, the maximum amount of money that can be withdrawn at the end of every year for *eternity* is $2000, which is the interest accumulated each year. This leaves the original $20,000 to earn interest so that another $2000 will be accumulated the next year.

The cash flows (costs, revenues, and savings) in a capitalized cost calculation are usually of two types: *recurring,* also called periodic, and *nonrecurring.* An annual operating cost of $50,000 and a rework cost estimated at $40,000 every 12 years are examples of recurring cash flows. Examples of nonrecurring cash flows are the initial investment amount in year 0 and one-time cash flow estimates at future times, for example, $500,000 in fees 2 years hence.

The procedure to determine the CC for an infinite sequence of cash flows is as follows:

1. Draw a cash flow diagram showing all nonrecurring (one-time) cash flows and at least two cycles of all recurring (periodic) cash flows.
2. Find the present worth of all nonrecurring amounts. This is their CC value.
3. Find the A value through *one life cycle* of all recurring amounts. (This is the same value in all succeeding life cycles, as explained in Chapter 6.) Add this to all other uniform amounts (A) occurring in years 1 through infinity. The result is the total equivalent uniform annual worth (AW).
4. Divide the AW obtained in step 3 by the interest rate *i* to obtain a CC value. This is an application of Equation [5.1].
5. Add the CC values obtained in steps 2 and 4.

Drawing the cash flow diagram (step 1) is more important in CC calculations than elsewhere, because it helps separate nonrecurring and recurring amounts. In step 5 the present worths of all component cash flows have been obtained; the total capitalized cost is simply their sum.

EXAMPLE 5.6

The Haverty County Transportation Authority (HCTA) has just installed new software to charge and track toll fees. The director wants to know the total equivalent cost of all future costs incurred to purchase the software system. If the new system will be used for the indefinite future, find the equivalent cost (*a*) now, a CC value, and (*b*) for each year hereafter, an AW value.

The system has an installed cost of $150,000 and an additional cost of $50,000 after 10 years. The annual software maintenance contract cost is $5000 for the first 4 years and $8000 thereafter. In addition, there is expected to be a recurring major upgrade cost of $15,000 every 13 years. Assume that $i = 5\%$ per year for county funds.

Solution

(*a*) The five-step procedure to find CC now is applied.
   1. Draw a cash flow diagram for two cycles (Figure 5–5).
   2. Find the present worth of the nonrecurring costs of $150,000 now and $50,000 in year 10 at $i = 5\%$. Label this CC_1.

$$CC_1 = -150,000 - 50,000(P/F,5\%,10) = \$-180,695$$

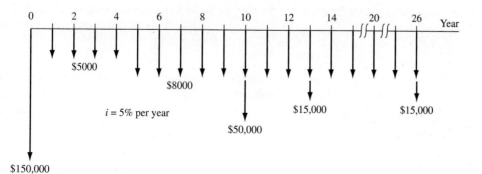

Figure 5–5
Cash flows for two cycles of recurring costs and all nonrecurring amounts, Example 5.6.

3 and 4. Convert the $15,000 recurring cost to an A value over the first cycle of 13 years, and find the capitalized cost CC_2 at 5% per year using Equation [5.1].

$$A = -15,000(A/F,5\%,13) = \$-847$$

$$CC_2 = -847/0.05 = \$-16,940$$

There are several ways to convert the annual software maintenance cost series to A and CC values. A straightforward method is to, first, consider the $-5000 an A series with a capitalized cost of

$$CC_3 = -5000/0.05 = \$-100,000$$

Second, convert the step-up maintenance cost series of $-3000 to a capitalized cost CC_4 in year 4, and find the present worth in year 0. (Refer to Figure 5–5 for cash flow timings.)

$$CC_4 = \frac{-3,000}{0.05}(P/F,5\%,4) = \$-49,362$$

5. The total capitalized cost CC_T for Haverty County Transportation Authority is the sum of the four component CC values.

$$CC_T = -180,695 - 16,940 - 100,000 - 49,362$$
$$= \$-346,997$$

(*b*) Equation [5.2] determines the AW value forever.

$$AW = Pi = CC_T(i) = \$346,997(0.05) = \$17,350$$

Correctly interpreted, this means Haverty County officials have committed the equivalent of $17,350 forever to operate and maintain the toll management software.

For the comparison of **two alternatives on the basis of capitalized cost,** use the procedure above to find the A value and CC_T for each alternative. Since the capitalized cost represents the total present worth of financing and maintaining a given alternative forever, the alternatives will automatically be compared for the same number of years (i.e., infinity). The alternative with the smaller capitalized cost will represent the more economical one. This evaluation is illustrated in Example 5.7 using the progressive example for this chapter.

EXAMPLE 5.7 Water for Semiconductor Manufacturing Case

Our case study has progressed (in Example 5.4) to the point that the life of the seawater option can be extended to 10 years with a major refurbishment cost after 5 years. This extension is possible only one time, after which a new life cycle would commence. In $1 million units, the estimates and PW values (from Figure 5–3) are as follows:

Seawater: $P_S = \$-20$; $AOC_S = \$-1.94$; $n_S = 10$ years; refurbishment, year 5 $= \$-10$; $S_S = 0.05(20) = \$1.00$; $PW_S = \$-36.31$

Groundwater: $P_G = \$-22$; $AOC_G = \$-2.10$; $n_G = 10$ years; $S_G = 0.10(22) = \$2.2$; $PW_G = \$-33.16$

If we assume that the UPW (ultrapure water) requirement will continue for the foreseeable future, a good number to know is the present worth of the long-term options at the selected MARR of 12% per year. What are these capitalized costs for the two options using the estimates made thus far?

Solution

Find the equivalent A value for each option over its respective life, then determine the CC value using the relation $CC = A/i$. Select the option with the lower CC. This approach satisfies the equal-service requirement because the time horizon is infinity when the CC is determined.

Seawater: $A_S = PW_S(A/P,12\%,10) = -36.31(0.17698) = \$-6.43$

$\quad\quad CC_S = -6.43/0.12 = \$-53.58$

Groundwater: $A_G = PW_G(A/P,12\%,10) = -33.16(0.17698) = \$-5.87$

$\quad\quad CC_G = -5.87/0.12 = \$-48.91$

In terms of capitalized cost, the *groundwater alternative is cheaper.*

Comment

If the seawater-life extension is not considered a viable option, the original alternative of 5 years could be used in this analysis. In this case, the equivalent A value and CC computations in $1 million units are as follows.

$$A_{S,5\text{ years}} = -20(A/P,12\%,5) - 1.94 + 0.05(20)(A/F,12\%,5)$$
$$= \$-7.33$$
$$CC_{S,5\text{ years}} = -7.33/0.12 = \$-61.08$$

Now, the economic advantage of the groundwater option is even larger.

If a finite-life alternative (for example, 5 years) is compared to one with an indefinite or very long life, capitalized costs can be used. To determine capitalized cost for the finite life alternative, calculate the equivalent A value for one life cycle and divide by the interest rate (Equation [5.1]). This procedure is illustrated in Example 5.8 using a spreadsheet.

EXAMPLE 5.8

The State Legislature has mandated a statewide recycling program to include all types of plastic, paper, metal, and glass refuse. The goal is zero landfill by 2020. Two options for the materials separation equipment are outlined below. The interest rate for state-mandated projects is 5% per year.

Contractor option (C): $8 million now and $25,000 per year will provide separation services at a maximum of 15 sites. No contract period is stated; thus the contract and services are offered for as long as the State needs them.

Purchase option (P): Purchase equipment at each site for $275,000 per site and expend an estimated $12,000 in annual operating costs (AOC). Expected life of the equipment is 5 years with no salvage value.

(*a*) Perform a capitalized cost analysis for a total of 10 recycling sites.
(*b*) Determine the maximum number of sites at which the equipment can be purchased and still have a capitalized cost less than that of the contractor option.

Solution

(*a*) Figure 5–6, column B, details the solution. The contract, as proposed, has a long life. Therefore, the $8 million is already a capitalized cost. The annual charge of $A = \$25,000$ is divided by $i = 0.05$ to determine its CC value. Summing the two values results in $CC_C = \$-8.5$ million.

For the finite, 5-year purchase alternative, column B shows the first cost ($-275,000 per site), AOC ($-12,000), and equivalent A value of $-755,181, which is determined via the PMT function (cell tag). Divide A by the interest rate of 5% to determine $CC_P = \$-15.1$ million.

The *contractor option is by far more economical* for the anticipated 10 sites.

(*b*) A quick way to find the maximum number of sites for which $CC_P < CC_C$ is to use Excel's **Goal Seek** tool, introduced in Chapter 2, Example 2.10. (See Appendix A for details on how to use this tool.) The template is set up in Figure 5–6 to make the two CC values equal as the number of sites is altered (decreased). The result, shown in column C, indicates that 5.63 sites make the options economically equivalent. Since the number of sites

	A	B	C
1	Interest rate	5%	
2			
3		(a) Evaluation	(b) Sensitivity
4	**Number of machines**	**10**	**5.63**
5			
6	**Alternative 1: Contractor**		
7	Initial contract cost	-8,000,000	-8,000,000
8	Annual cost	-25,000	-25,000
9	**CC for contractor (CC = A/i)**	-8,500,000	-8,500,000
10			
11	**Alternative 2: Purchase**		
12	Initial cost per machine	-275,000	-275,000
13	Expected life, years	5	5
14	AOC per machine	-12,000	-12,000
15	Equivalent annual value (A)	-755,181	-425,000
16	**CC for purchase (CC = A/i)**	-15,103,614	-8,500,000
17			

$$= \text{PMT}(\$B\$1,B13,-B12*B4) + B14*B4$$

Goal Seek

Set cell: $B$16
To value: -8500000
By changing cell: B4

Figure 5–6
Spreadsheet solution of Example 5.8 using capitalized cost (*a*) for 10 recycling sites and (*b*) to determine the number of sites to make the alternatives economically equal.

must be an integer, *5 or fewer sites will favor purchasing the equipment* and 6 or more sites will favor contracting the separation services.

This approach to problem solution will be called *breakeven analysis* in later chapters of the text. By the way, another way to determine the number of sites is by trial and error. Enter different values in cell B4 until the CC values favor the purchase alternative.

CHAPTER SUMMARY

The present worth method of comparing alternatives involves converting all cash flows to present dollars at the MARR. The alternative with the numerically larger (or largest) PW value is selected. When the alternatives have different lives, the comparison must be made for equal-service periods. This is done by performing the comparison over either the LCM of lives or a specific study period. Both approaches compare alternatives in accordance with the equal-service requirement. When a study period is used, any remaining value in an alternative is recognized through the estimated future market value.

If the life of the alternatives is considered to be very long or infinite, capitalized cost is the comparison method. The CC value is calculated as A/i, because the P/A factor reduces to $1/i$ in the limit of $n = \infty$.

PROBLEMS

Types of Projects

5.1 What is the difference between mutually exclusive alternatives and independent projects?

5.2 (*a*) What is meant by the do-nothing alternative?
  (*b*) When is the do-nothing alternative *not* an option?

5.3 (*a*) How many alternatives are possible from four independent projects identified as W, X, Y, and Z?
  (*b*) List all of the possibilities.

5.4 What is the difference between a revenue and a cost alternative?

5.5   What is meant by the term *equal service?*

5.6   What two approaches can be used to satisfy the equal-service requirement?

Alternative Comparison—Equal Lives

5.7   A company that manufactures magnetic membrane switches is investigating two production options that have the estimated cash flows shown ($1 million units). Which one should be selected on the basis of a present worth analysis at 10% per year?

	In-house	Contract
First cost, $	−30	0
Annual cost, $ per year	−5	−2
Annual income, $ per year	14	3.1
Salvage value, $	2	—
Life, years	5	5

5.8   The manager of a canned food processing plant must decide between two different labeling machines. Machine A will have a first cost of $42,000, an annual operating cost of $28,000, and a service life of 4 years. Machine B will cost $51,000 to buy and will have an annual operating cost of $17,000 during its 4-year life. At an interest rate of 10% per year, which should be selected on the basis of a present worth analysis?

5.9   A metallurgical engineer is considering two materials for use in a space vehicle. All estimates are made. (*a*) Which should be selected on the basis of a present worth comparison at an interest rate of 12% per year? (*b*) At what first cost for the material *not selected above* will it become the more economic alternative?

	Material X	Material Y
First cost, $	−15,000	−35,000
Maintenance cost, $ per year	−9,000	−7,000
Salvage value, $	2,000	20,000
Life, years	5	5

5.10   To retain high-performing engineers, a large semiconductor company provides corporate stock as part of the compensation package. In one particular year, the company offered 1000 shares of either class A or class B stock. The class A stock was selling for $30 per share at the time, and stock market analysts predicted that it would increase at a rate of 6% per year for the next 5 years. Class B stock was selling for $20 per share, but its price was expected to increase by 12% per year. At an interest rate of 8% per year, which stock should the engineers select on the basis of a present worth analysis and a 5-year planning horizon?

5.11   The Murphy County Fire Department is considering two options for upgrading its aging physical facilities. Plan A involves remodeling the fire stations on Alameda Avenue and Trowbridge Boulevard that are 57 and 61 years old, respectively. (The industry standard is about 50 years of use for a station.) The cost for remodeling the Alameda station is estimated at $952,000 while the cost of redoing the Trowbridge station is $1.3 million. Plan B calls for buying 5 acres of land somewhere between the two stations, building a new fire station, and selling the land and structures at the previous sites. The cost of land in that area is estimated to be $366,000 per acre. The size of the new fire station would be 9000 square feet with a construction cost of $151.18 per square foot. Contractor fees for overhead, profit, etc. are expected to be $340,000, and architect fees will be $81,500. (Assume all of the costs for plan B occur at time 0.) If plan A is adopted, the extra cost for personnel and equipment will be $126,000 per year. Under plan B, the sale of the old sites is anticipated to net a positive $500,000 five years in the future. Use an interest rate of 6% per year and a 50-year useful life for the remodeled and new stations to determine which plan is better on the basis of a present worth analysis.

5.12   Delcon Properties is a commercial developer of shopping centers and malls in various places around the country. The company needs to analyze the economic feasibility of rainwater drains in a 60-acre area that it plans to develop. Since the development won't be started for 3 years, this large open space will be subject to damage from heavy thunderstorms that cause soil erosion and heavy rutting. If no drains are installed, the cost of refilling and grading the washed out area is expected to be $1500 per thunderstorm. Alternatively, a temporary corrugated steel drainage pipe could be installed that will prevent the soil erosion. The cost of the pipe will be $3 per foot for the total length of 7000 feet required. Some of the pipe will be salvageable for $4000 at the end of the 3-year period between now and when the construction begins. Assuming that thunderstorms occur regularly at 3-month intervals, starting 3 months from now, which alternative should be selected on the basis of a present worth comparison using an interest rate of 4% per quarter?

5.13   A public water utility is trying to decide between two different sizes of pipe for a new water main. A 250-mm line will have an initial cost of $155,000, whereas a 300-mm line will cost $210,000. Since there is more head loss through the 250-mm pipe, the pumping cost is expected to be $3000 more per year than for the 250-mm line. If the lines are expected to last for 30 years, which size should be selected on the basis of a present worth analysis using an interest rate of 10% per year?

5.14 The supervisor of a community swimming pool has developed two methods for chlorinating the pool. If gaseous chlorine is added, a chlorinator will be required that has an initial cost of $8000 and a useful life of 5 years. The chlorine will cost $650 per year, and the labor cost will be $800 per year. Alternatively, dry chlorine can be added manually at a cost of $1000 per year for chlorine and $1900 per year for labor. Which method should be used on the basis of a present worth analysis if the interest rate is 10% per year?

5.15 Anion, an environmental engineering consulting firm, is trying to be eco-friendly in acquiring an automobile for general office use. It is considering a gasoline-electric hybrid and a gasoline-free all-electric hatchback. The hybrid under consideration is GM's Volt, which will cost $35,000 and have a range of 40 miles on the electric battery and several hundred more miles when the gasoline engine kicks in. Nissan's Leaf, on the other hand, is a pure electric that will have a range of only 100 miles, after which its lithium-ion battery will have to be recharged. The Leaf's relatively limited range creates a psychological effect known as *range anxiety*. This fact alone has caused the company to lean toward purchasing the Volt, which is assumed to have a salvage value of $15,000 in 5 years. The Leaf could be leased for $349 per month (end-of-month payments) for 5 years after an initial $1500 down payment for "account activation." If the consulting company plans to ignore the range anxiety effect in making its decision, which automobile is the better option on the basis of a present worth analysis at an interest rate of 0.75% per month? Assume the operating cost will be the same for both vehicles.

5.16 A pipeline engineer working in Kuwait for the oil giant BP wants to perform a present worth analysis on alternative pipeline routings—the first predominately by land and the second primarily undersea. The undersea route is more expensive initially due to extra corrosion protection and installation costs, but cheaper security and maintenance reduces annual costs. Perform the analysis for the engineer at 15% per year.

	Land	Undersea
Installation cost, $ million	−215	−350
Pumping, operating, security, $ million per year	−22	−2
Replacement of valves and appurtenances in year 25, $ million	−30	−70
Expected life, years	50	50

Alternative Comparison—Different Lives

5.17 An electric switch manufacturing company has to choose one of three different assembly methods. Method A will have a first cost of $40,000, an annual operating cost of $9000, and a service life of 2 years. Method B will cost $80,000 to buy and will have an annual operating cost of $6000 over its 4-year service life. Method C will cost $130,000 initially with an annual operating cost of $4000 over its 8-year life. Methods A and B will have no salvage value, but method C will have some equipment worth an estimated $12,000. Which method should be selected? Use present worth analysis at an interest rate of 10% per year.

5.18 Midwest Power and Light operates 14 coal-fired power plants in several states around the United States. The company recently settled a lawsuit by agreeing to pay $60 million in mitigation costs related to acid rain. The settlement included $21 million to reduce emissions from barges and trucks in the Ohio River Valley, $24 million for projects to conserve energy and produce alternative energy, $3 million for Chesapeake Bay, $2 million for Shenandoah National Park, and $10 million to acquire ecologically sensitive lands in Appalachia. The question of how to distribute the money over time has been posed. Plan A involves spending $5 million now and the remaining $55 million equally over a 10-year period (that is, $5.5 million in each of years 1 through 10). Plan B requires expenditures of $5 million now, $25 million 2 years from now, and $30 million 7 years from now. Determine which plan is more economical on the basis of a present worth analysis over a 10-year period at an interest rate of 10% per year.

5.19 Machines that have the following costs are under consideration for a robotized welding process. Using an interest rate of 10% per year, determine which alternative should be selected on the basis of a present worth analysis. Show (*a*) hand and (*b*) spreadsheet solutions.

	Machine X	Machine Y
First cost, $	−250,000	−430,000
Annual operating cost, $ per year	−60,000	−40,000
Salvage value, $	70,000	95,000
Life, years	3	6

5.20 **Water for Semiconductor Manufacturing Case**

Throughout the present worth analyses, the decision between seawater and groundwater switched multiple times in Examples 5.2 and 5.4. A summary is given here in $1 million units.

	Seawater (S)				Groundwater (G)		
Life n, years	First cost, $	PW at 12%, $	Selected	Life n, years	First cost, $	PW at 12%, $	Selected
10	−20	−30.64	Yes	10	−22	−33.16	No
5	−20, plus −10 after 5 years	−36.31	No	10	−22	−33.16	Yes
5 (study period)	−20	−26.43	Yes	5 (study period)	−22	−28.32	No

The confusion about the recommended source for UPW has not gone unnoticed by the general manager. Yesterday, you were asked to settle the issue by determining the first cost X_S of the seawater option to ensure that it is the economic choice over groundwater. The study period is set by the manager as 10 years, simply because that is the time period on the lease agreement for the building where the fab will be located. Since the seawater equipment must be refurbished or replaced after 5 years, the general manager told you to assume that the equipment will be purchased anew after 5 years of use. What is the maximum first cost that Angular Enterprises should pay for the seawater option?

5.21 Accurate airflow measurement requires straight unobstructed pipe for a minimum of 10 diameters upstream and 5 diameters downstream of the measuring device. In a field application, physical constraints compromise the pipe layout, so the engineer is considering installing the airflow probes in an elbow, knowing that flow measurement will be less accurate but good enough for process control. This is plan 1, which will be in place for only 3 years, after which a more accurate flow measurement system with the same costs as plan 1 will be available. This plan will have a first cost of $26,000 with an annual maintenance cost estimated at $5000.

Plan 2 involves installation of a recently designed submersible airflow probe. The stainless steel probe can be installed in a drop pipe with the transmitter located in a waterproof enclosure on the handrail. The first cost of this system is $83,000, but because it is accurate and more durable, it will not have to be replaced for at least 6 years. Its maintenance cost is estimated to be $1400 per year plus $2500 in year 3 for replacement of signal processing software. Neither system will have a salvage value. At an interest rate of 10% per year, which one should be selected on the basis of a present worth comparison?

5.22 An engineer is considering two different liners for an evaporation pond that will receive salty concentrate from a brackish water desalting plant. A plastic liner will cost $0.90 per square foot initially and will require replacement in 15 years when precipitated solids will have to be removed from the pond using heavy equipment. This removal will cost $500,000. A rubberized elastomeric liner is tougher and, therefore, is expected to last 30 years, but it will cost $2.20 per square foot. If the size of the pond is 110 acres (1 acre = 43,560 square feet), which liner is more cost effective on the basis of a present worth comparison at an interest rate of 8% per year?

5.23 A sports mortgage is the brainchild of Stadium Capital Financing Group, a company headquartered in Chicago, Illinois. It is an innovative way to finance cash-strapped sports programs by allowing fans to sign up to pay a "mortgage" over a certain number of years for the right to buy good seats at football games for several decades with season ticket prices locked in at current prices. In California, the locked-in price period is 50 years. Assume UCLA fan X purchases a $130,000 mortgage and pays for it now to get season tickets for $290 each for 50 years, while fan Y buys season tickets at $290 in year 1, with prices increasing by $20 per year for 50 years. (a) Which fan made the better deal if the interest rate is 8% per year? (b) What should fan X be willing to pay up front for the mortgage to make the two plans exactly equivalent economically? (Assume he has no reason to give extra money to UCLA at this point.)

5.24 A chemical processing corporation is considering three methods to dispose of a non-hazardous chemical sludge: land application, fluidized-bed incineration, and private disposal contract. The estimates for each method are shown. Determine which has the least cost on the basis of a present worth comparison at 10% per year for the following scenarios:
(a) The estimates as shown
(b) The contract award cost increases by 20% every 2-year renewal

	Land Application	Incineration	Contract
First cost, $	−130,000	−900,000	0
Annual operating cost, $ per year	−95,000	−60,000	−120,000
Salvage value, $	25,000	300,000	0
Life, years	3	6	2

5.25 An assistant to Stacy gave her the PW values for four alternatives they are comparing for the development of a remote control vibration control system for offshore platform application. The results in the table use a MARR of 14% per year. Determine which alternative(s) should be selected (*a*) if the alternatives are exclusive, and (*b*) if the projects are independent.

	I	J	K	L
Life *n*, years	3	4	12	6
PW over *n* years, $	16.08	31.12	−257.46	140.46
PW over 6 years, $	26.94	15.78	−653.29	140.46
PW over 12 years, $	39.21	60.45	−257.46	204.46

Future Worth Comparison

5.26 An industrial engineer is considering two robots for purchase by a fiber-optic manufacturing company. Robot *X* will have a first cost of $80,000, an annual maintenance and operation (M&O) cost of $30,000, and a $40,000 salvage value. Robot *Y* will have a first cost of $97,000, an annual M&O cost of $27,000, and a $50,000 salvage value. Which should be selected on the basis of a future worth comparison at an interest rate of 15% per year? Use a 3-year study period.

5.27 Two processes can be used for producing a polymer that reduces friction loss in engines. Process T will have a first cost of $750,000, an operating cost of $60,000 per year, and a salvage value of $80,000 after its 2-year life. Process W will have a first cost of $1,350,000, an operating cost of $25,000 per year, and a $120,000 salvage value after its 4-year life. Process W will also require updating at the end of year 2 at a cost of $90,000. Which process should be selected on the basis of a future worth analysis at an interest rate of 12% per year?

5.28 Compare the alternatives shown below on the basis of a future worth analysis, using an interest rate of 8% per year.

	P	Q
First cost, $	−23,000	−30,000
Annual operating cost, $ per year	−4,000	−2,500
Salvage value, $	3,000	1,000
Life, years	3	6

5.29 Two manufacturers supply MRI systems for medical imaging. St. Jude's Hospital wishes to replace its current MRI equipment that was purchased 8 years ago with the newer technology and clarity of a state-of-the-art system. System K will have a first cost of $1,600,000, an operating cost of $70,000 per year, and a salvage value of $400,000 after its 4-year life. System L will have a first cost of $2,100,000, an operating cost of $50,000 the first year with an expected increase of $3000 per year thereafter, and no salvage value after its 8-year life. Which system should be selected on the basis of a future worth analysis at an interest rate of 12% per year?

5.30 A retail shopping center developer signed a contract to build a $100 million high-end shopping center in City Center, because the city and county governments agreed to sales and tax rebates totaling $18.7 million over 10 years. The contract called for the developer to raze existing buildings 2 years from the date the contract was signed and to have the shopping center built by the end of year 3. However, due to a real estate–induced recession in the United States, the developer sought and was granted a new contract. The new contract required the developer to raze the existing buildings at the end of year 1, but the shopping center would not have to be completed for 7 years from the date the contract was signed. Assume that the cost for razing the existing buildings is $1.3 million and the developer does not build the shopping center until 7 years from now (at a cost of $100 million). Determine the difference in the future worth cost in year 7 of the two contracts at an interest rate of 10% per year.

Capitalized Cost

5.31 A wealthy businessman wants to start a permanent fund for supporting research directed toward sustainability. The donor plans to give equal amounts of money for each of the next 5 years, plus one now (i.e., six donations) so that $100,000 per year can be withdrawn each year forever, beginning in year 6. If the fund earns interest at a rate of 8% per year, how much money must be donated each time?

5.32 Bob, a philanthropist, is not sure what rate of return his gifts may realize once donated to his favorite charity. Determine the capitalized cost of $10,000 every 5 years *forever,* starting 5 years from now at an interest rate of (*a*) 3% and (*b*) 8% per year. (*c*) Explain the significant difference between the two capitalized costs.

5.33 Find the capitalized cost of a present cost of $300,000, annual costs of $35,000, and periodic costs every 5 years of $75,000. Use an interest rate of 12% per year.

5.34 **Water for Semiconductor Manufacturing Case**

It is anticipated that the needs for UPW (ultrapure water) at the new Angular Enterprises site will continue for a long time, as long as 50 years. This is the rationale for using capitalized cost as a basis for the economic decision between desalinated seawater (S) and purified groundwater (G). These costs were determined (Example 5.7) to be $CC_S = \$-53.58$ million and $CC_G = \$-48.91$ million. Groundwater is the clear economic choice.

Yesterday, the general manager had lunch with the president of Brissa Water, who offered to supply the needed UPW at a cost of $5 million per year for the indefinite future. It would mean a dependence upon a contractor to supply the water, but the equipment, treatment, and other costly activities to obtain UPW on-site would be eliminated. The manager asks you to make a recommendation about this seemingly attractive alternative under the following conditions at the same MARR of 12% per year as used for the other analyses:

(a) The annual cost of $5 million remains constant throughout the time it is needed.

(b) The annual cost starts at $5 million for the first year only, and then it increases 2% per year. (This increase is above the cost of providing UPW by either of the other two methods.)

5.35 Compare the alternatives shown on the basis of their capitalized costs using an interest rate of 10% per year.

	Alternative M	Alternative N
First cost, $	−150,000	−800,000
Annual operating cost, $ per year	−50,000	−12,000
Salvage value, $	8,000	1,000,000
Life, years	5	∞

5.36 The cost of maintaining a certain permanent monument in Washington, DC occurs as periodic outlays of $1000 every year and $5000 every 4 years. Calculate the capitalized cost of the maintenance using an interest rate of 10% per year.

5.37 Because you are thankful for what you learned in engineering economy, you plan to start a permanent scholarship fund in the name of the professor who taught the course. You plan to deposit money now with the stipulation that the scholarships be awarded beginning 12 years from now (which happens to be the exact time that your daughter plans to begin college). The interest that is accumulated between now and year 12 is to be added to the principal of the endowment. After that, the interest that is earned each year will be awarded as scholarship money. If you want the amount of the scholarships to be $40,000 per year, how much must you donate now if the fund earns interest at a rate of 8% per year?

5.38 A patriotic group of firefighters is raising money to erect a permanent (i.e., infinite life) monument in New York City to honor those killed in the line of duty. The initial cost of the monument will be $150,000, and the annual maintenance will cost $5000. There will be an additional one-time cost of $20,000 in 2 years to add names of those who were missed initially. At an interest rate of 6% per year, how much money must they raise now in order to construct and maintain the monument forever?

ADDITIONAL PROBLEMS AND FE EXAM REVIEW QUESTIONS

5.39 One assumption inherent in the present worth method of analysis is that:

(a) The alternatives will be used only through the life of the *shortest-lived* alternative.

(b) The alternatives will be used only through the life of the *longest-lived* alternative.

(c) The cash flows of each alternative will change only by the inflation or deflation rate in succeeding life cycles.

(d) At least one of the alternatives will have a finite life.

5.40 When only one alternative can be selected from two or more, the alternatives are said to be:

(a) Mutually exclusive

(b) Independent alternatives

(c) Cost alternatives

(d) Revenue alternatives

5.41 For the mutually exclusive alternatives shown, the one(s) that should be selected are:

Alternative	PW, $
A	−25,000
B	−12,000
C	10,000
D	15,000

(a) Only C

(b) Only A

(c) C and D

(d) Only D

5.42 The alternatives shown are to be compared on the basis of their present worth values. At an interest rate of 10% per year, the values of *n* that you should use in the uniform series factors to make a correct comparison by the present worth method are:

	Alternative A	Alternative B
First cost, $	−50,000	−90,000
Annual operating cost, $ per year	−10,000	−4000
Salvage value, $	13,000	15,000
Life, years	3	6

    (a) $n = 3$ years for A and $n = 3$ years for B
    (b) $n = 3$ years for A and $n = 6$ years for B
    (c) $n = 6$ years for A and $n = 6$ years for B
    (d)   None of the above

5.43   The value of the *future worth* for alternative P at an interest rate of 8% per year is closest to:

	P	Q
First cost, $	−23,000	−30,000
Annual operating cost, $ per year	−4,000	−2,500
Salvage value, $	3,000	1,000
Life, years	3	6

    (a) $FW_P = \$-88{,}036$
    (b) $FW_P = \$-86{,}026$
    (c) $FW_P = \$-81{,}274$
    (d) $FW_P = \$-70{,}178$

5.44   The present worth of $50,000 now, $10,000 per year in years 1 through 15, and $20,000 per year in years 16 through infinity at 10% per year is closest to:
    (a)   Less than $−169,000
    (b) $−169,580
    (c) $−173,940
    (d) $−195,730

5.45   A donor (you) wishes to start an endowment that will provide scholarship money of $40,000 per year beginning in year 5 and continuing indefinitely. If the university earns 10% per year on the endowment, the amount you must donate now is closest to:
    (a) $−225,470
    (b) $−248,360
    (c) $−273,200
    (d) $−293,820

Problems 5.46 through 5.48 are based on the following information.

	Alternative I	Alternative J
Initial cost, $	−150,000	−250,000
Annual income, $ per year	20,000	40,000
Annual expenses, $ per year	−9,000	−14,000
Salvage value, $	25,000	35,000
Life, years	3	6

The interest rate is 15% per year.

5.46   In comparing alternatives I and J by the present worth method, the value of n that must be used in $11{,}000(P/A,i,n)$ for alternative I is:
    (a)   3
    (b)   6
    (c)   18
    (d)   36

5.47   In comparing alternatives I and J by the present worth method, the equation that yields the present worth of alternative J is:
    (a) $PW_J = -250{,}000 + 40{,}000(P/A,15\%,6) + 35{,}000(P/F,15\%,6)$
    (b) $PW_J = -250{,}000 + 26{,}000(P/A,15\%,6) + 35{,}000(P/F,15\%,6)$
    (c) $PW_J = -250{,}000 - 26{,}000(P/A,15\%,6) + 35{,}000(P/F,15\%,6)$
    (d) $PW_J = -250{,}000 - 26{,}000(P/A,15\%,6) - 35{,}000(P/F,15\%,6)$

5.48   In comparing alternatives I and J by the present worth method, the equation that yields the present worth of alternative I is:
    (a) $PW_I = -150{,}000 + 11{,}000(P/A,15\%,3) + 25{,}000(P/F,15\%,3)$
    (b) $PW_I = -150{,}000 + 11{,}000(P/A,15\%,6) + 25{,}000(P/F,15\%,6)$
    (c) $PW_I = -150{,}000 + 11{,}000(P/A,15\%,6) + 175{,}000(P/F,15\%,3) + 25{,}000(P/F,15\%,6)$
    (d) $PW_I = -150{,}000 + 11{,}000(P/A,15\%,6) - 125{,}000(P/F,15\%,3) + 25{,}000(P/F,15\%,6)$

Problems 5.49 and 5.50 are based on the following information.

	Machine X	Machine Y
Initial cost, $	−80,000	−95,000
Annual operating cost, $ per year	−20,000	−15,000
Salvage value, $	10,000	30,000
Life, years	2	4

The interest rate is 10% per year.

5.49   The equation that will calculate the present worth of machine X is:
    (a) $PW_X = -80{,}000 - 15{,}000(P/A,10\%,4) + 30{,}000(P/F,10\%,4)$
    (b) $PW_X = -80{,}000 - 20{,}000(P/A,10\%,4) - 80{,}000(P/F,10\%,2) + 10{,}000(P/F,10\%,4)$
    (c) $PW_X = -80{,}000 - 20{,}000(P/A,10\%,2) + 10{,}000(P/F,10\%,2)$
    (d) $PW_X = -80{,}000 - 20{,}000(P/A,10\%,4) - 70{,}000(P/F,10\%,2) + 10{,}000(P/F,10\%,4)$

5.50   In comparing the machines on a present worth basis, the present worth of machine Y is closest to:
    (a) $−112,320
    (b) $−122,060
    (c) $−163,040
    (d) $−175,980

5.51 The capitalized cost of $10,000 every 5 years *forever,* starting *now* at an interest rate of 10% per year, is closest to:
(*a*) $-13,520
(*b*) $-16,380
(*c*) $-26,380
(*d*) $-32,590

5.52 At an interest rate of 10% per year, the capitalized cost of $10,000 in year 0, $5000 per year in years 1 through 5, and $1000 per year thereafter forever is closest to:
(*a*) $-$29,652
(*b*) $-$35,163
(*c*) $-$38,954
(*d*) $-$43,221

CASE STUDY

COMPARING SOCIAL SECURITY BENEFITS

Background

When Sheryl graduated from Northeastern University in 2000 and went to work for BAE Systems, she did not pay much attention to the monthly payroll deduction for social security. It was a "necessary evil" that may be helpful in retirement years. However, this was so far in the future that she fully expected this government retirement benefit system to be broke and gone by the time she could reap any benefits from her years of contributions.

This year, Sheryl and Brad, another engineer at BAE, got married. Recently, they both received notices from the Social Security Administration of their potential retirement amounts, were they to retire and start social security benefits at preset ages. Since both of them hope to retire a few years early, they decided to pay closer attention to the predicted amount of retirement benefits and to do some analysis on the numbers.

Information

They found that their projected benefits are substantially the same, which makes sense since their salaries are very close to each other. Although the numbers were slightly different in their two mailings, the similar messages to Brad and Sheryl can be summarized as follows:

If you stop working and start receiving benefits . . .

At age 62, your payment would be about	$1400 per month
At you full retirement age (67 years), your payment would be about	$2000 per month
At age 70, your payment would be about	$2480 per month

These numbers represent a reduction of 30% for early retirement (age 62) and an increase of 24% for delayed retirement (age 70).

This couple also learned that it is possible for a spouse to take spousal benefits at the time that one of them is at full retirement age. In other words, if Sheryl starts her $2000 benefit at age 67, Brad can receive a benefit equal to 50% of hers. Then, when Brad reaches 70 years of age, he can discontinue spousal benefits and start his own. In the meantime, his benefits will have increased by 24%. Of course, this strategy could be switched with Brad taking his benefits and Sheryl receiving spousal benefits until age 70.

All these options led them to define four alternative plans.

A: Each takes early benefits at age 62 with a 30% reduction to $1400 per month.

B: Each takes full benefits at full retirement age of 67 and receives $2000 per month.

C: Each delays benefits until age 70 with a 24% increase to $2480 per month.

D: One person takes full benefits of $2000 per month at age 67, and the other person receives spousal benefits ($1000 per month at age 67) and switches to delayed benefits of $2480 at age 70.

They realize, of course, that the numbers will change over time, based on their respective salaries and number of years of contribution to the social security system by them and by their employers.

Case Study Exercises

Brad and Sheryl are the same age. Brad determined that most of their investments make an average of 6% per year. With this as the interest rate, the analysis for the four alternatives is possible. Sheryl and Brad plan to answer the following questions, but don't have time this week. Can you please help them? (Do the analysis for one person at a time, not the couple, and stop at the age of 85.)

1. How much in total (without the time value of money considered) will each plan A through D pay through age 85?
2. What is the future worth at 6% per year of each plan at age 85?
3. Plot the future worth values for all four plans on one spreadsheet graph.
4. Economically, what is the best combination of plans for Brad and Sheryl, assuming they both live to be 85 years old?
5. Develop at least one additional question that you think Sheryl and Brad may have. Answer the question.

CHAPTER 6

Annual Worth Analysis

LEARNING OUTCOMES

Purpose: Utilize different annual worth techniques to evaluate and select alternatives.

SECTION	TOPIC	LEARNING OUTCOME
6.1	Advantages of AW	• Demonstrate that the AW value is the same for each life cycle.
6.2	CR and AW values	• Calculate and interpret the capital recovery (CR) and AW amounts.
6.3	AW analysis	• Select the best alternative using an annual worth analysis.
6.4	Perpetual life	• Evaluate alternatives with very long lives using AW analysis.
6.5	LCC analysis	• Perform a life-cycle cost (LCC) analysis using AW methods.

I n this chapter, we add to our repertoire of alternative comparison tools. In Chapter 5 we learned the PW method. Here we learn the equivalent annual worth, or AW, method. AW analysis is commonly considered the more desirable of the two methods because the AW value is easy to calculate; the measure of worth—AW in monetary units per year—is understood by most individuals; and its assumptions are essentially identical to those of the PW method.

Annual worth is also known by other titles. Some are *equivalent annual worth* (EAW), *equivalent annual cost* (EAC), *annual equivalent* (AE), and *equivalent uniform annual cost* (EUAC). The alternative selected by the AW method will always be the same as that selected by the PW method, and all other alternative evaluation methods, provided they are performed correctly.

An additional application of AW analysis treated here is **life-cycle cost (LCC) analysis.** This method considers all costs of a product, process, or system from concept to phaseout.

6.1 Advantages and Uses of Annual Worth Analysis ●●●

Online Presentation

For many engineering economic studies, the AW method is the best to use, when compared to PW, FW, and rate of return (Chapters 7 and 8). Since the AW value is the equivalent uniform annual worth of all estimated receipts and disbursements during the life cycle of the project or alternative, AW is easy to understand by any individual acquainted with annual amounts, for example, dollars per year. The AW value, which has the same interpretation as A used thus far, is the economic equivalent of the PW and FW values at the MARR for n years. All three can be easily determined from each other by the relation

$$\text{AW} = \text{PW}(A/P,i,n) = \text{FW}(A/F,i,n) \qquad [6.1]$$

The n in the factors is the number of years for equal-service comparison. This is the LCM or the stated study period of the PW or FW analysis.

When all cash flow estimates are converted to an AW value, this value applies for every year of the life cycle and for *each additional life cycle.*

> The annual worth method offers a prime computational and interpretation advantage because the AW value needs to be calculated for **only one life cycle.** The AW value determined over one life cycle is the AW for all future life cycles. Therefore, it is **not necessary to use the LCM** of lives to satisfy the equal-service requirement.

Equal-service requirement and LCM

As with the PW method, there are three fundamental assumptions of the AW method that should be understood. When alternatives being compared have different lives, the AW method makes the assumptions that

1. The services provided are needed for at least the LCM of the lives of the alternatives.
2. The selected alternative will be repeated for succeeding life cycles in exactly the same manner as for the first life cycle.
3. All cash flows will have the same estimated values in every life cycle.

In practice, no assumption is precisely correct. If, in a particular evaluation, the first two assumptions are not reasonable, a study period must be established for the analysis. Note that for assumption 1, the length of time may be the indefinite future (forever). In the third assumption, all cash flows are expected to change exactly with the inflation (or deflation) rate. If this is not a reasonable assumption, new cash flow estimates must be made for each life cycle, and again a study period must be used. AW analysis for a stated study period is discussed in Section 6.3.

EXAMPLE 6.1

In Example 5.3, National Homebuilders, Inc. evaluated cut-and-finish equipment from vendor A (6-year life) and vendor B (9-year life). The PW analysis used the LCM of 18 years. Consider only the vendor A option now. The diagram in Figure 6–1 shows the cash flows for all three life cycles (first cost $\$-15,000$; annual M&O costs $\$-3500$; salvage value $\$1000$). Demonstrate the equivalence at $i = 15\%$ of PW over three life cycles and AW over one cycle. In Example 5.3, present worth for vendor A was calculated as PW = $\$-45,036$.

Figure 6–1
PW and AW values
for three life cycles,
Example 6.1.

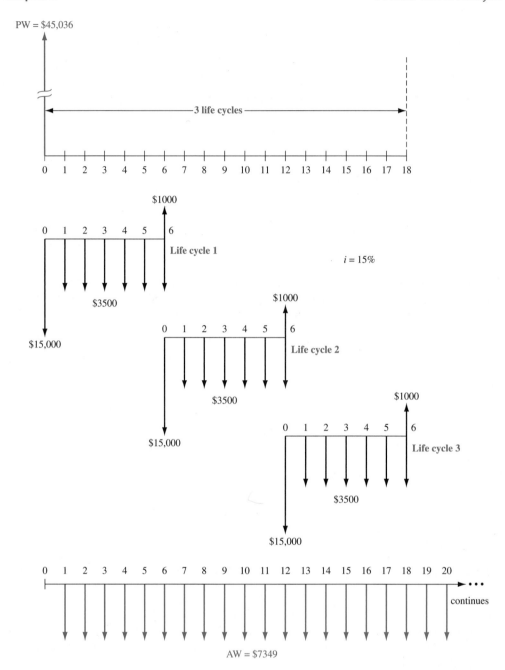

Solution

Calculate the equivalent uniform annual worth value for all cash flows in the first life cycle.

$$AW = -15,000(A/P,15\%,6) + 1000(A/F,15\%,6) - 3500 = \$-7349$$

When the same computation is performed on each succeeding life cycle, the AW value is $\$-7349$. Now Equation [6.1] is applied to the PW value for 18 years.

$$AW = -45,036(A/P,15\%,18) = \$-7349$$

The one-life-cycle AW value and the AW value based on 18 years are equal.

Not only is annual worth an excellent method for performing engineering economy studies, but also it is applicable in any situation where PW (and FW and benefit/cost) analysis can be utilized. The AW method is especially useful in certain types of studies: asset replacement and retention time studies to minimize overall annual costs (both covered in Chapter 11), breakeven studies and make-or-buy decisions (Chapter 13), and all studies dealing with production or manufacturing costs where a cost/unit or profit/unit measure is the focus.

If income taxes are considered, a slightly different approach to the AW method is used by some large corporations and financial institutions. It is termed *economic value added,* or EVA. This approach, covered in Chapter 17, concentrates upon the wealth-increasing potential that an alternative offers a corporation. The resulting EVA values are the equivalent of an AW analysis of after-tax cash flows.

6.2 Calculation of Capital Recovery and AW Values ●●◐

An alternative should have the following cash flow estimates:

Initial investment P. This is the total first cost of all assets and services required to initiate the alternative. When portions of these investments take place over several years, their present worth is an equivalent initial investment. Use this amount as *P*.

Salvage value S. This is the terminal estimated value of assets at the end of their useful life. The *S* is zero if no salvage is anticipated; *S* is negative when it will cost money to dispose of the assets. For study periods shorter than the useful life, *S* is the estimated market value or trade-in value at the end of the study period.

Salvage/market value

Annual amount A. This is the equivalent annual amount (costs only for cost alternatives; costs and receipts for revenue alternatives). Often this is the annual operating cost (AOC) or M&O cost, so the estimate is already an equivalent *A* value.

The annual worth (AW) value for an alternative is comprised of two components: **capital recovery** for the initial investment *P* at a stated interest rate (usually the MARR) and the equivalent annual amount *A*. The symbol CR is used for the capital recovery component. In equation form,

$$AW = CR + A \qquad [6.2]$$

Both CR and *A* represent costs. The total annual amount *A* is determined from uniform recurring costs (and possibly receipts) and nonrecurring amounts. The *P/A* and *P/F* factors may be necessary to first obtain a present worth amount; then the *A/P* factor converts this amount to the *A* value in Equation [6.2]. (If the alternative is a revenue project, there will be positive cash flow estimates present in the calculation of the *A* value.)

The recovery of an amount of capital *P* committed to an asset, **plus** the time value of the capital at a particular interest rate, is a **fundamental principle of economic analysis.**

Capital recovery (CR) is the equivalent annual amount that the asset, process, or system must earn (new revenue) each year to just **recover the initial investment plus a stated rate of return** over its expected life. Any expected salvage value is considered in the computation of CR.

Capital recovery

The *A/P* factor is used to convert *P* to an equivalent annual cost. If there is some anticipated positive salvage value *S* at the end of the asset's useful life, its equivalent annual value is recovered using the *A/F* factor. This action reduces the equivalent annual cost of the asset. Accordingly, CR is calculated as

$$CR = -P(A/P,i,n) + S(A/F,i,n) \qquad [6.3]$$

EXAMPLE 6.2

Lockheed Martin is increasing its booster thrust power in order to win more satellite launch contracts from European companies interested in opening up new global communications markets. A piece of earth-based tracking equipment is expected to require an investment of $13 million, with $8 million committed now and the remaining $5 million expended at the end of year 1 of the project. Annual operating costs for the system are expected to start the first year and continue at $0.9 million per year. The useful life of the tracker is 8 years with a salvage value of $0.5 million. Calculate the CR and AW values for the system, if the corporate MARR is 12% per year.

Solution

Capital recovery: Determine *P* in year 0 of the two initial investment amounts, followed by the use of Equation [6.3] to calculate the capital recovery. In $1 million units,

$$P = 8 + 5(P/F,12\%,1) = \$12.46$$

$$CR = -12.46(A/P,12\%,8) + 0.5(A/F,12\%,8)$$
$$= -12.46(0.20130) + 0.5(0.08130)$$
$$= \$-2.47$$

The correct interpretation of this result is very important to Lockheed Martin. It means that each and every year for 8 years, the equivalent total net revenue from the tracker must be at least \$2,470,000 just to recover the initial present worth investment plus the required return of 12% per year. This does not include the AOC of \$0.9 million each year.

Annual worth: To determine AW, the cash flows in Figure 6–2a must be converted to an equivalent AW series over 8 years (Figure 6–2b). Since CR = \$−2.47 million is an *equivalent annual cost,* as indicated by the minus sign, total AW is determined.

$$AW = -2.47 - 0.9 = \$-3.37 \text{ million per year}$$

This is the AW for all future life cycles of 8 years, provided the costs rise at the same rate as inflation, and the same costs and services are expected to apply for each succeeding life cycle.

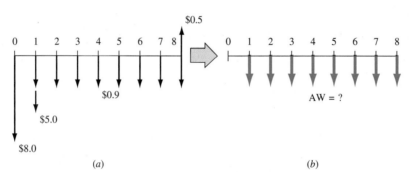

Figure 6–2

(*a*) Cash flow diagram for satellite tracker costs, and (*b*) conversion to an equivalent AW (in \$1 million), Example 6.2.

There is a second, equally correct way to determine CR. Either method results in the same value. There is a relation between the A/P and A/F factors.

$$(A/F,i,n) = (A/P,i,n) - i$$

Both factors are present in the CR Equation [6.3]. Substitute for the A/F factor to obtain

$$CR = -P(A/P,i,n) + S[(A/P,i,n) - i]$$
$$= -[(P - S)(A/P,i,n) + S(i)]$$

There is a basic logic to this formula. Subtracting S from the initial investment P before applying the A/P factor recognizes that the salvage value will be recovered. This reduces CR, the annual cost of asset ownership. However, the fact that S is not recovered until year n of ownership is compensated for by charging the annual interest $S(i)$ against the CR. Although either CR relation results in the same amount, it is better to consistently use the same method. The first method, Equation [6.3], will be used in this text.

For solution by spreadsheet, use the PMT function to determine CR only in a single spreadsheet cell. The general function = PMT($i\%,n,P,F$) is rewritten using the initial investment as P and $-S$ for the salvage value. The format is

$$= PMT(i\%,n,P,-S) \tag{6.4}$$

As an illustration, determine only the CR in Example 6.2. The equivalent initial investment in year 0 is \$12.46 million. The complete function for the CR amount (in \$1 million units) is = PMT(12%,8,12.46,−0.5). The answer of \$−2.47 (million) will be displayed in the spreadsheet cell.

As we learned in Section 3.1, one spreadsheet function can be embedded in another function. In the case of Example 6.2, the initial investment is distributed over a 2-year period. The one-cell PMT function, with the PV function embedded (in bold), can be written as = PMT(12%,8, **8+PV(12%,1,−5)**,−0.5) to display the same value CR = \$−2.47.

6.3 Evaluating Alternatives by Annual Worth Analysis ● ● ●

Online Presentation

The annual worth method is typically the easiest to apply of the evaluation techniques when the MARR is specified. The AW is calculated over the respective life of each alternative, and the selection guidelines are the same as those used for the PW method. For **mutually exclusive** alternatives, whether cost- or revenue-based, the guidelines are as follows:

Project evaluation

ME alternative selection

> **One alternative:** If AW ≥ 0, the requested MARR is met or exceeded and the alternative is economically justified.
>
> **Two or more alternatives:** Select the alternative with the AW that is **numerically largest,** that is, less negative or more positive. This indicates a lower AW of cost for cost alternatives or a larger AW of net cash flows for revenue alternatives.

If any of the three assumptions in Section 6.1 is not acceptable for an alternative, a study period analysis must be used. Then the cash flow estimates over the study period are converted to AW amounts. The following two examples illustrate the AW method for one project and two alternatives.

EXAMPLE 6.3

Heavenly Pizza, which is located in Toronto, fares very well with its competition in offering fast delivery. Many students at the area universities and community colleges work part-time delivering orders made via the web. The owner, Jerry, a software engineering graduate, plans to purchase and install five portable, in-car systems to increase delivery speed and accuracy. The systems provide a link between the web order-placement software and the On-Star system for satellite-generated directions to any address in the area. The expected result is faster, friendlier service to customers and larger income.

Each system costs $4600, has a 5-year useful life, and may be salvaged for an estimated $300. Total operating cost for all systems is $1000 for the first year, increasing by $100 per year thereafter. The MARR is 10%. Perform an annual worth evaluation for the owner that answers the following questions. Perform the solution by hand and by spreadsheet.

(a) How much new annual net income is necessary to recover the investment at the MARR of 10% per year?

(b) Jerry estimates increased net income of $6000 per year for all five systems. Is this project financially viable at the MARR?

(c) Based on the answer in part (b), determine how much new net income Heavenly Pizza must have to economically justify the project. Operating costs remain as estimated.

Solution by Hand

(a) The capital recovery amount calculated by Equation [6.3] answers the first question.

$$CR = -5[4600(A/P,10\%,5)] + 5[300(A/F,10\%,5)]$$
$$= -5[4600(0.26380)] + 5[300(0.16380)]$$
$$= \$-5822$$

The five systems must generate an equivalent annual new revenue of $5822 to recover the initial investment plus a 10% per year return.

(b) Figure 6–3 presents the cash flows over 5 years. The annual operating cost series, combined with the estimated $6000 annual income, forms an arithmetic gradient series with a base amount of $5000 and $G = \$-100$. The project is financially viable if AW ≥ 0 at $i = 10\%$ per year. Apply Equation [6.2], where A is the equivalent annual net income series.

$$AW = CR + A = -5822 + 5000 - 100(A/G,10\%,5)$$
$$= \$-1003$$

The system is not financially justified at the net income level of $6000 per year.

Figure 6–3
Cash flow diagram used to
compute AW, Example 6.3.

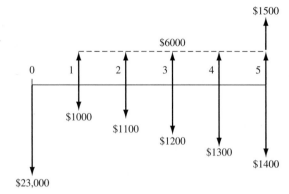

(*c*) Let the required income equal *R,* and set the AW relation equal to zero to find the minimum income to justify the system.

$$0 = -5822 + (R - 1000) - 100(A/G,10\%,5)$$
$$R = -5822 - 1000 - 100(1.8101)$$
$$= \$7003 \text{ per year}$$

Solution by Spreadsheet

The spreadsheet in Figure 6–4 summarizes the estimates and answers the questions posed for Heavenly Pizza with the same values that were determined in the hand solution.

Figure 6–4
Spreadsheet solution
of Example 6.3.
(*a*) Capital recovery
in cell B16, (*b*) AW in
cell E17, and (*c*) Goal
Seek template and
outcome in cell B5.

	A	B	C	D	E
1					
2	MARR	10%			
3	# of systems	5			
4	Cost/system	-4,600			
5	Income/year	$ 6,000			
6					
7		Investment	Annual	Annual	
8	Year	and salvage	costs	income	Net income
9	0	-23,000	0	0	-23,000
10	1	0	-1,000	6,000	5,000
11	2	0	-1,100	6,000	4,900
12	3	0	-1,200	6,000	4,800
13	4	0	-1,300	6,000	4,700
14	5	1,500	-1,400	6,000	6,100
15					
16	(a) CR value	-5,822			
17	(b) AW value				-1,003
18					
19	= −PMT($B$2, 5, NPV($B$2,B10:B14)+B9)				
20					
21	= −PMT($B$2, 5, NPV($B$2,E10:E14)+E9)				
22					

(*a*) and (*b*)

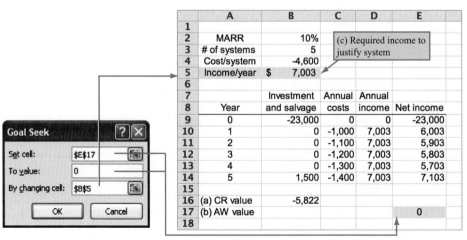

	A	B	C	D	E
1					
2	MARR	10%	(c) Required income to		
3	# of systems	5	justify system		
4	Cost/system	-4,600			
5	Income/year	$ 7,003			
6					
7		Investment	Annual	Annual	
8	Year	and salvage	costs	income	Net income
9	0	-23,000	0	0	-23,000
10	1	0	-1,000	7,003	6,003
11	2	0	-1,100	7,003	5,903
12	3	0	-1,200	7,003	5,803
13	4	0	-1,300	7,003	5,703
14	5	1,500	-1,400	7,003	7,103
15					
16	(a) CR value	-5,822			
17	(b) AW value				0
18					

Goal Seek

Set cell: $E$17
To value: 0
By changing cell: $B$5

OK   Cancel

(*c*)

Cell references are used in the spreadsheet functions to accommodate changes in estimated values.

(a) The capital recovery CR = \$–5822 is displayed in column B using the PMT function with an embedded NPV function, as shown in the cell tag.

(b) The annual worth AW = \$–1003 is displayed in column E using the PMT function shown. The arithmetic gradient series of costs and estimated income of \$6000 in columns C and D, respectively, are added to obtain the net income necessary for the PMT function.

(c) The minimum required income is determined in the lower part of Figure 6–4. This is easily accomplished by setting AW = 0 (column E) in the Goal Seek tool and letting it find the income per year of \$7003 to balance the AW equation.

EXAMPLE 6.4

Luby's Cafeterias is in the process of forming a separate business unit that provides meals to facilities for the elderly, such as assisted care and long-term care centers. Since the meals are prepared in one central location and distributed by trucks throughout the city, the equipment that keeps food and drink cold and hot is very important. Michele is the general manager of this unit, and she wishes to choose between two manufacturers of temperature retention units that are mobile and easy to sterilize after each use. Use the cost estimates below to select the more economic unit at a MARR of 8% per year.

	Hamilton (H)	Infinity Care (IC)
Initial cost P, \$	−15,000	−20,000
Annual M&O, \$/year	−6,000	−9,000
Refurbishment cost, \$	0	−2,000 every 4 years
Trade-in value S, % of P	20	40
Life, years	4	12

Solution

The best evaluation technique for these different-life alternatives is the annual worth method, where AW is taken at 8% per year over the respective lives of 4 and 12 years.

$$\text{AW}_H = \text{annual equivalent of } P - \text{annual M\&O} + \text{annual equivalent of } S$$
$$= -15{,}000(A/P,8\%,4) - 6000 + 0.2(15{,}000)(A/F,8\%,4)$$
$$= -15{,}000(0.30192) - 6000 + 3000(0.22192)$$
$$= \$-9{,}863$$

$$\text{AW}_{IC} = \text{annual equivalent of } P - \text{annual M\&O} - \text{annual equivalent of refurbishment}$$
$$+ \text{annual equivalent of } S$$
$$= -20{,}000(A/P,8\%,12) - 9000 - 2000[(P/F,8\%,4) + (P/F,8\%,8)](A/P,8\%,12)$$
$$+ 0.4(20{,}000)(A/F,8\%,12)$$
$$= -20{,}000(0.13270) - 9000 - 2000[0.7350 + 0.5403](0.13270) + 8000(0.05270)$$
$$= \$-11{,}571$$

The Hamilton unit is considerably less costly on an annual equivalent basis.

If the projects are **independent**, the AW at the MARR is calculated. All projects with AW ≥ 0 are acceptable.

Independent project selection

6.4 AW of a Permanent Investment ●●●

This section discusses the annual worth equivalent of the capitalized cost introduced in Section 5.5. Evaluation of public sector projects, such as flood control dams, irrigation canals, bridges, or other large-scale projects, requires the comparison of alternatives that have such long lives that they may be considered infinite in economic analysis terms. For this type of analysis, the annual worth (and capital recovery amount) of the initial investment is the perpetual annual interest on the initial investment, that is, $A = Pi = (CC)i$. This is Equation [5.2].

Cash flows recurring at regular or irregular intervals are handled exactly as in conventional AW computations; **convert them to equivalent uniform annual amounts A for one cycle**. This automatically annualizes them for each succeeding life cycle. Add all the A values to the CR amount to find total AW, as in Equation [6.2].

EXAMPLE 6.5

The U.S. Bureau of Reclamation is considering three proposals for increasing the capacity of the main drainage canal in an agricultural region of Nebraska. Proposal A requires dredging the canal to remove sediment and weeds that have accumulated during previous years' operation. The capacity of the canal will have to be maintained in the future near its design peak flow because of increased water demand. The Bureau is planning to purchase the dredging equipment and accessories for $650,000. The equipment is expected to have a 10-year life with a $17,000 salvage value. The annual operating costs are estimated to total $50,000. To control weeds in the canal itself and along the banks, environmentally safe herbicides will be sprayed during the irrigation season. The yearly cost of the weed control program is expected to be $120,000.

Proposal B is to line the canal with concrete at an initial cost of $4 million. The lining is assumed to be permanent, but minor maintenance will be required every year at a cost of $5000. In addition, lining repairs will have to be made every 5 years at a cost of $30,000.

Proposal C is to construct a new pipeline along a different route. Estimates are an initial cost of $6 million, annual maintenance of $3000 for right-of-way, and a life of 50 years.

Compare the alternatives on the basis of annual worth, using an interest rate of 5% per year.

Solution

Since this is an investment for a permanent project, compute the AW for one cycle of all recurring costs. For proposals A and C, the CR values are found using Equation [6.3], with $n_A = 10$ and $n_C = 50$, respectively. For proposal B, the CR is simply $P(i)$.

Proposal A	
CR of dredging equipment:	
$-650,000(A/P,5\%,10) + 17,000(A/F,5\%,10)$	$ -82,824
Annual cost of dredging	$-50,000
Annual cost of weed control	$-120,000
	$-252,824
Proposal B	
CR of initial investment: $-4,000,000(0.05)$	$-200,000
Annual maintenance cost	$-5,000
Lining repair cost: $-30,000(A/F,5\%,5)$	$-5,429
	$-210,429
Proposal C	
CR of pipeline: $-6,000,000(A/P,5\%,50)$	$-328,680
Annual maintenance cost	$-3,000
	$-331,680

Proposal B, which is a permanent solution, is selected due to its lowest AW of costs.

Comment

Note the use of the A/F factor for the lining repair cost in proposal B. The A/F factor is used instead of A/P, because the lining repair cost begins in year 5, not year 0, and continues indefinitely at 5-year intervals.

If the 50-year life of proposal C is considered infinite, $CR = P(i) = \$-300,000$, instead of $\$-328,680$ for $n = 50$. This is a small economic difference. How long lives of 40 or more years are treated economically is a matter of "local" practice.

EXAMPLE 6.6

At the end of each year, all owners and employees at Bell County Utility Cooperative are given a bonus check based on the net profit of the Coop for the previous year. Bart just received his bonus in the amount of $8530. He plans to invest it in an annuity program that returns 7% per year. Bart's long-term plans are to quit the Coop job some years in the future when he is still young enough to start his own business. Part of his future living expenses will be paid from the proceeds that this year's bonus accumulates over his remaining years at the Coop.

(*a*) Use a spreadsheet to determine the amount of annual year-end withdrawal that he can anticipate (starting 1 year after he quits) that will continue forever. He is thinking of working 15 or 20 more years.

(*b*) Determine the amount Bart must accumulate after 15 and 20 years to generate $3000 per year forever.

Solution by Spreadsheet

(*a*) Figure 6–5 presents the cash flow diagram for $n = 15$ years of accumulation at 7% per year on the $8530 deposited now. The accumulated amount after $n = 15$ years is indicated as $F_{\text{after } 15} = ?$ and the withdrawal series starts at the end of year 16. The diagram for $n = 20$ would appear the same, except there is a 20-year accumulation period.

The spreadsheet in Figure 6–6 shows the functions and answers for $n = 15$ years in columns C and D. The FV function displays the total accumulated at 7% after 15 years as $23,535. The perpetual withdrawal is determined by viewing this accumulated amount as a P value and by applying the formula

$$A = P(i) = 23,535(0.07) = \$1647 \text{ per year}$$

The spreadsheet function = D9*B7 performs this same computation in cell reference format in column D.

Answers for $n = 20$ years are displayed in column E. At a consistent 7% per year return, Bart's perpetual income is estimated as $1647 after 15 years, or $2311 per year if he waits for 20 years.

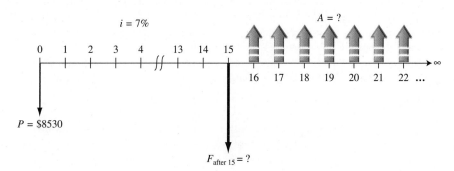

Figure 6–5
Diagram for a perpetual series starting after 15 years of accumulation, Example 6.6.

$P = \$8530$

$F_{\text{after } 15} = ?$

◇	A	B	C	D	E
1					
2					
3			Functions	Number of years, n	
4		Estimates	for n = 15 years	15	20
5					
6					
7	Interest rate, % per year	7%			
8	Amount deposited today	$8,530			
9	Amount accumulated after n years		'=-FV(B7,D4,,B8)	$ 23,535	$ 33,008
10	(a) Withdrawal forever, $ per year		'=D9*B7	$ 1,647	$ 2,311
11					
12					
13	(b) Required accumulation for $3000/year	7%	'= 3000/B13	$ 42,857	$ 42,857
14					
15	Accumulation years for $3000 per year	23.86	'= NPER(7%,,-8530,42857)		
16					

Figure 6–6
Spreadsheet solution, Example 6.6.

(*b*) To obtain a perpetual annual withdrawal of $3000, it is necessary to determine how much must be accumulated 1 year before the first withdrawal of $3000. This is an application of the relation $A = P(i)$ solved for P, or

$$P = \frac{A}{i} = \frac{3000}{0.07} = \$42,857$$

This *P* value is independent of how long Bart works at the Coop, because he must accumulate this amount to achieve his goal. Figure 6–6, row 13, shows the function and result. Note that the number of years *n* does not enter into the function = 3000/B13.

Comment

The NPER function can be used to determine how many years it will take for the current amount of $8530 to accumulate the required $42,857 at 7% per year. The row 15 function indicates that Bart will have to work at the Coop for just under 24 additional years.

6.5 Life-Cycle Cost Analysis ●●●

The PW and AW analysis techniques discussed thus far have concentrated on estimates for first cost *P*, annual operating and maintenance costs (AOC or M&O), salvage value *S*, and predictable periodic repair and upgrade costs, plus any revenue estimates that may favor one alternative over another. There are usually a host of additional costs involved when the complete project life costs are evaluated. A life-cycle cost analysis includes these additional estimates to the extent that they can be reliably determined.

Life-cycle cost (LCC) analysis utilizes AW or PW methods to evaluate cost estimates for the entire life cycle of one or more projects. Estimates will cover the **entire life span** from the early conceptual stage, through the design and development stages, throughout the operating stage, and even the phaseout and disposal stages. Both **direct and indirect costs** are included to the extent possible, and differences in revenue and savings projections between alternatives are included.

Some typical LCC applications are life-span analysis for military and commercial aircraft, new manufacturing plants, new automobile models, new and expanded product lines, and government systems at federal and state levels. For example, the U.S. Department of Defense requires that a government contractor include an LCC budget and analysis in the originating proposal for most defense systems. Most commonly the LCC analysis includes costs, and the AW method is used for the analysis, especially if only one alternative is evaluated. If there are expected revenue or other benefit differences between alternatives, a PW analysis is recommended. Public sector projects are usually evaluated using a benefit/cost analysis (Chapter 9), rather than LCC analysis, because estimates to the citizenry are difficult to make with much accuracy. The *direct costs* mentioned above include material, human labor, equipment, supplies, and other costs directly related to a product, process, or system. Some examples of *indirect cost components* are taxes, management, legal, warranty, quality, human resources, insurance, software, purchasing, etc. Direct and indirect costs are discussed further in Chapter 15.

LCC analysis is most effectively applied when a substantial percentage of the life span (postpurchase) costs, relative to the initial investment, will be expended in direct and indirect operating, maintenance, and similar costs once the system is operational. For example, the evaluation of two equipment purchase alternatives with expected useful lives of 5 years and M&O costs of 5% to 10% of the initial investment does not require an LCC analysis. However, let's assume that Exxon-Mobil wants to evaluate the design, construction, operation, and support of a new type and style of tanker that can transport oil over long distances of ocean. If the initial costs are in the $100 millions with support and operating costs ranging from 25% to 35% of this amount over a 25-year life, the logic of an LCC analysis will offer a better understanding of the economic viability of the project.

To understand how a LCC analysis works, first we must understand the phases and stages of systems engineering or systems development. Many books and manuals are available on systems development and analysis. Generally, the LCC estimates may be categorized into a simplified format for the major phases of *acquisition, operation,* and *phaseout/disposal,* and their respective stages.

Acquisition phase: all activities prior to the delivery of products and services.
- Requirements definition stage—Includes determination of user/customer needs, assessing them relative to the anticipated system, and preparation of the system requirements documentation.
- Preliminary design stage—Includes feasibility study, conceptual, and early-stage plans; final go–no go decision is probably made here.
- Detailed design stage—Includes detailed plans for resources—capital, human, facilities, information systems, marketing, etc.; there is some acquisition of assets, if economically justifiable.

Operation phase: all activities are functioning, products and services are available.
- Construction and implementation stage—Includes purchases, construction, and implementation of system components; testing; preparation, etc.
- Usage stage—Uses the system to generate products and services; the largest portion of the life cycle.

Phaseout and disposal phase: covers all activities to transition to a new system; removal/recycling/disposal of old system.

EXAMPLE 6.7

In the 1860s, General Mills Inc. and Pillsbury Inc. both started in the flour business in the Twin Cities of Minneapolis–St. Paul, Minnesota. In the decade of 2000 to 2010, General Mills purchased Pillsbury for a combination cash and stock deal worth more than $10 billion and integrated the product lines. Food engineers, food designers, and food safety experts made many cost estimates as they determined the needs of consumers and the combined company's ability to technologically and safely produce and market new food products. At this point only cost estimates have been addressed—no revenues or profits.

Assume that the major cost estimates below have been made based on a 6-month study about two new products that could have a 10-year life span for the company. Use LCC analysis at the industry MARR of 18% to determine the size of the commitment in AW terms. (Time is indicated in product-years. Since all estimates are for costs, they are not preceded by a minus sign.)

Consumer habits study (year 0)	$0.5 million
Preliminary food product design (year 1)	0.9 million
Preliminary equipment/plant design (year 1)	0.5 million
Detail product designs and test marketing (years 1, 2)	1.5 million each year
Detail equipment/plant design (year 2)	1.0 million
Equipment acquisition (years 1 and 2)	$2.0 million each year
Current equipment upgrades (year 2)	1.75 million
New equipment purchases (years 4 and 8)	2.0 million (year 4) + 10% per purchase thereafter
Annual equipment operating cost (AOC) (years 3–10)	200,000 (year 3) + 4% per year thereafter
Marketing, year 2	$8.0 million
years 3–10	5.0 million (year 3) and −0.2 million per year thereafter
year 5 only	3.0 million extra
Human resources, 100 new employees for 2000 hours per year (years 3–10)	$20 per hour (year 3) + 5% per year
Phaseout and disposal (years 9 and 10)	$1.0 million each year

Solution

LCC analysis can get complicated rapidly due to the number of elements involved. Calculate the PW by phase and stage, add all PW values, then find the AW over 10 years. Values are in $1 million units.

Acquisition phase:
Requirements definition: consumer study

$$PW = \$0.5$$

Preliminary design: product and equipment

$$PW = 1.4(P/F,18\%,1) = \$1.187$$

Detailed design: product and test marketing, and equipment

$$PW = 1.5(P/A,18\%,2) + 1.0(P/F,18\%,2) = \$3.067$$

Operation phase:
Construction and implementation: equipment and AOC

$$PW = 2.0(P/A,18\%,2) + 1.75(P/F,18\%,2) + 2.0(P/F,18\%,4) + 2.2(P/F,18\%,8)$$

$$+ 0.2\left[\frac{1 - \left(\frac{1.04}{1.18}\right)^8}{0.14}\right](P/F,18\%,2) = \$6.512$$

Use: marketing

$$PW = 8.0(P/F,18\%,2) + [5.0(P/A,18\%,8) - 0.2(P/G,18\%,8)](P/F,18\%,2)$$
$$\quad + 3.0(P/F,18\%,5)$$

$$= \$20.144$$

Use: human resources: (100 employees)(2000 h/yr)($20/h) = $4.0 million in year 3

$$PW = 4.0\left[\frac{1 - \left(\frac{1.05}{1.18}\right)^8}{0.13}\right](P/F,18\%,2) = \$13.412$$

Phaseout phase:

$$PW = 1.0(P/A,18\%,2)(P/F,18\%,8) = \$0.416$$

The sum of all PW of costs is PW = $45.238 million. Finally, determine the AW over the expected 10-year life span.

AW = 45.238 million($A/P,18\%,10$) = \$10.066 million per year

This is the LCC estimate of the equivalent annual commitment to the two proposed products.

Often the alternatives compared by LCC do not have the same level of output or amount of usage. For example, if one alternative will produce 20 million units per year and a second alternative will operate at 35 million per year, the AW values should be compared on a currency unit/unit produced basis, such as dollar/unit or euro/hour operated.

Figure 6–7 presents an overview of how costs may be distributed over an entire life cycle. For some systems, typically defense systems, operating and maintenance costs rise fast after acquisition and remain high until phaseout occurs.

The total LCC for a system is established or locked in early in the life cycle. It is not unusual to have 75% to 85% of the entire life span LCC committed during the preliminary and detail

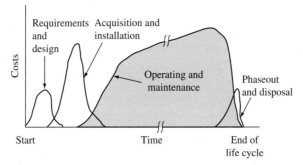

Figure 6–7

Typical distribution of life-cycle costs of the phases for one life cycle.

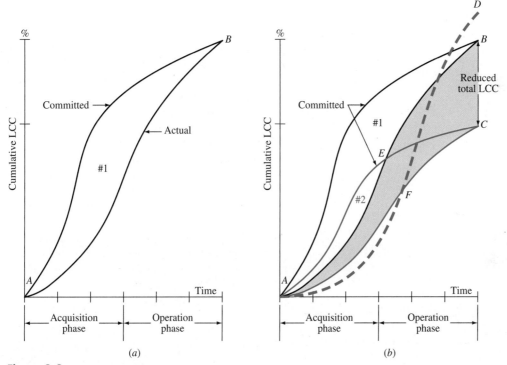

Figure 6–8

LCC envelopes for committed and actual costs: (*a*) design 1, (*b*) improved design 2.

design stages. As shown in Figure 6–8*a*, the actual or observed LCC (bottom curve *AB*) will trail the committed LCC throughout the life span (unless some major design flaw increases the total LCC of design #1 above point *B*).

The potential for significantly reducing total LCC occurs primarily during the early stages. A more effective design and more efficient equipment can reposition the envelope to design #2 in Figure 6–8*b*. Now the committed LCC curve *AEC* is below *AB* at all points, as is the actual LCC curve *AFC*. It is this lower envelope #2 we seek. The shaded area represents the reduction in actual LCC.

Even though an effective LCC envelope may be established early in the acquisition phase, it is not uncommon that unplanned cost-saving measures are introduced during the acquisition phase and early operation phase. These apparent "savings" may actually increase the total LCC, as shown by curve *AFD*. This style of ad hoc cost savings, often imposed by management early in the design stage and/or construction stage, can substantially increase costs later, especially in the after-sale portion of the use stage. For example, the use of inferior-strength concrete and steel has been the cause of structural failures many times, thus increasing the overall life span LCC.

CHAPTER SUMMARY

The annual worth method of comparing alternatives is often preferred to the present worth method, because the AW comparison is performed for only one life cycle. This is a distinct advantage when comparing different-life alternatives. AW for the first life cycle is the AW for the second, third, and all succeeding life cycles, under certain assumptions. When a study period is specified, the AW calculation is determined for that time period, regardless of the lives of the alternatives.

For infinite-life (perpetual) alternatives, the initial cost is annualized simply by multiplying P by i. For finite-life alternatives, the AW through one life cycle is equal to the perpetual equivalent annual worth.

Life-cycle cost analysis is appropriate for systems that have a large percentage of costs in operating and maintenance. LCC analysis helps in the analysis of all costs from design to operation to disposal phases.

PROBLEMS

Annual Worth Calculations

6.1 If you are asked to provide an annual worth (AW) comparison of alternatives after a present worth (PW) comparison has already been done, what factor multiplied by the PW values provide the correct AW values?

6.2 List three assumptions that are inherent in the annual worth method of comparing alternatives.

6.3 In the annual worth method of comparing alternatives that have different lives, why do you calculate the AW of the alternatives *over their respective life cycles* instead of over the least common multiple of their lives?

6.4 James developed the two cash flow diagrams shown at the bottom of this page. The cash flows for alternative B represent two life cycles of A. Calculate the annual worth value of each over the respective life cycles to demonstrate that they are the same. Use an interest rate of 10% per year.

6.5 An asset with a first cost of $20,000 has an annual operating cost of $12,000 and a $4000 salvage value

after its 4-year life. If the project will be needed for 6 years, what would the market (salvage) value of the 2-year-old asset have to be for the annual worth to be the same as it is for one life cycle of the asset? Use an interest rate of 10% per year.

6.6 A sports mortgage is an innovative way to finance cash-strapped sports programs by allowing fans to sign up to pay a "mortgage" for the right to buy good seats at football games for several decades with season tickets locked in at current prices. At Notre Dame, the locked-in price period is 50 years. If a fan pays a $130,000 "mortgage" fee now (i.e., in year 0) when season tickets are selling for $290 each, what is the equivalent annual cost of the football tickets over the 50-year period at an interest rate of 8% per year?

6.7 If a fan buys a sports mortgage to USC football games by paying $130,000 in 10 equal payments *starting now*, and then pays a fixed price of $290 per year for 50 years (starting 1 year from now) for season tickets, what is the AW in years 1 through 50 of the season tickets at 8% per year interest?

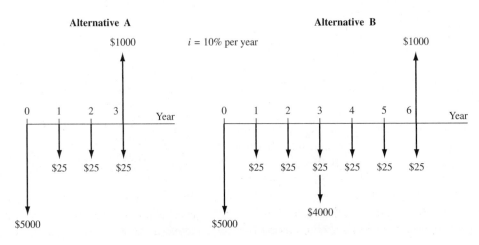

6.8   Eight years ago, Ohio Valley Trucking purchased a large-capacity dump truck for $115,000 to provide short-haul earthmoving services. The company sold it today for $45,000. Operating and maintenance costs averaged $10,500 per year. A complete overhaul at the end of year 4 cost an extra $3600. Calculate the annual cost of the truck at 8% per year interest.

6.9   A major repair on the suspension system of a 5-year-old car cost $2000 because the warranty expired after 3 years of ownership. The cost of periodic maintenance has been $800 every 2 years. If the owner donates the car to charity after 8 years of ownership, what is the equivalent annual cost of the repair and maintenance in the 8-year period of ownership? Use an interest rate of 8% per year, and assume that the owner paid the $800 maintenance cost immediately before donating the car in year 8.

Capital Recovery

6.10   Ten years ago, Jacobson Recovery purchased a wrecker for $285,000 to move disabled 18-wheelers. He anticipated a salvage value of $50,000 after 10 years. During this time his average annual revenue totaled $52,000. (*a*) Did he recover his investment and a 12% per year return? (*b*) If the annual M&O cost was $10,000 the first year and increased by a constant $1000 per year, was the AW positive or negative at 12% per year? Assume the $50,000 salvage was realized.

6.11   Sylvia has received a $500,000 inheritance from her favorite, recently deceased aunt in Hawaii. Sylvia is planning to purchase a condo in Hawaii in the same area where her aunt lived all her life and to rent it to vacationers. She hopes to make 8% per year on this purchase over an ownership period of 20 years. The condo's total first cost is $500,000, and she conservatively expects to sell it for 90% of the purchase price. No annual M&O costs are considered in the analysis. (*a*) What is the capital recovery amount? (*b*) If there is a real boom in rental real estate 10 years in the future, what sales price (as a percentage of original purchase price) is necessary at that time (year 10) to realize the same amount as the 8% return expected over the 20-year ownership period?

6.12   Humana Hospital Corporation installed a new MRI machine at a cost of $750,000 this year in its medical professional clinic in Cedar Park. This state-of-the-art system is expected to be used for 5 years and then sold for $75,000. Humana uses a return requirement of 24% per year for all of its medical diagnostic equipment. As a bioengineering student

currently serving a coop semester on the management staff of Humana Corporation in Louisville, Kentucky, you are asked to determine the minimum revenue required each year to realize the expected recovery and return. Also, you are asked to draw two cash flow diagrams, one showing the MRI purchase and sale cash flow and a second depicting the required capital recovery each year.

Alternative Comparison

6.13   Polypropylene wall caps, used for covering exterior vents for kitchen cooktops, bathroom fans, dryers, and other building air exhausts, can be made by two different methods. Method X will have a first cost of $75,000, an operating cost of $32,000 per year, and a $9000 salvage value after 4 years. Method Y will have a first cost of $140,000, an operating cost of $24,000 per year, and a $19,000 salvage value after its 4-year life. At an interest rate of 10% per year, which method should be used on the basis of an annual worth analysis?

6.14   Nissan's all-electric car, the Leaf, has a base price of $32,780 in the United States, but it is eligible for a $7500 federal tax credit. A consulting engineering company wants to evaluate the purchase or lease of one of the vehicles for use by its employees traveling to job sites in the local area. The cost for leasing the vehicle will be $4200 per year (payable at the end of each year) after an initialization charge of $2500 paid now. If the company purchases the vehicle, it will also purchase a home charging station for $2200 that will be partially offset by a 50% tax credit. If the company expects to be able to sell the car and charging station for 40% of the base price of the car alone at the end of 3 years, should the company purchase or lease the car? Use an interest rate of 10% per year and annual worth analysis.

6.15   A new structural design software package is available for analyzing and designing three-sided guyed towers and three- and four-sided self-supporting towers. A single-user license will cost $6000 per year. A site license has a one-time cost of $22,000. A structural engineering consulting company is trying to decide between two alternatives: buy a single-user license *now* and one each year for the next 3 years (which will provide 4 years of service), or buy a site license now. Determine which strategy should be adopted at an interest rate of 10% per year for a 4-year planning period using the annual worth method of evaluation.

6.16   The city council in a certain southwestern city is considering whether to construct permanent restrooms in 22 of its smaller parks (i.e., parks of less than 12 acres) or pay for portable toilets on a year-round basis. The cost of constructing the 22 permanent restrooms will be $3.8 million. The 22 portable restrooms can be rented for $7500 each for 1 year. The service life of a permanent restroom is 20 years. Using an interest rate of 6% per year and an annual worth analysis, determine if the city should build the permanent restrooms or lease the portable ones.

6.17   A remotely located air sampling station can be powered by solar cells or by running an above ground electric line to the site and using conventional power. Solar cells will cost $16,600 to install and will have a useful life of 5 years with no salvage value. Annual costs for inspection, cleaning, etc., are expected to be $2400. A new power line will cost $31,000 to install, with power costs expected to be $1000 per year. Since the air sampling project will end in 5 years, the salvage value of the line is considered to be zero. At an interest rate of 10% per year, (a) which alternative should be selected on the basis of an annual worth analysis and (b) what must be the first cost of the above ground line to make the two alternatives equally attractive economically?

6.18   The cash flows for two small raw water treatment systems are shown. Determine which should be selected on the basis of an annual worth analysis at 10% per year interest.

	MF	UF
First cost, $	−33,000	−51,000
Annual cost, $ per year	−8,000	−3,500
Salvage value, $	4,000	11,000
Life, years	3	6

6.19   PGM Consulting is under contract to Montgomery County for evaluating alternatives that use a robotic, liquid-propelled "pig" to periodically inspect the interior of buried potable water pipes for leakage, corrosion, weld strength, movement over time, and a variety of other parameters. Two equivalent robot instruments are available. Robot Joeboy will have a first cost of $85,000, annual M&O costs of $30,000, and a $40,000 salvage value after 3 years. Robot Watcheye will have a first cost of $125,000, annual M&O costs of $27,000, and a $33,000 salvage value after its 5-year life. Assume an interest rate of 8% per year.
(a)   Which robot is the better economic option?
(b)   Using the spreadsheet Goal Seek tool, determine the first cost of the robot not selected in (a) so that it will be the economic selection.

6.20   TT Racing and Performance Motor Corporation wishes to evaluate two alternative CNC machines for NHRA engine building. Use the AW method at 10% per year to select the better alternative.

	Machine R	Machine S
First cost, $	−250,000	−370,500
Annual operating cost, $ per year	−40,000	−50,000
Salvage value, $	20,000	30,000
Life, years	3	5

6.21   The maintenance and operation (M&O) cost of front-end loaders working under harsh environmental conditions tends to increase by a constant $1200 per year for the first 5 years of operation. For a loader that has a first cost of $39,000 and first-year M&O cost of $17,000, compare the equivalent annual worth of a loader kept for 4 years with one kept for 5 years at an interest rate of 12% per year. The salvage value of a used loader is $23,000 after 4 years and $18,000 after 5 years.

6.22   You work for Midstates Solar Power. A manager asked you to determine which of the following two machines will have the lower (a) capital recovery and (b) equivalent annual total cost. Machine Semi2 has a first cost of $80,000 and an operating cost of $21,000 in year 1, increasing by $500 per year through year 5, after which time it will have a salvage value of $13,000. Machine Auto1 has a first cost of $62,000 and an operating cost of $21,000 in year 1, increasing by 8% per year through year 5, after which time it will have a scavenge value of $2000. Utilize an interest rate of 10% per year to determine both estimates.

Permanent Investments

6.23   The State of Chiapas, Mexico, decided to fund a program for literacy. The first cost is $200,000 now, and an update budget of $100,000 every 7 years forever is requested. Determine the perpetual equivalent annual cost at an interest rate of 10% per year.

6.24   Calculate the perpetual equivalent annual cost (years 1 through infinity) of $5 million in year 0, $2 million in year 10, and $100,000 in years 11 through infinity. Use an interest rate of 10% per year.

6.25   A Pennsylvania coal mining operation has installed an in-shaft monitoring system for oxygen tank and gear readiness for emergencies. Based on maintenance patterns for previous systems, there are no maintenance costs for the first 2 years, they increase for a time period, and then they level off. Maintenance costs are expected to be $150,000 in year 3, $175,000 in year 4, and amounts increasing by $25,000 per year through year 6 and remain constant thereafter for the expected 10-year life of the system. If similar systems with similar costs will replace the current one, determine the perpetual equivalent annual maintenance cost at $i = 10\%$ per year.

6.26 Compare two alternatives for a security system surrounding a power distribution substation using annual worth analysis and an interest rate of 10% per year.

	Condi	Torro
First cost, $	−25,000	−130,000
Annual cost, $ per year	−9,000	−2,500
Salvage value, $	3,000	150,000
Life, years	3	∞

6.27 A new bridge across the Allegheny River in Pittsburgh is expected to be permanent and will have an initial cost of $30 million. This bridge must be resurfaced every 5 years at a cost of $1 million. The annual inspection and operating costs are estimated to be $50,000. Determine its equivalent annual worth at an interest rate of 10% per year.

6.28 For the cash flows shown, use an annual worth comparison and an interest rate of 10% per year.
(a) Determine the alternative that is economically best.
(b) Determine the first cost required for each of the two alternatives not selected in (a) so that all alternatives are equally acceptable. Use a spreadsheet to answer this question.

	X	Y	Z
First cost, $	−90,000	−400,000	−650,000
Annual cost, $ per year	−40,000	−20,000	−13,000
Overhaul every 10 years, $	—	—	−80,000
Salvage value, $	7,000	25,000	200,000
Life, years	3	10	∞

Life-Cycle Cost

6.29 Blanton Agriculture of Santa Monica, California, offers different types and levels of irrigation water conservation systems for use in areas where groundwater depletion is a serious concern. A large farming corporation in India, where depletion is occurring at an alarming rate of 1.6 inches (4 centimeters) per year due to exponential growth

in usage for crop irrigation, is considering the purchase of one of the Blanton systems. There are three options with two levels of automation for the first two options. The estimated costs and associated cash flow diagrams over a 10-year period are summarized below and on the next page, respectively, for each of the five alternatives. Costs are categorized as design (Des), development (Dev), and operation (Oper). For alternatives A and B, there is an extra cost of $15,000 per installation year to maintain the manual system in place now. Level 2 development costs are distributed equally over a 2-year period. Alternative C is a retrofit of the current manual system with no design or development costs, and there is no level 1 option. At an interest rate of 10% per year and a 10-year study period, determine which alternative and level has the lowest LCC.

Alternative	Cost Component	Level 1	Level 2
A	Des	$100,000	$200,000
	Dev	175,000	350,000
	Oper	60,000	55,000
	Install time	1 year	2 years
B	Des	$ 50,000	$100,000
	Dev	200,000	500,000
	Oper	45,000	30,000
	Install time	1 year	2 years
C	Oper		$100,000

6.30 The Pentagon asked a defense contractor to estimate the life-cycle cost for a proposed light-duty support vehicle. The list of items included the following categories: R&D costs (R&D), nonrecurring investment costs (NRI), recurring upgrade costs (RU), scheduled and unscheduled maintenance costs (Maint), equipment usage costs (Equip), and phaseout/disposal costs (Po/D). Use the cost estimates (shown in $1 million) for the 20-year life cycle to calculate the annual LCC at an interest rate of 7% per year.

Year	R&D	NRI	RU	Maint	Equip	Po/D
0	5.5	1.1				
1	3.5					
2	2.5					
3	0.5	5.2	1.3	0.6	1.5	
4		10.5	3.1	1.4	3.6	
5		10.5	4.2	1.6	5.3	
6–10			6.5	2.7	7.8	
11 on			2.2	3.5	8.5	
18–20						2.7

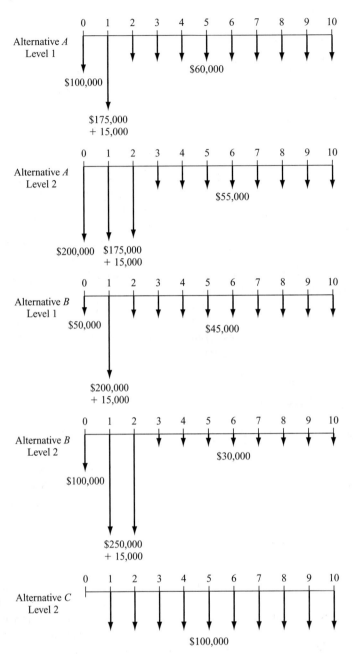

Cash flow diagrams for Problem 6.29

6.31 A manufacturing software engineer at a major aerospace corporation has been assigned the management responsibility of a project to design, build, test, and implement AREMSS, a new-generation automated scheduling system for routine and expedited maintenance. Reports on the disposition of each service will also be entered by field personnel, then filed and archived by the system. The initial application will be on existing Air Force in-flight refueling aircraft. The system is expected to be widely used over time for other aircraft maintenance scheduling. Once it is fully implemented, enhancements will have to be made, but the system is expected to serve as a worldwide scheduler for up to 15,000 separate aircraft. The engineer, who must make a presentation next week of the best estimates of costs over a 20-year life period, has decided to use the life-cycle cost approach of cost estimations. Use the following information to determine the current annual LCC at 6% per year for the AREMSS scheduling system.

Cost Category	Cost in Year ($ millions)							
	1	2	3	4	5	6 on	10	18
Field study	0.5							
Design of system	2.1	1.2	0.5					
Software design		0.6	0.9					
Hardware purchases			5.1					
Beta testing		0.1	0.2					
Users manual development		0.1	0.1	0.2	0.2	0.06		
System implementation				1.3	0.7			
Field hardware				0.4	6.0	2.9		
Training trainers				0.3	2.5	2.5	0.7	
Software upgrades						0.6	3.0	3.7

6.32   The U.S. Army received two proposals for a turnkey design-build project for barracks for infantry unit soldiers in training. Proposal A involves an off-the-shelf "bare-bones" design and standard grade construction of walls, windows, doors, and other features. With this option, heating and cooling costs will be greater, maintenance costs will be higher, and replacement will be sooner than for proposal B. The initial cost for A will be $750,000. Heating and cooling costs will average $72,000 per year, with maintenance costs averaging $24,000 per year. Minor remodeling will be required in years 5, 10, and 15 at a cost of $150,000 each time in order to render the units usable for 20 years. They will have no salvage value.

Proposal B will include tailored design and construction costs of $1.1 million initially, with estimated heating and cooling costs of $36,000 per year and maintenance costs of $12,000 per year. There will be no salvage value at the end of the 20-year life. Which proposal should be accepted on the basis of an annual life-cycle cost analysis, if the interest rate is 6% per year?

6.33   A medium-size municipality plans to develop a software system to assist in project selection during the next 10 years. A life-cycle cost approach has been used to categorize costs into development, programming, operating, and support costs for each alternative. There are three alternatives under consideration, identified as M, N, and O. The costs are summarized below. Use an annual life-cycle cost approach to identify the best alternative at 8% per year.

Alternative	Component	Estimated Cost
M	Development	$250,000 now, $150,000 years 1–4
	Programming	$45,000 now, $35,000 years 1, 2
	Operation	$50,000 years 1–10
	Support	$30,000 years 1–5
N	Development	$10,000 now
	Programming	$45,000 year 0, $30,000 years 1–3
	Operation	$80,000 years 1–10
	Support	$40,000 years 1–10
O	Operation	$175,000 years 1–10

ADDITIONAL PROBLEMS AND FE EXAM REVIEW QUESTIONS

6.34   All of the following are fundamental assumptions for the annual worth method of analysis *except*:
(*a*)   The alternatives will be needed for only *one* life cycle.
(*b*)   The services provided are needed for at least the LCM of the lives of the alternatives.
(*c*)   The selected alternative will be repeated for the succeeding life cycles in exactly the same manner as for the first life cycle.
(*d*)   All cash flows will have the same estimated values in every life cycle.

6.35   When comparing five alternatives that have different lives by the AW method, you must:
(*a*)   Find the AW of each over the life of the longest-lived alternative.
(*b*)   Find the AW of each over the life of the shortest-lived alternative.
(*c*)   Find the AW of each over the LCM of all of the alternatives.
(*d*)   Find the AW of each alternative over its life without considering the life of the other alternatives.

6.36   The annual worth of an alternative can be calculated from the alternative's:
(*a*)   Present worth by multiplying by ($A/P,i,n$)
(*b*)   Future worth by multiplying by ($F/A,i,n$)
(*c*)   Either (*a*) or (*b*)
(*d*)   Neither (*a*) nor (*b*)

6.37   The alternatives shown are to be compared on the basis of annual worth. At an interest rate of 10% per year, the values of *n* that you could use in the ($A/P,i,n$) factors to make a correct comparison by the annual worth method are:

	A	B
First cost, $	−50,000	−90,000
Annual cost, $ per year	−10,000	−4,000
Salvage value, $	13,000	15,000
Life, years	3	6

(*a*) $n = 3$ years for A and 3 years for B
(*b*) $n = 3$ years for A and 6 years for B
(*c*)   Either (*a*) or (*b*)
(*d*)   Neither (*a*) nor (*b*)

6.38 The alternatives shown are to be compared on the basis of a perpetual (i.e., forever) equivalent annual worth. At an interest rate of 10% per year, the equation that represents the perpetual AW of X1 is:

	X1	Y1
First cost, $	−50,000	−90,000
Annual cost, $ per year	−10,000	−4,000
Salvage value, $	13,000	15,000
Life, years	3	6

(a) $AW_{X1} = -50,000(0.10) - 10,000 + 13,000(0.10)$

(b) $AW_{X1} = -50,000(0.10) - 10,000 + 13,000(A/F,10\%,3)$

(c) $AW_{X1} = -50,000(0.10) - 10,000 - 37,000(P/F,10\%,3)(0.10) + 13,000(0.10)$

(d) $AW_{X1} = -50,000(A/P,10\%,3) - 10,000 + 13,000(A/F,10\%,3)$

6.39 To get the AW of a cash flow of $10,000 that occurs every 10 years forever, with the first one occurring 10 years from now, you should:
(a) Multiply $10,000 by $(A/P,i,10)$.
(b) Multiply $10,000 by $(A/F,i,10)$.
(c) Multiply $10,000 by i.
(d) Multiply $10,000 by $(A/F,i,n)$ and then multiply by i.

Problems 6.40 through 6.43 refer to the following estimates.

The alternatives are mutually exclusive and the MARR is 6% per year.

	Vendor 1	Vendor 2	Vendor 3
First cost, $	−200,000	−550,000	−1,000,000
Annual cost, $ per year	−50,000	−20,000	−10,000
Revenue, $ per year	120,000	120,000	110,000
Salvage value, $	25,000	0	500,000
Life, years	10	15	∞

6.40 The annual worth of vendor 2 cash flow estimates is closest to:
(a) $−63,370
(b) $43,370
(c) $−43,370
(d) $63,370

6.41 Of the following three relations, the correct one or ones to calculate the annual worth of vendor 1 cash flow estimates is (note: all dollar values are in thousands):
Relation 1: $AW_1 = -200(A/P,6\%,10) + 70 + 25(A/F,6\%,10)$
Relation 2: $AW_1 = [-200 - 50(P/A,6\%,10) + 120(P/A,6\%,10) + 25(P/F,6\%,10)](A/P,6\%,10)$

Relation 3: $AW_1 = -200(F/P,6\%,10) + 25 + (-50 + 120)(A/P,6\%,10)$

(a) 1 and 3
(b) Only 1
(c) 1 and 2
(d) Only 3

6.42 The AW values for the alternatives are listed below. The vendor or vendors that should be recommended is:
$AW_1 = \$44,723$ $AW_2 = \$43,370$
$AW_3 = \$40,000$

(a) 1 and 2
(b) 3
(c) 2
(d) 1

6.43 The capital recovery amount for vendor 3 is:
(a) $40,000 per year
(b) $60,000 per year
(c) $43,370 per year
(d) $100,000 per year

6.44 If a revenue alternative has a negative AW value and it was correctly calculated, it means the following:
(a) The equivalent annual worth of revenues does not exceed that of the costs.
(b) The estimates are wrong somewhere.
(c) A minus or plus sign of a cash flow was entered incorrectly into the PMT spreadsheet function.
(d) The alternative should have a longer life so revenues will exceed costs.

6.45 Estimates for one of two process upgrades are as follows: first cost of $40,000, annual cost of $5000 per year, market value that decreases by $2000 per year to the salvage value of $20,000 after the expected life of 10 years. If a 4-year study period is used for AW analysis at 15% per year, the correct AW value is closest to:
(a) $−15,000
(b) $−11,900
(c) $−7600
(d) $−12,600

6.46 The perpetual annual worth of investing $50,000 now and $20,000 per year starting in year 16 and continuing forever at 12% per year is closest to:
(a) $−4200
(b) $−8650
(c) $−9655
(d) $−10,655

6.47 All the following statements about the capital recovery amount for an alternative are false *except:*
  (*a*) Annual revenue can be no more than this amount, if the alternative is selected.
  (*b*) A monetary estimate of new capital funds required each year for the life of the alternative.
  (*c*) An amount of revenue required to recover the first cost plus a stated return over the life of the alternative.
  (*d*) Does not consider the salvage value, since it is returned at the end of the alternative's life.

CASE STUDY

THE CHANGING SCENE OF AN ANNUAL WORTH ANALYSIS

Background and Information

Harry, owner of an automobile battery distributorship in Atlanta, Georgia, performed an economic analysis 3 years ago when he decided to place surge protectors in-line for all his major pieces of testing equipment. The estimates used and the annual worth analysis at MARR = 15% are summarized below. Two different manufacturers' protectors were compared.

	PowrUp	Lloyd's
Cost and installation, $	−26,000	−36,000
Annual maintenance cost, $ per year	−800	−300
Salvage value, $	2,000	3,000
Equipment repair savings, $	25,000	35,000
Useful life, years	6	10

The spreadsheet in Figure 6–9 is the one Harry used to make the decision. Lloyd's was the clear choice due to its substantially larger AW value. The Lloyd's protectors were installed.

During a quick review this last year (year 3 of operation), it was obvious that the maintenance costs and repair savings have not followed (and will not follow) the estimates made 3 years ago. In fact, the maintenance contract cost (which includes quarterly inspection) is going from $300 to $1200 per year next year and will then increase 10% per year for the next 10 years. Also, the repair savings for the last 3 years were $35,000, $32,000, and $28,000, as best as Harry can determine. He believes savings will decrease by $2000 per year hereafter. Finally, these 3-year-old protectors are worth nothing on the market now, so the salvage in 7 years is zero, not $3000.

Case Study Exercises

1. Plot a graph of the newly estimated maintenance costs and repair savings projections, assuming the protectors last for 7 more years.
2. With these new estimates, what is the recalculated AW for the Lloyd's protectors? Use the old first cost and maintenance cost estimates for the first 3 years. If these estimates had been made 3 years ago, would Lloyd's still have been the economic choice?
3. How has the capital recovery amount changed for the Lloyd's protectors with these new estimates?

◇	A	B	C	D	E	F	G
1	MARR =	15%					
2							
3			PowrUp			Lloyd's	
4		Investment	Annual	Repair	Investment	Annual	Repair
5	Year	and salvage	maintenance	savings	and salvage	maintenance	savings
6	0	-26,000	0	0	-36,000	0	0
7	1	0	-800	25,000	0	-300	35,000
8	2	0	-800	25,000	0	-300	35,000
9	3	0	-800	25,000	0	-300	35,000
10	4	0	-800	25,000	0	-300	35,000
11	5	0	-800	25,000	0	-300	35,000
12	6	2,000	-800	25,000	0	-300	35,000
13	7				0	-300	35,000
14	8				0	-300	35,000
15	9				0	-300	35,000
16	10				3,000	-300	35,000
17	AW element	-6,642	-800	25,000	-7,025	-300	35,000
18	**Total AW**			$ 17,558			$ 27,675

Figure 6–9
Annual worth analysis of surge protector alternatives, case study.

CHAPTER 7

Rate of Return Analysis: One Project

LEARNING OUTCOMES

Purpose: Understand the meaning of rate of return and perform an ROR evaluation of a single project.

SECTION	TOPIC	LEARNING OUTCOME
7.1	Definition	• State and understand the meaning of rate of return.
7.2	Calculate ROR	• Use a PW or AW relation to determine the ROR of a series of cash flows.
7.3	Cautions	• State the difficulties of using the ROR method, relative to the PW and AW methods.
7.4	Multiple RORs	• Determine the maximum number of possible ROR values and their values for a specific cash flow series.
7.5	Calculate EROR	• Determine the external rate of return using the techniques of modified ROR and return on invested capital.
7.6	Bonds	• Calculate the nominal and effective rate of return for a bond investment.

T he most commonly quoted measure of economic worth for a project or alternative is its rate of return (ROR). Whether it is an engineering project with cash flow estimates or an investment in a stock or bond, the rate of return is a well-accepted way of determining if the project or investment is economically acceptable. Compared to the PW or AW value, the ROR is a generically different type of measure of worth, as is discussed in this chapter. Correct procedures to calculate a rate of return using a PW or AW relation are explained here, as are some cautions necessary when the ROR technique is applied to a single project's cash flows.

The ROR is known by other names such as the *internal rate of return* (IROR), which is the technically correct term, and *return on investment* (ROI). We will discuss the computation of ROI in the latter part of this chapter.

In some cases, more than one ROR value may satisfy the PW or AW equation. This chapter describes how to recognize this possibility and an approach to find the **multiple values.** Alternatively, one reliable ROR value can be obtained by using additional information established separately from the project cash flows. Two of the techniques are covered: the modified ROR technique and the ROIC (return on invested capital) technique.

Only one alternative is considered here; Chapter 8 applies these same principles to multiple alternatives. Finally, the rate of return for a bond investment is discussed.

7.1 Interpretation of a Rate of Return Value ●●●

From the perspective of someone who has borrowed money, the interest rate is applied to the *unpaid balance* so that the total loan amount and interest are paid in full exactly with the last loan payment. From the perspective of a lender of money, there is an *unrecovered balance* at each time period. The interest rate is the return on this unrecovered balance so that the total amount lent and the interest are recovered exactly with the last receipt. *Rate of return* describes both of these perspectives.

> Rate of return (ROR) is the rate paid on the **unpaid balance of borrowed money,** or the rate earned on the **unrecovered balance of an investment,** so that the final payment or receipt brings the **balance to exactly zero** with interest considered.

Rate of return

The rate of return is expressed as a percent per period, for example, $i = 10\%$ per year. It is stated as a positive percentage; the fact that interest paid on a loan is actually a negative rate of return from the borrower's perspective is not considered. The numerical value of i can range from -100% to infinity, that is, $-100\% \leq i < \infty$. In terms of an investment, a return of $i = -100\%$ means the entire amount is lost.

The definition above does not state that the rate of return is on the initial amount of the investment; rather it is on the **unrecovered balance,** which changes each time period. Example 7.1 illustrates this difference.

EXAMPLE 7.1

To get started in a new telecommuting position with AB Hammond Engineers, Jane took out a $1000 loan at $i = 10\%$ per year for 4 years to buy home office equipment. From the lender's perspective, the investment in this young engineer is expected to produce an equivalent net cash flow of $315.47 for each of 4 years.

$$A = \$1000(A/P,10\%,4) = \$315.47$$

This represents a 10% per year rate of return on the unrecovered balance. Compute the amount of the unrecovered investment for each of the 4 years using (*a*) the rate of return on the unrecovered balance (the correct basis) and (*b*) the return on the initial $1000 investment. (*c*) Explain why all of the initial $1000 amount is not recovered by the final payment in part (*b*).

Solution

(*a*) Table 7–1 shows the unrecovered balance at the end of each year in column 6 using the 10% rate on the *unrecovered balance at the beginning of the year.* After 4 years the total $1000 is recovered, and the balance in column 6 is exactly zero.

TABLE 7–1	Unrecovered Balances Using a Rate of Return of 10% on the Unrecovered Balance				
(1)	(2)	(3) = 0.10 × (2)	(4)	(5) = (4) − (3)	(6) = (2) + (5)
Year	Beginning Unrecovered Balance	Interest on Unrecovered Balance	Cash Flow	Recovered Amount	Ending Unrecovered Balance
0	—	—	$−1000.00	—	$−1000.00
1	$−1000.00	$100.00	+315.47	$215.47	−784.53
2	−784.53	78.45	+315.47	237.02	−547.51
3	−547.51	54.75	+315.47	260.72	−286.79
4	−286.79	28.68	+315.47	286.79	0
		$261.88		$1000.00	

TABLE 7–2	Unrecovered Balances Using a 10% Return on the Initial Amount				
(1)	(2)	(3) = 0.10 × (2)	(4)	(5) = (4) − (3)	(6) = (2) + (5)
Year	Beginning Unrecovered Balance	Interest on Initial Amount	Cash Flow	Recovered Amount	Ending Unrecovered Balance
0	—	—	$−1000.00	—	$−1000.00
1	$−1000.00	$100	+315.47	$215.47	−784.53
2	−784.53	100	+315.47	215.47	−569.06
3	−569.06	100	+315.47	215.47	−353.59
4	−353.59	100	+315.47	215.47	−138.12
		$400		$861.88	

(b) Table 7–2 shows the unrecovered balance if the 10% return is always figured on the *initial $1000*. Column 6 in year 4 shows a remaining unrecovered amount of $138.12, because only $861.88 is recovered in the 4 years (column 5).

(c) As shown in column 3, a total of $400 in interest must be earned if the 10% return each year is based on the initial amount of $1000. However, only $261.88 in interest must be earned if a 10% return on the unrecovered balance is used. There is more of the annual cash flow available to reduce the remaining loan when the rate is applied to the unrecovered balance as in part (*a*) and Table 7–1. Figure 7–1 illustrates the correct interpretation of rate of return in Table 7–1.

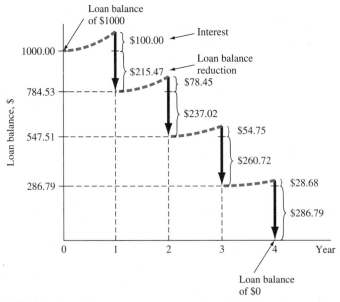

Figure 7–1
Plot of unrecovered balances and 10% per year rate of return on a $1000 amount, Table 7–1.

Each year the $315.47 receipt represents 10% interest on the unrecovered balance in column 2 plus the recovered amount in column 5.

Because rate of return is the interest rate on the unrecovered balance, the computations in *Table 7–1 for part (a) present a correct interpretation of a 10% rate of return.* Clearly, an interest rate applied only to the principal represents a higher rate than is stated. In practice, a so-called add-on interest rate is frequently based on principal only, as in part (*b*). This is sometimes referred to as the *installment financing* problem.

Installment financing can be discovered in many forms in everyday finances. One popular example is a "no-interest program" offered by retail stores on the sale of major appliances, audio and video equipment, furniture, and other consumer items. Many variations are possible, but in most cases, if the purchase is not paid for in full by the time the promotion is over, usually 6 months to 1 year later, *finance charges are assessed from the original date of purchase.* Further, the program's fine print may stipulate that the purchaser use a credit card issued by the retail company, which often has a higher interest rate than that of a regular credit card, for example, 24% per year compared to 15% per year. In all these types of programs, the one common theme is more interest paid over time by the consumer. Usually, the correct definition of *i* as interest on the unpaid balance does not apply directly; *i* has often been manipulated to the financial disadvantage of the purchaser. This was demonstrated by Example 4.4 using the Credit Card Case in Chapter 4.

7.2 Rate of Return Calculation Using a PW or AW Relation ●●●

Online Presentation

The ROR value is determined in a generically different way compared to the PW or AW value for a series of cash flows. For a moment, consider only the present worth relation for a cash flow series. Using the MARR, which is established independent of any particular project's cash flows, a mathematical relation determines the PW value in actual monetary units, say, dollars or euros. For the ROR values calculated in this and later sections, **only the cash flows themselves** are used to determine an interest rate that balances the present worth relation. Therefore, ROR may be considered a relative measure, while PW and AW are absolute measures. Since the resulting interest rate depends only on the cash flows themselves, the correct term is **internal rate of return (IROR);** however, the term *ROR* is used interchangeably. Another definition of rate of return is based on our previous interpretations of PW and AW.

Rate of return

> The *rate of return* is the interest rate that makes the present worth or annual worth of a cash flow series exactly equal to 0.

To determine the rate of return, develop the ROR equation using either a PW or AW relation, set it equal to 0, and solve for the interest rate. Alternatively, the present worth of cash outflows (costs and disbursements) PW_O may be equated to the present worth of cash inflows (revenues and savings) PW_I. That is, solve for *i* using either of the relations

$$0 = PW \qquad\qquad [7.1]$$

or $$PW_O = PW_I$$

The annual worth approach utilizes the AW values in the same fashion to solve for *i*.

$$0 = AW \qquad\qquad [7.2]$$

or $$AW_O = AW_I$$

The *i* value that makes these equations numerically correct is called i^*. It is the root of the ROR relation. To determine if the investment project's cash flow series is viable, compare i^* with the established MARR.

Figure 7–2
Cash flow for which
a value of i is to be
determined.

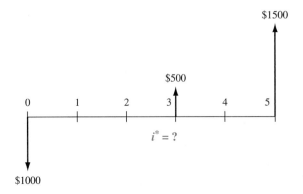

The guideline is as follows:

Project evaluation

> If $i^* \geq$ MARR, accept the project as economically viable.
> If $i^* <$ MARR, the project is not economically viable.

The purpose of engineering economy calculations is *equivalence* in PW or AW terms for a stated $i \geq 0\%$. In rate of return calculations, the objective is to *find the interest rate i^* at which the cash flows are equivalent*. The calculations are the reverse of those made in previous chapters, where the interest rate was known. For example, if you deposit $1000 now and are promised payments of $500 three years from now and $1500 five years from now, the rate of return relation using PW factors and Equation [7.1] is

$$1000 = 500(P/F,i^*,3) + 1500(P/F,i^*,5) \tag{7.3}$$

The value of i^* that makes the equality correct is to be determined (see Figure 7–2). If the $1000 is moved to the right side of Equation [7.3], we have the form $0 = $ PW.

$$0 = -1000 + 500(P/F,i^*,3) + 1500(P/F,i^*,5)$$

The equation is solved for $i^* = 16.9\%$ by hand using trial and error or using a spreadsheet function. The rate of return will always be greater than zero if the total amount of cash inflow is greater than the total amount of outflow, when the time value of money is considered. Using $i^* = 16.9\%$, a graph similar to Figure 7–1 can be constructed. It will show that the unrecovered balances each year, starting with -1000 in year 1, are exactly recovered by the $500 and $1500 receipts in years 3 and 5.

It should be evident that rate of return relations are merely a rearrangement of a present worth equation. That is, if the above interest rate is known to be 16.9%, and it is used to find the present worth of $500 three years from now and $1500 five years from now, the PW relation is

$$\text{PW} = 500(P/F,16.9\%,3) + 1500(P/F,16.9\%,5) = \$1000$$

This illustrates that rate of return and present worth equations are set up in exactly the same fashion. The only differences are what is given and what is sought.

There are several ways to determine i^* once the PW relation is established: solution via trial and error by hand, using a programmable calculator, and solution by spreadsheet function. The spreadsheet is faster; the first helps in understanding how ROR computations work. We summarize two methods here and in Example 7.2. Refer to Appendix D for the discussion about solution by calculator.

i^* Using Trial and Error The general procedure of using a PW-based equation is as follows:

1. Draw a cash flow diagram.
2. Set up the rate of return equation in the form of Equation [7.1].
3. Select values of i by trial and error until the equation is balanced.

When the trial-and-error method is applied to determine i^*, it is advantageous in step 3 to get fairly close to the correct answer on the first trial. If the cash flows are combined in such a manner that the income and disbursements can be represented by a *single factor* such as P/F or P/A, it is possible to look up the interest rate (in the tables) corresponding to the value of that factor for n years. The problem, then, is to combine the cash flows into the format of only one of the factors. This may be done through the following procedure:

1. Convert all *disbursements* into either single amounts (P or F) or uniform amounts (A) by neglecting the time value of money. For example, if it is desired to convert an A to an F value, simply multiply the A by the number of years n. The scheme selected for movement of cash flows should be the one that minimizes the error caused by neglecting the time value of money. That is, if most of the cash flow is an A and a small amount is an F, convert the F to an A rather than the other way around.
2. Convert all *receipts* to either single or uniform values.
3. Having combined the disbursements and receipts so that a P/F, P/A, or A/F format applies, use the interest tables to find the approximate interest rate at which the P/F, P/A, or A/F value is satisfied. The rate obtained is a good estimate for the first trial.

It is important to recognize that this first-trial rate is only an *estimate* of the actual rate of return, because the time value of money is neglected. The procedure is illustrated in Example 7.2.

i* by Spreadsheet   The fastest way to determine an i^* value when there is a series of equal cash flows (A series) is to apply the RATE function. This is a powerful one-cell function, where it is acceptable to have a separate P value in year 0 and a separate F value in year n. The format is

$$= \text{RATE}(n,A,P,F) \qquad\qquad [7.4]$$

When cash flows vary from year to year (period to period), the best way to find i^* is to enter the net cash flows into contiguous cells (including any \$0 amounts) and apply the IRR function in any cell. The format is

$$= \text{IRR(first\_cell:last\_cell,guess)} \qquad\qquad [7.5]$$

where "guess" is the i value at which the function starts searching for i^*.

The PW-based procedure for sensitivity analysis and a graphical estimation of the i^* value is as follows:

1. Draw the cash flow diagram.
2. Set up the ROR relation in the form of Equation [7.1], PW = 0.
3. Enter the cash flows onto the spreadsheet in contiguous cells.
4. Develop the IRR function to display i^*.
5. Use the NPV function to develop a PW graph (PW versus i values). This graphically shows the i^* value at which PW = 0.

EXAMPLE 7.2

Applications of green, lean manufacturing techniques coupled with value stream mapping can make large financial differences over future years while placing greater emphasis on environmental factors. Engineers with Monarch Paints have recommended to management an investment of \$200,000 now in novel methods that will reduce the amount of wastewater, packaging materials, and other solid waste in their consumer paint manufacturing facility. Estimated savings are \$15,000 per year for each of the next 10 years and an additional savings of \$300,000 at the end of 10 years in facility and equipment upgrade costs. Determine the rate of return using hand and spreadsheet solutions.

Solution by Hand

Use the trial-and-error procedure based on a PW equation.

1. Figure 7–3 shows the cash flow diagram.
2. Use Equation [7.1] format for the ROR equation.

$$0 = -200{,}000 + 15{,}000(P/A,i^*,10) + 300{,}000(P/F,i^*,10) \qquad [7.6]$$

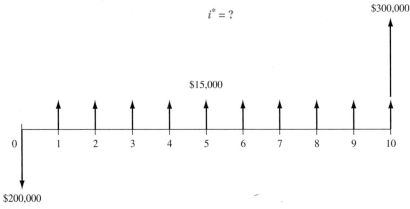

Figure 7–3
Cash flow diagram, Example 7.2.

3. Use the estimation procedure to determine i for the first trial. All income will be regarded as a single F in year 10 so that the P/F factor can be used. The P/F factor is selected because most of the cash flow ($300,000) already fits this factor and errors created by neglecting the time value of the remaining money will be minimized. Only for the first estimate of i, define $P = \$200,000$, $n = 10$, and $F = 10(15,000) + 300,000 = \$450,000$. Now we can state that

$$200,000 = 450,000(P/F,i,10)$$
$$(P/F,i,10) = 0.444$$

The roughly estimated i is between 8% and 9%. Use 9% as the first trial because this approximate rate for the P/F factor will be lower than the true value when the time value of money is considered.

$$0 = -200,000 + 15,000(P/A,9\%,10) + 300,000(P/F,9\%,10)$$
$$0 < \$22,986$$

The result is positive, indicating that the return is more than 9%. Try $i = 11\%$.

$$0 = -200,000 + 15,000(P/A,11\%,10) + 300,000(P/F,11\%,10)$$
$$0 > \$-6002$$

Since the interest rate of 11% is too high, linearly interpolate between 9% and 11%.

$$i^* = 9.00 + \frac{22,986 - 0}{22,986 - (-6002)}(2.0)$$
$$= 9.00 + 1.58 = 10.58\%$$

Solution by Spreadsheet

The fastest way to find i^* is to use the RATE function (Equation [7.4]). The entry = RATE(10,15000,−200000,300000) displays $i^* = 10.55\%$ per year. It is equally correct to use the IRR function. Figure 7–4, column B, shows the cash flows and = IRR(B2:B12) function to obtain i^*.

For a complete spreadsheet analysis, use the procedure outlined above.

1. Figure 7–3 shows cash flows.
2. Equation [7.6] is the ROR relation.
3. Figure 7–4 shows the net cash flows in column B.
4. The IRR function in cell B14 displays $i^* = 10.55\%$.
5. To graphically observe $i^* = 10.55\%$, column D displays the PW graph for different i values. The NPV function is used repeatedly to calculate PW for the xy scatter chart.

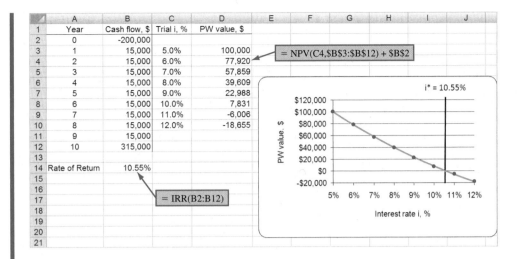

Figure 7–4
Spreadsheet to determine i^* and develop a PW graph, Example 7.2.

Just as i^* can be found using a PW equation, it may equivalently be determined using an AW relation. This method is preferred when uniform annual cash flows are involved. Solution by hand is the same as the procedure for a PW-based relation, except Equation [7.2] is used. In the case of Example 7.2, $i^* = 10.55\%$ is determined using the AW-based relation.

$$0 = -200{,}000(A/P,i^*,10) + 15{,}000 + 300{,}000(A/F,i^*,10)$$

The procedure for solution by spreadsheet is exactly the same as outlined above using the IRR function. Internally, IRR calculates the NPV function at different i values until NPV = 0. (There is no equivalent way to utilize the PMT function, since it requires a fixed value of i to calculate an A value.)

7.3 Special Considerations When Using the ROR Method ● ● ●

The rate of return method is commonly used in engineering and business settings to evaluate one project, as discussed in this chapter, and to select one alternative from two or more, as explained in the next chapter. As mentioned earlier, an ROR analysis is performed using a different basis than PW and AW analyses. The cash flows themselves determine the (internal) rate of return. As a result, there are some assumptions and special considerations with ROR analysis that must be made when calculating i^* and in interpreting its real-world meaning. A summary is provided below.

- *Multiple i^* values.* Depending upon the sequence of net cash inflows and outflows, there may be more than one real-number root to the ROR equation, resulting in *more than one i^* value*. This possibility is discussed in Section 7.4.
- *Reinvestment at i^*.* Both the PW and AW methods assume that any net positive investment (i.e., net positive cash flows once the time value of money is considered) is reinvested at the MARR. However, the ROR method assumes reinvestment at the i^* rate. When i^* is not close to the MARR (e.g., if i^* is substantially larger than MARR), this is an unrealistic assumption. In such cases, the i^* value is not a good basis for decision making. This situation is discussed in Section 7.5.
- *Different procedure for multiple alternative evaluations.* To correctly use the ROR method to choose from two or more mutually exclusive alternatives requires an *incremental analysis* procedure that is significantly more involved than PW and AW analysis. Chapter 8 explains this procedure.

If possible, from an engineering economic study perspective, the **AW or PW method at a stated MARR should be used in lieu of the ROR method**. However, there is a strong appeal for the ROR method because rate of return values are very commonly quoted. And it is easy to compare a proposed project's return with that of in-place projects.

> **When it is important to know the exact value of i^*, a good approach is to determine PW or AW at the MARR, then determine the specific i^* for the selected alternative.**

As an illustration, if a project is evaluated at MARR = 15% and has PW < 0, there is no need to calculate i^*, because $i^* < 15\%$. However, if PW is positive, but close to 0, calculate the exact i^* and report it along with the conclusion that the project is financially justified.

7.4 Multiple Rate of Return Values ● ● ●

In Section 7.2 a unique rate of return i^* was determined. In the cash flow series presented thus far, the algebraic signs on the *net cash flows* changed only once, usually from minus in year 0 to plus at some time during the series. This is called a *conventional* (or *simple*) *cash flow series*. However, for some series the net cash flows switch between positive and negative from one year to another, so there is more than one sign change. Such a series is called *nonconventional (non-simple)*. As shown in the examples of Table 7–3, each series of positive or negative signs may be one or more in length. Relatively large net cash flow (NCF) changes in amount and sign can occur in projects that require significant spending at the end of the expected life. Nuclear plants, open-pit mines, petroleum well sites, refineries, and the like often require environmental restoration, waste disposal, and other expensive phaseout costs. The cash flow diagram will appear similar to Figure 7–5a. Plants and systems that have anticipated major refurbishment costs or upgrade investments in future years may have considerable swings in cash flow and sign changes over the years, as shown by the pattern in Figure 7–5b.

TABLE 7–3	Examples of Conventional and Nonconventional Net Cash Flow for a 6-year Project							
	Sign on Net Cash Flow by Year							**Number of**
Type of Series	**0**	**1**	**2**	**3**	**4**	**5**	**6**	**Sign Changes**
Conventional	−	+	+	+	+	+	+	1
Conventional	−	−	−	+	+	+	+	1
Conventional	+	+	+	+	+	−	−	1
Nonconventional	−	+	+	+	−	−	−	2
Nonconventional	+	+	−	−	−	+	+	2
Nonconventional	−	+	−	−	+	+	+	3

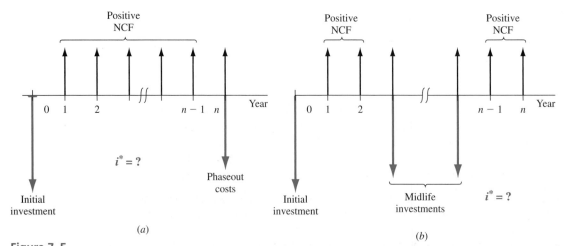

Figure 7–5

Typical cash flow diagrams for projects with (*a*) large restoration or remediation costs, and (*b*) upgrade or refurbishment costs.

When there is more than one sign change in the net cash flows, it is possible that there will be multiple i^* values in the -100% to plus infinity range. There are two tests to perform in sequence on the nonconventional series to determine if there is one unique value or possibly multiple i^* values that are real numbers.

Test 1: (Descartes') rule of signs states that the total number of real-number roots is always less than or equal to the number of sign changes in the series.

This rule is derived from the fact that the relation set up by Equation [7.1] or [7.2] to find i^* is an nth-order polynomial. (It is possible that imaginary values or infinity may also satisfy the equation.)

Test #2: Cumulative cash flow sign test, also known as *Norstrom's criterion,* states that only one sign change in a series of cumulative cash flows which *starts negatively* indicates that there is **one positive root** to the polynomial relation.

Zero values in the series are neglected when applying Norstrom's criterion. This is a more discriminating test that determines if there is one, real-number, positive i^* value. There may be negative roots that satisfy the ROR relation, but these are not useful i^* values. To perform the test, determine the series

$$S_t = \text{cumulative cash flows through period } t$$

Observe the sign of S_0 and count the sign changes in the series $S_0, S_1, \ldots, S_n$. Only if $S_0 < 0$ and signs change one time in the series is there a single, real-number, positive i^*.

With the results of these two tests, the ROR relation is solved for either the unique i^* or the multiple i^* values, using trial and error by hand, using a programmable calculator, or by spreadsheet using an IRR function that incorporates the "guess" option. Development of the PW graph is recommended, especially when using a spreadsheet. Examples 7.3 and 7.4 illustrate the tests and solution for i^*.

EXAMPLE 7.3

Sept-Îles Aluminum Company operates a bauxite mine to supply its aluminum smelter located about 2 km from the current open pit. A new branch for the pit is proposed that will supply an additional 10% of the bauxite currently available over the next 10-year period. The lease for the land will cost $400,000 immediately. The contract calls for the restoration of the land and development as part of a state park and wildlife area at the end of the 10 years. This is expected to cost $300,000. The increased production capacity is estimated to net an additional $75,000 per year for the company. Perform an ROR analysis that will provide the following information:

(*a*) Type of cash flow series and possible number of ROR values
(*b*) PW graph showing all i^* values
(*c*) Actual i^* values determined using the ROR relation and spreadsheet function
(*d*) Conclusions that can be drawn about the correct rate of return from this analysis

Solution

(*a*) The net cash flows will appear like those in Figure 7–5*a* with an initial investment of $\$-400,000$, annual net cash flow (NCF) of $75,000 for years 1 through 10, and a phase-out cost of $\$-300,000$ in year 10. Figure 7–6 details the NCF series (column B) and cumulative NCF (column C) for use in the two tests for unique and multiple i^* values. The **series is nonconventional** based on the sign changes throughout the series.

Test #1: There are two sign changes in the NCF series, which indicates a **possible maximum of two roots** to the polynomial equation or i^* values for the ROR equation.
Test #2: There is one sign change in the cumulative NCF series, which indicates a unique positive root or **one positive i^* value.**

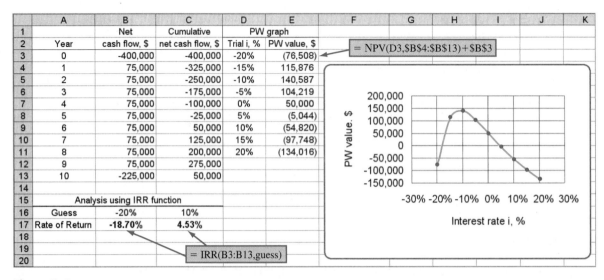

Figure 7–6
Spreadsheet determination of multiple i^* values and PW graph, Example 7.3.

(b) Columns D and E of the spreadsheet in Figure 7–6 use i values ranging from -20% to $+20\%$ per year to plot the PW vs i curve via the NPV function. There are two times that the parabolic-shaped curve crosses the PW = 0 line; these are approximately $i_1^* = -18\%$ and $i_2^* = 5\%$.

(c) The ROR equation based on PW computations is

$$0 = -400,000 + 75,000(P/A,i^*\%,10) - 300,000(P/F,i^*\%,10) \qquad [7.7]$$

i^* values by hand   If hand solution is chosen, the same procedure used in Example 7.2 can be applied here. However, the technique to estimate the initial i value will not work as well in this case since the majority of the cash flows do not fit either the P/F or the F/P factor. In fact, using the P/F factor, the initial i value is indicated to be 1.25%. Trial-and-error solution of Equation [7.7] with various i values will approximate the correct answer of about 4.5% per year. This complies with the test results of one positive i^* value.

i^* by spreadsheet function   Use the = IRR(B3:B13,guess) function to determine the i^* value for the NCF series in column B, Figure 7–6. Entering different values in the optional "guess" field will force the function to find multiple i^* values, if they exist. As shown in row 17, two are found.

$$i_1^* = -18.70\% \qquad i_2^* = +4.53\%$$

This result does not conflict with test results, as there is one positive value, but a negative value also balances the ROR equation.

(d) The positive $i^* = 4.53\%$ is accepted as the correct internal rate of return (IROR) for the project. The negative value is not useful in economic conclusions about the project.

EXAMPLE 7.4

The engineering design and testing group for Honda Motor Corp. does contract-based work for automobile manufacturers throughout the world. During the last 3 years, the net cash flows for contract payments have varied widely, as shown below, primarily due to a large manufacturer's inability to pay its contract fee.

Year	0	1	2	3
Cash Flow ($1000)	+2000	−500	−8100	+6800

(a) Determine the maximum number of i^* values that may satisfy the ROR equation.

(b) Write the PW equation and approximate the i^* value(s) by plotting PW vs i.

(c) What do the i^* values mean?

Solution

(a) Table 7–4 shows the annual cash flows and cumulative cash flows. Since there are two sign changes in the cash flow sequence, the rule of signs indicates a maximum of two i^* values. The cumulative cash flow sequence starts with a positive number $S_0 = +2000$, indicating that test #2 is inconclusive. As many as two i^* values can be found.

TABLE 7–4	Cash Flow and Cumulative Cash Flow Sequences, Example 7.4		
Year	Cash Flow ($1000)	Sequence Number	Cumulative Cash Flow ($1000)
0	+2000	S_0	+2000
1	−500	S_1	+1500
2	−8100	S_2	−6600
3	+6800	S_3	+200

(b) The PW relation is

$$PW = 2000 - 500(P/F,i,1) - 8100(P/F,i,2) + 6800(P/F,i,3)$$

The PW values are shown below and plotted in Figure 7–7 for several i values. The characteristic parabolic shape for a second-degree polynomial is obtained, with PW crossing the i axis at approximately $i_1^* = 8\%$ and $i_2^* = 41\%$.

i%	5	10	20	30	40	50
PW ($1000)	+51.44	−39.55	−106.13	−82.01	−11.83	+81.85

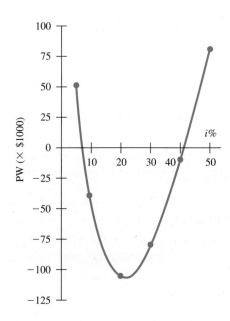

Figure 7–7
Present worth of cash flows at several interest rates, Example 7.4.

Figure 7–8 presents the spreadsheet PW graph with the PW curve crossing the x axis at PW = 0 two times. Also, the solution for **two positive i^* values** using the IRR function with different guess values is displayed. The values are

$$i_1^* = 7.47\% \qquad i_2^* = 41.35\%$$

(c) Since both i^* values are positive, they are not of much value, because neither can be considered the true ROR of the cash flow series. This result indicates that additional

	A	B	C	D	E
2		Cash flow	Cum. cash	PW graph	
3	Year	($1,000)	flow ($1000)	i value	PW ($1000)
4	0	2,000	2,000	5%	50.97
5	1	-500	1,500	10%	-39.82
6	2	-8,100	-6,600	20%	-106.48
7	3	6,800	200	30%	-82.39
8				40%	-11.66
9				50%	81.48
10					
11	Guess	0%	10%	30%	50%
12	i* value	**7.47%**	**7.47%**	**41.35%**	**41.35%**

IRR function format

= IRR($B4:$B7,guess)

Figure 7–8
Spreadsheet solution, Example 7.4.

information is needed to calculate a more useful project ROR, namely, some information about the anticipated return on funds invested external to the project and the cost of capital to borrow money to continue the project. This problem is a good example of when an approach discussed in the next section should be taken.

If the guess option is not used in the IRR function, the starting point is 10% per year. The function will find the one ROR closest to 10% that satisfies the PW relation. Entering various guess values will allow IRR to find multiple $i*$ values in the range of -100% to ∞, if they exist. Often, the results are unbelievable or unacceptable values that are rejected. Some helpful guidelines can be developed. Assume there are two $i*$ values for a particular cash flow series.

If the Results Are	What to Do
Both $i* < 0$	Discard both values.
Both $i* > 0$	Discard both values.
One $i* > 0$; one $i* < 0$	Use $i* > 0$ as ROR.

If both $i*$ values are discarded, proceed to the approach discussed in the next section to determine one rate of return value for the project. However, remember the prime recommendation.

> Always determine the PW or AW at the MARR first for a reliable measure of economic justification. If the PW or AW is greater than zero and the ROR is needed, then find the actual $i*$ of the project cash flows.

This recommendation is not to dissuade you from using the ROR method. Rather it is a recommendation that the use of the ROR technique be reserved for times when the actual $i*$ value is essential to the decision-making process.

7.5 Techniques to Remove Multiple Rates of Return ● ● ◐

The techniques developed here are used under the following conditions:

- The PW or AW value at the MARR is determined and could be used to make the decision, but information on the ROR is deemed necessary to finalize the economic decision, and
- The two tests of cash flow sign changes (Descartes' and Norstrom's) indicate multiple roots ($i*$ values) are possible, and

- More than one positive i^* value or all negative i^* values are obtained when the PW graph and IRR function are developed, and
- A single, reliable rate of return value is required by management or engineers to make a clear economic decision.

We will present a couple of ways to remove multiple i^* values. The selected approach depends upon what estimates are the most reliable for the project being evaluated. An important fact to remember is the following.

> The result of follow-up analysis to obtain a single ROR value when multiple, nonuseful i^* values are present does **not determine the internal rate of return (IROR)** for nonconventional net cash flow series. The resulting rate is a function of the additional information provided to make the selected technique work, and the accuracy is further dependent upon the reliability of this information.

We will refer to the resulting value as the **external rate of return (EROR)** as a reminder that it is different from the IROR obtained in all previous sections. First, it is necessary to identify the **perspective about the annual net cash flows** of a project. Take the following view: You are the project manager and the project generates cash flows each year. Some years produce positive NCF, and you want to invest the excess money at a good rate of return. We will call this the **investment rate i_i**. This can also be called the *reinvestment rate*. Other years, the net cash flow will be negative and you must borrow funds from some source to continue. The interest rate you pay should be as small as possible; we will call this the **borrowing rate i_b**, also referred to as the *finance rate*. Each year, you must consider the time value of money, which must utilize either the investment rate or the borrowing rate, depending upon the sign on the NCF of the preceding year. With this perspective, it is now possible to outline two approaches that rectify the multiple i^* situation. The resulting ROR value will not be the same for each method, because slightly different additional information is necessary and the cash flows are treated in slightly different fashions from the time value of money viewpoint.

Modified ROR (MIRR) Approach This is the easier approach to apply, and it has a spreadsheet function that can find the single EROR value quickly. However, the investment and borrowing rates must be reliably estimated, since the results may be quite sensitive to them. The symbol i' will identify the result.

Return on Invested Capital (ROIC) Approach Though more mathematically rigorous, this technique provides a more reliable estimate of the EROR and it requires only the investment rate i_i. The symbol i'' is used to indentify the result.

Before covering the techniques, it would be good to review the material in Section 7.1, including Example 7.1. Though the i' or i'' value determined here is not the ROR defined earlier in the chapter, the concepts used to make the ending cash flow balance equal to zero are used.

Modified ROR Approach

The technique requires that two rates external to the project net cash flows be estimated.

- **Investment rate i_i** is the rate at which extra funds are invested in some source external to the project. This applies to all positive annual NCF. It is reasonable that the MARR is used for this rate.
- **Borrowing rate i_b** is the rate at which funds are borrowed from an external source to provide funds to the project. This applies to all negative annual NCF. The weighted average cost of capital (WACC) can be used for this rate.

It is possible to make all rates the same, that is, $i_i = i_b = $ MARR $= $ WACC. However, this is not a good idea as it implies that the company is willing to borrow funds and invest in projects at the same rate. This implies no profit margin over time, so the company can't survive for long using this strategy. Commonly MARR $>$ WACC, so usually $i_i > i_b$. (See Section 1.9 for a quick review of MARR and WACC and Chapter 10 for a more detailed discussion of WACC.)

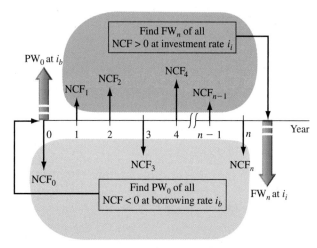

Figure 7–9
Typical cash flow diagram to determine modified rate of return i'.

Figure 7–9 is a reference diagram that has multiple $i*$ values, since the net cash flows change sign multiple times. The modified ROR method uses the following procedure to determine a single external rate of return i' and to evaluate the economic viability of the project.

1. Determine the PW value in year 0 of *all negative NCF at the borrowing rate i_b* (lightly shaded area and resulting PW_0 value in Figure 7–9).
2. Determine the FW value in year n of *all positive NCF at the investment rate i_i* (darker shaded area and resulting FW_n value in Figure 7–9).
3. Calculate the modified rate of return i' at which the *PW and FW values are equivalent* over the n years using the following relation, where i' is to be determined.

$$FW_n = PW_0 \,(F/P,i'\%,n) \qquad\qquad [7.8]$$

If using a spreadsheet rather than hand computation, the MIRR function displays i' directly with the format

$$= \text{MIRR(first\_cell:last\_cell, } i_b, i_i) \qquad\qquad [7.9]$$

4. The guideline for economic decision making compares the EROR or i' to MARR.

   If $i' >$ MARR, the project is economically justified.

   If $i' <$ MARR, the project is not economically justified.

   As in other situations, on the rare occasion that $i' =$ MARR, there is indifference to the project's economic acceptability; however, acceptance is the usual decision.

EXAMPLE 7.5

The cash flows experienced by Honda Motors in Example 7.4 are repeated below. There are two positive $i*$ values that satisfy the PW relation, 7.47% and 41.35% per year. Use the modified ROR method to determine the EROR value. Studies indicate that Honda has a WACC of 8.5% per year and that projects with an estimated return of less than 9% per year are routinely rejected. Due to the nature of this contract business, any excess funds generated are expected to earn at a rate of 12% per year.

Year	0	1	2	3
Net Cash Flow ($1000)	+2000	−500	−8100	+6800

Solution by Spreadsheet

Using the information in the problem statement, the rate estimates are as follows:

MARR:	9% per year
Investment rate, i_i:	12% per year
Borrowing rate, i_b:	8.5% per year

The fast way to find i' is with the MIRR function. Figure 7–10 shows the result of $i' = 9.39\%$ per year. Since $9.39\% > $ MARR, the project is economically justified.

	A	B	C	D	E	F
1	Year	NCF, $				
2	0	2000		Borrowing rate, i_b	8.5%	
3	1	-500		Investment rate, i_i	12.0%	
4	2	-8100				
5	3	6800		EROR value	**9.39%**	
6						
7				= MIRR(B2:B5,E2,E3)		
8						

Figure 7–10
Spreadsheet application of MIRR function, Example 7.5.

It is vital that the interpretation be correct. The 9.39% is not the internal rate of return (IROR); it is the external ROR (EROR) based on the two external rates for investing and borrowing money.

Solution by Hand

Figure 7–9 can serve as a reference as the procedure to find i' manually is applied.

Step 1. Find PW_0 of all negative NCF at $i_b = 8.5\%$.

$$PW_0 = -500(P/F,8.5\%,1) - 8100(P/F,8.5\%,2)$$
$$= \$-7342$$

Step 2. Find FW_3 of all positive NCF at $i_i = 12\%$.

$$FW_3 = 2000(F/P,12\%,3) + 6800$$
$$= \$9610$$

Step 3. Find the rate i' at which the PW and FW are equivalent.

$$PW_0(F/P,i',3) + FW_3 = 0$$
$$-7342(1 + i')^3 + 9610 = 0$$
$$i' = \left(\frac{9610}{7342}\right)^{1/3} - 1$$
$$= 0.939 \quad (9.39\%)$$

Step 4. Since $i' > $ MARR of 9%, the project is economically justified using this EROR approach.

Return on Invested Capital Approach

The definition of ROIC should be understood before we discuss the approach.

Return on invested capital (ROIC) is a rate-of-return measure of how effectively a project utilizes the funds invested in it, that is, funds that remain internal to the project. For a corporation, ROIC is a measure of how effectively it utilizes the funds invested in its operations, including facilities, equipment, people, systems, processes, and all other assets used to conduct business.

The technique requires that the investment rate i_i be estimated for excess funds generated in any year that they are not needed by the project. The ROIC rate, which has the symbol i'', is determined using an approach called the **net-investment procedure.** It involves developing a series of future worth (F) relations moving forward 1 year at a time. In those years that the net balance of the project cash flows is positive (extra funds generated by the project), the funds are invested at the i_i rate. Usually, i_i *is set equal to the MARR.* When the net balance is negative, the ROIC rate is used, since the project keeps all of its funds internal to itself. The ROIC method uses the following procedure to determine a single external rate of return i'' and to evaluate the economic viability of the project. Remember that the perspective is that you are the project manager and when the project generates extra cash flows, they are invested external to the project at the investment rate i_i.

1. Develop a series of future worth relations by setting up the following relation for each year t ($t = 1, 2, \ldots, n$ years).

$$F_t = F_{t-1}(1 + k) + \text{NCF}_t \qquad [7.10]$$

where F_t = **future worth in year t based on previous year and time value of money**

   NCF_t = **net cash flow in year t**

   $k = \begin{cases} i_i & \text{if } F_{t-1} > 0 \quad \textbf{(extra funds available)} \\ i'' & \text{if } F_{t-1} < 0 \quad \textbf{(project uses all available funds)} \end{cases}$

2. Set the future worth relation for the last year n equal to 0, that is, $F_n = 0$, and solve for i'' to balance the equation. The i'' value is the ROIC for the specified investment rate i_i.

   The F_t series and solution for i'' in the $F_n = 0$ relation can become involved mathematically. Fortunately, the Goal Seek spreadsheet tool can assist in the determination of i'' because there is only one unknown in the F_n relation and the target value is zero. (Both the hand and spreadsheet solutions are demonstrated in Example 7.6.)

3. The guideline for economic decision making is the same as above, namely,

   If ROIC $\geq$ MARR, the project is economically justified.

   If ROIC $<$ MARR, the project is not economically justified.

   It is important to remember that the **ROIC is an external rate of return dependent upon the investment rate choice.** It is not the same as the internal rate of return discussed at the beginning of this chapter, nor is it the multiple rates, nor is it the MIRR rate found by the previous method. This is a separate technique to find a single rate for the project.

EXAMPLE 7.6

Once again, we will use the cash flows experienced by Honda Motors in Example 7.4 (repeated below). Use the ROIC method to determine the EROR value. The MARR is 9% per year, and any excess funds generated by the project can earn at a rate of 12% per year.

Year	0	1	2	3
Net Cash Flow ($1000)	+2000	−500	−8100	+6800

Solution by Hand

The hand solution is presented first to provide the logic of the ROIC method. Use MARR = 9% and i_i = 12% per year in the procedure to determine i'', which is the ROIC. Figure 7–11 details the cash flows and tracks the progress as each F_t is developed. Equation [7.10] is applied to develop each F_t.

Step 1. Year 0: $F_0 = \$+2000$

                Since $F_0 > 0$, externally invest in year 1 at i_i = 12%.

   Year 1: $F_1 = 2000(1.12) - 500 = \$+1740$

                Since $F_1 > 0$, use i_i = 12% for year 2.

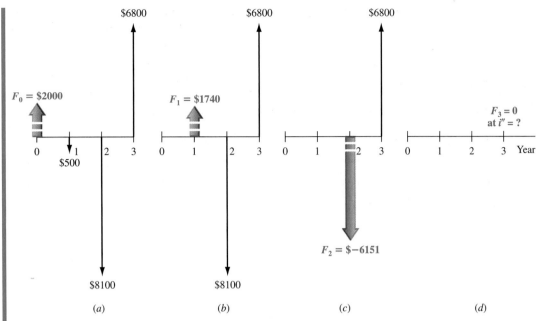

Figure 7–11
Application of ROIC method at $i_i = 12\%$ per year: (a) original cash flow; equivalent form in (b) year 1, (c) year 2, and (d) year 3.

Year 2: $F_2 = 1740(1.12) - 8100 = \$-6151$

Now $F_2 < 0$, use i'' for year 3, according to Equation [7.10].

Year 3: $F_3 = -6151(1 + i'') + 6800$

This is the last year. See Figure 7–11 for equivalent net cash flow diagrams. Go to step 2.

Step 2. Solve for $i'' = $ ROIC from $F_3 = 0$.

$$-6151(1 + i'') + 6800 = 0$$
$$i'' = 6800/6151 - 1$$
$$= 0.1055 \quad (10.55\%)$$

Step 3. Since ROIC > MARR = 9%, the project is economically justified.

Solution by Spreadsheet

Figure 7–12 provides a spreadsheet solution. The future worth values F_1 through F_3 are determined by the conditional IF statements in rows 3 through 5. The functions are shown in column D. In each year, Equation [7.10] is applied. If there are surplus funds generated by the project, $F_{t-1} > 0$ and the investment rate i_i (in cell E7) is used to find F_t. For example, because the F_1 value (in cell C3) of $1740 > 0$, the time value of money for the next year is calculated at the investment rate of 12% per year, as shown in the hand solution above for year 2.

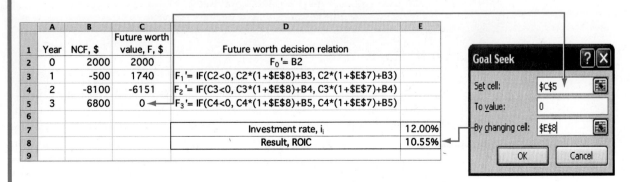

Figure 7–12
Spreadsheet application of ROIC method using Goal Seek, Example 7.6.

The Goal Seek template sets the F_3 value to zero by changing the ROIC value (cell E8). The result is $i'' = $ ROIC $= 10.55\%$ per year. As before, since $10.55\% > 9\%$, the MARR, the project is economically justified.

Comment

Note that the rate by the ROIC method (10.55%) is different than the MIRR rate (9.39%). Also these are both different than the multiple rates determined earlier (7.47% and 41.35%). This shows how dependent the different methods are upon the additional information provided when multiple $i*$ rates are indicated by the two sign tests.

Now that we have learned two techniques to remove multiple $i*$ values, there are some connections between the multiple $i*$ values, the external rate estimates, and the resulting external rates (i' and i'') obtained by the two methods.

Modified ROR technique   When both the borrowing rate i_b and the investment rate i_i are exactly equal to any one of the multiple $i*$ values, the rate i' found by the MIRR function, or by hand solution, will equal the $i*$ value. That is, all four parameters have the same value.

$$\text{If any } i* = i_b = i_i, \quad \text{then } i' = i*$$

ROIC technique   Similarly, if the investment rate i_i is exactly equal to any one of the multiple $i*$ values, the rate found by the Goal Seek tool, or by hand when the equation $F_n = 0$ is solved, will be $i'' = i*$ value.

Finally, it is very important to remember the following fact.

None of the details of the modified ROR (MIRR) technique or the return on invested capital (ROIC) technique are necessary if the PW or AW method of project evaluation is applied at a specific MARR. When the MARR is established, this is, in effect, fixing the $i*$ value. Therefore, a definitive economic decision can be made directly from the PW or AW value.

7.6 Rate of Return of a Bond Investment ●●○

A time-tested method of raising capital funds is through the issuance of an IOU, which is financing through debt, not equity (see Chapter 1). One very common form of IOU is a bond—a long-term note issued by a corporation or a government entity (the borrower) to finance major projects. The borrower receives money now in return for a promise to pay the *face value V* of the bond on a stated maturity date. Bonds are usually issued in face value amounts of $1000, $5000, or $10,000. *Bond dividend I,* also called *bond interest,* is paid periodically between the time the money is borrowed and the time the face value is repaid. The bond dividend is paid c times per year. Expected payment periods are usually semiannually or quarterly. The amount of interest is determined using the stated dividend or interest rate, called the *bond coupon rate b.*

$$I = \frac{(\text{face value})(\text{bond coupon rate})}{\text{number of payment periods per year}}$$

$$I = \frac{Vb}{c} \quad [7.11]$$

There are many types or classifications of bonds. Four general classifications are summarized in Table 7–5 according to their issuing entity, fundamental characteristics, and example names or purposes. For example, *Treasury securities* are issued in different monetary amounts ($1000 and up) with varying periods of time to the maturity date (Bills up to 1 year; Notes for 2 to 10 years). In the United States, Treasury securities are considered a very safe bond purchase because they are backed with the "full faith and credit of the U.S. government." The safe investment rate indicated in Figure 1−6 as the lowest level for establishing a MARR is the coupon rate on a U.S. Treasury security. Funds obtained through corporate bond issues are used for new product development, facilities upgrade, expansion into international markets, and similar business ventures.

TABLE 7–5	Classification and Characteristics of Bonds		
Classification	**Issued by**	**Characteristics**	**Examples**
Treasury securities	Federal government	Backed by faith and credit of the federal government	Bills ($\leq$ 1 year) Notes (2–10 years) Bonds (10–30 years)
Municipal	Local governments	Federal tax-exempt Issued against taxes received	General obligation Revenue Zero coupon Put
Mortgage	Corporation	Backed by specified assets or mortgage Low rate/low risk on first mortgage Foreclosure, if not repaid	First mortgage Second mortgage Equipment trust
Debenture	Corporation	Not backed by collateral, but by reputation of corporation Bond rate may "float" Higher interest rates and higher risks	Convertible Subordinated Junk or high yield

EXAMPLE 7.7

General Electric just released $10 million worth of $10,000 ten-year bonds. Each bond pays dividends semiannually at a rate of 6% per year. (*a*) Determine the amount a purchaser will receive each 6 months and after 10 years. (*b*) Suppose a bond is purchased at a time when it is discounted by 2% to $9800. What are the dividend amounts and the final payment amount at the maturity date?

Solution

(*a*) Use Equation [7.11] for the dividend amount.

$$I = \frac{10,000\,(0.06)}{2} = \$300 \text{ per 6 months}$$

The face value of $10,000 is repaid after 10 years.

(*b*) Purchasing the bond at a discount from face value does not change the dividend or final repayment amounts. Therefore, $300 per 6 months and $10,000 after 10 years remain the amounts.

The cash flow series for a bond investment is conventional and has one unique i^*, which is best determined by solving a PW-based rate of return equation in the form of Equation [7.1], that is, $0 = $ PW.

EXAMPLE 7.8

Allied Materials needs $3 million in debt capital for expanded composites manufacturing. It is offering small-denomination bonds at a discount price of $800 for a 4% $1000 bond that matures in 20 years with a dividend payable semiannually. What nominal and effective interest rates per year, compounded semiannually, will Allied Materials pay an investor?

Solution

The income that a purchaser will receive from the bond purchase is the bond dividend $I = \$20$ every 6 months plus the face value in 20 years. The PW-based equation for calculating the rate of return is

$$0 = -800 + 20(P/A,i^*,40) + 1000(P/F,i^*,40)$$

Solve by the IRR function or by hand to obtain $i^* = 2.8435\%$ semiannually. The nominal interest rate per year is computed by multiplying i^* by 2.

$$\text{Nominal } i = (2.8435)(2) = 5.6870\% \text{ per year, compounded semiannually}$$

Using Equation [4.5], the effective annual rate is

$$i_a = (1.028435)^2 - 1 = 5.7678\%$$

EXAMPLE 7.9

Gerry is a project engineer. He took a financial risk and bought a bond from a corporation that had defaulted on its interest payments. He paid $4240 for an 8% $10,000 bond with dividends payable quarterly. The bond paid no interest for the first 3 years after Gerry bought it. If interest was paid for the next 7 years and then Gerry was able to resell the bond for $11,000, what rate of return did he make on the investment? Assume the bond is scheduled to mature 18 years after he bought it. Perform hand and spreadsheet analysis.

Solution by Hand

The bond interest received in years 4 through 10 was

$$I = \frac{(10{,}000)(0.08)}{4} = \$200 \text{ per quarter}$$

The effective rate of return *per quarter* can be determined by solving the PW equation developed on a per quarter basis.

$$0 = -4240 + 200(P/A,i^* \text{ per quarter},28)(P/F,i^* \text{ per quarter},12)$$
$$+ 11{,}000(P/F,i^* \text{ per quarter},40)$$

The equation is correct for $i^* - 4.1\%$ per quarter, which is a nominal 16.4% per year, compounded quarterly.

Solution by Spreadsheet

Once all the cash flows are entered into contiguous cells, the function $= $ IRR(B2:B42) is used in Figure 7–13, row 43, to display the answer of a nominal rate of return of 4.10% per quarter. (Note that many of the row entries have been hidden to conserve space.) This is the same as the nominal annual rate of

$$i^* = 4.10\%(4) = 16.4\% \text{ per year, compounded quarterly}$$

Gerry did well on his bond investment.

	A	B
1	Quarter	Cash flow, $
2	0	-4,240
3	1	0
4	2	0
8	6	0
13	11	0
14	12	0
15	13	200
21	19	200
22	20	200
29	27	200
41	39	200
42	40	11,200
43	i*/qtr	4.10%

Figure 7–13
Spreadsheet solution for a bond investment, Example 7.9.

If a bond investment is being considered and a required rate of return is stated, the same PW-based relation used to find i^* can be used to determine the maximum amount to pay for the bond now to ensure that the rate is realized. The stated rate is the MARR, and the PW computations are performed exactly as they were in Chapter 5. As an illustration, in the last example, if 12% per year, compounded quarterly, is the target MARR, the PW relation is used to find the maximum that Gerry should pay now; P is determined to be $6004. The quarterly MARR is 12%/4 = 3%.

$$0 = -P + 200(P/A,3\%,28)(P/F,3\%,12) + 11{,}000(P/F,3\%,40)$$
$$P = \$6004$$

CHAPTER SUMMARY

The rate of return of a cash flow series is determined by setting a PW-based or AW-based relation equal to zero and solving for the value of i^*. The ROR is a term used and understood by almost everybody. Most people, however, can have considerable difficulty in calculating a rate of return correctly for anything other than a conventional cash flow series. For some types of series, more than one ROR possibility exists. The maximum number of i^* values is equal to the number of changes in the sign of the net cash flow series (Descartes' rule of signs). Also, a single positive rate can be found if the cumulative net cash flow series starts negatively and has only one sign change (Norstrom's criterion).

When multiple i^* values are indicated, either of the two techniques covered in this chapter can be applied to find a single, reliable rate for the nonconventional net cash flow series. In the case of the ROIC technique, additional information is necessary about the *investment rate* that excess project funds will realize, while the modified ROR technique requires this same information, plus the *borrowing rate* for the organization considering the project. Usually, the investment rate is set equal to the MARR, and the borrowing rate takes on the historical WACC rate. Each technique will result in slightly different rates, but they are reliable for making the economic decision, whereas the multiple rates are often not useful to decision making.

> **If an exact ROR is not necessary, it is strongly recommended that the PW or AW method at the MARR be used to decide upon economic justification.**

PROBLEMS

Understanding ROR

7.1 Under what circumstances would the rate of return be (*a*) -100%, and (*b*) infinite?

7.2 A shrewd investor loaned $1,000,000 to a start-up company at 10% per year interest for 3 years, but the terms of the agreement were such that interest would be charged on the principal rather than on the unpaid balance. How much extra interest did the company pay?

7.3 What is the *nominal* rate of return per year on an investment that increases in value by 8% every 3 months?

7.4 Assume you borrow $50,000 at 10% per year interest and you agree to repay the loan in five equal annual payments. What is the amount of the unrecovered balance immediately after you make the third payment?

7.5 International Potash got a $50 million loan amortized over a 10-year period at 10% per year interest. The loan agreement stipulates that the loan will be repaid in 10 equal annual payments with interest charged on the principal amount of the loan (not on the unrecovered balance).
(*a*) What is the amount of each payment?
(*b*) What is the total amount of interest paid? How does the total interest paid compare with the principal of the loan?

Determination of ROR

7.6 In 2010, the city of Houston, Texas, collected $24,112,054 in fines from motorists because of traffic violations caught by red-light cameras. The cost of operating the system was $8,432,372. The net profit, that is, profit after operating costs, is split equally (that is, 50% each) between the city and the operator of the camera system. What will be the rate of return over a 3-year period to the contractor that paid for, installed, and operates the system, if its initial cost was $9,000,000 and the profit for each of the 3 years is the same as it was in 2010?

7.7 P&G sold its prescription drug business to Warner-Chilcott, Ltd. for $3.1 billion. If income from product sales is $2 billion per year and net profit is 20% of sales, what rate of return will the company make over a 10-year planning horizon?

7.8 Water damage from a major flood in a Midwestern city resulted in damages estimated at $108 million. As a result of the claimant payouts, insurance companies raised homeowners' insurance rates by an average of $59 per year for each of the 160,000 households in the affected city. If a 20-year study period is considered, what was the rate of return on the $108 million paid by the insurance companies?

7.9 Determine the rate of return for the cash flows shown in the diagram. (If requested by your instructor, show both hand and spreadsheet solutions.)

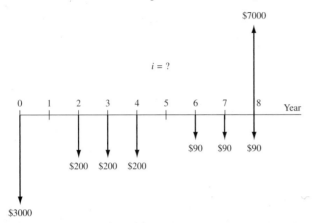

7.10 The Office of Naval Research sponsors a contest for college students to build underwater robots that can perform a series of tasks without human intervention. The University of Florida, with its SubjuGator robot, won the $7000 first prize (and serious bragging rights) over 21 other universities. If the team spent $2000 for parts (at time 0) and the project took 2 years, what annual rate of return did the team make?

7.11 For the cash flows shown, determine the rate of return.

Year	0	1	2	3	4	5
Expense, $	−17,000	−2,500	−2,500	−2,500	−2,500	−2,500
Revenue, $	0	5,000	6,000	7,000	8,000	12,000

7.12 In an effort to avoid foreclosure proceedings on struggling mortgage customers, Bank of America proposed an allowance that a jobless customer make no payment on their mortgage for up to 9 months. If the customer did not find a job within that time period, they would have to sign over their house to the bank. The bank would give them $2000 for moving expenses.

Assume John and his family had a mortgage payment of $2900 per month and he was not able to find a job within the 9-month period. If the bank saved $40,000 in foreclosure costs, what rate of return per month did the bank make on the allowance? Assume the first payment that was skipped was due at the end of month 1 and the $40,000 foreclosure savings and $2000 moving expense occurred at the end of the 9-month forbearance period.

7.13 The Closing the Gaps initiative by the Texas Higher Education Coordinating Board established the goal of increasing the number of students in higher education in Texas from 1,064,247 in 2000 to 1,694,247 in 2015. If the increase were to occur uniformly, what rate of increase would be required each year to meet the goal?

7.14 U.S. Census Bureau statistics show that the annual earnings for persons with a high school diploma are $35,220 versus $57,925 for someone with a bachelor's degree. If the cost of attending college is assumed to be $30,000 per year for 4 years and the forgone earnings during those years are assumed to be $35,220 per year, what rate of return does earning a bachelor's degree represent? Use a 35-year study period. (*Hint:* The investment in years 1 through 4 is the cost of college plus the foregone earnings, and the income in years 5 through 35 is the difference in income between a high school diploma and a bachelor's degree.)

7.15 The Ester Municipal Water Utility issued 20-year bonds in the amount of $53 million for several high-priority flood control improvement projects. The bonds carried a 5.38% dividend rate with the dividend payable annually. The U.S. economy was in a recession at that time, so as part of the federal stimulus program, the Utility gets a 35% reimbursement on the dividend it pays.
(a) What is the effective dividend rate that the Utility is paying on the bonds?
(b) What is the total dollar amount the Utility will *save* in dividends over the life of the bonds?
(c) What is the future worth in year 20 of the dividend savings, if the interest rate is 6% per year?

7.16 A contract between BF Goodrich and the Steelworkers Union of America called for the company to spend $100 million in capital investment to keep the facilities competitive. The contract also required the company to provide buyout packages for 400 workers. If the average buyout package is $100,000 and the company is able to reduce costs by $20 million per year, what rate of return will the company make over a 10-year period? Assume all of the company's expenditures occur at time 0 and the savings begin 1 year later.

7.17 Rubber sidewalks made from ground up tires are said to be environmentally friendly and easier on people's knees. Rubbersidewalks, Inc. of Gardena, California, manufactures the small rubberized squares that are being installed where tree roots, freezing weather, and snow removal have required sidewalk replacement or major repairs every 3 years. The District of Columbia spent $60,000 for a rubber sidewalk to replace broken concrete in a residential neighborhood lined with towering willow oaks. If a concrete sidewalk costs $28,000 and lasts only 3 years versus a 9-year life for the rubber sidewalks, what rate of return does this represent?

7.18 Efficient light jets (ELJs) are smaller aircraft that may revolutionize the way people travel by plane. They cost between $1.5 and $3 million, seat 5 to 7 people, and can fly up to 1100 miles at cruising speeds approaching 425 mph. Eclipse Aerospace was founded in 2009, and its sole business is making ELJs. The company invested $500 million (time 0) and began taking orders 2 years later. If the company accepted orders for 2500 planes and received 10% down (in year 2) on planes having an average cost of $1.8 million, what rate of return will the company make over a 10-year planning period? Assume 500 of the planes are delivered each year in years 6 through 10 and that the company's M&O costs average $10 million per year in years 1 through 10. (If requested by your instructor, show both hand and spreadsheet solutions.)

7.19 Betson Enterprises distributes and markets the Big Buck video game which allows players to "hunt" for elk, antelope, moose, and bucks without shivering outside in the cold. E-sports entertainment in New York City purchased five machines for $6000 each and took in an average of $600 total per week in sales. What rate of return does this represent (a) per week and (b) per year (nominal)? Use a 3-year study period with 52 weeks per year.

7.20 A 473-foot, 7000-ton World War II troop carrier (once commissioned as the *USS Excambion*) was sunk in the Gulf of Mexico to serve as an underwater habitat and diving destination. The project took 10 years of planning and cost $4 million. Assume the $4 million was expended equally in years 1 through 10. What rate of return does the venture represent, if increased fishing and recreation activities are valued at $270,000 per year beginning in year 11 and they continue in perpetuity? (If assigned by your instructor, show both hand and spreadsheet solutions.)

Multiple ROR Values

7.21 What is meant by a *nonconventional* cash flow series?

7.22 Explain at least three types of projects in which large net cash flow changes may cause sign changes during the life of the project, thus indicating the possible presence of multiple ROR values.

7.23 Explain a situation with which you are personally familiar for which the net cash flows have changed signs in a fashion similar to those in Figure 7–5.

7.24 According to Descartes' rule of signs, what is the maximum number of real-number values that will balance a rate of return equation?

7.25 According to Descartes' rule of signs, how many possible i^* values are there for net cash flows that have the following signs?
(a) $+-++-+-+$
(b) $---+++++$
(c) $++++---+----+-+---$

7.26 According to Norstrom's criterion, there are two requirements regarding the cumulative cash flows that must be satisfied to ensure that there is only one positive root in a rate of return equation. What are they?

7.27 According to Descartes' rule of signs, how many possible i^* values are there for the cash flows shown?

Year	1	2	3	4	5	6
Net Cash Flow, $	+4100	−2000	−7000	+12,000	−700	+800

7.28 According to Descartes' rule of signs, how many i^* values are possible for the cash flows shown?

Year	1	2	3	4
Revenue, $	25,000	13,000	4,000	70,000
Costs, $	−30,000	−7,000	−6,000	−12,000

7.29 According to Descartes' rule *and* Norstrom's criterion, how many i^* values are possible for the cash flow (CF) sequence shown?

Year	1	2	3	4	5
Net Cash Flow, $	+16,000	−32,000	−25,000	+50,000	−8,000
Cumulative CF, $	+16,000	−16,000	−41,000	+9,000	+1,000

7.30 For the cash flows shown, determine the sum of the cumulative cash flows.

Year	0	1	2	3	4
Revenue, $		25,000	15,000	4,000	18,000
Costs, $	−6,000	−30,000	−7,000	−6,000	−12,000

7.31 Stan-Rite Corp of Manitowoc, Wisconsin, is a B to B company that manufactures many types of industrial products, including portable measuring arms with absolute encoders, designed to perform 3D inspections of industrial parts. If the company's cash flow (in millions) for one of its product divisions is as shown on the next page, determine (a) the number of possible i^* values and (b) all rate of return values between 0% and 100%.

Year	Expenses	Revenues
0	$-30	$ 0
1	-20	18
2	-25	19
3	-15	36
4	-22	52
5	-20	38
6	-30	70

7.32 Julie received a $50 bill for her birthday at the end of January. At the end of February, she spent this $50 and an additional $150 to buy clothes. Her parents then gave her $50 and $125 at the end of March and April, respectively, as she prepared to go to summer school and needed the clothes. Her conclusion was that over the 4 months, she had received $25 more than she spent. Determine if Julie has a multiple rate of return situation for these cash flows. If so, determine the multiple rates and comment on their validity. The cash flow values are as follows:

Month	Jan (0)	Feb (1)	Mar (2)	Apr (3)
Cash Flow, $	50	-200	50	125

7.33 Veggie Burger Boy sells franchises to individuals who want to start small in the sandwiches-for-vegeterians business and grow in net cash flow over the years. A franchisee in Mississippi provided the $5000 up-front money, but did very poorly the first year. He was allowed to borrow at the end of his first year from the corporation's capital incentive fund with a promise to repay the loan in addition to the annual share that the corporation contractually receives from annual sales. The net cash flows from the corporation's perspective are shown below.

Year	0	1	2	3	4	5	6
NCF, $	5000	-10,100	500	2000	2000	2000	2000

The corporate chief financial officer (CFO) has some questions concerning this NCF series. Help her by doing the following, using a spreadsheet.
(a) Plot the PW versus i graph and estimate the rate of return for this franchise.
(b) Use the IRR function on a spreadsheet to find the corresponding return.
(c) Basing your conclusions on Descartes' and Norstrom's rules, provide the CFO with some advice on what ROR value is the most reliable for this franchise over the 6-year period. The normal corporate MARR used for franchisee evaluation is 30% per year.

7.34 In 2011, Vaught Industries closed its plant in Marionsville following labor, environmental, and safety problems. The plant was built in 2005 based on an older technology to produce meat products. It had positive NCF until 2010 and discontinued operation in 2011 due to labor and safety problems. In 2012, prior to the sale of the facility and property, Vaught spent $1 million to make the site environmentally acceptable to a potential buyer. The net cash flows in $100,000 over the years are listed below. Use a spreadsheet to do the following.
(a) Check for multiple rates of return.
(b) Find all rates that are real numbers between -25% and +50%, and calculate the PW value for interest rates in this range.
(c) Indicate which is the best and correct i^* value to use in a PW analysis.

Year	2005	2006	2007	2008	2009	2010	2011	2012
NCF, $	-25	+10	+10	+15	+15	-5	-6	-10

7.35 Five years ago, VistaCare spent $5 million to develop and introduce a new service in home health care for people who require frequent blood dialysis treatments. The service was not well received after the first year and was removed from the market. When reintroduced 4 years after its initial launch, it was much more profitable. Now, in year 5, VistaCare has spent a large sum on research to broaden the application of this service. Use the NCF series below to plot the PW versus i graph and estimate the ROR over the 5 years. NCF values are in $1 million units.

Year	0	1	2	3	4	5
Net Cash Flow, $	-5000	5000	0	0	15,000	-15,000

Removing Multiple i^* Values

7.36 In calculating the external rate of return by the modified rate of return approach, it is necessary to use two different rates of return, the investment rate i_i and the borrowing rate i_b. When is each used?

7.37 In the modified rate of return approach for determining a single interest rate from net cash flows, state which interest rate is usually higher, the investment rate i_i or the borrowing rate i_b. State why.

7.38 Use the modified rate of return approach with an investment rate of 18% per year and a borrowing rate of 10% to find the external rate of return for the following cash flows.

Year	0	1	2	3
Net Cash Flow, $	+16,000	-32,000	-25,000	+70,000

7.39 Harley worked for many years to save enough money to start his own residential landscape design business. The cash flows shown are those he recorded for the first 6 years as his own boss. Find the external rate of return using the modified rate of return approach, an investment rate of 15% per year, and a borrowing rate of 8%. (After using the procedure, use the MIRR function to confirm your answer.)

Year	0	1	2	3	4	5	6
NCF, $	−9000	+4100	−2000	−7000	+12,000	+700	+800

7.40 Samara, an engineer working for GE, invested her bonus money each year in company stock. Her bonus has been $8000 each year for the past 6 years (i.e., at the end of years 1 to 6). At the end of year 7, she sold the stock for $52,000 to buy a condo; she purchased no stock that year. In years 8 to 10, she again invested the $8000 bonus. Samara sold all of the remaining stock for $28,000 immediately after the investment at the end of year 10.
   (a) Determine the number of possible rate of return values in the net cash flow series using the two sign tests.
   (b) Determine the external rate of return by hand, using the modified rate of return approach with an investment rate of 12% per year and a borrowing rate of 8%.
   (c) Find the external rate of return by spreadsheet using the ROIC approach with an investment rate of 12% per year.
   (d) Enter the cash flows into a spreadsheet, and use the IRR function to find the i^* value. You should get the same value as the ROIC in part (c). Explain why this is so, given that the investment rate is 12% per year. (*Hint:* Look carefully at the column labeled "Future worth, F, $" when you solved part (c) using the spreadsheet.)

7.41 Swagelok Co. of Solon, Ohio, makes variable area flowmeters (VAFs) that measure liquid and gas flow rates by means of a tapered tube and float. If tooling and setup costs were $400,000 in year 0 and an additional $190,000 in year 3, determine the external rate of return using the modified rate of return approach. The revenue was $160,000 per year in years 1 through 10. Assume the company's MARR is 20% per year and its cost of capital is 9% per year.

7.42 A company that makes clutch disks for race cars has the cash flows shown for one department.

Year	Cash Flow, $1000
0	−65
1	30
2	84
3	−10
4	−12

   (a) Determine the number of positive roots to the rate of return relation.
   (b) Calculate the internal rate of return.

(c) Calculate the external rate of return using the return on invested capital (ROIC) approach with an investment rate of 15% per year. (As assigned by your instructor, solve by hand and/or spreadsheet.)

7.43 For the cash flow series below, calculate the external rate of return, using the return on invested capital approach with an investment rate of 14% per year.

Year	Cash Flow, $
0	3000
1	−2000
2	1000
3	−6000
4	3800

7.44 Five years ago, a company made a $500,000 investment in a new high-temperature material. The product did poorly after only 1 year on the market. However, with a new name and advertising campaign 4 years later it did much better. New development funds have been expended this year (year 5) at a cost of $1.5 million. Determine the external rate of return using the ROIC approach and an investment rate of 15% per year. The i^* rate is 44.1% per year.

Year	Cash Flow, $
0	−500,000
1	400,000
2	0
3	0
4	2,000,000
5	−1,500,000

Bonds

7.45 What is the bond coupon rate on a $25,000 mortgage bond that has semiannual interest payments of $1250 and a 20-year maturity date?

7.46 An equipment trust bond with a face value of $10,000 has a bond coupon rate of 8% per year, payable quarterly. What are the amount and frequency of the dividend payments?

7.47 What is the face value of a municipal bond that matures in 20 years and has a bond coupon rate of 6% per year with semiannual payments of $900?

7.48 What is the present worth of a $50,000 debenture bond that has a bond coupon rate of 8% per year, payable quarterly? The bond matures in 15 years. The interest rate in the marketplace is 6% per year, compounded quarterly.

7.49 Best Buy issued collateral bonds *4 years ago* that have a face value of $20,000 each and a coupon rate of 8% per year, payable semiannually. If the bond maturity date is 20 years *from the date they were issued* and the interest rate in the marketplace is now 12% per year, compounded semiannually, what is the present worth (now) of one bond?

7.50 In 2011, El Paso Water Utilities (EPWU) issued bonds worth $9.125 million to improve the Van Buren dam in central El Paso and to finance three other drainage projects. The bonds were purchased by the Texas Water Development Board under the federal stimulus program wherein EPWU did not have to pay any dividend on the bonds. If the bond dividend rate would have been 4% per year, payable quarterly, with a bond maturity date 18 years after issuance, what is the present worth of the dividend savings to EPWU rate payers? Assume the market interest rate is 6% per year.

7.51 A recently issued industrial bond with a face value of $10,000 has a coupon rate of 8% per year, payable annually. The bond matures 20 years from now. Jeremy is interested in buying one bond. If he pays $10,000 for the bond and plans to hold it to maturity, what rate of return per year will he realize?

7.52 Due to a significant troop buildup at the local military base, a school district issued $10,000,000 in bonds to build new schools. The bond coupon rate is 6% per year, payable semiannually, with a maturity date of 20 years. If an investor is able to purchase one of the bonds that has a face value of $5000 for $4800, what rate of return per 6 months will the investor realize? Assume the bond is kept to maturity.

7.53 As the name implies, a zero-coupon bond pays no dividend, only the face value when it matures. If a zero coupon bond that has a face value of $10,000 and a maturity date of 15 years is for sale for $2000, what rate of return will the purchaser make, provided the bond is held to maturity?

7.54 To provide infrastructure in the outlying areas of Morgantown, West Virginia, the city council

issued 30-year bonds with a face value of $25 million. The bond coupon rate was set at 5% per year, payable semiannually. Because the market interest rate increased immediately before the bonds were sold, the city received only $23.5 million from the bond sale. What was the semiannual interest rate when the bonds were sold?

7.55 An investor who purchased a $10,000 mortgage bond today paid only $6000 for it. The bond coupon rate is 8% per year, payable quarterly, and the maturity date is 18 years from the year of issuance. Because the bond is in default, it will pay no dividend for the next 2 years. If the bond dividend is in fact paid for the following 5 years (after the 2 years) and the investor then sells the bond for $7000, what rate of return will be realized (*a*) per quarter and (*b*) per year (nominal)?

7.56 Five years ago, GSI, an oil services company headquartered in Texas, issued $10 million worth of 12% 30-year bonds with the dividend payable quarterly. The bonds have a call date of this year if GSI decides to take advantage of it. The interest rate in the marketplace decreased enough that the company is considering calling the bonds since the coupon rate is relatively high. If the company buys the bonds back now for $11 million, determine the rate of return that the company will make (*a*) per quarter and (*b*) per year (nominal). (*Hint:* The "call" option means the following: By spending $11 million now, the company will not have to make the quarterly bond dividend payments or pay the face value of the bonds when they come due 25 years from now.)

7.57 Four years ago, Chevron issued $5 million worth of debenture bonds with a coupon rate of 10% per year, payable semiannually. Market interest rates dropped, and the company called the bonds (i.e., paid them off in advance) at a 10% premium on the face value. Therefore, it cost the corporation $5.5 million to retire the bonds. What semiannual rate of return did an investor make who purchased a $5000 bond 4 years ago and held it until it was called 4 years later?

ADDITIONAL PROBLEMS AND FE EXAM REVIEW QUESTIONS

7.58 All of the following mean the same as *rate of return* except:
    (*a*) Internal rate of return
    (*b*) Time for return of capital
    (*c*) Interest rate
    (*d*) Return on investment

7.59 The numerical value of *i* in a rate of return equation can range from:
    (*a*) 0% to 100%
    (*b*) 0 to ∞
    (*c*) −100% to +100%
    (*d*) −100% to ∞

7.60 The internal rate of return on an investment refers to the interest rate earned on the:
   (a) Initial investment
   (b) Unrecovered balance of the investment
   (c) Money recovered from an investment
   (d) Income from an investment

7.61 A conventional (or simple) cash flow series is one wherein:
   (a) The algebraic signs on the net cash flows change only once.
   (b) The interest rate you get is a simple interest rate.
   (c) The total of the net cash flows is equal to 0.
   (d) The total of the cumulative cash flows is equal to 0.

7.62 According to Descartes' rule of signs, for a net cash flow sequence of $- -++-+$, the number of possible i values is:
   (a) 2
   (b) 3
   (c) 4
   (d) 5

7.63 According to Norstrom's criterion, the one statement below that is correct is:
   (a) The cumulative cash flow must start out positively.
   (b) The cumulative cash flow must start out negatively.
   (c) The cumulative cash flow must equal 0.
   (d) The net cash flow must start out positively.

7.64 According to Descartes' rule *and* Norstrom's criterion, the number of positive i^* values for the following cash flow sequence is:

Year	1	2	3	4
Revenue, $	25,000	15,000	4,000	18,000
Cost, $	30,000	7,000	6,000	12,000

   (a) 1
   (b) 2
   (c) 3
   (d) 4

7.65 For the net cash flows and cumulative cash flows shown, the value of x is:

Year	1	2	3	4	5
NCF, $	+13,000	−29,000	−25,000	x	−8000
Cumulative NCF, $	+13,000	−16,000	−41,000	+9000	+1000

   (a) $7000
   (b) $16,000
   (c) $36,000
   (d) $50,000

7.66 A company that uses a minimum attractive rate of return of 10% per year is evaluating new processes to improve operational efficiency. The estimates associated with candidate processes are shown.

	Alternative I	Alternative J
First cost, $	−40,000	−50,000
Annual cost, $ per year	−15,000	−12,000
Salvage value, $	5,000	5,000
Life, years	3	6

The statement that is most correct is:
   (a) The alternatives are revenue alternatives.
   (b) The alternatives are cost alternatives.
   (c) The alternatives are revenue alternatives and DN is an option.
   (d) The alternatives are cost alternatives and DN is an option.

7.67 Scientific Instruments, Inc. uses a MARR of 8% per year. The company is evaluating a new process to reduce water effluents from its manufacturing processes. The estimate associated with the process follows. In evaluating the process on the basis of a rate of return analysis, the correct equation to use is:

	New Process
First cost, $	−40,000
NCF, $ per year	13,000
Salvage value, $	5,000
Life, years	3

   (a) $0 = -40,000 + 13,000(P/A,i,3) + 5000(P/F,i,3)$
   (b) $0 = -40,000(A/P,i,3) + 13,000 + 5000(A/F,i,3)$
   (c) $0 = -40,000(F/P,i,3) + 13,000(F/A,i,3) + 5000$
   (d) Any of the above

7.68 When one is using the modified ROR method to remove multiple ROR values, an additional estimate needed besides the cash flows and their timings is:
   (a) The ROIC value
   (b) External rate of return
   (c) Investment rate
   (d) Internal rate of return

7.69 For the following cash flows, the modified rate of return method uses a borrowing rate of 10%, and an investment rate is 12% per year. The correct computation for the present worth in year 0 is:

Year	1	2	3	4	5
NCF, $	−10,000	0	0	−19,000	+25,000

   (a) $-10,000 - 19,000(P/F,12\%,4)$
   (b) $-10,000 - 19,000(P/F,12\%,4) + 25,000(P/F,10\%,5)$

    (c) $25,000(P/F,10\%,5)$

    (d) $-10,000 - 19,000(P/F,10\%,4)$

7.70   The return on invested capital (ROIC) method removes multiple ROR values from a cash flow sequence. If the future worth computation in year t is $F_t < 0$, the ROIC rate i'' is used. The interpretation of $F_t < 0$ most closely means:

    (a)   The net balance of project cash flows in year t is negative.

    (b)   The resulting external rate of return will be positive.

    (c)   The net balance of project cash flows is positive in year t.

    (d)   The sequence has nonremovable negative ROR values.

7.71   The meaning of *return on invested capital* for a corporation is best stated as:

    (a)   A rate-of-return measure that equates the internal and external ROR

    (b)   A measure of how effectively the corporation uses capital funds invested in it

    (c)   The value at which borrowing ROR and investing ROR are equal

    (d)   The external rate of return value is based on total capital invested

7.72   A corporate $10,000 bond has a coupon rate of 8% per year, payable semiannually. The bond matures 20 years from now. At an interest rate of 6% per year, compounded semiannually, the amount and frequency of the bond dividend payments are:

    (a) $600 every 6 months

    (b) $800 every 6 months

    (c) $300 every 6 months

    (d) $400 every 6 months

7.73   A $20,000 mortgage bond that is due in 1 year pays interest of $500 every 3 months. The bond's coupon rate is:

    (a)   2.5% per year, payable quarterly

    (b)   5% per year, payable quarterly

    (c)   5% per year, payable semiannually

    (d)   10% per year, payable quarterly

7.74   A $10,000 municipal bond due in 10 years pays interest of $400 every 6 months. If an investor purchases the bond now for $9000 and holds it to maturity, the rate of return received can be determined by the following equation:

    (a) $0 = -9000 + 400(P/A,i,10)$
              $+ 10,000(P/F,i,10)$

    (b) $0 = -9000 + 400(P/A,i,20)$
              $+ 10,000(P/F,i,20)$

    (c) $0 = -10,000 + 400(P/A,i,20)$
              $+ 10,000(P/F,i,20)$

    (d) $0 = -9000 + 800(P/A,i,10)$
              $+ 10,000(P/F,i,10)$

7.75   A debenture bond issued 3 years ago has a face value of $5000, a coupon rate of 4% per year, payable semiannually, and a maturity date of 20 years *from the date it was issued*. The bond is for sale now for $4500. If the interest rate in the marketplace is compounded quarterly, the value of n that must be used in the P/A factor to calculate the rate of return for the bond is:

    (a)   34

    (b)   40

    (c)   68

    (d)   80

CASE STUDY

DEVELOPING AND SELLING AN INNOVATIVE IDEA

Background

Three engineers who worked for Mitchell Engineering, a company specializing in public housing development, went to lunch together several times a week. Over time they decided to work on solar energy production ideas. After a lot of weekend time over several years, they had designed and developed a prototype of a low-cost, scalable solar energy plant for use in multifamily dwellings on the low end and medium-sized manufacturing facilities on the upper end. For residential applications, the collector could be mounted along side a TV dish and be programmed to track the sun. The generator and additional equipment are installed in a closet-sized area in an apartment or on a floor for multiple-apartment supply.

The system serves as a supplement to the electricity provided by the local power company. After some 6 months of testing, it was agreed that the system was ready to market and reliably state that an electricity bill in high-rises could be reduced by approximately 40% per month. This was great news for low-income dwellers on government subsidy that are required to pay their own utility bills.

Information

With a hefty bank loan and $200,000 of their own capital, they were able to install demonstration sites in three cities in the sunbelt. Net cash flow after all expenses, loan repayment, and taxes for the first 4 years was acceptable; $55,000 at the end of the first year, increasing by 5% each year thereafter. A business acquaintance introduced them to a potential buyer of the patent rights and current subscriber base with an estimated $500,000 net cashout after only these 4 years of ownership. However, after serious discussion replaced the initial excitement of the sales offer, the trio decided to not sell at this time. They wanted to stay in the business for a while longer to develop some enhancement ideas and to see how much revenue may increase over the next few years.

During the next year, the fifth year of the partnership, the engineer who had received the patents upon which the collector and generator designs were based became very displeased with the partnering arrangements and left the trio to go into partnership with an international firm in the energy business. With new research and development funds and the patent rights, a competing design was soon on the market and took much of the business away from the original two developers. Net cash flow dropped to $40,000 in year 5 and continued to decrease by $5000 per year. Another offer to sell in year 8 was presented, but it was only for $100,000 net cash. This was considered too much of a loss, so the two owners did not accept. Instead, they decided to put $200,000 more of their own savings into the company to develop additional applications in the housing market.

It is now 12 years since the system was publicly launched. With increased advertising and development, net cash flow has been positive the last 4 years, starting at $5000 in year 9 and increasing by $5000 each year until now.

Case Study Exercises

It is now 12 years after the products were developed, and the engineers invested most of their savings in an innovative idea. However, the question of "When do we sell?" is always present in these situations. To help with the analysis, determine the following:

1. The rate of return at the end of year 4 for two situations: (*a*) The business is sold for the net cash amount of $500,000 and (*b*) no sale.
2. The rate of return at the end of year 8 for two situations: (*a*) The business is sold for the net cash amount of $100,000 and (*b*) no sale.
3. The rate of return now at the end of year 12.
4. Consider the cash flow series over the 12 years. Is there any indication that multiple rates of return may be present? If so, use the spreadsheet already developed to search for ROR values in the range ±100% other than the one determined in exercise 3 above.
5. Assume you are an investor with a large amount of ready cash, looking for an innovative solar energy product. What amount would you be willing to offer for the business at this point (end of year 12) if you require a 12% per year return on all your investments and, if purchased, you plan to own the business for 12 additional years? To help make the decision, assume the current NCF series continues increasing at $5000 per year for the years you would own it. **Explain your logic for offering this amount.**

CHAPTER 8

Rate of Return Analysis: Multiple Alternatives

LEARNING OUTCOMES

Purpose: Select the best alternative on the basis of incremental rate of return analysis.

SECTION	TOPIC	LEARNING OUTCOME
8.1	Incremental analysis	• State why the ROR method of comparing alternatives requires an incremental cash flow analysis.
8.2	Incremental cash flows	• Calculate the incremental cash flow series for two alternatives.
8.3	Meaning of Δi^*	• Interpret the meaning of the incremental ROR (Δi^*) determined from the incremental cash flow series.
8.4	Δi^* from PW relation	• Based on a PW relation, select the better of two alternatives using incremental ROR analysis or a breakeven ROR value.
8.5	Δi^* from AW relation	• Select the better of two alternatives using incremental ROR analysis based on an AW relation.
8.6	More than two alternatives	• Select the best from several alternatives using incremental ROR analysis.
8.7	All-in-one spreadsheet	• Use a single spreadsheet to perform PW, AW, ROR, and incremental ROR analyses for mutually exclusive and independent alternatives.

his chapter presents the methods by which two or more alternatives can be evaluated using a rate of return (ROR) comparison based on the methods of the previous chapter. The ROR evaluation, correctly performed, will result in the same selection as the PW and AW analyses, but the computational procedure is considerably different for ROR evaluations. The ROR analysis evaluates the increments between two alternatives in pairwise comparisons. As the cash flow series becomes more complex, spreadsheet functions help speed computations.

8.1 Why Incremental Analysis Is Necessary ● ● ●

Online Presentation

When two or more mutually exclusive alternatives are evaluated, engineering economy can identify the one alternative that is the best economically. As we have learned, the PW and AW techniques can be used to do so, and are the recommended methods. Now the procedure using ROR to identify the best is presented.

Let's assume that a company uses a MARR of 16% per year, that the company has $90,000 available for investment, and that two alternatives (A and B) are being evaluated. Alternative A requires an investment of $50,000 and has an internal rate of return i_A^* of 35% per year. Alternative B requires $85,000 and has an i_B^* of 29% per year. Intuitively we may conclude that the better alternative is the one that has the larger return, A in this case. However, this is not necessarily so. While A has the higher projected return, its initial investment ($50,000) is much less than the total money available ($90,000). What happens to the investment capital that is left over? It is generally assumed that excess funds will be invested at the company's MARR, as we learned in previous chapters. Using this assumption, it is possible to determine the consequences of the two alternative investments. If alternative A is selected, $50,000 will return 35% per year. The $40,000 left over will be invested at the MARR of 16% per year. The rate of return on the total capital available, then, will be the weighted average. Thus, if alternative A is selected,

$$\text{Overall ROR}_A = \frac{50{,}000(0.35) + 40{,}000(0.16)}{90{,}000} = 26.6\%$$

If alternative B is selected, $85,000 will be invested at 29% per year, and the remaining $5000 will earn 16% per year. Now the weighted average is

$$\text{Overall ROR}_B = \frac{85{,}000(0.29) + 5000(0.16)}{90{,}000} = 28.3\%$$

These calculations show that even though the i^* for alternative A is higher, alternative B presents the better overall ROR for the $90,000. If either a PW or AW comparison is conducted using the MARR of 16% per year as i, alternative B will be chosen.

This simple example illustrates a major fact about the rate of return method for ranking and comparing alternatives:

Under some circumstances, project ROR values do not provide the same ranking of alternatives as do PW and AW analyses. This situation does not occur if we conduct an **incremental ROR analysis** (discussed below).

When **independent projects** are evaluated, no incremental analysis is necessary between projects. Each project is evaluated separately from others, and more than one can be selected. Therefore, the only comparison is with the do-nothing alternative for each project. The project ROR can be used to accept or reject each one.

Independent project selection

8.2 Calculation of Incremental Cash Flows for ROR Analysis ● ● ●

To conduct an incremental ROR analysis, it is necessary to calculate the **incremental cash flow** series over the lives of the alternatives. Based upon the equivalence relations (PW and AW), ROR evaluation makes the equal-service assumption.

TABLE 8–1	Format for Incremental Cash Flow Tabulation		
	Cash Flow		Incremental Cash Flow
Year	Alternative A (1)	Alternative B (2)	(3) = (2) − (1)
0			
1			
.			
.			
.			

Equal-service requirement

The incremental ROR method requires that the equal-service requirement be met. Therefore, the **LCM (least common multiple) of lives for each pairwise comparison** must be used. All the assumptions of equal service present for PW analysis are necessary for the incremental ROR method.

A format for hand or spreadsheet solutions is helpful (Table 8–1). Equal-life alternatives have n years of incremental cash flows, while unequal-life alternatives require the LCM of lives for analysis. At the end of each life cycle, the salvage value and initial investment for the next cycle must be included for the LCM case.

When a **study period** is established, only this number of years is used for the evaluation. All incremental cash flows outside the period are neglected. As we learned earlier, using a study period, especially one shorter than the life of either alternative, can change the economic decision from that rendered when the full lives are considered.

Only for the purpose of simplification, use the convention that between two alternatives, the one with the *larger initial investment* will be regarded as *alternative B*. Then, for each year in Table 8–1,

$$\text{Incremental cash flow} = \text{cash flow}_B - \text{cash flow}_A \qquad [8.1]$$

The initial investment and annual cash flows for each alternative (excluding the salvage value) are one of the types identified in Chapter 5:

Revenue alternative, where there are both negative and positive cash flows

Cost alternative, where all cash flow estimates are negative

Revenue or cost alternative

In either case, Equation [8.1] is used to determine the incremental cash flow series with the sign of each cash flow carefully determined.

EXAMPLE 8.1

A tool and die company in Hanover is considering the purchase of a drill press with fuzzy-logic software to improve accuracy and reduce tool wear. The company has the opportunity to buy a slightly used machine for $15,000 or a new one for $21,000. Because the new machine is a more sophisticated model, its operating cost is expected to be $7000 per year, while the used machine is expected to require $8200 per year. Each machine is expected to have a 25-year life with a 5% salvage value. Tabulate the incremental cash flow.

Solution

Incremental cash flow is tabulated in Table 8–2. The subtraction performed is (new – used) since the new machine has a larger initial cost. The salvage values in year 25 are separated from ordinary cash flow for clarity. When disbursements are the same for a number of consecutive years, *for hand solution only,* it saves time to make a single cash flow listing, as is done for years 1 to 25. However, remember that several years were combined when performing the analysis. This approach *cannot be used for spreadsheets,* when the IRR or NPV function is used, as each year must be entered separately.

TABLE 8–2   Cash Flow Tabulation for Example 8.1

	Cash Flow		Incremental Cash Flow
Year	Used Press	New Press	(New − Used)
0	$−15,000	$−21,000	$−6,000
1–25	−8,200	−7,000	+1,200
25	+750	+1,050	+300

EXAMPLE 8.2

A sole-source vendor can supply a new industrial park with large transformers suitable for underground utilities and vault-type installation. Type A has an initial cost of $70,000 and a life of 8 years. Type B has an initial cost of $95,000 and a life expectancy of 12 years. The annual operating cost for type A is expected to be $9000, while the AOC for type B is expected to be $7000. If the salvage values are $5000 and $10,000 for type A and type B, respectively, tabulate the incremental cash flow using their LCM for hand and spreadsheet solutions.

Solution by Hand

The LCM of 8 and 12 is 24 years. In the incremental cash flow tabulation for 24 years (Table 8–3), note that the reinvestment and salvage values are shown in years 8 and 16 for type A and in year 12 for type B.

Solution by Spreadsheet

Figure 8–1 shows the incremental cash flows for the LCM of 24 years. As in the hand tabulation, reinvestment is made in the last year of each intermediate life cycle. The incremental values in column D are the result of subtractions of column B from C.

    Note that the final row includes a summation check. The total incremental cash flow should agree in both the column D total and the subtraction C29 − B29. Also note that the incremental values change signs three times, indicating the possibility of multiple i^* values, per Descartes' rule of signs. This possible dilemma is discussed later in the chapter.

TABLE 8–3   Incremental Cash Flow Tabulation, Example 8.2

	Cash Flow		Incremental Cash Flow
Year	Type A	Type B	(B − A)
0	$ −70,000	$ −95,000	$−25,000
1–7	−9,000	−7,000	+2,000
8	⎰ −70,000 ⎨ −9,000 ⎱ +5,000	−7,000	+67,000
9–11	−9,000	−7,000	+2,000
12	−9,000	⎰ −95,000 ⎨ −7,000 ⎱ +10,000	−83,000
13–15	−9,000	−7,000	+2,000
16	⎰ −70,000 ⎨ −9,000 ⎱ +5,000	−7,000	+67,000
17–23	−9,000	−7,000	+2,000
24	⎰ −9,000 ⎱ +5,000	⎰ −7,000 ⎱ +10,000	+7,000
	$−411,000	$−338,000	$+73,000

Figure 8–1
Spreadsheet computation of incremental cash flows for unequal-life alternatives, Example 8.2.

	A	B	C	D	E	F	G
1				Incremental			
2		Net cash flows		cash flow			
3	Year	Type A	Type B	(B - A)			
4	0	-70,000	-95,000	-25,000			
5	1	-9,000	-7,000	2,000			
6	2	-9,000	-7,000	2,000			
7	3	-9,000	-7,000	2,000			
8	4	-9,000	-7,000	2,000			
9	5	9,000	-7,000	2,000			
10	6	-9,000	-7,000	2,000			
11	7	-9,000	-7,000	2,000	**Starting new life cycle for A**		
12	8	-74,000	-7,000	67,000	= initial cost + AOC + salvage		
13	9	-9,000	-7,000	2,000	= − 70,000 − 9,000 + 5,000		
14	10	-9,000	-7,000	2,000			
15	11	-9,000	-7,000	2,000			
16	12	-9,000	-92,000	-83,000			
17	13	-9,000	-7,000	2,000			
18	14	-9,000	-7,000	2,000			
19	15	-9,000	-7,000	2,000			
20	16	-74,000	-7,000	67,000			
21	17	-9,000	-7,000	2,000			
22	18	-9,000	-7,000	2,000			
23	19	-9,000	-7,000	2,000			
24	20	-9,000	7,000	2,000			
25	21	-9,000	-7,000	2,000			
26	22	-9,000	-7,000	2,000			
27	23	-9,000	-7,000	2,000			
28	24	-4,000	3,000	7,000	**Check on summations**		
29	Totals	-411,000	-338,000	73,000	Incremental column should		
30					equal difference of columns		
31							

Online
Presentation

8.3 Interpretation of Rate of Return
on the Extra Investment ● ● ○

The incremental cash flows in year 0 of Tables 8–2 and 8–3 reflect the *extra investment* or *cost* required if the alternative with the larger first cost is selected. This is important in an incremental ROR analysis in order to determine the ROR earned on the extra funds expended for the larger-investment alternative. If the incremental cash flows of the larger investment don't justify it, we must select the cheaper one. In Example 8.1 the new drill press requires an extra investment of $6000 (Table 8–2). If the new machine is purchased, there will be a "savings" of $1200 per year for 25 years, plus an extra $300 in year 25. The decision to buy the used or new machine can be made on the basis of the profitability of investing the extra $6000 in the new machine. If the equivalent worth of the savings is greater than the equivalent worth of the extra investment at the MARR, the extra investment should be made (i.e., the larger first-cost proposal should be accepted). On the other hand, if the extra investment is not justified by the savings, select the lower-investment proposal.

It is important to recognize that the rationale for making the selection decision is the same as if only *one alternative* were under consideration, that alternative being the one represented by the incremental cash flow series. When viewed in this manner, it is obvious that unless this investment yields a rate of return equal to or greater than the MARR, the extra investment should not be made. As further clarification of this extra investment rationale, consider the following: The rate of return attainable through the incremental cash flow is an alternative to investing at the MARR. Section 8.1 states that any excess funds not invested in the alternative are assumed to be invested at the MARR. The conclusion is clear:

ME alternative selection

If the rate of return available through the incremental cash flow equals or exceeds the MARR, the alternative associated with the extra investment should be selected.

Not only must the return on the extra investment meet or exceed the MARR, but also the return on the investment that is common to both alternatives must meet or exceed the MARR. Accordingly, prior to performing an incremental ROR analysis, it is advisable to determine the internal rate of return i^* for each alternative. This can be done only for *revenue alternatives,* because cost alternatives have only cost (negative) cash flows and no i^* can be determined. The guideline is as follows:

> **For multiple revenue alternatives,** calculate the internal rate of return i^* for each alternative, and eliminate all alternatives that have an $i^* <$ MARR. Compare the remaining alternatives incrementally.

As an illustration, if the MARR $= 15\%$ and two alternatives have i^* values of 12% and 21%, the 12% alternative can be eliminated from further consideration. With only two alternatives, it is obvious that the second one is selected. If both alternatives have $i^* <$ MARR, no alternative is justified and the do-nothing alternative is the best economically. When three or more alternatives are evaluated, it is usually worthwhile, but not required, to calculate i^* for each alternative for preliminary screening. Alternatives that cannot meet the MARR may be eliminated from further evaluation using this option. This option is especially useful when performing the analysis by spreadsheet. The IRR function applied to each alternative's cash flow estimates can quickly indicate unacceptable alternatives, as demonstrated in Section 8.6.

When **independent projects** are evaluated, there is no comparison on the extra investment. The ROR value is used to accept all projects with $i^* \geq$ MARR, assuming there is no budget limitation. For example, assume MARR $= 10\%$, and three independent projects are available with ROR values of

$$i_A^* = 12\% \qquad i_B^* = 9\% \qquad i_C^* = 23\%$$

Independent project selection

Projects A and C are selected, but B is not because $i_B^* <$ MARR.

8.4 Rate of Return Evaluation Using PW: Incremental and Breakeven ●●●

In this section we discuss the primary approach to making mutually exclusive alternative selections by the incremental ROR method. A PW-based relation is developed for the incremental cash flows and set equal to zero. Use hand solution or spreadsheet functions to find Δi_{B-A}^*, the internal ROR for the series. Placing Δ (delta) before i_{B-A}^* distinguishes it from the overall ROR values i_A^* and i_B^*. (Δi^* may replace Δi_{B-A}^* when only two alternatives are present.)

Since incremental ROR requires equal-service comparison, the LCM of lives must be used in the PW formulation. Because of the reinvestment requirement for PW analysis for different-life assets, the incremental cash flow series may contain several sign changes, indicating multiple Δi^* values. Though incorrect, this indication is usually neglected in actual practice. The correct approach is to follow one of the techniques of Section 7.5. This means that the single external ROR ($\Delta i'$ or $\Delta i''$) for the incremental cash flow series is determined.

These three elements—**incremental cash flow series, LCM, and multiple roots**—are the primary reasons that the ROR method is often applied incorrectly in engineering economy analyses of multiple alternatives. As stated earlier, it is always possible, and generally advisable, to use a PW or AW analysis *at an established MARR* in lieu of the ROR method when multiple rates are indicated.

> The complete procedure for hand or spreadsheet solution for an incremental ROR analysis for two alternatives is as follows:
>
> 1. Order the alternatives by initial investment or cost, starting with the smaller one, called A. The one with the larger initial investment is in the column labeled B in Table 8–1.
> 2. Develop the cash flow and incremental cash flow series using the LCM of years, assuming reinvestment in alternatives.
> 3. Draw an incremental cash flow diagram, if needed.

> **4.** Count the number of sign changes in the incremental cash flow series to determine if multiple rates of return may be present. If necessary, use Norstrom's criterion to determine if a single positive root exists.
> **5.** Set up the PW = 0 equation and determine Δi^*_{B-A}.
> **6.** Select the economically better alternative as follows:
>
> > **If Δi^*_{B-A} < MARR, select alternative A.**
> >
> > **If Δi^*_{B-A} ≥ MARR, the extra investment is justified; select alternative B.**
>
> If Δi^* is exactly equal to or very near the MARR, noneconomic considerations help in the selection of the "better" alternative.

ME alternative
selection

In step 5, if trial and error by hand is used, time may be saved if the Δi^*_{B-A} value is bracketed, rather than approximated by a point value using linear interpolation, provided that a single ROR value is not needed. For example, if the MARR is 15% per year and you have established that Δi^*_{B-A} is in the 15% to 20% range, an exact value is not necessary to accept B since you already know that $\Delta i^*_{B-A} \geq$ MARR.

The IRR function on a spreadsheet will normally determine one Δi^* value. Multiple guess values can be input to find multiple roots in the range -100% to ∞ for a nonconventional series, as illustrated in Example 7.4. If this is not the case, to be correct, the indication of multiple roots in step 4 requires that one of the techniques of Section 7.5 be applied to find an EROR.

EXAMPLE 8.3

As Ford Motor Company retools an old truck assembly plant in Michigan to produce a fuel-efficient economy model, the Ford Focus. Ford and its suppliers are seeking additional sources for light, long-life transmissions. Automatic transmission component manufacturers use highly finished dies for precision forming of internal gears and other moving parts. Two United States–based vendors make the required dies. Use the per unit estimates below and a MARR of 12% per year to select the more economical vendor bid. Show both hand and spreadsheet solutions.

	A	B
Initial cost, $	−8,000	−13,000
Annual costs, $ per year	−3,500	−1,600
Salvage value, $	0	2,000
Life, years	10	5

Solution by Hand

These are cost alternatives, since all cash flows are costs. Use the procedure described above to determine Δi^*_{B-A}.

1. Alternatives A and B are correctly ordered with the higher first-cost alternative in column 2 of Table 8–4.
2. The cash flows for the LCM of 10 years are tabulated.

TABLE 8–4	Incremental Cash Flow Tabulation, Example 8.3		
Year	**Cash Flow A** **(1)**	**Cash Flow B** **(2)**	**Incremental Cash Flow (3) = (2) − (1)**
0	$ −8,000	$−13,000	$ −5,000
1–5	−3,500	−1,600	+1,900
5	—	{ +2,000 −13,000	−11,000
6–10	−3,500	−1,600	+1,900
10	—	+2,000	+2,000
	$−43,000	$−38,000	$ +5,000

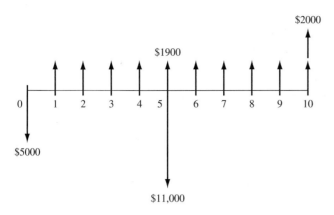

Figure 8–2
Diagram of incremental cash flows, Example 8.3.

3. The incremental cash flow diagram is shown in Figure 8–2.
4. There are three sign changes in the incremental cash flow series, indicating as many as three roots. There are also three sign changes in the cumulative incremental series, which starts negatively at $S_0 = \$-5000$ and continues to $S_{10} = \$+5000$, indicating that more than one positive root may exist.
5. The rate of return equation based on the PW of incremental cash flows is

$$0 = -5000 + 1900(P/A,\Delta i^*,10) - 11{,}000(P/F,\Delta i^*,5) + 2000(P/F,\Delta i^*,10) \qquad [8.2]$$

In order to resolve any multiple-root problem, we can assume that the investment rate i_i in the ROIC technique will equal the Δi^* found by trial and error. Solution of Equation [8.2] for the first root discovered results in Δi^* between 12% and 15%. By interpolation $\Delta i^* = 12.65\%$.
6. Since the rate of return of 12.65% on the extra investment is greater than the 12% MARR, the higher-cost vendor B is selected.

Comment

In step 4, the presence of up to three i^* values is indicated. The preceding analysis finds one of the roots at 12.65%. When we state that the incremental ROR is 12.65%, we assume that any positive net cash flows are reinvested at 12.65%. If this is not a reasonable assumption, the ROIC or modified ROR technique (Section 7.5) must be applied to find a different single $\Delta i'$ or $\Delta i''$ to compare with MARR = 12%.

The other two roots are very large positive and negative numbers, as the IRR function of Excel reveals. So they are not useful to the analysis.

Solution by Spreadsheet

Steps 1 through 4 are the same as above.

5. Figure 8–3 includes the same incremental net cash flows from Table 8–4 calculated in column D. Cell D15 displays the Δi^* value of 12.65% using the IRR function.

	A	B	C	D	E
1					
2		Vendor A	Vendor B	Incremental	
3	Year	cash flow, $	cash flow, $	cash flow, $	
4	0	-8,000	-13,000	-5,000	
5	1	-3,500	-1,600	1,900	
6	2	-3,500	-1,600	1,900	
7	3	-3,500	-1,600	1,900	
8	4	-3,500	-1,600	1,900	
9	5	-3,500	-12,600	-9,100	
10	6	-3,500	-1,600	1,900	
11	7	-3,500	-1,600	1,900	
12	8	-3,500	-1,600	1,900	
13	9	-3,500	-1,600	1,900	
14	10	-3,500	400	3,900	
15	Incremental i*			**12.65%**	
16					
17		Check: PW @ 12%		$137.67	
18					
19			= NPV(12%,D5:D14) + D4		
20					

Figure 8–3
Spreadsheet solution using LCM of lives and IRR function, Example 8.3.

6. Since the rate of return on the extra investment is greater than the 12% MARR, the higher-cost vendor B is selected.

Comment

Once the spreadsheet is set up, there are a wide variety of analyses that can be performed. For example, row 17 uses the NPV function to verify that the present worth is positive at MARR=12%. Charts such as PW versus Δi and PW versus i help graphically interpret the situation.

Online Presentation

Breakeven ROR

The rate of return determined for the incremental cash flow series or the actual cash flows can be interpreted as a **breakeven rate of return** value.

> The **breakeven rate of return** is the incremental i^* value, Δi^*, at which the PW (or AW) value of the *incremental cash flows* is exactly zero. Equivalently, the breakeven ROR is the i value, i^*, at which the PW (or AW) values of two alternatives' actual cash flows are exactly equal to each other.

If the incremental cash flow ROR (Δi^*) is greater than the MARR, the larger-investment alternative is selected. For example, if the PW versus Δi graph for the incremental cash flows in Table 8–4 (and spreadsheet Figure 8–3) is plotted for various interest rates, the graph shown in Figure 8–4 is obtained. It shows the Δi^* breakeven at 12.65%. The conclusions are that

- For MARR < 12.65%, the extra investment for B is justified.
- For MARR > 12.65%, the opposite is true—the extra investment in B should not be made, and vendor A is selected.
- If MARR is exactly 12.65%, the alternatives are equally attractive.

Figure 8–5, which is a breakeven graph of PW versus i for the cash flows (not incremental) of each alternative in Example 8.3, provides the same results. Since all net cash flows are negative

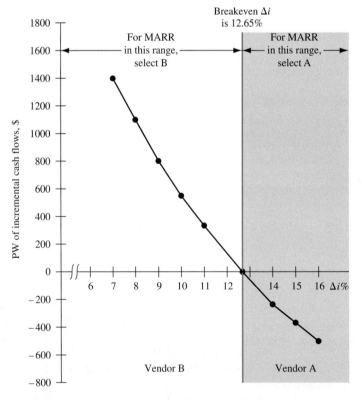

Figure 8–4
Plot of present worth of incremental cash flows for Example 8.3 at various Δi values.

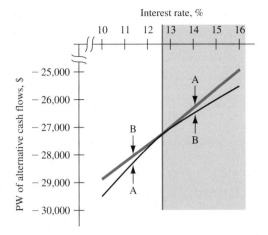

Figure 8–5
Breakeven graph of
Example 8.3 cash flows
(not incremental).

(cost alternatives), the PW values are negative. Now, the same conclusions are reached using the following logic:

- If MARR < 12.65%, select B since its PW of cost cash flows is smaller (numerically larger).
- If MARR > 12.65%, select A since its PW of costs is smaller.
- If MARR is exactly 12.65%, either alternative is equally attractive.

Example 8.4 illustrates incremental ROR evaluation and breakeven rate of return graphs for revenue alternatives. More of breakeven analysis is covered in Chapter 13.

EXAMPLE 8.4

New filtration systems for commercial airliners are available that use an electric field to remove up to 99.9% of infectious diseases and pollutants from aircraft air. This is vitally important, as many of the flu germs, viruses, and other contagious diseases are transmitted through the systems that recirculate aircraft air many times per hour. Investments in the new filtration equipment can cost from $100,000 to $150,000 per aircraft, but savings in fuel, customer complaints, legal actions, etc., can also be sizable. Use the estimates below (in $100 units) from two suppliers provided to an international carrier to do the following, using a spreadsheet and an MARR of 15% per year.

- Plot two graphs: PW versus i values for both alternatives' cash flows and PW versus Δi values for incremental cash flows.
- Estimate the breakeven ROR values from both graphs, and use this estimate to select one alternative.

	Air Cleanser (Filter 1)	Purely Heaven (Filter 2)
Initial cost per aircraft, $	−1000	−1500
Estimated savings, $ per year	375	700 in year 1, decreasing by 100 per year thereafter
Estimated life, years	5	5

Solution by Spreadsheet

Refer to Figure 8–6 as the solution is explained. For information, row 10 shows the ROR values calculated using the IRR function for filter 1 and filter 2 cash flows and the incremental cash flow series (filter 2 − filter 1).

The cash flow sign tests for each filter indicate no multiple rates. The incremental cash flow sign test does not indicate the presence of a unique positive root; however, the second rate is an extremely large and useless value. The PW values of filter 1 and filter 2 cash flows are plotted on the right side for i values ranging from 0% to 60%. Since the PW curves cross each other at approximately $i^* = 17\%$, the rate does exceed the MARR of 15%. The higher-cost filter 2 (Purely Heaven) is selected.

The PW curve for the incremental cash flows series (column G) is plotted at the bottom left of Figure 8–6. As expected, the curve crosses the PW = 0 line at approximately 17%, indicating the same economic conclusion of filter 2.

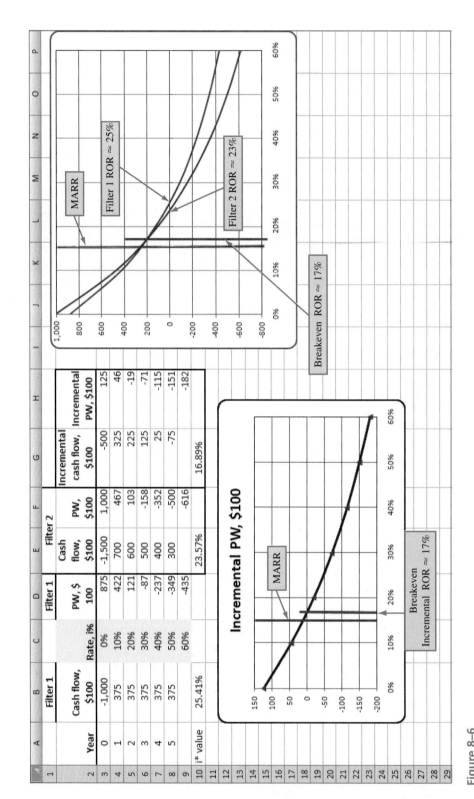

The data table shown in the figure:

	A	B	C	D	E	F	G	H
1		Filter 1		Filter 1	Filter 2			
2	Year	Cash flow, $100	Rate, i%	PW, $100	Cash flow, $100	PW, $100	Incremental cash flow, $100	Incremental PW, $100
3	0	-1,000	0%	875	-1,500	1,000	-500	125
4	1	375	10%	422	700	467	325	46
5	2	375	20%	121	600	103	225	-19
6	3	375	30%	-87	500	-158	125	-71
7	4	375	40%	-237	400	-352	25	-115
8	5	375	50%	-349	300	-500	-75	-151
9			60%	-435		-616		-182
10	i* value	25.41%				23.57%	16.89%	

Chart annotations: MARR; Filter 1 ROR ≈ 25%; Filter 2 ROR ≈ 23%; Breakeven ROR ≈ 17%

Incremental PW, $100 chart annotations: MARR; Breakeven Incremental ROR ≈ 17%

Figure 8–6
PW versus *i* graph and PW versus incremental *i* graph, Example 8.4.

Figure 8–6 provides an excellent opportunity to see why the ROR method can result in selecting the wrong alternative when only $i*$ values are used to select between two alternatives. This is sometimes called the **ranking inconsistency problem** of the ROR method. *The inconsistency occurs when the MARR is set less than the breakeven rate between two revenue alternatives.* Since the MARR is established based on conditions of the economy and market, MARR is established external to any particular alternative evaluation. In Figure 8–6 the incremental breakeven rate is 16.89%, and the MARR is 15%. The MARR is lower than breakeven; therefore, the *incremental ROR analysis* results in correctly selecting filter 2. But if only the $i*$ values were used, filter 1 would be wrongly chosen, because its $i*$ exceeds that of filter 2 (25.41% > 23.57%). This error occurs because the rate of return method assumes reinvestment at the alternative's ROR value, while PW and AW analyses use the MARR as the reinvestment rate. The conclusion is simple:

If the ROR method is used to evaluate two or more alternatives, use the **incremental cash flows and $\Delta i*$** to make the decision between alternatives.

8.5 Rate of Return Evaluation Using AW ●●◦

Comparing alternatives by the ROR method (correctly performed) always leads to the same selection as PW and AW analyses, whether the ROR is determined using a PW-based or an AW-based relation. However, for the AW-based technique, there are two equivalent ways to perform the evaluation: (1) using the *incremental cash flows* over the LCM of alternative lives, just as for the PW-based relation (Section 8.4), or (2) finding the AW for each alternative's *actual cash flows* and setting the difference of the two equal to zero to find the $\Delta i*$ value. There is no difference between the two approaches if the alternative lives are equal. Both methods are summarized here.

Equal-service requirement

Since the ROR method requires comparison for equal service, the **incremental cash flows must be evaluated over the LCM of lives.** There may be no real computational advantage to using AW, as was found in Chapter 6. The same six-step procedure of the previous section (for PW-based calculation) is used, except in step 5 the AW-based relation is developed.

The second AW-based method mentioned above takes advantage of the AW technique's assumption that the equivalent AW value is the same for each year of the first and all succeeding life cycles. Whether the lives are equal or unequal, set up the *AW relation for the cash flows of each alternative,* form the relation below, and solve for $i*$.

$$0 = AW_B - AW_A \qquad\qquad [8.3]$$

For both methods, all equivalent values are on an AW basis, so the $i*$ that results from Equation [8.3] is the same as the $\Delta i*$ found using the first approach. Example 8.5 illustrates ROR analysis using AW-based relations for unequal lives.

EXAMPLE 8.5

Compare the alternatives of vendors A and B for Ford in Example 8.3, using an AW-based incremental ROR method and the same MARR of 12% per year.

Solution

For reference, the PW-based ROR relation, Equation [8.2], for the incremental cash flow in Example 8.3 shows that vendor B should be selected with $\Delta i* = 12.65\%$.

For the AW relation, there are two equivalent solution approaches. Write an AW-based relation on the *incremental* cash flow series over the *LCM of 10 years,* or write Equation [8.3] for the *two actual* cash flow series over *one life cycle* of each alternative.

For the incremental method, the AW equation is

$0 = -5000(A/P,\Delta i*,10) - 11{,}000(P/F,\Delta i*,5)(A/P,\Delta i*,10) + 2000(A/F,\Delta i*,10) + 1900$

It is easy to enter the incremental cash flows onto a spreadsheet, as in Figure 8–3, column D, and use the = IRR(D4:D14) function to display $\Delta i^* = 12.65\%$.

For the second method, the ROR is found using the actual cash flows and the respective lives of 10 years for A and 5 years for B.

$$AW_A = -8000(A/P,i,10) - 3500$$
$$AW_B = -13{,}000(A/P,i,5) + 2000(A/F,i,5) - 1600$$

Now develop $0 = AW_B - AW_A$.

$$0 = -13{,}000(A/P,i^*,5) + 2000(A/F,i^*,5) + 8000(A/P,i^*,10) + 1900$$

Solution again yields $i^* = 12.65\%$.

Comment

It is very important to remember that when an incremental ROR analysis using an AW-based equation is made on the *incremental cash flows,* the LCM must be used.

Online
Presentation

8.6 Incremental ROR Analysis of Multiple Alternatives ●●●

This section treats selection from multiple alternatives that are mutually exclusive, using the incremental ROR method. Acceptance of one alternative automatically precludes acceptance of any others. The analysis is based upon PW (or AW) relations for incremental cash flows between two alternatives at a time.

When the incremental ROR method is applied, the entire investment must return at least the MARR. When the i^* values on several alternatives exceed the MARR, incremental ROR evaluation is required. (For revenue alternatives, if not even one $i^* \geq$ MARR, the do-nothing alternative is selected.) For all alternatives (revenue or cost), the incremental investment must be separately justified. If the return on the extra investment equals or exceeds the MARR, then the extra investment should be made in order to maximize the total return on the money available, as discussed in Section 8.1.

For ROR analysis of multiple, mutually exclusive alternatives, the following criteria are used.

ME alternative
selection

> Select the one alternative
>
> That requires the **largest investment**, and
>
> Indicates that the **extra investment** over another acceptable alternative **is justified.**

An important rule to apply when evaluating multiple alternatives by the incremental ROR method is that *an alternative should never be compared with one for which the incremental investment is not justified.*

The incremental ROR evaluation procedure for multiple, equal-life alternatives is summarized below. Step 2 applies only to revenue alternatives, because the first alternative is compared to DN only when revenue cash flows are estimated. The terms **defender** and **challenger** are dynamic in that they refer, respectively, to the alternative that is currently selected (the defender) and the one that is challenging it for acceptance based on Δi^*. In every pairwise evaluation, there is one of each. The steps for solution by hand or by spreadsheet are as follows:

1. Order the alternatives from **smallest to largest initial investment.** Record the annual cash flow estimates for each equal-life alternative.
2. *Revenue alternatives only:* Calculate i^* for the first alternative. In effect, this makes DN the defender and the first alternative the challenger. If $i^* <$ MARR, eliminate the alternative and go to the next one. Repeat this until $i^* \geq$ MARR for the first time, and define that alternative as the defender. The next alternative is now the challenger. Go to step 3. (*Note:* This is where solution by spreadsheet can be a quick assist. Calculate the i^* for all alternatives first using the IRR function, and select as the defender the first one for which $i^* \geq$ MARR. Label it the defender and go to step 3.)
3. Determine the incremental cash flow between the challenger and defender, using the relation

    Incremental cash flow = challenger cash flow − defender cash flow

    Set up the ROR relation.

4. Calculate Δi^* for the incremental cash flow series using a PW- or AW-based equation. (PW is most commonly used.)
5. If $\Delta i^* \geq$ MARR, the challenger becomes the defender and the previous defender is eliminated. Conversely, if $\Delta i^* <$ MARR, the challenger is removed, and the defender remains against the next challenger.
6. Repeat steps 3 to 5 until only one alternative remains. It is the selected one.

Note that only two alternatives are compared at any one time. It is vital that the correct alternatives be compared, or the wrong alternative may be selected.

EXAMPLE 8.6

Caterpillar Corporation wants to build a spare parts storage facility in the Phoenix, Arizona, vicinity. A plant engineer has identified four different location options. The initial cost of earthwork and prefab building and the annual net cash flow estimates are detailed in Table 8–5. The annual net cash flow series vary due to differences in maintenance, labor costs, transportation charges, etc. If the MARR is 10%, use incremental ROR analysis to select the one economically best location.

TABLE 8–5	Estimates for Four Alternative Building Locations, Example 8.6			
	A	B	C	D
Initial cost, $	−200,000	−275,000	−190,000	−350,000
Annual cash flow, $ per year	+22,000	+35,000	+19,500	+42,000
Life, years	30	30	30	30

Solution

All sites have a 30-year life, and they are revenue alternatives. The procedure outlined above is applied.

1. The alternatives are ordered by increasing initial cost in Table 8–6.
2. Compare location C with the do-nothing alternative. The ROR relation includes only the P/A factor.

$$0 = -190{,}000 + 19{,}500(P/A{,}i^*{,}30)$$

Table 8–6, column 1, presents the calculated $(P/A{,}\Delta i^*{,}30)$ factor value of 9.7436 and $\Delta i_c^* = 9.63\%$. Since 9.63% < 10%, location C is eliminated. Now the comparison is A to DN, and column 2 shows that $\Delta i_A^* = 10.49\%$. This eliminates the do-nothing alternative; the defender is now A and the challenger is B.

TABLE 8–6	Computation of Incremental Rate of Return for Four Alternatives, Example 8.6			
	C (1)	A (2)	B (3)	D (4)
Initial cost, $	−190,000	−200,000	−275,000	−350,000
Cash flow, $ per year	+19,500	+22,000	+35,000	+42,000
Alternatives compared	C to DN	A to DN	B to A	D to B
Incremental cost, $	−190,000	−200,000	−75,000	−75,000
Incremental cash flow, $	+19,500	+22,000	+13,000	+7,000
Calculated $(P/A{,}\Delta i^*{,}30)$	9.7436	9.0909	5.7692	10.7143
$\Delta i^*{,}\%$	9.63	10.49	17.28	8.55
Increment justified?	No	Yes	Yes	No
Alternative selected	DN	A	B	B

3. The incremental cash flow series, column 3, and Δi^* for *B-to-A comparison* are determined from

$$0 = -275{,}000 - (-200{,}000) + (35{,}000 - 22{,}000)(P/A,\Delta i^*,30)$$
$$= -75{,}000 + 13{,}000(P/A,\Delta i^*,30)$$

4. From the interest tables, look up the P/A factor at the MARR, which is $(P/A,10\%,30) =$ 9.4269. Now, any P/A value greater than 9.4269 indicates that the Δi^* will be less than 10% and is unacceptable. The P/A factor is 5.7692, so B is acceptable. For reference purposes, $\Delta i^* = 17.28\%$.
5. Alternative B is justified incrementally (new defender), thereby eliminating A.
6. Comparison D-to-B (steps 3 and 4) results in the PW relation $0 = -75{,}000 +$ $7000(P/A,\Delta i^*,30)$ and a P/A value of 10.7143 ($\Delta i^* = 8.55\%$). Location D is eliminated, and **only alternative B remains; it is selected.**

Comment

An alternative must *always* be incrementally compared with an acceptable alternative, and the do-nothing alternative can end up being the only acceptable one. Since C was not justified in this example, location A was not compared with C. Thus, *if* the B-to-A comparison had not indicated that B was incrementally justified, then the D-to-A comparison would be correct instead of D-to-B.

To demonstrate how important it is to apply the ROR method correctly, consider the following. If the i^* of each alternative is computed initially, the results by ordered alternatives are

Location	C	A	B	D
i^*, %	9.63	10.49	12.35	11.56

Now apply *only* the first criterion stated earlier; that is, make the largest investment that has a MARR of 10% or more. Location D is selected. But, as shown above, this is the wrong selection, because the extra investment of $75,000 over location B will not earn the MARR. In fact, it will earn only 8.55%. This is another example of the *ranking inconsistency problem of the ROR method* mentioned in Section 8.4.

For cost alternatives, the incremental cash flow is the difference between costs for two alternatives. There is no do-nothing alternative and no step 2 in the solution procedure. Therefore, the lowest-investment alternative is the initial defender against the next-lowest investment (challenger). This procedure is illustrated in Example 8.7 using a spreadsheet solution.

EXAMPLE 8.7

The complete failure of an offshore platform and the resulting spillage of up to 800,000 to 1,000,000 gallons per day into the Gulf of Mexico in the spring of 2010 have made major oil producers and transporters very conscious of the harm done to people's livelihood and all forms of aquatic life by spills of this magnitude. To address the specific danger to birds that are shoreline feeders and dwellers, environmental engineers from several international petroleum corporations and transport companies—Exxon-Mobil, BP, Shell, and some transporters for OPEC producers—have developed a plan to strategically locate throughout the world newly developed equipment that is substantially more effective than manual procedures in cleaning crude oil residue from bird feathers. The Sierra Club, Greenpeace, and other international environmental interest groups are in favor of the initiative. Alternative machines from manufacturers in Asia, America, Europe, and Africa are available with the cost estimates in Table 8–7. Annual cost estimates are expected to be high to ensure readiness at any time. The company representatives have agreed to use the average of the corporate MARR values, which results in MARR = 13.5%. Use a spreadsheet and incremental ROR analysis to determine which manufacturer offers the best economic choice.

TABLE 8–7	Costs for Four Alternative Machines, Example 8.7			
	Machine 1	**Machine 2**	**Machine 3**	**Machine 4**
First cost, $	−5,000	−6,500	−10,000	−15,000
Annual operating cost, $	−3,500	−3,200	−3,000	−1,400
Salvage value, $	+500	+900	+700	+1,000
Life, years	8	8	8	8

Solution by Spreadsheet

Follow the procedure for incremental ROR analysis. The spreadsheet in Figure 8–7 contains the complete solution.

1. The alternatives are already ordered by increasing first costs.
2. These are cost alternatives, so there is no comparison to DN, since i^* values cannot be calculated.
3. Machine 2 is the first challenger to machine 1; the incremental cash flows for the 2-to-1 comparison are in column D.
4. The 2-to-1 comparison results in $\Delta i^* = 14.57\%$ by applying the IRR function.
5. This return exceeds MARR = 13.5%, so machine 2 is the new defender (cell D17).

The comparison continues for 3-to-2 in column E, where the return is negative at $\Delta i^* = -18.77\%$; machine 2 is retained as the defender. Finally the 4-to-2 comparison has an incremental ROR of 13.60%, which is slightly larger than MARR = 13.5%. The conclusion is to purchase machine 4 because the extra investment is (marginally) justified.

	A	B	C	D	E	F
1		Year	Machine 1	Machine 2	Machine 3	Machine 4
2	Initial investment, $		-5,000	-6,500	-10,000	-15,000
3	Annual cost, $ per year		-3,500	-3,200	-3,000	-1,400
4	Salvage value, $		500	900	700	1,000
5	Incremental comparison			2 to 1	3 to 2	4 to 2
6	Incremental investment, $	0		-1,500	-3,500	-8,500
7	Incremental cash flow, $ per year	1		300	200	1,800
8		2		300	200	1,800
9		3		300	200	1,800
10		4		300	200	1,800
11		5		300	200	1,800
12		6		300	200	1,800
13		7		300	200	1,800
14		8		700	0	1,900
15	Incremental i* (Δi*)			14.57%	-18.77%	13.60%
16	Increment justified?			Yes	No	Yes
17	Alternative selected			2	2	4
18			= IRR(D6:D14)			
19						
20						

Figure 8–7
Spreadsheet solution to select from multiple cost alternatives, Example 8.7.

Comment

As mentioned earlier, it is not possible to generate a PW versus i graph for each cost alternative because all cash flows are negative. However, it is possible to generate PW versus Δi graphs for the incremental series in the same fashion as we have done previously. The curves will cross the PW = 0 line at the Δi^* values determined by the IRR functions.

Selection from multiple, mutually exclusive alternatives with **unequal lives** using Δi^* values requires that the incremental cash flows be evaluated over the LCM of the two alternatives being compared. This is another application of the principle of equal-service comparison. The spreadsheet application in the next section illustrates the computations.

It is always possible to rely on PW or AW analysis of the incremental cash flows at the MARR to make the selection. In other words, don't find Δi^* for each pairwise comparison; find PW or

AW at the MARR instead. However, it is still necessary to make the comparison over the LCM number of years for an incremental analysis to be performed correctly.

8.7 All-in-One Spreadsheet Analysis (Optional) ●●○

For professors and students who like to pack a spreadsheet, Example 8.8 combines many of the economic analysis techniques we have learned so far—(internal) ROR analysis, incremental ROR analysis, PW analysis, and AW analysis. Now that the IRR, NPV, and PV functions are mastered, it is possible to perform a wide variety of evaluations for multiple alternatives on a single spreadsheet. No cell tags are provided in this example. A nonconventional cash flow series for which multiple ROR values may be found, and selection from both mutually exclusive alternatives and independent projects, are included in this example.

EXAMPLE 8.8

In-flight texting, phone, and Internet connections provided at airline passenger seats are an expected service by many customers. Singapore Airlines knows it will have to replace 15,000 to 24,000 units in the next few years on its Boeing 757, 777, and A300 aircraft. Four optional data handling features that build upon one another are available from the manufacturer, but at an added cost per unit. Besides costing more, the higher-end options (e.g., satellite-based plug-in video service) are estimated to have longer lives before the next replacement is forced by new, advanced features expected by flyers. All four options are expected to boost annual revenues by varying amounts. Figure 8–8 spreadsheet rows 2 through 6 include all the estimates for the four options.

(a) Using MARR = 15%, perform ROR, PW, and AW evaluations to select the one level of options that is the most promising economically.

(b) If more than one level of options can be selected, consider the four that are described as independent projects. If no budget limitations are considered at this time, which options are acceptable if the MARR is increased to 20% when more than one option may be implemented?

	A	B	C	D	E	F	G	H
1	MARR =	15%						
2	**Alternative**		**A**	**B**	**C**		**D**	
3	Initial cost, $		-6,000	-7,000	-9,000		-17,000	
4	Annual cash flow, $ per year		2,000	3,000	3,000		3,500	
5	Salvage value, $		0	200	300		1,000	
6	Life, years	Year	3	4	6		12	
7	**Incr. ROR comparison**		**Actual CF**	**Actual CF**	**Actual CF**	**C to B**	**Actual CF**	**D to C**
8	Incremental investment, $	0	-6,000	-7,000	-9,000	-2,000	-17,000	-8,000
9	Incremental cash flow	1	2,000	3,000	3,000	0	3,500	500
10	over the LCM, $ per year	2	2,000	3,000	3,000	0	3,500	500
11		3	2,000	3,000	3,000	0	3,500	500
12		4		3,200	3,000	6,800	3,500	500
13		5			3,000	0	3,500	500
14		6			3,300	-8,700	3,500	9,200
15		7				0	3,500	500
16		8				6,800	3,500	500
17		9				0	3,500	500
18		10				0	3,500	500
19		11				0	3,500	500
20		12				100	4,500	1,200
21	Overall i*		0.00%	26.32%	24.68%		17.87%	
22	Retain or eliminate?		Eliminate	Retain	Retain		Retain	
23	Incremental i* ($\Delta i*$)					19.42%		11.23%
24	Increment justified?					Yes		No
25	**Alternative selected**					**C**		**C**
26	AW at MARR		-628	588	656		398	
27	PW at MARR		-3,403	3,188	3,557		2,159	
28	Alternative selected		No	No	Yes		No	

Figure 8–8
Spreadsheet analysis using ROR, PW, and AW methods for unequal-life, revenue alternatives, Example 8.8.

Solution by Spreadsheet

(*a*) The spreadsheet (Figure 8–8) is divided into six sections:

Section 1 (rows 1, 2): MARR value and the alternative names (A through D) are in increasing order of initial cost.

Section 2 (rows 3 to 6): Per-unit net cash flow estimates for each alternative. These are revenue alternatives with unequal lives.

Section 3 (rows 7 to 20): Actual and incremental cash flows are displayed here.

Section 4 (rows 21, 22): Because these are all revenue alternatives, i^* values are determined by the IRR function. If an alternative passes the MARR test ($i^* > 15\%$), it is retained and a column is added to the right of its actual cash flows so the incremental cash flows can be determined. Columns F and H were inserted to make space for the incremental evaluations. Alternative A does not pass the i^* test.

Section 5 (rows 23 to 25): The IRR functions display the Δi^* values in columns F and H. Comparison of C to B takes place over the LCM of 12 years. Since $\Delta i^*_{C-B} = 19.42\% > 15\%$, eliminate B; alternative C is the new defender and D is the next challenger. The final comparison of D to C over 12 years results in $\Delta i^*_{D-C} = 11.23\% < 15\%$, so D is eliminated. Alternative C is the chosen one.

Section 6 (rows 26 to 28): These include the AW and PW analyses. The AW value over the life of each alternative is calculated using the PMT function at the MARR with an embedded NPV function. Also, the PW value is determined from the AW value for 12 years using the PV function. For both measures, alternative C has the numerically largest value, as expected.

Conclusion: All methods result in the same, correct choice of alternative C.

(*b*) Since each option is independent of the others, and there is no budget limitation at this time, each i^* value in row 21 of Figure 8–8 is compared to MARR = 20%. This is a comparison of each option with the do-nothing alternative. Of the four, options B and C have $i^* > 20\%$. They are acceptable; the other two are not.

Comment

In part (*a*), we should have applied the two multiple-root sign tests to the incremental cash flow series for the C-to-B comparison. The series itself has three sign changes, and the cumulative cash flow series starts negatively and also has three sign changes. Therefore, up to three real-number roots may exist. The IRR function is applied in cell F23 to obtain $\Delta i^*_{C-B} = 19.42\%$ without using a supplemental (Section 7.5) procedure. This means that the investment assumption of 19.42% for positive cash flows is a reasonable one. If the MARR = 15%, or some other earning rate were more appropriate, the ROIC procedure could be applied to determine a single rate, which would be different from 19.42%. Depending upon the investment rate chosen, alternative C may or may not be incrementally justified against B. Here, the assumption is made that the Δi^* value is reasonable, so C is justified.

CHAPTER SUMMARY

Just as present worth and annual worth methods find the best alternative from among several, incremental rate of return calculations can be used for the same purpose. In using the ROR technique, it is necessary to consider the **incremental cash flows** when selecting between mutually exclusive alternatives. The incremental investment evaluation is conducted between only two alternatives at a time, beginning with the lowest initial investment alternative. Once an alternative has been eliminated, it is not considered further.

Rate of return values have a natural appeal to management, but the ROR analysis is often more difficult to set up and complete than the PW or AW analysis using an established MARR. Care must be taken to perform a ROR analysis correctly on the incremental cash flows; otherwise it may give incorrect results.

If there is no budget limitation when independent projects are evaluated, the ROR value of each project is compared to the MARR. Any number, or none, of the projects can be accepted.

PROBLEMS

Understanding Incremental ROR

8.1 In a tabulation of cash flow, the column entitled "Rate of return on the incremental cash flow" represents the rate of return on what?

8.2 If the rate of return on the incremental cash flow between two alternatives is less than the minimum attractive rate of return, which alternative should be selected, if any?

8.3 An engineer is comparing three projects by the incremental ROR method. There are revenue and cost cash flow estimates. He used the IRR function of Excel to determine the ROR values for each project. The first one is 3.5% above the MARR, the second is 1.2% below the MARR, and the third is 2.4% above the MARR. Which alternatives, if any, must he include in the incremental ROR analysis?

8.4 The rates of return on alternatives X and Y are 15% and 12%, respectively. Alternative Y requires a larger investment than alternative X.
  (a) What is known about the rate of return on the increment of investment between the two alternatives?
  (b) If the MARR is 12%, which alternative should be selected and why?

8.5 Victoria is comparing two mutually exclusive alternatives, A and B. The overall ROR on alternative A is greater than the MARR, and the overall ROR on alternative B, which requires the larger investment, is exactly equal to the MARR.
  (a) What is known about the ROR on the increment between A and B?
  (b) Which alternative should be selected?

8.6 A food processing company is considering two types of moisture analyzers. Only one can be selected. The company expects an infrared model to yield a rate of return of 27% per year. A more expensive microwave model will yield a rate of return of 22% per year. If the company's MARR is 19% per year, can you determine which model should be purchased solely on the basis of the overall rate of return information provided? Why or why not?

8.7 If $80,000 is invested at 30% and another $50,000 is invested at 20% per year, what is the overall rate of return on the entire $130,000?

8.8 A total of $100,000 was invested in two different projects identified as Z1 and Z2. If the overall rate

of return on the $100,000 was 30% and the rate of return on the $30,000 invested in Z1 was 15%, what was the rate of return on Z2?

8.9 Tuggle, Inc., which manufactures rigid shaft couplings, has $600,000 to invest. The company is considering three different projects that will yield the following rates of return.

Project X	$i_X = 24\%$
Project Y	$i_Y = 18\%$
Project Z	$i_Z = 30\%$

The initial investment required for each project is $100,000, $300,000, and $200,000, respectively. If Tuggle's MARR is 15% per year and it invests in all three projects, what rate of return will the company make?

8.10 Two options are available for setting up a wireless meter scanner and controller. A simple setup is good for 2 years and has an initial cost of $12,000, no salvage value, and an operating cost of $27,000 per year. A more permanent system has a higher first cost of $73,000, but it has an estimated life of 6 years and a salvage value of $15,000. It costs only $14,000 per year to operate and maintain. If the two options are compared using an incremental rate of return, what are the incremental cash flows in (a) year 0 and (b) year 2?

8.11 Prepare a tabulation of incremental cash flows for the two machine alternatives below.

	Machine X	Machine Y
First cost, $	−35,000	−90,000
Annual operating cost, $ per year	−31,600	−19,400
Salvage value, $	0	8,000
Life, years	2	4

8.12 For the alternatives shown, determine the sum of the incremental cash flows for Q − P.

	Alternative P	Alternative Q
First cost, $	−50,000	−85,000
Annual operating cost, $ per year	−8,600	−2,000
Annual revenue, $ per year	22,000	45,000
Salvage value, $	3,000	8,000
Life, years	3	6

8.13 The tabulation of the incremental cash flows between alternatives A and B is shown on the next page. Alternative A has a 3-year life and alternative B a 6-year life. If neither alternative has a salvage value, what is (a) the first cost of alternative A and (b) the first cost of alternative B?

Year	Incremental Cash Flow (B − A), $
0	−20,000
1	5,000
2	5,000
3	12,000
4	5,000
5	5,000
6	5,000

8.14 Standby power for pumps at water distribution booster stations can be provided by either gasoline- or diesel-powered engines. The costs for the gasoline engines are as follows:

	Gasoline
First cost, $	−150,000
Annual M&O, $ per year	−41,000
Salvage value, $	23,000
Life, years	15

The incremental PW cash flow equation associated with (diesel − gasoline) is

$$0 = -40,000 + 11,000(P/A,i,15) + 16,000(P/F,i,15)$$

Determine the following:
(a) First cost of the diesel engines
(b) Annual M&O cost of the diesel engines
(c) Salvage value of the diesel engines

8.15 Several high-value parts for NASA's reusable space exploration vehicle can be either anodized or powder-coated. Some of the costs for each process are shown below:

	Anodize	Powder Coat
First cost, $	?	−65,000
Annual cost, $ per year	−21,000	?
Resale value, $	?	6,000
Life, years	3	3

The incremental AW cash flow equation associated with (powder coat − anodize) is

$$0 = -14,000(A/P,i,3) + 5000 + 2000(A/F,i,3)$$

What is (a) the first cost for anodizing, (b) the annual cost for powder coating, and (c) the resale (salvage) value of the anodized parts?

Incremental ROR Comparison (Two Alternatives)

8.16 Specialty Gases & Chemicals manufactures nitrogen trifluoride, a highly specialized gas used as an industrial cleansing agent for flat panels installed in laptop computers, televisions, and desktop monitors. The incremental cash flow associated with two alternatives for chemical storage and handling systems

(identified as P3 and X3) has been calculated in $1000 units. Determine (a) the rate of return on the incremental cash flows and (b) which one should be selected if the company's MARR is 25% per year. Alternative X3 requires the larger initial investment.

Year	Incremental Cash Flow (X3 − P3), $1000
0	$−4600
1–9	1100
10	2000

8.17 As groundwater wells age, they sometimes begin to pump sand (and they become known as "sanders"), and this can cause damage to downstream desalting equipment. This situation can be dealt with by drilling a new well at a cost of $1,000,000 or by installing a tank and self-cleaning screen ahead of the desalting equipment. The tank and screen will cost $230,000 to install and $61,000 per year to operate and maintain. A new well will have a pump that is more efficient than the old one, and it will require almost no maintenance, so its operating cost will be only $18,000 per year. If the salvage values are estimated at 10% of the first cost, use a present worth relation to (a) calculate the incremental rate of return and (b) determine which alternative is better at a MARR of 6% per year over a 20-year study period.

8.18 Konica Minolta plans to sell a copier that prints documents on both sides simultaneously, cutting in half the time it takes to complete big commercial jobs. The costs associated with producing chemically-treated vinyl rollers and fiber-impregnated rubber rollers are shown below. Determine which of the two types should be selected by calculating the rate of return on the incremental investment. Assume the company's MARR is 21% per year. (Solve by hand and/or spreadsheet, as instructed.)

	Treated	Impregnated
First cost, $	−50,000	−95,000
Annual cost, $ per year	−100,000	−85,000
Salvage value, $	5,000	11,000
Life, years	3	6

8.19 The Texas Department of Transportation (TxDOT) is considering two designs for crash barriers along a reconstructed portion of I-10. Design 2B will cost $3 million to install and $135,000 per year to maintain. Design 4R will cost $3.7 million to install and $70,000 per year to maintain. Calculate the rate of return and determine which design is preferred if TxDOT uses a MARR of 6% per year and a 20-year project period.

8.20 Chem-Tex Chemical is considering two additives for improving the dry-weather stability of its low-cost acrylic paint. Additive A has a first cost of $110,000 and an annual operating cost of $60,000. Additive B has a first cost of $175,000 and an annual operating cost of $35,000. If the company uses a 3-year recovery period for paint products and a MARR of 20% per year, which process is economically favored? Use an incremental ROR analysis.

8.21 The manager of Liquid Sleeve, Inc., a company that makes a sealing solution for machine shaft surfaces that have been compromised by abrasion, high pressures, or inadequate lubrication, is considering adding metal-based nanoparticles of either type Al or Fe to its solution to increase the product's performance at high temperatures. The costs associated with each are shown below. The company's MARR is 20% per year. Do the following using a PW-based rate of return analysis and a spreadsheet:
(a) Determine which nanoparticle type the company should select using the Δi^* value.
(b) On the same graph, plot the PW versus different i values for each alternative. Indicate the breakeven i^* value and the MARR value on the plot.
(c) Use the plot of PW versus Δi values to select the better alternative with MARR = 20% per year. Is the answer the same as in part (a)?

	Type Fe	Type Al
First cost, $	−150,000	−280,000
Annual operating cost, $ per year	−92,000	−74,000
Salvage value, $	30,000	70,000
Life, years	2	4

8.22 A chemical company is considering two processes for isolating DNA material. The incremental cash flows between the two alternatives, J and S, have an incremental rate of return that is less than 40%, which is the MARR of the company. However, the company CEO prefers the more expensive process S. She believes the company can implement cost controls to reduce the annual cost of the more expensive process. By how much would she have to reduce the annual operating cost of alternative S (in $ per year) for it to have an incremental rate of return of exactly 40%?

Year	Incremental Cash Flow (S − J), $
0	−900,000
1	400,000
2	400,000
3	400,000

8.23 The incremental cash flows for two alternative electrode setups are shown. The MARR is 12% per year, and alternative Dryloc requires a larger initial investment compared to NPT.
(a) Determine which should be selected using an AW-based rate of return analysis.
(b) Use a graph of incremental values to determine the largest MARR value that will justify the NPT alternative.

Year	Incremental Cash Flow (Dryloc − NPT), $
0	−56,000
1−8	+8,900
9	+12,000

8.24 Hewett Electronics manufactures amplified pressure transducers. It must decide between two machines for a finishing operation. Select one for them on the basis of AW-based rate of return analysis. The company's MARR is 18% per year.

	Variable Speed	Dual Speed
First cost, $	−270,000	−245,000
Annual operating cost, $ per year	−135,000	−139,000
Salvage value, $	75,000	35,000
Life, years	6	6

8.25 A manufacturer of hydraulic equipment is trying to determine whether it should use monoflange double block and bleed (DBB) valves or a multi-valve system (MVS) for chemical injection. The costs are shown below. Use an AW-based rate of return analysis and a MARR of 18% per year to determine the better of the two options.

	DBB	MVS
First cost, $	−40,000	−71,000
Annual cost, $ per year	−60,000	−65,000
Salvage value, $	0	18,000
Life, years	2	4

8.26 Poly-Chem Plastics is considering two types of injection molding machines—hydraulic and electric. The hydraulic press (HP) will have a first cost of $600,000, annual costs of $200,000, and a salvage value of $70,000 after 5 years. Electric machine technology (EMT) will have a first cost of $800,000, annual costs of $150,000, and a salvage value of $130,000 after 5 years.
(a) Use an AW-based rate of return equation to determine the ROR on the increment of investment between the two.

(b) Determine which machine the company should select, if the MARR = 16% per year.

(c) Plot the AW versus i graph for each alternative's cash flows, and utilize it to determine the largest MARR that will justify the EMT extra investment of $200,000.

8.27 Last week Eduardo calculated the overall project ROR values for two alternatives A and B using the estimates below. He calculated $i_A^* = 34.2\%$ and $i_B^* = 31.2\%$ and recommended acceptance of A since its rate of return exceeded the established MARR of 30% by a greater amount than project B. Yesterday, the general manager of the company announced a major capital investment program, which includes a large drop in the MARR from 30% to 20% per year. Do the following to help Eduardo better understand the rate of return method and what this reduction in MARR means.

(a) Explain the error that Eduardo made in performing the rate of return analysis.

(b) Perform the correct analysis using each MARR value.

(c) Illustrate the ranking inconsistency problem using the two MARR values, and determine the maximum MARR that will justify alternative B.

	Alternative A	Alternative B
First cost, $	−40,000	−85,000
Annual operating cost, $ per year	−5,500	−15,000
Annual revenue, $ per year	22,000	45,000
Salvage value, $	0	20,000
Life, years	6	6
i^*, %	34.2	29.2

Multiple Alternative Comparison

8.28 Four mutually exclusive revenue alternatives are under consideration to automate a baking and packaging process at Able Bakery Products. The alternatives are ranked in order of increasing initial investment and compared by incremental rate of return analysis. The rate of return on each increment of investment was less than the MARR. Which alternative should be selected?

8.29 A WiMAX wireless network integrated with a satellite network can provide connectivity to any location within 10 km of the base station. The number of sectors per base station can be varied to increase the bandwidth. An independent cable operator is considering three bandwidth alternatives. Assume a life of 20 years and a MARR of 10% per year to determine which alternative is best using an incremental rate of return analysis.

Bandwidth, Mbps	First Cost, $1000	Operating Cost, $1000 per Year	Annual Income, $1000 per Year
44	−40,000	−2000	+5000
55	−46,000	−1000	+5000
88	−61,000	−500	+8000

8.30 Xerox's iGenX high-speed commercial printers cost $1.5 billion to develop. The machines cost $500,000 to $750,000 depending on what options the client selects. Spectrum Imaging Systems is considering the purchase of a new printer based on recent contracts it received for printing weekly magazine and mail-out advertising materials. The operating costs and revenues generated are related to a large extent to the speed and other capabilities of the copier. Spectrum is considering the four machines shown below. The company uses a 3-year planning period and a MARR of 15% per year. Determine which copier the company should acquire on the basis of an incremental rate of return analysis.

Copier	Initial Investment, $	Operating Cost, $ per Year	Annual Revenue, $ per Year	Salvage Value, $
iGen-1	−500,000	−350,000	+450,000	+70,000
iGen-2	−600,000	−300,000	+460,000	+85,000
iGen-3	−650,000	−275,000	+480,000	+95,000
iGen-4	−750,000	−200,000	+510,000	+120,000

8.31 Ashley Foods, Inc. has determined that any one of five machines can be used in one phase of its chili canning operation. The costs of the machines are estimated below, and all machines are estimated to have a 4-year useful life. If the minimum attractive rate of return is 20% per year, determine which machine should be selected on the basis of a rate of return analysis.

Machine	First Cost, $	Annual Operating Cost, $ per Year
1	−31,000	−16,000
2	−29,000	−19,300
3	−34,500	−17,000
4	−49,000	−12,200
5	−41,000	−15,500

8.32 Five revenue projects are under consideration by General Dynamics for improving material flow through an assembly line. The initial cost in

$1000 and the life of each project are as follows (revenue estimates are not shown):

	Project				
	A	B	C	D	E
Initial cost, $1000	−700	−2300	−900	−300	−1600
Life, years	5	8	5	5	6

An engineer made the comparisons shown below. From the calculations, determine which project, if any, should be undertaken if the company's MARR is (a) 11.5% per year and (b) 13.5% per year. If other calculations are necessary to make a decision, state which ones.

Comparison	Incremental Rate of Return, %
B vs DN	13%
A vs B	19%
D vs DN	11%
E vs B	15%
E vs D	24%
E vs A	21%
C vs DN	7%
C vs A	19%
E vs DN	12%
A vs DN	10%
E vs C	33%
D vs C	33%
D vs B	29%

8.33 The five alternatives shown here are being evaluated by the rate of return method.

	Initial Investment, $	ROR versus DN, %	Incremental Rate of Return, %				
Alternative			A	B	C	D	E
A	−25,000	9.6	—	27.3	9.4	35.3	25.0
B	−35,000	15.1		—	0	38.5	24.4
C	−40,000	13.4			—	46.5	27.3
D	−60,000	25.4				—	6.8
E	−75,000	20.2					—

(a) If the alternatives are *mutually exclusive* and the MARR is 26% per year, which alternative should be selected?

(b) If the alternatives are *mutually exclusive* and the MARR is 15% per year, which alternative should be selected?

(c) If the alternatives are *independent* and the MARR is 15% per year, which alternative(s) should be selected?

8.34 Five mutually exclusive revenue alternatives that have *infinite lives* are under consideration for increasing productivity in a manufacturing operation. The initial costs and cash flows of each project are shown. If the MARR is 14.9% per year, which alternative should be selected?

	Alternative				
	A	B	C	D	E
Initial cost, $	−7,000	−23,000	−9,000	−3,000	−16,000
Cash flow, $ per year	1,000	3,500	1,400	500	2,200
Rate of return (vs DN), %	14.3	15.2	15.6	16.7	13.8

8.35 The plant manager at Automaton Robotics is looking at the summarized incremental rate of return information shown below for five mutually exclusive alternatives, one of which must be chosen. The table includes the overall ROR and the incremental comparison of alternatives. Which alternative is best if the minimum attractive rate of return is (a) 15% per year and (b) 12% per year?

	First	Overall	Incremental Rate of Return, %				
Alternative	Cost, $	ROR, %	A	B	C	D	E
A	−80,000	14	—	12	11	17	24
B	−60,000	16	12	—	14	23	21
C	−40,000	17	11	14	—	35	29
D	−30,000	12	17	23	35	—	17
E	−20,000	8	24	21	29	17	—

8.36 Only one of four different machines can be purchased for Glass Act Products. An engineer performed the following analysis to select the best machine, all of which have a 10-year life. Which machine, if any, should the company select at a MARR of 22% per year?

	Machine			
	1	2	3	4
Initial cost, $	−44,000	−60,000	−72,000	−98,000
Annual cost, $ per year	−70,000	−64,000	−61,000	−58,000
Annual savings, $ per year	+80,000	+80,000	+80,000	+82,000
Overall ROR, %	18.6	23.4	23.1	20.8
Machines compared		2 to 1	3 to 2	4 to 3
Incremental investment, $		−16,000	−12,000	−26,000
Incremental cash flow, $ per year		+6,000	+3,000	+5,000
ROR on increment, %		35.7	21.4	14.1

8.37 The U.S. Bureau of Reclamation is considering five national park projects shown below, all of which can be considered to last indefinitely. At a MARR of 7.5% per year, determine which should be selected, if they are (a) independent and (b) mutually exclusive.

Project ID	First Cost, $1000	Annual Income, $1000	Rate of Return, %
A	−20,000	+2000	10.0
B	−10,000	+1300	13.0
C	−15,000	+1000	6.6
D	−70,000	+4000	5.7
E	−50,000	+2600	5.2

ADDITIONAL PROBLEMS AND FE EXAM REVIEW QUESTIONS

8.38 When conducting a rate of return (ROR) analysis involving multiple mutually exclusive alternatives, the first step is to:
(a) Rank the alternatives according to decreasing initial investment cost
(b) Rank the alternatives according to increasing initial investment cost
(c) Calculate the present worth of each alternative using the MARR
(d) Find the LCM between all of the alternatives

8.39 In comparing mutually exclusive alternatives by the ROR method, you should:
(a) Find the ROR of each alternative and pick the one with the *highest* ROR
(b) Select the alternative whose *incremental* ROR is the highest
(c) Select the alternative with ROR ≥ MARR that has the lowest initial investment cost
(d) Select the alternative with the largest initial investment that has been incrementally justified

8.40 When comparing independent projects by the ROR method, you should:
(a) Find the ROR of each project and pick the ones with the highest ROR
(b) Select all projects that have an overall ROR ≥ MARR
(c) Select the project with an overall ROR ≥ MARR that involves the lowest initial investment cost
(d) Select the project with the largest initial investment that has been incrementally justified

8.41 Of the following scenarios, alternative Y requires a higher initial investment than alternative X, and the MARR is 20% per year. The only scenario that requires an incremental investment analysis to select an alternative is that:
(a) X has an overall ROR of 22% per year, and Y has an overall ROR of 24% per year
(b) X has an overall ROR of 19% per year, and Y has an overall ROR of 23% per year
(c) X has an overall ROR of 18% per year, and Y has an overall ROR of 19% per year
(d) X has an overall ROR of 28% per year, and Y has an overall ROR of 26% per year

8.42 Alternatives whose cash flows (excluding the salvage value) are all negative are called:
(a) Revenue alternatives
(b) Nonconventional alternatives
(c) Cost alternatives
(d) Independent alternatives

8.43 For these alternatives, the sum of the incremental cash flows is:

Year	A	B
0	−10,000	−14,000
1	+2,500	+4,000
2	+2,500	+4,000
3	+2,500	+4,000
4	+2,500	+4,000
5	+2,500	+4,000

(a) $2500
(b) $3500
(c) $6000
(d) $8000

8.44 Helical Systems, Inc. uses a minimum attractive rate of return of 8% per year, compounded annually. The company is evaluating two new processes for improving the efficiency of its manufacturing operation. The cash flow estimates associated with each process are shown below. A correct equation to use for an incremental rate of return analysis is:

	Process I	Process J
First cost, $	−40,000	−50,000
Annual cost, $ per year	−15,000	−12,000
Salvage value, $	5,000	6,000
Life, years	3	3

(a) $0 = -10,000 + 3000(P/A,\Delta i^*,3) + 1000(P/F,\Delta i^*,3)$
(b) $0 = -40,000(A/P,\Delta i^*,3) - 15,000 + 5000(A/F,\Delta i^*,3)$
(c) $0 = -50,000(A/P,\Delta i^*,3) - 12,000 + 6000(A/F,\Delta i^*,3)$
(d) $0 = -10,000 - 3000(P/A,\Delta i^*,3) + 1000(P/F,\Delta i^*,3)$

8.45 For the four independent projects shown, the one or ones to select using a MARR of 14% per year are:

Project	Rate of Return, % per Year
A	14
B	12
C	15
D	10

(a) Only C
(b) Only A and C
(c) Only A
(d) Can't tell; need to conduct incremental analysis

Problems 8.46 through 8.48 are based on the following information.

Five alternatives are being evaluated by the incremental rate of return method.

Alternative	Initial Investment, $	Overall ROR versus DN, %	Incremental Rate of Return, %				
			A	B	C	D	E
A	−25,000	9.6	—	27.3	9.4	35.3	25.0
B	−35,000	15.1		—	0	38.5	24.4
C	−40,000	13.4			—	46.5	27.3
D	−60,000	25.4				—	6.8
E	−75,000	20.2					—

8.46 If the projects are mutually exclusive and the minimum attractive rate of return is 14% per year, the best alternative is:

(a) B   (b) C   (c) D   (d) E

8.47 If the projects are mutually exclusive and the MARR is 20% per year, the best alternative is:
(a) B
(b) C
(c) D
(d) E

8.48 If the projects are independent, instead of mutually exclusive, the one or ones to select at an MARR of 18% per year are:
(a) B and C
(b) B, D, and E
(c) D and E
(d) B, C, and E

CASE STUDY

ROR ANALYSIS WITH ESTIMATED LIVES THAT VARY

Background

Make-to-Specs is a software system under development by ABC Corporation. It will be able to translate digital versions of three-dimensional computer models, containing a wide variety of part shapes with machined and highly finished (ultra-smooth) surfaces. The product of the system is the numerically controlled (NC) machine code for the part's manufacturing. Additionally, Make-to-Specs will build the code for superfine finishing of surfaces with continuous control of the finishing machines.

Information

There are two alternative computers that can provide the server function for the software interfaces and shared database updates on the manufacturing floor while Make-to-Specs is operating in parallel mode. The server first cost and estimated contribution to annual net cash flow are summarized below.

	Server 1	Server 2
First cost, $	100,000	200,000
Net cash flow, $/year	35,000	50,000 year 1, plus 5000 per year for years 2, 3, and 4 (gradient) 70,000 maximum for years 5 on, even if the server is replaced
Life, years	3 or 4	5 or 8

The life estimates were developed by two different individuals: a design engineer and a manufacturing manager. They have asked that, at this stage of the project, all analyses be performed using both life estimates for each system.

Case Study Exercises

Use spreadsheet analysis to determine the following:

1. If the MARR = 12%, which server should be selected? Use the PW or AW method to make the selection.
2. Use incremental ROR analysis to decide between the servers at MARR = 12%.
3. Use any method of economic analysis to display on the spreadsheet the value of the incremental ROR between server 2 with a life estimate of 5 years and a life estimate of 8 years.

CASE STUDY

HOW A NEW ENGINEERING GRADUATE CAN HELP HIS FATHER[1]

Background

"I don't know whether to sell it, expand it, lease it, or what. But I don't think we can keep doing the same thing for many more years. What I really want to do is to keep it for 5 more years, then sell it for a bundle," Elmer Kettler said to his wife, Janise, their son, John Kettler, and new daughter-in-law, Suzanne Gestory, as they were gathered around the dinner table. Elmer was sharing thoughts on Gulf Coast Wholesale Auto Parts, a company he has owned and operated for 25 years on the southern outskirts of Houston, Texas. The business has excellent contracts for parts supply with several national retailers operating in the area—NAPA, AutoZone, O'Reilly, and Advance. Additionally, Gulf Coast operates a rebuild shop serving these same retailers for major automobile components, such as carburetors, transmissions, and air conditioning compressors.

At his home after dinner, John decided to help his father with an important and difficult decision: What to do with his business? John graduated just last year with an engineering degree from a major state university in Texas, where he completed a course in engineering economy. Part of his job at Energcon Industries is to perform basic rate of return and present worth analyses on energy management proposals.

Information

Over the next few weeks, John outlined five options, including his dad's favorite of selling in 5 years. John summarized all the estimates over a 10-year horizon. The options and estimates were given to Elmer, and he agreed with them.

Option 1: *Remove rebuild.* Stop operating the rebuild shop and concentrate on selling wholesale parts. The removal of the rebuild operations and the switch to an "all-parts house" are expected to cost $750,000 in the first year. Overall revenues will drop to $1 million the first year with an expected 4% increase per year thereafter. Expenses are projected at $0.8 million the first year, increasing 6% per year thereafter.

Option 2: *Contract rebuild operations.* To get the rebuild shop ready for an operations contractor to take over will cost $400,000 immediately. If expenses stay the same for 5 years, they will average $1.4 million per year, but they can be expected to rise to $2 million per year in year 6 and thereafter. Elmer thinks revenues under a contract arrangement can be $1.4 million the first year and can rise 5% per year for the duration of a 10-year contract.

Option 3: *Maintain status quo and sell out after 5 years* (Elmer's personal favorite). There is no cost now, but

the current trend of negative net profit will probably continue. Projections are $1.25 million per year for expenses and $1.15 million per year in revenue. Elmer had an appraisal last year, and the report indicated Gulf Coast Wholesale Auto Parts is worth a net $2 million. Elmer's wish is to sell out completely after 5 more years at this price, and to make a deal that the new owner pay $500,000 per year at the end of year 5 (sale time) and the same amount for the next 3 years.

Option 4: *Trade-out.* Elmer has a close friend in the antique auto parts business who is making a "killing," so he says, with e-commerce. Although the possibility is risky, it is enticing to Elmer to consider a whole new line of parts, but still in the basic business that he already understands. The trade-out would cost an estimated $1 million for Elmer immediately. The 10-year horizon of annual expenses and revenues is considerably higher than for his current business. Expenses are estimated at $3 million per year and revenues at $3.5 million each year.

Option 5: *Lease arrangement.* Gulf Coast could be leased to some turnkey company with Elmer remaining the owner and bearing part of the expenses for building, delivery trucks, insurance, etc. The first-cut estimates for this option are $1.5 million to get the business ready now, with annual expenses at $500,000 per year and revenues at $1 million per year for a 10-year contract.

Case Study Exercises

Help John with the analysis by doing the following:

1. Develop the actual cash flow series and incremental cash flow series (in $1000 units) for all five options in preparation for an incremental ROR analysis.
2. Discuss the possibility of multiple rate of return values for all the actual and incremental cash flow series. Find any multiple rates in the range of 0% to 100%.
3. If John's father insists that he make 25% per year or more on the selected option over the next 10 years, what should he do? Use all the methods of economic analysis you have learned so far (PW, AW, ROR) so John's father can understand the recommendation in one way or another.
4. Prepare plots of the PW versus *i* for each of the five options. Estimate the breakeven rate of return between options.
5. What is the minimum amount that must be received in each of years 5 through 8 for option 3 (the one Elmer wants) to be best economically? Given this amount, what does the sale price have to be, assuming the same payment arrangement as presented above?

[1]Based upon a study by Alan C. Stewart, Consultant, Communications and High Tech Solutions Engineering, Accenture LLP.

CHAPTER 9

Benefit/Cost Analysis and Public Sector Economics

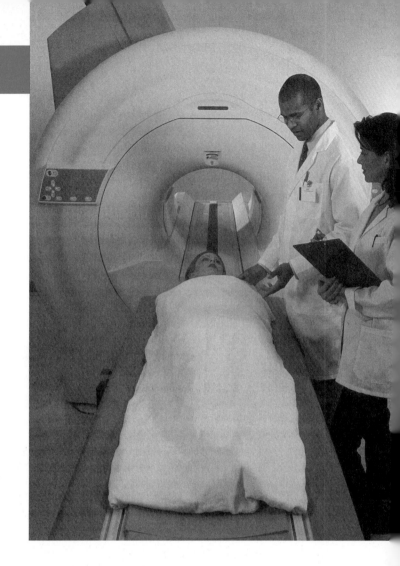

T he evaluation methods of previous chapters are usually applied to alternatives in the private sector, that is, for-profit and not-for-profit corporations and businesses. This chapter introduces **public sector and service sector alternatives** and their economic consideration. In the case of public projects, the owners and users (beneficiaries) are the citizens and residents of a government unit—city, county, state, province, or nation. Government units provide the mechanisms to raise capital and operating funds. Public-private partnerships have become increasingly common, especially for large infrastructure projects such as major highways, power generation plants, water resource developments, and the like.

The benefit/cost (B/C) ratio introduces objectivity into the economic analysis of public sector evaluation, thus reducing the effects of politics and special interests. The different formats of B/C analysis, and associated disbenefits of an alternative, are discussed here. The B/C analysis can use equivalency computations based on PW, AW, or FW values. Performed correctly, the benefit/cost method will always select the same alternative as PW, AW, and ROR analyses.

This chapter also introduces **service sector projects** and discusses how their economic evaluation is different from that for other projects. Finally, there is a discussion on **professional ethics** and ethical dilemmas in the public sector.

PE

Water Treatment Facility #3 Case: Allen Water Utilities has planned for the last 25 years to construct a new drinking water treatment facility that will supply the rapidly growing north and northwest areas of the city. An expectation of over 100,000 new residents in the next several years and 500,000 by 2040 prompted the development of the plant starting in 2012. The supply is from a large surface lake currently used to provide water to all of Allen and the surrounding communities. The project is termed WTF3, and its initial capital investment is $540 million for the treatment plant and two large steel-pipe transmission mains (84- and 48-inch) that will be installed via tunneling approximately 100 to 120 feet under suburban areas of the city to reach current reservoirs.

Tunneling was selected after geotechnical borings indicated that open trenching was not supportable by the soil and based upon a large public outcry against trenching in the living areas along the selected transmission routes. Besides the treatment plant construction on the 95-acre site, there must be at least three large vertical shafts (25 to 50 feet in diameter) bored along each transmission main to gain underground access for equipment and debris removal during the tunneling operations.

The stated criteria used to make decisions for WTF3 and the transmission mains were economics, environment, community impact, and constructability.

There are major long-term benefits for the new facility. These are some mentioned by city engineers:

- It will meet projected water needs of the city for the next 50 years.

- The new treatment plant is at a higher elevation than the current two plants, allowing gravity flow to replenish reservoirs, thereby using little or no electric pumping.

- There will be an increase in the diversity and reliability of supply as other plants age.

- It will provide a water quality that is more consistent due to the location of the raw water intakes.

- The facility uses water supplies already purchased; therefore, there is no need to negotiate additional allowances.

The disbenefits are mostly short-term during the construction of WTF3 and transmission mains. Some of these are mentioned by citizen groups and one retired city engineer:

- There will be disruption of habitat for some endangered species of birds, lizards, and trees not found in any other parts of the country.

- Large amounts of dust and smoke will enter the atmosphere in a residential area during the 3½ years of construction, tunneling, and transmission main completion.

- Noise pollution and traffic congestion will result during an estimated 26,000 truck trips to remove debris from the plant site and tunnel shafts, in addition to the problems from regular construction traffic.

- Natural landscape in plant and tunnel shaft sites will be destroyed.

- Safety will be compromised for children in a school where large trucks will pass about every 5 minutes for approximately 12 hours per day, 6 days per week for 2½ years.

- There may be delays in fire and ambulance services in emergencies, since many neighborhood streets are country-road width and offer only single ingress/egress streets for neighborhoods along the indicated routes.

- The need for the facility has not been proved, as the water will be sold to developers outside the city limits, not provided to residences within Allen.

- Newly generated revenues will be used to pay off the capital funding bonds approved for the plant's construction.

Last year, the city engineers did a benefit/cost analysis for this massive public sector project; none of the results were publicized. Public and elected official intervention has now caused some of the conclusions using the criteria mentioned above to be questioned by the general manager of Allen Water Utilities.

This case is used in the following topics of this chapter:

Public sector projects (Section 9.1)

Incremental B/C analysis, two alternatives (Section 9.3)

Incremental B/C analysis, more than two alternatives (Section 9.4)

Online Presentation

9.1 Public Sector Projects ● ● ●

Virtually all the examples and problems of previous chapters have involved the private sector, where products, systems, and services are developed and offered by corporations and businesses for use by individual customers and clients, the government, or other companies. (Notable exceptions are the long-life alternatives discussed in Chapters 5 (PW) and 6 (AW) where capitalized cost analysis was applied.) Now we will explore projects that concentrate on government units and the citizens they serve. These are called public sector projects.

A **public sector project** is a product, service, or system used, financed, and owned by the citizens of any government level. The primary purpose is to **provide service to the citizenry for the public good at no profit.** Areas such as public health, criminal justice, safety, transportation, welfare, and utilities are publically owned and require economic evaluation.

Upon reflection, it is surprising how much of what we use on a daily or as-needed basis is publicly owned and financed to serve us—the citizenry. These are some public sector examples:

Hospitals and clinics

Parks and recreation

Utilities: water, electricity, gas, sewer, sanitation

Schools: primary, secondary, community colleges, universities

Economic development projects

Convention centers

Sports arenas

Transportation: highways, bridges, waterways

Police and fire protection   Public housing

Courts and prisons   Emergency relief

Food stamp and rent relief programs   Codes and standards

Job training

There are significant differences in the characteristics of private and public sector alternatives. They are summarized here.

Characteristic	Public sector	Private sector
Size of investment	Large	Some large; more medium to small

Often alternatives developed to serve public needs require large initial investments, possibly distributed over several years. Modern highways, public transportation systems, universities, airports, and flood control systems are examples.

Characteristic	Public sector	Private sector
Life estimates	Longer (30–50+ years)	Shorter (2–25 years)

The long lives of public projects often prompt the use of the capitalized cost method, where infinity is used for n and annual costs are calculated as $A = P(i)$. As n gets larger, especially over 30 years, the differences in calculated A values become small. For example, at $i = 7\%$, there will be a very small difference in 30 and 50 years, because $(A/P,7\%,30) = 0.08059$ and $(A/P,7\%,50) = 0.07246$.

Characteristic	Public sector	Private sector
Annual cash flow estimates	No profit; costs, benefits, and disbenefits are estimated	Revenues contribute to profits; costs are estimated

Public sector projects (also called publicly owned) do not have profits; they do have costs that are paid by the appropriate government unit; and they benefit the citizenry. Public sector projects often have undesirable consequences, as interpreted by some sectors of the public. It is these consequences that can cause public controversy about the projects. The economic analysis should consider these consequences in monetary terms to the degree estimable. (Often in private sector analysis, undesirable consequences are not considered, or they may be directly addressed as costs.) To perform a benefit/cost economic analysis of public alternatives, the costs (initial and annual), the benefits, and the disbenefits, if considered, must be estimated as accurately as possible in *monetary units.*

Costs—estimated expenditures *to the government entity* for construction, operation, and maintenance of the project, less any expected salvage value.

Benefits—advantages to be experienced *by the owners, the public.*

Disbenefits—expected undesirable or negative consequences *to the owners* if the alternative is implemented. Disbenefits may be indirect economic disadvantages of the alternative.

It is difficult to estimate and agree upon the economic impact of benefits and disbenefits for a public sector alternative. For example, assume a short bypass around a congested area in town is recommended. How much will it benefit a driver in *dollars per driving minute* to be able to bypass five traffic lights while averaging 35 miles per hour, as compared to currently driving through the lights averaging 20 miles per hour and stopping at an average of two lights for an average of 45 seconds each? The bases and standards for benefits estimation are always difficult to establish and verify. Relative to revenue cash flow estimates in the private sector, benefit estimates are much harder to make, and vary more widely around uncertain averages. (The inability to make economic estimates for benefits *may be overcome* by using the evaluation technique discussed in Section 9.5.) And the disbenefits that accrue from an alternative are even harder to estimate. In fact, the disbenefit itself may not be known at the time the evaluation is performed.

Characteristic	Public sector	Private sector
Funding	Taxes, fees, bonds, private funds	Stocks, bonds, loans, individual owners

The capital used to finance public sector projects is commonly acquired from taxes, bonds, and fees. Taxes are collected from those who are the owners—the citizens (e.g., federal gasoline taxes for highways are paid by all gasoline users, and health care costs are covered by insurance premiums). This is also the case for fees, such as toll road fees for drivers. Bonds are often issued: U.S. Treasury bonds, municipal bond issues, and special-purpose bonds, such as utility district bonds. Private lenders can provide up-front financing. Also, private donors may provide funding for museums, memorials, parks, and garden areas through gifts.

Characteristic	Public sector	Private sector
Interest rate	Lower	Higher, based on cost of capital

Because many of the financing methods for public sector projects are classified as *low-interest,* the interest rate is virtually always lower than for private sector alternatives. Government agencies are exempt from taxes levied by higher-level units. For example, municipal projects do not have to pay state taxes. (Private corporations and individual citizens do pay taxes.) Many loans are very low-interest, and grants with no repayment requirement from federal programs may share project costs. This results in interest rates in the 4% to 8% range. It is common that a government agency will direct that all projects be evaluated at a specific rate. As a matter of standardization, directives to use a specific interest rate are beneficial because different government agencies are able to obtain varying types of funding at different rates. This can result in projects of the same type being rejected in one state or city but accepted in another. Standardized rates tend to increase the consistency of economic decisions and to reduce gamesmanship.

The determination of the interest rate for public sector evaluation is as important as the determination of the MARR for a private sector analysis. The public sector interest rate is identified as i; however, it is referred to by other names to distinguish it from the private sector rate. The most common terms are *discount rate* and *social discount rate.*

Characteristic	Public sector	Private sector
Alternative selection criteria	Multiple criteria	Primarily based on rate of return

Multiple categories of users, economic as well as noneconomic interests, and special-interest political and citizen groups make the selection of one alternative over another much more difficult in public sector economics. Seldom is it possible to select an alternative on the sole basis of a criterion such as PW or ROR. It is important to describe and itemize the criteria and selection method prior to the analysis. This helps determine the perspective or viewpoint when the evaluation is performed. Viewpoint is discussed below.

Characteristic	Public sector	Private sector
Environment of the evaluation	Politically inclined	Primarily economic

There are often public meetings and debates associated with public sector projects to accommodate the various interests of citizens (owners). Elected officials commonly assist with the selection, especially when pressure is brought to bear by voters, developers, environmentalists, and others. The selection process is not as "clean" as in private sector evaluation.

The **viewpoint of the public sector analysis** must be determined before cost, benefit, and disbenefit estimates are made and before the evaluation is formulated and performed. There are several viewpoints for any situation, and the different perspectives may alter how a cash flow estimate is classified.

Some example perspectives are the citizen; the city tax base; number of students in the school district; creation and retention of jobs; economic development potential; a particular industry interest (agriculture, banking, electronics manufacturing); even the reelection of a public officeholder (often termed *pork projects*). In general, the viewpoint of the analysis should be as broadly defined as those who will bear the costs of the project and reap its benefits. Once established, the viewpoint assists in categorizing the costs, benefits, and disbenefits of each alternative, as illustrated in Example 9.1.

EXAMPLE 9.1 Water Treatment Facility #3 Case

(PE)

The situation with the location and construction of the new WTF3 and associated transmission mains described in the chapter's introduction has reached a serious level because of recent questions posed by some city council members and citizen groups. Before going public to the city council with the analysis performed last year, the director of Allen Water Utilities has asked an engineering management consultant to review it and determine if it was an acceptable analysis and correct economic decision, then and now. The lead consultant, Joel Whiterson, took engineering economy as a part of his B.S. education and has previously worked on economic studies in the government sector, but never as the lead person.

Within the first hour of checking background notes, Joel found several initial estimates (shown below) from last year for expected consequences if WTF3 were built. He realized that no viewpoint of the study was defined, and, in fact, the estimates were never classified as costs, benefits, or disbenefits. He did determine that disbenefits were considered at some point in the analysis, though the estimates for them are very sketchy.

Joel defined two viewpoints: a *citizen* of Allen and the Allen Water Utilities *budget*. He wants to identify each of the estimates as a cost, benefit, or disbenefit from each viewpoint. Please help with this classification.

Economic Dimension	Monetary Estimate
1. Cost of water: 10% annual increase to Allen households	Average of $29.7 million (years 1–5, steady thereafter)
2. Bonds: Annual debt service at 3% per year on $540 million	$16.2 million (years 1–19); $516.2 million (year 20)
3. Use of land: Payment to Parks and Recreation for shaft sites and construction areas	$300,000 (years 1–4)
4. Property values: Loss in value, sales price, and property taxes	$4 million (years 1–5)
5. Water sales: Increases in sales to surrounding communities	$5 million (year 4) plus 5% per year (years 5–20)
6. M&O: Annual maintenance and operations costs	$300,000 plus 4% per year increase (years 1–20)
7. Peak load purchases: Savings in purchases of treated water from secondary sources	$500,000 (years 5–20)

Solution

The perspective of each viewpoint is identified and estimates are classified. (How this classification is done will vary depending upon who does the analysis. This solution offers only one logical answer.)

Viewpoint 1: Citizen of the city of Allen. Goal: Maximize the quality of life and wellness of citizens with family and neighborhood as prime concerns.

Costs: 1, 2, 4, 6   Benefits: 5, 7   Disbenefits: 3

Viewpoint 2: Allen Water Utilities budget. Goal: Ensure the budget is balanced and of sufficient size to fund rapidly growing city service demands.

Costs: 2, 3, 6   Benefits: 1, 5, 7   Disbenefits: 4

Citizens view costs in a different light than a city budget employee does. For example, the loss of property values (item 4) is considered a real cost to a citizen, but is an unfortunate disbenefit from the city budget perspective. Similarly, the Allen Water Utilities budget interprets estimate 3 (payment for use of land to Parks and Recreation) as a real cost; but a citizen might interpret this as merely a movement of funds between two municipal budgets—therefore, it is a disbenefit, not a real cost.

Comment

The inclusion of disbenefits can easily change the economic decision. However, agreement on the disbenefits and their monetary estimates is difficult (to impossible) to develop, often resulting in the exclusion of any disbenefits from the economic analysis. Unfortunately, this usually transfers the consideration of disbenefits to the noneconomic (i.e., political) realm of public project decision making.

Most of the large public sector projects are developed through public-private partnerships (PPPs). A partnership is advantageous in part because of the greater efficiency of the private sector and in part because of the sizable cost to design, construct, and operate such projects. Full funding by the government unit may not be possible using traditional means—fees, taxes, and bonds. Some examples of the projects are as follows:

Project	Some Purposes of the Project
Mass transportation	Reduce transit time; reduce congestion; improve environment; decrease road accidents
Bridges and tunnels	Speed traffic flows; reduce congestion; improve safety
Ports and harbors	Increase cargo capacity; support industrial development; increase tourism
Airports	Increase capacity; improve passenger safety; support development
Water resources	Desalination for drinking water; meet irrigation and industrial needs; improve wastewater treatment

In these joint ventures, the public sector (government) is responsible for the funding and service to the citizenry, and the private sector partner (corporation) is responsible for varying aspects of the projects as detailed below. The government unit cannot make a profit, but the corporation(s) involved can realize a reasonable profit; in fact, the profit margin is usually written into the contract that governs the design, construction, and operation of the project.

Traditional methods of contracting were *fixed-price* (historically called lump-sum) and *cost reimbursable* (also called cost-plus). In these formats, a government unit took responsibility for funding and possibly some of the design elements, and later all operation activities, while the contractor did not share in the risks involved—liability, natural disasters, funding shortfalls, etc. More recently, the PPP has become the arrangement of choice for most large public projects. Commonly these are called **design-build contracts**, under which contractors take on more and more of the functions from design to operation. Details are explained in publications such as *Design-Build Contracting Handbook* (Cushman and Loulakis). The most reliance is placed upon a contractor or contractors with a DBOMF contract, as described below.

The **Design-Build-Operate-Maintain-Finance (DBOMF) contract** is considered a turnkey approach to a project. It requires the contractor(s) to perform all the DBOMF activities with collaboration and approval of the owner (the government unit). The activity of **financing** is the *management of cash flow to support project implementation* by a contracting firm. Although a contractor may assist in some instances, the **funding** (obtaining the capital funds) remains the government's responsibility through bonding, commercial loans, taxation, grants, and gifts.

When the financing activity is not managed by a contractor, the contract is a DBOM; it is also common to develop a design-build contract. In virtually all cases, some forms of design-build arrangements for public projects are made because they offer several advantages to the government and citizens served:

- Cost and time savings in the design, build, and operate phases
- Earlier and more reliable (less variable) cost estimates

- Reduced administrative responsibilities for the owner
- Better efficiency of resource allocation of private enterprise
- Environmental, liability, and safety issues addressed by the private sector, where there usually is greater expertise

Many of the projects in international settings and in developing countries utilize the public-private partnership. There are, of course, disadvantages to this arrangement. One risk is that the amount of funding committed to the project may not cover the actual build cost because it is considerably higher than estimated. Another risk is that a reasonable profit may not be realized by the private corporation due to low usage of the facility during the operate phase. To prevent such problems, the original contract may provide for special subsidies and loans guaranteed by the government unit. The subsidy may cover costs plus (contractually agreed-to) profit if usage is lower than a specified level. The level used may be the breakeven point with the agreed-to profit margin considered.

9.2 Benefit/Cost Analysis of a Single Project ●●●

Online Presentation

The benefit/cost ratio is relied upon as a fundamental analysis method for public sector projects. The B/C analysis was developed to introduce greater objectivity into public sector economics, and as one response to the U.S. Congress approving the Flood Control Act of 1936. There are several variations of the B/C ratio; however, the fundamental approach is the same. All cost and benefit estimates must be converted to a common equivalent monetary unit (PW, AW, or FW) at the discount rate (interest rate). The B/C ratio is then calculated using one of these relations:

$$\text{B/C} = \frac{\text{PW of benefits}}{\text{PW of costs}} = \frac{\text{AW of benefits}}{\text{AW of costs}} = \frac{\text{FW of benefits}}{\text{FW of costs}} \qquad [9.1]$$

Present worth and annual worth equivalencies are preferred to future worth values. The sign convention for B/C analysis is positive signs; **costs are preceded by a + sign.** Salvage values and additional revenues to the government, when they are estimated, are subtracted from costs in the denominator. Disbenefits are considered in different ways depending upon the model used. Most commonly, **disbenefits are subtracted from benefits** and placed in the numerator. The different formats are discussed below.

The decision guideline is simple:

If B/C ≥ 1.0, accept the project as economically justified for the estimates and discount rate applied.

If B/C < 1.0, the project is not economically acceptable.

Project evaluation

If the B/C value is exactly or very near 1.0, noneconomic factors will help make the decision.

The *conventional B/C ratio,* probably the most widely used, is calculated as follows:

$$\text{B/C} = \frac{\text{benefits} - \text{disbenefits}}{\text{costs}} = \frac{B - D}{C} \qquad [9.2]$$

In Equation [9.2] disbenefits are subtracted from benefits, not added to costs. The B/C value could change considerably if disbenefits are regarded as costs. For example, if the numbers 10, 8, and 5 are used to represent the PW of benefits, disbenefits, and costs, respectively, the correct procedure results in B/C = (10 − 8)/5 = 0.40. The incorrect placement of disbenefits in the denominator results in B/C = 10/(8 + 5) = 0.77, which is approximately twice the correct B/C value of 0.40. Clearly, then, the method by which disbenefits are handled affects the magnitude of the B/C ratio. However, regardless of whether disbenefits are (correctly) subtracted from the numerator or (incorrectly) added to costs in the denominator, a B/C ratio of less than 1.0 by the first method will always yield a B/C ratio less than 1.0 by the second method, and vice versa.

The *modified B/C ratio* includes all the estimates associated with the project, once operational. Maintenance and operation (M&O) costs are placed in the numerator and treated in a manner similar to disbenefits. The denominator includes only the initial investment. Once all amounts are expressed in PW, AW, or FW terms, the modified B/C ratio is calculated as

$$\text{Modified B/C} = \frac{\text{benefits} - \text{disbenefits} - \text{M\&O costs}}{\text{initial investment}} \qquad [9.3]$$

Salvage value is usually included in the denominator as a negative cost. The modified B/C ratio will obviously yield a different value than the conventional B/C method. However, as with disbenefits, *the modified procedure can change the magnitude of the ratio but not the decision to accept or reject the project.*

The *benefit and cost difference* measure of worth, which does not involve a ratio, is based on the difference between the PW, AW, or FW of benefits and costs, that is, $B - C$. If $(B - C) \geq 0$, the project is acceptable. This method has the advantage of eliminating the discrepancies noted above when disbenefits are regarded as costs, because B represents *net benefits*. Thus, for the numbers 10, 8, and 5, the same result is obtained regardless of how disbenefits are treated.

Subtracting disbenefits from benefits: $B - C = (10 - 8) - 5 = -3$
Adding disbenefits to costs: $B - C = 10 - (8 + 5) = -3$

Before calculating the B/C ratio by any formula, check whether the alternative with the larger AW or PW of costs also yields a larger AW or PW of benefits. It is possible for one alternative with larger costs to generate lower benefits than other alternatives, thus making it unnecessary to further consider the larger-cost alternative.

By the very nature of benefits and especially disbenefits, monetary estimates are difficult to make and will vary over a wide range. The extensive use of **sensitivity analysis** on the more questionable parameters helps determine how sensitive the economic decision is to estimate variation. This approach assists in determining the **economic and public acceptance risk** associated with a defined project. Also, the use of sensitivity analysis can alleviate some of the public's concerns commonly expressed that people (managers, engineers, consultants, contractors, and elected officials) designing (and promoting) the public project are narrowly receptive to different approaches to serving the public's interest.

EXAMPLE 9.2

In the past, the Afram Foundation has awarded many grants to improve the living and medical conditions of people in war-torn and poverty-stricken countries throughout the world. In a proposal for the foundation's board of directors to construct a new hospital and medical clinic complex in a deprived central African country, the project manager has developed some estimates. These are developed, so she states, in a manner that does not have a major negative effect on prime agricultural land or living areas for citizens.

Award amount:	$20 million (end of) first year, decreasing by $5 million per year for 3 additional years; local government will fund during the first year only
Annual costs:	$2 million per year for 10 years, as proposed
Benefits:	Reduction of $8 million per year in health-related expenses for citizens
Disbenefits:	$0.1 to $0.6 million per year for removal of arable land and commercial districts

Use the conventional and modified B/C methods to determine if this grant proposal is economically justified over a 10-year study period. The foundation's discount rate is 6% per year.

Solution
Initially, determine the AW for each parameter over 10 years. In $1 million units,

Award:	$20 - 5(A/G,6\%,4) = \$12.864$ per year
Annual costs:	$2 per year
Benefits:	$8 per year
Disbenefits:	Use $0.6 for the first analysis

The *conventional B/C analysis* applies Equation [9.2].

$$B/C = \frac{8.0 - 0.6}{12.864 + 2.0} = 0.50$$

The *modified B/C analysis* uses Equation [9.3].

$$\text{Modified B/C} = \frac{8.0 - 0.6 - 2.0}{12.864} = 0.42$$

The proposal is not justified economically since both measures are less than 1.0. If the low disbenefits estimate of $0.1 million per year is used, the measures increase slightly, but not enough to justify the proposal.

It is possible to develop a direct formula connection between the B/C of a public sector and B/C of a private sector project that is a **revenue alternative;** that is, both revenues and costs are estimated. Further, we can identify a direct correspondence between the modified B/C relation in Equation [9.3] and the PW method we have used repeatedly. (The following development also applies to AW or FW values.) Let's neglect the initial investment in year 0 for a moment, and concentrate on the cash flows of the project for year 1 through its expected life. For the private sector, the PW for project cash flows is

$$\text{PW of project} = \text{PW of revenue} - \text{PW of costs}$$

Since private sector revenues are approximately the same as public sector benefits minus disbenefits (B − D), the modified B/C relation in Equation [9.3] may be written as

$$\text{Modified B/C} = \frac{\text{PW of (B − D)} - \text{PW of C}}{\text{PW of initial investment}}$$

This relation can be slightly rewritten to form the **profitability index (PI)**, which can be used to evaluate revenue projects in the public or private sector. For years $t = 1, 2, \ldots, n$,

$$PI = \frac{\text{PW of NCF}_t}{\textbf{PW of initial investment}} \qquad [9.4]$$

Note that the denominator includes only first cost (initial investment) items, while the numerator has only cash flows that result from the project for years 1 through its life. The PI measure of worth provides a sense of getting the most for the investment dollar (euro, yen, etc.). That is, the result is in PW units per PW of money invested at the beginning. This is a "bang for the buck" measure. When used solely for a private sector project, the disbenefits are usually omitted, whereas they should be estimated and included in the modified B/C version of this measure for a public project.

The evaluation guideline for a single project using the PI is the same as for the conventional B/C or modified B/C.

Project evaluation

If PI ≥ 1.0, the project is economically acceptable at the discount rate.
If PI < 1.0, the project is not economically acceptable at the discount rate.

Remember, the computations for PI and modified B/C are essentially the same, except the PI is usually applied without disbenefits estimated. The PI has another name: the **present worth index (PWI)**. It is often used to rank and assist in the selection of independent projects when the capital budget is limited. This application is discussed in Chapter 12, Section 12.5.

EXAMPLE 9.3

The Georgia Transportation Directorate is considering a public-private partnership with Young Construction as the prime contractor using a DBOMF contract for a new 22.51-mile toll road on the outskirts of Atlanta's suburban area. The design includes three 4-mile-long commercial/retail corridors on both sides of the toll road. Highway construction is expected to require

5 years at an average cost of \$3.91 million per mile. The discount rate is 4% per year, and the study period is 30 years. Evaluate the economics of the proposal using (*a*) the modified B/C analysis from the State of Georgia perspective and (*b*) the profitability index from the Young corporate viewpoint in which disbenefits are not included.

Initial investment: \$88 million distributed over 5 years; \$4 million now and in year 5 and \$20 million in each of years 1 through 4.

Annual M&O cost: \$1 million per year, plus an additional \$3 million each fifth year, including year 30.

Annual revenue/benefits: Include tolls and retail/commercial growth; start at \$2 million in year 1, increasing by a constant \$0.5 million annually through year 10, and then increasing by a constant \$1 million per year through year 20 and remaining constant thereafter.

Estimable disbenefits: Include loss of business income, taxes, and property value in surrounding areas; start at \$10 million in year 1, decrease by \$0.5 million per year through year 21, and remain at zero thereafter.

Solution

The PW values in year 0 for all estimates must be developed initially usually by hand, calculator, or spreadsheet computations. If the 30 years of estimates are entered into a spreadsheet and NPV functions at 4% are applied, the results in \$1 million units are obtained. All values are positive because of the sign convention for B/C and PI measures.

$$\text{PW of investment} = \$71.89 \qquad \text{PW of benefits} = \$167.41$$
$$\text{PW of costs} = \$26.87 \qquad \text{PW of disbenefits} = \$80.12$$

(*a*) From the public project perspective, the State will apply Equation [9.3].

$$\text{Modified B/C} = \frac{167.41 - 80.12 - 26.87}{71.89} = 0.84$$

The toll road proposal is *not economically acceptable*, since B/C < 1.0.

(*b*) From the private corporation viewpoint, Young Construction will apply Equation [9.4].

$$\text{PI} = \frac{167.41 - 26.87}{71.89} = 1.95$$

The proposal is clearly justified without the disbenefits, since PI > 1.0. The private project perspective predicts that every investment dollar will return an equivalent of \$1.95 over 30 years at a 4% per year discount rate.

Comment

The obvious question that arises concerns the correct measure to use. When PI is used in the private project setting, there is no problem, since disbenefits are virtually never considered in the economic analysis. The public project setting will commonly use some form of the B/C ratio with the disbenefit considered. When a public-private partnership is initiated, there should be some agreement beforehand that establishes the economic measure acceptable for analysis and decision making throughout the project. Then the numerical dilemma presented above should not occur.

Online Presentation

ME alternative selection

9.3 Alternative Selection Using Incremental B/C Analysis ● ● ○

The technique to compare two mutually exclusive alternatives using benefit/cost analysis is virtually the same as that for incremental ROR in Chapter 8. The incremental (conventional) B/C ratio, which is identified as ΔB/C, is determined using PW, AW, or FW calculations. The higher-cost alternative is justified if ΔB/C is equal to or larger than 1.0. The selection rule is as follows:

If ΔB/C ≥ 1.0, choose the higher-cost alternative, because its extra cost is economically justified.

If ΔB/C < 1.0, choose the lower-cost alternative.

To perform a correct incremental B/C analysis, it is required that each alternative be compared only with another alternative for which the incremental cost is already justified. This same rule was used for incremental ROR analysis.

There are two dimensions of an incremental B/C analysis that differ from the incremental ROR method in Chapter 8. We already know the first, all costs have a positive sign in the B/C ratio. The second, and significantly more important, concerns the ordering of alternatives prior to incremental analysis.

> Alternatives are **ordered by increasing equivalent total costs,** that is, PW or AW of all cost estimates that will be utilized in the denominator of the B/C ratio. When not done correctly, the incremental B/C analysis may reject a justified higher-cost alternative.

If two alternatives, A and B, have equal initial investments and lives, but B has a larger equivalent annual cost, then B must be incrementally justified against A. (This is illustrated in Example 9.4 below.) If this convention is not correctly followed, it is possible to get a negative cost value in the denominator, which can incorrectly make B/C < 1 and reject a higher-cost alternative that is actually justified.

Follow these steps to correctly perform a conventional B/C ratio analysis of two alternatives. Equivalent values can be expressed in PW, AW, or FW terms.

1. Determine the equivalent total costs for both alternatives.
2. Order the alternatives by equivalent total cost: first smaller, then larger. Calculate the incremental cost (ΔC) for the larger-cost alternative. This is the denominator in ΔB/C.
3. Calculate the equivalent total benefits and any disbenefits estimated for both alternatives. Calculate the incremental benefits (ΔB) for the larger-cost alternative. This is $\Delta(B - D)$ if disbenefits are considered.
4. Calculate the ΔB/C ratio using Equation [9.2], $(B - D)/C$.
5. Use the selection guideline to select the higher-cost alternative if ΔB/C ≥ 1.0.

When the B/C ratio is determined for the lower-cost alternative, it is a comparison with the do-nothing (DN) alternative. If B/C < 1.0, then DN should be selected and compared to the second alternative. If neither alternative has an acceptable B/C value and one of the alternatives does not have to be selected, the DN alternative must be selected. In public sector analysis, the DN alternative is usually the current condition.

EXAMPLE 9.4

The city of Garden Ridge, Florida, has received designs for a new patient room wing to the municipal hospital from two architectural consultants. One of the two designs must be accepted in order to announce it for construction bids. The costs and benefits are the same in most categories, but the city financial manager decided that the estimates below should be considered to determine which design to recommend at the city council meeting next week and to present to the citizenry in preparation for an upcoming bond referendum next month.

	Design A	Design B
Construction cost, $	10,000,000	15,000,000
Building maintenance cost, $/year	35,000	55,000
Patient usage copay, $/year	450,000	200,000

The patient usage copay is an estimate of the amount paid by patients over the insurance coverage generally allowed for a hospital room. The discount rate is 5%, and the life of the building is estimated at 30 years.

(*a*) Use incremental B/C analysis to select design A or B.

(*b*) Once the two designs were publicized, the privately owned hospital in the directly adjacent city of Forest Glen lodged a complaint that design A will reduce its own municipal hospital's income by an estimated $500,000 per year because some of the day-surgery features of design A duplicate its services. Subsequently, the Garden Ridge merchants' association argued that design B could reduce its annual revenue by an estimated $400,000, because it will eliminate an entire parking lot used by their patrons for short-term parking. The city financial manager stated that these concerns would be entered into the evaluation as disbenefits of the respective designs. Redo the B/C analysis to determine if the economic decision is still the same as when disbenefits were not considered.

Solution

(*a*) Since most of the cash flows are already annualized, the incremental B/C ratio will use AW values. No disbenefit estimates are considered. Follow the steps of the procedure above:

1. The AW of costs is the sum of construction and maintenance costs.

$$AW_A = 10,000,000(A/P,5\%,30) + 35,000 = \$685,500$$
$$AW_B = 15,000,000(A/P,5\%,30) + 55,000 = \$1,030,750$$

2. Design B has the larger AW of costs, so it is the alternative to be incrementally justified. The incremental cost is

$$\Delta C = AW_B - AW_A = \$345,250 \text{ per year}$$

3. The AW of benefits is derived from the patient usage copays, since these are consequences to the public. The benefits for the ΔB/C analysis are not the estimates themselves, but the *difference* if design B is selected. The lower usage copay is a positive benefit for design B.

$$\Delta B = \text{usage}_A - \text{usage}_B = \$450,000 - \$200,000 = \$250,000 \text{ per year}$$

4. The incremental B/C ratio is calculated by Equation [9.2].

$$\Delta B/C = \frac{\$250,000}{\$345,250} = 0.72$$

5. The B/C ratio is less than 1.0, indicating that the extra costs associated with design B are not justified. Therefore, design A is selected for the construction bid.

(*b*) The revenue loss estimates are considered disbenefits. Since the disbenefits of design B are $100,000 less than those of A, this positive difference is added to the $250,000 benefits of B to give it a total benefit of $350,000. Now

$$\Delta B/C = \frac{\$350,000}{\$345,250} = 1.01$$

Design B is slightly favored. In this case the inclusion of disbenefits has reversed the previous economic decision. This has probably made the situation more difficult politically. New disbenefits will surely be claimed in the near future by other special-interest groups.

Like other methods, incremental B/C analysis requires **equal-service comparison** of alternatives. Usually, the expected useful life of a public project is long (25 or 30 or more years), so alternatives generally have equal lives. However, when alternatives do have unequal lives, the use of PW or AW to determine the equivalent costs and benefits requires that the LCM of lives be used to calculate ΔB/C. As with ROR analysis of two alternatives, this is an excellent opportunity to use the AW equivalency of estimated (not incremental) costs and benefits, if the implied assumption that the project could be repeated is reasonable. Therefore, use AW-based analysis of actual costs and benefits for B/C ratios when different-life alternatives are compared.

EXAMPLE 9.5   Water Treatment Facility #3 Case   PE

As our case unfolds, the consultant, Joel Whiterson, has pieced together some of the B/C analysis estimates for the 84-inch Jolleyville transmission main study completed last year. The two options for constructing this main were open trench (OT) for the entire 6.8-mile distance or a combination of trenching and bore tunneling (TT) for a shorter route of 6.3 miles. One of the two options had to be selected to transport approximately 300 million gallons per day (gpd) of treated water from the new WTF3 to an existing aboveground reservoir.

The general manager of Allen Water Utilities has stated publicly several times that the trench-tunnel combination option was selected over the open-trench alternative based on analysis of both quantitative and nonquantitative data. He stated the equivalent annual costs in an internal e-mail some months ago, based on the expected construction periods of 24 and 36 months, respectively, as equivalent to

$$AW_{OT} = \$1.20 \text{ million per year}$$
$$AW_{TT} = \$2.37 \text{ million per year}$$

This analysis indicated that the open-trench option was economically better, at that time. The planning horizon for the transmission mains is 50 years; this is a reasonable study period, Joel concluded. Use the estimates below that Joel has unearthed to perform a correct incremental B/C analysis and comment on the results. The interest (discount) rate is 3% per year, compounded annually, and 1 mile is 5280 feet.

	Open trench (OT)	Trench-tunnel (TT)
Distance, miles	6.8	6.3
First cost, $ per foot	700	Trench for 2.0 miles: 700
		Tunnel for 4.3 miles: 2100
Time to complete, months	24	36
Construction support costs, $ per month	250,000	175,000
Ancillary expenses, $ per month:		
Environmental	150,000	20,000
Safety	140,000	60,000
Community interface	20,000	5,000

Solution

One of the alternatives must be selected, and the construction lives are unequal. Since it is not reasonable to assume that this construction project will be repeated many cycles in the future, it is incorrect to conduct an AW analysis over the respective completion periods of 24 and 36 months, or the LCM of these time periods. However, the study period of 50 years is a reasonable evaluation time frame, since the mains are considered permanent installations. We can assume that the construction first costs are a present worth value in year 0, but the equivalent PW and 50-year AW of other monthly costs must be determined.

$$
\begin{aligned}
PW_{OT} &= PW \text{ of construction} + PW \text{ of construction support costs} \\
&= 700(6.8)(5280) + 250,000(12)(P/A,3\%,2) \\
&= \$30,873,300
\end{aligned}
$$

$$
\begin{aligned}
AW_{OT} &= 30,873,300(A/P,3\%,50) \\
&= \$1.20 \text{ million per year}
\end{aligned}
$$

$$
\begin{aligned}
PW_{TT} &= [700(2.0) + 2100(4.3)](5280) + 175,000(12)(P/A,3\%,3) \\
&= \$61,010,460
\end{aligned}
$$

$$
\begin{aligned}
AW_{TT} &= 61,010,460(A/P,3\%,50) \\
&= \$2.37 \text{ million per year}
\end{aligned}
$$

The *trench-tunnel (TT) alternative has a larger equivalent cost*; it must be justified against the OT alternative. The incremental cost is

$$\Delta C = AW_{TT} - AW_{OT} = 2.37 - 1.20 = \$1.17 \text{ million per year}$$

The difference between ancillary expenses defines the incremental benefit for TT.

$$
\begin{aligned}
PW_{OT\text{-anc}} &= 310,000(12)(P/A,3\%,2) \\
&= \$7,118,220
\end{aligned}
$$

$$
\begin{aligned}
AW_{OT\text{-anc}} &= 7,118,220(A/P,3\%,50) \\
&= \$276,685 \text{ per year}
\end{aligned}
$$

$$PW_{\text{TT-anc}} = 85,000(12)(P/A,3\%,3)$$
$$= \$2,885,172$$

$$AW_{\text{TT-anc}} = 2,885,172(A/P,3\%,50)$$
$$= \$112,147 \text{ per year}$$

$$\Delta B = AW_{\text{OT-anc}} - AW_{\text{TT-anc}} = 276,685 - 112,147 = \$164,538 \text{ per year} \quad (\$0.16 \text{ million})$$

Calculate the incremental B/C ratio.

$$\Delta B/C = 0.16/1.17 = 0.14$$

Since $\Delta B/C \ll 1.0$, the trench-tunnel option is not economically justified. Joel can now conclude that the general manager's earlier comment that the TT option was selected based on quantitative *and* nonquantitative data must have had heavy dependence on nonquantitative information not yet discovered.

9.4 Incremental B/C Analysis of Multiple, Mutually Exclusive Alternatives ● ● ●

The procedure necessary to select one from three or more mutually exclusive alternatives using incremental B/C analysis is essentially the same as that of Section 9.3. The procedure also parallels that for incremental ROR analysis in Section 8.6. The selection guideline is as follows:

> Choose the largest-cost alternative that is justified with an incremental B/C ≥ 1.0 when this selected alternative has been compared with another justified alternative.

There are two types of benefit estimates—estimation of **direct benefits,** and **implied benefits based on usage cost** estimates. The previous two examples (9.4 and 9.5) are good illustrations of the second type of implied benefit estimation. *When direct benefits are estimated,* the B/C ratio for each alternative may be calculated first as an initial screening mechanism to eliminate unacceptable alternatives. At least one alternative must have B/C ≥ 1.0 to perform the incremental B/C analysis. If all alternatives are unacceptable, the DN alternative is the choice. (This is the same approach as that of step 2 for "revenue alternatives only" in the ROR procedure of Section 8.6. However, the term *revenue alternative* is not applicable to public sector projects.)

As in the previous section when comparing two alternatives, selection from multiple alternatives by incremental B/C ratio utilizes equivalent total costs to initially order alternatives from smallest to largest. Pairwise comparison is then undertaken. Also, remember that all costs are considered positive in B/C calculations. The terms *defender* and *challenger alternative* are used in this procedure, as in a ROR-based analysis. The procedure for incremental B/C analysis of multiple alternatives is as follows:

1. Determine the equivalent total cost for all alternatives. Use AW, PW, or FW equivalencies.
2. Order the alternatives by equivalent total cost, smallest first.
3. Determine the equivalent total benefits (and any disbenefits estimated) for each alternative.
4. *Direct benefits estimation only:* Calculate the B/C for the first ordered alternative. If B/C < 1.0, eliminate it. By comparing each alternative to DN in order, we eliminate all that have B/C < 1.0. The lowest-cost alternative with B/C ≥ 1.0 becomes the defender and the next higher-cost alternative is the challenger in the next step. (For analysis by spreadsheet, determine the B/C for all alternatives initially and retain only acceptable ones.)
5. Calculate incremental costs (ΔC) and benefits (ΔB) using the relations

$$\Delta C = \text{challenger cost} - \text{defender cost}$$
$$\Delta B = \text{challenger benefits} - \text{defender benefits}$$

If relative **usage costs** are estimated for each alternative, rather than direct benefits, ΔB may be found using the relation

$$\Delta B = \textbf{defender usage costs} - \textbf{challenger usage costs} \qquad [9.5]$$

6. Calculate the ΔB/C for the first challenger compared to the defender.

$$\Delta B/C = \Delta B/\Delta C \qquad [9.6]$$

If $\Delta B/C \geq 1.0$ in Equation [9.6], the challenger becomes the defender and the previous defender is eliminated. Conversely, if $\Delta B/C < 1.0$, remove the challenger and the defender remains against the next challenger.

7. Repeat steps 5 and 6 until only one alternative remains. It is the selected one.

In all the steps above, incremental disbenefits may be considered by replacing ΔB with $\Delta(B - D)$.

EXAMPLE 9.6

Schlitterbahn Waterparks of Texas, a very popular water and entertainment park headquartered in New Braunfels, has been asked by four different cities outside of Texas to consider building a park in their area. All the offers include some version of the following incentives:

- Immediate cash incentive (year 0)
- A 10% of first-year incentive as a direct property tax reduction for 8 years
- Sales tax rebate sharing plan for 8 years
- Reduced entrance (usage) fees for area residents for 8 years

Table 9–1 (top section) summarizes the estimates for each proposal, including the present worth of the initial construction cost and anticipated annual revenue. The annual M&O costs are expected to be the same for all locations. Use incremental B/C analysis at 7% per year and an 8-year study period to advise the board of directors if they should consider any of the offers to be economically attractive.

Solution

The viewpoint is that of Schlitterbahn, and the benefits are direct estimates. Develop the AW equivalents over 8 years, and use the procedure detailed above. The results are presented in Table 9–1.

1. AW of total costs and an example for city 1 are determined in $1 million units.

$$\begin{aligned}
\text{AW of costs} &= \text{first cost}(A/P,7\%,8) + \text{entrance fee reduction to residents} \\
&= 38.5(0.16747) + 0.5 \\
&= \$6.948 \quad (\$6,948,000 \text{ per year})
\end{aligned}$$

2. The four alternatives are correctly ordered by increasing equivalent total cost in Table 9–1.

TABLE 9–1   Incremental B/C Analysis of Water Park Proposals, Example 9.6

	City 1	City 2	City 3	City 4
First cost, $ million	38.5	40.1	45.9	60.3
Entrance fee costs, $/year	500,000	450,000	425,000	250,000
Annual revenue, $ million/year	7.0	6.2	10.0	10.4
Initial cash incentive, $	250,000	350,000	500,000	800,000
Property tax reduction, $/year	25,000	35,000	50,000	80,000
Sales tax sharing, $/year	310,000	320,000	320,000	340,000
AW of total costs, $ million/year	6.948	7.166	8.112	10.348
AW of total benefits, $ million/year	7.377	6.614	10.454	10.954
Overall B/C	1.06	0.92	1.29	1.06
Alternatives compared	1-to-DN	B/C < 1.0	3-to-1	4-to-3
Incremental costs ΔC, $/year	6.948		1.164	2.236
Incremental benefits ΔB, $/year	7.377		3.077	0.50
$\Delta B/C$	1.06		2.64	0.22
Increment justified?	Yes	Eliminated	Yes	No
City selected	1		3	3

3. AW of total benefits and an example for city 1 are also determined in $1 million units.

$$\text{AW of benefits} = \text{revenue} + \text{initial incentive}(A/P,7\%,8)$$
$$+ \text{ property tax reduction} + \text{sales tax sharing}$$
$$= 7.0 + 0.25(0.16747) + 0.025 + 0.31$$
$$= \$7.377 \quad (\$7,377,000 \text{ per year})$$

4. Since benefits are directly estimated (and no disbenefits are included), determine the overall B/C for each alternative using Equation [9.1]. In the case of city 1,

$$\text{B/C}_1 = 7.377/6.948 = 1.06$$

City 2 is eliminated with $\text{B/C}_2 = 0.92$; the rest are initially acceptable.

5. The ΔC and ΔB values are the actual estimates for the 1-to-DN comparison.
6. The overall B/C is the same as $\Delta \text{B/C} = 1.06$, using Equation [9.6]. City 1 is economically justified and becomes the defender.
7. Repeat steps 5 and 6. Since city 2 is eliminated, the 3-to-1 comparison results in

$$\Delta C = 8.112 - 6.948 = 1.164$$
$$\Delta B = 10.454 - 7.377 = 3.077$$
$$\Delta \text{B/C} = 3.077/1.164 = 2.64$$

City 3 is well justified and becomes the defender against city 4. From Table 9–1, $\Delta \text{B/C} = 0.22$ for the 4-to-3 comparison. City 4 falls out easily, and **city 3** is the one to recommend to the board. Note that the DN alternative could have been selected had no proposal met the B/C or $\Delta \text{B/C}$ requirements.

Independent project selection

When two or more *independent projects* are evaluated using B/C analysis and there is no budget limitation, no incremental comparison is necessary. The only comparison is between each project separately with the do-nothing alternative. The project B/C values are calculated, and those with $\text{B/C} \geq 1.0$ are accepted.

This is the same procedure as that used to select from independent projects using the ROR method (Chapter 8). When a budget limitation is imposed, the capital budgeting procedure discussed in Chapter 12 must be applied.

When the lives of mutually exclusive alternatives are so long that they can be considered infinite, the capitalized cost is used to calculate the equivalent PW or AW values for costs and benefits. Equation [5.3], $A = P(i)$, is used to determine the equivalent AW values in the incremental B/C analysis. Example 9.7 illustrates this using the progressive example and a spreadsheet.

EXAMPLE 9.7 Water Treatment Facility #3 Case

Land for Water Treatment Facility #3 was initially purchased in the year 2010 for $19.3 million; however, when it was publicized, influential people around Allen spoke strongly against the location. We will call this location 1. Some of the plant design had already been completed when the general manager announced that this site was not the best choice anyway, and that it would be sold and a different, better site (location 2) would be purchased for $28.5 million. This was well over the budget amount of $22.0 million previously set for land acquisition. As it turns out, there was a third site (location 3) available for $35.0 million that was never seriously considered.

In his review and after much resistance from Allen Water Utilities staff, the consultant, Joel, received a copy of the estimated costs and benefits for the three plant location options. The revenues, savings, and sale of bulk water rights to other communities are estimated as increments from a base amount for all three locations. Using the assumption of a very long life for the WTF3 facility and the established discount rate of 3% per year, determine what Joel discovered when he did the B/C analysis. Was the general manager correct in concluding that location 2 was the best, all said and done?

	Location 1	Location 2	Location 3
Land cost, $ million	19.3	28.5	35.0
Facility first cost, $ million	460.0	446.0	446.0
Benefits, $ per year:			
Pumping cost savings	5	3	0
Sales to area communities	12	10	8
Added revenue from Allen	6	6	6
Total benefits, $ per year	23	19	14

Solution

A spreadsheet can be very useful when performing an incremental B/C analysis of three or more alternatives. Figure 9–1a presents the analysis with the preliminary input of AW values for costs using the relation $A = P(i)$ and annual benefits. Figure 9–1b details all the functions used in the analysis. Logical IF statements indicate alternative elimination and selection decisions. In $1 million units,

$$\text{Location 1:} \quad \text{AW of costs} = A \text{ of land cost} + A \text{ of facility first cost}$$
$$= (19.3 + 460.0)(0.03)$$
$$= \$14.379 \text{ per year}$$
$$\text{AW of benefits} = \$23$$

$$\text{Location 2:} \quad \text{AW of costs} = \$14.235 \quad \text{AW of benefits} = \$19$$

$$\text{Location 3:} \quad \text{AW of costs} = \$14.430 \quad \text{AW of benefits} = \$14$$

Though the AW of cost values are close to one another, the increasing order is locations 2, 1, and 3 to determine ΔB/C values. The benefits are direct estimates; therefore, the overall B/C ratios indicate that location 3 (row 5; B/C$_3$ = 0.97) is not economically justified at the outset. It is eliminated, and one of the remaining locations must be selected. Location 2 is justified against the DN alternative (B/C$_2$ = 1.33); the only remaining comparison is 1-to-2 as detailed in column C of Figure 9–1. **Location 1** is a clear winner with ΔB/C = 27.78.

In conclusion, Joel has learned that location 1 is indeed the best and that, from the economic perspective, the general manager was incorrect in stating that location 2 was better. However, given the original evaluation criteria listed in the introduction—*economics, environment, community impact, and constructability*—location 2 is likely a good compromise selection.

	A	B	C	D
1				
2	**Order of analysis**	**Location 2**	**Location 1**	**Location 3**
3	AW of cost, $M/year	14.235	14.379	14.430
4	Annual benefits, $M/year	19.0	23.0	14.0
5	Overall B/C	1.33	1.60	0.97
6	Acceptable	Yes	Yes	No
7				
8	Comparison		1-to-2	Eliminated
9	ΔC $/year		0.144	
10	ΔB, $/year		4.0	
11	ΔB/C		27.78	
12	Increment justified?		Yes	
13	Selection		Location 1	

(a)

	A	B	C	D
1				
2	**Order of analysis**	**Location 2**	**Location 1**	**Location 3**
3	AW of cost, $M/year	14.235	14.379	14.43
4	Annual benefits, $M/year	19	23	14
5	Overall B/C	= B4/B3	= C4/C3	= D4/D3
6	Acceptable	= IF(B5<1,"No","Yes")	= IF(C5<1,"No","Yes")	= IF(D5<1,"No","Yes")
7				
8	Comparison		1-to-2	Eliminated
9	ΔC $/year		= C3-B3	
10	ΔB, $/year		= C4-B4	
11	ΔB/C		= C10/C9	
12	Increment justified?		= IF(C11<1,"No","Yes")	
13	Selection		= IF(C12="Yes",C2,"No")	

(b)

Figure 9–1
Incremental B/C analysis for WTF3 case: (a) numerical results and (b) functions developed for the analysis.

Comment

This is an actual situation with changed names and values. Location 1 was initially purchased and planned for WTF3. However, the presence of political, community, and environmental stress factors changed the decision to location 2, when all was said and done.

9.5 Service Sector Projects and Cost-Effectiveness Analysis ● ● ●

Much of the GDP of the United States and some countries in Europe and Asia is generated by what has become known as the service sector of the economy. A large percentage of service sector projects are generated by and dependent upon the private sector (corporations, businesses, and other for-profit institutions). However, many projects in the public sector are also service sector projects.

> A **service sector project** is a process or system that provides services to individuals, businesses, or government units. The economic value is developed primarily by the **intangibles** of the process or system, **not the physical entities,** such as buildings, machines, and equipment. Manufacturing and construction activities are commonly not considered a service sector project, though they may support the theme of the service provided.

Service projects have a tremendous range of variety and purpose; to name a few: health care systems, health and life insurance, airline reservation systems, credit card services, police and court systems, security programs, safety training programs, and all types of consulting projects. The intangible and intellectual work done by engineers and other professionals is often a part of a service sector project.

The economic evaluation of a service project is difficult to a great degree because the cost and benefit estimates are not accurate and often not within an acceptable degree of error. In other words, undue risk may be introduced into the decision because of poor monetary estimates. For example, consider the decision to place red-light cameras at stop lights to ticket drivers who run the red light. This is a public and a service project, but its (*economic*) *benefits are quite difficult to estimate*. Depending upon the viewpoint, benefits could be in terms of accidents averted, deaths prevented, police personnel released from patrolling the intersection, or, from a more mercenary viewpoint, amount of fines collected. In all but the last case, benefits in monetary terms will be poor estimates. These are examples where B/C analysis does not work well and a different form of analysis is needed.

In service and public sector projects, as expected, it is the benefits that are the more difficult to estimate. An evaluation method that combines monetary cost estimates with nonmonetary benefit estimates is **cost-effectiveness analysis (CEA)**. The CEA approach utilizes a **cost-effectiveness measure** or the **cost-effectiveness ratio (CER)** as a basis of ranking projects and selecting the best of independent projects or mutually exclusive alternatives. The CER ratio is defined as

$$\text{CER} = \frac{\text{equivalent total costs}}{\text{total of effectiveness measure}} = \frac{C}{E} \qquad [9.7]$$

In the red-light camera example, the effectiveness measure (the benefit) may be one of the samples mentioned earlier, accidents averted or deaths prevented. Different from the B/C ratio of costs to benefits, CER places the PW or AW of total costs in the numerator and the effectiveness measure in the denominator. (The reciprocal of Equation [9.7] can also be used as the measure of worth, but we will use CER as defined above.) With costs in the numerator, smaller ratio values are more desirable for the same value of the denominator, since smaller ratio values indicate a lower cost for the same level of effectiveness.

Like ROR and B/C analysis, cost-effectiveness analysis requires the ordering (ranking) of alternatives prior to selection and the use of incremental analysis for mutually exclusive alternative selection. Cost-effectiveness analysis utilizes a different ranking criterion than ROR or BC analysis. The *ordering criteria* are as follows:

> **Independent projects:** Initially rank projects by **CER value.**
>
> **Mutually exclusive alternatives:** Initially rank alternatives by **effectiveness measure,** then perform an incremental CER analysis.

Return again to the public/service project of red-light cameras. If the CER is defined as "*cost per total accidents averted*" and the projects are independent, increasing CER value is the ranking basis. If the projects are mutually exclusive, "*total accidents averted*" is the correct ranking basis and an incremental analysis is necessary.

    There are significantly different analysis procedures for independent and mutually exclusive proposals. To select some from several (independent) projects, a budget limit, termed *b,* is inherently necessary once ordering is complete. However, for selecting one from several (mutually exclusive) alternatives, a pairwise incremental analysis is necessary and selection is made on the basis of ΔC/E ratios. The procedures and examples follow.

> For **independent projects,** the procedure is as follows:
>
> 1. Determine the equivalent total cost *C* and effectiveness measure *E*, and calculate the CER measure for each project.
> 2. Order projects from the *smallest to the largest CER value.*
> 3. Determine cumulative cost for each project and compare with the budget limit *b*.
> 4. The selection criterion is to fund all projects such that *b* is not exceeded.

Independent project selection

EXAMPLE 9.8

Recent research indicates that corporations throughout the world need employees who demonstrate creativity and innovation for new processes and products. One measure of these talents is the number of patents approved each year through the R&D efforts of a company. Rollings Foundation for Innovative Thinking has allocated $1 million in grant funds to award to corporations that enroll their top R&D personnel in a 1- to 2-month professional training program in their home state that has a historically proven track record over the last 5 years in helping individuals earn patents.

    Table 9–2 summarizes data for six corporations that submitted proposals. Columns 2 and 3 give the proposed number of attendees and cost per person, respectively, and column 4 provides the historical track record of program graduates in patents per year. Use cost-effectiveness analysis to select the corporations and programs to fund.

Solution

We assume that across all programs and all patent awards there is equal quality. Use the procedure for independent projects and *b* = $1 million to select from the proposals.

TABLE 9–2	Data for Programs to Increase Patents Used for CEA		
Program (1)	Total Personnel (2)	Cost/Person, $ (3)	5-Year History, Patents/Graduate/Year (4)
1	50	5000	0.5
2	35	4500	3.1
3	57	8000	1.9
4	24	2500	2.1
5	12	5500	2.9
6	87	3800	0.6

1. Using Equation [9.7], the effectiveness measure E is patents per year, and the CER is

$$\text{CER} = \frac{\text{program cost per person}}{\text{patents per graduate}} = \frac{C}{E}$$

The program cost C is a PW value, and the E values are obtained from the proposals.
2. The CER values are shown in Table 9–3 in increasing order, column 5.
3. Cost per course, column 6, and cumulative costs, column 7, are determined.
4. Programs 4, 2, 5, 3, and 6 (68 of the 87 people) are selected to not exceed $1 million.

TABLE 9–3	Programs Ordered by CER Value, Example 9.8					
Program (1)	Total Personnel (2)	Cost/Person C, $ (3)	Patents per Year E (4)	CER, $ per Patent (5) = (3)/(4)	Program Cost, $ (6) = (2)(3)	Cumulative Cost, $ (7) = Σ(6)
4	24	2,500	2.1	1,190	60,000	60,000
2	35	4,500	3.1	1,452	157,500	217,500
5	12	5,500	2.9	1,897	66,000	283,500
3	57	8,000	1.9	4,211	456,000	739,500
6	87	3,800	0.6	6,333	330,600	1,070,100
1	50	5,000	0.5	10,000	250,000	1,320,100

Comment

This is the first time that a budget limit has been imposed for the selection among independent projects. This is often referred to as *capital budgeting*, which is discussed further in Chapter 12.

For **mutually exclusive** alternatives and no budget limit, the alternative with the highest effectiveness measure E is selected without further analysis. Otherwise, an incremental CER analysis is necessary and the budget limit is applied to the selected alternative(s). The analysis is based on the incremental ratio $\Delta C/E$, and the procedure is similar to that we have applied for incremental ROR and B/C, except now the concept of *dominance* is utilized.

ME alternative
selection

Dominance occurs when the incremental analysis indicates that the challenger alternative offers an improved incremental CER measure compared to the defender's CER, that is,

$$(\Delta C/E)_{\text{challenger}} < (C/E)_{\text{defender}}$$

Otherwise, no dominance is present, and both alternatives remain in the analysis.

For **mutually exclusive alternatives,** the selection procedure is as follows:

1. Order the alternatives from *smallest to largest effectiveness measure E*. Record the cost for each alternative.
2. Calculate the CER measure for the first alternative. This, in effect, makes DN the defender and the first alternative the challenger. This CER is a baseline for the next incremental comparison, and the first alternative becomes the new defender.
3. Calculate incremental costs (ΔC) and effectiveness (ΔE) and the incremental measure $\Delta C/E$ for the new challenger using the relation

$$\Delta C/E = \frac{\text{cost of challenger} - \text{cost of defender}}{\text{effectiveness of challenger} - \text{effectiveness of defender}} = \frac{\Delta C}{\Delta E}$$

4. If $\Delta C/E < C/E_{\text{defender}}$, the challenger dominates the defender and it becomes the new defender; the previous defender is eliminated. Otherwise, no dominance is present and both alternatives are retained for the next incremental evaluation.
5. *Dominance present:* Repeat steps 3 and 4 to compare the next ordered alternative (challenger) and new defender. Determine if dominance is present.

Dominance not present: The current challenger becomes the new defender, and the next alternative is the new challenger. Repeat steps 3 and 4 to compare the new challenger and new defender. Determine if dominance is present.

6. Continue steps 3 through 5 until only one alternative or only nondominated alternatives remain.
7. Apply the budget limit (or other criteria) to determine which of the remaining alternative(s) can be funded.

EXAMPLE 9.9

One of the corporations not selected for funding in Example 9.8 decided to fund its 50 R&D personnel to attend one of the innovation and creativity programs at its own expense. One criterion is that the program must have a historical average for a graduate of at least 2.0 patents per year. Use the data in Table 9–2 to select the best program.

TABLE 9–4   Mutually Exclusive Alternatives Evaluated by Cost-Effectiveness Analysis, Example 9.9

Program (1)	Total Personnel (2)	Cost/Person C, $ (3)	Patents per Year E (4)	CER, $ per Patent (5) = (3)/(4)	Program Cost, $ (6) = (2)(3)
4	50	2,500	2.1	1190	125,000
5	50	5,500	2.9	1897	275,000
2	50	4,500	3.1	1452	225,000

Solution

From Table 9–2, three programs—2, 4, and 5—have a historical record of at least two patents per graduate per year. Since only one program will be selected, these are now *mutually exclusive alternatives.* Use the procedure to perform the incremental analysis.

1. The alternatives are ranked by increasing patents per year in Table 9–4, column 4.
2. The CER measure for program 4 is compared to the DN alternative.

$$C/E_4 = \frac{\text{program cost per person}}{\text{patents per graduate}} = \frac{2500}{2.1} = 1190$$

3. Program 5 is now the challenger.

$$\text{5-to-4 comparison: } \Delta C/E = \frac{\Delta C}{\Delta E} = \frac{5500 - 2500}{2.9 - 2.1} = 3750$$

4. In comparison to $C/E_4 = 1190$, it costs $3750 per additional patent if program 5 is chosen over 4. Program 5 is more expensive for more patents; however, clear dominance is not present; both programs are retained for further evaluation.
5. *Dominance not present:* Program 5 becomes the new defender, and program 2 is the new challenger. Perform the 2-to-5 comparison.

$$\text{2-to-5 comparison: } \Delta C/E = \frac{\Delta C}{\Delta E} = \frac{4500 - 5500}{3.1 - 2.9} = -5000$$

Compared to $C/E_5 = 1897$, this increment is much cheaper—more patents for less money per person. Dominance is present; eliminate program 5 and compare 2 to 4.
6. Repeat steps 3 through 5 and compare $\Delta C/E$ to $C/E_4 = 1190$.

$$\text{2-to-4 comparison: } \Delta C/E = \frac{\Delta C}{\Delta E} = \frac{4500 - 2500}{3.1 - 2.1} = 2000$$

This does not represent dominance of program 2 over 4. The conclusion is that both programs are eligible for funding, that is, CEA in this case does not indicate only one program. This occurs when there is not lower cost and higher effectiveness of one alternative over another; that is, one alternative does not dominate all the others.
7. Now the budget and other considerations (probably noneconomic) are brought to bear to make the final decision. The fact that program 4 costs $125,000—significantly less than program 2 at $225,000—will likely enter into the decision.

Cost-effectiveness analysis is a form of multiattribute decision-making in which economic and noneconomic dimensions are integrated to evaluate alternatives from several perspectives by different decision makers. See Chapter 10 for further discussion of alternative analysis using multiple attributes.

9.6 Ethical Considerations in the Public Sector ● ● ○

Usual expectations by citizens of their elected officials—locally, nationally, and internationally—are that they make decisions for the good of the public, ensuring safety and minimizing risk and cost to the public. Above these is the long-standing expectation that public servants have **integrity.**

Similarly, the expectations of engineers employed by government departments, and those serving as consultants to government agencies, are held to high standards. **Impartiality,** consideration of a **wide range of circumstances,** and the use of **realistic assumptions** are but three of the foundation elements upon which engineers should base their recommendations to decision makers. This implies that engineers in public service avoid

- Self-serving, often greedy individuals and clients with goals of excessive profits and future contract awards
- Using a politically favorable perspective that compromises the results of a study
- Narrowly-defined assumptions that serve special interest groups and subcommunities potentially affected by the findings

Many people are disappointed and discouraged with government when elected officials and public employees (engineers and others, alike) do not have real commitment to integrity and unbiasedness in their work.

Engineers are routinely involved in two of the major aspects of public sector activities:

Public policy making—the **development of strategy** for public service, behavior, fairness, and justice. This may involve literature study, background discovery, data collection, opinion giving, and hypothesis testing. An example is *transportation management.* Engineers make virtually all the recommendations based on data and long-standing decision algorithms for policy items such as capacity of roads, expansion of highways, planning and zoning rules, traffic signal usage, speed limit corridors, and many related topics in transportation policy. Public officials use these findings to establish public transportation policy.

Public planning—the **development of projects** that implement strategy and affect people, the environment, and financial resources in a variety of ways. Consider *traffic control,* where the use and placement of traffic control signs, signals, speed limits, parking restrictions, etc. are detailed based upon established policy and current data. (In effect, this is systems engineering, that is, an application of the life-cycle phases and stages explained in Section 6.5.)

Whether in the arena of policy making or public planning, engineers can find ethical compromise a possibility when working with the public sector. A few circumstances are summarized here.

- *Use of technology*   Many public projects involve the use of new technology. The public risks and safety factors are not always known for these new advances. It is common and expected that engineers make every attempt to apply the latest technology while ensuring that the public is not exposed to undue risk.
- *Scope of study*   A client may pressure the engineers to limit the range of options, the assumption base, or the breadth of alternative solutions. These restrictions may be based on financial reasons, politically-charged topics, client-favored options, or a wide variety of other reasons. To remain impartial, it is the responsibility of the engineer to submit a fully unbiased analysis, report, and recommendation, even though it may jeopardize future contract possibilities, promote public disfavor, or generate other negative consequences.
- *Negative community impact*   It is inevitable that public projects will adversely affect some groups of people, or the environment, or businesses. The intentional silencing of these projected effects is often the cause of strong public outcries against what may be a project that is in the best interest of the community at large. Engineers who find (stumble onto) such negative impacts may be pressured by clients, managers, or public figures to overlook them, though

TABLE 9–5 Some Ethical Considerations When Performing B/C and CEA Analysis

What the Study Includes	Ethical Dimension	Example
Audience for study	Is it ethical to select a specific group of people affected by the project and neglect possible effects on other groups?	Construct children's health care clinics for city dwellers, but neglect rural families with poor transportation means.
Impact time of decision	Is it ethical to decide now for future generations who may be adversely and economically affected by the current project decision?	Accomplish financial bailouts of corporations when future generations' taxes will be significantly higher to recover the costs, plus interest and inflation effects.
Greater good for community as a whole	Vulnerable minority groups, especially economically deprived ones, may be disproportionally affected. Is this ethical if the impact is predictable?	Allow a chemical plant that is vital to the community's employment to pollute a waterway when a minority group is known to eat fish from the water that is predictably contaminated.
Reliance on economic measures only	Is it acceptable to reduce all costs and benefits to monetary estimates for a decision, then subjectively impute nonquantified factors in the final decision?	Softening of building codes can improve the financial outlook for home builders; however, increased risks of fire loss, storm and water damage to structures, and reduced future resale values are considered only in passing as a new subdivision is approved by the planning and zoning committee.
Scope of disbenefits estimated and evaluated	Is it ethical to disregard any disbenefits in the B/C study or use indirect effectiveness measures in a CEA study based on the difficulty to estimate some of them?	Noise and air pollution caused by a planned open-pit quarry will have a negative effect on area ranchers, residents, wildlife, and plant life; but the effectiveness measure considers only suburban residents due to estimation difficulty of effects on other constituencies.

the Code of Ethics for Engineers dictates a full and fair analysis and report. For example, a planned rerouting of a city street may effectively cut off a section of international citizens' businesses, thus resulting in a clearly predictable economic downturn. Considering this outcome in the recommendation to the transportation department should be a goal of the analyzing engineers, yet pressure to bias the results may be quite high.

The results of a B/C or CEA analysis are routinely depended upon by public officials and government staff members to assist in making public planning decisions. As discussed earlier, estimations for benefits, disbenefits, effectiveness measures, and costs can be difficult and inaccurate, but these analysis tools are often the best available to structure a study. Some example ethically-oriented challenges that may be confronted during B/C and CEA analyses are summarized in Table 9–5.

CHAPTER SUMMARY

The benefit/cost method is used primarily to evaluate alternatives in the public sector. When one is comparing mutually exclusive alternatives, the incremental B/C ratio must be greater than or equal to 1.0 for the incremental equivalent total cost to be economically justified. The PW, AW, or FW of the initial costs and estimated benefits can be used to perform an incremental B/C analysis. For independent projects, no incremental B/C analysis is necessary. All projects with B/C ≥ 1.0 are selected provided there is no budget limitation. It is usually quite *difficult to make accurate estimates of benefits* for public sector projects. The characteristics of public sector projects are substantially different from those of the private sector: initial costs are larger; expected life is longer; additional sources of capital funds include taxation, user fees, and government grants; and interest (discount) rates are lower.

Service projects develop economic value largely based on the intangibles of the services provided to users, not the physical items associated with the process or system. Evaluation by B/C analysis can be difficult with no good way to make monetary estimates of benefits. Cost-effectiveness analysis (CEA) combines cost estimates and a nonmonetary effectiveness measure (the benefit) to evaluate independent or mutually exclusive projects using procedures that are similar to incremental ROR and B/C analysis. The concept of dominance is incorporated into the procedure for comparing mutually exclusive alternatives.

As a complement to the discussion on professional ethics in Chapter 1, some potential ethical challenges in the public sector for engineers, elected officials, and government consultants are discussed here. Examples are included.

PROBLEMS

Public Projects and B/C Concepts

9.1 What is the difference between disbenefits and costs?

9.2 Identify the following as primarily public or private sector undertakings: eBay, farmer's market, state police department, car racing facility, social security, EMS, ATM, travel agency, amusement park, gambling casino, swap meet, football stadium.

9.3 State whether the following characteristics are primarily associated with public or private sector projects: large initial investment, park user fees, short-life projects, profit, disbenefits, tax-free bonds, subsidized loans, low interest rate, income tax, water quality regulations.

9.4 Identify the following cash flows as a benefit, disbenefit, or cost.
  (a) Loss of income to local businesses because of a new freeway
  (b) Less travel time because of a loop bypass
  (c) $400,000 annual income to local businesses because of tourism created by a national park
  (d) Cost of fish from a hatchery to stock a lake at the state park
  (e) Less tire wear because of smoother road surfaces
  (f) Decrease in property values due to the closure of a government research lab
  (g) School overcrowding because of a military base expansion
  (h) Revenue to local motels because of an extended weekend holiday

9.5 What is a fundamental difference between DBOM and DBOMF contracts?

9.6 Buster County has proposed a strict water conservation policy for all industrial plants within the county limits. Enforcement is proposed to be a public-private partnership between the sheriff's office and a private security company. In preparing for a B/C analysis of the proposal, estimates must be categorized as a benefit, cost, or disbenefit. The categories chosen will vary depending upon a person's viewpoint. Assume that possible viewpoints include sales revenues, politics, service to the public, the environment, contract obligations, future generations of residents, legal matters, revenue and budget (in general), and customers.
  (a) Select the *top two viewpoints* (in your opinion) for each of the following individuals as they would categorize estimates as a cost, benefit, or disbenefit.
    1. An industrial plant manager in the county
    2. County sheriff's deputy (appointed office)
    3. County commissioner (elected office)
    4. Security company president
  (b) Explain your answers by writing a short description of why you selected as you did.

Project B/C Value

9.7 If an alternative has a salvage value, how is it handled in the calculation of a B/C ratio relative to benefits, disbenefits, costs, or savings?

9.8 The cost of grading and spreading gravel on a short rural road is expected to be $300,000. The road will have to be maintained at a cost of $25,000 per year. Even though the new road is not very smooth, it allows access to an area that previously could only be reached with off-road vehicles. The improved accessibility has led to a 150% increase in the property values along the road. If the previous market value of a property was $900,000, calculate the B/C ratio using an interest rate of 6% per year and a 20-year study period.

9.9 Arsenic enters drinking water supplies from natural deposits in the earth or from agricultural and industrial practices. Since it has been linked to cancer of the bladder, kidney, and other internal organs, the EPA has lowered the arsenic standard for drinking water from 0.050 parts per million to 0.010 parts per million (10 parts per billion). The annual cost to public water utilities to meet the new standard is estimated to be $200 per household. If it is estimated that there are 90 million households in the United States and that the lower standard can save 50 lives per year valued at $4,000,000 per life, what is the benefit/cost ratio of the regulation?

9.10 From the following estimates, determine the B/C ratio for a project that has a 20-year life. Use an interest rate of 8% per year.

Consequences to the People		Consequences to the Government	
Annual benefits	$90,000 per year	First cost	$750,000
Annual disbenefits	$10,000 per year	Annual cost	$50,000 per year
		Annual savings	$30,000 per year

9.11 A project to extend irrigation canals into an area that was recently cleared of mesquite trees (a nuisance tree in Texas) and large weeds is projected to have a capital cost of $2,000,000. Annual maintenance and operation costs will be $100,000 per year. Annual favorable consequences to the general public of $820,000 per year will be offset to some extent by annual adverse consequences of $400,000 to a portion of the general public. If the project is assumed to have a 20-year life, what is the B/C ratio at an interest rate of 8% per year?

9.12 Calculate the B/C ratio for the following cash flow estimates at a discount rate of 7% per year.

Item	Cash Flow
FW of benefits, $	30,800,000
AW of disbenefits, $ per year	105,000
First cost, $	1,200,000
M&O costs, $ per year	400,000
Life of project, years	20

9.13 The benefits associated with a nuclear power plant cooling water filtration project located on the Ohio River are $10,000 per year forever, starting in year 1. The costs are $50,000 in year 0 and $50,000 at the end of year 2. Calculate the B/C ratio at $i = 10\%$ per year.

9.14 A privately funded wind-based electric power generation company in the southern part of the country has developed the following estimates

(in $1000) for a new turbine farm. The MARR is 10% per year, and the project life is 25 years. Calculate (*a*) the profitability index and (*b*) the modified B/C ratio.

Benefits:	$20,000 in year 0 and $30,000 in year 5
Savings:	$2000 in years 1–20
Cost:	$50,000 in year 0
Disbenefits:	$3000 in years 1–10

9.15 For the values shown, calculate the conventional B/C ratio at $i = 10\%$ per year.

	PW, $	AW, $/Year	FW, $
First cost	100,000	—	259,370
M&O cost	61,446	10,000	159,374
Benefits	—	40,000	637,496
Disbenefits	30,723	5,000	—

9.16 A proposal to reduce traffic congestion on I-5 has a B/C ratio of 1.4. The annual worth of benefits minus disbenefits is $560,000. What is the first cost of the project if the interest rate is 6% per year and the project is expected to have a 20-year life?

9.17 Oil spills in the Gulf of Mexico have been known to cause extensive damage to both public and private oyster grounds along the Louisiana and Mississippi shores. One way to protect shellfish along the shoreline is to release large volumes of freshwater from the Mississippi River to flush oil out to sea. This procedure inevitably results in death to some of the saltwater shellfish while preventing more widespread destruction to public reefs. Oil containment booms and other temporary structures can also be used to intercept floating oil before it damages sensitive fishing grounds. If the Fish and Wildlife Service spent $110 million in year 0 and $50 million in years 1 and 2 to minimize environmental damage from one particular oil spill, what is the benefit-to-cost ratio provided the efforts resulted in saving 3000 jobs valued at a total of $175 million per year? Assume disbenefits associated with oyster deaths amounted to $30 million in year 0. Use a 5-year study period and an interest rate of 8% per year.

9.18 On-site granular ferric hydroxide (GFH) systems can be used to remove arsenic from water when daily flow rates are relatively low. If the operating cost is $600,000 per year and the public health benefits are assumed to be $800,000 per year, what initial investment in the GFH system is necessary to guarantee a modified B/C ratio of at least 1.0? Assume that the equipment life is 10 years and the interest rate is 6% per year.

9.19 The Parks and Recreation Department of Burkett County has estimated that the initial cost of a "bare-bones" permanent river park will be $2.3 million. Annual upkeep costs are estimated at $120,000. Benefits of $340,000 per year and disbenefits of $40,000 per year have also been identified. Using a discount rate of 6% per year, calculate (a) the conventional B/C ratio and (b) the modified B/C ratio.

9.20 From the following data, calculate the (a) conventional and (b) modified benefit/cost ratios using an interest rate of 6% per year and an infinite project period.

	To the People		To the Government
Benefits:	$300,000 now and $100,000 per year thereafter	Costs:	$1.5 million now and $200,000 three years from now
Disbenefits:	$40,000 per year	Savings:	$70,000 per year

9.21 In 2010, Brazil began construction of the Belo Monte hydroelectric dam on the Xingu River (which feeds the Amazon River). The project is funded by a consortium of investors and is expected to cost $11 billion. It will begin producing electricity in 2015. Even though the dam will provide clean energy for millions of people, environmentalists are sharply opposed. They say it will devastate wildlife and the livelihoods of 40,000 people who live in the area to be flooded.

Assume that the funding will occur evenly over the 5-year period from 2010 through 2014 at $2.2 billion per year. The disbenefits are estimated to be $100,000 for each displaced person and $1 billion for wildlife destruction. Assume that the disbenefits will occur evenly through the 5-year construction period and anticipated benefits will begin at the end of 2015 and continue indefinitely. Use an interest rate of 8% per year to determine what the equivalent annual benefits must be to ensure a B/C ratio of at least 1.0.

9.22 In the United States, the average number of airplanes in the sky on an average morning is 4000. There are another 16,000 planes on the ground. Aerospace company Rockwell Collins developed what it calls a *digital parachute*—a panic-button technology that will land any plane in a pinch at the closest airport, no matter what the weather or geography and without the help of a pilot. The technology can be applied if a pilot is no longer capable of flying the plane or is panicked and confused about what to do in an emergency. Assume that the cost of retrofitting 20,000 commercial air-

planes is $100,000 each and the plane stays in service for 15 years. If the technology saves an average of 30 lives per year, with the value of a human life placed at $4,000,000, what is the B/C ratio? Use an interest rate of 10% per year.

9.23 Although the lower Rio Grande is regulated by the Elephant Butte Dam and the Caballo Reservoir, serious flooding has occurred in El Paso and other cities located along the river. This has required homeowners living in valley areas near the river to purchase flood insurance costing between $145 and $2766 per year. To alleviate the possibility of flooding, the International Boundary and Water Commission undertook a project costing $220 million to raise the levees along flood-prone portions of the river. As a result, 13,000 properties were freed of the federal mandate to purchase flood insurance. In addition, historical records indicate that damage to infrastructure will be avoided, which amounts to an average benefit of $8,200,000 per year. If the average cost of flood insurance is $460 per household per year, calculate the benefit-to-cost ratio of the levee-raising project. Use an interest rate of 6% per year and a 30-year study period.

9.24 For the data shown, calculate the *conventional* B/C ratio at $i = 6\%$ per year.

Benefits:	$20,000 in year 0 and $30,000 in year 5
Disbenefits:	$7000 in year 3
Savings (to government):	$25,000 in years 1–4
Cost:	$100,000 in year 0
Project life:	5 years

9.25 Explain a fundamental difference between the modified B/C ratio and the profitability index.

9.26 Gerald Corporation entered a public-private partnership using a DBOM contract with the state of Massachusetts 10 years ago for railroad system upgrades. Determine the profitability index for the financial results listed below using a MARR of 8% per year.

Investments:	Year 0	$-4.2 million
	Year 5	$-3.5 million
Net savings:	Years 1–5	$1.2 million per year
	Years 6–10	$2.5 million per year

9.27 A project had a staged investment distributed over the 6-year contract period. For the cash flows shown (next page) and an interest rate of 10% per year, determine the profitability index and determine if the project was economically justified.

Year	0	1	2	3	4	5	6
Investment, $1000	−25	0	−10	0	−5	0	0
NCF, $1000 per year	0	5	7	9	11	13	15

Two Alternative Comparison

9.28 In comparing two alternatives by the B/C method, if the overall B/C ratio for both alternatives is calculated to be exactly 1.0, which alternative should you select?

9.29 In comparing alternatives X and Y by the B/C method, if $B/C_X = 1.6$ and $B/C_Y = 1.8$, what is known about the B/C ratio on the increment of investment between X and Y?

9.30 Logan Well Services Group is considering two sites for storage and recovery of reclaimed water. The mountain site (MS) will use injection wells that cost $4.2 million to develop and $280,000 per year for M&O costs. This site will be able to accommodate 150 million gallons per year. The valley site (VS) will involve recharge basins that cost $11 million to construct and $400,000 to operate and maintain. At this site, 890 million gallons can be injected each year. If the value of the injected water is $3.00 per 1000 gallons, which alternative, if either, should be selected according to the B/C ratio method? Use an interest rate of 8% per year and a 20-year study period.

9.31 The estimates shown are for a bridge under consideration for a river crossing in Wheeling, West Virginia. Use the B/C ratio method at an interest rate of 6% per year to determine which bridge, if either, should be built.

	East Location	West Location
Initial cost, $	11×10^6	27×10^6
Annual M&O, $/year	100,000	90,000
Benefits, $/year	990,000	2,400,000
Disbenefits, $/year	120,000	100,000
Life, years	∞	∞

9.32 Select the better of two proposals to improve street safety and lighting in a colonia in south central New Mexico. Use a B/C analysis and an interest rate of 8% per year.

	Proposal 1	Proposal 2
Initial cost, $	900,000	1,700,000
Annual M&O cost, $/year	120,000	60,000
Annual benefits, $/year	530,000	650,000
Annual disbenefits, $/year	300,000	195,000
Life, years	10	20

9.33 Conventional and solar alternatives are available for providing energy at a remote radar site. Use the B/C ratio to determine which method should be selected at an interest rate of 8% per year and a 5-year study period.

	Conventional	Solar
Initial cost, $	300,000	2,500,000
Annual cost, $ per year	700,000	5,000

9.34 The two alternatives shown are under consideration for improving security at a county jail in Travis County, New York. Determine which one should be selected, based on a B/C analysis, an interest rate of 7% per year and a 10-year study period.

	Extra Cameras (EC)	New Sensors (NS)
First cost, $	38,000	87,000
Annual M&O, $/year	49,000	64,000
Benefits, $/year	110,000	160,000
Disbenefits, $/year	26,000	—

9.35 The U.S. government recently released an RFP to construct a second-story floor on an existing building at the Pentagon Complex. Separate contractors proposed two methods. Method 1 will use lightweight expanded shale on a metal deck with open web joists and steel beams. For this method, the costs will be $14,100 for concrete, $6000 for metal decking, $4300 for joists, and $2600 for beams. Method 2 will construct a reinforced concrete slab costing $5200 for concrete, $1400 for rebar, $2600 for equipment rental, and $1200 for expendable supplies. Special additives will be included in the lightweight concrete that will improve the heat-transfer properties of the floor. If the energy costs for method 1 will be $600 per year lower than for method 2, which one is more attractive? Use an interest rate of 7% per year, a 20-year study period, and the B/C method.

9.36 A project to control flooding from rare, but sometimes heavy rainfalls in the arid southwest will have the cash flows shown below. Determine which project should be selected on the basis of a B/C analysis at $i = 8\%$ per year and a 20-year study period.

	Sanitary Sewers	Open Channels
First cost, $	26 million	53 million
M&O cost, $ per year	400,000	30,000
Homeowner cleanup costs, $ per year	60,000	0

9.37 Two routes are under consideration for a new interstate highway. The long intervalley route would be 25 kilometers in length and would have an initial cost of $25 million. The short transmountain route would be 10 kilometers long and would have an initial cost of $45 million. Maintenance costs are estimated at $150,000 per year for the long route and $35,000 per year for the short route. Regardless of which route is selected, the volume of traffic is expected to be 400,000 vehicles per year. If the vehicle operating expenses are assumed to be $0.30 per kilometer, determine which route should be selected on the basis of (a) conventional B/C analysis and (b) modified B/C analysis. Assume an infinite life for each road, and use an interest rate of 8% per year.

9.38 The Idaho Department of Fish and Wildlife (IDFW) is considering two locations for a new state park. Location E would require an investment of $3 million and $50,000 per year to maintain. Location W would cost $7 million to construct, but the IDFW would receive an additional $25,000 per year in park fees. The operating cost of location W will be $65,000 per year. The revenue to park concessionaires will be $500,000 per year at location E and $700,000 at location W. The disbenefits associated with each location are $30,000 per year for location E and $40,000 per year for location W. Assume the park will be maintained indefinitely. Use an interest rate of 12% per year to determine which location, if either, should be selected on the basis of (a) the B/C method and (b) the modified B/C method.

9.39 Three engineers made the estimates shown below for two optional methods by which new construction technology would be implemented at a site for public housing. Either one of the two options or the current method may be selected. Set up a spreadsheet for B/C sensitivity analysis and determine if option 1, option 2 or the do-nothing option is selected by each of the three engineers. Use a life of 5 years and a discount rate of 10% per year for all analyses.

	Engineer Bob		Engineer Judy		Engineer Chen	
	Option 1	Option 2	Option 1	Option 2	Option 1	Option 2
Initial cost, $	50,000	90,000	75,000	90,000	60,000	70,000
Cost, $/year	3,000	4,000	3,800	3,000	6,000	3,000
Benefits, $/year	20,000	29,000	30,000	35,000	30,000	35,000
Disbenefits, $/year	500	1,500	1,000	0	5,000	1,000

Multiple Alternatives

9.40 A group of engineers responsible for developing advanced missile detection and tracking technologies, such as shortwave infrared, thermal infrared detection, target tracking radar, etc., recently came up with six proposals for consideration. The present worth (in $ billions) of the capital requirements and benefits is shown for each alternative in the table. Determine which one(s) should be undertaken, if they are (a) independent and (b) mutually exclusive.

Alternative	A	B	C	D	E	F
PW of capital, $	80	50	72	43	89	81
PW of benefits, $	70	55	76	52	85	84

9.41 Comparison of five mutually exclusive alternatives is shown. One *must* be accepted. According to the B/C ratio, which alternative should be selected (costs increase from A to E).

Comparison	ΔB/C Ratio
A versus B	0.75
B versus C	1.4
C versus D	1.3
A versus C	1.1
A versus D	0.2
B versus D	1.9
C versus E	1.2
D versus E	0.9

9.42 A consulting engineer is currently evaluating four different projects for the Department of Housing and Urban Development. The future worth of costs, benefits, disbenefits, and cost savings is shown. The interest rate is 10% per year, compounded continuously. Determine which of the projects, if any, should be selected, if the projects are (a) independent and (b) mutually exclusive.

	Project ID			
	Good	Better	Best	Best of All
FW of first costs, $	10,000	8,000	20,000	14,000
FW of benefits, $	15,000	11,000	25,000	42,000
FW of disbenefits, $	6,000	1,000	20,000	32,000
FW of cost savings, $	1,500	2,000	16,000	3,000

9.43 From the data shown below for six mutually exclusive projects, determine which project, if any, should be selected.

	Project ID					
	A	B	C	D	E	F
Annual cost, $ per year	8000	25,000	15,000	32,000	17,000	20,000
Annual benefits, $ per year	?	?	?	?	?	?
B/C ratio (alternative vs. DN)	1.23	1.12	0.87	0.97	0.71	1.10

Selected Incremental B/C Ratios

A versus B = 1.07
A versus C = 0.46
A versus F = 1.02
B versus D = 0.43
B versus E = 2.00
B versus F = 1.20
C versus D = 1.06
C versus F = 1.80

9.44 Four mutually exclusive revenue alternatives are being compared using the B/C method. Which alternative, if any, should be selected?

Alternative	Initial Cost, $ Millions	B/C Ratio vs. DN	Incremental B/C When Compared with Alternative			
			A	B	C	D
A	30	0.87	—	2.38	1.30	1.38
B	38	1.18		—	0.58	1.13
C	52	1.04			—	1.45
D	81	1.16				—

9.45 The city of St. Louis, Missouri, is considering various proposals regarding the disposal of used tires. All of the proposals involve shredding, but the charges for the service and the handling of the tire shreds differ in each plan. An incremental B/C analysis was initiated but never completed.
(a) Fill in all the missing blanks in the table.
(b) Determine which alternative should be selected.

Alternative	PW of Costs, $	PW of Benefits, $	PW of Disbenefits, $	B/C Ratio	Incremental B/C When Compared with Alternative			
					J	K	L	M
J	20	?	1	1.05	—	?	?	?
K	23	28	?	1.13		—	?	?
L	28	35	3	?			—	?
M	?	51	4	1.34				—

Cost-Effectiveness Analysis

9.46 In an effort to improve productivity in a large semiconductor manufacturing plant, the plant manager decided to undertake on a trial basis a series of actions directed toward improving employee morale. Six different strategies were implemented, such as increased employee autonomy, flexible work schedules, improved training, company picnics, electronic suggestion box, and better work environment. Periodically, the company surveyed the employees to measure the change in morale. The measure of effectiveness is the difference between the number of employees who rate their job satisfaction as very high and those who rate it very low. The per-employee cost of each strategy (identified as A through F) and the resultant measurement score are shown in the next column.

The manager has a maximum of $50 per employee to spend on the permanent implementation of as many of the strategies as are justified from both the effectiveness and economic viewpoints. Determine which strategies are the best to implement. (Hand or spreadsheet solution is acceptable, as you are instructed.)

Strategy	Cost/Employee, $	Measurement Score
A	5.20	50
B	23.40	182
C	3.75	40
D	10.80	75
E	8.65	53
F	15.10	96

9.47 There are a number of techniques to help people stop smoking, but their cost and effectiveness vary widely. One accepted measure of effectiveness of a program is the percentage of enrollees quitting. The table below shows several techniques touted as effective stop-smoking methods, some historical data on the approximate cost of each program per person, and the percentage of people smoke-free 3 months after the program ended.

The Cancer Society provides annual cost-offset funding to cancer patients so more people can afford these programs. A large clinic in St. Louis has the capacity to treat each year the number of people shown. If the clinic plans to place a proposal with the Cancer Society to treat a specified number of people annually, estimate the amount of money the clinic should ask for in its proposal to do the following:
(a) Conduct programs at the capacity level for the technique with the lowest cost-effectiveness ratio.
(b) Offer programs using as many techniques as possible to treat up to 1300 people per year using the most cost-effective techniques.

Technique	Cost, $/ Enrollee	% Quitting	Treatment Capacity per Year
Acupuncture	700	9	250
Subliminal message	150	1	500
Aversion therapy	1700	10	200
Outpatient clinic	2500	39	400
In-patient clinic	1800	41	550
Nicotine replacement therapy (NRT)	1300	20	100

9.48 The cost in $ per year and effectiveness measure in items salvaged per year for four mutually exclusive service sector alternatives have been collected. (a) Calculate the cost effectiveness ratio for each alternative, and (b) use the CER to identify the best alternative.

Alternative	Cost C, $/Year	Salvaged Items/Year E
W	355	20
X	208	17
Y	660	41
Z	102	7

9.49 An engineering student has only 30 minutes before the final exam in statics and dynamics. He wants to get help solving a type of problem that he knows will be on the test from the professor's review during the previous class. There is time for using only one method of assistance before the exam; he must select well. In a rapid process of estimation, he determines how many minutes it would take for each method of assistance and how many points it might gain for him on the final. The method and estimates follow. Where should he seek help to be most effective?

Assistance from	Minutes for Assistance	Points Gained
Teaching assistant	15	15
Course slides on Web	20	10
Friend in class	10	5
Professor	20	15

Public Sector Ethics

9.50 During the design and specifications development stages of a remote meter reading system for residential electricity use (a system that allows monthly usage to be transmitted via phone lines with no need to physically view meters), the two engineers working on the project for the city of Forest Ridge noted something different from what they expected. The first, an electrical/software engineer, noted that the city liaison staff member provided all the information on the software options, but only one option, the one from Lorier Software, was ever discussed and detailed. The second designer, an industrial/systems engineer, further noted that all the hardware specifications provided to them by this same liaison came from the same distributor, namely, Delsey Enterprises. Coincidently, at a weekend family picnic for city employees, to which the engineers had been invited, they met a couple named Don Delsey and Susan Lorier. Upon review, they learned that Don is the son-in-law of the city liaison and Susan is his stepdaughter. Based on these observations and before they complete the system design and specifications, what should the two engineers do, if anything, about their suspicions that the city liaison person is trying to bias the design to favor of the use of his relatives' software and hardware businesses?

9.51 Explain the difference between public policy making and public planning.

9.52 Since transportation via automobile was introduced, drivers throughout the country of Yalturia in eastern Europe have driven on the left side of the road. Recently the Yalturian National Congress passed a law that within 3 years, a right-hand driving convention will be adopted and implemented throughout the country. This is a major policy change for the country and will require significant public planning and project development to implement successfully and safely. Assume you are the lead engineering consultant from Halcrow Engineers, responsible for developing and describing many of these major projects. Identify six of the projects you deem necessary. State a name and provide a one- or two-sentence description of each project.

ADDITIONAL PROBLEMS AND FE EXAM REVIEW QUESTIONS

9.53 All of the following are examples of public sector projects, except:
  (a) Bridges   (b) Emergency relief
  (c) Prisons   (d) Oil wells

9.54 All of the following are usually associated with public sector projects except:
  (a) Funding from taxes   (b) Profit
  (c) Disbenefits   (d) Infinite life

9.55 All of the following would be examples of public projects except:
  (a) Air traffic control system
  (b) Establishing a dot.com company
  (c) Dam with irrigation canals
  (d) Mass transit system

9.56 In a conventional B/C ratio, revenue received by the government from admission to national parks should be:
  (a) Added to benefits in the numerator
  (b) Subtracted from benefits in the numerator

  (c) Subtracted from costs in the denominator
  (d) Added to costs in the denominator

9.57 In a conventional B/C ratio:
  (a) Disbenefits and M&O costs are subtracted from benefits.
  (b) Disbenefits are subtracted from benefits and M&O costs are included in costs.
  (c) Disbenefits and M&O costs are added to costs.
  (d) Disbenefits are added to costs and M&O costs are subtracted from benefits.

9.58 In a modified B/C ratio:
  (a) Disbenefits are put in the denominator.
  (b) Benefits are subtracted from costs.
  (c) M&O costs are put in the denominator.
  (d) M&O costs are put in the numerator.

9.59 If two mutually exclusive alternatives have B/C ratios of 1.4 and 1.5 for the lower- and higher-cost alternatives, respectively, the following is correct:

(a) The B/C ratio on the increment between them is equal to 1.5.
(b) The B/C ratio on the increment between them is between 1.4 and 1.5.
(c) The B/C ratio on the increment between them is less than 1.5.
(d) The higher-cost alternative is the better one economically.

9.60 In evaluating three independent alternatives by the B/C method, the alternatives were ranked A, B, and C, respectively, in terms of increasing cost, and the following results were obtained for *overall* B/C ratios: 1.1, 0.9, and 1.3. On the basis of these results, you should:
(a) Select only alternative A
(b) Select only alternative C
(c) Compare A and C incrementally
(d) Select alternatives A and C

9.61 An alternative has the following cash flows:

Benefits of $50,000 per year
Disbenefits of $27,000 per year
Initial cost of $250,000
M&O costs of $10,000 per year

If the alternative has an infinite life and the interest rate is 10% per year, the B/C ratio is closest to:
(a) 0.52   (b) 0.66
(c) 0.91   (d) 1.16

9.62 At the interest rate of 10% per year, an alternative with the following estimates has a modified B/C ratio that is closest to:

Benefits of $60,000 per year
Disbenefits of $29,000 per year
Amortized first cost of $20,000 per year
M&O costs of $15,000 per year

(a) 0.65   (b) 0.72
(c) 0.80   (d) 1.04

9.63 In evaluating three mutually exclusive alternatives by the B/C method, the alternatives are ranked A, B, and C, respectively, in terms of increasing cost, and the following results are obtained for the *overall B/C ratios*: 1.1, 0.9, and 1.06. On the basis of these results, you should:
(a) Select A
(b) Select C

(c) Select A and C
(d) Compare A and C incrementally

9.64 An alternative with an infinite life has a B/C ratio of 1.5. The alternative has benefits of $50,000 per year and annual maintenance costs of $10,000 per year. The first cost of the alternative at an interest rate of 10% per year is closest to:
(a) $23,300   (b) $85,400
(c) $146,100   (d) $233,000

9.65 Cost-effectiveness analysis (CEA) differs from cost-benefit (B/C) analysis in that:
(a) CEA cannot handle multiple alternatives.
(b) CEA expresses outcomes in natural units rather than in currency units.
(c) CEA cannot handle independent alternatives.
(d) CEA is more time-consuming and resource-intensive.

9.66 Several private colleges claim to have programs that are very effective at teaching enrollees how to become entrepreneurs. Two programs, identified as program X and program Y, have produced 4 and 6 persons per year, respectively, who were recognized as entrepreneurs. If the total cost of the programs is $25,000 and $33,000, respectively, the incremental cost-effectiveness ratio is closest to:
(a) 6250   (b) 5500
(c) 4000   (d) 1333

9.67 The statements contained in a code of ethics are variously known as all of the following except:
(a) Canons   (b) Laws
(c) Standards   (d) Norms

9.68 Of the following, the word *not* related to ethics is:
(a) Virtuous   (b) Honest
(c) Lucrative   (d) Proper

9.69 All of the following are examples of unethical behavior except:
(a) Offering services at prices lower than the competition
(b) Price fixing
(c) Bait and switch
(d) Selling on the *black market*

CASE STUDY

COMPARING B/C ANALYSIS AND CEA OF TRAFFIC ACCIDENT REDUCTION

Background

This case study compares benefit/cost analysis and cost-effectiveness analysis on the same information about highway lighting and its role in accident reduction.

Poor highway lighting may be one reason that proportionately more traffic accidents occur at night. Traffic accidents are categorized into six types by severity and value. For example, an accident with a *fatality* is valued at approximately $4 million, while an accident in which there is *property damage* (to the car and contents) is valued at $6000. One method by which the impact of lighting is measured compares day and night accident rates for lighted and unlighted highway sections with similar characteristics. Observed reductions in accidents seemingly caused by too low lighting can be translated into either monetary estimates of the benefits *B* of lighting or used as the effectiveness measure *E* of lighting.

Information

Freeway accident data were collected in a 5-year study. The property damage category is commonly the largest based on the accident rate. The number of accidents recorded on a section of highway is presented here.

| | Number of Accidents Recorded[1] | | | |
| | Unlighted | | Lighted | |
Accident Type	Day	Night	Day	Night
Property damage	379	199	2069	839

The ratios of night to day accidents involving property damage for the unlighted and lighted freeway sections are $199/379 = 0.525$ and $839/2069 = 0.406$, respectively. These results indicate that the lighting was beneficial. To quantify the benefit, the accident rate ratio from the unlighted section will be applied to the lighted section. This will yield the number of accidents that were prevented. Thus, there would have been $(2069)(0.525) = 1086$ accidents instead of 839 if there had not been lights on the freeway. This is a difference of 247 accidents. At a cost of $6000 per accident, this results in a net annual benefit of

$$B = (247)(\$6000) = \$1,482,000$$

For an effectiveness measure of number of accidents prevented, this results in $E = 247$.

To determine the cost of the lighting, it will be assumed that the light poles are center poles 67 meters apart with 2 bulbs each. The bulb size is 400 watts, and the installation cost is $3500 per pole. Since these data were collected over 87.8 kilometers of lighted freeway, the installed cost of the lighting is (with number of poles rounded off):

$$\text{Installation cost} = \$3500 \left(\frac{87.8}{0.067} \right)$$
$$= 3500(1310)$$
$$= \$4,585,000$$

There are a total of $87.8/0.067 = 1310$ poles, and electricity costs $0.10 per kWh. Therefore, the annual power cost is

Annual power cost
$$= 1310 \text{ poles}(2 \text{ bulbs/pole})(0.4 \text{ kilowatt/bulb})$$
$$\times (12 \text{ hours/day})(365 \text{ days/year})$$
$$\times (\$0.10/\text{kilowatt-hour})$$
$$= \$459,024 \text{ per year}$$

The data were collected over a 5-year period. Therefore, the annualized cost *C* at $i = 6\%$ per year is

$$\text{Total annual cost} = \$4,585,000(A/P,6\%,5)$$
$$+ 459,024$$
$$= \$1,547,503$$

If a benefit/cost analysis is the basis for a decision on additional lighting, the B/C ratio is

$$\text{B/C} = \frac{1,482,000}{1,547,503} = 0.96$$

Since B/C < 1.0, the lighting is not justified. Consideration of other categories of accidents is necessary to obtain a better basis for decisions. If a cost-effectiveness analysis (CEA) is applied, due to a judgment that the monetary estimates for lighting's benefit is not accurate, the *C/E* ratio is

$$C/E = \frac{1,547,503}{247} = 6265$$

This can serve as a base ratio for comparison when an incremental CEA is performed for additional accident reduction proposals.

These preliminary B/C and *C/E* analyses prompted the development of four lighting options:

W)   Implement the plan as detailed above; light poles every 67 meters at a cost of $3500 per pole.

X)   Install poles at twice the distance apart (134 meters). This is estimated to cause the accident prevention benefit to decrease by 40%.

Y)   Install cheaper poles and surrounding safety guards, plus slightly lowered lumen bulbs (350 watts) at a cost of $2500 per pole; place the poles 67 meters apart. This is estimated to reduce the benefit by 25%.

Z)   Install cheaper equipment for $2500 per pole with 350-watt lightbulbs and place them 134 meters apart. This plan is estimated to reduce the accident prevention measure by 50% from 247 to 124.

Case Study Exercises

Determine if a definitive decision on lighting can be determined by doing the following:

1. Use a *benefit/cost analysis* to compare the four alternatives to determine if any are economically justified.
2. Use a cost-effectiveness analysis to compare the four alternatives.

From an understanding viewpoint, consider the following:

3. How many property-damage accidents could be prevented on the unlighted portion if it were lighted?
4. What would the lighted, night-to-day accident ratio have to be to make alternative Z economically justified by the B/C ratio?
5. Discuss the analysis approaches of B/C and C/E. Does one seem more appropriate in this type of situation than the other? Why? Can you think of other bases that might be better for decisions for public projects such as this one?

[1]Portion of data reported in Michael Griffin, "Comparison of the Safety of Lighting on Urban Freeways," *Public Roads,* vol. 58, pp. 8–15, 1994.

LEARNING STAGE 2: EPILOGUE

Selecting the Basic Analysis Tool

In the previous five chapters, several equivalent evaluation techniques have been discussed. Any method—PW, AW, FW, ROR, or B/C—can be used to select one alternative from two or more and obtain the same, correct answer. Only one method is needed to perform the engineering economy analysis, because any method, correctly performed, will select the same alternative. Yet different information about an alternative is available with each different method. The selection of a method and its correct application can be confusing.

Table LS2–1 gives a recommended evaluation method for different situations, if it is not specified by the instructor in a course or by corporate practice in professional work. The primary criteria for selecting a method are speed and ease of performing the analysis. Interpretation of the entries in each column follows.

Evaluation period: Most private sector alternatives (revenue and cost) are compared over their equal or unequal estimated lives, or over a specific period of time. Public sector projects are commonly evaluated using the B/C ratio and usually have long lives that may be considered infinite for economic computation purposes.

Type of alternatives: Private sector alternatives have cash flow estimates that are revenue-based (includes income and cost estimates) or cost-based (cost estimates only). For cost alternatives, the revenue cash flow series is assumed to be equal for all alternatives. For public sector projects, the difference between costs and timing is used to select one alternative over another. *Service sector projects* for which benefits are estimated using a nonmonetary effectiveness measure are usually evaluated with a method such as cost-effectiveness analysis. This applies to all evaluation periods.

Recommended method: Whether an analysis is performed by hand, calculator, or spreadsheet, the method(s) recommended in Table LS2–1 will correctly select one alternative from two or more as

TABLE LS2–1	Recommended Method to Compare Mutually Exclusive Alternatives, Provided the Method Is Not Preselected		
Evaluation Period	**Type of Alternatives**	**Recommended Method**	**Series to Evaluate**
Equal lives of alternatives	Revenue or cost	AW or PW	Cash flows
	Public sector	B/C, based on AW or PW	Incremental cash flows
Unequal lives of alternatives	Revenue or cost	AW	Cash flows
	Public sector	B/C, based on AW	Incremental cash flows
Study period	Revenue or cost	AW or PW	Updated cash flows
	Public sector	B/C, based on AW or PW	Updated incremental cash flows
Long to infinite	Revenue or cost	AW or PW	Cash flows
	Public sector	B/C, based on AW	Incremental cash flows

rapidly as possible. Any other method can be applied subsequently to obtain additional information and, if needed, verification of the selection. For example, if lives are unequal and the rate of return is needed, it is best to first apply the AW method at the MARR and then determine the selected alternative's i^* using the same AW relation with i as the unknown.

Series to evaluate: The estimated cash flow series for one alternative and the incremental series between two alternatives are the only two options for present worth or annual worth evaluation. For spreadsheet analyses, this means that the NPV or PV functions (for present worth) or the PMT function (for annual worth) is applied. The word *updated* is added as a reminder that a study period analysis requires that cash flow estimates (especially salvage/market values) be reexamined and updated before the analysis is performed.

Once the evaluation method is selected, a specific procedure must be followed. These procedures were the primary topics of the last five chapters. Table LS2–2 summarizes the important

TABLE LS2–2	Characteristics of an Economic Analysis of Mutually Exclusive Alternatives Once the Evaluation Method Is Determined					
Evaluation Method	Equivalence Relation	Lives of Alternatives	Time Period for Analysis	Series to Evaluate	Rate of Return; Interest Rate	Decision Guideline: Select[1]
	PW	Equal	Lives	Cash flows	MARR	Numerically largest PW
	PW	Unequal	LCM	Cash flows	MARR	Numerically largest PW
Present worth	PW	Study period	Study period	Updated cash flows	MARR	Numerically largest PW
	CC	Long to infinite	Infinity	Cash flows	MARR	Numerically largest CC
Future worth	FW	Same as present worth for equal lives, unequal lives, and study period				Numerically largest FW
	AW	Equal or unequal	Lives	Cash flows	MARR	Numerically largest AW
Annual worth	AW	Study period	Study period	Updated cash flows	MARR	Numerically largest AW
	AW	Long to infinite	Infinity	Cash flows	MARR	Numerically largest AW
	PW or AW	Equal	Lives	Incremental cash flows	Find Δi^*	Last $\Delta i^* \geq$ MARR
	PW or AW	Unequal	LCM of pair	Incremental cash flows	Find Δi^*	Last $\Delta i^* \geq$ MARR
Rate of return	AW	Unequal	Lives	Cash flows	Find Δi^*	Last $\Delta i^* \geq$ MARR
	PW or AW	Study period	Study period	Updated incremental cash flows	Find Δi^*	Last $\Delta i^* \geq$ MARR
	PW	Equal or unequal	LCM of pairs	Incremental cash flows	Discount rate	Last ΔB/C $\geq$ 1.0
Benefit/cost	AW	Equal or unequal	Lives	Incremental cash flows	Discount rate	Last ΔB/C $\geq$ 1.0
	AW or PW	Long to infinite	Infinity	Incremental cash flows	Discount rate	Last ΔB/C $\geq$ 1.0

[1]Lowest equivalent cost or largest equivalent income.

elements of the procedure for each method—PW, AW, ROR, and B/C. FW is included as an extension of PW. The meaning of the entries in Table LS2–2 follows.

Equivalence relation   The basic equation written to perform any analysis is either a PW or an AW relation. The capitalized cost (CC) relation is a PW relation for infinite life, and the FW relation is likely determined from the PW equivalent value. Additionally, as we learned in Chapter 6, AW is simply PW times the *A/P* factor over the LCM or study period.

Lives of alternatives and time period for analysis   The length of time for an evaluation (the *n* value) will always be one of the following: equal lives of the alternatives, LCM of unequal lives, specified study period, or infinity because the lives are very long.

- PW analysis always requires the LCM of compared alternatives.
- Incremental ROR and B/C methods require the LCM of the two alternatives being compared.
- The AW method allows analysis over the respective alternative lives.
- CC analysis has an infinite time line and uses the relation $P = A/i$.

The one exception is for the incremental ROR method for unequal-life alternatives using an AW relation for *incremental cash flows*. The LCM of the two alternatives compared must be used. This is equivalent to using an AW relation for the *actual cash flows* over the respective lives. Both approaches find the incremental rate of return Δi^*.

Series to evaluate   Either the estimated cash flow series or the incremental series is used to determine the PW value, the AW value, the i^* value, or the B/C ratio.

Rate of return (interest rate)   The MARR value must be stated to complete the PW, FW, or AW method. This is also correct for the discount rate for public sector alternatives analyzed by the B/C ratio. The ROR method requires that the incremental rate be found in order to select one alternative. It is here that the dilemma of multiple rates appears, if the sign tests indicate that a unique, real number root does not necessarily exist for a nonconventional series.

Decision guideline   The selection of one alternative is accomplished using the general guideline in the rightmost column. Always select the alternative with the **numerically largest PW, FW, or AW value.** This is correct for both revenue and cost alternatives. The incremental cash flow methods—ROR and B/C—require that the largest initial cost and incrementally justified alternative be selected, provided it is justified against an alternative that is itself justified. This means that the Δi^* exceeds MARR, or the ΔB/C exceeds 1.0.

EXAMPLE LS2–1

Read through the problem statement of the following examples, neglecting the evaluation method used in the example. Determine which evaluation method is probably the fastest and easiest to apply. Is this the method used in the example? (*a*) 8.6, (*b*) 6.5, (*c*) 5.8, (*d*) 5.4.

Solution

Referring to the contents of Table LS2–1, the following methods should be applied first.

(*a*) Example 8.6 involves four revenue alternatives with equal lives. Use the AW or PW value at the MARR of 10%. The incremental ROR method was applied in the example.

(*b*) Example 6.5 requires selection between three public sector alternatives with unequal lives, one of which is 50 years and another is infinite. The B/C ratio of AW values is the best choice. This is how the problem was solved.

(c) Since Example 5.8 involves two cost alternatives with one having a long life, either AW or PW can be used. Since one life is long, capitalized cost, based on $P = A/i$, is best in this case. This is the method applied in the example.

(d) Example 5.4 is in the series of progressive examples. It involves 5-year and 10-year cost alternatives. The AW method is the best to apply in this case. The PW method for the LCM of 10 years and a study period of 5 years were both presented in the example.

Making Better Decisions

Most of the evaluations in the real world involve more than a simple economic selection of new assets or projects. The chapters in this stage introduce information-gathering and techniques that make decisions better. For example, **noneconomic parameters** can be introduced into the project analysis study through multiple attribute evaluation, and the **appropriate MARR** for a corporation or type of alternative can tailor and improve the economic decision.

The future is certainly not exact. However, techniques such as **replacement/retention** studies, **breakeven analysis,** and **payback analysis** help make informed decisions about future uses of existing assets and systems.

After completing these chapters, you will be able to go beyond the basic alternative analysis tools of the previous chapters. The techniques covered in this learning stage take into consideration the moving targets of change over time.

Important note: If asset depreciation and taxes are to be considered by an *after-tax analysis,* Chapters 16 and 17 should be covered before or in conjunction with these chapters.

CHAPTER 10

Project Financing and Noneconomic Attributes

T his chapter discusses the different ways to finance a project through debt and equity sources and explains how the MARR is established. The descriptions here complement the introductory material of Chapter 1 on the same topics. Some of the parameters specified earlier are unspecified here, and in future chapters. As a result, some of the textbook aspects apparent in previous chapters are removed, thus coming closer to treating the more complex, real-world situations in which professional practice and decision making occur.

Until now, only one dimension—the economic one—has been the basis for judging the economic viability of one project, or the selection basis from two or more alternatives. In this chapter, guidelines and techniques explain the determination and use of multiple (noneconomic) attributes helpful in selecting between alternatives.

10.1 MARR Relative to the Cost of Capital ● ● ◌

The MARR value used in alternative evaluation is one of the most important parameters of a study. In Chapter 1 the MARR was described relative to the weighted costs of debt and equity capital. This and the next four sections explain how to establish a MARR under varying conditions.

To form the basis for a realistic MARR, the types and cost of each source of project financing should be understood and estimated. There is a strong connection between the costs of debt and equity capital and the MARR used to evaluate one or more alternatives, whether they are mutually exclusive or independent. There are several terms and relationships important to the understanding of project financing and the MARR that is specified to evaluate projects using PW, AW, FW, or B/C methods. (Reference to Section 1.9 will complement the following material.)

> The **cost of capital** is the weighted average interest rate paid based on the proportion of investment capital from *debt* and *equity sources*.
>
> The **MARR** is then set relative to the cost of capital. The MARR can be set for one project, a series of projects, a division of a corporation, or the entire company. MARR values *change over time* due to changing circumstances.
>
> When no specific MARR is established, the estimated net cash flows and available capital establish an inherent MARR. This rate is determined by finding the ROR (i^*) value of the project cash flows. This rate is utilized as the **opportunity cost**, which is the ROR of the first project not funded due to the lack of capital funds.

Cost of capital
_____

MARR
_____

Opportunity cost

Before we discuss cost of capital, we review the two primary sources of capital.

Debt capital represents borrowing from outside the company, with the principal repaid at a stated interest rate following a specified time schedule. Debt financing includes borrowing via *bonds, loans,* and *mortgages*. The lender does not share in the profits made using the debt funds, but there is **risk** in that the borrower could default on part of or all the borrowed funds. The amount of outstanding debt financing is indicated in the liabilities section of the corporate balance sheet.

Equity capital is corporate money comprised of the *funds of owners* and *retained earnings*. Owners' funds are further classified as common and preferred stock proceeds or owners' capital for a private (non-stock-issuing) company. Retained earnings are funds previously retained in the corporation for capital investment. The amount of equity is indicated in the net worth section of the corporate balance sheet.

To illustrate the relation between cost of capital and MARR, assume a new greenhouse gas emission control system will be completely financed by a $25,000,000 bond issue (100% debt financing), and assume the dividend rate on the bonds is 8%. Therefore, the cost of debt capital is 8% as shown in Figure 10–1. This 8% is the minimum for MARR. Management may increase this MARR in increments that reflect its desire for added return and its perception of risk. For example, management may add an amount for all capital commitments in this area. Suppose this amount is 2%. This increases the expected return to 10% (Figure 10–1). Also, if the risk associated with the investment is considered substantial enough to warrant an additional 1% return requirement, the final MARR is 11%.

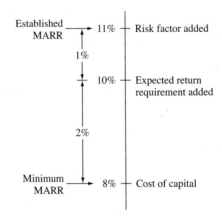

Figure 10–1
A fundamental relation between cost of capital and MARR used in practice.

The recommended approach does not follow the logic presented above. Rather, the cost of capital (8% here) should be the established MARR. Then the i^* value is determined from the estimated net cash flows. Using this approach, suppose the control system is estimated to return 11%. Now, additional return requirements and risk factors are considered to determine if 3% above the MARR of 8% is sufficient to justify the capital investment. After these considerations, if the project is not funded, the effective MARR is now 11%. This is the opportunity cost discussed previously—the unfunded project i^* has established the effective MARR for emission control system alternatives at 11%, not 8%.

The setting of the MARR for an economy study is not an exact process. The debt and equity capital mix changes over time and between projects. Also, the MARR is not a fixed value established corporatewide. It is altered for different opportunities and types of projects. For example, a corporation may use a MARR of 10% for evaluating the purchase of assets (equipment, cars) and a MARR of 20% for expansion investments, such as acquiring smaller companies.

The effective MARR varies from one project to another and through time because of factors such as the following:

Project risk. Where there is greater risk (perceived or actual) associated with proposed projects, the tendency is to set a higher MARR. This is encouraged by the higher cost of debt capital for projects considered risky. This usually means that there is some concern that the project will not realize its projected revenue requirements.

Investment opportunity. If management is determined to expand in a certain area, the MARR may be lowered to encourage investment with the hope of recovering lost revenue in other areas. This common reaction to investment opportunity can create havoc when the guidelines for setting a MARR are too strictly applied. Flexibility becomes very important.

Government intervention. Depending upon the state of the economy, international relations, and a host of other factors, the federal government (and possibly lower levels) can dictate the forces and direction of the free market. This may occur through price limits, subsidies, import tariffs, and limitation on availability. Both short-term and long-term government interventions are commonly present in different areas of the economy. Examples are steel imports, foreign capital investment, car imports, and agricultural product exports. During the time that such government actions are in force, there is a strong impact to increase or decrease taxes, prices, etc., thus tending to move the MARR up or down.

Tax structure. If corporate taxes are rising (due to increased profits, capital gains, local taxes, etc.), pressure to increase the MARR is present. Use of after-tax analysis may assist in eliminating this reason for a fluctuating MARR, since accompanying business expenses will tend to decrease taxes and after-tax costs.

Limited capital. As debt and equity capital become limited, the MARR is increased. If the demand for limited capital exceeds supply, the MARR may tend to be set even higher. The opportunity cost has a large role in determining the MARR actually used.

Market rates at other corporations.   If the MARR increases at other corporations, especially competitors, a company may alter its MARR upward in response. These variations are often based on changes in interest rates for loans, which directly impact the cost of capital.

If the details of after-tax analysis are not of interest, but the effects of income taxes are important, the MARR may be increased by incorporating an effective tax rate using the formula

$$\text{Before-tax MARR} = \frac{\text{after-tax MARR}}{1 - \text{tax rate}} \qquad [10.1]$$

The total or effective tax rate, including federal, state, and local taxes, for most corporations is in the range of 30% to 50%. If an after-tax rate of return of 10% is required and the effective tax rate is 35%, the MARR for the before-tax economic analysis is $10\%/(1 - 0.35) = 15.4\%$.

EXAMPLE 10.1

Twin brother and sister, Carl and Christy graduated several years ago from college. Carl, an architect, has worked in home design with Bulte Homes since graduation. Christy, a civil engineer, works with Butler Industries in structural components and analysis. They both reside in Richmond, Virginia. They have started a creative e-commerce network through which Virginia-based builders can buy their "spec home" plans and construction materials much more cheaply. Carl and Christy want to expand into a regional e-business corporation. They have gone to the Bank of America (BA) in Richmond for a business development loan. Identify some factors that might cause the loan rate to vary when BA provides the quote. Also, indicate any impact on the established MARR when Carl and Christy make economic decisions for their business.

Solution

In all cases the direction of the loan rate and the MARR will be the same. Using the six factors mentioned above, some loan rate considerations are as follows:

Project risk: The loan rate may increase if there has been a noticeable downturn in housing starts, thus reducing the need for the e-commerce connection.

Investment opportunity: The rate could increase if other companies offering similar services have already applied for a loan at other BA branches regionally or nationwide.

Government intervention: The loan rate may decrease if the federal government has recently offered Federal Reserve loan money at low rates to banks. The intervention may be designed to boost the housing economic sector in an effort to offset a significant slowdown in new home construction.

Taxes: If the state recently removed house construction materials from the list of items subject to sales tax, the rate might be lowered slightly.

Capital limitation: Assume the computer equipment and software rights held by Carl and Christy were bought with their own funds and there are no outstanding loans. If additional equity capital is not available for this expansion, the rate for the loan (debt capital) should be lowered.

Market loan rates: The local BA branch probably obtains its development loan money from a large national pool. If market loan rates to this BA branch have increased, the rate for this loan will likely increase, because money is becoming "tighter."

10.2 Debt-Equity Mix and Weighted
Average Cost of Capital ● ● ●

Online
Presentation

The **debt-to-equity (D-E) mix** identifies the percentages of debt and equity financing for a corporation. A company with a 40–60 D-E mix has 40% of its capital originating from debt capital sources (bonds, loans, and mortgages) and 60% derived from equity sources (stocks and retained

Figure 10–2
General shape of different cost of capital curves.

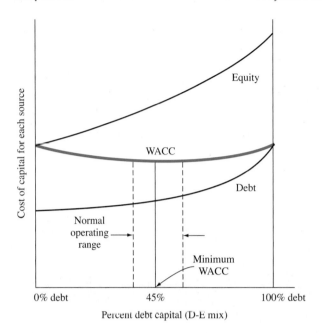

earnings). Most projects are funded with a combination of debt and equity capital made available specifically for the project or taken from a corporate *pool of capital*. The **weighted average cost of capital (WACC)** of the pool is estimated by the relative fractions from debt and equity sources. If known exactly, these fractions are used to estimate WACC; otherwise the historical fractions for each source are used in the relation

$$\textbf{WACC = (equity fraction)(cost of equity capital)}$$
$$\textbf{+ (debt fraction)(cost of debt capital)} \qquad \textbf{[10.2]}$$

The two *cost* terms are expressed as percentage interest rates.

Since virtually all corporations have a mixture of capital sources, the WACC is a value between the debt and equity costs of capital. If the fraction of each type of equity financing—common stock, preferred stock, and retained earnings—is known, Equation [10.2] is expanded.

$$\textbf{WACC = (common stock fraction)(cost of common stock capital)}$$
$$\textbf{+ (preferred stock fraction)(cost of preferred stock capital)}$$
$$\textbf{+ (retained earnings fraction)(cost of retained earnings capital)}$$
$$\textbf{+ (debt fraction)(cost of debt capital)} \qquad \textbf{[10.3]}$$

Figure 10–2 indicates the usual shape of cost of capital curves. If 100% of the capital is derived from equity or 100% is from debt sources, the WACC equals the cost of capital of that source of funds. There is virtually always a mixture of capital sources involved for any capitalization program. As an illustration only, Figure 10–2 indicates a minimum WACC at about 45% debt capital. Most firms operate over a range of D-E mixes. For example, a range of 30% to 50% debt financing for some companies may be very acceptable to lenders, with no increases in risk or MARR. However, another company may be considered "risky" with only 20% debt capital. It takes knowledge about management ability, current projects, and the economic health of the specific industry to determine a reasonable operating range of the D-E mix for a particular company.

EXAMPLE 10.2

Historically, Hong Kong has imported over 95% of its fresh vegetables each day. In an effort to develop sustainable and renewable vegetable sources, a new commercial vertical crop technology is being installed through a public-private partnership with Valcent Products.[1] For

---

[1] "Valcent Announces Agreement to Supply Verticrop™ Vertical Farming Technology to Hong Kong's VF Innovations Ltd.," www.valcent.net, June 16, 2010 news release.

illustration purposes, assume that the present worth of the total system cost is $20 million with financing sources and costs as follows.

Commercial loan for debt financing	$10 million at 6.8% per year
Retained earnings from partnering corporations	$4 million at 5.2% per year
Sale of stock (common and preferred)	$6 million at 5.9% per year

There are three existing international vertical farming projects with capitalization and WACC values as follows:

Project 1:	$5 million with $\text{WACC}_1 = 7.9\%$
Project 2:	$30 million with $\text{WACC}_2 = 10.2\%$
Project 3:	$7 million with $\text{WACC}_3 = 4.8\%$

Compare the WACC for the Hong Kong (HK) project with the WACC of the existing projects.

Solution

To apply Equation [10.3] to this new project, the fraction of equity (stock and retained earnings) and debt financing is needed. These are 0.3 for stock ($6 out of $20 million), 0.2 for retained earnings, and 0.5 for debt ($10 out of $20 million).

$$\text{WACC}_{HK} = 0.3(5.9\%) + 0.2(5.2\%) + 0.5(6.8\%) = 6.210\%$$

To correctly weight the other three project WACCs by size, determine the fraction in each one of the $42 million in total capital: project 1 has $5 million/$42 million = 0.119; project 2 has 0.714; project 3 has 0.167. The WACC weighted by project size is WACC_W.

$$\text{WACC}_W = 0.119(7.9\%) + 0.714(10.2\%) + 0.167(4.8\%) = 9.025\%$$

The Hong Kong project has a considerably lower cost of capital than the weighted average of other projects, considering all sources of funding.

The WACC value can be computed using before-tax or after-tax values for cost of capital. The after-tax method is the correct one since debt financing has a distinct tax advantage, as discussed in Section 10.3 below. Approximations of after-tax or before-tax cost of capital are made using the effective tax rate T_e in the relation

> **After-tax cost of debt capital = (before-tax cost)$(1 - T_e)$**   [10.4]

The effective tax rate is a combination of federal, state, and local tax rates. They are reduced to a single number T_e to simplify computations. Equation [10.4] may be used to approximate the cost of debt capital separately or inserted into Equation [10.2] for an after-tax WACC rate. Chapter 17 treats taxes and after-tax economic analysis in detail.

10.3 Determination of the Cost of Debt Capital ● ● ●

Debt financing includes borrowing, primarily via bonds and loans. (We learned about bonds in Section 7.6.) In most industrialized countries, bond dividends and loan interest payments are tax-deductible as a corporate expense. This reduces the taxable income base upon which taxes are calculated, with the end result of less taxes paid. The cost of debt capital is, therefore, reduced because there is an annual *tax savings* of the expense cash flow times the effective tax rate T_e. This tax savings is subtracted from the debt-capital expense in order to calculate the cost of debt capital. In formula form,

> **Tax savings = (expenses) (effective tax rate) = expenses (T_e)**   [10.5]
> **Net cash flow = expenses − tax savings = expenses $(1 - T_e)$**   [10.6]

To find the cost of debt capital, develop a PW- or AW-based relation of the net cash flow (NCF) series with i^* as the unknown. Find i^* by trial and error, by calculator, or by the RATE or IRR function on a spreadsheet. This is the cost of debt capital used in the WACC computation, Equation [10.2].

EXAMPLE 10.3

AT&T will generate $5 million in debt capital by issuing five thousand $1000 8% per year 10-year bonds. If the effective tax rate of the company is 30% and the bonds are discounted 2%, compute the cost of debt capital (*a*) before taxes and (*b*) after taxes from the company perspective. Obtain the answers by hand and spreadsheet.

Solution by Hand

(*a*) The annual bond dividend is $1000(0.08) = $80, and the 2% discounted sales price is $980 now. Using the company perspective, find the i^* in the PW relation

$$0 = 980 - 80(P/A, i^*,10) - 1000(P/F, i^*,10)$$
$$i^* = 8.3\%$$

The before-tax cost of debt capital is $i^* = 8.3\%$, which is slightly higher than the 8% bond interest rate, because of the 2% sales discount.

(*b*) With the allowance to reduce taxes by deducting the bond dividend, Equation [10.5] shows a tax savings of $80(0.3) = $24 per year. The bond dividend amount for the PW relation is now $80 - 24 = $56. Solving for i^* after taxes reduces the cost of debt capital to 5.87%.

Solution by Spreadsheet

Figure 10–3 is a spreadsheet image for both before-tax (column B) and after-tax (column C) analysis using the IRR function. The after-tax net cash flow is calculated using Equation [10.6] with $T_e = 0.3$.

	A	B	C	D	E	F	G
1		Before-tax	After-tax				
2	Year	cash flow	cash flow				
3	0	980	980				
4	1	-80	-56				
5	2	-80	-56	Bond dividend before taxes			
6	3	-80	-56	= −1000*0.08			
7	4	-80	-56				
8	5	-80	-56				
9	6	-80	-56				
10	7	-80	-56	Bond dividend after taxes			
11	8	-80	-56	= (−1000*0.08)*(1 − 0.3)			
12	9	-80	-56				
13	10	-1,080	-1,056				
14	**Cost of debt capital**	8.30%	5.87%	= IRR(C3:C13)			
15							
16							
17							

Figure 10–3
Use of IRR function to determine cost of debt capital before taxes and after taxes, Example 10.3.

EXAMPLE 10.4

LST Trading Company will purchase a $20,000 ten-year-life asset. Company managers have decided to put $10,000 down now from retained earnings and borrow $10,000 at an interest rate of 6%. The simplified loan repayment plan is $600 in interest each year, with the entire $10,000 principal paid in year 10. (*a*) What is the after-tax cost of debt capital if the effective tax rate is 42%? (*b*) How are the interest rate and cost of debt capital used to calculate WACC?

Solution

(*a*) The after-tax net cash flow for interest on the $10,000 loan is an annual amount of 600(1 − 0.42) = $348 by Equation [10.6]. The loan repayment is $10,000 in year 10. PW is used to estimate a cost of debt capital of 3.48%.

$$0 = 10,000 - 348(P/A, i^*,10) - 10,000(P/F, i^*,10)$$

(*b*) The 6% annual interest on the $10,000 loan is not the WACC because 6% is paid only on the borrowed funds. Nor is 3.48% the WACC, since it is only the cost of debt capital. The cost of the $10,000 equity capital is needed to determine the WACC.

10.4 Determination of the Cost of Equity Capital and the MARR ●●●

Equity capital is usually obtained from the following sources:

> Sale of preferred stock
>
> Sale of common stock
>
> Use of retained earnings
>
> Use of owner's private capital

The cost of each type of financing is estimated separately and entered into the WACC computation. A summary of one commonly accepted way to estimate each source's cost of capital is presented here. One additional method for estimating the cost of equity capital via common stock is presented. *There are no tax savings for equity capital, because dividends paid to stockholders and owners are not tax-deductible.*

Issuance of **preferred stock** carries with it a commitment to pay a stated dividend annually. The cost of capital is the stated dividend percentage, for example, 10%, or the dividend amount divided by the price of the stock. Preferred stock may be sold at a discount to speed the sale, in which case the actual proceeds from the stock should be used as the denominator. For example, if a 10% dividend preferred stock with a value of $200 is sold at a 5% discount for $190 per share, there is a cost of equity capital of ($20/$190) × 100% = 10.53%.

Estimating the cost of equity capital for **common stock** is more involved. The dividends paid are not a true indication of what the stock issue will actually cost in the future. Usually a valuation of the common stock is used to estimate the cost. If R_e is the cost of equity capital (in decimal form),

$$R_e = \frac{\text{first-year dividend}}{\text{price of stock}} + \text{expected dividend growth rate}$$

$$= \frac{DV_1}{P} + g \qquad\qquad [10.7]$$

The growth rate g is an estimate of the annual increase in returns that the shareholders receive. Stated another way, it is the compound growth rate on dividends that the company believes is required to attract stockholders. For example, assume a multinational corporation plans to raise capital through its U.S. subsidiary for a new plant in South America by selling $2,500,000 worth of common stock valued at $20 each. If a 5% or $1 dividend is planned for the first year and an appreciation of 4% per year is anticipated for future dividends, the cost of capital for this common stock issue from Equation [10.7] is 9%.

$$R_e = \frac{1}{20} + 0.04 = 0.09$$

The **retained earnings** and **owner's funds** cost of equity capital is usually set equal to the common stock cost, since it is the shareholders and owners who will realize any returns from projects in which these funds are invested.

Figure 10–4
Expected return on com-
mon stock issue using
CAPM.

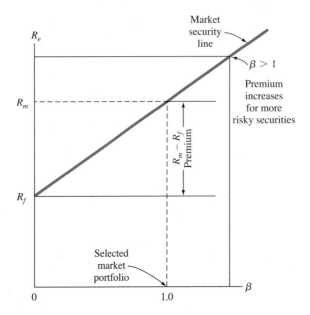

Once the cost of capital for all planned equity sources is estimated, the WACC is calculated using Equation [10.3].

A second method used to estimate the **cost of common stock capital** is the **capital asset pricing model (CAPM)**. Because of the fluctuations in stock prices and the higher return demanded by some corporations' stocks compared to others, this valuation technique is commonly applied. The cost of equity capital from common stock R_e, using CAPM, is

$$R_e = \text{risk-free return} + \text{premium above risk-free return}$$
$$= R_f + \beta(R_m - R_f) \qquad\qquad\qquad [10.8]$$

where β = volatility of a company's stock relative to other stocks in the market ($\beta = 1.0$ is
              the norm)
       R_m = return on stocks in a defined market portfolio measured by a prescribed index
       R_f = return on a "safe investment," which is usually the U.S. Treasury bill rate

The term $(R_m - R_f)$ is the premium paid above the safe or risk-free rate. The coefficient β (beta) indicates how the stock is expected to vary compared to a selected portfolio of stocks in the same general market area, often the Standard and Poor's 500 stock index. If $\beta < 1.0$, the stock is less volatile, so the resulting premium can be smaller; when $\beta > 1.0$, larger price movements are expected, so the premium is increased.

Security is a word that identifies a stock, bond, or any other instrument used to develop capital. To better understand how CAPM works, consider Figure 10–4. This is a plot of a market security line, which is a linear fit by regression analysis to indicate the expected return for different β values. When $\beta = 0$, the risk-free return R_f is acceptable (no premium). As β increases, the premium return requirement grows. Beta values are published periodically for most stock-issuing corporations. Once complete, this estimated cost of common stock equity capital can be included in the WACC computation in Equation [10.3].

EXAMPLE 10.5

The lead software engineer at SafeSoft, a food industry service corporation, has convinced the president to develop new software technology for the meat and food safety industry. It is envisioned that processes for prepared meats can be completed more safely and faster using this automated control software. A common stock issue is a possibility to raise capital if the cost of equity capital is below 9%. SafeSoft, which has a historical beta value of 1.09, uses CAPM to determine the premium of its stock compared to other software corporations. The security market line indicates that a 5% premium above the risk-free rate is desirable. If U.S. Treasury bills are paying 2%, estimate the cost of common stock capital.

Solution

The premium of 5% represents the term $R_m - R_f$ in Equation [10.8].

$$R_e = 2.0 + 1.09(5.0) = 7.45\%$$

Since this cost is lower than 9%, SafeSoft should issue common stock to finance this new venture.

In theory, a correctly performed engineering economy study uses a MARR equal to the cost of the capital committed to the specific alternatives in the study. Of course, such detail is not known. For a combination of debt and equity capital, the calculated WACC sets the minimum for the MARR. The most rational approach is to set MARR between the cost of equity capital and the corporation's WACC. The risks associated with an alternative should be treated separately from the MARR determination, as stated earlier. This supports the guideline that the MARR should not be arbitrarily increased to account for the various types of risk associated with the cash flow estimates. Unfortunately, the MARR is often set above the WACC because management does want to account for risk by increasing the MARR.

EXAMPLE 10.6

The Engineering Products Division of 4M Corporation has two mutually exclusive alternatives A and B with ROR values of $i_A^* = 9.2\%$ and $i_B^* = 5.9\%$. The financing scenario is yet unsettled, but it will be one of the following: plan 1—use all equity funds, which are currently earning 8% for the corporation; plan 2—use funds from the corporate capital pool which is 25% debt capital costing 14.5% and the remainder from the same equity funds mentioned above. The cost of debt capital is currently high because the company has narrowly missed its projected revenue on common stock for the last two quarters, and banks have increased the borrowing rate for 4M. Make the economic decision on alternative A versus B under each financing scenario. The MARR is set by the calculated WACC.

Solution

The capital is available for one of the two mutually exclusive alternatives. For plan 1, 100% equity, the financing is specifically known, so the cost of equity capital is the MARR, that is, 8%. Only alternative A is acceptable; alternative B is not since the estimated return of 5.9% does not exceed this MARR.

Under financing plan 2, with a D-E mix of 25–75,

$$WACC = 0.25(14.5) + 0.75(8.0) = 9.625\%$$

Now, neither alternative is acceptable since both ROR values are less than MARR = WACC = 9.625%. The selected alternative should be to do nothing, unless one alternative absolutely must be selected, in which case noneconomic attributes must be considered.

10.5 Effect of Debt-Equity Mix on Investment Risk ●●●

Online Presentation

The D-E mix was introduced in Section 10.2. As the proportion of debt capital increases, the calculated cost of capital decreases due to the tax advantages of debt capital.

The leverage offered by larger debt capital percentages increases the riskiness of projects undertaken by the company. When large debts are already present, additional financing using debt (or equity) sources gets more difficult to justify, and the corporation can be placed in a situation where it owns a smaller and smaller portion of itself. This is sometimes referred to as a **highly leveraged corporation**.

Inability to obtain operating and investment capital means increased difficulty for the company and its projects. Thus, a reasonable balance between debt and equity financing is important for the financial health of a corporation. Example 10.7 illustrates the disadvantages of unbalanced D-E mixes.

EXAMPLE 10.7

Three auto parts manufacturing companies have the following debt and equity capital amounts and D-E mixes. Assume all equity capital is in the form of common stock.

	Amount of Capital		
Company	Debt ($ in Millions)	Equity ($ in Millions)	D-E Mix (%–%)
A	10	40	20–80
B	20	20	50–50
C	40	10	80–20

Assume the annual revenue is $15 million for each one and that after interest on debt is considered, the net incomes are $14.4, $13.4, and $10.0 million, respectively. Compute the return on common stock for each company, and comment on the return relative to the D-E mixes.

Solution

Divide the net income by the stock (equity) amount to compute the common stock return. In million dollars,

$$\text{Return}_A = \frac{14.4}{40} = 0.36 \quad (36\%)$$

$$\text{Return}_B = \frac{13.4}{20} = 0.67 \quad (67\%)$$

$$\text{Return}_C = \frac{10.0}{10} = 1.00 \quad (100\%)$$

As expected, the return is by far the largest for highly leveraged C, where only 20% of the company is in the hands of the ownership. The return is excellent, but the risk associated with this firm is high compared to A, where the D-E mix is only 20% debt.

The use of *large percentages of debt financing greatly increases the risk* taken by lenders and stock owners. Long-term confidence in the corporation diminishes, no matter how large the short-term return on stock.

The leverage of large D-E mixes does increase the return on *equity capital,* as shown in previous examples; but it can also work against the owners and investors. A decrease in asset value will more negatively affect a highly debt-leveraged company compared to one with small leveraging. Example 10.8 illustrates this fact.

EXAMPLE 10.8

During the last several years, the U.S. airline industry has had financial problems, in part due to high fuel costs, fewer customers, security problems, government regulations, aging aircraft, and union dissatisfaction. As a consequence, the D-E mixes of the so-called traditional companies (American, United, Delta, and others) have become larger on the debt side than is historically acceptable. Meanwhile, the D-E mixes of so-called low-cost airlines (Southwest, JetBlue, and others) have suffered, but not to the same degree. In an effort to reduce costs, assume that three airlines joined forces to cooperate on a range of services (baggage handling, onboard food preparation, ticket services, and software development) by forming a new company called FullServe, Inc. This required $5 billion ($5 B) up-front funding from each airline.

Table 10–1 summarizes the D-E mixes and the total equity capitalization for each airline *after* its share of $5 B was removed from available equity funds. The percentage of the $5 B obtained as debt capital was the same proportion as the debt in the company's D-E mix. For

TABLE 10–1	Debt and Equity Statistics, Example 10.8		
Airline Company	Corporate D-E Mix, %	Amount Borrowed, $ B	Equity Capital Available, $ B
National	30–70	1.50	5.0
Global	65–35	3.25	3.7
PanAm	91–9	4.55	6.7

example, National had 30% of its capitalization in debt capital; therefore, 30% of $5 B was borrowed, and 70% was provided from National's equity fund.

Unfortunately, after a short time, it was clear that the three-way collaborative effort was a complete failure, and FullServe was dissolved and its assets were distributed or sold for a total of $3.0 billion, only 20% of its original value. A total of $1.0 billion in equity capital was returned to each airline. The commercial banks that provided the original loans then required that the airlines each pay back the entire borrowed amount now, since FullServe was dissolved and no profit from the venture could be realized. Assuming the loan and equity amounts are the same as shown in Table 10–1, determine the resulting equity capital situation for each airline after it pays off the loan from its own equity funds. Also, describe one impact on each company as a result of this expensive and risky venture that failed.

Solution

Determine the level of post-FullServe equity capital using the following relation, in $ billions.

$$\text{Equity capital} = \text{pre-FullServe level} + \text{returned capital} - \text{loan repayment}$$

National: Equity capital = 5.0 + 1.0 − 1.50 = $4.50

Global: Equity capital = 3.7 + 1.0 − 3.25 = $1.45

PanAm: Equity capital = 6.7 + 1.0 − 4.55 = $3.15

Comparing the equity capital levels (Table 10–1) with the levels above indicates that the FullServe effort *reduced equity amounts* by 10% for National, 60% for Global, and 53% for PanAm. The debt capital to fund the failed FullServe effort has affected National airlines the least, in large part due to its low D-E mix of 30%–70%. However, Global and PanAm are in much worse shape financially, and they now must maintain business with a significantly lower ownership level and much reduced ability to obtain future capital—debt or equity.

The same principles discussed above for corporations are applicable to individuals. The person who is highly leveraged has large debts in terms of credit card balances, personal loans, and house mortgages. As an example, assume two engineers each have a take-home amount of $40,000 after all income tax, social security, and insurance premiums are deducted from their annual salaries. Further, assume that the cost of the debt (money borrowed via credit cards and loans) averages 15% per year and that the total debt is being repaid in equal amounts over 20 years. If Jamal has a total debt of $25,000 and Barry owes $100,000, the remaining amount of the annual take-home pay may be calculated as follows:

		Amount Paid, $ per Year		
Person	Total Debt, $	Cost of Debt at 15%, $	Repayment of Debt, $	Amount Remaining from $40,000, $
Jamal	25,000	3,750	1,250	35,000
Barry	100,000	15,000	5,000	20,000

Jamal has 87.5% of his base available while Barry has only 50% available.

Online
Presentation

10.6 Multiple Attribute Analysis: Identification and Importance of Each Attribute ● ● ●

In Chapter 1 the role and scope of engineering economy in decision making were outlined. The decision-making process explained in that chapter (Figure 1–1) included the seven steps listed on the right side of Figure 10–5. Step 4 is to identify the one or multiple attributes (criteria) upon which the selection will be based. In all prior evaluations, only one attribute—the economic one—has been identified and used to select the best alternative. The criterion has been the maximization of the equivalent value of PW, AW, FW, ROR, B/C ratio, or the CER for service projects. As we are all aware, most evaluations do and should take into account multiple attributes in decision making. These are the factors labeled as noneconomic in step 5 of Figure 1–1. However, these noneconomic dimensions tend to be intangible and often difficult, if not impossible, to quantify with economic and other scales. Nonetheless, among the many attributes that can be identified, there are key ones that must be considered in earnest before the alternative selection process is complete. This section and the next describe some of the techniques that accommodate multiple attributes in an engineering study.

Multiple attributes enter into the decision-making process in many studies. Public and service sector projects are excellent examples of multiple-attribute problem solving. For example, the proposal to construct a dam to form a lake in a low-lying area or to widen the catch basin of a river usually has several purposes, such as flood control; drinking water; industrial use; commercial development; recreation; nature conservation for fish, plants, and birds; and possibly other less obvious purposes. High levels of complexity are introduced into the selection process by the multiple attributes thought to be important in selecting an alternative for the dam's location, design, environmental impact, etc.

The left side of Figure 10–5 expands steps 4 and 5 to consider multiple attributes. The discussion below concentrates on the expanded step 4 and the next section focuses on the evaluation measure and alternative selection of step 5.

4-1 Attribute Identification   Attributes to be considered in the evaluation methodology can be identified and defined by several methods, some much better than others depending upon the situation surrounding the study itself. To seek input from individuals other than the analyst is important; it helps focus the study on key attributes. The following is an incomplete listing of ways in which key attributes are identified.

- Comparison with similar studies that include multiple attributes
- Input from experts with relevant past experience
- Surveys of constituencies (customers, employees, managers) impacted by the alternatives
- Small group discussions using approaches such as focus groups, brainstorming, or nominal group technique
- Delphi method, which is a progressive procedure to develop reasoned consensus from different perspectives and opinions

Figure 10–5

Expansion of the decision-making process to include multiple attributes.

Consider multiple attributes

Emphasis on one attribute

1. Understand the problem; define the objective.
2. Collect relevant information; define alternatives.
3. Make estimates.

4-1. Identify the attributes for decision making.
4-2. Determine the relative importance (weights) of attributes.
4-3. For each alternative, determine each attribute's value rating.

4. Identify the selection criteria (one or more attributes).

5. Evaluate each alternative using a multiple-attribute technique. Use sensitivity analysis for key attributes.

5. Evaluate each alternative; use sensitivity and risk analysis.

6. Select the best alternative.
7. Implement the solution and monitor results.

As an illustration, assume that Delta Airlines has decided to purchase five new Boeing 787s for overseas flights, primarily between the North American west coast and Asian cities, principally Hong Kong, Tokyo, and Singapore. There are approximately 8000 options for each plane that must be decided upon by Delta's engineering, purchasing, maintenance, and marketing personnel before the order to Boeing is placed. Options range *in scope* from the material and color of the plane's interior to the type of latching devices used on the engine cowlings, and *in function* from maximum engine thrust to pilot instrument design. An economic study based on the equivalent AW of the estimated passenger income per trip has determined that 150 of these options are clearly advantageous. But other noneconomic attributes are to be considered before some of the more expensive options are specified. A Delphi study was performed using input from 25 individuals. Concurrently, option choices for another, unidentified airline's recent order were shared with Delta personnel. From these two studies it was determined that there are 10 key attributes for options selection. Four of the most important attributes are

- *Repair time:* mean time to repair or replace (MTTR) if the option is or affects a flight-critical component.
- *Safety:* mean time to failure (MTTF) of flight-critical components.
- *Economic:* estimated extra revenue for the option. (Basically, this is the attribute evaluated by the economic study already performed.)
- *Crewmember needs:* some measure of the necessity and/or benefits of the option as judged by representative crewmembers—pilots and attendants.

The economic attribute of extra revenue may be considered an indirect measure of customer satisfaction, one that is more quantitative than customer opinion/satisfaction survey results. Of course, there are many other attributes that can be, and are, used. However, the point is that the economic study may directly address only one or a few of the key attributes vital to alternative decision making.

An attribute routinely identified by individuals and groups is **risk**.

Risk

> Risk is a possible **variation** in a parameter from an expected, desired, or predicted value that may be detrimental to observing the intended outcome(s) of the product, process, or system. It represents the absence of or deviation from certainty. Risk is present when there are two or more observable values of a parameter and it is possible to assume or estimate the **chance** that each value may occur.

Actually, risk is not a stand-alone attribute, because it is a part of every attribute in one form or another. Considerations of variation, probabilistic estimates, etc., in the decision-making process are treated in Chapters 18 and 19. Formalized sensitivity analysis, expected values, simulation, and decision trees are some of the techniques useful in handling risk.

4-2 Importance (Weights) for the Attributes   Determination of the **extent of importance** for each attribute *i* results in a weight W_i that is incorporated into the final evaluation measure. The weight, a number between 0 and 1, is based upon the experienced opinion of one individual or a group of persons familiar with the attributes, and possibly the alternatives. If a group is utilized to determine the weights, there must be consensus among the members for each weight. Otherwise, some averaging technique must be applied to arrive at one weight value for each attribute.

Table 10–2 is a tabular layout of attributes and alternatives used to perform a multiple attribute evaluation. Weights W_i for each attribute are entered on the left side. The remainder of the table is discussed as we proceed through steps 4 and 5 of the expanded decision-making process.

Attribute weights are usually normalized such that their sum over all the alternatives is 1.0. This normalizing implies that each attribute's importance score is divided by the sum *S* over all attributes. Expressed in formula form, these two properties of weights for attribute *i* ($i = 1, 2, \ldots, m$) are

$$\text{Normalized weights:} \sum_{i=1}^{m} W_i = 1.0 \qquad [10.9]$$

$$\text{Weight calculation:} \ W_i = \frac{\text{importance score}_i}{\sum\limits_{i=1}^{m} \text{importance score}_i} = \frac{\text{importance score}_i}{S} \qquad [10.10]$$

TABLE 10–2	Tabular Layout of Attributes and Alternatives Used for Multiple Attribute Evaluation						
		Alternatives					
Attributes	**Weights**	**1**	**2**	**3**	**...**		***n***
1	W_1						
2	W_2						
3	W_3		Value ratings V_{ij}				
⋮	⋮						
m	W_m						

Of the many procedures developed to assign weights to an attribute, an analyst is likely to rely upon one that is relatively simple, such as equal weighting, rank order, or weighted rank order. Pairwise comparison is another technique. Each is briefly presented below.

Equal Weighting   All attributes are considered to be of approximately the **same importance**, or there is no rationale to distinguish the more important from the less important attribute. This is the default approach. Each weight in Table 10–2 will be $1/m$, according to Equation [10.10]. Alternatively, the normalizing can be omitted, in which case each weight is 1 and their sum is *m*. In this case, the final evaluation measure for an alternative will be the sum over all attributes.

Rank Order   The *m* attributes are ordered (ranked) by increasing importance with a score of 1 assigned to the least important and *m* assigned to the most important. By Equation [10.10], the weights follow the pattern $1/S, 2/S, \ldots, m/S$. With this method, the difference in weights between attributes of **increasing importance is constant**.

Weighted Rank Order   The *m* attributes are again placed in the order of increasing importance. However, now **differentiation between attributes** is possible. The most important attribute is assigned a score, usually 100, and all other attributes are scored relative to it between 100 and 0. Now, define the score for each attribute as s_i, and Equation [10.10] takes the form

$$W_i = \frac{s_i}{\sum\limits_{i=1}^{m} s_i} \qquad\qquad [10.11]$$

This is a very practical method to determine weights because one or more attributes can be heavily weighted if they are significantly more important than the remaining ones, and Equation [10.11] automatically normalizes the weights. For example, suppose the four key attributes in the previous aircraft purchase example are ordered: safety, repair time, crewmember needs, and economic. If repair time is only one-half as important as safety, and the last two attributes are each one-half as important as repair time, the scores and weights are as follows.

Attribute	Score	Weights
Safety	100	100/200 = 0.50
Repair time	50	50/200 = 0.25
Crewmember needs	25	25/200 = 0.125
Economic	25	25/200 = 0.125
Sum of scores and weights	200	1.000

Pairwise Comparison   Each attribute is compared to each other attribute in a pairwise fashion using a rating scale that indicates the importance of one attribute over the other. Assume the

TABLE 10–3	Pairwise Comparison of Three Attributes to Determine Weights		
Attribute i	1 = Cost	2 = Constructability	3 = Environment
Cost	—	0	1
Constructability	2	—	1
Environment	1	1	—
Sum of scores	3	1	2
Weight W_i	0.500	0.167	0.333

three criteria (attributes) upon which a public works project decision is based are *cost, constructability,* and *environmental impact.* Define the importance comparison scale as follows:

0   if attribute is *less important* than one compared to

1   if attribute is *equally important* as one compared to

2   if attribute is *more important* than one compared to

Set up a table listing attributes across the top and down the side, and perform the pairwise comparison for each column attribute with each row attribute. Table 10–3 presents a comparison with importance scores included. The arrow to the right of the table indicates the direction of comparison, i.e., column with row attribute. For example, cost is judged more important than constructability, thus a score of 2. The complement score of 0 is placed in the reverse comparison of constructability with cost. The weights are determined by normalizing the scores using Equation [10.11], where the sum for each column is s_i. For the first attribute, cost $i = 1$.

$$s_1 = 3$$
$$\sum s_i = 3 + 1 + 2 = 6$$
$$\text{Cost weight } W_1 = 3/6 = 0.500$$

Similarly, the other weights are $W_2 = 1/6 = 0.167$ and $W_3 = 2/6 = 0.333$.

There are other attribute weighting techniques, especially for group processes, such as utility functions, and the Dunn-Rankin procedure. These become increasingly sophisticated, but they are able to provide an advantage that these simple methods do not afford the analyst: *consistency of ranks and scores* between attributes and between individuals. If this consistency is important in that several decision makers with diverse opinions about attribute importance are involved in a study, a more sophisticated technique may be warranted. There is substantial literature on this topic.

4-3 Value Rating of Each Alternative by Attribute   This is the final step prior to calculating the evaluation measure. Each alternative j is awarded a value rating V_{ij} for each attribute i. These are the entries within the cells in Table 10–2. The ratings are appraisals by decision makers of how well an alternative will perform as each attribute is considered.

The scale for the value rating can vary depending upon what is easiest to understand for those who do the valuation. A scale of 0 to 100 can be used for attribute importance scoring. However, the most popular is a scale of 4 or 5 gradations about the perceived ability of an alternative to accomplish the intent of the attribute. This is called a *Likert scale,* which can have descriptions for the gradations (e.g., very poor, poor, good, very good), or numbers assigned between 0 and 10, or -1 to $+1$, or -2 to $+2$. The last two scales can give a negative impact to the evaluation measure for poor alternatives. An example numerical scale of 0 to 10 is as follows:

If You Value the Alternative as	Give It a Rating between the Numbers
Very poor	0–2
Poor	3–5
Good	6–8
Very good	9–10

It is preferable to have a Likert scale with four choices (an even number) so that the central tendency of "fair" is not overrated.

TABLE 10–4	Completed Layout for Four Attributes and Three Alternatives for Multiple Attribute Evaluation			
		Alternatives		
Attributes	Weights	1	2	3
Safety	0.50	6	4	8
Repair	0.25	9	3	1
Crew needs	0.125	5	6	6
Economic	0.125	5	9	7

If we now build upon the aircraft purchase illustration to include value ratings, the cells are filled with ratings awarded by a decision maker. Table 10–4 includes example ratings V_{ij} and the weights W_i determined above. Initially, there will be one such table for each decision maker. Prior to calculating the final evaluation measure R_j, the ratings can be combined in some fashion; or a different R_j can be calculated using each decision maker's ratings. Determination of this evaluation measure is discussed below.

10.7 Evaluation Measure for Multiple Attributes ● ● ●

Online Presentation

The need for an evaluation measure that accommodates multiple attributes is indicated in step 5 of Figure 10–5. The measure can be one that attempts to retain all of the ratings, values, and complexity of previous assessments by multiple decision makers, or the measure can reduce these inputs to a single-dimension measure. This section introduces a single-dimension measure that is widely accepted.

A **single-dimension measure** effectively combines the different aspects addressed by the attribute importance weights W_i and the alternative value ratings V_{ij}. The resulting evaluation measure is a formula that calculates an aggregated measure for use in selecting from two or more alternatives. The approach applied in this process is called the *rank-and-rate method*.

This reduction process removes much of the complexity of trying to balance the different attributes; however, it also eliminates much of the robust information captured by the process of ranking attributes for their importance and rating each alternative's performance against each attribute.

There are additive, multiplicative, and exponential measures, but by far the most commonly applied is the additive model. The most used additive model is the **weighted attribute method**, also called the **additive weight technique**. The evaluation measure, symbolized by R_j for each alternative j, is defined as

$$R_j = \text{sum of (weight} \times \text{value rating)}$$
$$= \sum_{i=1}^{m} W_i \times V_{ij} \qquad [10.12]$$

The W_i numbers are the attribute importance weights, and V_{ij} is the value rating by attribute i for each alternative j. If the attributes are of equal weight (also called *unweighted*), all $W_i = 1/m$, as determined by Equation [10.10]. This means that W_i can be moved outside of the summation in the formula for R_j. (If an equal weight of $W_i = 1.0$ is used for all attributes, in lieu of $1/m$, then the R_j value is simply the sum of all ratings for the alternative.)

The selection guideline is as follows:

Choose the alternative with the **largest R_j value.** This measure assumes that increasing weights W_i mean more important attributes and increasing ratings V_{ij} mean better performance of an alternative.

ME alternative selection

Sensitivity analysis for any score, weight, or value rating is used to determine sensitivity of the decision to it. The Chapter 18 case study includes an example of multiple attribute sensitivity analysis.

EXAMPLE 10.9

The Island of Niue in the South Pacific Ocean (www.niueisland.com) released a request for proposal (RFP)[1] for a new or reconditioned workboat as it upgrades the infrastructure of its port services. The spreadsheet in Figure 10–6, left two columns, presents the attributes and normalized weights W_i published in the RFP for use in selecting one of the tenders presenting proposals. Four acceptable proposals were received. The next four columns (C through F) include value ratings between 0 and 100 developed by a group of decision makers when the details of each proposal were evaluated against each attribute. For example, proposal 2 received a perfect score of 100 on delivery time, but lifetime costs were considered too high (rating of 20) and the price was considered relatively high (rating of 55). Use these weights and ratings to determine which proposal to pursue first.

Solution

Assume an additive weighting model is appropriate and apply the weighted attribute method. Equation [10.12] determines the R_j measure for the four alternatives. As an illustration, for proposal 3,

$$R_3 = 0.30(95) + 0.20(60) + 0.05(90) + 0.35(85) + 0.10(100)$$
$$= 28.5 + 12.0 + 4.5 + 29.8 + 10.0$$
$$= 84.8$$

The four totals in Figure 10–6 (columns G through J, row 8) indicate that proposal 3 is the overall best choice for the attributes and weights published in the RFP.

Comment

Any economic measure can be incorporated into a multiple attribute evaluation using this method. All measures of worth—PW, AW, ROR, B/C, and C/E—can be included; however, their impact on the final selection will vary relative to the importance placed on the noneconomic attributes.

	A	B	C	D	E	F	G	H	I	J
1			Value rating V_{ij} (0 to 100 basis)				Evaluation measure R_j			
2	Attribute i	Normalized weight, W_i	Proposal 1	Proposal 2	Proposal 3	Proposal 4	Proposal 1	Proposal 2	Proposal 3	Proposal 4
3	Construction*	0.30	95	60	95	95	28.5	18.0	28.5	28.5
4	Delivery time	0.20	60	100	60	60	12.0	20.0	12.0	12.0
5	Contract form‡	0.05	100	80	90	100	5.0	4.0	4.5	5.0
6	Price	0.35	20	55	85	65	7.0	19.3	29.8	22.8
7	Lifetime costs	0.10	100	20	100	100	10.0	2.0	10.0	10.0
8	Total	1.00					62.5	63.3	84.8	78.3
9										
10	* Includes skills and experience of designer and builder, previous projects, and references.									
11	‡ Includes warranties and payment schedule based on stated milestones.									

Figure 10–6

Attributes, weights, ratings, and evaluation measure for Niue workboat proposals, Example 10.9.

[1] Used with permission of Government of Niue, Infrastructure Department, "Request for Proposal: Supply of Workboat," released March 30, 2010, www.gov.nu/Documents/workboattender3022.pdf.

CHAPTER SUMMARY

The interest rate at which the MARR is established depends principally upon the cost of capital and the mix between debt and equity financing. The MARR is strongly influenced by the *weighted average* cost of capital (WACC). Risk, profit, and other factors can be considered after the AW, PW, or ROR analysis is completed and prior to final alternative selection. A high debt-to-equity mix can significantly increase the riskiness of a project and make further debt financing difficult to acquire for the corporation.

If multiple attributes, which include more than the economic dimension of a study, are to be considered in making the alternative decision, first the attributes must be identified and their relative importance assessed. Then each alternative can be value-rated for each attribute. The evaluation measure is determined using a model such as the weighted attribute method, where the measure is calculated by Equation [10.12]. The largest value indicates the best alternative.

PROBLEMS

Working with MARR

10.1 List at least three factors that affect the MARR, and discuss how each one affects it.

10.2 State whether each of the following involves debt financing or equity financing.
   (a) $10,000 taken from one partner's savings account to pay for equipment repair
   (b) Issuance of preferred stock worth $1.3 million
   (c) Short-term loan of $75,000 from a local bank
   (d) Issuance of $3 million worth of 20-year bonds
   (e) Del Engineering buyback of $8 million of its own stock using internal funds

10.3 Helical Products, Inc. uses an after-tax MARR of 12% per year. If the company's effective tax rate (federal, state, and local taxes) is 40%, determine the company's before-tax MARR.

10.4 The owner of a small pipeline construction company is trying to figure out how much he should bid in his attempt to win his first "big" contract. He estimates that his cost to complete the project will be $7.2 million. He wants to bid an amount that will give him an after-tax rate of return of 15% per year if he gets the job, but he doesn't know how much he should bid on a before-tax basis. He told you that his effective state tax rate is 7% and his effective federal tax rate is 22%.
   (a) The expression for determining the overall effective tax rate is

   state rate + (1 − state rate)(federal rate)

   What should his before-tax MARR be in order for him to make an after-tax MARR of 15% per year?
   (b) How much should he bid on the job?

10.5 A group of private equity investors provided $16 million to a start-up company involved in making high-technology detection systems for drugs and other types of contraband. Immediately after the investment was made, another investment opportunity came up for which the investors didn't have enough capital. That project would have yielded an estimated rate of return of 29% per year before taxes. If the group's effective tax rate is 32%, what after-tax rate of return would the forgone project have yielded?

10.6 Five projects were ranked in decreasing order by two measures—rate of return (ROR) and present worth (PW)—to determine which ones should be funded, with the total initial investment not to exceed $18 million. Use the results below to determine the opportunity cost for each measure.

				Ranking by ROR		Ranking by PW	
Project	ROR, %	PW at 15%, $1000	Initial Investment, $1000	Project	Cumulative Investment, $1000	Project	Cumulative Investment, $1000
A	44.5	7,138	8,000	A	8,000	A	8,000
B	12.8	−1,162	15,000	E	13,000	C	16,000
C	20.4	1,051	8,000	C	21,000	E	21,000
D	9.6	−863	8,000	B	36,000	D	29,000
E	26.0	936	5,000	D	44,000	B	44,000

10.7 Tom, the owner of Burger Palace, determined that his weighted average cost of capital is 8%. He expects a return of 4% per year on all of his investments. A proposal presented by the owner of the Dairy Choice next door seems quite risky to Tom, but is an intriguing partnership opportunity. Tom has determined that the proposal's "risk factor" will require an additional 3% per year return for him to accept it.
   (a) Use the recommended approach to determine the MARR that Tom should use, and explain how the 3% risk factor is compensated for in this MARR.
   (b) Determine the effective MARR for his business if Tom turns down the proposal.

D-E Mix and WACC

10.8 Electrical generators produce not only electricity, but also heat from conductor resistance and from friction losses in bearings. A company that manufactures generator coolers for nuclear and gas turbine power plants undertook a plant expansion through financing that had a debt-equity mix of 45–55. If $18 million came from mortgages and bond sales, what was the total amount of the financing?

10.9 Determine the debt-to-equity mix when Applied Technology bought out Southwest Semiconductor using financing as follows: $12 million from mortgages, $5 million from retained earnings, $10 million from cash on hand, and $20 million from bonds.

10.10 Business and engineering seniors are comparing methods of financing their college education during their senior year. The business student has $30,000 in student loans that come due at graduation. Interest is an effective 4% per year. The engineering senior owes $50,000, 50% from his parents with no interest due and 50% from a credit union loan. This latter amount is also due at graduation with an effective rate of 7% per year.

(a) What is the D-E mix for each student?

(b) If their grandparents pay the loans in full at graduation, what are the amounts on the checks they write for each graduate?

(c) When grandparents pay the full amount at graduation, what percent of the principal does the interest represent?

10.11 Two public corporations, First Engineering and Midwest Development, each show capitalization of $175 million in their annual reports. The balance sheet for First Engineering indicates a total debt of $87 million, and that of Midwest Development indicates a net worth of $62 million. Determine the D-E mix for each company.

10.12 Forest Products, Inc. invested $50 million. The company's overall D-E mix is 60–40. What is the return on the company's equity, if the net income is $5 million on a revenue base of $6 million?

10.13 Determine the weighted average cost of capital for a company that manufactures miniature triaxial accelerometers for space-restricted applications. The financing profile, with interest rates, is as follows: $3 million in stock sales at 15% per year, $4 million in bonds at 9%, and $6 million in retained earnings at 7% per year.

10.14 Growth Transgenics Enterprises (GTE) is contemplating the purchase of its rival. One of GTE's genetics engineers got interested in the financing strategy of the buyout. He learned there are two plans being considered. Plan 1 requires 50% equity funds from GTE's retained earnings that currently earn 9% per year, with the balance borrowed externally at 6%, based on the company's excellent stock rating. Plan 2 requires only 20% equity funds with the balance borrowed at a higher rate of 8% per year.

(a) Which plan has the lower average cost of capital?

(b) If GTE's owners decide that the current corporate WACC of 8.2% will not be exceeded, what is the maximum cost of debt capital allowed for each plan? Are these rates higher or lower than the current estimates?

10.15 Midac Corporation wants to arrange for $50 million in capital to finance the manufacturing of a new consumer product. The current plan is 60% equity capital and 40% debt financing. Calculate the WACC for the following scenario:

Equity capital: 60%, or $35 million, via common stock sales for 40% of this amount that will pay

dividends at a rate of 5% per year, and the remaining 60% from retained earnings, which currently earn 9% per year.

Debt capital: 40%, or $15 million, obtained through two sources—bank loans for $10 million borrowed at 8% per year, and the remainder in convertible bonds at an estimated 10% per year bond dividend rate.

10.16 A public corporation in which you own common stock reported a WACC of 11.1% for the year in its report to stockholders. The common stock that you own has averaged a total return of 7% per year over the last 3 years. The annual report also mentions that projects within the corporation are 75% funded by its own capital. Estimate the company's cost of debt capital.

10.17 BASF will invest $14 million this year to upgrade its ethylene glycol processes. This chemical is used to produce polyester resins to manufacture products varying from construction materials to aircraft, and from luggage to home appliances. Equity capital costs 14.5% per year and will supply 65% of the capital funds. Debt capital costs 10% per year before taxes. The effective tax rate for BASF is 36%.

(a) Determine the amount of annual revenue after taxes that is consumed in covering the interest on the project's initial cost.

(b) If the corporation does not want to use 65% of its own funds, the financing plan may include 75% debt capital. Determine the amount of annual revenue needed to cover the interest with this plan, and explain the effect it may have on the corporation's ability to borrow in the future.

10.18 A couple planning for their child's college education can fund part of or all the expected $100,000 tuition cost from their own funds (through an education IRA) or borrow all or part of it. The average return for their own funds is 7% per year, but the loan is expected to have a higher interest rate as the loan amount increases. Use a spreadsheet to generate a plot of the WACC curve with the estimated loan interest rates below and determine the best D-E mix for the couple.

Loan Amount, $	Interest Rate, % per year
10,000	5.0
30,000	6.0
50,000	8.0
60,000	9.0
75,000	11.0
100,000	13.0

10.19 Over the last few years, Carol's Fashion Store, a statewide franchise, has experienced the D-E mixes and costs of debt and equity capital on several projects summarized below.

(a) Plot debt, equity, and weighted average cost of capital.

(b) Determine what mix of debt and equity capital provided the lowest WACC.

	Debt capital		Equity capital	
Project	Percent	Rate	Percent	Rate
A	100%	14.5%		
B	70	13.0	30%	7.8%
C	65	12.0	35	7.8
D	50	11.5	50	7.9
E	35	9.9	65	9.8
F	20	12.4	80	12.5
G			100	12.5

10.20 For Problem 10.19, use a spreadsheet to (a) determine the best D-E mix and (b) determine the best D-E mix if the cost of debt capital increases by 10% per year, for example, 13.0% increases to 14.3%.

Cost of Debt Capital

10.21 The cash flow plan associated with a debt financing transaction allowed a company to receive $2,800,000 now in lieu of future interest payments of $196,000 per year for 10 years plus a lump sum of $2,800,000 in year 10. If the company's effective tax rate is 33%, determine the company's cost of debt capital (a) before taxes and (b) after taxes.

10.22 A company that makes several different types of skateboards, Jennings Outdoors, incurred interest expenses of $1,200,000 per year from various types of debt financing. The company borrowed $19,000,000 in year 0 and repaid the principal of the loans in year 15 in a lump-sum payment of $20,000,000. If the company's effective tax rate is 29%, what was the company's cost of debt capital (a) before taxes and (b) after taxes?

10.23 Molex Inc., a manufacturer of cable assemblies for polycrystalline photovoltaic solar modules, requires $3.1 million in debt capital. The company plans to sell 15-year bonds that carry a dividend of 6% per year, payable semiannually. Molex has an effective tax rate of 32% per year. Determine (a) the nominal annual after-tax cost of debt capital and (b) the effective annual after-tax cost of debt capital.

10.24 Tri-States Gas Producers expects to borrow $800,000 for field engineering improvements. Two methods of debt financing are possible—borrow it all from a bank or issue debenture bonds. The company will pay an effective 8% per year to the bank for 8 years. The principal on the loan will be reduced uniformly over the 8 years, with the remainder of each annual payment going toward interest. The bond issue will be for 800 10-year bonds of $1000 each that require a 6% per year dividend payment.

(a) Which method of financing is cheaper after an effective tax rate of 40% is considered?

(b) Which is the cheaper method using a before-tax analysis?

10.25 The Sullivan Family Partnership plans to purchase a refurbished condo in their hometown for investment purposes. The negotiated $200,000 purchase price will be financed with 20% of savings (retained earnings) which consistently makes 6.5% per year after all relevant income taxes are paid. Eighty percent will be borrowed at a before-tax rate of 9% per year for 15 years with the principal repaid in equal annual installments. If the effective tax rate is 22% per year, based only on these data, answer the following.

(a) What is the partnership's annual loan payment for each of the 15 years?

(b) What is the net present worth difference between the $200,000 now and the PW of the cost of the 80–20 D-E mix series of cash flows necessary to finance the purchase? What does this PW value mean?

(c) What is the after-tax WACC for this purchase?

Cost of Equity Capital

10.26 Determine the cost of equity capital to Hy-Lok USA if the company sells 500,000 shares of its preferred stock at a 5% discount from its price of $130. The stock carries a $10 per year dividend.

10.27 The initial public offering price for the common stock of SW Refining is $23 per share, and it will pay a first-year dividend of $0.92 per share. If the appreciation rate in dividends is anticipated to be 3.2% per year, determine the cost of equity capital for the stock offering.

10.28 The cost of debt capital is lower after taxes than before taxes. The cost of equity capital is more difficult to estimate using the dividend method or the CAPM model, for example, yet the after-tax and before-tax cost of equity capital is the same. Why are the after-tax rates not the same for both types of financing?

10.29 H2W Technologies is considering raising capital to expand its offerings of 2-phase and 4-phase linear stepper motors. The beta value for its stock is 1.41. Use the capital asset pricing model and a 3.8% premium above the risk-free return to determine the cost of equity capital if the risk-free return is 3.2%.

10.30 Management at Hirschman Engineering has asked you to determine the cost of equity capital based on the company's common stock. The company wants you to use two methods: the dividend method and the CAPM. Last year, the first year for dividends, the stock paid $0.75 per share on the average of $11.50 on the New York Stock Exchange. Management hopes to grow the dividend rate at 3% per year. Hirschman Engineering stock has a volatility that is higher than the norm at 1.3. If safe investments are returning 5.5% and the 3% growth on common stocks is also the premium above the risk-free investments that Hirschman Engineering plans to pay, calculate the cost of equity capital using the two methods.

10.31 Common stocks issued by Meggitt Sensing Systems paid stockholders $0.93 per share on an average price of $18.80 last year. The company expects to grow the dividend rate at a maximum of 1.5% per year. The stock volatility is 1.19, and other stocks in the same industry are paying an average of 4.95% per year dividend. U.S Treasury bills are returning 4.5%. Determine the company's cost of equity capital last year using (a) the dividend method and (b) the CAPM.

10.32 Last year a Japanese engineering materials corporation, Yamachi Inc., purchased some U.S. Treasury bonds that return an average of 4% per year. Now, Euro bonds are being purchased with a realized average return of 3.9% per year. The volatility factor of Yamachi stock last year was 1.10 and has increased this year to 1.18. Other publicly traded stocks in this same business are paying an average of 5.1% dividends per year. Determine the cost of equity capital for each year, and explain why the increase or decrease seems to have occurred.

10.33 The engineering manager at FXO Plastics wants to complete an alternative evaluation study. She asked the finance manager for the corporate MARR. The finance manager gave her some data on the project and stated that all projects must clear their average (pooled) cost by at least 4%.

Funds Source	Amount, $	Average Cost, %
Retained earnings	4,000,000	7.4
Stock sales	6,000,000	4.8
Long-term loans	5,000,000	9.8

(a) Use the data to determine the minimum MARR.
(b) The study is after taxes and part (a) provided the before-tax MARR. Determine the correct MARR to use if the tax rate was 32% last year and the finance manager meant that the 4% above the cost is for after-tax evaluations.

Different D-E Mixes

10.34 Why is it financially unhealthy for an individual to maintain a large percentage of debt financing over a long time, that is, to be highly leveraged?

10.35 In a leveraged buyout of one company by another, the purchasing company usually obtains borrowed money and inserts as little of its own equity as possible into the purchase. Explain some circumstances under which such a buyout may put the purchasing company at economic risk.

10.36 Grainger and Company has an opportunity to invest $500,000 in a new line of direct-drive rotary screw compressors. Financing will be equally split between common stock ($250,000) and a loan with an 8% after-tax interest rate. The estimated annual NCF after taxes is $48,000 for the next 7 years. The effective tax rate is 50%. Grainger uses the capital asset pricing model for evaluation of its common stock. Recent analysis shows that it has a volatility rating of 0.95 and is paying a premium of 5% above a safe return on its common stock. Nationally, the safest investment is currently paying 3% per year. Is the investment financially attractive if Grainger uses as the MARR its (a) equity cost of capital and (b) WACC?

10.37 Fairmont Industries primarily relies on 100% equity financing to fund projects. A good opportunity is available that will require $250,000 in capital. The Fairmont owner can supply the money from personal investments that currently earn an average of 7.5% per year. The annual net cash flow from the project is estimated at $30,000 for the next 15 years. Alternatively, 60% of the required amount can be borrowed for 15 years at 7% per year. If the MARR is the WACC, determine which plan, if either, should be undertaken. This is a before-tax analysis.

10.38 Omega Engineering Inc. has an opportunity to invest $10,000,000 in a new engineering remote control system for offshore drilling platforms. Financing will be split between common stock sales ($5,000,000) and a loan with an 8% per year interest rate. Omega's share of the annual net cash flow is estimated to be $1.35 million for each of the next 5 years. Omega is about to initiate CAPM as its common stock evaluation model. Recent analysis shows that it has a volatility rating of 1.22 and is paying a premium of 5% on its common stock dividend. The U.S. Treasury bills are currently paying 4% per year. Is the venture financially attractive if the MARR equals (*a*) the cost of equity capital and (*b*) the WACC?

10.39 A new annular die process is to be installed for extruding pipes, tubes, and tubular films. The phase I installed price for the dies and machinery is $2,000,000. The manufacturer has not decided how to finance the system. The WACC over the last 5 years has averaged 10% per year.

(*a*) Two financing alternatives have been defined. The first requires an investment of 40% equity funds at 9% and a loan for the balance at an interest rate of 10% per year. The second alternative requires only 25% equity funds and the balance borrowed at 10.5% per year. Which approach will result in the smaller average cost of capital?

(*b*) Yesterday, the corporate finance committee decided that the WACC for all new projects must not exceed the 5-year historical average of 10% per year. With this restriction, what is the maximum loan interest rate that can be incurred for each of the financing alternatives?

10.40 Shadowland, a manufacturer of air-freightable pet crates, has identified two projects that, though having a relatively high risk, are expected to move the company into new revenue markets. Utilize a spreadsheet solution to (*a*) select any combination of the projects if the MARR is equal to the after-tax WACC and (*b*) determine if the same projects should be selected if the risk factors are enough to require an additional 2% per year for the investment to be made.

Project	Initial Investment, $	After-Tax Cash Flow, $/Year	Life, Years
Wildlife (W)	−250,000	48,000	10
Reptiles (R)	−125,000	30,000	5

Financing will be developed using a D-E mix of 60–40 with equity funds costing 7.5% per year. Debt financing will be developed from $10,000, 5% per year, paid quarterly, 10-year bonds. The effective tax rate is 30% per year.

10.41 Two friends each invested $20,000 of their own (equity) funds. Stan, being more conservative, purchased utility and manufacturing corporation stocks. Theresa, being a risk taker, *leveraged* the $20,000 and purchased a $100,000 condo for rental

property. Considering no taxes, dividends, or revenues, analyze these two purchases by doing the following for one year after the funds were invested.

(*a*) Determine the year-end values of their equity funds if there was a 10% *increase* in the value of the stocks and the condo.

(*b*) Determine the year-end values of their equity funds if there was a 10% *decrease* in the value of the stocks and the condo.

(*c*) Use your results to explain why *leverage* can be financially risky.

Multiple, Noneconomic Attributes

10.42 In multiple attribute analysis, if three different alternatives are to be evaluated on the basis of eight attributes that are considered of equal importance, what is the weight of each attribute?

10.43 A consulting engineer asked a company manager to assign importance values (0 to 100) to five attributes that will be included in an alternative evaluation process. Determine the weight of each attribute using the importance scores.

Attribute	Importance Score
1. Safety	60
2. Cost	40
3. Impact	80
4. Environmental	30
5. Acceptability	20

10.44 Ten attributes were rank-ordered in terms of increasing importance and were identified as A, B, C, . . . , and J. Determine the weight of (*a*) attribute C and (*b*) attribute J.

10.45 A team of three people submitted the following statements about the attributes to be used in a weighted attribute evaluation. Use the statements to determine the normalized weights if assigned scores are between 0 and 100.

Attribute	Comment
1. Flexibility (F)	The most important factor
2. Safety (S)	70% as important as uptime
3. Uptime (U)	One-half as important as flexibility
4. Rate of return (R)	Twice as important as safety

10.46 Different types and capacities of crawler hoes are being considered for use in a major excavation on a pipe-laying project. Several supervisors who served on similar projects in the past have identified some of the attributes and their view of relative importance. For the information that follows, determine the weighted rank order, using a 0-to-10 scale and the normalized weights.

Attribute	Comment
1. Truck versus hoe height	90% as important as trenching speed
2. Type of topsoil	Only 10% of most important attribute
3. Type of subsoil	30% as important as trenching speed
4. Hoe cycle time	Twice as important as type of subsoil
5. Hoe trenching speed	Most important attribute
6. Pipe laying speed	80% as important as hoe cycle time

10.47 John, who works at Swatch, has decided to use the weighted attribute method to compare three systems for manufacturing a watchband. The vice president and her assistant VP have evaluated each of three attributes in terms of importance to them, and John has placed an evaluation from 0 to 100 on each alternative for the three attributes. John's *ratings* for each alternative are as follows:

	Alternative		
Attribute	1	2	3
Economic return > MARR	50	70	100
High throughput	100	60	30
Low scrap rate	100	40	50

Use the *weights* below to evaluate the alternatives. Are the results the same for both persons' weights? Why?

Importance Score	VP	Assistant VP
Economic return > MARR	20	100
High throughput	80	80
Low scrap rate	100	20

10.48 The Athlete's Shop has evaluated two proposals for weight lifting and exercise equipment. A present worth analysis at $i = 15\%$ per year of estimated revenues and costs resulted in $PW_A = \$440,000$ and $PW_B = \$390,000$. In addition to this economic measure, three more attributes were independently assigned a relative importance score from 0 to 100 by the shop manager and the lead trainer.

	Importance Score	
Attribute	Manager	Trainer
Economics	80	80
Durability	35	80
Flexibility	30	100
Maintainability	20	50

Separately, you have used the four attributes to rate the two equipment proposals on a scale of 0 to 1.0 as shown in the following table. The economic attribute was rated using the PW values.

Attribute	Proposal A	Proposal B
Economics	1.00	0.90
Durability	0.35	1.00
Flexibility	1.00	0.90
Maintainability	0.25	1.00

Select the better proposal using each of the following methods.

(*a*) Present worth
(*b*) Weighted evaluations of the shop manager
(*c*) Weighted evaluations of the lead trainer

ADDITIONAL PROBLEMS AND FE EXAM REVIEW QUESTIONS

10.49 The term *opportunity cost* refers to:
(*a*) The first cost of an alternative that has been accepted for funding
(*b*) The total cost of an alternative that has been accepted for funding
(*c*) The rate of return or profit available on the next-best alternative that had to be forgone due to lack of capital funds
(*d*) The cost of an alternative that was not recognized as an alternative that actually represented a good opportunity

10.50 The cost of capital is established on the basis of:
(*a*) The cost of debt financing
(*b*) The weighted average of debt and equity financing

(*c*) The cost of equity financing
(*d*) The cost of debt financing plus the expected inflation rate

10.51 All of the following are examples of debt capital except:
(*a*) Retained earnings
(*b*) Long-term bonds
(*c*) Loan from a local bank
(*d*) Purchase of equipment using a credit card

10.52 All of the following are examples of equity capital except:
(*a*) Sale of preferred stock
(*b*) Long-term bonds
(*c*) Company cash on hand
(*d*) Use of retained earnings

10.53 If a public utility expands its capacity to generate electricity by obtaining $41 million from retained earnings and $30 million from municipal bond sales, the utilities' debt-to-equity mix is closest to:
(a) 58% debt and 42% equity
(b) 73% debt and 27% equity
(c) 27% debt and 73% equity
(d) 42% debt and 58% equity

10.54 Gentech, Inc. financed a new product as follows: $5 million in stock sales at 13.7% per year, $2 million in retained earnings at 8.9% per year, and $3 million through convertible bonds at 7.8% per year. The company's WACC is closest to:
(a) 9% per year
(b) 10% per year
(c) 11% per year
(d) 12% per year

10.55 If the after-tax rate of return for a cash flow series is 11.2% and the corporate effective tax rate is 39%, the approximated before-tax rate of return is closest to:
(a) 6.8%
(b) 5.4%
(c) 18.4%
(d) 28.7%

10.56 Medzyme Pharmaceuticals has maintained a 50-50 D-E mix for capital investments. Equity capital has cost 11%; however, debt capital that historically cost 9% has now increased by 20% per year.

If Medzyme does not want to exceed its historical weighted average cost of capital (WACC), and it is forced to go to a D-E mix of 75–25, the maximum acceptable cost of equity capital is closest to:
(a) 7.6%
(b) 9.2%
(c) 9.8%
(d) 10.9%

10.57 The importance values (0 to 100) for five attributes are shown below. The weight to assign to attribute 1 is:
(a) 0.16
(b) 0.20
(c) 0.22
(d) 0.25

Attribute	Importance Score
1	55
2	45
3	85
4	30
5	60

10.58 For eight attributes rank-ordered in terms of increasing importance, the weighting of the sixth attribute is closest to:
(a) 0.17
(b) 0.14
(c) 0.08
(d) 0.03

CASE STUDY

WHICH IS BETTER—DEBT OR EQUITY FINANCING?

Background

Pizza Hut Corporation has decided to enter the catering business in three states within its Southeastern U.S. Division, using the name Pizza Hut At-Your-Place. To deliver the meals and serving personnel, it is about to purchase 200 vans with custom interiors for a total of $1.5 million. Each van is expected to be used for 10 years and have a $1000 salvage value.

A feasibility study completed last year indicated that the At-Your-Place business venture could realize an estimated annual net cash flow of $300,000 before taxes in the three states. After-tax considerations would have to take into account the effective tax rate of 35% paid by Pizza Hut.

An engineer with Pizza Hut's Distribution Division has worked with the corporate finance office to determine how to best develop the $1.5 million capital needed for the purchase of vans. There are two viable financing plans.

Information

Plan A is debt financing for 50% of the capital ($750,000) with the 8% per year compound interest loan repaid over 10 years with uniform year-end payments. (A simplifying assumption that $75,000 of the principal is repaid with each annual payment can be made.)

Plan B is 100% equity capital raised from the sale of $15 per share common stock. The financial manager informed the engineer that stock is paying $0.50 per share in dividends and that this dividend rate has been increasing at an average of 5% each year. This dividend pattern is expected to continue, based on the current financial environment.

Case Study Exercises

1. What values of MARR should the engineer use to determine the better financing plan?

2. The engineer must make a recommendation on the financing plan by the end of the day. He does not know how to consider all the tax angles for the debt financing in plan A. However, he does have a handbook that gives these relations for equity and debt capital about taxes and cash flows:

Equity capital: no income tax advantages

   After-tax net cash flow
   = (before-tax net cash flow)(1 − tax rate)

Debt capital: income tax advantage comes from interest paid on loans

   After-tax net cash flow = before-tax net cash flow
                 − loan principal
                 − loan interest − taxes

   Taxes = (taxable income)(tax rate)
   Taxable income = net cash flow
                 − loan interest

He decides to forget any other tax consequences and use this information to prepare a recommendation. Is A or B the better plan?

3. The division manager would like to know how much the WACC varies for different D-E mixes, especially about 15% to 20% on either side of the 50% debt financing option in plan A. Plot the WACC curve and compare its shape with that of Figure 10–2.

CHAPTER 11

Replacement and Retention Decisions

LEARNING OUTCOMES

Purpose: Perform a replacement/retention study between an in-place asset, process, or system and one that could replace it.

One of the most common and important issues in industrial practice is that of replacement or retention of an asset, process, or system that is currently installed. This differs from previous situations where all the alternatives were new. The fundamental question answered by a replacement study (also called a replacement/retention study) about a currently installed system is, *Should it be replaced now or later?* When an asset is currently in use and its function is needed in the future, it will be replaced at some time. In reality, a replacement study answers the question of **when, not if,** to replace.

A replacement study is usually designed to first make the economic decision to retain or replace *now*. If the decision is to replace, the study is complete. If the decision is to retain, the cost estimates and decision can be revisited each year to ensure that the decision to retain is still economically correct. This chapter explains how to perform the initial-year and follow-on year replacement studies.

A replacement study is an application of the AW method of comparing unequal-life alternatives, first introduced in Chapter 6. In a replacement study with no specified study period, the AW values are determined by a technique called the **economic service life (ESL)** analysis. If a study period is specified, the replacement study procedure is different from that used when no study period is set.

If asset depreciation and taxes are to be considered in an *after-tax replacement analysis,* Chapters 16 and 17 should be covered before or in conjunction with this chapter. After-tax replacement analysis is included in Chapter 17.

PE

Keep or Replace the Kiln Case: B&T Enterprises manufactures and sells high-melting-temperature ceramics and high-performance metals to other corporations. The products are sold to a wide range of industries from the nuclear and solar power industry to sports equipment manufacturers of specialty golf and tennis gear, where kiln temperatures up to approximately 1700°C are needed. For years, B&T has owned and been very satisfied with Harper International pusher-plate tunnel kilns. Two are in use currently at plant locations on each coast of the country; one kiln is 10 years old, and the second was purchased only 2 years ago and serves, primarily, the ceramics industry needs on the west coast. This newer kiln can reach temperatures of 2800°C.

During the last two or three quarterly maintenance visits, the Harper team leader and the head of B&T quality have discussed the ceramic and metal industry needs for higher temperatures. In some cases the temperatures are as high as 3000°C for emerging nitride, boride, and carbide transition metals that form very high-melting-temperature oxides. These may find use in hypersonic vehicles, engines, plasma arc electrodes, cutting tools, and high-temperature shielding.

A looming question on the mind of the senior management and financial officers of B&T revolves around the need to seriously consider a new graphite hearth kiln, which can meet higher temperature and other needs of the current and projected customer base. This unit will have lower operating costs and significantly greater furnace efficiency in heat time, transit, and other crucial parameters. Since virtually all of this business is on the west coast, the graphite hearth kiln would replace the newer of the two kilns currently in use.

For identification, let

PT identify the currently installed pusher-plate tunnel kiln (defender)

GH identify the proposed new graphite hearth kiln (challenger)

Relevant estimates follow in $ millions for monetary units.

	PT	GH
First cost, $ M	$25; 2 years ago	$38; with no trade-in
AOC, $ M per year	year 1: $5.2; year 2: $6.4	starts at $3.4, increases 10%/year
Life, years	6 (remaining)	12 (estimated)
Heating element, $ M	—	$2.0 every 6 years

This case is used in the following topics of this chapter:

Economic service life (Section 11.2)

Replacement study (Section 11.3)

Replacement study with study period (Section 11.5)

Replacement value (Section 11.6)

Problems 11.18 and 11.41

11.1 Basics of a Replacement Study ● ● ●

The need for a replacement study can develop from several sources:

Reduced performance. Because of physical deterioration, the ability to perform at an expected level of *reliability* (being available and performing correctly when needed) or *productivity* (performing at a given level of quality and quantity) is not present. This usually results in increased costs of operation, higher scrap and rework costs, lost sales, reduced quality, diminished safety, and larger maintenance expenses.

Altered requirements. New requirements of accuracy, speed, or other specifications cannot be met by the existing equipment or system. Often the choice is between complete replacement or enhancement through retrofitting or augmentation.

Obsolescence. International competition and rapidly changing technology make currently-used systems and assets perform acceptably but less productively than equipment coming available. The ever-decreasing development cycle time to bring new products to market is often the reason for *premature replacement* studies, that is, studies performed before the estimated useful or economic life is reached.

Replacement studies use some terminology that is closely related to terms in previous chapters.

Defender and **challenger** are the names for two mutually exclusive alternatives. The defender is the currently installed asset, and the challenger is the potential replacement. A replacement study compares these two alternatives. The challenger is the "best" challenger because it has been selected as the best one to possibly replace the defender. (This is the same terminology used earlier for incremental ROR and B/C analysis, but both alternatives were new).

Salvage/market value

Market value is the current value of the installed asset if it were sold or traded on the open market. Also called *trade-in value*, this estimate is obtained from professional appraisers, resellers, or liquidators familiar with the industry. As in previous chapters, **salvage value** is the estimated value at the end of the expected life. In replacement analysis, the salvage value at the end of one year is used as the market value at the beginning of the next year.

AW values are used as the primary economic measure of comparison between the defender and challenger. The term *equivalent uniform annual cost* (*EUAC*) may be used in lieu of AW, because often only costs are included in the evaluation; revenues generated by the defender or challenger are assumed to be equal. (Since EUAC calculations are exactly the same as for AW, we use the term AW.) Therefore, all values will be negative when only costs are involved. Salvage or market value is an exception; it is a cash inflow and carries a plus sign.

Economic service life

Economic service life (ESL) for an alternative is the *number of years at which the lowest AW of cost occurs*. The equivalency calculations to determine ESL establish the life n for the best challenger and the lowest cost life for the defender in a replacement study. The next section explains how to find the ESL.

Defender first cost is the initial investment amount P used for the defender. The *current market value* (*MV*) is the correct estimate to use for P for the defender in a replacement study. The estimated salvage value at the end of one year becomes the market value at the beginning of the next year, provided the estimates remain correct as the years pass. It is incorrect to use the following as MV for the defender first cost: trade-in value that *does not represent a fair market value*, or the depreciated book value taken from accounting records. If the defender must be upgraded or augmented to make it equivalent to the challenger (in speed, capacity, etc.), this cost is added to the MV to obtain the estimated defender first cost.

Challenger first cost is the amount of capital that must be recovered (amortized) when replacing a defender with a challenger. This amount is almost always equal to P, the first cost of the challenger.

If an unrealistically high trade-in value is offered for the defender compared to its fair market value, the *net* cash flow required for the challenger is reduced, and this fact should be considered in the analysis. The correct amount to recover and use in the economic analysis for the challenger is its first cost minus the difference between the trade-in value (TIV) and market value (MV) of the defender. In equation form, this is $P - (TIV - MV)$. This amount represents the actual cost to the company because it includes both the opportunity cost (i.e., market value of the defender) and the out-of-pocket cost (i.e., first cost − trade-in) to acquire the challenger. Of course, when the trade-in and market values are the same, the challenger P value is used in all computations.

The challenger first cost is the estimated initial investment necessary to acquire and install it. Sometimes, an analyst or manager will attempt to *increase* this first cost by an amount equal to the *unrecovered capital* remaining in the defender, as shown on the accounting records for the asset. This incorrect treatment of capital recovery is observed most often when the defender is working well and in the early stages of its life, but technological obsolescence, or some other reason, has forced consideration of a replacement. This leads us to identify two additional characteristics of replacement analysis, in fact, of any economic analysis: *sunk costs* and *nonowner's viewpoint*.

> A **sunk cost** is a prior expenditure or loss of capital (money) that cannot be recovered by a decision about the future. The replacement alternative for an asset, system, or process that has incurred a nonrecoverable cost should not include this cost in any direct fashion; sunk costs should be handled in a realistic way using tax laws and write-off allowances.

Sunk cost

A sunk cost should never be added to the challenger's first cost, because it will make the *challenger appear to be more costly than it actually is*. For example, assume an asset costing $100,000 two years ago has a depreciated value of $80,000 on the corporate books. It must be replaced prematurely due to rapidly advancing technology. If the replacement alternative (challenger) has a first cost of $150,000, the $80,000 from the current asset is a sunk cost were the challenger purchased. For the purposes of an economic analysis, it is *incorrect to increase* the challenger's first cost to $230,000 or any number between this and $150,000.

The second characteristic is the perspective taken when conducting a replacement study. You, the analyst, are a consultant from outside the company.

> The **nonowner's viewpoint**, also called the *outsider's viewpoint* or *consultant's viewpoint*, provides the greatest objectivity in a replacement study. This viewpoint performs the analysis without bias; it means the analyst owns neither the defender nor the challenger. Additionally, it assumes the services provided by the defender can be purchased now by making an "initial investment" equal to the market value of the defender.

Besides being unbiased, this perspective is correct because the defender's market value is a forgone opportunity of cash inflow were the replacement not selected, and the defender chosen.

As mentioned in the introduction, a replacement study is an application of the annual worth method. As such, the fundamental assumptions for a replacement study parallel those of an AW analysis. If the *planning horizon is unlimited*, that is, a study period is not specified, the assumptions are as follows:

1. The services provided are needed for the indefinite future.
2. The challenger is the best challenger available now and in the future to replace the defender. When this challenger replaces the defender (now or later), it will be repeated for succeeding life cycles.
3. Cost estimates for every life cycle of the defender and challenger will be the same as in their first cycle.

As expected, none of these assumptions is precisely correct. We discussed this previously for the AW method (and the PW method). When the intent of one or more of the assumptions becomes incorrect, the estimates for the alternatives must be updated and a new replacement study conducted. The replacement procedure discussed in Section 11.3 explains how to do this. When the planning horizon is limited to a specified study period, the assumptions above do not hold. The procedure of Section 11.5 discusses replacement analysis over a fixed study period.

EXAMPLE 11.1

Only 2 years ago, Techtron purchased for $275,000 a fully loaded SCADA (supervisory control and data acquisition) system including hardware and software for a processing plant operating on the Houston ship channel. When it was purchased, a life of 5 years and salvage of 20% of first cost were estimated. Actual M&O costs have been $25,000 per year, and the book value is $187,000. There has been a series of insidious malware infections targeting Techtron's command and control software, plus next-generation hardware marketed only recently could greatly reduce the competitiveness of the company in several of its product lines. Given these factors, the system is likely worth nothing if kept in use for the final 3 years of its anticipated useful life.

Model K2-A1, a new replacement turnkey system, can be purchased for $300,000 net cash, that is, $400,000 first cost and a $100,000 trade-in for the current system. A 5-year life, salvage value of 15% of stated first cost or $60,000, and an M&O cost of $15,000 per year are good estimates for the new system. The current system was appraised this morning, and a market value of $100,000 was confirmed for today; however, with the current virus discovery, the appraiser anticipates that the market value will fall rapidly to the $80,000 range once the virus problem and new model are publicized.

Using the above values as the best possible today, state the correct defender and challenger estimates for P, M&O, S, and n in a replacement study to be performed today.

Solution

Defender: Use the current market value of $100,000 as the first cost for the defender. All others—original cost of $275,000, book value of $187,000, and trade-in value of $100,000—are irrelevant to a replacement study conducted today. The estimates are as follows:

First cost	$P = \$-100,000$
M&O cost	$A = \$-25,000$ per year
Expected life	$n = 3$ years
Salvage value	$S = 0$

Challenger: The $400,000 stated first cost is the correct one to use for P, because the trade-in and market values are equal.

First cost	$P = \$-400,000$
M&O cost	$A = \$-15,000$ per year
Expected life	$n = 5$ years
Salvage value	$S = \$60,000$

Comment

If the replacement study is conducted next week when estimates will have changed, the defender's first cost will be $80,000, the new market value according to the appraiser. The challenger's first cost will be $380,000, that is, $P - (\text{TIV} - \text{MV}) = 400,000 - (100,000 - 80,000)$.

Online Presentation

11.2 Economic Service Life ● ● ○

Until now the estimated life n of an alternative or asset has been stated. In reality, the best life estimate to use in the economic analysis is not known initially. When a replacement study or an analysis between new alternatives is performed, the best value for n should be determined using current cost estimates. The best life estimate is called the *economic service life*.

Economic service life

The **economic service life (ESL)** is the number of years n at which the equivalent uniform annual worth (AW) of costs is the minimum, considering the most current cost estimates over all possible years that the asset may provide a needed service.

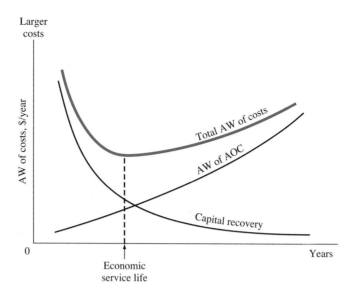

Figure 11–1
Annual worth curves of
cost elements that deter-
mine the economic
service life.

The ESL is also referred to as the *economic life* or *minimum cost life*. Once determined, the ESL should be the estimated life for the asset used in an engineering economy study, if only economics are considered. When n years have passed, the ESL indicates that the asset should be replaced to minimize overall costs. To perform a replacement study correctly, it is important that the ESL of the challenger and the ESL of the defender be determined, since their n values are usually not preestablished.

The ESL is determined by calculating the total AW of costs if the asset is in service 1 year, 2 years, 3 years, and so on, up to the last year the asset is considered useful. Total AW of costs is the sum of capital recovery (CR), which is the AW of the initial investment and any salvage value, and the AW of the estimated annual operating cost (AOC), that is,

> **Total AW = capital recovery − AW of annual operating costs**
> **= CR − AW of AOC**   [11.1]

The ESL is the n value for the smallest total AW of costs. (Remember: These AW values are *cost* estimates, so the AW values are negative numbers. Therefore, $-200 is a lower cost than $-500.) Figure 11–1 shows the characteristic shape of a total AW of cost curve. The CR component of total AW decreases, while the AOC component increases, thus forming the concave shape. The two AW components are calculated as follows.

Decreasing cost of capital recovery. The capital recovery is the AW of investment; it decreases with each year of ownership. Capital recovery is calculated by Equation [6.3], which is repeated here. The salvage value S, which usually decreases with time, is the estimated market value (MV) in that year.

Capital recovery

$$\text{Capital recovery} = -P(A/P,i,n) + S(A/F,i,n) \qquad [11.2]$$

Increasing cost of AW of AOC. Since the AOC (or M&O) estimates usually increase over the years, the AW of AOC increases. To calculate the AW of the AOC series for 1, 2, 3, ... years, determine the present worth of each AOC value with the P/F factor, then redistribute this P value over the years of ownership, using the A/P factor.

The complete equation for total AW of costs over k years ($k = 1, 2, 3, \dots$) is

$$\text{Total AW}_k = -P(A/P,i,k) + S_k(A/F,i,k) - \left[\sum_{j=1}^{j=k} \text{AOC}_j(P/F,i,j) \right](A/P,i,k) \qquad [11.3]$$

where P = initial investment or current market value
             S_k = salvage value or market value after k years
      AOC_j = annual operating cost for year j ($j = 1$ to k)

The current MV is used for P when the asset is the defender, and the estimated future MV values are substituted for the S values in years 1, 2, 3, ... Plotting the AW_k series as in Figure 11–1 clearly indicates where the ESL is located and the trend of the AW_k curve on each side of the ESL.

To determine ESL by spreadsheet, the PMT function (with embedded NPV functions as needed) is used repeatedly for each year to calculate capital recovery and the AW of AOC. Their sum is the total AW for k years of ownership. The PMT function formats for the capital recovery and AOC components for each year k ($k = 1, 2, 3, ...$) are as follows:

Capital recovery for the challenger:	PMT(i%,years,P,−MV_in_year_k)
Capital recovery for the defender:	PMT(i%,years,current_MV,−MV_in_ year_k) [11.4]
AW of AOC:	−PMT(i%,years,NPV(i%,year_1_AOC: year_k_AOC)+0)

When the spreadsheet is developed, it is recommended that the PMT functions in year 1 be developed using cell-reference format; then drag down the function through each column. A final column summing the two PMT results displays total AW. Augmenting the table with an Excel xy scatter chart graphically displays the cost curves in the general form of Figure 11–1, and the ESL is easily identified. Example 11.2 illustrates ESL determination by hand and by spreadsheet.

EXAMPLE 11.2

A 3-year-old backup power system is being considered for early replacement. Its current market value is $20,000. Estimated future market values and annual operating costs for the next 5 years are given in Table 11–1, columns 2 and 3. What is the economic service life of this defender if the interest rate is 10% per year? Solve by hand and by spreadsheet.

Solution by Hand

Equation [11.3] is used to calculate total AW_k for $k = 1, 2, ..., 5$. Table 11–1, column 4, shows the capital recovery for the $20,000 current market value ($j = 0$) plus 10% return. Column 5 gives the equivalent AW of AOC for k years. As an illustration, the computation of total AW for $k = 3$ from Equation [11.3] is

$$\text{Total AW}_3 = -P(A/P,i,3) + MV_3(A/F,i,3) - [\text{PW of AOC}_1, \text{AOC}_2, \text{and AOC}_3](A/P,i,3)$$
$$= -20,000(A/P,10\%,3) + 6000(A/F,10\%,3) - [5000(P/F,10\%,1)$$
$$+ 6500(P/F,10\%,2) + 8000(P/F,10\%,3)](A/P,10\%,3)$$
$$= -6230 - 6405 = \$-12,635$$

A similar computation is performed for each year 1 through 5. The lowest equivalent cost (numerically largest AW value) occurs at $k = 3$. Therefore, the defender ESL is $n = 3$ years, and the AW value is $\$-12,635$. In the replacement study, this AW will be compared with the best challenger AW determined by a similar ESL analysis.

Solution by Spreadsheet

See Figure 11–2 for the spreadsheet screen shot and chart that shows the ESL is $n = 3$ years and AW = $\$-12,634$. (This format is a template for any ESL analysis; simply change the

TABLE 11–1	Computation of Economic Service Life				
Year j (1)	MV$_j$, $ (2)	AOC$_j$, $ (3)	Capital Recovery, $ (4)	AW of AOC, $ (5)	Total AW$_k$, $ (6) = (4) + (5)
1	10,000	−5,000	−12,000	−5,000	−17,000
2	8,000	−6,500	−7,714	−5,714	−13,428
3	6,000	−8,000	−6,230	−6,405	−12,635
4	2,000	−9,500	−5,878	−7,072	−12,950
5	0	−12,500	−5,276	−7,961	−13,237

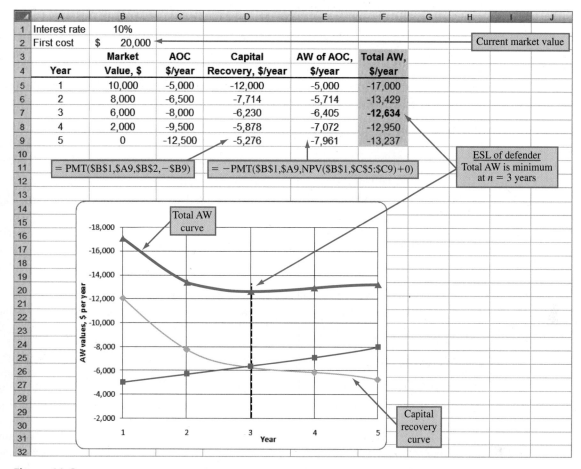

Figure 11–2
Determination of ESL and plot of curves, Example 11.2

estimates and add rows for more years.) Contents of columns D and E are described below. The PMT functions apply the formats as described in Equation [11.4]. Cell tags show detailed cell-reference format for year 5. The $ symbols are included for absolute cell referencing, needed when the entry is dragged down through the column.

Column D: Capital recovery is the AW of the $20,000 investment in year 0 for each year 1 through 5 with the estimated MV in that year. For example, in actual numbers, the cell-reference PMT function in year 5 shown on the spreadsheet reads $= PMT(10\%,5,20000,-0)$, resulting in $\$-5276$. This series is plotted in Figure 11–2.

Column E: The NPV function embedded in the PMT function obtains the present worth in year 0 of all AOC estimates through year k. Then PMT calculates the AW of AOC over k years. For example, in year 5, the PMT in numbers is $= -PMT(10\%,5,NPV(10\%,C5:C9)+0)$. The 0 is the AOC in year 0; it is optional. The graph plots the AW of AOC curve, which constantly increases in cost because the AOC estimates increase each year.

Comment

The *capital recovery curve* in Figure 11–2 (middle curve) is not the expected shape (see year 4) because the estimated market value changes each year. If the same MV were estimated for each year, the curve would appear like Figure 11–1. When several total AW values are approximately equal, the curve will be flat over several periods. This indicates that the ESL is relatively insensitive to costs.

It is reasonable to ask about the difference between the ESL analysis above and the AW analyses performed in previous chapters. Previously we had a *specific life estimated to be n years* with associated other estimates: first cost in year 0, possibly a salvage value in year *n*, and an AOC that remained constant or varied each year. For all previous analyses, the calculation of AW using these estimates determined the AW over *n* years. This is the economic service life when *n* is fixed. Also, in all previous cases, there were no year-by-year market value estimates. Therefore, we can conclude the following:

> When the **expected life n is known and specified** for the challenger or defender, no ESL computations are necessary. Determine the AW over *n* years, using the first cost or current market value, estimated salvage value after *n* years, and AOC estimates. This AW value is the correct one to use in the replacement study.

However, when *n* is not fixed, the following is useful. First the market/salvage series is needed. It is not difficult to estimate this series for a new or current asset. For example, an asset with a first cost of *P* can lose market value of, say, 20% per year, so the market value series for years 0, 1, 2, . . . is *P*, 0.8*P*, 0.64*P*, . . . , respectively. If it is reasonable to predict the MV series on a year-by-year basis, it can be combined with the AOC estimates to produce what is called the *marginal costs* for the asset.

> **Marginal costs (MC)** are year-by-year estimates of the costs to own and operate an asset for that year. Three components are added to determine the marginal cost:
>
> - Cost of ownership (loss in market value is the best estimate of this cost)
> - Forgone interest on the market value at the beginning of the year
> - AOC for each year

Once the marginal costs are estimated for each year, their equivalent AW value is calculated. *The sum of the AW values of the first two of these components is the capital recovery amount.* Now, it should be clear that the total AW of all three marginal cost components over *k* years is the same value as the total annual worth for *k* years calculated in Equation [11.3]. That is, the following relation is correct.

$$\text{AW of marginal costs} = \text{total AW of costs} \qquad [11.5]$$

Therefore, there is no need to perform a separate, detailed marginal cost analysis when yearly market values are estimated. The ESL analysis presented in Example 11.2 is sufficient in that it results in the same numerical values. This is demonstrated in Example 11.3 using the progressive example.

EXAMPLE 11.3 Keep or Replace the Kiln Case   PE

In our progressive example, B&T Enterprises is considering the replacement of a 2-year-old kiln with a new one to meet emerging market needs. When the current tunnel kiln was purchased 2 years ago for $25 million, an ESL study indicated that the minimum cost life was between 3 and 5 years of the expected 8-year life. The analysis was not very conclusive because the total AW cost curve was flat for most years between 2 and 6, indicating insensitivity of the ESL to changing costs. Now, the same type of question arises for the proposed graphite hearth model that costs $38 million new: What are the ESL and the estimated total AW of costs? The Manager of Critical Equipment at B&T estimates that the market value after only 1 year will drop to $25 million and then retain 75% of the previous year's value over the 12-year expected life. Use this market value series and *i* = 15% per year to illustrate that an ESL analysis and marginal cost analysis result in exactly the same total AW of cost series.

Solution

Figure 11–3 is a spreadsheet screen shot of the two analyses in $ million units. The market value series is detailed in column B starting at $25 (million) and decreasing by 25% per year. A brief description of each analysis follows.

	A	B	C	D	E	F	G
1	Interest rate	15%	**ESL analysis**		First cost, $	38.00	
2		**Market**	**AOC**	**Capital**	**AW of AOC,**	**Total AW,**	
3	**Year**	**Value, $**	**$/year**	**Recovery, $/year**	**$/year**	**$/year**	
4	1	25.00	-3.40	-18.70	-3.40	-22.10	
5	2	18.75	-3.74	-14.65	-3.56	-18.21	
6	3	14.06	-4.11	-12.59	-3.72	-16.31	
7	4	10.55	-4.53	-11.20	-3.88	-15.08	
8	5	7.91	-4.98	-10.16	-4.04	-14.21	
9	6	5.93	-7.48	-9.36	-4.43	-13.80	Two AW series are identical
10	7	4.45	-6.02	-8.73	-4.58	-13.31	
11	8	3.34	-6.63	-8.23	-4.73	-12.95	
12	9	2.50	-7.29	-7.81	-4.88	-12.69	
13	10	1.88	-8.02	-7.48	-5.03	-12.51	
14	11	1.41	-8.82	-7.20	-5.19	-12.39	
15	12	1.06	-9.70	-6.97	-5.35	-12.32	
16	Formulas, year 12	=B14*0.75	=C14*1.1	=PMT($B$1, $A15,$F$1,-$B15)	=-PMT($B$1,$A15, NPV($B$1,$C$4:$C15)+0)	=E15+D15	
17							
18	Interest rate	15%	**Marginal cost analysis**		Current MV, $	38.00	
19		**Market**	**Loss in**	**Lost Interest**		**Marginal Cost**	**AW of Marginal**
20	**Year**	**Value, $**	**Market Value, $**	**on MV for Year, $**	**AOC, $**	**for the Year, $**	**Cost, $/year**
21	1	25.00	-13.00	-5.70	-3.40	-22.10	-22.10
22	2	18.75	-6.25	-3.75	-3.74	-13.74	-18.21
23	3	14.06	-4.69	-2.81	-4.11	-11.61	-16.31
24	4	10.55	-3.52	-2.11	-4.53	-10.15	-15.08
25	5	7.91	-2.64	-1.58	-4.98	-9.20	-14.21
26	6	5.93	-1.98	-1.19	-7.48	-10.64	-13.80
27	7	4.45	-1.48	-0.89	-6.02	-8.40	-13.31
28	8	3.34	-1.11	-0.67	-6.63	-8.41	-12.95
29	9	2.50	-0.83	-0.50	-7.29	-8.62	-12.69
30	10	1.88	-0.63	-0.38	-8.02	-9.02	-12.51
31	11	1.41	-0.47	-0.28	-8.82	-9.57	-12.39
32	12	1.06	-0.35	-0.21	-9.70	-10.26	-12.32
33	Formulas, year 12	=B31*0.75	=B32-B31	=-$B$1*$B31	=E31*1.1	=$C32+$D32+$E32	=-PMT($B$1,$A32, NPV($B$1,$F$21:$F32)+0)
34							

Figure 11–3

Comparison of annual worth series resulting from ESL analysis and marginal cost analysis, Example 11.3.

ESL analysis: Equation [11.4] is applied repeatedly for $k = 1, 2, \ldots, 12$ years (columns C, D, and E) in the top of Figure 11–3. Row 16 details the spreadsheet functions for year 12. The result in column F is the total AW series that is of interest now.

Marginal cost (MC): The functions in the bottom of Figure 11–3 (columns C, D, and E) develop the three components added to obtain the MC series. Row 33 details the functions for year 12. The resulting AW of marginal costs (column G) is the series to compare with the corresponding ESL series above (column F).

The two AW series are identical, thus demonstrating that Equation [11.5] is correct. Therefore, either an ESL or a marginal cost analysis will provide the same information for a replacement study. In this case, the results show that the new kiln will have a minimum AW of costs of $-12.32 million at its full 12-year life.

We can draw two important conclusions about the n and AW values to be used in a replacement study. These conclusions are based on the extent to which detailed annual estimates are made for the market value.

1. **Year-by-year market value estimates are made.** Use them to perform an ESL analysis, and determine the n value with the lowest total AW of costs. These are the best n and AW values for the replacement study.

2. **Yearly market value estimates are not available.** The only estimate available is market value (salvage value) in year n. Use it to calculate the AW over n years. These are the n and AW values to use; however, they may not be the "best" values in that they may not represent the best equivalent total AW of cost value found if an ESL analysis were performed.

Upon completion of the ESL analysis (item 1 above), the replacement study procedure in Section 11.3 is applied using the values

Challenger alternative (C): AW_C for n_C years

Defender alternative (D): AW_D for n_D years

Online Presentation

11.3 Performing a Replacement Study ● ● ●

Replacement studies are performed in one of two ways: without a study period specified or with one defined. Figure 11–4 gives an overview of the approach taken for each situation. The procedure discussed in this section applies when no study period (planning horizon) is specified. If a specific number of years is identified for the replacement study, for example, over the next 5 years, with no continuation considered after this time period in the economic analysis, the procedure in Section 11.5 is applied.

A replacement study determines when a challenger replaces the in-place defender. The complete study is finished if the challenger (C) is selected to replace the defender (D) now. However, if the defender is retained now, the study may extend over a number of years equal to the life of the defender n_D, after which a challenger replaces the defender. Use the annual worth and life values for C and D determined in the ESL analysis in the following procedure. Assume the services provided by the defender could be obtained at the AW_D amount.

The replacement study procedure is:

New replacement study:

1. On the basis of the better AW_C or AW_D value, select the challenger C or defender D. When the challenger is selected, replace the defender now, and expect to keep the challenger for n_C years. This replacement study is complete. If the defender is selected, plan to retain it for up to n_D more years. (This is the leftmost branch of Figure 11–4.) Next year, perform the following steps.

One-year-later analysis:

2. Determine if all estimates are still current for both alternatives, especially first cost, market value, and AOC. If not, proceed to step 3. If yes and this is year n_D, replace the defender. If this is not year n_D, retain the defender for another year and repeat this same step. This step may be repeated several times.

3. Whenever the estimates have changed, update them and determine new AW_C and AW_D values. Initiate a new replacement study (step 1).

Figure 11–4

Overview of replacement study approaches.

If the defender is selected initially (step 1), estimates may need updating after 1 year of retention (step 2). Possibly there is a new best challenger to compare with D. Either significant changes in defender estimates or availability of a new challenger indicates that a new replacement study is to be performed. In actuality, a replacement study can be performed each year or more frequently to determine the advisability of replacing or retaining any defender, provided a competitive challenger is available.

Example 11.4 illustrates the application of ESL analysis for a challenger and defender, followed by the use of the replacement study procedure. The planning horizon is unspecified in this example.

EXAMPLE 11.4

Two years ago, Toshiba Electronics made a $15 million investment in new assembly line machinery. It purchased approximately 200 units at $70,000 each and placed them in plants in 10 different countries. The equipment sorts, tests, and performs insertion-order kitting on electronic components in preparation for special-purpose circuit boards. This year, new international industry standards will require a $16,000 retrofit on each unit, in addition to the expected operating cost. Due to the new standards, coupled with rapidly changing technology, a new system is challenging the retention of these 2-year-old machines. The chief engineer at Toshiba USA realizes that the economics must be considered, so he has asked that a replacement study be performed this year and each year in the future, if need be. The i is 10% and the estimates are below.

Challenger: First cost: $50,000
  Future market values: decreasing by 20% per year
  Estimated retention period: no more than 10 years
  AOC estimates: $5000 in year 1 with increases of $2000 per year
    thereafter

Defender: Current international market value: $15,000
  Future market values: decreasing by 20% per year
  Estimated retention period: no more than 3 more years
  AOC estimates: $4000 next year, increasing by $4000 per year
    thereafter, plus the $16,000 retrofit next year

(a) Determine the AW values and economic service lives necessary to perform the replacement study.
(b) Perform the replacement study now.
(c) After 1 year, it is time to perform the follow-up analysis. The challenger is making large inroads to the market for electronic components assembly equipment, especially with the new international standards features built in. The expected market value for the defender is still $12,000 this year, but it is expected to drop to virtually nothing in the future—$2000 next year on the worldwide market and zero after that. Also, this prematurely outdated equipment is more costly to keep serviced, so the estimated AOC next year has been increased from $8000 to $12,000 and to $16,000 two years out. Perform the follow-up replacement study analysis.

Solution

(a) The results of the ESL analysis, shown in Figure 11–5, include all the MV and AOC estimates in columns B and C. For the challenger, note that $P = \$50,000$ is also the MV in year 0. The total AW of costs is for each year, should the challenger be placed into service for that number of years. As an example, the year $k = 4$ amount of $\$-19{,}123$ is determined using Equation [11.3], where the A/G factor accommodates the arithmetic gradient series in the AOC.

$$
\begin{aligned}
\text{Total AW}_4 &= -50{,}000(A/P,10\%,4) + 20{,}480(A/F,10\%,4) \\
&\quad - [5000 + 2000(A/G,10\%,4)] \\
&= \$-19{,}123
\end{aligned}
$$

	A	B	C	D	E	F	G H	I	J	K
1	Interest rate	10%	**ESL Analysis of Challenger**							
2	First cost	$ 50,000								
3		**Market**				**Total AW**				
4	**Year**	**Value**	**AOC**	**Capital Recovery**	**AW of AOC**	**of costs**				
5	1	40,000	-5,000	-15,000	-5,000	-20,000				
6	2	32,000	-7,000	-13,571	-5,952	-19,524				
7	3	25,600	-9,000	-12,372	-6,873	-19,245				
8	4	20,480	-11,000	-11,361	-7,762	-19,123				
9	5	16,384	-13,000	-10,506	-8,620	-19,126				
10	6	13,107	-15,000	-9,782	-9,447	-19,229				
11	7	10,486	-17,000	-9,165	-10,243	-19,408				
12	8	8,389	-19,000	-8,639	-11,009	-19,648				
13	9	6,711	-21,000	-8,188	-11,745	-19,933				
14	10	5,369	-23,000	-7,800	-12,451	-20,251				
15	Formulas, year 10	= B13*0.8	= $C13-2000	= PMT($B$1,$A14,$B$2,-$B14)	= -PMT($B$1,$A14,NPV($B$1,$C$5:$C14)+0)	= E14+D14	ESL results			
16										
17	Interest rate	10%	**ESL Analysis of Defender**							
18	Market value	$ 15,000								
19		**Market**				**Total AW**				
20	**Year**	**Value**	**AOC**	**Capital Recovery**	**AW of AOC**	**of costs**				
21	1	12,000	-20,000	-4,500	-20,000	-24,500				
22	2	9,600	-8,000	-4,071	-14,286	-18,357				
23	3	7,680	-12,000	-3,711	-13,595	-17,307				
24										
25										

Figure 11–5
ESL analysis of challenger and defender, Example 11.4.

For spreadsheet-based ESL analysis, this same result is achieved in cell F8 using Equation [11.4]. The functions are

$$\text{Total AW}_4 = \text{PMT}(10\%,4,50000,-20480) - \text{PMT}(10\%,4,\text{NPV}(10\%,C5:C8)+0)$$
$$= -11,361 - 7,762$$
$$= -19,123$$

The defender costs are analyzed in the same way up to the maximum retention period of 3 years.

The lowest AW cost (numerically largest) values for the replacement study are as follows:

Challenger: $AW_C = \$-19,123$ for $n_C = 4$ years
Defender: $AW_D = \$-17,307$ for $n_D = 3$ years

The challenger total AW of cost curve (Figure 11–5) is classically shaped and relatively flat between years 3 and 6; there is virtually no difference in the total AW for years 4 and 5. For the defender, note that the estimated AOC values change substantially over 3 years, and they do not constantly increase or decrease.

(b) To perform the replacement study now, apply only the first step of the procedure. Select the defender because it has the better AW of costs ($-17,307$), and expect to retain it for 3 more years. Prepare to perform the one-year-later analysis 1 year from now.
(c) One year later, the situation has changed significantly for the equipment Toshiba retained last year. Apply the steps for the one-year-later analysis:

   2. After 1 year of defender retention, the challenger estimates are still reasonable, but the defender market value and AOC estimates are substantially different. Go to step 3 to perform a new ESL analysis for the defender.
   3. The defender estimates in Figure 11–5 are updated below for the ESL analysis. New AW values are calculated using Equation [11.3]. There is now a maximum of 2 more years of retention, 1 year less than the 3 years determined last year.

Year k	Market Value, $	AOC, $	Total AW If Retained k More Years, $
0	12,000	—	—
1	2,000	-12,000	-23,200
2	0	-16,000	-20,819

The AW and n values for the new replacement study are as follows:

          Challenger: unchanged at $AW_C = \$-19{,}123$ for $n_C = 4$ years
          Defender: new $AW_D = \$-20{,}819$ for $n_D = 2$ more years

Now select the challenger based on its favorable AW value. Therefore, replace the defender now, not 2 years from now. Expect to keep the challenger for 4 years, or until a better challenger appears on the scene.

EXAMPLE 11.5   Keep or Replace the Kiln Case   (PE)

We continue with the progressive example of possibly replacing a kiln at B&T Enterprises. A marketing study revealed that the improving business activity on the west coast implies that the revenue profile between the installed kiln (PT) and the proposed new one (GH) would be the same, with the new kiln possibly bringing in new revenue within the next couple of years. The president of B&T decided it was time to do a replacement study. Assume you are the lead engineer and that you previously completed the ESL analysis on the challenger (Example 11.3). It indicates that for the GH system the ESL is its expected useful life.

Challenger: ESL $n_{GH} = 12$ years with total equivalent annual cost $AW_{GH} = \$-12.32$ million

The president asked you to complete the replacement study, stipulating that, due to the rapidly rising annual operating costs (AOC), the defender would be retained a maximum of 6 years. You are expected to make the necessary estimates for the defender (PT) and perform the study at a 15% per year return.

Solution

After some data collection, you have good evidence that the market value for the PT system will stay high, but that the increasing AOC is expected to continue rising about $1.2 million per year. The best estimates for the next 6 years in $ million units are these:

Year	1	2	3	4	5	6
Market Value, $ M	22.0	22.0	22.0	20.0	18.0	18.0
AOC, $ M per year	−5.2	−6.4	−7.6	−8.8	−10.0	−11.2

You developed a spreadsheet and performed the analysis in Figure 11–6. As an illustration, total AW computation for 3 years of retention, in $ million units, is

$$
\begin{aligned}
\text{Total AW}_3 &= -22.0(A/P,15\%,3) + 22.0(A/F,15\%,3) - [5.2(P/F,15\%,1) + 6.4(P/F,15\%,2) \\
&\quad + 7.6(P/F,15\%,3)](A/P,15\%,3) \\
&= -9.63 + 6.34 - [14.36](0.43798) \\
&= \$-9.59 \text{ per year}
\end{aligned}
$$

Though the system could be retained up to 6 years, the ESL is much shorter at 1 year.

Defender: ESL $n_{PT} = 1$ year with total equivalent annual cost $AW_{PT} = \$-8.50$ million

	A	B	C	D	E	F
1			**ESL Analysis of PT**			
2	Interest rate	15%			Market value	$ 22.00
3						
4		**Market**				**Total AW**
5	Year	**Value**	**AOC**	**Capital Recovery**	**AW of AOC**	**of costs**
6	1	22.00	-5.20	-3.30	-5.20	-8.50
7	2	22.00	-6.40	-3.30	-5.76	-9.06
8	3	22.00	-7.60	-3.30	-6.29	-9.59
9	4	20.00	-8.80	-3.70	-6.79	-10.49
10	5	18.00	-10.00	-3.89	-7.27	-11.16
11	6	18.00	-11.20	-3.76	-7.72	-11.47
12	Formula, year 3	22.00	-7.60	`= PMT($B$2, $A8,$F$2,-$B8)`	`= -PMT($B$2,$A8, NPV($B$2,$C$6:$C8)+0)`	`= D8+E8`
13						

Figure 11–6
ESL analysis of defender kiln PT for progressive example, Example 11.5.

To make the replacement/retention decision, apply step 1 of the procedure. Since $AW_{PT} = \$-8.50$ million per year is considerably less than $AW = \$-12.32$ million, you should recommend keeping the current kiln only 1 more year and doing another study during the year to determine if the current estimates are still reliable.

Comment

An observation of the trends of the two final AW series in this problem is important. A comparison of Figure 11–3 (top), column F, and Figure 11–6, column F, shows us that the largest total AW of the current system ($-11.47 M for 6 years) is still below the smallest total AW of the proposed system ($-12.32 M for 12 years). This indicates that the graphite hearth system (the challenger) will not be chosen on an economic basis, if the decision to consider the installed kiln as a defender in the future were made. It would take some significant estimate changes to justify the challenger.

11.4 Additional Considerations in a Replacement Study ● ● ●

There are several additional aspects of a replacement study that may be introduced. Three of these are identified and discussed in turn.

- Future-year replacement decisions at the time of the initial replacement study
- Opportunity cost versus cash flow approaches to alternative comparison
- Anticipation of improved future challengers

In most cases when management initiates a replacement study, the question is best framed as, "Replace now, 1 year from now, 2 years from now, etc.?" The procedure above does answer this question provided the estimates for C and D do not change as each year passes. In other words, *at the time it is performed, step 1 of the procedure does answer the replacement question for multiple years.* It is only when estimates change over time that the decision to retain the defender may be prematurely reversed (prior to n_D years) in favor of the then-best challenger.

The first costs (P values) for the challenger and defender have been correctly taken as the initial investment for the challenger C and current market value for the defender D. This is called the **opportunity cost approach** because it recognizes that a cash inflow of funds equal to the market value is forgone if the defender is selected. This approach, also called the conventional approach, is correct for every replacement study. A second approach, called the **cash flow approach,** recognizes that when C is selected, the market value cash inflow for the defender is received and, in effect, immediately reduces the capital needed to invest in the challenger. *Use of the cash flow approach is strongly discouraged* for at least two reasons: possible violation of the equal-service requirement and incorrect capital recovery value for C. As we are aware, all economic evaluations must compare alternatives with equal service. Therefore, the cash flow approach can work only when challenger and defender lives are exactly equal. This is commonly not the case; in fact, the ESL analysis and the replacement study procedure are designed to compare two mutually exclusive, *unequal-life* alternatives via the annual worth method. If this equal-service comparison reason is not enough to avoid the cash flow approach, consider what happens to the challenger's capital recovery amount when its first cost is decreased by the market value of the defender. The capital recovery (CR) terms in Equation [11.3] will decrease, resulting in a falsely low value of CR for the challenger, were it selected. From the vantage point of the economic study itself, the decision for C or D will not change; but when C is selected and implemented, this CR value is not reliable. The conclusion is simple:

Use the initial investment of C and the market value of D as the first costs in the ESL analysis and in the replacement study.

A basic premise of a replacement study is that some challenger will replace the defender at a future time, provided the service continues to be needed and a worthy challenger is available. The expectation of ever-improving challengers can offer strong encouragement to retain the defender until some situational elements—technology, costs, market fluctuations, contract negotiations, etc.—

stabilize. This was the case in the previous two examples. A large expenditure on equipment when the standards changed soon after purchase forced an early replacement consideration and a large loss of invested capital. The replacement study is no substitute for forecasting challenger availability. *It is important to understand trends, new advances, and competitive pressures that can complement the economic outcome of a good replacement study.* It is often better to compare a challenger with an augmented defender in the replacement study. Adding needed features to a currently installed defender may prolong its useful life and productivity until challenger choices are more appealing.

It is possible that a significant tax impact may occur when a defender is traded early in its expected life. If taxes should be considered, proceed now, or after the next section, to Chapter 17 and the after-tax replacement analysis in Section 17.7.

11.5 Replacement Study over a Specified Study Period ● ● ◐

Online Presentation

When the time period for the replacement study is limited to a specified study period or planning horizon, for example, 6 years, the ESL analysis is not performed.

> The AW values for the challenger and for the remaining life of the defender are not based on the economic service life; the AW is calculated over the study period only. What happens to the alternatives after the study period is not considered in the replacement analysis.

This means that the defender or challenger is not needed beyond the study period. In fact, a study period of fixed duration does not comply with the three assumptions stated in Section 11.1—service needed for indefinite future, best challenger available now, and estimates will be identical for future life cycles.

When performing a replacement study over a fixed study period, it is crucial that the estimates used to determine the AW values be accurate and used in the study. This is especially important for the defender. Failure to do the following violates the requirement of equal-service comparison.

> When the defender's remaining life is **shorter than the study period,** the cost of providing the defender's services from the end of its expected remaining life to the end of the study period must be estimated as accurately as possible and included in the replacement study.

Study period

The right branch of Figure 11–4 presents an overview of the replacement study procedure for a stated study period.

1. *Succession options and AW values.* Develop all the viable ways to use the defender and challenger during the study period. There may be only one option or many options; the longer the study period, the more complex this analysis becomes. The AW values for the challenger and defender cash flows are used to build the equivalent cash flow values for each option.

2. *Selection of the best option.* The PW or AW for each option is calculated over the study period. Select the option with the lowest cost, or highest income if revenues are estimated. (As before, the best option will have the numerically largest PW or AW value.)

The following examples use this procedure and illustrate the importance of making cost estimates for the defender alternative when its remaining life is less than the study period.

EXAMPLE 11.6

Claudia works with Lockheed-Martin (LMCO) in the aircraft maintenance division. She is preparing for what she and her boss, the division chief, hope to be a new 10-year defense contract with the U.S. Air Force on C-5A cargo aircraft. A key piece of equipment for maintenance operations is an avionics circuit diagnostics system. The current system was purchased 7 years ago on an earlier contract. It has no capital recovery costs remaining, and the following are reliable estimates: current market value = $70,000, remaining life of 3 more years, no salvage value, and AOC = $30,000 per year. The only options for this system are to replace it now or retain it for the full 3 additional years.

Claudia has found that there is only one good challenger system. Its cost estimates are: first cost = $750,000, life = 10 years, $S = 0$, and AOC = $50,000 per year.

Realizing the importance of accurate defender alternative cost estimates, Claudia asked the division chief what system would be a logical follow-on to the current one 3 years hence, if LMCO wins the contract. The chief predicted LMCO would purchase the very system she had identified as the challenger, because it is the best on the market. The company would keep it for the entire 10 additional years for use on an extension of this contract or some other application that could recover the remaining 3 years of invested capital. Claudia interpreted the response to mean that the last 3 years would also be capital recovery years, but on some project other than this one. Claudia's estimate of the first cost of this same system 3 years from now is $900,000. Additionally, the $50,000 per year AOC is the best estimate at this time.

The division chief mentioned any study had to be conducted using the interest rate of 10%, as mandated by the U.S. Office of Management and Budget (OMB). Perform a replacement study for the fixed contract period of 10 years.

Solution

The study period is fixed at 10 years, so the intent of the replacement study assumptions is not present. This means the defender follow-on estimates are very important to the analysis. Further, *any analyses to determine the ESL values are unnecessary* since alternative lives are already set and no projected annual market values are available. The first step of the replacement study procedure is to define the options. Since the defender will be replaced now or in 3 years, there are only two options:

1. Challenger for all 10 years.
2. Defender for 3 years, followed by the challenger for 7 years.

Cash flows are diagrammed in Figure 11–7. For option 1, the challenger is used for all 10 years. Equation [11.3] is applied to calculate AW using the following estimates:

Challenger: $P = \$-750,000$ $AOC = \$-50,000$
$n = 10$ years $S = 0$

$$AW_C = -750,000(A/P,10\%,10) - 50,000 = \$-172,063$$

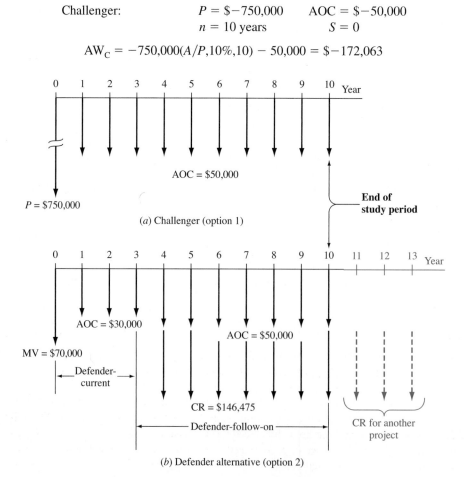

(a) Challenger (option 1)

(b) Defender alternative (option 2)

Figure 11–7
Cash flow diagrams for a 10-year study period replacement study, Example 11.6.

The second option has more complex cost estimates. The AW for the in-place system is calculated over the first 3 years. Added to this is the capital recovery for the defender follow-on for the next 7 years. *However in this case, the CR amount is determined over its full 10-year life.* (It is not unusual for the recovery of invested capital to be moved between projects, especially for contract work.) Refer to the AW components as AW_{DC} (subscript DC for defender current) and AW_{DF} (subscript DF for defender follow-on). The final cash flows are shown in Figure 11–7b.

Defender current: market value $= \$-70,000$ $AOC = \$-30,000$
$n = 3$ years $S = 0$

$$AW_{DC} = [-70,000 - 30,000(P/A,10\%,3)](A/P,10\%,10) = \$-23,534$$

Defender follow-on: $P = \$-900,000$, $n = 10$ years for capital recovery calculation only, $AOC = \$-50,000$ for years 4 through 10, $S = 0$.

The CR and AW for all 10 years are

$$CR_{DF} = -900,000(A/P,10\%,10) = \$-146,475 \qquad\qquad [11.6]$$
$$AW_{DF} = (-146,475 - 50,000)(F/A,10\%,7)(A/F,10\%,10) = \$-116,966$$

Total AW_D for the defender is the sum of the two annual worth values above. This is the AW for option 2.

$$AW_D = AW_{DC} + AW_{DF} = -23,534 - 116,966 = \$-140,500$$

Option 2 has a lower cost ($\$-140,500$ versus $\$-172,063$). Retain the defender now and expect to purchase the follow-on system 3 years hence.

Comment

The capital recovery cost for the defender follow-on will be borne by some yet-to-be-identified project for years 11 through 13. If this assumption were not made, its capital recovery cost would be calculated over 7 years, not 10, in Equation [11.6], increasing CR to $\$-184,869$. This raises the annual worth to $AW_D = \$-163,357$. The defender alternative (option 2) is still selected.

EXAMPLE 11.7

Three years ago Chicago's O'Hare Airport purchased a new fire truck. Because of flight increases, new fire-fighting capacity is needed once again. An additional truck of the same capacity can be purchased now, or a double-capacity truck can replace the current fire truck. Estimates are presented below. Compare the options at 12% per year using (*a*) a 12-year study period and (*b*) a 9-year study period.

	Presently Owned	New Purchase	Double Capacity
First cost P, $	−151,000 (3 years ago)	−175,000	−190,000
AOC, $	−1,500	−1,500	−2,500
Market value, $	70,000	—	—
Salvage value, $	10% of P	12% of P	10% of P
Life, years	12	12	12

Solution

Identify option 1 as retention of the presently owned truck and augmentation with a new same-capacity vehicle. Define option 2 as replacement with the double-capacity truck.

	Option 2		
	Presently Owned	Augmentation	Double Capacity
P, $	−70,000	−175,000	−190,000
AOC, $	−1,500	−1,500	−2,500
S, $	15,100	21,000	19,000
n, years	9	12	12

(*a*) For a full-life 12-year study period of option 1,

$$AW_1 = \text{(AW of presently owned)} + \text{(AW of augmentation)}$$
$$= [-70,000(A/P,12\%,9) + 15,100(A/F,12\%,9) - 1500]$$
$$+ [-175,000(A/P,12\%,12) + 21,000(A/F,12\%,12) - 1500]$$
$$= -13,616 - 28,882$$
$$= \$-42,498$$

This computation assumes the equivalent services provided by the current fire truck can be purchased at $\$-13,616$ per year for years 10 through 12.

$$AW_2 = -190,000(A/P,12\%,12) + 19,000(A/F,12\%,12) - 2500$$
$$= \$-32,386$$

Replace now with the double-capacity truck (option 2) at an advantage of $10,112 per year.

(*b*) The analysis for an abbreviated 9-year study period is identical, except that $n = 9$ in each factor; that is, 3 fewer years are allowed for the augmentation and double-capacity trucks to recover the capital investment plus a 12% per year return. The salvage values remain the same since they are quoted as a percentage of P for all years.

$$AW_1 = \$-46,539 \qquad AW_2 = \$-36,873$$

Option 2 is again selected.

The previous two examples indicate an important consideration for setting the length of the study period for a replacement analysis. It involves the capital recovery amount for the challenger, when the strict definition of a study period is applied.

Study period

Capital recovery

> When a study period shorter than the life of the challenger is defined, the challenger's capital recovery amount **increases** in order to recover the initial investment plus a return in this **shortened time period**. Highly abbreviated study periods tend to disadvantage the challenger because no consideration of time beyond the end of the study period is made in calculating the challenger's capital recovery amount.

If there are several options for the number of years that the defender may be retained before replacement with the challenger, the first step of the replacement study—succession options and AW values—must include all the viable options. For example, if the study period is 5 years and the defender will remain in service 1 year, or 2 years, or 3 years, cost estimates must be made to determine AW values for each defender retention period. In this case, there are four options; call them W, X, Y, and Z.

Option	Defender Retained, Years	Challenger Serves, Years
W	3	2
X	2	3
Y	1	4
Z	0	5

The respective AW values for defender retention and challenger use define the cash flows for each option. Example 11.8 illustrates the procedure using the progressive example.

EXAMPLE 11.8 Keep or Replace the Kiln Case

We have progressed to the point that the replacement study between the defender PT and challenger GH was completed (Example 11.5). The defender was the clear choice with a much smaller AW value ($-8.50 M$) than that of the challenger ($-12.32 M$). Now the management of B&T is in a dilemma. They know the current tunnel kiln is much cheaper than the new graphite hearth, but the prospect of future new business should not be dismissed.

TABLE 11–2		Replacement Study Options and Total AW Values, Example 11.8		
	Defender PT		**Challenger GH**	
Option	Years Retained	AW, $ M/Year	Years Retained	AW, $ M/Year
A	1	−8.50	5	−14.21
B	2	−9.06	4	−15.08
C	3	−9.59	3	−16.31
D	4	−10.49	2	−18.21
E	5	−11.16	1	−22.10
F	6	−11.47	0	—

The president asked, "Is it possible to determine when it is economically the cheapest to purchase the new kiln, provided the current one is kept at least 1 year, but no more than 6 years, its remaining expected life?" The chief financial officer answered, yes, of course. (*a*) Determine the answer for the president. (*b*) Discuss the next step in the analysis based on the conclusion reached here.

Solution

(*a*) Actually, this is a quite easy question to answer, because all the information has been determined previously. We know the MARR is 15% per year, the study period has been established at 6 years, and the defender PT will stay in place between 1 and 6 years. Therefore, the challenger GH will be considered for 0 to 5 years of service. The total AW values were determined for the defender in Example 11.5 (Figure 11–6) and for the challenger in Example 11.3 (Figure 11–3). They are repeated in Table 11–2 for convenience. Use the procedure for a replacement study with a fixed study period.

Step 1: Succession options and AW values. There are six options in this case; the defender is retained from 1 to 6 years while the challenger is installed from 0 to 5 years. We will label them A through F. Figure 11–8 presents the options and the AW series for each option from Table 11–2. No consideration of the fact that the challenger has an expected life of 12 years is made since the study period is fixed at 6 years.

Step 2: Selection of the best option. The PW value for each option is determined over the 6-year study period in column J of Figure 11–8. The conclusion is clearly to **keep the defender** in place for 6 more years.

(*b*) Every replacement analysis has indicated that the defender should be retained for the near future. If the analysis is to be carried further, the possibility of increased revenue based on services of the challenger's high-temperature and operating efficiency should be considered next. In the introductory material, new business opportunities were mentioned. A revenue increase for the challenger will reduce its AW of costs and possibly make it more economically viable.

	A	B	C	D	E	F	G	H	I	J
1		Time in Service, Years								Option
2		Defender	Challenger	AW Cash Flows for Each Option, $M per year						PW at 15%,
3	Option	PT	GH	1	2	3	4	5	6	$M
4	A	1	5	-8.50	-14.21	-14.21	-14.21	-14.21	-14.21	-48.81
5	B	2	4	-9.06	-9.06	-15.08	-15.08	-15.08	-15.08	-47.28
6	C	3	3	-9.59	-9.59	-9.59	-16.31	-16.31	-16.31	-46.38
7	D	4	2	-10.49	-10.49	-10.49	-10.49	-18.21	-18.21	-46.87
8	E	5	1	-11.16	-11.16	-11.16	-11.16	-11.16	-22.10	-46.96
9	F	6	0	-11.47	-11.47	-11.47	-11.47	-11.47	-11.47	**-43.41**
10										
11				Conclusion: Keep defender all 6 years						
12										
13										

Figure 11–8
PW values for 6-year study period replacement analysis, Example 11.8.

11.6 Replacement Value ● ● ○

Often it is helpful to know the minimum market value of the defender necessary to make the challenger economically attractive. If a realizable market value or trade-in of at least this amount can be obtained, from an economic perspective the challenger should be selected immediately. This is a *breakeven value* between AW_C and AW_D; it is referred to as the **replacement value (RV)**. Set up the relation $AW_C = AW_D$ with the market value for the defender identified as RV, which is the unknown. The AW_C is known, so RV can be determined. The selection guideline is as follows:

> If the actual market trade-in **exceeds** the breakeven **replacement value**, the challenger is the better alternative and should replace the defender now.

Determination of the RV for a defender is an excellent opportunity to utilize the Goal Seek tool in Excel. The target cell is the current market value, and the AW_D value is forced to equal the AW_C amount. Example 11.9 discusses a replacement value of the progressive example.

EXAMPLE 11.9   Keep or Replace the Kiln Case   PE

As one final consideration of the challenger kiln, you decide to determine what the trade-in amount would have to be so that the challenger is the economic choice next year. This is based on the ESL analysis that concluded the following (Examples 11.3 and 11.5):

Defender: ESL $n_{PT} = 1$ year with $AW_{PT} = \$-8.50$ million
Challenger: ESL $n_{GH} = 12$ years with $AW_{GH} = \$-12.32$ million

The original defender price was $25 million, and a current market value of $22 million was estimated earlier (Figure 11–6). Since the installed kiln is known for retention of its market value (MV), you are hopeful the difference between RV and estimated MV may not be so significant. What will you discover RV to be? The MARR is 15% per year.

Solution

Set the AW relation for the defender for the ESL time of 1 year equal to $AW_{GH} = \$-12.32$ and solve for RV. The estimates for AOC and MV next year are in Figure 11–6; they are, in $ million,

Year 1:   AOC = $-5.20   MV = $22.0

$$-12.32 = -RV(A/P,15\%,1) + 22.00(A/F,15\%,1) - 5.20$$
$$1.15RV = 12.32 + 22.00 - 5.20$$
$$RV = \$25.32$$

Though the RV is larger than the defender's estimated MV of $22 million, some flexibility in the trade-in offer or the challenger's first cost may cause the challenger to be economically justifiable.

Comment

To find RV using a spreadsheet, return to Figure 11–6. In the Goal Seek template, the "set" cell is the AW for 1 year (currently -8.50), and the required value is -12.32, the AW_{GH} for its ESL of 12 years. The "changing" cell is the current market value (cell F2), currently $22.00. When "OK" is touched, $25.32 is displayed as the breakeven market value. This is the RV.

CHAPTER SUMMARY

It is important in a replacement study to compare the best challenger with the defender. *Best (economic) challenger is described as the one with the lowest annual worth (AW) of costs for some number of years.* If the expected remaining life of the defender and the estimated life of the challenger are specified, the AW values over these years are determined and the replacement study proceeds. However, if reasonable estimates of the expected market value (MV) and AOC for each year of ownership can be made, these year-by-year (marginal) costs help determine the best challenger.

The economic service life (ESL) analysis is designed to determine the best challenger's years of service and the resulting lowest total AW of costs. The resulting n_C and AW_C values are used in the replacement study. The same analysis can be performed for the ESL of the defender.

Replacement studies in which no study period (planning horizon) is specified utilize the annual worth method of comparing two unequal-life alternatives. The better AW value determines how long the defender is retained before replacement.

When a study period is specified for the replacement study, it is vital that the market value and cost estimates for the defender be as accurate as possible. When the defender's remaining life is shorter than the study period, it is critical that the cost for continuing service be estimated carefully. All the viable options for using the defender and challenger are enumerated, and their AW equivalent cash flows are determined. For each option, the PW or AW value is used to select the best option. This option determines how long the defender is retained before replacement.

PROBLEMS

Foundations of Replacement

11.1 In a replacement study, what is meant by "taking the nonowner's viewpoint"?

11.2 An asset that was purchased 3 years ago for $100,000 is becoming obsolete faster than expected. The company thought the asset would last 5 years and that its book value would decrease by $20,000 each year and, therefore, be worthless at the end of year 5. In considering a more versatile, more reliable high-tech replacement, the company discovered that the presently owned asset has a market value of only $15,000. If the replacement is purchased immediately at a first cost of $75,000 and if it will have a lower annual worth, what is the amount of the sunk cost? Assume the company's MARR is 15% per year.

11.3 As a muscle car aficionado, a friend of yours likes to restore cars of the 60s and 70s and sell them for a profit. He started his latest project (a 1965 Shelby GT350) four months ago and has a total of $126,000 invested so far. Another opportunity has come up (a 1969 Dodge Charger) that he is thinking of buying because he believes he could sell it for a profit of $60,000 after it is completely restored. To do so, however, he would have to sell the unfinished Shelby first. He thought that the completely restored Shelby would be worth $195,000, resulting in a tidy profit of $22,000, but in its half-restored condition, the most he could get now is $115,000. In discussing the situation with you, he stated that if he could sell the Shelby now and buy the Charger at a reduced price, he would make up for the money he will lose in selling the Shelby at a lower-than-desired price.
(a) What is wrong with this thinking?
(b) What is his sunk cost in the Shelby?

11.4 In conducting a replacement study wherein the planning horizon is unspecified, list three assumptions that are inherent in an annual worth analysis of the defender and challenger.

11.5 A civil engineer who owns his own design/build/operate company purchased a small crane 3 years ago at a cost of $60,000. At that time, it was expected to be used for 10 years and then traded in for its salvage value of $10,000. Due to increased construction activities, the company would prefer to trade for a new, larger crane now that will cost $80,000. The company estimates that the old crane can be used, if necessary, for another 3 years, at which time it would have a $23,000 estimated market value. Its current market value is estimated to be $39,000, and if it is used for another 3 years, it will have M&O costs (exclusive of operator costs) of $17,000 per year. Determine the values of P, n, S, and AOC that should be used for the existing crane in a replacement analysis.

11.6 Equipment that was purchased by Newport Corporation for making pneumatic vibration isolators cost $90,000 two years ago. It has a market value that can be described by the relation $90,000 - 8000k$, where k is the years from time of purchase. Experience with this type of equipment has shown that the operating cost for the first 4 years is $65,000 per year, after which it increases by $6300 per year. The asset's salvage value was originally estimated to be $7000 after a predicted 10-year useful life. Determine the values of P, S, and AOC if a replacement study is done (a) now and (b) 1 year from now.

11.7 A piece of equipment that was purchased 2 years ago by Toshiba Imaging for $50,000 was expected to have a useful life of 5 years with a $5000 salvage value. Its performance was less than expected, and it was upgraded for $20,000 one year ago. Increased demand now requires that the equipment be upgraded again for another $17,000 so that it can be used for 3 more years. If upgraded, its annual operating cost will be $27,000 and it will have a $12,000 salvage after 3 years. Alternatively, it can be replaced with new equipment priced at $65,000 with operating costs of $14,000 per year and a salvage value of $23,000 after 6 years. If replaced now, the existing equipment will be sold for $7000. Determine the values of P, S, AOC, and n for the defender in a replacement study.

Economic Service Life

11.8 For equipment that has a first cost of $10,000 and the estimated operating costs and year-end salvage values shown below, determine the economic service life at $i = 10\%$ per year.

Year	Operating Cost, $ per Year	Salvage Value, $
1	−1000	7000
2	−1200	5000
3	−1300	4500
4	−2000	3000
5	−3000	2000

11.9 To improve package tracking at a UPS transfer facility, conveyor equipment was upgraded with RFID sensors at a cost of $345,000. The operating cost is expected to be $148,000 per year for the first 3 years and $210,000 for the next 3 years. The salvage value of the equipment is expected to be $140,000 for the first 3 years, but due to obsolescence, it won't have a significant value after that. At an interest rate of 10% per year, determine
(a) The economic service life of the equipment and associated annual worth
(b) The percentage increase in the AW of cost if the equipment is retained 2 years longer than the ESL

11.10 Economic service life calculations for an asset are shown below. If an interest rate of 10% per year was used in making the calculations, determine the values of P and S that were used in calculating the AW for year 3.

Years Retained	AW of First Cost, $	AW of Operating Cost, $ per Year	AW of Salvage Value, $
1	−51,700	−15,000	35,000
2	−27,091	−17,000	13,810
3	−18,899	−19,000	6,648
4	−14,827	−21,000	4,309
5	−12,398	−23,000	2,457

11.11 The initial cost of a bridge that is expected to be in place forever is $70 million. Maintenance can be done at 1-, 2-, 3-, or 4-year intervals, but the longer the interval between servicing, the higher the cost. The costs of servicing are estimated at $83,000, $91,000, $125,000, and $183,000 for intervals of 1 through 4 years, respectively. What interval should be scheduled for maintenance to minimize the overall equivalent annual cost? The interest rate is 8% per year.

11.12 A construction company bought a 180,000 metric ton earth sifter at a cost of $65,000. The company expects to keep the equipment a maximum of 7 years. The operating cost is expected to follow the series described by $40,000 + 10,000k$, where k is the number of years since it was purchased ($k = 1$, $2, \ldots, 7$). The salvage value is estimated to be $30,000 for years 1 and 2 and $20,000 for years 3 through 7. At an interest rate of 10% per year, determine the economic service life and the associated equivalent annual cost of the sifter.

11.13 An engineer determined the ESL of a new $80,000 piece of equipment and recorded the calculations shown below. [Note that the numbers are annual worth values associated with various years of retention; that is, if the equipment is kept for, say, 3 years, the AW (years 1 through 3) of the first cost is $32,169, the AW of the operating cost is $51,000, and the AW of the salvage value is $6042.] The engineer forgot to enter the AW of the salvage value for 2 years of retention. From the information available, determine the following:
(a) The interest rate used in the ESL calculations.
(b) The salvage value after 2 years, if the total AW of the equipment in year 2 was $78,762. Use the interest rate determined in part (a).

Years Retained	AW of First Cost, $	AW of Operating Cost, $	AW of Salvage Value, $
1	−88,000	−45,000	50,000
2	−46,095	−46,000	?
3	−32,169	−51,000	6,042

11.14 A large, standby electricity generator in a hospital operating room has a first cost of $70,000 and may be used for a maximum of 6 years. Its salvage value, which decreases by 15% per year, is described by the equation $S = 70,000(1 - 0.15)^n$, where n is the number of years after purchase. The operating cost of the generator will be constant at $75,000 per year. At an interest rate of 12% per year, what are the economic service life and the associated AW value?

11.15 A piece of equipment has a first cost of $150,000, a maximum useful life of 7 years, and a market (salvage) value described by the relation $S = 120,000 - 20,000k$, where k is the number of years since it was purchased. The salvage value cannot go below zero. The AOC series is estimated using $AOC = 60,000 + 10,000k$. The interest rate is 15% per year. Determine the economic service life (*a*) by hand solution, using regular AW computations, and (*b*) by spreadsheet, using annual marginal cost estimates.

11.16 Determine the economic service life and corresponding AW value for a machine that has the following cash flow estimates. Use an interest rate of 14% per year and hand solution.

Year	Salvage Value, $	Operating Cost, $ per Year
0	100,000	—
1	75,000	−28,000
2	60,000	−31,000
3	50,000	−34,000
4	40,000	−34,000
5	25,000	−34,000
6	15,000	−45,000
7	0	−49,000

11.17 Use the annual marginal costs to find the economic service life for Problem 11.16 on a spreadsheet. Assume the salvage values are the best estimates of future market value. Develop an Excel chart of annual marginal costs (MC) and AW of MC over 7 years.

11.18 Keep or Replace the Kiln Case

In Example 11.3, the market value (salvage value) series of the proposed $38 million replacement kiln (GH) dropped to $25 million in only 1 year and then retained 75% of the previous year's market value through the remainder of its 12-year expected life. Based on the experience with the current kiln, and the higher-temperature capability of the replacement, the new kiln is actually expected to retain only 50% of the previous year's value starting in year 5. Additionally, the heating element replacement in year 6 will probably cost $4 million, not $2 million. And finally, the maintenance costs will be considerably higher as the kiln ages. Starting in year 5, the AOC is expected to increase by 25% per year, not 10% as predicted earlier. The Manager of Critical Equipment is now very concerned that the ESL will be significantly decreased from the 12 years calculated earlier (Example 11.3).
(*a*) Determine the new ESL and associated AW value.
(*b*) In percentage changes, estimate how much these new cost estimates may affect the minimum-cost life and AW of cost estimate.

Replacement Study

11.19 In a one-year-later analysis, what action should be taken if (*a*) all estimates are still current and the year is n_D, (*b*) all estimates are still current and the year is not n_D, and (*c*) the estimates have changed?

11.20 Based on company records of similar equipment, a consulting aerospace engineer at Aerospatiale estimated AW values for a presently owned, highly accurate steel rivet inserter.

If Retained This Number of Years	AW Value, $ per Year
1	−62,000
2	−51,000
3	−49,000
4	−53,000
5	−70,000

A challenger has ESL = 2 years and $AW_C = \$-48,000$ per year. (*a*) If the consultant must recommend a replace/retain decision today, should the company keep the defender or purchase the challenger? Why? The MARR is 15% per year. (*b*) When should the next replacement evaluation take place?

11.21 In planning a plant expansion, MedImmune has an economic decision to make—upgrade the existing controlled-environment rooms or purchase new ones. The presently owned ones were purchased 4 years ago for $250,000. They have a current "quick sale" value of $20,000, but for an investment of $100,000 now, they would be adequate for another 4 years, after which they would be sold for $40,000. Alternatively, new controlled-environment rooms could be purchased at a cost of $270,000. They are expected to have a 10-year life with a $50,000 salvage value at that time. Determine whether the company should upgrade or replace. Use a MARR of 20% per year.

11.22 Three years ago, Witt Gas Controls purchased equipment for $80,000 that was expected to have a useful life of 5 years with a $9000 salvage value. Increased demand necessitated an upgrade costing $30,000 one year ago. Technology changes now require that the equipment be upgraded again for another $25,000 so that it can be used for 3 more years. Its annual operating cost will be $47,000, and it will have a $22,000 salvage after 3 years. Alternatively, it can be replaced with new equipment that will cost $68,000 with operating costs of $35,000 per year and a salvage value of $21,000 after 3 years. If replaced now, the existing equipment will be sold for $9000. Calculate the annual worth of the defender at an interest rate of 10% per year.

11.23 A recent environmental engineering graduate is trying to decide whether he should keep his presently owned car or purchase a more environmentally friendly hybrid. A new car will cost $26,000 and have annual operation and maintenance costs of $1200 per year with an $8000 salvage value in 5 years (which is its estimated economic service life).

The presently owned car has a resale value *now* of $5000; one year from now it will be $3000, two years from now $2500, and 3 years from now $2200. Its operating cost is expected to be $1900 this year, with costs increasing by $200 per year. The presently owned car will definitely not be kept longer than 3 more years. Assuming used cars like the one presently owned *will always* be available, should the presently owned car be sold now, 1 year from now, 2 years from now, or 3 years from now? Use annual worth calculations at $i = 10\%$ per year and show your work.

11.24 A pulp and paper company is evaluating whether it should retain the current bleaching process that uses chlorine dioxide or replace it with a proprietary "oxypure" process. The relevant information for each process is shown. Use an interest rate of 15% per year to perform the replacement study.

	Current Process	Oxypure Process
Original cost 6 years ago, $	−450,000	—
Investment cost now, $	—	−700,000
Current market value, $	25,000	—
Annual operating cost, $/year	−180,000	−70,000
Salvage value, $	0	50,000
Remaining life, years	5	10

11.25 A machine that is critical to the Phelps-Dodge copper refining operation was purchased 7 years ago for $160,000. Last year a replacement study was performed with the decision to retain it for 3 more years. The situation has changed. The equipment is estimated to have a value of $8000 if "scavenged" for parts now or anytime in the future. If kept in service, it can be minimally upgraded at a cost of $43,000, which will make it usable for up to 2 more years. Its operating cost is expected to be $22,000 the first year and $25,000 the second year. Alternatively, the company can purchase a new system that will have an equivalent annual worth of $−47,063 per year over its ESL. The company uses a MARR of 10% per year. Calculate the relevant annual worth values, and determine when the company should replace the machine.

11.26 A crushing machine that is a basic component of a metal recycling operation is wearing out faster than expected. The machine was purchased 2 years ago for $400,000. At that time, the buyer thought the machine would serve its needs for at least 5 years, at which time the machine would be sold to a smaller independent recycler for $80,000. Now, however, the company thinks the market value of the diminished machine is only $50,000. If it is kept, the machine's operating cost will be $37,000 per year for the next 2 years, after which it will be scrapped for $1000. If it is kept for only 1 year, the market value is estimated to be $10,000. Alternatively, the company can outsource the process now for a fixed cost of $56,000 per year. At an interest rate of 10% per year, should the company continue with the current machine or outsource the process?

11.27 The data associated with operating and maintaining an asset are shown below. The company manager has already decided to keep the machine for 1 more year (i.e., until the end of year 1), but you have been asked to determine the cost of keeping it 1 more year after that. At an interest rate of 10% per year, estimate the AW of keeping the machine from year 1 to year 2.

Year	Market Value, $	Operating Cost, $ per Year
0	30,000	−15,000
1	25,000	−15,000
2	14,000	−15,000
3	10,000	−15,000

11.28 A machine that cost $120,000 three years ago can be sold now for $54,000. Its market value for the next 2 years is expected to be $40,000 and $20,000 one year and 2 years from now, respectively. Its operating cost was $18,000 for the first 3 years of its life, but the M&O cost is expected to be $23,000 for the next 2 years. A new improved machine that can be purchased for $138,000 will have an economic life of 5 years, an operating cost of $9000 per year, and a salvage value of $32,000 after 5 years. At an interest rate of 10% per year, determine if the presently owned machine should be replaced now, 1 year from now, or 2 years from now.

11.29 The projected market value and M&O costs associated with a presently owned machine are shown (next page). An outside vendor of services has offered to provide the service of the existing machine at a fixed price per year. If the presently owned machine is replaced now, the cost of the fixed-price contract will be $33,000 for each of the next 3 years. If the presently owned machine is replaced next year or the year after that, the contract price will be $35,000 per year. Determine if and when the defender should be replaced with the outside vendor using an interest rate of 10% per year. Assume used equipment similar to the defender will always be available.

Year	Market Value, $	M&O Cost, $ per Year
0	30,000	—
1	25,000	−24,000
2	14,000	−25,000
3	10,000	−26,000
4	8,000	—

11.30 BioHealth, a biodevice systems leasing company, is considering a new equipment purchase to replace a currently owned asset that was purchased 2 years ago for $250,000. It is appraised at a current market value of only $50,000. An upgrade is possible for $200,000 now that would be adequate for another 3 years of lease rights, after which the entire system could be sold on the international circuit for an estimated $40,000. The challenger, which can be purchased for $300,000, has an expected life of 10 years and a $50,000 salvage value. Determine whether the company should upgrade or replace at a MARR of 12% per year. Assume the AOC estimates are the same for both alternatives.

11.31 For the estimates in Problem 11.30, use a spreadsheet-based analysis to determine the first cost for the augmentation of the current system that will make the defender and challenger break even. Is this a maximum or minimum for the upgrade, if the current system is to be retained?

11.32 Herald Richter and Associates, 5 years ago, purchased for $45,000 a microwave signal graphical plotter for corrosion detection in concrete structures. It is expected to have the market values and annual operating costs shown below for its remaining useful life of up to 3 years. It could be traded now at an appraised market value of $8000.

Year	Market Value at End of Year, $	AOC, $ per Year
1	6000	−50,000
2	4000	−53,000
3	1000	−60,000

A replacement plotter with new Internet-based, digital technology costing $125,000 has an estimated $10,000 salvage value after its 5-year life and an AOC of $31,000 per year. At an interest rate of 15% per year, determine how many more years

Richter should retain the present plotter. Solve (a) by hand and (b) by using a spreadsheet.

11.33 In the opportunity cost approach to replacement analysis, what does the *opportunity* refer to?

11.34 State what is meant by the *cash flow approach* to replacement analysis, and list two reasons why it is not a good idea to use this method.

Replacement Study over a Specified Study Period

11.35 ABB Communications is considering replacing equipment that had a first cost of $300,000 five years ago. The company CEO wants to know if the equipment should be replaced now or at any other time over the next 3 years to minimize the cost of producing miniature background suppression sensors. Since the present equipment or the proposed equipment can be used for any or all of the 3-year period, one of the company's industrial engineers produced AW cost information for the defender and challenger as shown below. The values represent the annual costs of the respective equipment if used for the indicated number of years. Determine when the defender should be replaced to minimize the cost to ABB for the 3-year study period using an interest rate of 10% per year.

Number of Years Kept	AW If Kept Stated Number of Years, $ per Year	
	Defender	Challenger
1	−22,000	−29,000
2	−24,000	−26,000
3	−27,000	−25,000

11.36 The table below shows present worth calculations of the costs associated with using a presently owned machine (defender) and a possible replacement (challenger) for different numbers of years. Determine when the defender should be replaced using an interest rate of 10% per year and a 5-year study period. Show solutions (a) by hand and (b) by spreadsheet.

Number of Years Kept/Used	PW If Kept/Used Stated Number of Years, $	
	Defender	Challenger
1	−36,000	−89,000
2	−75,000	−96,000
3	−125,000	−102,000
4	−166,000	−113,000
5	−217,000	−149,000

11.37 Nano Technologies intends to use the newest and finest equipment in its labs. Accordingly, a senior engineer has recommended that a 2-year-old piece of precision measurement equipment be replaced immediately. This engineer believes it can be shown that the proposed equipment is economically advantageous at a 10% per year return and a planning horizon of 3 years.

(a) Perform the replacement analysis using the annual worth method for a specified 3-year study period.

(b) Determine the challenger's capital recovery amount for the 3-year study period and the expected full life. Comment on the effect made by the 3-year study period.

	Current	Proposed
Original purchase price, $	−30,000	−40,000
Current market value, $	17,000	—
Remaining life, years	5	15
Estimated value in 3 years, $	9,000	20,000
Salvage value after 15 years, $	—	5,000
Annual operating cost, $ per year	− 8,000	−3,000

11.38 An industrial engineer at a fiber-optic manufacturing company is considering two robots to reduce costs in a production line. Robot X will have a first cost of $82,000, an annual maintenance and operation (M&O) cost of $30,000, and salvage values of $50,000, $42,000, and $35,000 after 1, 2, and 3 years, respectively. Robot Y will have a first cost of $97,000, an annual M&O cost of $27,000, and salvage values of $60,000, $51,000, and $42,000 after 1, 2, and 3 years, respectively. Which robot should be selected if a 2-year study period is specified at an interest rate of 15% per year and replacement after 1 year is not an option?

11.39 A 3-year-old machine purchased for $140,000 is not able to meet today's market demands. The machine can be upgraded now for $70,000 or sold to a subcontracting company for $40,000. The current machine will have an annual operating cost of $85,000 per year and a $30,000 salvage value in 3 years. If upgraded, the presently owned machine will be retained for only 3 more years, then replaced with a machine to be used in the manufacture of several other product lines. The replacement machine, which will serve the company now and for at least 8 years, will cost $220,000. Its salvage value will be $50,000 for years 1 through 4; $20,000 after 5 years; and $10,000 thereafter. It will have an estimated operating cost of $65,000 per year. You want to perform an economic analysis at 15% per year using a 3-year planning horizon.

(a) Should the company replace the presently owned machine now, or do it 3 years from now?

(b) Compare the capital recovery requirements for the replacement machine (challenger) over the study period and an expected life of 8 years.

11.40 Two processes can be used for producing a polymer that reduces friction loss in engines. Process K, which is currently in place, has a market value of $165,000 now, an operating cost of $69,000 per year, and a salvage value of $50,000 after 1 more year and $40,000 after its maximum 2-year remaining life. Process L, the challenger, will have a first cost of $230,000, an operating cost of $65,000 per year, and salvage values of $100,000 after 1 year, $70,000 after 2 years, $45,000 after 3 years, and $26,000 after its maximum expected 4-year life. The company's MARR is 12% per year. You have been asked to determine which process to select when (a) a 2-year study period is used and (b) a 3-year study period is used.

11.41 **Keep or Replace the Kiln Case**

In Example 11.8, the in-place kiln and replacement kiln (GH) were evaluated using a fixed study period of 6 years. This is a significantly shortened period compared to the expected 12-year life of the challenger. Use the best estimates available throughout this case to determine the impact on the capital recovery amount for the GH kiln of shortening the evaluation time from 12 to 6 years.

11.42 Nabisco Bakers currently employs staff to operate the equipment used to sterilize much of the mixing, baking, and packaging facilities in a large cookie and cracker manufacturing plant in Iowa. The plant manager, who is dedicated to cutting costs but not sacrificing quality and hygiene, has the projected data shown in the table below if the current system were retained for up to its maximum expected life of 5 years. A contract company has proposed a turnkey sanitation system for $5.0 million per year if Nabisco signs on for 4 to 10 years, and $5.5 million per year for a shorter number of years.

Years Retained	AW, $ per Year	Close-Down Expense, $
0	—	−3,000,000
1	−2,300,000	−2,500,000
2	−2,300,000	−2,000,000
3	−3,000,000	−1,000,000
4	−3,000,000	−1,000,000
5	−3,500,000	−500,000

(a) At a MARR = 8% per year, perform a replacement study for the plant manager with a fixed study period of 5 years, when it is anticipated that the plant will be shut down due to the age of the facility and projected technological obsolescence. As you perform the study, take into account that regardless of the number of years that the current sanitation system is retained, a one-time close-down cost will be incurred for personnel and equipment during the last year of operation. (*Hint:* Calculate AW values for all combinations of defender/challenger options.)

(*b*) What is the percentage change in the AW amount each year for the 5-year period? If the decision to retain the current sanitation system for all 5 years is made, what is the economic disadvantage in AW compared to that of the most economic retention period?

Replacement Value

11.43 In 2008, Amphenol Industrial purchased a new quality inspection system for $550,000. The estimated salvage value was $50,000 after 8 years. Currently the expected remaining life is 3 years with an AOC of $27,000 per year and an estimated salvage value of $30,000. The new president has recommended early replacement of the system with one that costs $400,000 and has a 5-year economic service life, a $45,000 salvage value, and an estimated AOC of $50,000 per year. If the MARR for the corporation is 12% per year, find the minimum trade-in value now necessary to make the president's replacement economically advantageous.

11.44 A CNC milling machine purchased by Proto Tool and Die 10 years ago for $75,000 can be used for 3 more years. Estimates are an annual operating cost of $63,000 and a salvage value of $25,000. A challenger will cost $130,000 with an economic life of 6 years and an operating cost of $32,000 per year. Its salvage value will be $45,000. On the basis of

these estimates, what market value for the existing asset will render the challenger equally attractive? Use an interest rate of 12% per year.

11.45 Hydrochloric acid, which fumes at room temperatures, creates a very corrosive work environment, causing steel tools to rust and equipment to fail prematurely. A distillation system purchased 4 years ago for $80,000 can be used for only 2 more years, at which time it will be scrapped with no value. Its operating cost is $75,000 per year. A more corrosion-resistant challenger will cost $220,000 with an operating cost of $49,000 per year. It is expected to have a $30,000 salvage value after its 6-year ESL. At an interest rate of 10% per year, what minimum replacement value now for the present system will render the challenger attractive?

11.46 Engine oil purifier machines can effectively remove acid, pitch, particles, water, and gas from used oil. Purifier A was purchased 5 years ago for $90,000. Its operating cost is higher than expected, so if it is not replaced now, it will likely be used for only 3 more years. Its operating cost this year will be $140,000, increasing by $2000 per year through the end of its useful life, at which time it will be donated for its recyclable scrap value. A more efficient challenger, purifier B, will cost $150,000 with a $50,000 salvage value after its 8-year ESL. Its operating cost is expected to be $82,000 for year 1, increasing by $500 per year thereafter. What is the market value for A that will make the two purifiers equally attractive at an interest rate of 12% per year?

ADDITIONAL PROBLEMS AND FE EXAM REVIEW QUESTIONS

11.47 In conducting a replacement study, all of the following are correct viewpoints for the analyst except:
(*a*) Consultant's
(*b*) Owner's
(*c*) Outsider's
(*d*) Nonowner's

11.48 A sunk cost is the difference between:
(*a*) The first cost and the salvage value
(*b*) The market value and the salvage value
(*c*) The first cost and the market value
(*d*) The book value and the market value

11.49 A truck was purchased 3 years ago for $45,000 and can be sold today for $24,000. The operating costs are $9000 per year, and it is expected to last 4 more years with a $5000 salvage value. A new truck,

which will perform that same service, can be purchased for $50,000, and it will have a life of 10 years with operating costs of $28,000 per year and a $10,000 salvage value. The value that should be used as *P* for the presently owned vehicle in a replacement study is:
(*a*) $45,000
(*b*) $5000
(*c*) $50,000
(*d*) $24,000

11.50 The economic service life of an asset with the PW and AW values on the next page is:
(*a*) 1 year
(*b*) 2 years
(*c*) 3 years
(*d*) 4 years

Years n	Present Worth If Kept n Years, $	Annual Worth If Kept n Years, $/Year
1	−15,455	−17,000
2	−28,646	−16,286
3	−45,019	−18,102
4	−57,655	−18,188
5	−72,867	−19,222

11.51 In looking for ways to cut costs and increase profit (to make the company's stock go up), one of the company's industrial engineers (IEs) determined that the equivalent annual worth of an existing machine over its remaining useful life of 3 years will be $−70,000 per year. The IE also determined that a replacement with more advanced features will have an AW of $−80,000 per year if it is kept for 2 years or less, $−75,000 if it is kept between 3 and 4 years, and $−65,000 if it is kept for 5 to 10 years. If the engineer uses a 3-year study period and an interest rate of 15% per year, she should recommend that the existing machine be:
(a) Replaced now
(b) Replaced 1 year from now
(c) Replaced 2 years from now
(d) Not replaced

11.52 The equivalent annual worth of an existing machine at American Semiconductor is estimated to be $−70,000 per year over its remaining useful life of 3 years. It can be replaced now or later with a machine that will have an AW of $−90,000 per year if it is kept for 2 years or less, −$65,000 if it is kept between 3 and 5 years, and $−110,000 if it is kept for 6 to 8 years. The company wants an analysis of what it should do for a 3-year study period at an interest rate of 15% per year. The replacement must be made now or 3 years from now, according to the department supervisor. You should recommend that the existing machine be replaced:
(a) Now
(b) 1 year from now
(c) 2 years from now
(d) Can't tell; must find AW of different retention combinations for the two machines

11.53 The cost characteristics of a CO testing machine at Dytran Instruments are shown below. The cost of a new tester is $100,000. The equation for determining the AW of keeping the tester for 2 years is:
(a) $\text{AW} = -100,000(A/P,i,8) - [42,000(P/F,i,1) + 47,000(P/F,i,2)](A/P,i,8) + 40,000(A/F,i,8)$
(b) $\text{AW} = -100,000(A/P,i,2) - [42,000(P/F,i,1) + 47,000(P/F,i,2)](A/P,i,2) + 40,000(A/F,i,2)$
(c) $\text{AW} = -100,000(A/P,i,2) - 47,000 + 40,000(A/F,i,2)$
(d) $\text{AW} = -100,000(A/P,i,2) - 42,000 + 40,000(A/F,i,2)$

Machine Age, Years	M&O Costs, $ per Year	Salvage Value at End of Year, $
1	−42,000	60,000
2	−47,000	40,000
3	−49,000	31,000
4	−50,000	24,000
5	−52,000	15,000
6	−54,000	10,000
7	−63,000	10,000
8	−67,000	10,000

11.54 An engineer with Haliburton calculated the AW values shown for a presently owned machine by using estimates he obtained from the vendor and company records.

Retention Period, Years	AW Value, $ per Year
1	−92,000
2	−81,000
3	−87,000
4	−89,000
5	−95,000

A challenger has an economic service life of 7 years with an AW of $−86,000 per year. Assume that used machines like the one presently owned will always be available and that the MARR is 12% per year. If all future costs remain as estimated for the analysis, the company should purchase the challenger:
(a) Now
(b) After 2 years
(c) After 3 years
(d) Never

11.55 The annual worth values for a defender, which can be replaced with a similar used asset, and a challenger are estimated. The defender should be replaced:
(a) Now
(b) 1 year from now
(c) 2 years from now
(d) 3 years from now

Number of Years Retained	AW Value, $ per Year	
	Defender	Challenger
1	−14,000	−21,000
2	−13,700	−18,000
3	−16,900	−13,100
4	−17,000	−15,600
5	−18,000	−17,500

CASE STUDY

WILL THE CORRECT ESL PLEASE STAND?

Background

New pumper system equipment is under consideration by a Gulf Coast chemical processing plant. One crucial pump moves highly corrosive liquids from specially lined tanks on intercoastal barges into storage and preliminary refining facilities dockside. Because of the variable quality of the raw chemical and the high pressures imposed on the pump chassis and impellers, a close log is maintained on the number of hours per year that the pump operates. Safety records and pump component deterioration are considered critical control points for this system. As currently planned, rebuild and M&O cost estimates are increased accordingly when cumulative operating time reaches the 6000-hour mark.

Information

You are the safety engineer at the plant. Estimates made for this pump are as follows:

First cost: $-800,000

Rebuild cost: $-150,000 whenever 6000 cumulative hours are logged. Each rework will cost 20% more than the previous one. A maximum of 3 rebuilds is allowed.

M&O costs: $-25,000 for each year 1 through 4 $-40,000 per year starting the year after the first rebuild, plus 15% per year thereafter

MARR: 10% per year

Based on previous logbook data, the current estimates for number of operating hours per year are as follows:

Year	Hours per Year
1	500
2	1500
3 on	2000

Case Study Questions

1. Determine the economic service life of the pump. How does the ESL compare with the maximum allowed rebuilds?

2. The *plant superintendent* told you, the safety engineer, that only one rebuild should be planned for, because these types of pumps usually have their minimum-cost life before the second rebuild. Determine a market value for this pump that will force the ESL to be 6 years. Comment on the practicality of ESL = 6 years, given the MV calculated.

3. In a separate conversation, the *line manager* told you to not plan for a rebuild after 6000 hours, because the pump will be replaced after a total of 10,000 hours of operation. The line manager wants to know what the base AOC in year 1 can be to make the ESL 6 years. He also told you to assume now that the 15% growth rate applies from year 1 forward. How does this base AOC value compare with the rebuild cost after 6000 hours?

4. What do you think of these suggestions from the plant superintendent and the line manager?

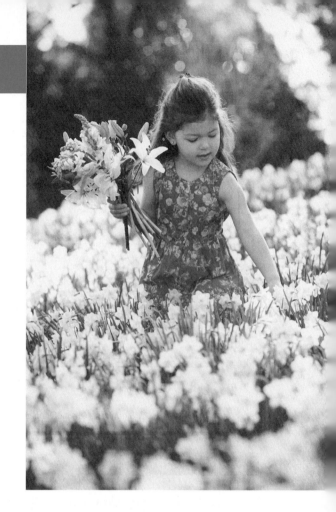

CHAPTER 12

Independent Projects with Budget Limitation

LEARNING OUTCOMES

Purpose: Select independent projects for funding when there is a limitation on the amount of capital available for investment.

SECTION	TOPIC	LEARNING OUTCOME
12.1	Capital rationing	• Explain how a capital budgeting problem is approached.
12.2	Equal-life projects	• Use PW-based capital budgeting to select from several equal-life independent projects.
12.3	Unequal-life projects	• Use PW-based capital budgeting to select from several unequal-life independent projects.
12.4	Linear programming	• Set up the linear programming model and use the Solver spreadsheet tool to select projects.
12.5	Ranking options	• Use the internal rate of return (IROR) and profitability index (PI) to rank and select from independent projects.

n most of the previous economic comparisons, the alternatives have been mutually exclusive; only one could be selected. If the projects are not mutually exclusive, they are categorized as independent of one another, as discussed at the beginning of Chapter 5. Now we learn techniques to select from several independent projects. It is possible to select any number of projects from none (do nothing) to all viable projects.

There is virtually always some upper limit on the amount of capital available for investment in new projects. This limit is considered as each independent project is economically evaluated. The techniques applied are called *capital budgeting methods,* also referred to as capital rationing. They determine the economically best rationing of initial investment capital among independent projects based upon different measures, such as PW, ROR, and the profitability index. These three are discussed here.

12.1 An Overview of Capital Rationing among Projects ●●●

Online Presentation

Investment capital is a scarce resource for all corporations; thus there is virtually always a limited amount to be distributed among competing investment opportunities. When a corporation has several options for placing investment capital, a "reject or accept" decision must be made for each project. Effectively, each option is independent of other options, so the evaluation is performed on a project-by-project basis. Selection of one project does not impact the selection decision for any other project. This is the fundamental difference between mutually exclusive alternatives and independent projects.

The term **project** is used to identify each independent option. We use the term **bundle** to identify a collection of independent projects. The term *mutually exclusive alternative* continues to identify a project when only one may be selected from several.

There are two exceptions to purely independent projects: A *contingent project* is one that has a condition placed upon its acceptance or rejection. Two examples of contingent projects A and B are as follows: A cannot be accepted unless B is accepted; and A can be accepted in lieu of B, but both are not needed. A *dependent project* is one that must be accepted or rejected based on the decision about another project(s). For example, B must be accepted if both A and C are accepted. In practice, these complicating conditions can be bypassed by forming packages of related projects that are economically evaluated themselves as independent projects along with the remaining, unconditioned projects.

A **capital budgeting study** has the following characteristics:

- Several independent projects are identified, and net cash flow estimates are available.
- Each project is either selected entirely or not selected; that is, partial investment in a project is not possible.
- A stated budgetary constraint restricts the total amount available for investment. Budget constraints may be present for the first year only or for several years. This investment limit is identified by the symbol *b*.
- The objective is to maximize the return on the investments using a measure of worth, such as the PW value.

Independent project selection

Limited budget

By nature, independent projects are usually quite different from one another. For example, in the public sector, a city government may develop several projects to choose from: drainage, city park, metro rail, and an upgraded public bus system. In the private sector, sample projects may be a new warehousing facility, expanded product base, improved quality program, upgraded information system, automation, and acquisition of another firm. The size of the study can range from only four or five projects to a complex study involving 50 to 100 projects. The typical capital budgeting problem is illustrated in Figure 12–1. For each independent project there is an initial investment, project life, and estimated net cash flows that can include a salvage value.

Present worth analysis using the capital budgeting process is the recommended method to select projects. The general selection guideline is as follows:

Accept projects with the **best PW values** determined at the MARR over the project life, provided the investment capital limit is not exceeded.

Figure 12–1
Basic characteristics of a
capital budgeting study.

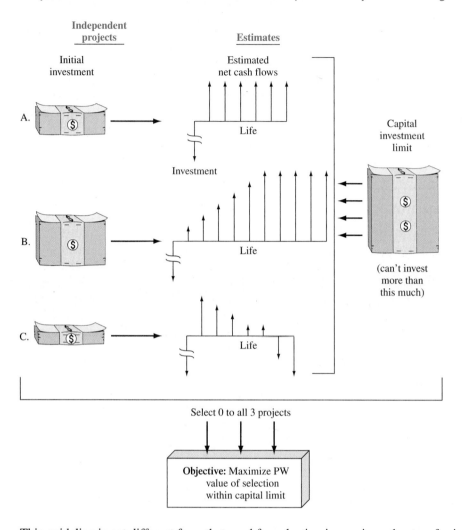

Select 0 to all 3 projects

Objective: Maximize PW
value of selection
within capital limit

This guideline is not different from that used for selection in previous chapters for independent projects. As before, each project is compared with the do-nothing project; that is, incremental analysis between projects is not necessary. The primary difference now is that the amount of money available to invest is limited, thus the title *capital budgeting* or *rationing*. A specific solution procedure that incorporates this constraint is needed.

Previously, PW analysis had the requirement of equal service between alternatives. This assumption is not necessary for capital rationing, because there is no life cycle of a project beyond its estimated life. Rather, the selection guideline has the following implied assumption.

Equal-service requirement

> When the present worth at the MARR over the respective project life is used to select projects, the reinvestment assumption is that all positive net cash flows are reinvested at the MARR from the time they are realized until the **end of the longest-lived project**.

This fundamental assumption is demonstrated to be correct at the end of Section 12.3, which treats PW-based capital rationing for unequal-life projects.

Another dilemma of capital rationing among independent projects concerns the flexibility of the capital investment limit *b*. The limit may marginally disallow an acceptable project that is next in line for acceptance. For example, assume project A has a positive PW value at the MARR. If A will cause the capital limit of $5,000,000 to be exceeded by only $1000, should A be included in the PW analysis? Commonly, a capital investment limit is somewhat flexible, so project A would usually be included. However, in the examples here, we will not exceed a stated investment limit.

Opportunity cost

As we learned earlier (Sections 1.9 and 10.1), the rate of return on the first unfunded project is an **opportunity cost**. The lack of capital to fund the next project defines the ROR level that is forgone. The opportunity cost will vary with each set of independent projects evaluated, but over time it provides information to fine-tune the MARR and other measures used by the company in future evaluations.

It is common to rank independent projects and select from them based on measures other than PW at the MARR. Two are the *internal rate of return (IROR),* discussed in Chapter 7, and the *profitability index (PI),* also called the *present worth index (PWI),* introduced in Chapter 9. Neither of these measures guarantees an optimal PW-based selection. The capital budgeting process, covered in the next 3 sections, does find the optimal solution for PW values. We recommend use of this PW-based technique; however, it should be recognized that both PI and IROR usually provide excellent, near-optimal selections, and they both work very well when the number of projects is large. Application of these two measures is presented in Section 12.5.

12.2 Capital Rationing Using PW Analysis of Equal-Life Projects ● ● ○

Online Presentation

To select from projects that have the same expected life while investing no more than the limit b, first formulate all **mutually exclusive bundles**—one project at a time, two at a time, etc. Each feasible bundle must have a total investment that does not exceed b. One of these bundles is the do-nothing (DN) project. The total number of bundles for m projects is 2^m. The number increases rapidly with m. For $m = 4$, there are $2^4 = 16$ bundles, and for $m = 16$, $2^{16} = 65,536$ bundles. Then the PW of each bundle is determined at the MARR. The bundle with the largest PW value is selected.

To illustrate the development of mutually exclusive bundles, consider these four projects with equal lives.

Project	Initial Investment, $
A	$-10,000$
B	$-5,000$
C	$-8,000$
D	$-15,000$

If the investment limit is $b = \$25,000$, of the 16 bundles, there are 12 feasible ones to evaluate. The bundles ABD, ACD, BCD, and ABCD have investment totals that exceed $25,000. The viable bundles are shown below.

Projects	Total Initial Investment, $	Projects	Total Initial Investment, $
A	$-10,000$	AD	$-25,000$
B	$-5,000$	BC	$-13,000$
C	$-8,000$	BD	$-20,000$
D	$-15,000$	CD	$-23,000$
AB	$-15,000$	ABC	$-23,000$
AC	$-18,000$	Do nothing	0

The procedure to conduct a capital budgeting study using PW analysis is as follows:

1. Develop all mutually exclusive bundles with a total initial investment that does not exceed the capital limit b.
2. Sum the net cash flows NCF_{jt} for all projects in each bundle j ($j = 1, 2, \ldots, 2^m$) and each year t ($t = 1, 2, \ldots, n_j$). Refer to the initial investment of bundle j at time $t = 0$ as NCF_{j0}.
3. Compute the present worth value PW_j for each bundle at the MARR.

$$PW_j = \text{PW of bundle net cash flows} - \text{initial investment}$$

$$= \sum_{t=1}^{t=n_j} NCF_{jt}(P/F,i,t) - NCF_{j0} \qquad [12.1]$$

4. Select the bundle with the (numerically) largest PW_j value.

Selecting the maximum PW_j means that this bundle has a PW value larger than any other bundle. Any bundle with $PW_j < 0$ is discarded, because it does not produce a return of at least the MARR.

EXAMPLE 12.1

The projects review committee of Microsoft has $20 million to allocate next year to new software product development. Any or all of five projects in Table 12–1 may be accepted. All amounts are in $1000 units. Each project has an expected life of 9 years. Select the project(s) if a 15% return is expected.

TABLE 12–1	Five Equal-Life Independent Projects ($1000 Units)		
Project	Initial Investment, $	Annual Net Cash Flow, $	Project Life, Years
A	−10,000	2870	9
B	−15,000	2930	9
C	−8,000	2680	9
D	−6,000	2540	9
E	−21,000	9500	9

Solution

Use the procedure above with $b = \$20,000$ to select one bundle that maximizes present worth. Remember the units are in $1000.

1. There are $2^5 = 32$ possible bundles. The eight bundles that require no more than $20,000 in initial investments are described in columns 2 and 3 of Table 12–2. The $21,000 investment for E eliminates it from all bundles.

TABLE 12–2	Summary of Present Worth Analysis of Equal-Life Independent Projects ($1000 Units)			
Bundle j (1)	Projects Included (2)	Initial Investment NCF_{j0}, $ (3)	Annual Net Cash Flow NCF_j, $ (4)	Present Worth PW_j, $ (5)
1	A	−10,000	2,870	+3,694
2	B	−15,000	2,930	−1,019
3	C	−8,000	2,680	+4,788
4	D	−6,000	2,540	+6,120
5	AC	−18,000	5,550	+8,482
6	AD	−16,000	5,410	+9,814
7	CD	−14,000	5,220	+10,908
8	Do nothing	0	0	0

2. The bundle net cash flows, column 4, are the sum of individual project net cash flows.
3. Use Equation [12.1] to compute the present worth for each bundle. Since the annual NCF and life estimates are the same for a bundle, PW_j reduces to

$$\text{PW}_j = \text{NCF}_j(P/A,15\%,9) - \text{NCF}_{j0}$$

4. Column 5 of Table 12–2 summarizes the PW_j values at $i = 15\%$. Bundle 2 does not return 15%, since $\text{PW}_2 < 0$. The largest is $\text{PW}_7 = \$10,908$; therefore, invest $14 million in C and D. This leaves $6 million uncommitted.

This analysis assumes that the $6 million not used in this initial investment will return the MARR by placing it in some other, unspecified investment opportunity.

12.3 Capital Rationing Using PW Analysis of Unequal-Life Projects ● ● ●

Online Presentation

Usually independent projects do not have the same expected life. As stated in Section 12.1, the PW method for solution of the capital budgeting problem assumes that each project will last for the period of the longest-lived project n_L. Additionally, reinvestment of any positive net cash flows is assumed to be at the MARR from the time they are realized until the end of the longest-lived project, that is, from year n_j through year n_L. Therefore, use of the **LCM of lives is not necessary**, and it is correct to use Equation [12.1] to select bundles of unequal-life projects by PW analysis using the procedure of the previous section.

EXAMPLE 12.2

For MARR = 15% per year and b = $20,000, select from the following independent projects. Solve by hand and by spreadsheet.

Project	Initial Investment, $	Annual Net Cash Flow, $	Project Life, Years
A	−8,000	3870	6
B	−15,000	2930	9
C	−8,000	2680	5
D	−8,000	2540	4

Solution by Hand

The unequal-life values make the net cash flows vary over a bundle's life, but the selection procedure is the same as above. Of the $2^4 = 16$ bundles, 8 are economically feasible. Their PW values by Equation [12.1] are summarized in Table 12–3. As an illustration, for bundle 7:

$$PW_7 = -16,000 + 5220(P/A,15\%,4) + 2680(P/F,15\%,5) = \$235$$

Select bundle 5 (projects A and C) for a $16,000 investment.

TABLE 12–3 Present Worth Analysis for Unequal-Life Independent Projects, Example 12.2

Bundle J (1)	Project (2)	Initial Investment, NCF_{j0}, $ (3)	Net Cash Flows Year t (4)	Net Cash Flows NCF_{jt}, $ (5)	Present Worth PW_j, $ (6)
1	A	−8,000	1–6	3,870	+6,646
2	B	−15,000	1–9	2,930	−1,019
3	C	−8,000	1–5	2,680	+984
4	D	−8,000	1–4	2,540	−748
5	AC	−16,000	1–5	6,550	+7,630
			6	3,870	
6	AD	−16,000	1–4	6,410	+5,898
			5–6	3,870	
7	CD	−16,000	1–4	5,220	+235
			5	2,680	
8	Do nothing	0		0	0

Solution by Spreadsheet

Figure 12–2 presents a spreadsheet with the same information as in Table 12–3. It is necessary to initially develop the mutually exclusive bundles manually and total net cash flows each year using each project's NCF. The NPV function is used to determine PW for each bundle j over its respective life. Bundle 5 (projects A and C) has the largest PW value (row 16).

	A	B	C	D	E	F	G	H	I
1	MARR =	15%							= D7+E7
2									
3	Bundle	1	2	3	4	5	6	7	8
4	Projects	A	B	C	D	AC	AD	CD	Do nothing
5	Year				Net cash flows, NCF, $				
6	0	-8,000	-15,000	-8,000	-8,000	-16,000	-16,000	-16,000	0
7	1	3,870	2,930	2,680	2,540	6,550	6,410	5,220	0
8	2	3,870	2,930	2,680	2,540	6,550	6,410	5,220	0
9	3	3,870	2,930	2,680	2,540	6,550	6,410	5,220	0
10	4	3,870	2,930	2,680	2,540	6,550	6,410	5,220	0
11	5	3,870	2,930	2,680		6,550	3,870	2,680	0
12	6	3,870	2,930			3,870	3,870		0
13	7		2,930						0
14	8		2,930						0
15	9		2,930						0
16	PW @ 15%	6,646	-1,019	984	-748	7,630	5,898	235	0
17									
18					= NPV($B$1,F7:F15) + F6				
19									

Figure 12–2
Computation of PW values for independent project selection, Example 12.2.

The rest of this section will help you understand why solution of the capital budgeting problem by PW evaluation using Equation [12.1] is correct. The following logic verifies the assumption of reinvestment at the MARR for all net positive cash flows when project lives are unequal. Refer to Figure 12–3, which uses the general layout of a two-project bundle. Assume each project has the same net cash flow each year. The P/A factor is used for PW computation. Define n_L as the life of the longer-lived project. At the end of the shorter-lived project, the bundle has a total future worth of $NCF_j(F/A,MARR,n_j)$ as determined for each project. Now, assume reinvestment at the MARR from year n_{j+1} through year n_L (a total of $n_L - n_j$ years). The assumption of the return at the MARR is important; this PW approach does not necessarily select the correct projects if the return is not at the MARR. The results are the two future worth arrows in year n_L in Figure 12–3. Finally,

Figure 12–3
Representative cash flows used to compute PW for a bundle of two independent unequal-life projects by Equation [12.1].

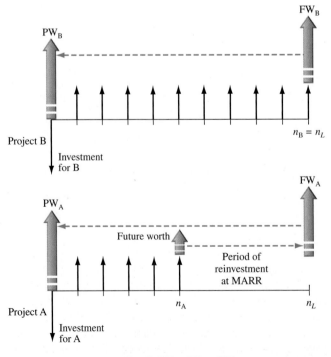

Bundle PW = PW$_A$ + PW$_B$

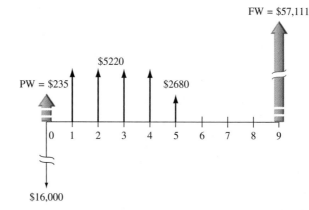

FW = \$57,111

Figure 12–4
Initial investment and cash flows for bundle 7, projects C and D, Example 17.2.

compute the bundle PW value in the initial year. This is the bundle PW = $PW_A + PW_B$. In general form, the bundle j present worth is

$$PW_j = NCF_j(F/A,MARR,n_j)(F/P,MARR,n_L-n_j)(P/F,MARR,n_L) \qquad [12.2]$$

Substitute the symbol i for the MARR, and use the factor formulas to simplify.

$$PW_j = NCF_j \frac{(1+i)^{n_j}-1}{i}(1+i)^{n_L-n_j}\frac{1}{(1+i)^{n_L}}$$

$$= NCF_j \left[\frac{(1+i)^{n_j}-1}{i(1+i)^{n_j}}\right] \qquad [12.3]$$

$$= NCF_j(P/A,i,n_j)$$

Since the bracketed expression in Equation [12.3] is the $(P/A,i,n_j)$ factor, computation of PW_j for n_j years assumes reinvestment at the MARR of all positive net cash flows until the longest-lived project is completed in year n_L.

To demonstrate numerically, consider bundle $j = 7$ in Example 12.2. The evaluation is in Table 12–3, and the net cash flow is pictured in Figure 12–4. Calculate the future worth in year 9, which is the life of the longest-lived project (B).

$$FW = 5220(F/A,15\%,4)(F/P,15\%,5) + 2680(F/P,15\%,4) = \$57,111$$

The present worth at the initial investment time is

$$PW = -16,000 + 57,111(P/F,15\%,9) = \$235$$

The PW value is the same as PW_7 in Table 12–3 and Figure 12–2. This demonstrates the reinvestment assumption for positive net cash flows. If this assumption is not realistic, the PW analysis must be conducted using the *LCM of all project lives.*

12.4 Capital Budgeting Problem Formulation Using Linear Programming ● ● ●

The procedure discussed above requires the development of mutually exclusive bundles one project at a time, two projects at a time, etc., until all 2^m bundles are developed and each one is compared with the capital limit b. As the number of independent projects increases, this process becomes prohibitively cumbersome and unworkable. Fortunately, the capital budgeting problem can be stated in the form of a linear programming model. The problem is formulated using the integer linear programming (ILP) model, which means simply that all relations are linear and that the variable x can take on only integer values. In this case, the variables can only take on the values 0 or 1, which makes it a special case called the 0-or-1 ILP model. The formulation in words follows.

Maximize: Sum of PW of net cash flows of independent projects.

Constraints:
- Capital investment constraint is that the sum of initial investments must not exceed a specified limit.
- Each project is completely selected or not selected.

For the math formulation, define b as the capital investment limit, and let x_k ($k = 1$ to m projects) be the variables to be determined. If $x_k = 1$, project k is completely selected; if $x_k = 0$, project k is not selected. Note that the subscript k *represents each independent project,* not a mutually exclusive bundle.

If the sum of PW of the net cash flows is Z, the math programming formulation is as follows:

$$\text{Maximize:} \quad \sum_{k=1}^{k=m} PW_k x_k = Z$$

$$\text{Constraints:} \quad \sum_{k=1}^{k=m} NCF_{k0} x_k \leq b \qquad \text{[12.4]}$$

$$x_k = 0 \text{ or } 1 \qquad \text{for } k = 1, 2, \ldots, m$$

The PW_k of each project is calculated using Equation [12.1] at MARR $= i$.

$$PW_k = \text{PW of project net cash flows for } n_k \text{ years}$$

$$= \sum_{t=1}^{t=n_k} NCF_{kt}(P/F,i,t) - NCF_{k0} \qquad \text{[12.5]}$$

Computer solution is accomplished by a linear programming software package which treats the ILP model. Also, Excel and its optimizing tool **Solver** can be used to develop the formulation and select the projects. The Solver tool is similar to Goal Seek with significantly more capabilities. For example, Solver allows the target cell to be maximized, minimized, or set to a specific value. This means that the function Z in Equation [12.4] can be maximized. Also, multiple changing cells can be identified, so the 0 or 1 value of the unknowns can be determined. Additionally, with the added capability to include constraints, the investment limit b and 0-or-1 requirement on the unknowns in Equation [12.4] can be accommodated. Solver is explained in detail in Appendix A, and Example 12.3 illustrates its use.

EXAMPLE 12.3

Review Example 12.2. (*a*) Formulate the capital budgeting problem using the math programming model presented in Equation [12.4]. (*b*) Select the projects using Solver.

Solution

(*a*) Define the subscript $k = 1$ through 4 for the four projects, which are relabeled as 1, 2, 3, and 4. The capital investment limit is $b = \$20,000$ in Equation [12.4].

$$\text{Maximize:} \quad \sum_{k=1}^{k=4} PW_k x_k = Z$$

$$\text{Constraints:} \quad \sum_{k=1}^{k=4} NCF_{k0} x_k \leq 20,000$$

$$x_k = 0 \text{ or } 1 \qquad \text{for } k = 1, 2, 3, 4$$

Now, substitute the PW_k and NCF_{k0} values from Table 12–3 into the model. Plus signs are used for all values in the budget constraint. We have the complete 0-or-1 ILP formulation.

$$\text{Maximize:} \quad 6646x_1 - 1019x_2 + 984x_3 - 748x_4 = Z$$

$$\text{Constraints:} \quad 8000x_1 + 15,000x_2 + 8000x_3 + 8000x_4 \leq 20,000$$

$$x_1, x_2, x_3, \text{ and } x_4 = 0 \text{ or } 1$$

The maximum PW is \$7630, and the solution from Example 12.2 is written as

$$x_1 = 1 \qquad x_2 = 0 \qquad x_3 = 1 \qquad x_4 = 0$$

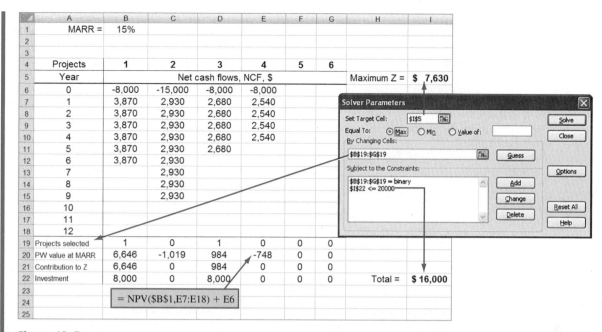

Figure 12–5
Spreadsheet and Solver template configured to solve a capital budgeting problem, Example 12.3.

(b) Figure 12–5 presents a spreadsheet template developed to select from six or fewer independent projects with 12 years or less of net cash flow estimates per project. The spreadsheet template can be expanded in either direction if needed. Figure 12–5 (inset) shows the Solver parameters set to solve this example for four projects and an investment limit of $20,000. The descriptions below and the cell tag identify the contents of the rows and cells in Figure 12–5, and their linkage to Solver parameters.

Rows 4 and 5: Projects are identified by numbers to distinguish them from spreadsheet column letters. Cell I5 is the expression for Z, the sum of the PW values for the projects. This is the target cell for Solver to maximize.

Rows 6 to 18: These are initial investments and net cash flow estimates for each project. Zero values that occur after the life of a project need not be entered; however, any $0 estimates that occur during a project's life must be entered.

Row 19: The entry in each cell is 1 for a selected project and 0 if not selected. These are the changing cells for Solver. Since each entry must be 0 or 1, a binary constraint is placed on all row 19 cells in Solver, as shown in Figure 12–5. When a problem is to be solved, it is best to initialize the spreadsheet with 0s for all projects. Solver will find the solution to maximize Z.

Row 20: The NPV function is used to find the PW for each net cash flow series. The NPV functions are developed for any project with a life up to 12 years at the MARR entered in cell B1.

Row 21: When a project is selected, the contribution to the Z function is shown by multiplying rows 19 and 20.

Row 22: This row shows the initial investment for the selected projects. Cell I22 is the total investment. This cell has the budget limitation placed on it by the constraint in Solver. In this example, the constraint is I22 <= $20,000.

To solve the example, set all values in row 19 to 0, set up the Solver parameters as described above, and click on Solve. (Since this is a linear model, the Solver options choice "Assume Linear Model" may be checked, if desired.) If needed, further directions on saving the solution, making changes, etc., are available in Appendix A, Section A.5, and on the Excel help function.

For this problem, the selection is projects 1 and 3 with Z = $7630, the same as determined previously, and $16,000 of the $20,000 limit is invested.

12.5 Additional Project Ranking Measures ● ● ●

The PW-based method of solving a capital budgeting problem covered in Sections 2.2 to 2.4 provides an optimal solution that maximizes the PW of the competing projects. However, it is very common in industrial, professional, and government settings to learn that the rate of return is the basis for ranking projects. The **internal rate of return (IROR),** as we learned in Section 7.2, is determined by setting a PW or AW relation equal to zero and solving for i^* — the IROR. Using a PW basis and the estimated net cash flow (NCF) series for each project j, solve for i^* in the relation

$$0 = \sum_{t=1}^{t=n_j} NCF_{jt}(P/F,i^*,t) - NCF_{j0} \qquad [12.6]$$

This is the same as Equation [12.1] set equal to zero with $i = i^*$ as the unknown. The spreadsheet function RATE or IRR will provide the same answer. The selection guideline is as follows:

Independent project
selection

> Once the project ranking by IROR is complete, select all projects in order without exceeding the investment limit b.
>
> If there is no budget limit, select all projects that have IROR ≥ MARR.

The ordering of projects using *IROR ranking may differ from the PW-based ranking* we used in previous sections. This can occur because IROR ranking maximizes the overall rate of return, not necessarily the PW value. The use of IROR ranking is illustrated in the next example.

Another common ranking method is the **profitability index (PI)** that we learned in Section 9.2. This is a "bang for the buck" measure that provides a sense of getting the most for the investment dollar over the life of the project. (Refer to Section 9.2 for more details.) When utilized for project ranking, it is often called the **present worth index (PWI);** however, we will use the PI term for consistency, and because there are other ways to mathematically define measures also referred to as a PW index. The PI measure is defined as

$$\frac{\text{PW of net cash flows}}{\text{PW of initial investment}} = \frac{\sum_{t=1}^{t=n_j} NCF_{jt}(P/F,i,t)}{|NCF_{j0}|} \qquad [12.7]$$

Note that the denominator includes only the initial investment, and it is its *absolute value* that is used. The numerator has only cash flows that result from the project for years 1 through its life n_j. Salvage value, if there is one estimated, is incorporated into the numerator. Similar to the previous case, the selection guideline is as follows:

Independent project
selection

> Once the project ranking by PI is complete, select all projects in order without exceeding the investment limit b.
>
> If there is no budget limit, select all projects that have PI ≥ 1.0.

Depending upon the project NCF estimates, the *PI ranking can differ from the IROR ranking.* Example 12.4 compares results using the different ranking methods—IROR, PI, and PW values. None of these results are incorrect; they simply maximize different measures, as you will see. The use of IROR, PI, or other measures is common when there are a large number of projects, because the PW basis (when solved by hand, Solver, or ILP software) becomes increasingly cumbersome as the number of possible mutually exclusive bundles grows using the formula 2^m. Additionally, greater complexity is introduced when dependent and contingent projects are involved.

EXAMPLE 12.4

Georgia works as a financial analyst in the Management Science Group of General Electronics. She has been asked to recommend which of the five projects detailed in Table 12–4 should be funded if the MARR is 15% per year and the investment budget limit for next year is a firm

TABLE 12–4	IROR, PI, and PW Values for Five Projects, Example 12.4				
Projects	**1**	**2**	**3**	**4**	**5**
Investment, NCF_0, $1000	−8,000	−15,000	−8,000	−8,000	−5,000
NCF, $1000 per year	4,000	2,900	2,700	2,500	2,600
Life n_j, years	6	9	5	4	3
IROR, %	44.5	12.8	20.4	9.6	26.0
PI	1.89	0.92	1.13	0.89	1.19
PW at 15%, $1000	7,138	−1,162	1,051	−863	936

$18 million. She has confirmed the computations and is ready to do the ranking and make the selection. Help her by doing the following:

(a) Use the IROR measure to rank and select projects.
(b) Use the PI measure to rank and select projects.
(c) Use the PW measure to rank and select projects.
(d) Compare the selected projects by the three methods and determine which one will maximize the overall ROR value of the $18 million budget.

Solution

Refer to Table 12–5 for the ranking, cumulative investment for each project, and selection based on the ranking and budget limit b = $18 million.

(a) Ranking by overall IROR values indicates that projects 1 and 5 should be selected with $13 million of the $18 million budget expended. The remaining $5 million is assumed to be invested at the MARR of 15% per year.
(b) As an example, the PI for project 1 is calculated using Equation [12.7].

$$PI_1 = \frac{4000(P/A,15\%,6)}{|-8000|}$$
$$= 15,138/8,000$$
$$= 1.89$$

Ranking and projects selected are the same as in the IROR-based analysis. Again, the remaining $5 million is assumed to generate a return of MARR = 15% per year.

(c) Ranking by PW value results in a different selection from that of IROR and PI ranking. Projects 1 and 3, rather than 1 and 5, are selected for a total PW = $8.189 million and an investment of $16 million. The remaining $2 million is assumed to earn 15% per year.
(d) IROR and PI rankings result in projects 1 and 5 being selected. PW ranking results in the selection of projects 1 and 3. With MARR = 15%, the $18 million will earn at the following ROR values. In $1000 units,

Projects 1 and 5   NCF, year 0: $−13,000
    NCF, years 1–3: $6,600
    NCF, years 4–6: $4,000

$$0 = -13,000 + 6600(P/A,i,3) + 4000(P/A,i,3)(P/F,i,3)$$

TABLE 12–5	Ranking of Projects by Different Measures, Example 12.4 (Monetary units in $1000)							
Ranking by IROR			**Ranking by PI**			**Ranking by PW**		
IROR, % (1)	Project (2)	Cumulative Investment, $ (3)	PI (4)	Project (5)	Cumulative Investment, $ (6)	PW, $ (7)	Project (8)	Cumulative Investment, $ (9)
44.5	1	8,000	1.89	1	8,000	7,138	1	8,000
26.0	5	13,000	1.19	5	13,000	1,051	3	16,000
20.4	3	21,000	1.13	3	21,000	936	5	21,000
12.8	2		0.92	2		−863	4	
9.6	4		0.89	4		−1,162	2	

By IRR function, the rate of return is 39.1%. The overall return on the entire budget is

$$ROR = [39.1(13,000) + 15.0(5000)]/18,000$$
$$= 32.4\%$$

Projects 1 and 3   NCF, year 0: $-16,000$
                          NCF, years 1–5: $6,700
                          NCF, year 6: $4,000

$$0 = -16,000 + 6700(P/A,i,5) + 4000(P/F,i,6)$$

By IRR function, the rate of return is 33.5%. The overall return on the entire budget is

$$ROR = [33.5(16,000) + 15.0(2000)]/18,000$$
$$= 31.4\%$$

In conclusion, the IROR and PI selections maximize the overall rate of return at 32.4%. The PW selection maximizes the PW value at $8.189 (i.e., 7.138 + 1.051) million, as determined from Table 12–5, column 7.

CHAPTER SUMMARY

Investment capital is always a scarce resource, so it must be rationed among competing projects using economic and noneconomic criteria. Capital budgeting involves proposed projects, each with an initial investment and net cash flows estimated over the life of the project. The fundamental capital budgeting problem has specific characteristics (Figure 12–1).

- Selection is made from among independent projects.
- Each project must be accepted or rejected as a whole.
- Maximizing the present worth of the net cash flows is the objective.
- The total initial investment is limited to a specified maximum.

The present worth method is used for evaluation. To start the procedure, formulate all mutually exclusive bundles that do not exceed the investment limit, including the do-nothing bundle. There are a maximum of 2^m bundles for m projects. Calculate the PW at MARR for each bundle, and select the bundle with the largest PW value. Reinvestment of net positive cash flows at the MARR is assumed for all projects with lives shorter than that of the longest-lived project.

The capital budgeting problem may be formulated as a linear programming problem to select projects directly in order to maximize the total PW. Excel's Solver tool solves this problem by spreadsheet.

Projects can be ranked and selected on bases other than PW values. Two measures are the internal rate of return (IROR) and the profitability index (PI), also called the PW index. The ordering of projects may differ between the various ranking bases since different measures are optimized. When there are a large number of projects, the IROR basis is commonly applied in industrial settings.

PROBLEMS

Understanding Capital Rationing

12.1 Define the following terms: bundle, contingent project, dependent project.

12.2 State two assumptions made when doing capital rationing using a PW analysis for unequal-life projects.

12.3 For independent projects identified as A, B, C, D, E, F, and G, how many mutually exclusive bundles can be formed?

12.4 For independent projects identified as W, X, Y, and Z, develop all of the mutually exclusive bundles. Projects X and Y perform the same function with different processes; both should not be selected.

12.5 Five projects have been identified for possible implementation by a company that makes dry ice blasters—machines that propel tiny dry ice pellets at supersonic speeds so they flash-freeze and then lift grime, paint, rust, mold, asphalt, and other contaminants off in-place machines and a wide range of surfaces. The total present worth of the initial investment for each project is shown. Determine which bundles are possible, provided the budget limitation is (a) $34,000 and (b) $45,000.

Project	PW of Investment at 15%, $
L	28,000
M	11,000
N	43,000
O	38,000
P	6,000

12.6 Four independent projects (1, 2, 3, and 4) are proposed for investment by Perfect Manufacturing, Inc. List all the acceptable mutually exclusive bundles based on the following selection restrictions developed by the department of engineering production:

Project 2 can be selected only if project 3 is selected.

Projects 1 and 4 should not both be selected; they are essentially duplicates.

12.7 Develop all acceptable mutually exclusive bundles for the four independent projects described below if the investment limit is $400 and the following project selection restriction applies: Project 1 can be selected only if both projects 3 and 4 are selected.

Project	Investment, $
1	−100
2	−150
3	−75
4	−235

Selection from Independent Projects

12.8 Listed below are bundles, each comprised of three independent proposals for which the PW has been estimated. Select the best bundle if the capital budget limit is $45,000 and the MARR is the cost of capital, which is 9% per year.

Proposal Bundle	Initial Investment, For Bundle, $	PW at 9%, $
1	−18,000	−1,400
2	−26,000	8,500
3	−34,000	7,100
4	−41,000	10,500

12.9 The general manager for Woodslome Appliance Company Plant #A14 in Mexico City has four independent projects that she can fund this year to

improve surface durability on stainless steel products. The project costs and 18% per year PW values are as shown. What projects should be accepted if the budget limit is (a) no limit and (b) $55,000?

Project	Initial Investment, $	PW at 18%, $
1	−15,000	−400
2	−25,000	8500
3	−20,000	500
4	−40,000	9600

12.10 The capital fund for research project investment at SummaCorp is limited to $100,000 for next year. Select any or all of the following proposals if a MARR of 15% per year is established by the board of directors.

Project	Initial Investment, $	Annual Net Cash Flow, $/Year	Life, Years	Salvage Value, $
I	−25,000	6,000	4	4,000
II	−30,000	9,000	4	−1,000
III	−50,000	15,000	4	20,000

12.11 An electrical engineer at GE is assigned the responsibility to determine how to invest up to $100,000 in none, some, or all of the following independent proposals. Use (a) hand and (b) spreadsheet-based PW analysis and a 15% per year return requirement to help this engineer make the best decision from a purely economic perspective.

Project	Initial Investment, $	Annual Net Cash Flow, $/Year	Life, Years	Salvage Value, $
A	−25,000	6,000	4	4,000
B	−20,000	9,000	4	0
C	−50,000	15,000	4	20,000

12.12 Dwayne has four independent vendor proposals to contract the nationwide oil recycling services for Ford Corporation manufacturing plants. All combinations are acceptable, except that vendors B and C cannot both be chosen. Revenue sharing of recycled oil sales with Ford is a part of the requirement. Develop all possible mutually exclusive bundles under the additional following restrictions and select the best projects. The corporate MARR is 10% per year.
(a) A maximum of $4 million can be spent.
(b) A larger budget of $5.5 million is allowed, but no more than two vendors can be selected.
(c) There is no limit on spending.

Vendor	Initial Investment, $	Life, Years	Annual Net Revenue, $ per Year
A	−1.5 million	8	360,000
B	−3.0 million	10	600,000
C	−1.8 million	5	620,000
D	−2.0 million	4	630,000

12.13 Use the PW method at 8% per year to select up to three projects from the four available ones if no more than $20,000 can be invested. Estimated lives and annual net cash flows vary.

Project	Initial Investment, $	Net Cash Flow, $ per Year					
		1	2	3	4	5	6
W	−5,000	1,000	1,700	1,800	2,500	2,000	
X	−8,000	900	950	1,000	1,050	10,500	
Y	−8,000	4,000	3,000	1,000	500	500	2,000
Z	−10,000	0	0	0	17,000		

12.14 Charlie's Garage has $70,000 to spend on new equipment that may increase revenue for his car repair shop. Use the PW method to determine which of these independent investments are financially acceptable at 6% per year, compounded monthly. All are expected to last 3 years.

Feature	Installed Cost, $	Estimated Added Revenue, $ per Month
Diagnostics system	−45,000	2200
Exhaust analyzer	−30,000	2000
Hybrid engine tester	−22,000	1500

12.15 (a) Determine which of the following independent projects should be selected for investment if $315,000 is available and the MARR is 10% per year. Use the PW method to evaluate mutually exclusive bundles to make the selection. (Solve by hand or spreadsheet as instructed.)

Project	Initial Investment, $	NCF, $ per Year	Life, Years
A	−100,000	50,000	8
B	−125,000	24,000	8
C	−120,000	75,000	8
D	−220,000	39,000	8
E	−200,000	82,000	8

(b) If the five projects are mutually exclusive alternatives, perform the present worth analysis and select the best alternative.

12.16 Use the following analysis of five independent projects to select the best, if the capital limitation is (a) $30,000, (b) $52,000, and (c) unlimited.

Project	Initial Investment, $	Life, Years	PW at 12% per Year, $
S	−15,000	6	8,540
A	−25,000	8	12,325
M	−10,000	6	3,000
E	−25,000	4	10
H	−40,000	12	15,350

12.17 The independent project estimates below have been developed by the engineering and finance managers. The corporate MARR is 8% per year, and the capital investment limit is $4 million. Select the economically best projects using the PW method and (a) hand solution and (b) spreadsheet solution.

Project	Project Cost, $ M	Life, Years	NCF, $ per Year
1	−1.5	8	360,000
2	−3.0	10	600,000
3	−1.8	5	520,000
4	−2.0	4	820,000

12.18 Use the PW method to evaluate four independent projects. Select as many as three of the four projects. The MARR is 12% per year, and up to $16,000 in capital investment funds are available.

	Project			
	1	2	3	4
Investment, $	−5000	−8,000	−9,000	−10,000
Life, years	5	5	3	4

Year	NCF Estimates, $ per Year			
1	1000	500	5000	0
2	1700	500	5000	0
3	2400	500	2000	0
4	3000	500		17,000
5	3800	10,500		

12.19 Work Problem 12.18 using a spreadsheet.

12.20 A capital rationing problem is defined for you as follows: Three projects are to be evaluated at a MARR of 12.5% per year. No more than $3.0 million can be invested.
   (a) Use a spreadsheet to select from the independent projects.
   (b) If the life of project 3 can be increased from 5 to 10 years for the same $1 million investment, use Goal Seek to determine the NCF in year 1 for project 3 alone to have the same PW as the best bundle in part (a). All other estimates remain the same. With these new NCF and life estimates, what are the best projects for investment?

Project	Investment, $ M	Life, Years	Year 1	Gradient after Year 1
			Estimated NCF, $ per Year	
1	−0.9	6	250,000	−5000
2	−2.1	10	485,000	+5000
3	−1.0	5	200,000	+20%

Linear Programming and Capital Budgeting

12.21 Formulate the linear programming model, develop a spreadsheet, and solve the capital rationing problem in Example 12.1 (*a*) as presented and (*b*) using an investment limit of $13 million.

12.22 Use linear programming and a spreadsheet-based solution technique to select from the independent unequal-life projects in Problem 12.17.

12.23 Solve the capital budgeting problem in Problem 12.20(*a*), using the linear programming model and a spreadsheet.

12.24 Johnson and Johnson is expanding its first-aid products line for individuals allergic to latex. The research team comprised of doctors, engineers, and chemists has proposed the four projects estimated in Problem 12.18. Develop the linear programming model and use a spreadsheet to select the projects to be funded. The MARR is 12% per year, and the budget limit is $16 million. All monetary values are in $1000 units.

12.25 Using the estimates in Problem 12.18 and repeated spreadsheet solution of the capital budgeting problem for budget limits ranging from $b = 5000 to $b = $25,000$, develop a spreadsheet chart that plots b versus the value of Z.

Other Ranking Measures

12.26 A newly proposed project has a first cost of $325,000 and estimated annual income of $60,000 per year for 8 years.
  (*a*) Determine the IROR, PI, and PW values if the MARR is 15% per year.
  (*b*) Is the project economically justified?

12.27 An engineer at Delphi Systems is considering the projects below, all of which can be considered to last indefinitely. The company's MARR is 13% per year.
  (*a*) Determine which projects should be selected on the basis of IROR if the budget limitation is $39,000.
  (*b*) What is the overall rate of return if the money not invested in projects is assumed to earn a rate of return equal to the MARR?

Project	First Cost, $	Net Income, $ per Year	Rate of Return, %
A	−20,000	4,000	20.0
B	−10,000	1,900	19.0
C	−15,000	2,600	17.3
D	−70,000	10,000	14.3
E	−50,000	6,000	12.0

12.28 The five independent projects shown below are under consideration for implementation by KNF Neuberger, Inc. The company's MARR is 15% per year.
  (*a*) Determine which projects should be undertaken on the basis of IROR if the budget limitation is $97,000. (Solve by hand or spreadsheet as instructed.)
  (*b*) Determine the overall rate of return if the money not invested in projects is assumed to earn a rate of return equal to the MARR.

Project	First Cost, $	Annual Income, $ per Year	Project Life, Years
L	−30,000	9,000	10
A	−15,000	4,900	10
N	−45,000	11,100	10
D	−70,000	9,000	10
T	−40,000	10,000	10

12.29 An estimated 6 billion gallons of clean drinking water disappear each day across the United States due to aging, leaky pipes and water mains before it gets to the consumer or industrial user. The American Society of Civil Engineers (ASCE) has teamed with municipalities, counties, and several excavation companies to develop robots that can travel through mains, detect leaks, and repair many of them immediately. Four proposals have been received for funding. There is a $100 million limit on capital funding, and the MARR is established at 12% per year.
  (*a*) Use the IROR method to rank and determine which of the four independent projects should be funded. Solve by spreadsheet, unless instructed to use hand solution.
  (*b*) Determine the rate of return for the combination of projects selected in part (*a*). Is it economically justified?
  (*c*) Determine the overall rate of return for the $100 million with the projects selected in part (*a*). Assume that excess funds are invested at the MARR.

Project	First Cost, $ M	Estimated Annual Savings, $M per Year	Project Life, Years
W	−12	5.0	3
X	−25	7.3	4
Y	−45	12.1	6
Z	−60	9.0	8

12.30 Determine the profitability index at 10% per year interest for a project that has a first cost of $200,000 in year 0 and $200,000 in year 2, an annual operating cost of $80,000 per year, revenue of $170,000 per year, and a salvage value of $60,000 after its 5-year life.

12.31 The budget limit is $120,000, and the interest rate is 10% per year. All projects have a 10-year life. Use (a) the PI method and (b) the IROR method to rank and select from the independent projects. (c) Are different projects selected using the two methods?

Project	First Cost, $	Net Income, $ per Year	IROR, %
A	−18,000	4,000	18.0
B	−15,000	2,800	13.3
C	−35,000	12,600	34.1
D	−60,000	13,000	17.3
E	−50,000	8,000	9.6

12.32 The six independent projects shown below are under consideration by Peyton Packing under budget-constrained conditions. The company always has more projects to engage in than it has capital to fund projects. Therefore, it uses a relatively high MARR of 25% per year. Since all projects are considered long-term ventures, the company uses an infinite period for their life. Determine which projects the company should fund and the total investment for a capital budget of $700,000 if the capital budgeting method used is (a) the IROR method, (b) the PI method, and (c) the PW method.

Project	First Cost, $	Estimated Annual Income, $ per Year
F	−200,000	54,000
G	−120,000	21,000
H	−250,000	115,000
I	−370,000	205,000
J	−50,000	26,000
K	−9,000	2,100

ADDITIONAL PROBLEMS AND FE EXAM REVIEW QUESTIONS

12.33 A project that has a condition associated with its acceptance or rejection is known as:
(a) A mutually exclusive alternative
(b) A contingent project
(c) A dependent project
(d) Both (b) and (c)

12.34 An assumption inherent in capital budgeting is that all positive net cash flows are reinvested at the MARR from the time they are realized until:
(a) The end of the shortest-lived project
(b) The end of the longest-lived project
(c) The end of the respective project that generated the cash flow
(d) The average of the lives of the projects included in the bundle

12.35 All of the following are correct when a capital budgeting problem is solved using the 0-1 integer linear programming model except:
(a) Partial investment in a project is acceptable.
(b) The objective is to maximize the present worth of the investments.
(c) Budget constraints may be present for the first year only or for several years.
(d) Contingent and dependent project restrictions may be considered.

12.36 All of the following are true when formulating mutually exclusive bundles of independent projects except:

(a) One of the bundles is the do-nothing project.
(b) A bundle may consist of only one project.
(c) The capital limit may be exceeded as long as it is exceeded by less than 3%.
(d) A bundle may include contingent and dependent projects.

12.37 When there are 5 projects involved in a capital budgeting study, the maximum number of bundles that can be formulated is:
(a) 6
(b) 10
(c) 31
(d) 32

12.38 For a capital limit of $25,000, the selected independent projects are:
(a) P only
(b) Q only
(c) R only
(d) P and R

Project	Initial Investment, $	Life, Years	PW at 12%, $
P	−15,000	6	8,540
Q	−25,000	8	12,325
R	−10,000	6	3,000
S	−25,000	4	10
T	−40,000	12	15,350

12.39 The independent projects shown below are under consideration for possible implementation by Renishaw Inc. of Hoffman Estates, Rhode Island. If the company's MARR is 14% per year and it uses the IROR method of capital budgeting, the projects it should select under a budget limitation of $105,000 are:

(*a*) A, B, and C
(*b*) A, B, and D
(*c*) B, C, and D
(*d*) A, C, and D

Project	First Cost, $	Annual Income, $ per Year	Rate of Return, %
A	−20,000	4,000	20.0
B	−10,000	1,900	19.0
C	−15,000	2,600	17.3
D	−70,000	10,000	14.3
E	−50,000	6,000	12.0

12.40 For a project that requires an initial investment of $26,000 and yields $10,000 per year for 4 years, the PI at an interest rate of 10% per year is closest to:

(*a*) 1.03
(*b*) 1.22
(*c*) 1.38
(*d*) 1.56

CHAPTER 13

Breakeven and Payback Analysis

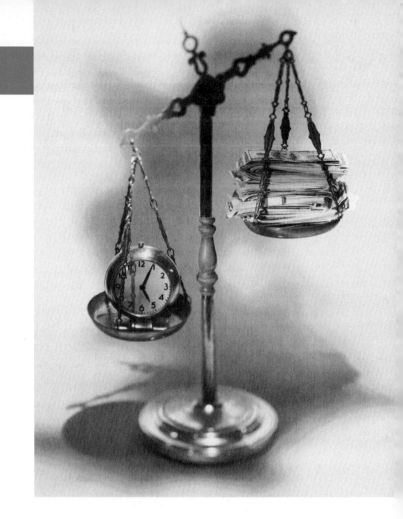

LEARNING OUTCOMES

Purpose: Determine the breakeven for one or two alternatives and calculate the payback period with and without a return required.

SECTION	TOPIC	LEARNING OUTCOME
13.1	Breakeven point	• Determine the breakeven point for one parameter.
13.2	Two alternatives	• Calculate the breakeven point of a parameter and use it to select between two alternatives.
13.3	Payback period	• Determine the payback period of a project at $i = 0\%$ and $i > 0\%$. Illustrate the cautions when using payback analysis.
13.4	Spreadsheet	• Answer breakeven and payback questions that are best resolved by spreadsheet and Goal Seek tools.

B reakeven analysis is performed to determine the value of a variable or parameter of a project or alternative that makes two elements equal, for example, the sales volume that will equate revenues and costs. A breakeven study is performed for two alternatives to determine when either alternative is equally acceptable. Breakeven analysis is commonly applied in **make-or-buy decisions** when a decision is needed about the source for manufactured components, services, etc.

Payback analysis determines the required minimum life of an asset, process, or system to recover the initial investment. There are two types of payback: return ($i > 0\%$) and no return ($i = 0\%$). Payback analysis should not be considered the final decision maker; it is used as a **screening tool** or to provide **supplemental information** for a PW, AW, or other analysis. These aspects are discussed in depth in this chapter.

Breakeven and payback studies use estimates that are considered to be certain; that is, if the estimated values are expected to vary enough to possibly change the outcome, another study is necessary using different estimates. If the variable of interest is allowed to vary, the approaches of sensitivity analysis (Chapter 18) should be used. Additionally, if probability and risk assessment are considered, the tools of simulation (Chapter 19) can be used to supplement the static nature of a breakeven or payback study.

13.1 Breakeven Analysis for a Single Project ● ● ○

Online
Presentation

When one of the engineering economy symbols—P, F, A, i, or n—is not known or not estimated, a breakeven quantity can be determined by setting an equivalence relation for PW or AW equal to zero. This form of breakeven analysis has been used many times so far. For example, we have solved for the rate of return i^*, found the replacement value for a defender, and determined the P, F, A, or salvage value S at which a series of cash flow estimates return a specific MARR. Methods used to determine the quantity include

Direct solution by hand if only one factor is present (say, P/A) or only single amounts are estimated (for example, P and F)

Trial and error by hand or calculator when multiple factors are present

Spreadsheet when cash flow and other estimates are entered into cells and used in resident functions (PV, FV, RATE, IRR, NPV, PMT, and NPER) or tools (Goal Seek and Solver).

We now concentrate on the determination of the **breakeven quantity Q_{BE} for one parameter or decision variable**. For example, the variable may be a design element to minimize cost or the production level needed to realize revenues that exceed costs by 10%.

Breakeven

> Breakeven analysis finds the value of a parameter that **makes two elements equal**. The breakeven point Q_{BE} is determined from mathematical relations, e.g., product revenue and costs or materials supply and demand or other parameters that involve the parameter Q. Breakeven analysis is fundamental to evaluations such as make-buy decisions.

The unit of the parameter Q may vary widely: units per year, cost per kilogram, hours per month, percentage of full plant capacity, etc.

Figure 13–1a presents different shapes of a revenue relation identified as R. A linear revenue relation is commonly assumed, but a nonlinear relation is often more realistic. It can model an increasing per unit revenue with larger volumes (curve 1 in Figure 13–1a) or a decreasing per unit revenue that usually prevails at higher quantities (curve 2).

Costs, which may be linear or nonlinear, usually include two components—fixed and variable—as indicated in Figure 13–1b.

Fixed costs (FC). These include costs such as buildings, insurance, fixed overhead, some minimum level of labor, equipment capital recovery, and information systems.

Variable costs (VC). These include costs such as direct labor, materials, indirect costs, contractors, marketing, advertisement, and warranty.

The **fixed-cost** component is essentially constant for all values of the variable, so it does not vary for a large range of operating parameters, such as production level or workforce size. Even if no units are produced, fixed costs are incurred at some threshold level. Of course, this situation

Figure 13–1
Linear and nonlinear
revenue and cost relations.

(a) Revenue relations—(1) increasing and
(2) decreasing revenue per unit

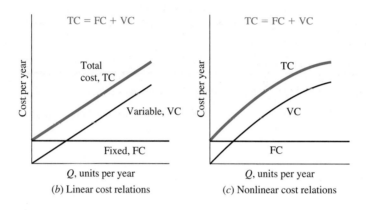

(b) Linear cost relations   (c) Nonlinear cost relations

cannot last long before the plant must shut down to reduce fixed costs. Fixed costs are reduced through improved equipment, information systems, and workforce utilization; and less costly fringe benefit packages, subcontracting specific functions, and so on.

Variable costs change with production level, workforce size, and other parameters. It is usually possible to decrease variable costs through better product design, manufacturing efficiency, improved quality and safety, and higher sales volume.

When FC and VC are added, they form the **total cost relation TC.** Figure 13–1b illustrates the TC relation for linear fixed and variable costs. Figure 13–1c shows a general TC curve for a nonlinear VC in which unit variable costs decrease as the quantity level rises.

At a specific but unknown value Q of the decision variable, the revenue R and total cost TC relations will intersect to identify the breakeven point Q_{BE} (Figure 13–2). If $Q > Q_{BE}$, there is a predictable **profit;** but if $Q < Q_{BE}$, there is a loss. For linear models of R and VC, the greater the quantity, the larger the profit. Profit is calculated as

$$\text{Profit} = \text{revenue} - \text{total cost}$$
$$= R - \text{TC}$$
$$= R - (\text{FC} + \text{VC}) \qquad [13.1]$$

A relation for the breakeven point may be derived when revenue and total cost are *linear functions of quantity Q* by setting the relations for R and TC equal to each other, indicating a profit of zero.

$$R = \text{TC}$$
$$rQ = \text{FC} + vQ$$

where r = revenue per unit
             v = variable cost per unit

Solve for the breakeven quantity $Q = Q_{BE}$ for linear R and TC functions.

$$Q_{BE} = \frac{\text{FC}}{r - v} \qquad [13.2]$$

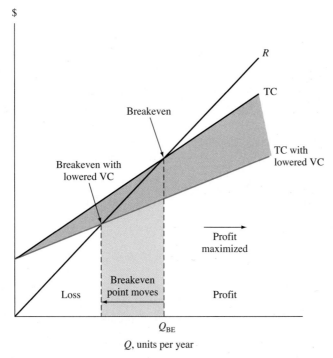

Figure 13–2
Effect on the breakeven point when the variable cost per unit is reduced.

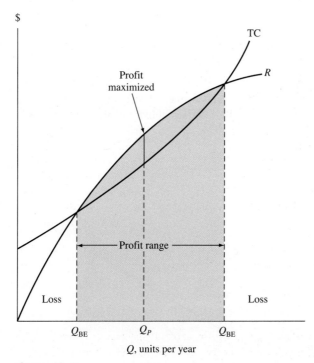

Figure 13–3
Breakeven points and maximum profit point for a nonlinear analysis.

The breakeven graph is an important management tool because it is easy to understand and may be used in decision making in a variety of ways. For example, if the variable cost per unit is reduced, then the TC line has a smaller slope (Figure 13–2) and the breakeven point will decrease. This is an advantage because the smaller the value of Q_{BE}, the greater the profit for a given amount of revenue. A similar analysis is possible for fixed VC and increased levels of production, as shown in the next example.

If nonlinear R or TC models are used, there may be more than one breakeven point. Figure 13–3 presents this situation for two breakeven points. The **maximum profit** occurs at Q_P between the two breakeven points where the distance between the R and TC relations is greatest.

Of course, no static R and TC relations—linear or nonlinear—are able to estimate exactly the revenue and cost amounts over an extended period of time. But the breakeven point is an excellent target for planning purposes.

EXAMPLE 13.1

Indira Industries is a major producer of diverter dampers used in the gas turbine power industry to divert gas exhausts from the turbine to a side stack, thus reducing the noise to acceptable levels for human environments. Normal production level is 60 diverter systems per month, but due to significantly improved economic conditions in Asia, production is at 72 per month. The following information is available.

Fixed costs	FC = \$2.4 million per month
Variable cost per unit	v = \$35,000
Revenue per unit	r = \$75,000

(a) How does the increased production level of 72 units per month compare with the current breakeven point?
(b) What is the current profit level per month for the facility?
(c) What is the difference between the revenue and variable cost per damper that is necessary to break even at a significantly reduced monthly production level of 45 units, if fixed costs remain constant?

Solution

(a) Use Equation [13.2] to determine the breakeven number of units. All dollar amounts are in \$1000 units.

$$Q_{BE} = \frac{FC}{r - v}$$

$$= \frac{2400}{75 - 35} = 60 \text{ units per month}$$

Figure 13–4 is a plot of R and TC lines. The breakeven value is 60 damper units. The increased production level of 72 units is above the breakeven value.

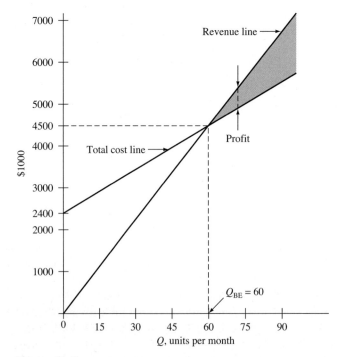

Figure 13–4
Breakeven graph, Example 13.1.

(b) To estimate profit (in $1000 units) at $Q = 72$ units per month, use Equation [13.1].

$$\begin{aligned}
\text{Profit} = R - \text{TC} &= rQ - (\text{FC} + vQ) \\
&= (r - v)Q - \text{FC} \\
&= (75 - 35)72 - 2400 \qquad\qquad [13.3] \\
&= \$480
\end{aligned}$$

There is a profit of $480,000 per month currently.

(c) To determine the required difference $r - v$, use Equation [13.3] with profit $= 0$, $Q = 45$, and FC $= \$2.4$ million. In $1000 units,

$$0 = (r - v)(45) - 2400$$

$$r - v = \frac{2400}{45} = \$53.33 \text{ per unit}$$

The spread between r and v must be $53,330. If v stays at $35,000, the revenue per damper must increase from $75,000 to $88,330 (i.e., $35,000 + 53,330) just to break even at a production level of $Q = 45$ per month.

In some circumstances, breakeven analysis performed on a per unit basis is more meaningful. The value of Q_{BE} is still calculated using Equation [13.2], but the TC relation is divided by Q to obtain an expression for cost per unit, also termed *average cost per unit* C_u.

$$C_u = \frac{\text{TC}}{Q} = \frac{\text{FC} + vQ}{Q} = \frac{\text{FC}}{Q} + v \qquad\qquad [13.4]$$

At the breakeven quantity $Q = Q_{\text{BE}}$, the revenue per unit is exactly equal to the cost per unit. If graphed, the FC per unit term in Equation [13.4] takes on the shape of a hyperbola.

The breakeven point for a project with one unknown variable can always be determined by equating revenue and total cost. This is the same as setting profit equal to zero in Equation [13.1]. It may be necessary to perform some dimensional analysis initially to obtain the correct revenue and total cost relations in order to use the same dimension for both relations, for example, $ per unit, miles per month, or units per year.

13.2 Breakeven Analysis Between Two Alternatives ● ● ●

Online Presentation

Breakeven

Now we consider breakeven analysis between two mutually exclusive alternatives.

> Breakeven analysis determines the value of a common variable or parameter between two alternatives. Equating the two PW or AW relations determines the breakeven point. Selection of the alternative is different depending upon two facts: **slope of the variable cost curve** and the **parameter value relative to the breakeven point.**

The parameter can be the interest rate i, first cost P, annual operating cost (AOC), or any parameter. We have already performed breakeven analysis between alternatives on several parameters. For example, the incremental ROR value (Δi^*) is the breakeven rate between alternatives. If the MARR is lower than Δi^*, the extra investment of the larger-investment alternative is justified. In Section 11.6, the replacement value (RV) of a defender was determined. If the market value is larger than RV, the decision should favor the challenger.

Often breakeven analysis involves revenue or cost variables common to both alternatives, such as price per unit, operating cost, cost of materials, or labor cost. Figure 13–5 illustrates the breakeven concept for two alternatives with linear cost relations. The fixed cost of alternative 2 is greater than that of alternative 1. However, alternative 2 has a smaller variable cost, as indicated by its lower slope. The intersection of the total cost lines locates the breakeven point, and the variable cost establishes the slope. Thus, if the number of units of the common variable is greater than the breakeven amount, alternative 2 is selected, since the total cost will be lower. Conversely, an anticipated level of operation below the breakeven point favors alternative 1.

Figure 13–5
Breakeven between two
alternatives with linear
cost relations.

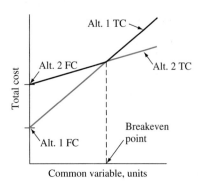

Instead of plotting the total costs of each alternative and estimating the breakeven point graphically, it may be easier to calculate the breakeven point numerically using engineering economy expressions for the PW or AW at the MARR. The AW is preferred when the variable units are expressed on a yearly basis, and AW calculations are simpler for alternatives with unequal lives. The following steps determine the breakeven point of the common variable and the slope of a linear total cost relation.

1. Define the common variable and its dimensional units.
2. Develop the PW or AW relation for each alternative as a function of the common variable.
3. Equate the two relations and solve for the breakeven value of the variable.

Selection between alternatives is based on this guideline:

> If the anticipated level of the common variable is **below** the breakeven value, select the alternative with the higher variable cost (larger slope).
>
> If the level is **above** the breakeven point, select the alternative with the lower variable cost. (Refer to Figure 13–5.)

EXAMPLE 13.2

A small aerospace company is evaluating two alternatives: the purchase of an automatic feed machine and a manual feed machine for a finishing process. The auto feed machine has an initial cost of $23,000, an estimated salvage value of $4000, and a predicted life of 10 years. One person will operate the machine at a rate of $12 per hour. The expected output is 8 tons per hour. Annual maintenance and operating cost is expected to be $3500.

The alternative manual feed machine has a first cost of $8000, no expected salvage value, a 5-year life, and an output of 6 tons per hour. However, three workers will be required at $8 per hour each. The machine will have an annual maintenance and operation cost of $1500. All projects are expected to generate a return of 10% per year. How many tons per year must be finished to justify the higher purchase cost of the auto feed machine?

Solution

Use the steps above to calculate the breakeven point between the two alternatives.

1. Let x represent the number of tons per year.
2. For the auto feed machine, the annual variable cost is

$$\text{Annual VC} = \frac{\$12}{\text{hour}} \frac{1 \text{ hour}}{8 \text{ tons}} \frac{x \text{ tons}}{\text{year}}$$
$$= 1.5x$$

The VC is developed in dollars per year. The AW expression for the auto feed machine is

$$AW_{auto} = -23{,}000(A/P,10\%,10) + 4000(A/F,10\%,10) - 3500 - 1.5x$$
$$= \$-6992 - 1.5x$$

Similarly, the annual variable cost and AW for the manual feed machine are

$$\text{Annual VC} = \frac{\$8}{\text{hour}}(3 \text{ operators})\frac{1 \text{ hour}}{6 \text{ tons}}\frac{x \text{ tons}}{\text{year}}$$

$$= 4x$$

$$\text{AW}_{\text{manual}} = -8000(A/P,10\%,5) - 1500 - 4x$$

$$= \$-3610 - 4x$$

3. Equate the two cost relations and solve for x.

$$\text{AW}_{\text{auto}} = \text{AW}_{\text{manual}}$$

$$-6992 - 1.5x = -3610 - 4x$$

$$x = 1353 \text{ tons per year}$$

If the output is expected to exceed 1353 tons per year, purchase the auto feed machine, since its VC slope of 1.5 is smaller than the manual feed VC slope of 4.

The breakeven analysis approach is commonly used for **make-or-buy decisions.** This means the company contracts to *buy* the product or service from the *outside,* or *makes* it *within* the company. The alternative to buy usually has no fixed cost and a larger variable cost than the option to make. Where the two cost relations cross is the make-buy decision quantity. Amounts above this indicate that the item should be made, not purchased outside.

EXAMPLE 13.3

Guardian is a national manufacturing company of home health care appliances. It is faced with a make-or-buy decision. A newly engineered lift can be installed in a car trunk to raise and lower a wheelchair. The steel arm of the lift can be purchased internationally for $3.50 per unit or made in-house. If manufactured on site, two machines will be required. Machine A is estimated to cost $18,000, have a life of 6 years, and have a $2000 salvage value; machine B will cost $12,000, have a life of 4 years, and have a $−500 salvage value (carry-away cost). Machine A will require an overhaul after 3 years costing $3000. The annual operating cost for machine A is expected to be $6000 per year and for machine B is $5000 per year. A total of four operators will be required for the two machines at a rate of $12.50 per hour per operator. In a normal 8-hour period, the operators and two machines can produce parts sufficient to manufacture 1000 units. Use a MARR of 15% per year to determine the following.

(*a*) Number of units to manufacture each year to justify the in-house (make) option.
(*b*) The maximum capital expense justifiable to purchase machine A, assuming all other estimates for machines A and B are as stated. The company expects to produce 10,000 units per year.

Solution

(*a*) Use steps 1 to 3 stated previously to determine the breakeven point.

1. Define x as the number of lifts produced per year.
2. There are variable costs for the operators and fixed costs for the two machines for the make option.

$$\text{Annual VC} = (\text{cost per unit})(\text{units per year})$$

$$= \frac{4 \text{ operators}}{1000 \text{ units}}\frac{\$12.50}{\text{hour}}(8 \text{ hours})x$$

$$= 0.4x$$

The annual fixed costs for machines A and B are the AW amounts.

$$\text{AW}_{\text{A}} = -18,000(A/P,15\%,6) + 2000(A/F,15\%,6)$$
$$-6000 - 3000(P/F,15\%,3)(A/P,15\%,6)$$

$$\text{AW}_{\text{B}} = -12,000(A/P,15\%,4) - 500(A/F,15\%,4) - 5000$$

Total cost is the sum of AW_{A}, AW_{B}, and VC.

3. Equating the annual costs of the buy option ($3.50x$) and the make option yields

$$-3.50x = AW_A + AW_B - VC$$
$$= -18{,}000(A/P{,}15\%{,}6) + 2000(A/F{,}15\%{,}6) - 6000$$
$$\qquad -3000(P/F{,}15\%{,}3)(A/P{,}15\%{,}6) - 12{,}000(A/P{,}15\%{,}4)$$
$$\qquad -500(A/F{,}15\%{,}4) - 5000 - 0.4x \qquad\qquad [13.5]$$
$$-3.10x = -20{,}352$$
$$x = 6565 \text{ units per year}$$

A minimum of 6565 lifts must be produced each year to justify the make option, which has the lower variable cost of $0.4x$.

(b) Substitute 10,000 for x and P_A for the to-be-determined first cost of machine A (currently $18,000) in Equation [13.5]. Solution yields $P_A = \$58{,}295$. This is approximately three times the estimated first cost of $18,000, because the production of 10,000 per year is considerably larger than the breakeven amount of 6565.

Even though the preceding examples treat only two alternatives, the same type of analysis can be performed for three or more alternatives. To do so, compare the alternatives in pairs to find their respective breakeven points. The results are the ranges through which each alternative is more economical. For example, in Figure 13–6, if the output is less than 40 units per hour, alternative 1 should be selected. Between 40 and 60, alternative 2 is more economical; and above 60, alternative 3 is favored.

If the variable cost relations are nonlinear, analysis is more complicated. If the costs increase or decrease uniformly, mathematical expressions that allow direct determination of the breakeven point can be developed.

13.3 Payback Analysis ● ● ○

Payback analysis is another use of the present worth technique. It is used to determine the amount of time, usually expressed in years, required to recover the first cost of an asset or project. Payback is allied with breakeven analysis; this is illustrated later in the section. The **payback period,** also called *payback* or *payout period,* has the following definition and types.

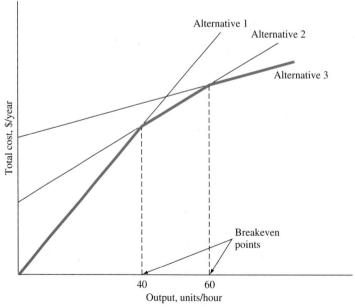

Figure 13–6
Breakeven points for three alternatives.

The **payback period** n_p is an estimated time for the revenues, savings, and any other monetary benefits to completely recover the **initial investment plus a stated rate of return i.**

There are two types of payback analysis as determined by the required return.

No return; $i=0\%$: Also called *simple payback,* this is the recovery of only the initial investment.

Discounted payback; $i > 0\%$: The time value of money is considered in that some return, for example, 10% per year, must be realized in addition to recovering the initial investment.

Payback period

An example application of payback may be a corporate senior manager who insists that every proposal return the initial cost and some stated return within 3 years. Using payback as an initial screening tool, no proposal with $n_p > 3$ years can become a viable alternative. The payback period should be determined using a required $i > 0\%$. Unfortunately in practice, no-return payback is used too often to make economic decisions. After the formulas are presented, a couple of cautions about payback usage are provided.

The equations used to determine n_p differ for each type of analysis. For both types, the terminology is P for the initial investment in the asset, project, contract, etc., and NCF for the estimated annual net cash flow. Using Equation [1.5], annual NCF is

$$NCF = \text{cash inflows} - \text{cash outflows}$$

To calculate the payback period for $i = 0\%$ or $i > 0\%$, determine the pattern of the NCF series. Note that n_p is usually not an integer. For $t = 1, 2, \ldots, n_p$,

No return, $i = 0\%$; NCF_t varies annually: $\qquad 0 = -P + \sum_{t=1}^{t=n_p} \text{NCF}_t \qquad$ [13.6]

No return, $i = 0\%$; annual uniform NCF: $\qquad n_p = \dfrac{P}{\text{NCF}} \qquad$ [13.7]

Discounted, $i > 0\%$; NCF_t varies annually: $\qquad 0 = -P + \sum_{t=1}^{t=n_p} \text{NCF}_t(P/F,i,t) \qquad$ [13.8]

Discounted, $i > 0\%$; annual uniform NCF: $\qquad 0 = -P + \text{NCF}(P/A,i,n_p) \qquad$ [13.9]

After n_p years, the cash flows will recover the investment in year 0 plus the required return of $i\%$. If the alternative is used more than n_p years, with the same or similar cash flows, a larger return results. If the estimated life is less than n_p years, there is not enough time to recover the investment and $i\%$ return. It is important to understand that payback analysis *neglects all cash flows after the payback period of n_p years.* Consequently, it is preferable to use payback as an **initial screening method** or **supplemental tool** rather than as the primary means to select an alternative. The reasons for this caution are that

- No-return payback **neglects the time value of money,** since no return on an investment is required.

- Either type of payback **disregards all cash flows occurring after the payback period.** These cash flows may increase the return on the initial investment.

Payback analysis utilizes a significantly different approach to alternative evaluation than the primary methods of PW, AW, ROR, and B/C. It is possible for payback analysis to select a different alternative than these techniques. However, the information obtained from discounted payback analysis performed at an appropriate $i > 0\%$ can be very useful in that a sense of the **risk** involved in undertaking an alternative is provided. For example, if a company plans to utilize a machine for only 3 years and payback is 6 years, indication is that the equipment should not be obtained. Even here, the 6-year payback is considered supplemental information and does not replace a complete economic analysis.

EXAMPLE 13.4

The board of directors of Halliburton International has just approved an $18 million worldwide engineering construction design contract. The services are expected to generate new annual net cash flows of $3 million. The contract has a potentially lucrative repayment clause to Halliburton of $3 million at any time that the contract is canceled by either party during the 10 years of the contract period. (*a*) If $i = 15\%$, compute the payback period. (*b*) Determine the no-return payback period and compare it with the answer for $i = 15\%$. This is an initial check to determine if the board made a good economic decision. Show both hand and spreadsheet solutions.

Solution by Hand

(*a*) The net cash flow each year is $3 million. The single $3 million payment (call it CV for cancellation value) could be received at any time within the 10-year contract period. Equation [13.9] is altered to include CV.

$$0 = -P + \text{NCF}(P/A,i,n) + \text{CV}(P/F,i,n)$$

In $1,000,000 units,

$$0 = -18 + 3(P/A,15\%,n) + 3(P/F,15\%,n)$$

The 15% payback period is $n_p = 15.3$ years, found by trail and error. During the period of 10 years, the contract will not deliver the required return.

(*b*) If Halliburton requires absolutely no return on its $18 million investment, Equation [13.6] results in $n_p = 5$ years, as follows (in $ million).

$$0 = -18 + 5(3) + 3$$

There is a very significant difference in n_p for 15% and 0%. At 15% this contract would have to be in force for 15.3 years, while the no-return payback period requires only 5 years. A longer time is always required for $i > 0\%$ for the obvious reason that the time value of money is considered.

Solution by Spreadsheet

Enter the function = NPER(15%,3,−18,3) to display 15.3 years. Change the rate from 15% to 0% to display the no-return payback period of 5 years.

If two or more alternatives are evaluated using payback periods to indicate that one may be better than the other(s), the second shortcoming of payback analysis (neglect of cash flows after n_p) may lead to an economically incorrect decision. When cash flows that occur after n_p are neglected, it is possible to favor short-lived assets even when longer-lived assets produce a higher return. In these cases, PW (or AW) analysis should always be the primary selection method. Comparison of short- and long-lived assets in Example 13.5 illustrates this situation.

EXAMPLE 13.5

Two equivalent pieces of quality inspection equipment are being considered for purchase by Square D Electric. Machine 2 is expected to be versatile and technologically advanced enough to provide net income longer than machine 1.

	Machine 1	Machine 2
First cost, $	12,000	8,000
Annual NCF, $	3,000	1,000 (years 1–5), 3,000 (years 6–14)
Maximum life, years	7	14

The quality manager used a return of 15% per year and software that incorporates Equations [13.8] and [13.9] to recommend machine 1 because it has a shorter payback period of 6.57 years at $i = 15\%$. The computations are summarized here.

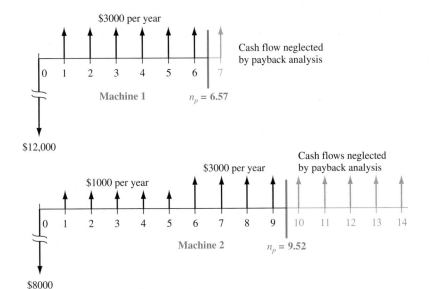

Figure 13–7
Illustration of payback periods and neglected net cash flows, Example 13.5.

Machine 1: $n_p = 6.57$ years, which is less than the 7-year life.

Equation used: $0 = -12{,}000 + 3000(P/A,15\%,n_p)$

Machine 2: $n_p = 9.52$ years, which is less than the 14-year life.

Equation used: $0 = -8000 + 1000(P/A,15\%,5)$
$$+ 3000(P/A,15\%,n_p-5)(P/F,15\%,5)$$

Recommendation: Select machine 1.

Now, use a 15% PW analysis to compare the machines and comment on any difference in the recommendation.

Solution

For each machine, consider the net cash flows for all years during the estimated (maximum) life. Compare them over the LCM of 14 years.

$$PW_1 = -12{,}000 - 12{,}000(P/F,15\%,7) + 3000(P/A,15\%,14) = \$663$$
$$PW_2 = -8000 + 1000(P/A,15\%,5) + 3000(P/A,15\%,9)(P/F,15\%,5)$$
$$= \$2470$$

Machine 2 is selected since its PW value is numerically larger than that of machine 1 at 15%. This result is the opposite of the payback period decision. The PW analysis accounts for the increased cash flows for machine 2 in the later years. As illustrated in Figure 13–7 (for one life cycle for each machine), payback analysis neglects all cash flow amounts that may occur after the payback time has been reached.

Comment

This is a good example of why payback analysis is best used for initial screening and supplemental risk assessment. Often a shorter-lived alternative evaluated by payback analysis may appear to be more attractive, when the longer-lived alternative has cash flows later in its life that make it more economically attractive.

As mentioned in the introduction to this section, breakeven and payback analyses are allied. They can be used in conjunction to determine the payback period when a desired level of break-even is specified. The reverse is also possible; when a desired payback period is established, the breakeven value with or without a return requirement can be determined. By working together in this fashion, better economic decisions can be made. Example 13.6 illustrates the second of the situations mentioned above.

EXAMPLE 13.6

The president of a local company expects a product to have a profitable life of between 1 and 5 years. Help her determine the breakeven number of units that must be sold annually (without any return) to realize payback for each of the time periods 1 year, 2 years, and so on up to 5 years. The cost and revenue estimates are as follows:

Fixed costs: Initial investment of $80,000 with $1000 annual operating cost.

Variable cost: $8 per unit.

Revenue: Twice the variable cost for the first 5 years and 50% of the variable cost thereafter.

Solution by Hand

Define X_{BE} as the breakeven quantity and n_p as the payback period. Since values of X_{BE} are sought for $n_p = 1, 2, 3, 4, 5$, solve for breakeven by substituting each payback period. First develop the FC, r, and v terms.

Fixed cost, FC	$\dfrac{80,000}{n_p} + 1000$
Revenue per unit, r	$16 (years 1 through 5 only)
Variable cost per unit, v	$8

The breakeven relation from Equation [13.2] is

$$X_{BE} = \frac{80,000/n_p + 1000}{8} \qquad\qquad [13.10]$$

Insert n_p values and solve for X_{BE}, the breakeven value.

n_p, payback years	1	2	3	4	5
X_{BE}, units per year	10,125	5125	3458	2625	2125

Solution by Spreadsheet

Figure 13–8 presents a spreadsheet solution for the breakeven values. Equation [13.10] is encoded to display the answers in column C. The breakeven values are the same as those above, e.g., sell 5125 units per year to pay back in 2 years. The breakeven curve rapidly flattens out as shown in the accompanying chart in Figure 13–8.

Figure 13–8
Breakeven number of units for different payback periods, Example 13.6.

13.4 More Breakeven and Payback Analysis on Spreadsheets ●●●

The Goal Seek tool that we have used previously is an excellent tool to perform breakeven and payback analysis. Examples 13.7 and 13.8 demonstrate the use of Goal Seek for both types of problems.

EXAMPLE 13.7

The Naruse brake-accelerator pedal (www.autoblog.com/tag/Masuyuki+Naruse) is designed to minimize the chances that a driver will accidently step on the accelerator pedal of the car when the brake pedal is the intended target. The design is based on the fact that a person naturally steps downward on his or her foot when surprised, shocked, or struck with a medical emergency. In this pedal design, downward motion of the foot will always engage the brake, never the accelerator. Assume that for the manufacture of pedal components, two equally qualified machines have been identified and estimates made.

	Machine 1	Machine 2
First cost, $	−80,000	−110,000
Net cash flow, $ per year	25,000	22,000
Salvage value, $	2,000	3,000
Life, years	4	6

Using an AW analysis at MARR = 10%, the spreadsheet screen shot in Figure 13–9 indicates that machine 1 is the economic choice with a positive AW value of $193. However, the automated controls, safety features, and ergonomic design of machine 2 make it a better choice for the plant in the opinion of the project engineer. Use breakeven analysis to find the threshold values for each of several parameters that will make machine 2 equally qualified economically. The parameters to concentrate on are (*a*) first cost, (*b*) net cash flow, and (*c*) life of machine 2, if all other estimates remain the same.

	A	B	C	D
1	MARR =	10%		
2		Net cash flows, $/year		
3	Year	Machine 1	Machine 2	
4	0	-80,000	-110,000	
5	1	25,000	22,000	
6	2	25,000	22,000	
7	3	25,000	22,000	
8	4	27,000	22,000	
9	5		22,000	
10	6		25,000	
11	AW @ MARR	193	-2,868	
12				
13	= −PMT($B$1,4,NPV($B1,B5:B8)+B4)			
14				

Figure 13–9
AW values for two machines, Example 13.7.

Solution

Figure 13–10 shows the spreadsheet and Goal Seek templates that determine breakeven values for first cost and NCF.

(*a*) Figure 13–10*a*: By forcing the AW for machine 2 to equal $193, Goal Seek finds a breakeven of $96,669. If the first cost can be negotiated down to this cost from $110,000, machine 2 will be economically equivalent to machine 1.

(*b*) Figure 13–10*b*: (Remember to reset the first cost to $−110,000 on the spreadsheet.) By setting all NCFs equal to the value in year 1 (using the function = $C$5), Goal Seek determines a breakeven of $25,061 per year. Therefore, if the NCF estimate can realistically be increased from $22,000 to $25,061, again machine 2 will be economically equivalent.

(*c*) Finding an extended life estimate for machine 2 is a payback question; Goal Seek is not needed. The easiest approach is to use the NPER function to find the payback period. Entering = NPER(10%,22000,−110000,3000) displays n_p = 7.13 years. Therefore, extending the estimated life from 6 to between 7 and 8 years and retaining the salvage value of $3000 will select machine 2 over 1.

Figure 13–10
Breakeven values for (a) first cost and (b) annual net cash flow using Goal Seek, Example 13.7.

EXAMPLE 13.8

Chris and her father just purchased a small office building for $160,000 that is in need of a lot of repairs, but is located in a prime commercial area of the city. The estimated costs each year for repairs, insurance, etc. are $18,000 the first year, increasing by $1000 per year thereafter. At an expected 8% per year return, use spreadsheet analysis to determine the payback period if the building is (a) kept for 2 years and sold for $290,000 sometime beyond year 2 or (b) kept for 3 years and sold for $370,000 sometime beyond 3 years.

Solution

Figure 13–11 shows the annual costs (column B) and the sales prices if the building is kept 2 or 3 years (columns C and E, respectively). The NPV function is applied (columns D and F) to determine when the PW changes sign from plus to minus. These results bracket the payback period for each retention period and sales price. When PW > 0, the 8% return is exceeded.

(a) The 8% return payback period is between 3 and 4 years (column D). If the building is sold after exactly 3 years for $290,000, the payback period was not exceeded; but after 4 years it is exceeded.

(b) At a sales price of $370,000, the 8% return payback period is between 5 and 6 years (column F). If the building is sold after 4 or 5 years, the payback is not exceeded; however, a sale after 6 years is beyond the 8%-return payback period.

	A	B	C	D	E	F	G	H	I
1		Costs,	If sold after 2 years		If sold after 3 years				
2	Year	$ per year	Price, $	PW at 8%, $	Price, $	PW at 8%, $			
3	0	-160,000							
4	1	-18,000							
5	2	-19,000							
6	3	-20,000	290,000	21,379					
7	4	-21,000	290,000	-11,110	370,000	47,693			
8	5	-22,000	290,000	-41,872	370,000	12,575			
9	6	-23,000			370,000	-20,572			
10	7	-24,000			370,000	-51,847			
11									
12		= NPV(8%,$B$4:B7)+$B$3 − PV(8%,A7,,290000)							
13									

If kept 2 years and sold, payback is between 3 and 4

If kept 3 years and sold, payback is between 5 and 6

Figure 13–11
Payback period analysis, Example 13.8

CHAPTER SUMMARY

The breakeven point for a variable for one project is expressed in terms such as units per year or hours per month. At the breakeven amount Q_{BE}, there is indifference to accept or reject the project. Use the following decision guideline:

Single Project   (Refer to Figure 13–2.)

             Estimated quantity is *larger* than $Q_{BE} \rightarrow$ accept project

             Estimated quantity is *smaller* than $Q_{BE} \rightarrow$ reject project

For two or more alternatives, determine the breakeven value of the common variable. Use the following guideline to select an alternative:

Two Alternatives   (Refer to Figure 13–5.)

         Estimated level is *below* breakeven $\rightarrow$ select alternative with higher
                                   variable cost (larger slope)

         Estimated level is *above* breakeven $\rightarrow$ select alternative with lower
                                   variable cost (smaller slope)

Payback analysis estimates the number of years necessary to recover the initial investment plus a stated rate of return. This is a supplemental analysis technique used primarily for initial screening prior to a full evaluation by PW or some other method. The technique has some drawbacks, especially for no-return payback analysis, where $i = 0\%$ is the stated return.

PROBLEMS

Breakeven Analysis for a Project

13.1 A design-to-cost approach to product pricing involves determining the selling price of the product and then figuring out if it can be made at a cost lower than that. Banner Engineering's QT50R radar-based sensor features frequency-modulated technology to accurately monitor or detect objects up to 15 miles away while resisting rain, wind, humidity, and extreme temperatures. It has a list price of $589, and the variable cost of manufacturing the unit is $340.
   (*a*) What could the company's fixed cost per year be in order for Banner to break even with sales of 9000 units per year?
   (*b*) If Banner's fixed cost is actually $750,000 per year, what is the profit at a sales level of 7000 units per year?

13.2 Handheld fiber-optic meters with white light polarization interferometry are useful for measuring temperature, pressure, and strain in electrically noisy environments. The fixed costs associated with manufacturing are $800,000 per year. If a base unit sells for $2950 and its variable cost is $2075, (*a*) how many units must be sold each year for breakeven and (*b*) what will the profit be for sales of 3000 units per year?

13.3 A metallurgical engineer has estimated that the capital investment cost for recovering valuable metals (nickel, silver, platinum, gold, etc.) from the copper refinery's wastewater stream will be $12 million. The equipment will have a useful life of 15 years with no salvage value. Its operating cost is represented by the relation $(\$2,600,000)E^{1.9}$, where E is the efficiency of the metal recovery operation (in decimal form). The amount of metal currently discharged is 2880 pounds per year prior to recovery operations, and the efficiency of recovery is estimated at 71%. What must the average selling price per pound be for the precious metals that are recovered and sold in order for the company to break even at its MARR of 15% per year?

Problems 13.4 through 13.7 are based on the following information.

Hambry Enterprises produces a component for recycling uranium used as a nuclear fuel in power plant generators in France and the United States. Use the following cost and revenue figures, quoted in U.S. dollars per hundredweight (cwt), recorded for this year to calculate the answers for each plant.

Location	Fixed Cost, $ million	Revenue, $ per cwt	Cost, $ per cwt
France	3.50	8,500	3,900
United States	2.65	12,500	9,900

13.4 Determine the breakeven point for each plant.

13.5 Estimate the minimum revenue per hundredweight required for next year if breakeven values and variable costs remain constant, but fixed costs increase by 10%.

13.6 During this year, the French plant sold 950 units in Europe, and the U.S. plant sold 850 units. Determine the year's profit (loss) for each plant.

13.7 Hambry's president has a goal of $1 million profit next year at each plant with no revenue or fixed cost increases. Determine the decreases *in dollar amounts and percentages* in variable cost necessary to meet this goal, if the number of units sold is the same as this year.

13.8 The National Highway Traffic Safety Administration raised the average fuel efficiency standard to 35.5 miles per gallon (mpg) for cars and light trucks by the year 2016. The rules will cost consumers an average of $926 extra per vehicle in the 2016 model year. Assume a person purchases a new car in 2016 that gets 35.5 mpg and keeps it for 5 years. If the person drives an average of 1000 miles per month and gets an extra 10 miles per gallon of gasoline, how much will the gasoline have to cost in order for the buyer to recover the extra investment in 5 years at an interest rate of 0.75% per month?

13.9 A call center in India used by U.S. and U.K. credit card holders has a capacity of 1,400,000 calls annually. The fixed cost of the center is $775,000 with an average variable cost of $1 and revenue of $2.50 per call.
  (a) Find the percentage of the call capacity that must be placed each year to break even.
  (b) The center manager expects to dedicate the equivalent of 500,000 of the 1,400,000 capacity to a new product line. This is expected to increase the center's fixed cost to $900,000 of which 50% will be allocated to the new product line. Determine the average revenue per call necessary to make 500,000 calls the breakeven point for only the new product. How does this required revenue compare with the current center revenue of $2.50 per call?

13.10 The addition of a turbocharger to a small V-6 engine that gets 18 miles per gallon of gasoline can boost its power to that of a V-8 engine and increase fuel efficiency at the same time. If Bill will pay $800 to turbocharge his engine and his fuel efficiency increases by 3 miles per gallon, how many miles will he have to drive each month for 3 years

in order to break even? Assume the cost of gasoline is $3.25 per gallon and the interest rate is 1% per month.

13.11 Transporting extremely heavy patients (people who weigh more than 500 pounds) is much more difficult than transporting normal-weight patients. Various cities in Colorado, Nebraska, and Kansas charge $1421 for an extremely obese patient compared to $758 for a typical patient. The extra fees are justified by the ambulance companies on the basis of the specialty equipment required and the extra personnel involved. If it is assumed that 50 extremely obese patients are transported every year, how much could the ambulance companies afford to spend on the specialty equipment *now* and break even on the initial cost in 5 years just from the extra charges? Assume the extra costs are $400 per patient and the company's MARR is 10% per year.

13.12 High-profile vehicles have poor fuel efficiency because of increased wind resistance from their large front area. A number of companies make devices that they claim will significantly increase a vehicle's fuel efficiency. One company claims that by spending $560 on its friction-reducing device, the fuel efficiency of a pickup truck will increase by 25%. Assuming the device works as claimed, for a vehicle that currently gets 20 miles per gallon (mpg), how many miles would the owner have to drive each year to break even in 5 years? Assume the cost of gasoline is $3.50 per gallon and the interest rate is 10% per year.

13.13 As the price of gasoline goes up, people are willing to drive farther to fill their tank in order to save money. Assume you had been buying gasoline for $2.90 per gallon and that it went up to $2.98 per gallon at the station where you usually go. If you drive an F-150 pickup that gets 18 miles per gallon, what is the round-trip distance you can drive to break even if it will take 20 gallons to fill your tank? Use an interest rate of 8% per year.

13.14 An automobile company is investigating the advisability of converting a plant that manufactures economy cars into one that will make retro sports cars. The initial cost for equipment conversion will be $200 million with a 20% salvage value anytime within a 5-year period. The cost of producing a car will be $21,000, but it is expected to have a selling price of $33,000 to dealers. The production capacity for the first year will be 4000 units. At an interest rate of 12% per year, by what uniform amount will production have to increase each year in order for the company to recover its investment in 3 years?

13.15 For the last 2 years, The Health Company has experienced a fixed cost of $850,000 per year and an $(r - v)$ value of $1.25 per unit for its multivitamin line of products. International competition has become severe enough that some financial changes must be made to keep market share at the current level.

(a) Perform a spreadsheet-based graphical analysis to estimate the effect on the breakeven point if the difference between revenue and variable cost per unit increases somewhere between 1% and 15% of its current value.

(b) If fixed costs and revenue per unit remain at their current values, what type of change must take place to make the breakeven point go down?

13.16 (This is an extension of Problem 13.15) Expand the analysis performed in Problem 13.15 by changing the variable cost per unit. The financial manager estimates that fixed costs will fall to $750,000 when the required production rate to break even is at or below 600,000 units. What happens to the breakeven points over the $(r - v)$ range of 1% to 15% increase as evaluated previously?

Breakeven Analysis Between Alternatives

13.17 Providing restrooms at parks, zoos, and other city-owned recreation facilities is a considerable expense for municipal governments. City councils usually opt for permanent restrooms in larger parks and portable restrooms in smaller ones. The cost of renting and servicing a portable restroom is $7500 per year. In one northeastern municipality, the parks director informed the city council that the cost of constructing a permanent restroom is $218,000 and the annual cost of maintaining it is $12,000. He remarked that the rather high cost is due to the necessity to use expensive materials and construction techniques that are tailored to minimize damage from vandalism that often occurs in unattended public facilities. If the useful life of a permanent restroom is assumed to be 20 years, how many portable restrooms could the city afford to rent each year and break even with the cost of one permanent facility? Let the interest rate be 6% per year.

13.18 A consulting engineer is considering two methods for lining ponds used for evaporating concentrate generated during reverse osmosis treatment of brackish groundwater for the Clay County Industrial Park. A geosynthetic bentonite clay liner (GCL) will cost $1.8 million to install, and if it is renovated after 4 years at a cost of $375,000, its life can be extended another 2 years. Alternatively,

a high-density polyethylene (HDPE) geomembrane can be installed that will have a useful life of 12 years. At an interest rate of 6% per year, how much money can be spent on the HDPE liner for the two methods to break even?

13.19 An irrigation canal contractor wants to determine whether he should purchase a used Caterpillar mini excavator or a Toro powered rotary tiller for servicing irrigation ditches in an agricultural area of California. The initial cost of the excavator is $26,500 with a $9000 salvage value after 10 years. Fixed costs for insurance, license, etc. are expected to be $18,000 per year. The excavator will require one operator at $15 per hour and maintenance at $1 per hour. In 1 hour, 0.15 mile of ditch can be prepared. Alternatively, the contractor can purchase a tiller and hire 2 workers at $11 per hour each. The tiller costs $1200 and has a useful life of 5 years with no salvage value. Its operating cost is expected to be $1.20 per hour, and with the tiller, the two workers can prepare 0.04 mile of ditch in 1 hour. The contractor's MARR is 10% per year. Determine the number of miles of ditch per year the contractor would have to service for the two options to break even.

13.20 An effective method to recover water used for regeneration of ion exchange resins is to use a reverse osmosis system in a batch treatment mode. Such a system involves recirculation of the partially treated water back into the feed tank, causing the water to heat up. The water can be cooled using one of two systems: a single-pass heat exchanger or a closed-loop heat exchange system. The single-pass system, good for 3 years, requires a small chiller costing $920 plus stainless steel tubing, connectors, valves, etc. costing $360. The cost of water, treatment charges, electricity, etc. will be $3.10 per hour. The closed-loop system will cost $3850 to buy, will have a useful life of 5 years, and will cost $1.28 per hour to operate. What is the minimum number of hours per year that the cooling system must be used in order to justify purchase of the closed-loop system? The MARR is 10% per year, and the salvage values are negligible.

13.21 Samsung Electronics is trying to reduce supply chain risk by making more responsible make-buy decisions through improved cost estimation. A high-use component (expected usage is 5000 units per year) can be purchased for $25 per unit with delivery promised within a week. Alternatively, Samsung can make the component in-house and have it readily available at a cost of $5 per unit, if equipment costing $150,000 is purchased. Labor and other operating costs are estimated to be $35,000 per year over the study period of 5 years. Salvage is estimated at 10% of first cost and $i = 12\%$ per year. Neglect the element of availability (*a*) to determine the breakeven quantity and (*b*) to recommend making or buying at the expected usage level.

13.22 A partner in a medium-size A/E (architectural/engineering) design firm is evaluating two alternatives for improving the exterior appearance of the building they occupy. The building can be completely painted at a cost of $6500. The paint is expected to remain attractive for 4 years, at which time repainting will be necessary. Every time the building is repainted, the cost will be 20% higher than the previous time. Alternatively, the building can be sandblasted now and every 6 years at a cost 40% greater than the previous time. If the company's MARR is 10% per year, what is the maximum amount that could be spent now on the sandblasting alternative that would render the two alternatives indifferent over a study period of 12 years?

13.23 A junior mechanical engineering student is co-oping this semester at Regency Aircraft, which customizes the interiors of private and corporate jets. Her first assignment is to develop the specifications for a new machine to cut, shape, and sew leather or vinyl covers and trims. The first cost is not easy to estimate due to many options, but the annual revenue and M&O costs should net out at $+15,000 per year over a 10-year life. Salvage is expected to be 20% of the first cost. Determine the breakeven first cost of the machine to just recover its first cost and a return of 8% per year under two scenarios:

   I: No outside revenue will be developed by the machine.

   II: Outside contracting will occur with estimated revenue of $10,000 the first year, increasing by $5000 per year thereafter.

   Solve using (*a*) hand and (*b*) spreadsheet solutions.

13.24 Ascarate Fishing Club (a nonprofit organization dedicated to teaching kids how to fish) is considering two options for providing a heavily stocked pond for kids who have never caught a fish before. Option 1 is an above-ground swimming pool made of heavy vinyl plastic that will be assembled and disassembled for each quarterly event. The purchase price will be $400. Leaks from hooks piercing the fabric will be repaired with a vinyl repair kit at a cost of $70 per year, but the pool will have to be replaced when too many repairs have been made.

Option 2 is an in-ground pond that will be excavated by club members at no cost and lined with fabric that costs $1 per square foot. The pond will be 15 ft in diameter and 3 ft deep. Assume 300 ft^2 of liner will be purchased. A chain link fence at $10 per lineal foot will be installed around the pond (100 ft of fence). Maintenance inside the fence is expected to cost $20 per year. The park where the pond will be constructed has committed the land for only 10 years. At an interest rate of 6% per year, how long would the above-ground pool have to last to break even?

13.25 A rural subdivision has several miles of access roads that need a new surface treatment. Alternative 1 is a gravel base and pavement with an initial cost of $500,000 that will last for 15 years and has an annual upkeep cost of $100 per mile. Alternative 2 is to enhance the gravel base now at a cost of $50,000 and immediately coat the surface with a durable hot oil mix, which costs $130 per barrel applied. Annual reapplication of the mix is required. A barrel covers 0.05 mile. (*a*) If the discount rate is 6% per year, determine the number of miles at which the two alternatives break even. (*b*) A drive in a pickup indicates a total of 12.5 miles of road. Which is the more economical alternative?

13.26 A waste-holding lagoon situated near the main plant receives sludge daily. When the lagoon is full, it is necessary to remove the sludge to a site located 8.2 kilometers from the main plant. Currently, when the lagoon is full, the sludge is removed by pump into a tank truck and hauled away. This process requires the use of a portable pump that initially costs $800 and has an 8-year life. The company pays a contract individual to operate the pump and oversee environmental and safety factors at a rate of $100 per day, plus the truck and driver must be rented for $200 per day. The company has the option to install a pump and pipeline to the remote site. The pump would have an initial cost of $1600 and a life of 10 years and will cost $3 per day to operate. The company's MARR is 10% per year.
   (*a*) If the pipeline will cost $12 per meter to construct and will have a 10-year life, how many days per year must the lagoon require pumping to justify construction of the pipeline?
   (*b*) If the company expects to pump the lagoon once per week every week of the year, how much money can it afford to spend now on the 10-year life pipeline to just break even?

13.27 Lorraine can select from two nutrient injection systems for her cottage industry of hydroponic tomato and lettuce greenhouses. (*a*) Use an AW relation to determine the minimum number of hours per year to operate the pumps that will justify the Auto Green system, if the MARR is 10% per year. (*b*) Which pump is economically better if it operates 7 hours per day, 365 days per year?

	Nutra Jet (N)	Auto Green (A)
Initial cost, $	−4,000	−10,300
Life, years	3	6
Rebuild cost, $	−1,000	−2,200
Time before rebuild, annually or minimum hours	2,000	8,000
Cost to operate, $ per hour	1.00	0.90

13.28 An engineering practitioner can lease a fully equipped computer and color printer system for $800 per month or purchase one for $8500 now and pay a $75 per month maintenance fee. If the nominal interest rate is 15% per year, determine the months of use necessary for the two to break even. Show both (*a*) hand and (*b*) spreadsheet solutions.

13.29 The office manager of an environmental engineering consulting firm was instructed to make an eco-friendly decision in acquiring an automobile for general office use. He is considering a gasoline-electric hybrid or a gasoline-free, all-electric hatchback. The hybrid under consideration is GM's Volt, which will cost $35,000, will have a salvage value of $15,000 after 5 years, and will have a range of 40 miles on the electric battery, plus several hundred more miles when the gasoline engine kicks in. Nissan's Leaf, on the other hand, is a pure electric that will have a range of only 100 miles, after which its lithium ion battery must be recharged. The Leaf's relatively limited range creates a psychological effect known as *range anxiety* (RA), which has the company leaning toward purchasing the Volt. The Leaf can be leased for $349 per month after an initial $500 down payment.

The accountant for the consulting firm told the office manager that the Leaf is the better economic option based on an evaluation she performed earlier. If the office manager purchases the Volt anyway (instead of leasing the Leaf), what is the monthly equivalent (AW value) of the extra amount of money the company will be paying to eliminate range anxiety? Assume the operating costs will be the same for both vehicles and the MARR is 0.75% per month.

Payback Analysis

13.30 How long will you have to sell a product that has an income of $5000 per month and expenses of $1500 per month if your initial investment is $28,000 and your MARR is (*a*) 0% and (*b*) 3% per month? Solve by formula. (*c*) Write the spreadsheet functions to display the payback period for both 0% and 3% per month.

13.31 (*a*) Determine the payback period at an interest rate of 8% per year for an asset that initially cost $28,000, has a scrap value of $1500 whenever it is sold, and generates cash flow of $2900 per year.

    (*b*) If the asset will be in service for 12 years, should it be purchased?

13.32 ABB purchased fieldbus communication equipment for a project in South Africa for $3.15 million. The net cash flow is estimated at $500,000 per year, and a salvage value of $400,000 is anticipated regardless of when it is sold. Determine the number of years the equipment must be used to obtain payback at MARR values of (*a*) 0% and 8% per year and (*b*) 15% and 16% per year. (*c*) Use a spreadsheet to plot the payback years for all four return values.

13.33 Sundance Detective Agency purchased new surveillance equipment with the following estimates. The year index is $k = 1, 2, 3, \ldots$.

First cost	$1050
Annual maintenance cost	$70 + 5k$ per year
Extra annual revenue	$200 + 50k$ per year
Salvage value	$600 for all years

    (*a*) Calculate the payback period to make a return of 10% per year.

    (*b*) For a preliminary conclusion, should the equipment be purchased if the actual useful life is 7 years?

13.34 Clarisa, an engineering manager, wants to purchase a resort accommodation to rent to skiers. She is considering the purchase of a three-bedroom lodge in upper Montana that will cost $250,000. The property in the area is rapidly appreciating in value because people anxious to get away from urban developments are bidding up the prices. If Clarisa spends an average of $500 per month for utilities and the investment increases at a rate of 2% per month, how long would it be before she could sell the property for $100,000 more than she has invested in it?

13.35 Laura's grandparents helped her purchase a small self-serve laundry business to make extra money during her 5 college years. When she completed her engineering management degree, she sold the business and her grandparents told her to keep the money as a graduation present. For the net cash flows listed below, determine the following:

(a) The percentage of the investment recovered during the 5 years

(b) The actual rate of return over the 5-year period

(c) How long it took to pay back the $75,000 investment in year 0, plus a 7% per year return

Year	0	1	2	3	4	5
NCF, $ per year	−75,000	−10,500	18,600	−2000	28,000	105,000

13.36 Buhler Tractor sold a tractor for $45,000 to Tom Edwards 10 years ago. (a) What is the uniform net cash flow that Tom must make each year to realize payback and a return of 5% per year on his investment over a period of 3 years? 5 years? 8 years? All 10 years? (b) If the net cash flow was actually $5000 per year, what is the amount Tom should have paid for the tractor to realize payback plus the 5% per year return over these 10 years?

13.37 National Parcel Service has historically owned and maintained its own delivery trucks. Leasing is an option being seriously considered because costs for maintenance, fuel, insurance, and some liability issues will be transferred to Pacific Leasing, the truck leasing company. The study period is no more than 24 months for either alternative. The annual lease cost is paid at the beginning of each year and is not refundable for partially used years. Use the first cost and net cash flow estimates to determine the payback in months with a nominal 9% per year return for the (a) purchase option and (b) lease option.

Purchase: $P = \$-30,000$ now
Monthly cost $= \$-1000$
Monthly revenue $= \$4500$

Lease: $P = \$-10,000$ at the beginning of each year (months 0 and 12)
Monthly cost $= \$-2500$
Monthly revenue $= \$4500$

13.38 Julian Browne, owner of Clear Interior Environments, purchased an air scrubber, HEPA vacuum, and other equipment for mold removal for $15,000 eight months ago. Net cash flows were $-2000 for each of the first 2 months, followed by $1000 per month for months 3 and 4. For the last 4 months, a contract generated a net $6000 per month. Julian sold the equipment yesterday for $3000 to a friend. Determine (a) the no-return payback period and (b) the nominal 18%-per-year payback period.

13.39 Explain why payback analysis may favor an alternative with a shorter payback period when it is not the better choice economically.

13.40 When comparing two alternatives, why is it best to use no-return payback analysis as a preliminary screening tool prior to conducting a complete PW or AW evaluation?

Spreadsheet Problems

13.41 Benjamin used regression analysis to fit quadratic relations to monthly revenue and cost data with the following results:

$$R = -0.007Q^2 + 32Q$$
$$TC = 0.004Q^2 + 2.2Q + 8$$

(a) Plot R and TC. Estimate the quantity Q_p at which the maximum profit should occur. Estimate the amount of profit at this quantity.

(b) The profit relation $P = R - TC$ and calculus can be used to determine the quantity Q_p at which the maximum profit will occur and the amount of the profit. The equations are

$$\text{Profit} = aQ^2 + bQ + c$$
$$Q_p = \frac{-b}{2a}$$
$$\text{Maximum profit} = \frac{-b^2}{4a} + c$$

Use these relations to confirm the graphical estimates you made in (a). (Your instructor may ask you to derive the relations above.)

13.42 The National Potato Cooperative purchased a de-skinning machine last year for $150,000. Revenue for the first year was $50,000. Over the total estimated life of 8 years, what must the remaining equivalent annual revenues (years 2 through 8) equal to break even by recovering the investment and a return of 10% per year? Costs are expected to be constant at $42,000 per year, and a salvage value of $20,000 is anticipated.

Problems 13.43 and 13.44 are based on the following information.

Wilson Partners manufactures thermocouples for electronics applications. The current system has a fixed cost of $300,000 per year and a variable cost of $10 per unit. Wilson sells the units for $14 each. A newly proposed process will add onboard features that allow the revenue to increase to $16 per unit, but the fixed cost will now be $500,000 per year. The variable cost of the new system will be based on a $48 per hour rate with 0.2 hour required to produce each unit.

13.43 Determine the annual breakeven quantity for (a) the current system and (b) the new system.

13.44 Plot the two profit relations and estimate graphically the breakeven quantity between the two alternatives.

Problems 13.45 through 13.48 are based on the following information.

Mid-Valley Industrial Extension Service, a state-sponsored agency, provides water quality sampling services to all business and industrial firms in a 10-county region. Just last month, the service purchased all necessary lab equipment for full in-house testing and analysis. Now an outsourcing agency has offered to take over this function on a per sample basis. Data and quotes for the two options have been collected. The MARR for government projects is 5% per year, and a study period of 8 years is chosen.

In-house: Equipment and supplies initially cost $125,000 for a life of 8 years, an AOC of $15,000, and annual salaries of $175,000. Sample costs average $25 each. There is no significant market value for the equipment and supplies currently owned.

Outsourced: Contractors quote sample cost averages of $100 for the first 5 years, increasing to $125 per sample for years 6 through 8.

13.45 Determine the breakeven number of tests between the two options.

13.46 Use a spreadsheet to graph the AW curves for both options for test loads between 0 and 4000 per year in increments of 1000 tests. What is the estimated breakeven quantity?

13.47 The service director has asked the outsource company to reduce the per sample costs by 25% across the board over the 8-year study period. What will this do to the breakeven point? (*Hint:* Look carefully at your graph from Problem 13.46 before answering.)

13.48 Assume the Extension Service can reduce its annual salaries from $175,000 to $100,000 per year and the per sample cost from $25 to $20. What will this do to the breakeven point? (*Hint:* Again, look carefully at your graph from the previous problem before answering.) What is the new annual breakeven test quantity?

ADDITIONAL PROBLEMS AND FE EXAM REVIEW QUESTIONS

13.49 In linear breakeven analysis, if a company expects to operate at a point *above* the breakeven point, it should select the alternative:
(*a*) With the lower fixed cost
(*b*) With the higher fixed cost
(*c*) With the lower variable cost
(*d*) With the higher variable cost

13.50 A company is considering two alternatives to automate the pH of process liquids. Alternative A will have fixed costs of $42,000 per year and will require 2 workers at $48 per day each. Together, these workers can generate 100 units of product per day. Alternative B will have fixed costs of $56,000 per year, but with this alternative, 3 workers will generate 200 units of product. If x is the number of units per year, the variable cost (VC) in $ per year for alternative B is represented by:
(*a*) $[2(48)/100]x$
(*b*) $[3(48)/200]x$
(*c*) $[3(48)/200]x + 56{,}000$
(*d*) $[2(48)/100]x + 42{,}000$

13.51 When the variable cost is reduced for linear total cost and revenue lines, the breakeven point decreases. This is an economic advantage because:
(*a*) The revenue per unit will increase.
(*b*) The two lines will now cross at zero.
(*c*) The profit will increase for the same revenue per unit.
(*d*) The total cost line becomes nonlinear.

13.52 $AW_1 = -23{,}000(A/P,10\%,10) + 4000(A/F,10\%,10)$
$\quad\quad - 3000 - 3x$
$AW_2 = -8{,}000(A/P,10\%,4) - 2000 - 6x$

For these two AW relations, the breakeven point x, in miles per year, is closest to:
(*a*) 1130
(*b*) 1224
(*c*) 1590
(*d*) 655

13.53 To make an item in-house, equipment costing $250,000 must be purchased. It will have a life of 4 years, an annual cost of $80,000, and each unit will cost $40 to manufacture. Buying the item externally will cost $100 per unit. At $i = 15\%$ per year, it is cheaper to make the item in-house if the number per year needed is:
(*a*) Above 1047 units
(*b*) Above 2793 units
(*c*) Equal to 2793 units
(*d*) Below 2793 units

13.54 A procedure at Mercy Hospital has fixed costs of $10,000 per year and variable costs of $50 per test. If the procedure is automated, its fixed cost will be $21,500 per year, but its variable cost will be only $10 per test. The number of tests that must be performed each year for the two operations to break even is closest to:
(*a*) 290   (*b*) 455   (*c*) 750   (*d*) Over 800

13.55 An assembly process can be completed using either alternative X or Y. Alternative X has fixed costs of $10,000 per year with a variable cost of $50 per unit. If the process is automated per alternative Y, its fixed cost will be $5000 per year and its variable cost will be only $10 per unit. The number of units that must be produced each year in order for alternative Y to be favored is closest to:
(a) Y will be favored for any level of production
(b) 125
(c) 375
(d) X will be favored for any level of production

13.56 Two different methods are under consideration for building a bypass road. Material C will cost $100,000 per mile and last for 10 years. Its annual maintenance cost will be $10,000 per year per mile. Material D will cost $30,000 per mile and last for 5 years. At an interest rate of 6% per year, the annual maintenance cost for material D that will make the two methods cost the same is closest to:
(a) Less than $14,000   (b) $14,270
(c) $16,470   (d) $19,510

13.57 A construction company can purchase a piece of equipment for $50,000 and spend $100 per day in operating costs. The equipment will have a 5-year life with no salvage value. Alternatively, the company can lease the equipment for $400 per day. The number of days per year the company must require the equipment to justify its purchase at an interest rate of 8% per year is closest to:
(a) 10 days   (b) 42 days
(c) 51 days   (d) 68 days

13.58 A tractor has a first cost of $40,000, a monthly operating cost of $1500, and a salvage value of $12,000 in 10 years. The MARR is 12% per year. An identical tractor can be rented for $3200 per month (operating cost not included). If n is the minimum number of months per year the tractor must be used in order to justify its purchase, the relation to find n is represented by:
(a) $-40,000(A/P,1\%,10) - 1500n + 12,000(A/F,1\%,10) = 3200n$
(b) $-40,000(A/P,12\%,10) - 1500n + 12,000(A/F,12\%,10) = 3200n$
(c) $-40,000(A/P,1\%,120) - 1500n + 12,000(A/F,1\%,120) = 3200n$
(d) $-40,000(A/P,11.4\%,10) - 1500n + 12,000(A/F,11.4\%,10) = 3200n$

13.59 An anticorrosive coating for a chemical storage tank will cost $5000 and last 5 years if touched up at the end of 3 years at a cost of $1000. If an oil-base enamel coating could be used that will last 2 years, the amount the enamel coating can cost for the two to breakeven at $i = 8\%$ per year is closest to:
(a) $2120   (b) $2390
(c) $2590   (d) $2725

13.60 The price of a car is $50,000 today. Its price is expected to increase by $2400 each year. You now have $25,000 in an investment that is earning 20% per year. The number of years before you have enough money to buy the car, without borrowing any money, is closest to:
(a) 3 years   (b) 5 years
(c) 7 years   (d) 9 years

13.61 Process A has a fixed cost of $16,000 per year and a variable cost of $40 per unit. For process B, 5 units can be produced in 1 day at a cost of $125. If the company's MARR is 10% per year, the fixed cost of process B that will make the two alternatives have the same annual cost at a production rate of 1000 units per year is closest to:
(a) Less than $10,000
(b) $18,000
(c) $27,000
(d) Over $30,000

13.62 The profit relation for the following estimates at a quantity that is 20% above breakeven is:

Fixed cost = $500,000 per year
Variable cost per unit = $200
Revenue per unit = $250

(a) Profit = 200(12,000) − 250(12,000) − 500,000
(b) Profit = 250(12,000) − 500,000 − 200(12,000)
(c) Profit = 250(12,000) − 200(12,000) + 500,000
(d) Profit = 250(10,000) − 200(10,000) − 500,000

13.63 Two methods of weed control in an irrigation canal are under consideration. Method A involves lining at a cost of $4000. The lining will last 20 years. The maintenance cost with this method will be $3 per mile per year. Method B involves spraying a chemical that costs $40 per gallon. One gallon will treat 8 miles, but the treatment must be applied 4 times per year. In determining the number of miles per year that would result in breakeven, the variable cost for method B is closest to:
(a) $5 per mile   (b) $15 per mile
(c) $20 per mile   (d) $40 per mile

13.64 How long will you have to maintain a business that has an income of $5000 per year and expenses of $1500 per year if your initial investment was $28,000 and your MARR is 10% per year?
(a) Less than 6 years   (b) 8 years
(c) 12 years   (d) 17 years

CASE STUDY

WATER TREATMENT PLANT PROCESS COSTS

Background

Aeration and sludge recirculation have been practiced for many years at municipal and industrial water treatment plants. Aeration is used primarily for the physical removal of gases or volatile compounds, while sludge recirculation can be beneficial for turbidity removal and hardness reduction.

When the advantages of aeration and sludge recirculation in water treatment were first recognized, energy costs were so low that such considerations were seldom of concern in treatment plant design and operation. With the huge increases in electricity cost that have occurred in some localities, however, it became necessary to review the cost-effectiveness of all water treatment processes that consume significant amounts of energy. This study was conducted at a municipal water treatment plant for evaluating the cost-effectiveness of the pre-aeration and sludge recirculation practices.

Information

This study was conducted at a 106 m³ per minute water-treatment plant where, under normal operating circumstances, sludge from the secondary clarifiers is returned to the aerator and subsequently removed in the primary clarifiers. Figure 13–12 is a schematic of the process.

To evaluate the effect of sludge recirculation, the sludge pump was turned off, but aeration was continued. Next, the sludge pump was turned back on, and aeration was discontinued. Finally, both processes were discontinued. Results obtained during the test periods were averaged and compared to the values obtained when both processes were operational.

The results obtained from the four operating modes showed that the hardness decreased by 4.7% when both processes were in operation (i.e., sludge recirculation and aeration). When only sludge was recirculated, the reduction was 3.8%. There was no reduction due to aeration only, or when

there was neither aeration nor recirculation. For turbidity, the reduction was 28% when both recirculation and aeration were used. The reduction was 18% when *neither* aeration nor recirculation was used. The reduction was also 18% when aeration alone was used, which means that aeration alone was of no benefit for turbidity reduction. With sludge recirculation alone, the turbidity reduction was only 6%, meaning that sludge recirculation alone actually resulted in an *increase* in turbidity—the difference between 18% and 6%.

Since aeration and sludge recirculation did cause readily identifiable effects on treated water quality (some good and others bad), the cost-effectiveness of each process for turbidity and hardness reduction was investigated. The calculations are based on the following data:

Aerator motor = 40 hp

Aerator motor efficiency = 90%

Sludge recirculation motor = 5 hp

Recirculation pump efficiency = 90%

Electricity cost = 9 ¢/kWh (previous analysis)

Lime cost = 7.9 ¢/kg

Lime required = 0.62 mg/L per mg/L hardness

Coagulant cost = 16.5 ¢/kg

Days/month = 30.5

As a first step, the costs associated with aeration and sludge recirculation were calculated. In each case, costs are independent of flow rate.

Aeration cost:

$$40 \text{ hp} \times 0.75 \text{ kW/hp} \times 0.09 \text{ \$/kWh} \times 24 \text{ h/day}$$
$$\div \, 0.90 = \$72 \text{ per day or } \$2196 \text{ per month}$$

Sludge recirculation cost:

$$5 \text{ hp} \times 0.75 \text{ kW/hp} \times 0.09 \text{ \$/kWh} \times 24 \text{ h/day}$$
$$\div \, 0.90 = \$9 \text{ per day or } \$275 \text{ per month}$$

Figure 13–12
Schematic of water treatment plant.

TABLE 13–1	Cost Summary in Dollars per Month						
		Savings from Discontinuation of			Extra Cost for Removal of		Total	Net
Alt. I.D.	Alternative Description	Aeration (1)	Recirculation (2)	Total Savings (3) = (1) + (2)	Hardness (4)	Turbidity (5)	Extra Cost (6) = (4) + (5)	Net Savings (7) = (3) − (6)
1	Sludge recirculation and aeration	Normal operating condition						
2	Aeration only	—	275	275	1380	469	1849	−1574
3	Sludge recirculation only	2196	—	2196	262	845	1107	+1089
4	Neither aeration nor sludge recirculation	2196	275	2471	1380	469	1849	+622

The estimates appear in columns 1 and 2 of the cost summary in Table 13–1.

Costs associated with turbidity and hardness removal are a function of the chemical dosage required and the water flow rate. The calculations below are based on a design flow of 53 m^3/minute.

As stated earlier, there was less turbidity reduction through the primary clarifier without aeration than there was with it (28% versus 6%). The extra turbidity reaching the flocculators could require further additions of the coagulating chemical. If it is assumed that, as a worst case, these chemical additions would be proportional to the extra turbidity, then 22 percent more coagulant would be required. Since the average dosage before discontinuation of aeration was 10 mg/L, the *incremental chemical cost* incurred because of the increased turbidity in the clarifier effluent would be

$$(10 \times 0.22) \text{ mg/L} \times 10^{-6} \text{kg/mg} \times 53 \text{ m}^3/\text{min}$$
$$\times 1000 \text{ L/m}^3 \times 0.165 \text{ \$/kg} \times 60 \text{ min/h}$$
$$\times 24 \text{ h/day} = \$27.70/\text{day or } \$845/\text{month}$$

Similar calculations for the other operating conditions (i.e., aeration only, and neither aeration nor sludge recirculation) reveal that the additional cost for turbidity removal would be $469 per month in each case, as shown in column 5 of Table 13–1.

Changes in hardness affect chemical costs by virtue of the direct effect on the amount of lime required for water softening. With aeration and sludge recirculation, the average hardness reduction was 12.1 mg/L (that is, 258 mg/L × 4.7%). However, with sludge recirculation only, the reduction was 9.8 mg/L, resulting in a difference of 2.3 mg/L attributed to aeration. The *extra cost of lime* incurred because of the discontinuation of aeration, therefore, was

$$2.3 \text{ mg/L} \times 0.62 \text{ mg/L lime} \times 10^{-6} \text{kg/mg}$$
$$\times 53 \text{m}^3/\text{min} \times 1000 \text{ L/m}^3 \times 0.079 \text{ \$/kg}$$
$$\times 60 \text{ min/h} \times 24 \text{ h/day} = \$8.60/\text{day or}$$
$$\$262/\text{month}$$

When sludge recirculation was discontinued, there was no hardness reduction through the clarifier, so that the extra lime cost would be $1380 per month.

The total savings and total costs associated with changes in plant operating conditions are tabulated in columns 3 and 6 of Table 13–1, respectively, with the net savings shown in column 7. Obviously, the optimum condition is represented by "sludge recirculation only." This condition would result in a net savings of $1089 per month, compared to a net savings of $622 per month when both processes are discontinued and a net *cost* of $1574 per month for aeration only. Since the calculations made here represent worst-case conditions, the actual savings that resulted from modifying the plant operating procedures were greater than those indicated.

In summary, the commonly applied water treatment practices of sludge recirculation and aeration can significantly affect the removal of some compounds in the primary clarifier. However, increasing energy and chemical costs warrant continued investigations on a case-by-case basis of the cost-effectiveness of such practices.

Case Study Exercises

1. What will be the monthly savings in electricity from discontinuation of aeration if the cost of electricity is now 12 ¢/kWh?

2. Does a decrease in the efficiency of the aerator motor make the selected alternative of sludge recirculation only more attractive, less attractive, or the same as before?

3. If the cost of lime were to increase by 50%, would the cost difference between the best alternative and second-best alternative increase, decrease, or remain the same?

4. If the efficiency of the sludge recirculation pump were reduced from 90% to 70%, would the net savings difference between alternatives 3 and 4 increase, decrease, or stay the same?

5. If hardness removal were to be discontinued at the treatment plant, which alternative would be the most cost-effective?

6. If the cost of electricity decreased to 8 ¢/kWh, which alternative would be the most cost-effective?

7. At what electricity cost would the following alternatives just break even? (*a*) Alternatives 1 and 2, (*b*) alternatives 1 and 3, (*c*) alternatives 1 and 4.

LEARNING STAGE 4

Rounding Out the Study

This stage includes topics to enhance your ability to perform a thorough engineering economic study of one project or several alternatives. The effects of **inflation, depreciation, income taxes** in all types of studies, and **indirect costs** are incorporated into the methods of previous chapters. Techniques of **cost estimation** to better predict cash flows are treated in order to base alternative selection on more accurate estimates. The last two chapters include additional material on the use of engineering economics in decision making. An expanded version of **sensitivity analysis** is developed to examine parameters that vary over a predictable range of values. The use of **decision trees** and an introduction to **real options** are included. Finally, the elements of **risk** and **probability** are explicitly considered using expected values, probabilistic analysis, and spreadsheet-based Monte Carlo simulation.

Several of these topics can be covered earlier in the text, depending on the objectives of the course. Use the chart in the Preface to determine appropriate points at which to introduce the material in Learning Stage 4.

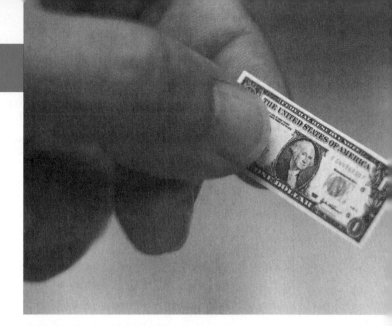

CHAPTER 14

Effects of Inflation

his chapter concentrates upon understanding and calculating the effects of inflation in time value of money computations. Inflation is a reality that we deal with
nearly everyday in our professional and personal lives.

The annual inflation rate is closely watched and historically analyzed by government
units, businesses, and industrial corporations. An engineering economy study can have different outcomes in an environment in which inflation is a serious concern compared to one
in which it is of minor consideration. In the first decade of the 21st century, inflation has not
been a major concern in the United States or most industrialized nations. But the inflation
rate is sensitive to real, as well as perceived, factors of the economy. Factors such as the cost
of energy, interest rates, availability and cost of skilled people, scarcity of materials, political stability, and other, less tangible factors have short-term and long-term impacts on the
inflation rate. In some industries, it is vital that the effects of inflation be integrated into an
economic analysis. The basic techniques to do so are covered here.

14.1 Understanding the Impact of Inflation ● ● ○

Online
Presentation

We are all very well aware that $20 now does not purchase the same amount as $20 did in 2005
and purchases significantly less than in 2000. Why? Primarily this is due to inflation and the
purchasing power of money.

Inflation

> **Inflation** is an increase in the amount of money necessary to obtain the **same amount** of
> goods or services before the inflated price was present.
>
> **Purchasing power**, or **buying power**, measures the value of a currency in terms of the quan-
> tity and quality of goods or services that one unit of money will purchase. Inflation decreases
> the purchasing ability of money in that **less** goods or services **can be purchased** for the same
> one unit of money.

Inflation occurs because the value of the currency has changed—it has gone down in value. The
value of money has decreased, and as a result, it takes more money for the same amount of
goods or services. This is a **sign of inflation**. To make comparisons between monetary amounts
that occur in different time periods, the different-valued money first must be converted to constant-value money in order to represent the same **purchasing power over time**. This is especially important when future sums of money are considered, as is the case with all alternative
evaluations.

Money in one period of time t_1 can be brought to the same value as money in another period
of time t_2 by using the equation

$$\text{Amount in period } t_1 = \frac{\text{amount in period } t_2}{\text{inflation rate between } t_1 \text{ and } t_2} \qquad [14.1]$$

Using dollars as the currency, dollars in period t_1 are called **constant-value dollars** or **today's
dollars.** Dollars in period t_2 are called **future dollars** or **then-current dollars** and have inflation
taken into account. If f represents the inflation rate per period (year) and n is the number of time
periods (years) between t_1 and t_2, Equation [14.1] is

$$\text{Constant-value dollars} = \frac{\text{future dollars}}{(1 + f)^n} \qquad [14.2]$$

$$\text{Future dollars} = \text{constant-value dollars}(1 + f)^n \qquad [14.3]$$

We can express future dollars in terms of constant-value dollars, and vice versa, by applying the
last two equations. This is how the Consumer Price Index (CPI) and cost estimation indices
(of Chapter 15) are determined. As an illustration, use the price of a cheese pizza.

$$\$8.99 \qquad \text{March 2011}$$

If inflation on food prices averaged 5% during the last year, in *constant-value 2010 dollars,* this
cost is last year's equivalent of

$$\$8.99/1.05 = \$8.56 \qquad \text{March 2010}$$

A predicted price in 2012, according to Equation [14.3], is

$$\$8.99(1.05) = \$9.44 \qquad \text{March 2012}$$

The price of $9.44 in 2012 buys exactly the same cheese pizza as $8.56 did in 2010. If inflation averages 5% per year over the next 10 years, Equation [14.3] is used to predict a price in 2020 based on 2010.

$$\$8.56(1.05)^{10} = \$13.94 \qquad \text{March 2020}$$

This is a 63% increase over the 2010 price at 5% inflation for prepared food prices, which is generally not considered excessive. In some areas of the world, hyperinflation may average 50% per year. In such an unfortunate economy, the cheese pizza in 10 years rises from the dollar equivalent of $8.99 to $518.44! This is why countries experiencing hyperinflation must devalue the currency by factors of 100 and 1000 when unacceptable inflation rates persist.

Placed into an industrial or business context, at a reasonably low inflation rate averaging 4% per year, equipment or services with a first cost of $209,000 will increase by 48% to $309,000 over a 10-year span. This is before any consideration of the rate of return requirement is placed upon the equipment's revenue-generating ability. **Make no mistake: Inflation is a formidable force in our economy**.

There are three different rates that are important to understanding inflation: the real interest rate (i), the market interest rate (i_f), and the inflation rate (f). Only the first two are interest rates.

Real or inflation-free interest rate i. This is the rate at which interest is earned when the effects of changes in the value of currency (inflation) have been removed. Thus, the real interest rate presents an actual gain in purchasing power. (The equation used to calculate i, with the influence of inflation removed, is derived later in Section 14.3.) The real rate of return that generally applies for individuals is approximately 3.5% per year. This is the "safe investment" rate. The required real rate for corporations (and many individuals) is set above this safe rate when a MARR is established without an adjustment for inflation.

Inflation-adjusted or market interest rate i_f. As its name implies, this is the interest rate that has been adjusted to take inflation into account. This is the interest rate we hear everyday. It is a combination of the real interest rate i and the inflation rate f, and, therefore, it changes as the inflation rate changes. It is also known as the *inflated interest rate*. A company's MARR adjusted for inflation is referred to as the inflation-adjusted or market MARR. The determination of this value is discussed in Section 14.3.

Inflation rate f. As described above, this is a measure of the rate of change in the value of the currency.

Deflation is the opposite of inflation in that when deflation is present, the purchasing power of the monetary unit is greater in the future than at present. That is, it will take fewer dollars in the future to buy the same amount of goods or services as it does today. Inflation occurs much more commonly than deflation, especially at the national economy level. In deflationary economic conditions, the market interest rate is always less than the real interest rate.

Temporary price deflation may occur in specific sectors of the economy due to the introduction of improved products, cheaper technology, or imported materials or products that force current prices down. In normal situations, prices equalize at a competitive level after a short time. However, deflation over a short time in a specific sector of an economy can be orchestrated through **dumping**. An example of dumping may be the importation of materials, such as steel, cement, or cars, into one country from international competitors at very low prices compared to current market prices in the targeted country. The prices will go down for the consumer, thus forcing domestic manufacturers to reduce their prices in order to compete for business. If domestic manufacturers are not in good financial condition, they may fail, and the imported items replace the domestic supply. Prices may then return to normal levels and, in fact, become inflated over time, if competition has been significantly reduced.

On the surface, having a moderate rate of deflation sounds good when inflation has been present in the economy over long periods. However, if deflation occurs at a more general level, say nationally, it is likely to be accompanied by the lack of money for new capital. Another result is that individuals and families have less money to spend due to fewer jobs, less credit, and fewer loans available; an *overall "tighter" money* situation prevails. As money gets tighter, less is available to be committed to industrial growth and capital investment. In the extreme case, this can evolve over time into a deflationary spiral that disrupts the entire economy. This has happened on occasion, notably in the United States during the Great Depression of the 1930s.

Engineering economy computations that consider deflation use the same relations as those for inflation. For basic equivalence between constant-value dollars and future dollars, Equations [14.2] and [14.3] are used, except the deflation rate is a $-f$ value. For example, if deflation is estimated to be 2% per year, an asset that costs $10,000 today would have a first cost 5 years from now determined by Equation [14.3].

$$10,000(1 - f)^n = 10,000(0.98)^5 = 10,000(0.9039) = \$9039$$

14.2 Present Worth Calculations Adjusted for Inflation ● ● ◦

When the dollar amounts in different time periods are to be expressed in *constant-value dollars,* the equivalent present and future amounts must be determined using the *real interest rate i.* The calculations involved in this procedure are illustrated in Table 14–1, where the inflation rate is 4% per year. Column 2 shows the inflation-driven increase for each of the next 4 years for an item that has a cost of $5000 today. Column 3 shows the cost in future dollars, and column 4 verifies the cost in constant-value dollars via Equation [14.2]. When the future dollars of column 3 are converted to constant-value dollars (column 4), the cost is always $5000, the same as the cost at the start. This is predictably true when the costs are increasing by an amount *exactly equal* to the inflation rate. The actual cost (in inflated dollars) of the item 4 years from now will be $5849, but in constant-value dollars the cost in 4 years will still amount to $5000. Column 5 shows the present worth of future amounts of $5000 at a real interest rate of $i = 10\%$ per year.

Two conclusions can be drawn. At $f = 4\%$, $5000 today inflates to $5849 in 4 years. And $5000 four years in the future has a PW of only $3415 now in constant-value dollars at a real interest rate of 10% per year.

Figure 14–1 graphs the differences over a 4-year period of the constant-value amount of $5000, the future-dollar costs at 4% inflation, and the present worth at 10% real interest with inflation considered. The effect of compounded inflation and interest rates can be large, as you can see by the shaded area.

An alternative, less complicated method of accounting for inflation in a present worth analysis involves adjusting the interest formulas themselves to account for inflation. Consider the P/F formula, where i is the real interest rate.

$$P = F \frac{1}{(1 + i)^n}$$

TABLE 14–1	Inflation Calculations Using Constant-Value Dollars ($f = 4\%$, $i = 10\%$)			
Year n (1)	Cost Increase due to 4% Inflation, $ (2)	Cost in Future Dollars, $ (3)	Future Cost in Constant-Value Dollars, $ (4) = (3)/1.04^n	Present Worth at Real $i = 10\%$, $ (5) = (4)(P/F,10%,n)
0		5000	5000	5000
1	5000(0.04) = 200	5200	$5200/(1.04)^1 = 5000$	4545
2	5200(0.04) = 208	5408	$5408/(1.04)^2 = 5000$	4132
3	5408(0.04) = 216	5624	$5624/(1.04)^3 = 5000$	3757
4	5624(0.04) = 225	5849	$5849/(1.04)^4 = 5000$	3415

Figure 14–1
Comparison of constant-value dollars, future dollars, and their present worth values.

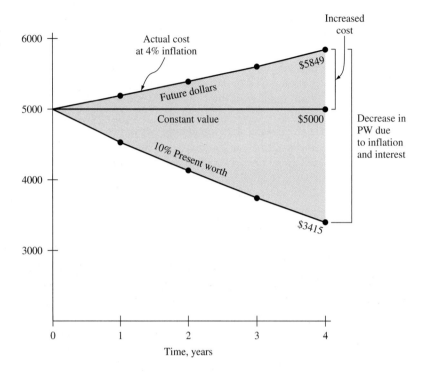

The F, which is a future-dollar amount with inflation built in, can be converted to constant-value dollars by using Equation [14.2].

$$P = \frac{F}{(1+f)^n} \frac{1}{(1+i)^n}$$

$$= F\frac{1}{(1+i+f+if)^n} \qquad [14.4]$$

If the term $i + f + if$ is defined as i_f, the equation becomes

$$P = F\frac{1}{(1+i_f)^n} = F(P/F,i_f,n) \qquad [14.5]$$

As described earlier, i_f is the **inflation-adjusted or market interest rate** and is defined as

$$i_f = i + f + if \qquad [14.6]$$

where i = real interest rate
f = inflation rate

For a real interest rate of 10% per year and an inflation rate of 4% per year, Equation [14.6] yields a market interest rate of 14.4%.

$$i_f = 0.10 + 0.04 + 0.10(0.04) = 0.144$$

Table 14–2 illustrates the use of $i_f = 14.4\%$ in PW calculations for $5000 now, which inflates to $5849 in future dollars 4 years hence. As shown in column 4, the present worth for each year is the same as column 5 of Table 14–1.

The present worth of any series of cash flows—uniform, arithmetic gradient, or geometric gradient—can be found similarly. That is, either i or i_f is introduced into the P/A, P/G, or P_g factors, depending upon whether the cash flow is expressed in constant-value (today's) dollars or future dollars, respectively.

If a cash flow series is expressed in today's (constant-value) dollars, then its PW is the discounted value using the real interest rate i.

If the cash flow is expressed in future dollars, the PW value is obtained using i_f.

It is always acceptable to first convert all future dollars to constant-value dollars using Equation [14.2] and then find the PW at the real interest rate i.

TABLE 14–2	Present Worth Calculation Using an Inflated Interest Rate		
Year n (1)	Cost in Future Dollars, \$ (2)	$(P/F,14.4\%,n)$ (3)	PW, \$ (4) = (2)(3)
0	5000	1	5000
1	5200	0.8741	4545
2	5408	0.7641	4132
3	5624	0.6679	3757
4	5849	0.5838	3415

EXAMPLE 14.1

Glyphosate is the active ingredient in the herbicide Roundup® marketed by Monsanto Co. Roundup has been a dependable product used by farmers, municipalities, and suburbanites alike to control weeds in fields, yards, gardens, streets, and parks. Contributions to Monsanto's revenue have been reduced significantly by international dumping of generic glyphosate, as announced in mid-2010.[1] Monsanto's sales price was decreased from \$16 to \$12 per gallon to compete with the highly competitive pricing, and it is expected that the international price will settle at approximately \$10 per gallon. Assume when the price was set at \$16 per gallon, there was a prediction that in 5 years the price would inflate to \$19 per gallon. Perform the following analysis.

(a) Determine the annual rate of inflation over 5 years to increase the price from \$16 to \$19.
(b) Using the same annual rate determined above as the rate at which the price continues to *decline* from the new \$12 price, calculate the expected price in 5 years. Compare this result with \$10 per gallon that Monsanto predicted would be the longer-term price.
(c) Provided Monsanto were somehow able to recover the same market share as it had previously, and the same inflation rate was applied to the reduced \$12 per gallon price, determine the price 5 years in the future and compare it with the pre-dumping price of \$16 per gallon.
(d) Determine the market interest rate that must be used in economic equivalence computations, if inflation is considered and an 8% per year real return is expected by Monsanto.

Solution

The first three parts involve inflation only—no return on investments.

(a) Solve Equation [14.2] for the annual inflation rate f with known constant-value and future amounts.

$$16 = 19(P/F,f,5) = \frac{19}{(1+f)^5}$$

$$1 + f = (1.1875)^{0.2}$$

$$f = 0.035 \quad (3.5\% \text{ per year})$$

(b) If the price deflation rate is 3.5% per year, find the F value in 5 years with $P = \$12$.

$$F = P(F/P,-3.5\%,5) = 12(1 - 0.035)^5$$
$$= 12(0.8368)$$
$$= \$10.04$$

The price will fall to exactly \$10 per gallon after 5 years, as Monsanto predicted.

[1]S. Kilman and I. Berry, "Monsanto Cuts Roundup Prices as Knockoffs Flood Farm Belt," *Wall Street Journal,* May 28, 2010.

(c) Five years in the future, at 3.5% per year inflation, the price will be

$$F = P(F/P,3.5\%,5) = 12(1.035)^5$$
$$= 12(1.1877)$$
$$= \$14.25$$

After 5 years of recovery at the same level as historically experienced, the price will still be considerably lower than it was at the pre-dumping point ($14.25 versus $16 per gallon).

(d) With inflation at 3.5% per year and a real return of 8% per year, Equation [14.6] results in a market rate of 11.78% per year.

$$i_f = 0.08 + 0.035 + (0.08)(0.035)$$
$$= 0.1178 \quad (11.78\% \text{ per year})$$

EXAMPLE 14.2

A 15-year $50,000 bond that has a dividend rate of 10% per year, payable semiannually, is currently for sale. If the expected rate of return of the purchaser is 8% per year, compounded semiannually, and if the inflation rate is expected to be 2.5% each 6-month period, what is the bond worth now (a) without an adjustment for inflation and (b) when inflation is considered? Show both hand and spreadsheet solutions.

Solution by Hand

(a) *Without inflation adjustment*: The semiannual dividend is $I = [(50,000)(0.10)]/2 = \$2500$. At a nominal 4% per 6 months for 30 periods,

$$PW = 2500(P/A,4\%,30) + 50,000(P/F,4\%,30) = \$58,645$$

(b) *With inflation*: Use the inflated rate i_f.

$$i_f = 0.04 + 0.025 + (0.04)(0.025) = 0.066 \text{ per semiannual period}$$
$$PW = 2500(P/A,6.6\%,30) + 50,000(P/F,6.6\%,30)$$
$$= 2500(12.9244) + 50,000(0.1470)$$
$$= \$39,660$$

Solution by Spreadsheet

Both (a) and (b) require simple, single-cell functions on a spreadsheet (Figure 14–2). Without an inflation adjustment, the PV function is developed at the nominal 4% rate for 30 periods; with inflation considered the rate is $i_f = 6.6\%$, as determined above.

Comment

The $18,985 difference in PW values illustrates the tremendous negative impact made by only 2.5% inflation each 6 months (5.06% per year). Purchasing the $50,000 bond means receiving $75,000 in dividends over 15 years and the $50,000 principal in year 15. Yet, this is worth only $39,660 in constant-value dollars.

	A	B	C
1		Function	PW value, $
2	(a) PW without inflation	`= PV(4%,30,-2500,-50000)	58,646
3			
4	(b) PW with inflation adjustment	`= PV(6.6%,30,-2500,-50000)	39,660
5			
6			

Figure 14–2
PW computation of a bond purchase (a) without and (b) with an inflation adjustment, Example 14.2.

EXAMPLE 14.3

A self-employed chemical engineer is on contract with Dow Chemical, currently working in a relatively high-inflation country in Central America. She wishes to calculate a project's PW with estimated costs of $35,000 now and $7000 per year for 5 years beginning 1 year from now with increases of 12% per year thereafter for the next 8 years. Use a real interest rate of 15% per year to make the calculations (*a*) without an adjustment for inflation and (*b*) considering inflation at a rate of 11% per year.

Solution

(*a*) Figure 14–3 presents the cash flows. The PW without an adjustment for inflation is found using $i = 15\%$ and $g = 12\%$ in Equations [2.34] and [2.35] for the geometric series.

$$PW = -35,000 - 7000(P/A,15\%,4)$$

$$-\left\{\frac{7000\left[1 - \left(\frac{1.12}{1.15}\right)^9\right]}{0.15 - 0.12}\right\}(P/F,15\%,4)$$

$$= -35,000 - 19,985 - 28,247$$

$$= \$-83,232$$

In the P/A factor, $n = 4$ because the $7000 cost in year 5 is the A_1 term in Equation [2.34].

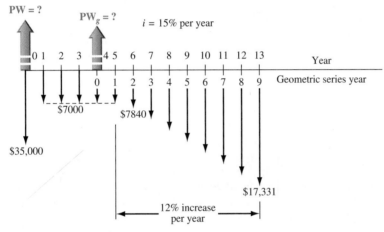

Figure 14–3
Cash flow diagram, Example 14.3.

(*b*) To adjust for inflation, calculate the inflated interest rate by Equation [14.6] and use it to calculate PW.

$$i_f = 0.15 + 0.11 + (0.15)(0.11) = 0.2765$$

$$PW = -35,000 - 7000(P/A,27.65\%,4)$$

$$-\left\{\frac{7000\left[1 - \left(\frac{1.12}{1.2765}\right)^9\right]}{0.2765 - 0.12}\right\}(P/F,27.65\%,4)$$

$$= -35,000 - 7000(2.2545) - 30,945(0.3766)$$

$$= \$-62,436$$

Comment

This result demonstrates that in a high-inflation economy, when negotiating the amount of the payments to repay a loan, it is economically advantageous for the borrower to use future (inflated) dollars whenever possible to make the payments. The present value of future inflated dollars is significantly less when the inflation adjustment is included. And the higher the inflation rate, the larger the discounting because the P/F and P/A factors decrease in size.

Examples 14.2 and 14.3 above add credence to the "buy now, pay later" philosophy. However, at some point, the debt-ridden company or individual will have to pay off the debts and the accrued interest with the inflated dollars. If cash is not readily available at that time, the debts cannot be repaid. This can happen, for example, when a company unsuccessfully launches a new product, when there is a serious downturn in the economy, or when an individual loses a salary. In the longer term, this buy now, pay later approach must be tempered with sound financial practices now, and in the future.

Online Presentation

14.3 Future Worth Calculations Adjusted for Inflation ● ● ●

In future worth calculations, a future amount F can have any one of four different interpretations:

Case 1. The *actual amount* of money that will be *accumulated* at time n.

Case 2. The *purchasing power* of the actual amount accumulated at time n, but stated in today's (constant-value) dollars.

Case 3. The number of *future dollars required* at time n to maintain the *same purchasing power* as today; that is, inflation is considered, but interest is not.

Case 4. The amount of money required at time n to *maintain purchasing power and earn a stated real interest rate.*

Depending upon which interpretation is intended, the F value is calculated differently, as described below. Each case is illustrated.

Case 1: Actual Amount Accumulated   It should be clear that F, the actual amount of money accumulated, is obtained using the inflation-adjusted (market) interest rate.

$$F = P(1 + i_f)^n = P(F/P,i_f,n) \tag{14.7}$$

For example, when we quote a market rate of 10%, the inflation rate is included. Over a 7-year period, $1000 invested at 10% per year will accumulate to

$$F = 1000(F/P,10\%,7) = \$1948$$

Case 2: Constant-Value Dollars with Purchasing Power   The purchasing power of future dollars is determined by first using the market rate i_f to calculate F and then deflating the future amount through division by $(1 + f)^n$.

$$F = \frac{P(1 + i_f)^n}{(1 + f)^n} = \frac{P(F/P, i_f, n)}{(1 + f)^n} \tag{14.8}$$

This relation, in effect, recognizes the fact that inflated prices mean $1 in the future purchases less than $1 now. The percentage loss in purchasing power is a measure of how much less. As an illustration, consider the same $1000 now, and a 10% per year *market rate,* which includes an inflation rate of 4% per year. In 7 years, the purchasing power has risen, but only to $1481.

$$F = \frac{1000 \, (F/P, 10\%, 7)}{(1.04)^7} = \frac{\$1948}{1.3159} = \$1481$$

This is $467 (or 24%) less than the $1948 actually accumulated at 10% (case 1). Therefore, we conclude that 4% inflation over 7 years reduces the purchasing power of money by 24%.

    Also for case 2, the future amount of money accumulated with today's buying power could equivalently be determined by calculating the real interest rate and using it in the F/P factor to compensate for the decreased purchasing power. This **real interest rate** is the i in Equation [14.6].

$$i_f = i + f + if$$
$$= i(1 + f) + f$$

$$i = \frac{i_f - f}{1 + f} \tag{14.9}$$

The real interest rate i represents the rate at which today's dollars expand with their *same purchasing power* into equivalent future dollars. An inflation rate larger than the market interest rate leads to a negative real interest rate. The use of this interest rate is appropriate for calculating the future worth of an investment (such as a savings account or money market fund) when the effect of inflation must be removed. For the example of $1000 in today's dollars from Equation [14.9]

$$i = \frac{0.10 - 0.04}{1 + 0.04} = 0.0577 \quad (5.77\%)$$

$$F = 1000(F/P,5.77\%,7) = \$1481$$

The market interest rate of 10% per year has been reduced to a real rate that is less than 6% per year because of the erosive effects of 4% per year inflation.

Case 3: Future Amount Required, No Interest   This case recognizes that prices increase when inflation is present. Simply put, future dollars are worth less, so more are needed. No interest rate is considered in this case—only inflation. This is the situation if someone asks, "How much will a car cost in 5 years if its current cost is $20,000 and its price will increase by the inflation rate of 6% per year?" (The answer is $26,765.) No interest rate—only inflation—is involved. To find the future cost, substitute f for the interest rate in the F/P factor.

$$F = P(1 + f)^n = P(F/P,f,n) \qquad\qquad [14.10]$$

Reconsider the $1000 used previously. If it is escalating at exactly the inflation rate of 4% per year, the amount 7 years from now will be

$$F = 1000(F/P,4\%,7) = \$1316$$

Case 4: Inflation and Real Interest   This is the case applied when a market MARR is established. Maintaining purchasing power and earning interest must account for both increasing prices (case 3) and the time value of money. If the growth of capital is to keep up, funds must grow at a rate equal to or above the real interest rate i plus the inflation rate f. Thus, to make a *real rate of return of 5.77%* when the inflation rate is 4%, i_f is the market (inflation-adjusted) rate that must be used. For the same $1000 amount,

$$i_f = 0.0577 + 0.04 + 0.0577(0.04) = 0.10$$
$$F = 1000(F/P,10\%,7) = \$1948$$

This calculation shows that $1948 seven years in the future will be equivalent to $1000 now with a real return of $i = 5.77\%$ per year and inflation of $f = 4\%$ per year.

    Table 14–3 summarizes which rate is used in the equivalence formulas for the different interpretations of F. The calculations made in this section explain the following:

- The amount of $1000 now at a market rate of 10% per year will accumulate to $1948 in 7 years.
- The $1948 will have the purchasing power of $1481 of today's dollars if $f = 4\%$ per year.
- An item with a cost of $1000 now will cost $1316 in 7 years at an inflation rate of 4% per year.
- It will take $1948 of future dollars to be equivalent to $1000 now at a real interest rate of 5.77% with inflation considered at 4% per year.

Most corporations evaluate alternatives at a MARR large enough to cover inflation plus some return greater than their cost of capital, and significantly higher than the safe investment return of approximately 3.5% mentioned earlier. Therefore, for case 4, the resulting MARR will normally be higher than the market rate i_f. Define the symbol MARR$_f$ as the inflation-adjusted or market MARR, which is calculated in a fashion similar to i_f.

$$\text{MARR}_f = i + f + i(f) \qquad\qquad [14.11]$$

TABLE 14–3	Calculation Methods for Various Future Worth Interpretations	
Future Worth Desired	**Method of Calculation**	**Example for $P = \$1000$, $n = 7$, $i_f = 10\%$, $f = 4\%$**
Case 1: Actual dollars accumulated	Use stated market rate i_f in equivalence formulas	$F = 1000(F/P,10\%,7)$
Case 2: Purchasing power of accumulated dollars in terms of constant-value dollars	Use market rate i_f in equivalence and divide by $(1 + f)^n$ or Use real i	$F = \dfrac{1000\,(F/P,10\%,7)}{(1.04)^7}$ or $F = 1000(F/P,5.77\%,7)$
Case 3: Dollars required for same purchasing power	Use f in place of i in equivalence formulas	$F = 1000(F/P,4\%,7)$
Case 4: Future dollars to maintain purchasing power **and** to earn a return	Calculate i_f and use in equivalence formulas	$F = 1000(F/P,10\%,7)$

The real rate of return i used here is the required rate for the corporation relative to its cost of capital. Now the future worth F, or FW, is calculated as

$$F = P(1 + \text{MARR}_f)^n = P(F/P,\text{MARR}_f,n)$$

For example, if a company has a WACC (weighted average cost of capital) of 10% per year and requires that a project return 3% per year above its WACC, the real return is $i = 13\%$. The inflation-adjusted MARR is calculated by including the inflation rate of, say, 4% per year. Then the project PW, AW, or FW will be determined at the rate obtained from Equation [14.11].

$$\text{MARR}_f = 0.13 + 0.04 + 0.13(0.04) = 0.1752 \qquad (17.52\%)$$

EXAMPLE 14.4

Abbott Mining Systems wants to determine whether it should upgrade a piece of equipment used in deep mining operations in one of its international operations now or later. If the company selects plan A, the upgrade will be purchased now for $200,000. However, if the company selects plan I, the purchase will be deferred for 3 years when the cost is expected to rise to $300,000. Abbott is ambitious; it expects a real MARR of 12% per year. The inflation rate in the country has averaged 3% per year. From only an economic perspective, determine whether the company should purchase now or later (*a*) when inflation is not considered and (*b*) when inflation is considered.

Solution

(*a*) *Inflation not considered*: The real rate, or MARR, is $i = 12\%$ per year. The cost of plan I is $300,000 three years hence. Calculate the FW value for plan A three years from now and select the lower cost.

$$\text{FW}_A = -200,000(F/P,12\%,3) = \$-280,986$$
$$\text{FW}_I = \$-300,000$$

Select plan A (purchase now).

(*b*) *Inflation considered*: This is case 4; the real rate (12%), and inflation of 3% must be accounted for. First, compute the inflation-adjusted MARR by Equation [14.11].

$$\text{MARR}_f = 0.12 + 0.03 + 0.12(0.03) = 0.1536$$

Use MARR$_f$ to compute the FW value for plan A in future dollars.

$$FW_A = -200,000(F/P,15.36\%,3) = \$-307,040$$
$$FW_I = \$-300,000$$

Purchase later (plan I) is now selected, because it requires fewer equivalent future dollars. The inflation rate of 3% per year has raised the equivalent future worth of costs by 9.3% from $280,986 to $307,040. This is the same as an increase of 3% per year, compounded over 3 years, or $(1.03)^3 - 1 = 9.3\%$.

Most countries have inflation rates in the range of 2% to 8% per year, but **hyperinflation** is a problem in countries where political instability, overspending by the government, weak international trade balances, etc., are present. Hyperinflation rates may be very high—10% to 100% *per month*. In these cases, the government may take drastic actions: redefine the currency in terms of the currency of another country, control banks and corporations, and control the flow of capital into and out of the country in order to decrease inflation.

In a hyperinflated environment, people usually spend all their money immediately since the cost of goods and services will be so much higher the next month, week, or day. To appreciate the disastrous effect of hyperinflation on a company's ability to keep up, we can rework Example 14.4b using an inflation rate of 10% per month, that is, a nominal 120% per year (not considering the compounding effect of inflation). The FW$_A$ amount skyrockets and plan I is a clear choice. Of course, in such an environment the $300,000 purchase price for plan I three years hence would obviously not be guaranteed, so the entire economic analysis is unreliable. Good economic decisions in a hyperinflated economy are very difficult to make using traditional engineering economy methods, since the estimated future values are totally unreliable and the future availability of capital is uncertain.

14.4 Capital Recovery Calculations Adjusted for Inflation ● ● ●

It is particularly important in capital recovery (CR) calculations used for AW analysis to include inflation because current capital dollars must be recovered with future inflated dollars. Since future dollars have less buying power than today's dollars, it is obvious that more dollars will be required to recover the present investment. This suggests the use of the inflated interest rate in the A/P formula. For example, if **$1000 is invested today** at a real interest rate of 10% per year when the inflation rate is 8% per year, the equivalent amount that must be recovered each year for 5 years in future dollars is

$$A = 1000(A/P,18.8\%,5) = \$325.59$$

On the other hand, the decreased value of dollars through time means that investors can spend fewer present (higher-value) dollars to accumulate a specified amount of future (inflated) dollars. This suggests the use of a higher interest rate, that is, the i_f rate, to produce a lower A value in the A/F formula. The annual equivalent (with adjustment for inflation) of F = **$1000 five years from now** in future dollars is

$$A = 1000(A/F,18.8\%,5) = \$137.59$$

For comparison, the equivalent annual amount to accumulate F = $1000 at a real i = 10% (without adjustment for inflation) is $1000(A/F,10\%,5) = \$163.80$. When F is a future known cost, uniformly distributed payments should be spread over as long a time period as possible so that the leveraging effect of inflation will reduce the effective annual payment ($137.59 versus $163.80 here).

EXAMPLE 14.5

What annual deposit is required for 5 years to accumulate an amount of money with the same purchasing power as $680.58 today, if the market interest rate is 10% per year and inflation is 8% per year?

Solution

First, find the actual number of inflated dollars required 5 years in the future that is equivalent to $680.58 today. This is case 3; Equation [14.10] applies.

$$F = (\text{present purchasing power})(1 + f)^5 = 680.58(1.08)^5 = \$1000$$

The actual amount of the annual deposit is calculated using the market interest rate of 10%. This is case 4 where A is calculated for a given F.

$$A = 1000(A/F,10\%,5) = \$163.80$$

Comment

The real interest rate is $i = 1.85\%$ as determined using Equation [14.9]. To put these calculations into perspective, if the inflation rate is zero when the real interest rate is 1.85%, the future amount of money with the same purchasing power as $680.58 today is obviously $680.58. Then the annual amount required to accumulate this future amount in 5 years is $A = 680.58(A/F,1.85\%,5) = \$131.17$. This is $32.63 lower than the $163.80 calculated above for $f = 8\%$. This difference is due to the fact that during inflationary periods, dollars deposited now have more purchasing power than the dollars returned at the end of the period. To make up the purchasing power difference, more higher-value dollars are required. That is, to maintain equivalent purchasing power at $f = 8\%$ per year, an extra $32.63 per year is required.

The logic discussed here explains why, in times of increasing inflation, lenders of money (credit card companies, mortgage companies, and banks) tend to further increase their market interest rates. People tend to pay off less of their incurred debt at each payment because they use any excess money to purchase additional items before the price is further inflated. Also, the lending institutions must have more money in the future to cover the expected higher costs of lending money. All this is due to the spiraling effect of increasing inflation. Breaking this cycle is difficult to do at the individual level and much more difficult to alter at a national level.

CHAPTER SUMMARY

Inflation, treated computationally as an interest rate, makes the cost of the same product or service increase over time due to the decreased value of money. There are several ways to consider inflation in engineering economy computations in terms of today's (constant-value) dollars and in terms of future dollars. Some important relations are the following:

Inflated interest rate: $i_f = i + f + if$

Real interest rate: $i = (i_f - f)/(1 + f)$

PW of a future amount with inflation considered: $P = F(P/F,i_f,n)$

Future worth in constant-value dollars of a present amount with the same purchasing power:
$F = P(F/P,i,n)$

Future amount to cover a current amount with inflation only: $F = P(F/P,f,n)$

Future amount to cover a current amount with inflation and interest: $F = P(F/P,i_f,n)$

Annual equivalent of a future amount: $A = F(A/F,i_f,n)$

Annual equivalent of a present amount in future dollars: $A = P(A/P,i_f,n)$

Hyperinflation implies very high f values. Available funds are expended immediately because costs increase so rapidly that larger cash inflows cannot offset the fact that the currency is losing value. This can, and usually does, cause a national financial disaster when it continues over extended periods of time.

PROBLEMS

Interest Rate and Currency Considerations with Inflation

14.1 What is the difference between today's dollars and constant-value dollars (*a*) when using *today* as the reference point in time and (*b*) when using 2 years ago as the reference point?

14.2 State the conditions under which the market interest rate is (*a*) higher than, (*b*) lower than, and (*c*) the same as the real interest rate.

14.3 What annual inflation rate is implied from an inflation-adjusted interest rate of 10% per year, when the real interest rate is 4% per year?

14.4 Determine the inflation-adjusted interest rate for a growth company that wants to earn a real rate of return of 20% per year when the inflation rate is 5% per year.

14.5 For a high-growth company that wants to make a real rate of return of 30% per year, compounded monthly, determine the inflation-adjusted nominal interest rate per year. Assume the inflation rate is 1.5% per month.

14.6 A high-tech company whose stock trades on the NASDAQ stock exchange uses a MARR of 35% per year. If the chief financial officer (CFO) said the company expects to make a *real rate of return* of 25% per year on its investments over the next 3-year period, what is the company expecting the inflation rate per year to be over that time period?

14.7 Calculate the inflation-adjusted interest rate per quarter when the real interest rate is 4% per quarter and the inflation rate is 1% per quarter.

14.8 Calculate the real interest rate per month if the nominal inflation-adjusted interest rate per year, compounded monthly, is 18% and the inflation rate per month is 0.5%.

14.9 A southwestern city has a contract with Firestone Vehicle Fleet Maintenance to provide maintenance services on its fleet of city-owned vehicles such as street sweepers, garbage trucks, backhoes, and carpool vehicles. The contract price is fixed at $45,000 per year for 4 years. If you were asked to convert the future amounts into constant-value amounts per today's dollars, what would the respective amounts be? Assume the current market interest rate of 10% per year and inflation rate of 5% per year are expected to remain the same over the next 4-year period.

14.10 When the inflation rate is 5% per year, how many inflated dollars will be required 10 years from now to buy the same things that $10,000 buys now?

14.11 Assume that you want to retire 30 years from now with an amount of money that will have the same value (same purchasing power) as $1.5 million today. If you estimate the inflation rate will be 4% per year, how many future (then-current) dollars will you need?

Adjusting for Inflation

14.12 If the inflation rate is 7% per year, how many years will it take for the cost of something to double when prices increase at exactly the same rate as inflation?

14.13 The inflation rate in a Central American country is 6% per year. What real rate of return will an investor make on a $100,000 investment in a copper mine stock that yields an overall internal rate of return of 28% per year?

14.14 During periods of hyperinflation, prices increase rapidly over short periods of time. In 1993, the Consumer Price Index (CPI) in Brazil was 113.6 billion. In 1994, the CPI was 2472.4 billion.
   (*a*)   What was the inflation rate per year between 1993 and 1994?
   (*b*)   Assuming the inflation rate calculated in part (*a*) occurred uniformly throughout the year and represented a nominal rate, what were the monthly and daily inflation rates over that time period?

14.15 A trust was set up by your grandfather that states you are to receive $250,000 exactly 5 years from today. Determine the buying power of the $250,000 in terms of today's dollars if the market interest rate is 10% per year and the inflation rate is 4% per year.

14.16 Assume the inflation rate is 4% per year and the market interest rate is 5% above the inflation rate. Determine (*a*) the number of constant-value dollars 5 years in the future that is equivalent to $30,000 now and (*b*) the number of future dollars that will be equivalent to $30,000 now.

14.17 The Pell Grant program of the federal government provides financial aid to needy college students. If the average grant is slated to increase from $4050 to $5400 over the next 5 years "to keep up with inflation," what is the average inflation rate per year expected to be?

14.18 Ford Motor Company announced that the price of its F-150 pickup trucks is going to increase by only the inflation rate for the next 3 years. If the current price of a well-equipped truck is $28,000 and the inflation rate averages 2.1% per year, what is the expected price of a comparably equipped truck next year? 3 years from now?

14.19 In an effort to reduce pipe breakage, water hammer, and product agitation, a chemical company plans to install several chemically resistant pulsation dampeners. The cost of the dampeners today is $120,000, but the company has to wait until a permit is approved for its bidirectional port-to-plant product pipeline. The permit approval process will take at least 2 years because of the time required for preparation of an environmental impact statement. Because of intense foreign competition, the manufacturer plans to increase the price only by the inflation rate each year. If the inflation rate is 2.8% per year and the company's MARR is 20% per year, estimate the cost of the dampeners in 2 years in terms of (*a*) today's dollars and (*b*) future dollars.

14.20 A machine currently under consideration by Holzmann Industries has a cost of $45,000. When the purchasing manager complained that a similar machine the company purchased 5 years ago was much cheaper, the salesman responded that the cost of the machine has increased solely in accordance with the inflation rate, which has averaged 3% per year. When the purchasing manager checked the invoice for the machine he purchased 5 years ago, he saw that the price was $29,000. Was the salesman telling the truth about the increase in the cost of the machine? If not, what should the machine cost now, provided the price increased by only the inflation rate?

14.21 A report by the National Center for Public Policy and Higher Education stated that tuition and fees (T&F) at public colleges and universities increased by 439% over the last 25 years. During this same time period, the median family income (MFI) rose 147%. When the report was written, T&F at a 4-year public university constituted 28% of the MFI of $52,000 (tuition and fees at a private university constituted 76% of MFI).
  (*a*) What was the tuition and fee cost per year when the report was written?
  (*b*) What was the T&F cost 25 years ago?
  (*c*) What percentage was T&F of the MFI 25 years ago?

14.22 The headline on a *Chronicle of Higher Education* article reads "College Costs Rise Faster than Inflation." The article states that tuition at public colleges and universities increased by 58% over the past 5 years. (*a*) What was the average annual percentage increase over that period of time? (*b*) If the real increase in tuition (i.e., without inflation) was 5% per year, what was the inflation rate per year?

14.23 Stadium Capital Financing Group helps cash-strapped sports programs through the marketing of its "sports mortgage." At the University of Kansas, Jayhawks fans can sign up to pay $105,000 over 10 years for the right to buy top seats for football during the next 30 years. In return, the season tickets will stay locked in at current-year prices. Season tickets in tier 1 are currently selling for $350. A fan plans to purchase the sports mortgage along with season tickets now and each year for the next 30 years (31 seasons). What is the dollar amount of the savings on the tickets (with no interest considered), if ticket prices rise at a rate of 3% per year for the next 30 years?

Present Worth Calculations with Inflation

14.24 There are two ways to account for inflation in present worth calculations. What are they?

14.25 An environmental testing company needs to purchase equipment 2 years from now and expects to pay $50,000 *at that time*. At a real interest rate of 10% per year and inflation rate of 4% per year, what is the present worth of the cost of the equipment?

14.26 Carlsbad Gas and Electric is planning to purchase a degassing tower for removing CO_2 from acidified saltwater. The supplier quoted a price of $125,000 if the unit is purchased within the next 3 years. Your supervisor has asked you to calculate the present worth of the tower when considering inflation. Assuming the tower will not be purchased for 3 years, calculate the present worth at an interest rate of 10% per year and an inflation rate of 4% per year.

14.27 How much can the manufacturer of superconducting magnetic energy storage systems afford to spend now on new equipment in lieu of spending $75,000 four years from now? The company's real MARR is 12% per year, and the inflation rate is 3% per year.

14.28 Find the present worth of the cash flows shown. Some are expressed as constant-value (CV) dollars, and others are inflated dollars. Assume a real interest rate of 8% per year and an inflation rate of 6% per year.

Year	1	2	3	4	5
Cash Flow, $	3000	6000	8000	4000	5000
Stated as	CV	Inflated	Inflated	CV	CV

14.29 A doctor is on contract to a medium-sized oil company to provide medical services at remotely located, widely separated refineries. The doctor is considering the purchase of a private plane to reduce the total travel time between refineries. The doctor can buy a used Lear jet now or wait for a new very light jet (VLJ) that will be available 3 years from now. The cost of the VLJ will be $1.9 million, payable when the plane is delivered in 3 years. The doctor has asked you, his friend, to determine the present worth of the VLJ so that he can decide whether to buy the used Lear now or wait for the VLJ. If the MARR is 15% per year and the inflation rate is projected to be 3% per year, what is the present worth of the VLJ with inflation considered?

14.30 A regional infrastructure building and maintenance contractor is trying to decide whether to buy a new compact horizontal directional drilling (HDD) machine now or wait to buy it 2 years from now (when a large pipeline contract will require the new equipment). The HDD machine will include an innovative pipe loader design and maneuverable undercarriage system. The cost of the system is $68,000 if purchased now or $81,000 if purchased 2 years from now. At a real MARR of 10% per year and an inflation rate of 5% per year, determine if the company should buy now or later (*a*) without any adjustment for inflation and (*b*) with inflation considered.

14.31 An engineer must recommend one of two rapid-prototyping machines for integration into an upgraded manufacturing line. She obtained estimates from salespeople from two companies. Salesman A gave her the estimates in constant-value (today's) dollars, while saleswoman B provided the estimates in future (then-current) dollars. The company's MARR is equal to the real rate of return of 20% per year, and inflation is estimated at 4% per year. Use PW analysis to determine which machine the engineer should recommend.

	A (in CV Dollars)	B (in Future Dollars)
First cost, $	−140,000	−155,000
AOC, $ per year	−25,000	−40,000
Life, years	10	10

14.32 A salesman from Industrial Water Services (IWS), who is trying to get his foot in the door for a large account in Fremont, offered water chlorination equipment for $2.1 million. This is $400,000 more than the price offered by a competing saleswoman from AG Enterprises. However, IWS said the company would not have to pay for the equipment until the warranty runs out. If the equipment has a 2-year warranty, determine which offer is better. The company's real MARR is 12% per year, and the inflation rate is 4% per year.

14.33 A chemical engineer is considering two sizes of pipes, small (S) and large (L), for moving distillate from a refinery to the tank farm. A small pipeline will cost less to purchase (including valves and other appurtenances) but will have a high head loss and, therefore, a higher pumping cost. In writing the report, the engineer compared the alternatives based on future worth values, but the company president wants the costs expressed as present dollars. Determine present worth values if future worth values are $FW_S = \$2.3$ million and $FW_L = \$2.5$ million. The company uses a real interest rate of 1% per month and an inflation rate of 0.4% per month. Assume the future worth values were for a 10-year project period.

14.34 As an innovative way to pay for various software packages, a new high-tech service company has offered to pay your company, Custom Computer Services (CCS), in one of three ways: (1) pay $480,000 now, (2) pay $1.1 million 5 years from now, or (3) pay an amount of money 5 years from now that will have the same purchasing power as $850,000 now. If you, as president of CCS, want to earn a real interest rate of 10% per year when the inflation rate is 6% per year, which offer should you accept?

Future Worth and Other Calculations with Inflation

14.35 If the inflation rate is 6% per year and a person wants to earn a true (real) interest rate of 10% per year, determine the number of future dollars she has to receive 10 years from now if the present investment is $10,000.

14.36 How many future dollars would you need 5 years from now just to have the same buying power as $50,000 now, if the *deflation* rate is 3% per year?

14.37 If a company deposits $100,000 into an account that earns a market interest rate of 10% per year at a time when the *deflation* rate is 1% per year, what will be the purchasing power of the accumulated amount (with respect to today's dollars) at the end of 15 years?

14.38 Harmony Corporation plans to set aside $60,000 per year beginning 1 year from now for replacing equipment 5 years from now. What will be the purchasing power (in terms of current-value dollars) of the amount accumulated, if the investment grows by 10% per year, but inflation averages 4% per year?

14.39 The strategic plan of a solar energy company that manufactures high-efficiency solar cells includes an expansion of its physical plant in 4 years. The engineer in charge of planning estimates the expenditure required now to be $8 million, but in 4 years, the cost will be higher by an amount equal to the inflation rate. If the company sets aside $7,000,000 now into an account that earns interest at 7% per year, what will the inflation rate have to be in order for the company to have exactly the right amount of money for the expansion?

14.40 Five years ago, an industrial engineer deposited $10,000 into an account and left it undisturbed until now. The account is now worth $25,000.
   (a) What was the overall ROR during the 5 years?
   (b) If the inflation over that period was 4% per year, what was the real ROR?
   (c) What is the buying power of the $25,000 now compared to thc buying power 5 years ago?

14.41 A Toyota Tundra can be purchased today for $32,350. A civil engineering firm is going to need three more trucks in 2 years because of a land development contract it just won. If the price of the truck increases exactly in accordance with an estimated inflation rate of 3.5% per year, determine how much the three trucks will cost in 2 years.

14.42 The Nobel Prize is administered by The Nobel Foundation, a private institution that was founded in 1900 based on the will of Alfred Nobel, the inventor of dynamite. In part, his will stated: "The capital shall be invested by my executors in safe securities and shall constitute a fund, the interest on which shall be annually distributed in the form of prizes to those who, during the preceding year, shall have conferred the greatest benefit on mankind." The will further stated that the prizes were to be awarded in physics, chemistry, peace, physiology or medicine, and literature. In addition to a gold medal and a diploma, each recipient receives a substantial sum of money that depends on the Foundation's income that year. The first Nobel Prize was awarded in 1901 in the amount of $150,000. In 1996, the award was $653,000; it was $1.4 million in 2009.
   (a) If the increase between 1996 and 2009 was strictly due to inflation, what was the average inflation rate per year during that 13-year period?
   (b) If the Foundation expects to invest money with a return of 5% above the inflation rate, how much will a laureate receive in 2020, provided the inflation rate averages 3% per year between 2009 and 2020?

14.43 Timken Roller Bearing is a manufacturer of seamless tubes for drill bit collars. The company is planning to add larger-capacity robotic arms to one of its assembly lines 3 years from now. If it is done now, the cost of the equipment with installation is $2.4 million. If the company's real MARR is 15% per year, determine the equivalent amount the company can spend 3 years from now in then-current dollars if the inflation rate is 2.8% per year.

14.44 The data below show two patterns of inflation that are exactly the opposite of each other over a 20-year time period.
   (a) If each machine costs $10,000 in year 0 and they both increase in cost exactly in accordance with the inflation rate, how much will each machine cost at the end of year 20?
   (b) What is the average inflation rate over the time period for machine A (that is, what single inflation rate would result in the same final cost for machine A)?
   (c) In which years will machine A cost more than machine B?

Year	Machine A, %	Machine B, %
1	10	2
2	10	2
3	2	10
4	2	10
5	10	2
6	10	2
7	2	10
8	2	10
⋮	⋮	⋮
19	2	10
20	2	10

14.45 Factors that increase costs and prices—especially for materials and manufacturing costs sensitive to market, technology, and labor availability—can be considered separately using the real interest rate i, the inflation rate f, and additional increases that grow at a geometric rate g. The future amount is calculated based on a current estimate by using the relation

$$F = P(1 + i)^n(1 + f)^n(1 + g)^n$$
$$= P[(1 + i)(1 + f)(1 + g)]^n$$

The product of the first two terms enclosed in parentheses results in the inflated interest rate i_f. The geometric rate is the same one used in the geometric series (Chapter 2). It commonly applies to maintenance and repair cost increases as machinery ages. This is over and above the inflation rate. If the current cost to manufacture an electronic subcomponent is $300,000 per year, what is the equivalent cost in 3 years, provided the average annual rates for i, f, and g are 10%, 3%, and 2%, respectively?

Capital Recovery with Inflation

14.46 An electric utility is considering two alternatives for satisfying state regulations regarding pollution control for one of its generating stations. This particular station is located at the outskirts of a major U.S. city and a short distance from a large city in a neighboring country. The station is currently producing excess VOCs and oxides of nitrogen. Two plans have been proposed for satisfying the regulators. Plan A involves replacing the burners and switching from fuel oil to natural gas. The cost of the option will be $300,000 initially and an extra $900,000 per year in fuel costs. Plan B involves going to the foreign city and running gas lines to many of the "backyard" brick-making sites that now use wood, tires, and other combustible waste materials for firing the bricks. The idea behind plan B is that by reducing the particulate pollution responsible for smog in the neighboring city, there would be greater benefit to U.S. citizens than would be achieved through plan A. The initial cost of plan B will be $1.2 million for installation of the lines. Additionally, the electric company would subsidize the cost of gas for the brick makers to the extent of $200,000 per year. Extra air monitoring associated with this plan will cost an additional $150,000 per year. For a 10-year project period and no salvage value for either plan, which one should be selected on the basis of an annual worth analysis at a real interest rate of 7% per year and an inflation rate of 4% per year?

14.47 A chemical engineer who is smitten with wanderlust wants to build a reserve fund that will be sufficient to permit her to take time off from work to travel around the world. Her goal is to save enough money over the next 3 years so that when she begins her trip, the amount she has accumulated will have the same buying power as $72,000 today. If she expects to earn a market rate of 12% per year on her investments and inflation averages 5% per year over the 3 years, how much must she save each year to reach her goal?

14.48 An entrepreneur engaged in wildcat oil well drilling is seeking investors who will put up $500,000 for an opportunity to reap high returns if the venture is successful. The prospectus states that a return of at least 22% per year for 5 years is likely. How much will the investors have to receive each year to recover their money if an inflation rate of 5% per year is to be included in the calculation?

14.49 Veri-Trol, Inc. manufactures in situ calibration verification systems that confirm flow measurement accuracies without removing the meters. The company has decided to modify the main assembly line with one of the enhancement processes shown below. If the company's real MARR is 15% per year, which process has the lower annual cost when inflation of 5% per year is considered?

	Process X	Process Y
First cost, $	−65,000	−90,000
Operating cost, $ per year	−40,000	−34,000
Salvage value, $	0	10,000
Life, years	5	5

14.50 Johnson Thermal Products used austenitic nickel-chromium alloys to manufacture resistance heating wire. The company is considering a new annealing-drawing process to reduce costs. If the new process will cost $3.7 million now, how much must be saved each year to recover the investment in 5 years if the company's MARR is a real 12% per year and the inflation rate is 3% per year?

14.51 The costs associated with a small X-ray inspection system are $40,000 now and $24,000 per year, with a $6000 salvage value after 3 years. Determine the equivalent cost of the system if the real interest rate is 10% per year and the inflation rate is 4% per year.

14.52 Maintenance costs for pollution control equipment on a pulverized coal cyclone furnace are expected to be $180,000 now and another $70,000 three years from now. The CFO of Monongahela Power wants to know the equivalent annual cost of the equipment in years 1 through 5. If the company uses a real interest rate of 9% per year and the inflation rate averages 3% per year, what is the equivalent annual cost of the equipment?

14.53 MetroKlean LLC, a hazardous waste soil cleaning company, borrowed $2.5 million for 5 years to finance start-up costs for a new project involving site reclamation. The company expects to earn a real rate of return of 20% per year. The average inflation rate is 5% per year.
   (a) Determine the capital recovery required each year with inflation considered.
   (b) Determine the capital recovery if the company is satisfied with accumulating $2.5 million at the end of the 5 years with inflation considered.
   (c) Determine the capital recovery in part (b) without considering inflation.

ADDITIONAL PROBLEMS AND FE EXAM REVIEW QUESTIONS

14.54 Inflation occurs when:
 (a) Productivity increases
 (b) The value of the currency decreases
 (c) The value of the currency increases
 (d) The price of gold decreases

14.55 "Then-current" (future) dollars can be converted into constant-value dollars by:
 (a) Multiplying them by $(1 + f)^n$
 (b) Multiplying them by $(1 + f)^n/(1 + i)^n$
 (c) Dividing them by $(1 + f)^n$
 (d) Dividing them by $(1 + i_f)^n$

14.56 To calculate how much something will cost if you expect its cost to increase by exactly the inflation rate, you should:
 (a) Multiply by $(1 + f)^n$
 (b) Multiply by $[(1 + f)^n/(1 + i)^n]$
 (c) Divide by $(1 + f)^n$
 (d) Multiply by $(1 + i_f)^n$

14.57 If the market interest rate is 16% per year when the inflation rate is 9% per year, the real interest rate is closest to:
 (a) 6.4%
 (b) 7%
 (c) 9%
 (d) 15.6%

14.58 The market interest rate i_f is 12% per year, compounded semiannually. For an inflation rate of 2% per 6 months, the effective semiannual interest rate is closest to:
 (a) 2%
 (b) 3%
 (c) 4%
 (d) 6%

14.59 Construction equipment has a cost today of $40,000. If its cost has increased only by the inflation rate of 6% per year, when the market interest rate was 10% per year, its cost 10 years ago was closest to:
 (a) $15,420
 (b) $22,335
 (c) $27,405
 (d) $71,630

14.60 The amount of money that would be accumulated now from an investment of $1000 25 years ago at a market rate of 5% per year and an inflation rate averaging 2% per year over that time period is closest to:
 (a) $1640
 (b) $3385
 (c) $5430
 (d) Over $5500

14.61 If $1000 is invested now, the number of future dollars required 10 years from now to earn a real interest rate of 6% per year, when the inflation rate is 4% per year, is closest to:
 (a) $1480
 (b) $1790
 (c) $2650
 (d) Over $2700

14.62 If you expect to receive a gift of $50,000 six years from now, the present worth of the gift at a real interest rate of 4% per year and an inflation rate of 3% per year is closest to:
 (a) $27,600
 (b) $29,800
 (c) $33,100
 (d) $37,200

14.63 New state-mandated emission testing equipment for annual inspection of automobiles at Charlie's Garage has a first cost of $30,000, an annual operating cost of $7000, and a $5000 salvage value after its 20-year life. For a real interest rate of 5% per year and an inflation rate of 4% per year, the annual capital recovery requirement for the equipment (in future dollars) is determined by:
 (a) $\text{AW} = -30,000(A/P,4\%,20) - 7000 + 5000(A/F,4\%,20)$
 (b) $\text{AW} = -30,000(A/P,5\%,20) - 7000 + 5000(A/F,5\%,20)$
 (c) $\text{AW} = -30,000(A/P,9\%,20) - 7000 + 5000(A/F,9\%,20)$
 (d) $\text{AW} = -30,000(A/P,9.2\%,20) - 7000 + 5000(A/F,9.2\%,20)$

14.64 Temporary price deflation may occur in specific sectors of the economy for all of the following reasons except:
 (a) Improved technology
 (b) Excessive demand
 (c) Dumping
 (d) Increased productivity

CASE STUDY

INFLATION VERSUS STOCK AND BOND INVESTMENTS

Background

The savings and investments that an individual maintains should have some balance between equity (corporate stocks that rely on market growth and dividend income) and fixed-income investments (bonds that pay dividends to the purchaser and a guaranteed amount upon maturity). When inflation is moderately high, bonds offer a low return relative to stocks, because the potential for market growth is not present with bonds. Additionally, the forces of inflation make the dividends worth less in future years, because there is no inflation adjustment made in the amount the dividend pays as time passes. However, bonds do offer a steady income that may be important to an individual, and they serve to preserve the principal invested in the bond, because the face value is returned at maturity.

Information

Earl is an engineer who wants a predictable flow of money for travel and vacations. He has a collection of stocks in his retirement portfolio, but no bonds. He has accumulated a total of $50,000 of his own funds in low-yielding savings accounts and wants to improve his long-term return from this nonretirement program "nest egg." He can choose additional stocks or bonds, but has decided to not split the $50,000 between the two forms of investments. There are two choices he has outlined, with the best estimates he can make at this time. He assumes the effects of federal and state income taxes will be the same for both forms of investment.

Stock purchase: Stocks purchased through a mutual fund would pay an estimated 2% per year dividend and appreciate in value at 5% per year.

Bond purchase: If he purchased a bond, he would have a predictable income of 5% per year and the $50,000 face value after the 12-year maturity period.

Case Study Questions

The analysis that Earl has laid out has the following questions. Can you answer them for him for both choices?

1. What is the overall rate of return after 12 years?
2. If he decided to sell the stock or bond immediately after the fifth annual dividend, what is his minimum selling price to realize a 7% real return? Include an adjustment of 4% per year for inflation.
3. If Earl needed some money in the future, say, immediately after the fifth dividend payment, what would be the minimum selling price in future dollars, if he were only interested in recovering an amount that maintained the purchasing power of the original price?
4. As a follow-on to question 3, what happens to the selling price (in future dollars) 5 years after purchase, if Earl is willing to remove (net out) the future purchasing power of each of the dividends in the computation to determine the required selling price 5 years hence?
5. Earl plans to keep the stocks or bonds for 12 years, that is, until the bond matures. However, he wants to make the 7% per year real return and make up for the expected 4% per year inflation. For what amount must he sell the stocks after 12 years, or buy the bonds now to ensure he realizes this return? Do these amounts seem reasonable to you, given your knowledge of the way that stocks and bonds are bought and sold?

CHAPTER 15

Cost Estimation and Indirect Cost Allocation

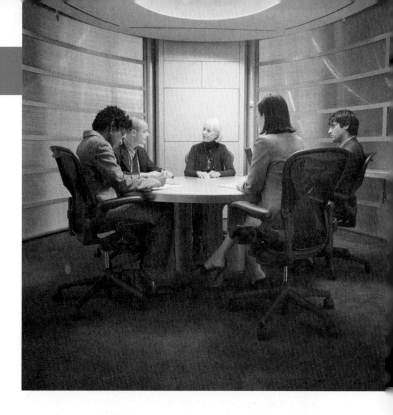

LEARNING OUTCOMES

Purpose: Make cost estimates using different methods; demonstrate the allocation of indirect costs using traditional and activity-based costing rates.

SECTION	TOPIC	LEARNING OUTCOME
15.1	Approach	• Explain the bottom-up and design-to-cost (top-down) approaches to cost estimation.
15.2	Unit method	• Use the unit method to make a preliminary cost estimate.
15.3	Cost index	• Use a cost index to estimate a present cost based on historical data.
15.4	Cost capacity	• Use a cost-capacity equation to estimate component, system, or plant costs.
15.5	Factor method	• Estimate total plant cost using the factor method.
15.6	Indirect cost rates	• Allocate indirect costs using traditional indirect cost rates.
15.7	ABC allocation	• Use the Activity-Based Costing (ABC) method to allocate indirect costs.
15.8	Ethics and profit	• Describe how biased estimation can become an ethical dilemma.

U p to this point, cost and revenue cash flow values have been stated or assumed as known. In reality, they are not; they must be estimated. This chapter explains what cost estimation involves and applies cost estimation techniques. **Cost estimation** is important in all aspects of a project, but especially in the stages of project conception, preliminary design, detailed design, and economic analysis. When a project is developed in the private or the public sector, questions about costs and revenues will be posed by individuals representing many different functions: management, engineering, construction, production, quality, finance, safety, environmental, legal, and marketing, to name some. In engineering practice, the estimation of costs receives much more attention than revenue estimation; costs are the topic of this chapter.

Unlike direct costs for labor and materials, indirect costs are not easily traced to a specific department, machine, or processing line. Therefore, **allocation of indirect costs** for functions such as utilities, safety, management and administration, purchasing, and quality is made using some rational basis. Both the traditional method of allocation and the Activity-Based Costing (ABC) method are covered in this chapter.

Online
Presentation

15.1 Understanding How Cost Estimation Is Accomplished ● ● ●

Cost estimation is a major activity performed in the initial stages of virtually every effort in industry, business, and government. In general, most cost estimates are developed for either a *project* or a *system;* however, combinations of these are very common. A **project** usually involves physical items, such as a building, bridge, manufacturing plant, or offshore drilling platform, to name just a few. A **system** is usually an operational design that involves processes, services, software, and other nonphysical items. Examples might be a purchase order system, a software package, an Internet-based remote-control system, or a health care delivery system. Of course, many projects will have major elements that are not physical, so estimates of both types must be developed. For example, consider a computer network system. There would be no operational system if only the costs of computer hardware plus wire and wireless connectors were estimated; it is equally important to estimate the software, personnel, and maintenance costs.

Thus far virtually all cash flow estimates in the examples, problems, progressive examples, and case studies were stated or assumed to be known. In real-world practice, the cash flows for costs and revenues must be estimated prior to the evaluation of a project or comparison of alternatives. We concentrate on cost estimation because costs are the primary values estimated for the economic analysis. Revenue estimates utilized by engineers are usually developed in marketing, sales, and other departments.

Costs are comprised of **direct costs** and **indirect costs.** Normally direct costs are estimated with some detail, then the indirect costs are added using standard rates and factors. However, direct costs in many industries, including manufacturing and assembly settings, have become a small percentage of overall product cost, while indirect costs have become much larger. Accordingly, many industrial settings require some estimating for indirect costs as well. Indirect cost allocation is discussed in detail in later sections of this chapter. Primarily, direct costs are discussed here.

Direct/Indirect costs

Because cost estimation is a complex activity, the following questions form a structure for the discussion that follows.

- What cost *components* must be estimated?
- What *approach* to cost estimation will be applied?
- How *accurate* should the estimates be?
- What estimation *techniques* will be utilized?

Costs to Estimate   If a project revolves around a single piece of equipment, for example, an industrial robot, the *cost components* will be significantly simpler and fewer than the components for a complete system such as the manufacturing and testing line for a new product. Therefore, it is important to know up front how much the cost estimation task will involve. Examples of cost components are the first cost P and the annual operating cost (AOC), also called the M&O costs (maintenance and operating) of equipment. Each component will have several *cost elements*. Listed below are sample elements of the first cost and AOC components.

> *First cost component P:*
>
> Elements: Equipment cost
>
> Delivery charges
>
> Installation cost
>
> Insurance coverage
>
> Initial training of personnel for equipment use

Delivered-equipment cost is the sum of the first two elements in the list above; **installed-equipment cost** adds the third element.

> *AOC component* (part of the equivalent annual cost A):
>
> Elements: Direct labor cost for operating personnel
>
> Direct materials
>
> Maintenance costs (daily, periodic, repairs, etc.)
>
> Rework and rebuild

Some of these elements, such as equipment cost, can be determined with high accuracy; others, such as maintenance costs, are harder to estimate. When costs for an entire system must be estimated, the number of cost components and elements is likely to be in the hundreds. It is then necessary to prioritize the estimation tasks.

For familiar projects (houses, office buildings, highways, and some chemical plants) there are standard cost estimation software packages available. For example, state highway departments utilize software that prompts for the cost components (bridges, pavement, cut-and-fill profiles, etc.) and estimates costs with time-proven, built-in relations. Once these components are estimated, exceptions for the specific project are added. However, there are no "canned" software packages for a large percentage of industrial, business, service and public sector projects.

Cost Estimation Approach   Historically in industry, business, and the public sector, a **bottom-up** approach to cost estimation was applied. For a simple rendition of this approach, see Figure 15–1 (left side). The progression is as follows: cost components and their elements are identified, cost elements are estimated, and estimates are summed to obtain total direct cost. The price is then determined by adding indirect costs and the profit margin, which is usually a percentage of the total cost.

> The **bottom-up approach** treats the required price as an **output variable** and the cost estimates as input variables. This approach works well when competition is not a dominant factor in pricing the product or service.

Figure 15–1 (right side) shows a simplistic progression for the **design-to-cost,** or top-down, approach. The competitive price establishes the target cost.

> The **design-to-cost,** or **top-down, approach** treats the competitive price as an **input variable** and the cost estimates as output variables. This approach is useful in encouraging innovation, new design, manufacturing process improvement, and efficiency. These are some of the essentials of *value engineering* and value-added systems engineering.

This approach places greater emphasis on the accuracy of the price estimation activity. The target cost must be realistic, or else it can be a disincentive to design and engineering staff. The design-to-cost approach is best applied in the early stages of a new or enhanced product design. The detailed design and specific equipment options are not yet known, but the price estimates assist in establishing target costs for different components.

Usually, the resulting approach is some combination of these two philosophies. However, it is helpful to understand up front what approach is to be emphasized. Historically, the bottom-up approach was more predominant in Western engineering cultures, especially in North America. The design-to-cost approach is considered routine in Eastern engineering cultures; however,

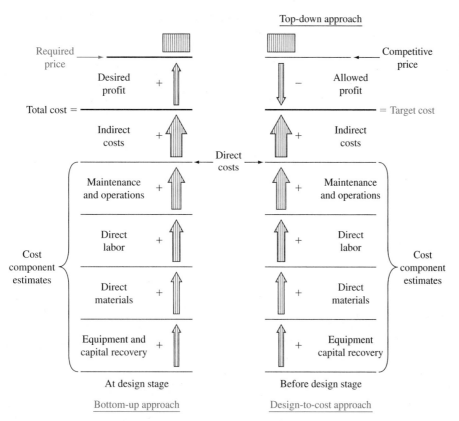

Figure 15–1
Simplified cost estimation processes for bottom-up and top-down approaches.

globalization of engineering design has speeded the adoption of the design-to-cost approach worldwide.

Accuracy of the Estimates   No cost estimates are expected to be exact; however, they are expected to be reasonable and accurate enough to support economic scrutiny. The accuracy required increases as the project progresses from preliminary design to detailed design and on to economic evaluation. Cost estimates made before and during the preliminary design stage are expected to be good "first-cut" estimates that serve as input to the project budget.

When utilized at early and conceptual design stages, estimates are referred to as **order-of-magnitude** estimates and generally range within ±20% of actual cost. At the detailed design stage, cost estimates are expected to be accurate enough to support economic evaluation for a go–no go decision. Every project setting has its own characteristics, but a range of ±5% of actual costs is expected at the detailed design stage. Figure 15–2 shows the general range of estimate accuracy for the construction cost of a building versus time spent in preparing the estimate. Obviously, the desire for better accuracy has to be balanced against the cost of obtaining it.

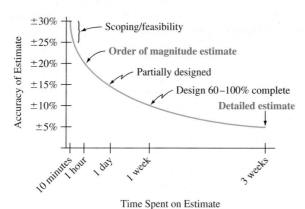

Figure 15–2
Characteristic curve of estimate accuracy versus time spent to estimate construction cost of a building.

Cost Estimation Techniques   Methods such as expert opinion and comparison with comparable installations serve as excellent estimators. The use of the *unit method* and *cost indexes* base the present estimate on past cost experiences, with inflation considered. Models such as *cost-capacity equations* and the *factor method* are simple mathematical techniques applied at the preliminary design stage. They are called **cost estimating relationships (CERs).** There are many additional methods discussed in the handbooks and publications of different industries.

Most cost estimates made in a professional setting are accomplished in part or wholly using software packages linked to updated databases that contain cost indexes and rates for the locations, products, or processes being studied. There are a wide variety of estimators, cost trackers, and cost compliance software systems, most of them developed for specific industries. Corporations usually standardize on one or two packages to ensure consistency over time and projects.

Online Presentation

15.2 Unit Method ● ● ○

The **unit method** is a popular preliminary estimation technique applicable to virtually all professions. The total estimated cost C_T is obtained by multiplying the number of units N by a per unit cost factor u.

$$C_T = u \times N \qquad\qquad [15.1]$$

Unit cost factors must be updated frequently to remain current with changing costs, areas, and inflation. Some sample unit cost factors (and values) are

Total average cost of operating an automobile (52¢ per mile)

Cost to bury fiber cable in a suburban area ($30,000 per mile)

Cost to construct a parking space in a parking garage ($4500 per space)

Cost of constructing interstate highway ($6.2 million per mile)

Cost of house construction per livable area ($225 per square foot)

Applications of the unit method to estimate costs are easily found. If house construction costs average $225 per square foot, a preliminary cost estimate for an 1800-square-foot house, using Equation [15.1], is $405,000. Similarly, a 200-mile trip should cost about $104 for the car only at 52¢ per mile.

When there are several components to a project or system, the unit cost factors for each component are multiplied by the amount of resources needed, and the results are **summed** to obtain the total cost C_T. This is illustrated in Example 15.1.

EXAMPLE 15.1

Justin, an ME with Dynamic Castings, has been asked to make a preliminary estimate of the total cost to manufacture 1500 sections of high-pressure gas pipe using an advanced centrifugal casting method. Since a ±20% estimate is acceptable at this preliminary stage, a unit method estimate is sufficient. Use the following resource and unit cost factor estimates to help Justin.

Materials: 3000 tons at $45.90 per ton

Machinery and tooling: 1500 hours at $120 per hour

Direct labor in plant:

    Casting and treating: 3000 hours at $55 per hour

    Finishing and shipping: 1200 hours at $45 per hour

Indirect labor: 400 hours at $75 per hour

Solution

Apply Equation [15.1] to each of the five areas and sum the results to obtain the total cost estimate of $566,700. Table 15–1 provides the details.

TABLE 15–1	Total Cost Estimate Using Unit Cost Factors for Several Resource Areas, Example 15.1		
Resource	**Amount N**	**Unit Cost Factor u, $**	**Cost Estimate, u × N, $**
Materials	3000 tons	45.90 per ton	137,700
Machinery, tooling	1500 hours	120 per hour	180,000
Labor, casting	3000 hours	55 per hour	165,000
Labor, finishing	1200 hours	45 per hour	54,000
Labor, indirect	400 hours	75 per hour	30,000
Total cost estimate			**566,700**

15.3 Cost Indexes ● ● ●

Online Presentation

This section explains indexes and their use in cost estimation. An *index* is a ratio or other number based on observation and used as an indicator or measure. A preliminary cost estimate is often based on a cost index.

A **cost index** is a ratio of the cost of something today to its cost sometime in the past. The index is **dimensionless** and **measures relative cost change over time.** Because these indexes are sensitive to technological change, the predefined quantity and quality of elements used to define the index may be hard to retain over time, thus causing "index creep." Timely updating of the index is very important.

One such index that most people are familiar with is the Consumer Price Index (CPI), which shows the relationship between present and past costs for many of the things that "typical" consumers must buy. This index includes such items as rent, food, transportation, and certain services. Other indexes track the costs of equipment, and goods and services that are more pertinent to the engineering disciplines. Table 15–2 is a listing of some of the more common indexes.

TABLE 15–2	Types and Sources of Various Cost Indexes
Type of Index	**Source**
Overall prices	
Consumer (CPI)	Bureau of Labor Statistics
Producer (wholesale)	U.S. Department of Labor
Construction	
Chemical plant overall	*Chemical Engineering*
Equipment, machinery, and supports	
Construction labor	
Buildings	
Engineering and supervision	
Engineering News Record overall	*Engineering News Record* (*ENR*)
Construction	
Building	
Common labor	
Skilled labor	
Materials	
EPA treatment plant indexes	Environmental Protection Agency
Large-city advanced treatment (LCAT)	and *ENR*
Small-city conventional treatment (SCCT)	
Federal highway	
Contractor cost	
Equipment	
Marshall and Swift (M&S) overall	Marshall & Swift
M&S specific industries	
Labor	
Output per worker-hour by industry	U.S. Department of Labor

	TABLE 15–3 Values for Selected Indexes		
Year	**CE** Plant Cost Index	**ENR** Construction Cost Index	**M&S** Equipment Cost Index
1995	381.1	5471	1027.5
1996	381.8	5620	1039.2
1997	386.5	5826	1056.8
1998	389.5	5920	1061.9
1999	390.6	6059	1068.3
2000	394.1	6221	1089.0
2001	394.3	6343	1093.9
2002	395.6	6538	1104.2
2003	401.7	6694	1123.6
2004	444.2	7115	1178.5
2005	468.2	7446	1244.5
2006	499.6	7751	1302.3
2007	525.4	7967	1373.3
2008	575.4	8310	1449.3
2009	521.9	8570	1468.6
2010 (midyear)	555.3	8837	1461.3

The general equation for updating costs through the use of a cost index over a period from time $t = 0$ (base) to another time t is

$$C_t = C_0 \left(\frac{I_t}{I_0} \right) \qquad [15.2]$$

where C_t = estimated cost at present time t
   C_0 = cost at previous time t_0
   I_t = index value at time t
   I_0 = index value at time t_0

Generally, the indexes for equipment and materials are made up of a mix of components that are assigned certain weights, with the components sometimes further subdivided into more basic items. For example, the equipment, machinery, and support component of the chemical plant cost index is subdivided into process machinery, pipes, valves and fittings, pumps and compressors, and so forth. These subcomponents, in turn, are built up from even more basic items such as pressure pipe, black pipe, and galvanized pipe. Table 15–3 shows the *Chemical Engineering* plant cost index (CEPCI), the *Engineering News Record (ENR)* construction cost index, and the Marshall and Swift (M&S) equipment cost index for several years. The base period of 1957 to 1959 is assigned a value of 100 for the CEPCI, 1913 = 100 for the *ENR* index, and 1926 = 100 for the M&S equipment cost index.

Current and past values of several of the indexes may be obtained from the Internet (usually for a fee). For example, the *CE* plant cost index is available at www.che.com/pci. The *ENR* construction cost index is found at www.construction.com. This latter site offers a comprehensive series of construction-related resources, including several *ENR* cost indexes and cost estimation systems. A website used by many engineering professionals in the form of a "technical chat room" for all types of topics, including estimation, is www.eng-tips.com.

EXAMPLE 15.2

In evaluating the feasibility of a major construction project, an engineer is interested in estimating the cost of skilled labor for the job. The engineer finds that a project of similar complexity and magnitude was completed 5 years ago at a skilled labor cost of $360,000. The *ENR* skilled labor index was 3496 then and is now 5127. What is the estimated skilled labor cost for the new project?

Solution

The base time t_0 is 5 years ago. Using Equation [15.2], the present cost estimate is

$$C_r = 360{,}000\left(\frac{5127}{3496}\right)$$
$$= \$527{,}952$$

In the manufacturing and service industries, tabulated cost indexes are not readily available. The cost index will vary, perhaps with the region of the country, the type of product or service, and many other factors. When estimating costs for a manufacturing system, for example, it is often necessary to develop the cost index for high-priority items such as subcontracted components, selected materials, and labor costs. The development of the cost index requires the actual cost at different times for a prescribed quantity and quality of the item. The **base period** is a selected time when the index is defined with a basis value of 100 (or 1). The index each year (period) is determined as the cost divided by the base-year cost and multiplied by 100 (or 1). Future index values may be forecast using simple extrapolation or more refined mathematical techniques, such as time-series analysis.

EXAMPLE 15.3

Sean, owner of Alamo Pictures, makes fact-based documentaries about the Old West and sells them in a variety of outlets, by mail, and online. He has decided to expand into new areas and wants to make cost estimates for three of the more significant labor costs involved in making these types of films. The Director of Finance of Alamo Pictures generated the annual average hourly costs (Table 15–4).

(a) Make 2008 the base year, and determine the cost indexes using a basis of 1.00. Comment on the trend of each index over the years.
(b) Sean expects to utilize a lot of graphics services in 2014 for a planned documentary; however, it is the most rapidly increasing cost component. The cost in 2010 was $78 per hour; assume a worst-case scenario is that the graphics index continues the same arithmetic trend it had from 2010 to 2011. Determine the hourly cost that he should budget for in 2014.

Solution

(a) For each type of service, calculate I_t/I_0 where $t = 2005, 2006, \ldots$ with 2008 as the base year 0. Table 15–5 presents the indexes. Observations about trend are as follows:

    Graphics labor cost: Constantly increasing over all years.

    Stuntmen labor cost: Rising until 2009, then stable.

    Actors labor cost: Higher in 2007 and 2008; comparatively lower and stable in other years.

TABLE 15–4   Average Hourly Costs for Three Services, Example 15.3

Type of Service	Cost of Service, Average $ per Hour						
	2005	2006	2007	2008	2009	2010	2011
Graphics	50	56	65	67	70	78	90
Stuntmen	50	55	60	70	87	83	85
Actors	80	80	90	90	80	75	85

TABLE 15–5   Index Values with 2008 as Base Year, Example 15.3

Type of Service	Index I_t/I_0						
	2005	2006	2007	2008	2009	2010	2011
Graphics	0.75	0.84	0.97	1.00	1.04	1.16	1.34
Stuntmen	0.71	0.79	0.86	1.00	1.24	1.19	1.21
Actors	0.89	0.89	1.00	1.00	0.89	0.83	0.94

(b) The index in 2010 is 1.16. The increase to 2011 is 0.18; the index value in 2014 will be 1.34 + 3(0.18) = 1.88. Equation [15.2] finds the expected, worst-case cost in 2014.

$$C_{2014} = C_{2010}(I_{2014}/I_{2010}) = 78(1.88/1.16)$$
$$= 78(1.62)$$
$$= \$126 \text{ per hour}$$

Online Presentation

15.4 Cost-Estimating Relationships: Cost-Capacity Equations ●●○

Design variables (speed, weight, thrust, physical size, etc.) for plants, equipment, and construction are determined in the early design stages. **Cost-estimating relationships (CERs)** use these design variables to predict costs. Thus, a CER is generically different from the cost index method, because the index is based on the cost history of a defined quantity and quality of a variable.

One of the most widely used CER models is a **cost-capacity equation.** As the name implies, an equation relates the cost of a component, system, or plant to its capacity. This is also known as the *power law and sizing model.* Since many cost capacity equations plot as a straight line on log-log paper, a common form is

$$C_2 = C_1\left(\frac{Q_2}{Q_1}\right)^x \qquad [15.3]$$

where C_1 = cost at capacity Q_1
C_2 = cost at capacity Q_2
x = correlating exponent

The value of the exponent for various components, systems, or entire plants can be obtained or derived from a number of sources, including *Plant Design and Economics for Chemical Engineers, Preliminary Plant Design in Chemical Engineering, Chemical Engineers' Handbook,* technical journals (especially *Chemical Engineering*), the U.S. Environmental Protection Agency, professional or trade organizations, consulting firms, handbooks, and equipment companies. Table 15–6 is a partial listing of typical values of the exponent for various units. When an exponent value for a particular unit is not known, it is common practice to use the value of $x = 0.6$. In fact, in the chemical processing industry, Equation [15.3] is referred to as the six-tenths model.

The exponent x in the cost-capacity equation is commonly in the range $0 < x \leq 1$.

If $x < 1$, *economies of scale* provide a cost advantage for larger sizes.
If $x = 1$, a *linear* relationship is present.
If $x > 1$, there are *diseconomies of scale* present in that a larger size is more costly than that of a linear relation.

It is especially powerful to combine the time adjustment of the cost index (I_t/I_0) from Equation [15.2] with a cost-capacity equation to estimate costs that change over time. If the index is embedded into the cost-capacity computation in Equation [15.3], the cost at time t and capacity level 2 may be written as the product of two independent terms.

$$C_{2,t} = \text{(cost at time 0 of level 2)} \times \text{(time adjustment cost index)}$$

$$= \left[C_{1,0}\left(\frac{Q_2}{Q_1}\right)^x\right]\left(\frac{I_t}{I_0}\right)$$

This is commonly expressed without the time subscripts. Thus,

$$C_2 = C_1\left(\frac{Q_2}{Q_1}\right)^x\left(\frac{I_t}{I_0}\right) \qquad [15.4]$$

The following example illustrates the use of this relation.

TABLE 15-6	Sample Exponent Values for Cost-Capacity Equations	
Component/System/Plant	Size Range	Exponent
Activated sludge plant	1–100 MGD	0.84
Aerobic digester	0.2–40 MGD	0.14
Blower	1000–7000 ft/min	0.46
Centrifuge	40–60 in	0.71
Chlorine plant	3000–350,000 tons/year	0.44
Clarifier	0.1–100 MGD	0.98
Compressor, reciprocating (air service)	5–300 hp	0.90
Compressor	200–2100 hp	0.32
Cyclone separator	20–8000 ft^3/min	0.64
Dryer	15–400 ft^2	0.71
Filter, sand	0.5–200 MGD	0.82
Heat exchanger	500–3000 ft^2	0.55
Hydrogen plant	500–20,000 scfd	0.56
Laboratory	0.05–50 MGD	1.02
Lagoon, aerated	0.05–20 MGD	1.13
Pump, centrifugal	10–200 hp	0.69
Reactor	50–4000 gal	0.74
Sludge drying beds	0.04–5 MGD	1.35
Stabilization pond	0.01–0.2 MGD	0.14
Tank, stainless	100–2000 gal	0.67

Note: MGD = million gallons per day; hp = horsepower; scfd = standard cubic feet per day.

EXAMPLE 15.4

The total design and construction cost for a digester to handle a flow rate of 0.5 million gallons per day (MGD) was $1.7 million in 2010. Estimate the cost today for a flow rate of 2.0 MGD. The exponent from Table 15–6 for the MGD range of 0.2 to 40 is 0.14. The cost index in 2010 of 131 has been updated to 225 for this year.

Solution

Equation [15.3] can estimate the cost of the larger system in 2010, but it must be updated by the cost index to today's dollars. Equation [15.4] performs both operations at once. The estimated cost is

$$C_2 = 1,700,000 \left(\frac{2.0}{0.5}\right)^{0.14} \left(\frac{225}{131}\right)$$
$$= 1,700,000(1.214)(1.718) = \$3.546 \text{ million}$$

15.5 Cost-Estimating Relationships: Factor Method ● ● ●

Another widely used model for preliminary cost estimates of process plants is called the **factor method.** While the methods discussed above can be used to estimate the costs of major items of equipment, processes, and the total plant costs, the factor method was developed specifically for **total plant costs.** The method is based on the premise that fairly reliable total plant costs can be obtained by multiplying the cost of the major equipment by certain factors. Since major equipment costs are readily available, rapid plant estimates are possible if the appropriate factors are known. These factors are commonly referred to as *Lang factors* after Hans J. Lang, who first proposed the method.

In its simplest form, the factor method is expressed in the same form as the unit method

$$C_T = hC_E \qquad\qquad [15.5]$$

where C_T = total plant cost
   h = overall cost factor or sum of individual cost factors
   C_E = total cost of major equipment

The h may be one overall cost factor or, more realistically, the sum of individual cost components such as construction, maintenance, direct labor, materials, and indirect cost elements. This follows the cost estimation approaches presented in Figure 15–1.

Direct/Indirect costs

In his original work, Lang showed that *direct cost factors* and *indirect cost factors* can be combined into one overall factor for some types of plants as follows: solid process plants, 3.10; solid-fluid process plants, 3.63; and fluid process plants, 4.74. These factors reveal that the total installed-plant cost is many times the first cost of the major equipment.

EXAMPLE 15.5

An engineer with Valero Petroleum has learned that an expansion of the solid-fluid process plant is expected to have a delivered equipment cost of $2.08 million. If the overall cost factor for this type of plant is 3.63, estimate the plant's total cost.

Solution

The total plant cost is estimated by Equation [15.5].

$$C_T = 3.63(2,080,000)$$
$$= \$7,550,400$$

Subsequent refinements of the factor method have led to the development of separate factors for direct and indirect cost components. Direct costs as discussed in Section 15.1 are specifically identifiable with a product, function, or process. Indirect costs are not directly attributable to a single function, but are shared by several because they are necessary to perform the overall objective. Examples of indirect costs are general administration, computer services, quality, safety, taxes, security, and a variety of support functions. The factors for both direct and indirect costs are sometimes developed for use with *delivered-equipment costs* and other times for *installed-equipment costs,* as defined in Section 15.1. In this text, we assume that all factors apply to delivered-equipment costs, unless otherwise specified.

For indirect costs, some of the factors apply to equipment costs only, while others apply to the total direct cost. In the former case, the simplest procedure is to add the direct and indirect cost factors before multiplying by the delivered-equipment cost. The overall cost factor h can be written as

$$h = 1 + \sum_{i=1}^{n} f_i \qquad\qquad [15.6]$$

where f_i = factor for each cost component
   i = 1 to n components, including indirect cost

If the indirect cost factor is applied to the total direct cost, only the direct cost factors are added to obtain h. Therefore, Equation [15.5] is rewritten as

$$C_T = \left[C_E \left(1 + \sum_{i=1}^{n} f_i \right) \right] (1 + f_I) \qquad\qquad [15.7]$$

where f_I = indirect cost factor
   f_i = factors for direct cost components

Examples 15.6 and 15.7 illustrate these equations.

EXAMPLE 15.6

The delivered-equipment cost for a small chemical process plant is expected to be $2 million. If the direct cost factor is 1.61 and the indirect cost factor is 0.25, determine the total plant cost.

Solution

Since all factors apply to the delivered-equipment cost, they are added to obtain h, the total cost factor in Equation [15.6].

$$h = 1 + 1.61 + 0.25 = 2.86$$

The total plant cost is

$$C_T = 2.86(2,000,000) = \$5,720,000$$

EXAMPLE 15.7

A new container-handling crane at the Port of Singapore is expected to have a delivered-equipment cost of $875,000. The cost factor for the installation of tracks, concrete, steel, noise abatement, supports, etc., is 0.49. The construction factor is 0.53, and the indirect cost factor is 0.21. Determine the total cost if (*a*) all cost factors are applied to the cost of the delivered equipment and (*b*) the indirect cost factor is applied to the total direct cost.

Solution

(*a*) Total equipment cost is $875,000. Since both the direct and indirect cost factors are applied to only the equipment cost, the overall cost factor from Equation [15.6] is

$$h = 1 + 0.49 + 0.53 + 0.21 = 2.23$$

The total cost is

$$C_T = 2.23(875,000) = \$1,951,250$$

(*b*) Now the total direct cost is calculated first, and Equation [15.7] is used to estimate total cost.

$$h = 1 + \sum_{i=1}^{n} f_i = 1 + 0.49 + 0.53 = 2.02$$
$$C_T = [875,000(2.02)](1.21) = \$2,138,675$$

Comment

Note the decrease in estimated cost when the indirect cost is applied to the equipment cost only in part (*a*). This illustrates the importance of determining exactly what the factors apply to before they are used.

15.6 Traditional Indirect Cost Rates and Allocation ● ● ●

Costs incurred in the production of an item or delivery of a service are tracked and assigned by a **cost accounting system.** For the manufacturing environment, it can be stated generally that the *statement of cost of goods sold* (discussed in Appendix B) is one end product of this system. The cost accounting system accumulates material costs, labor costs, and indirect costs (also called overhead costs or factory expenses) by using **cost centers.** All costs incurred in one department or process line are collected under a cost center title, for example, Department 3X. Since direct materials and direct labor are usually directly assignable to a cost center, the system need only identify and track these costs. Of course, this in itself is no easy chore, and the cost of the tracking system may prohibit collection of all direct cost data to the level of detail desired.

One of the primary and more difficult tasks of cost accounting is the allocation of **indirect costs** when it is necessary to allocate them separately to departments, processes, and product lines.

TABLE 15–7	Sample Indirect Cost Allocation Bases
Cost Category	**Possible Allocation Basis**
Taxes	Space occupied
Heat, light	Space, usage, number of outlets
Power	Space, direct labor hours, horsepower-hours, machine hours
Receiving, purchasing	Cost of materials, number of orders, number of items
Personnel, machine shop	Direct labor hours, direct labor cost
Building maintenance	Space occupied, direct labor cost
Software	Number of accesses
Quality control	Number of inspections

Indirect costs

Indirect costs are costs associated with property taxes, service and maintenance departments, personnel, legal, quality, supervision, purchasing, utilities, software development, etc. They must all be allocated to the using cost center. Detailed collection of these data is cost-prohibitive and often impossible; thus, allocation schemes are utilized to distribute the expenses on a reasonable basis.

A listing of possible allocation bases is included in Table 15–7. Historically, common bases have been direct labor cost, direct labor hours, machine-hours, number of employees, space, and direct materials.

Most allocation is accomplished utilizing a predetermined *indirect cost rate,* computed by using the general relation.

$$\text{Indirect cost rate} = \frac{\text{estimated total indirect costs}}{\text{estimated basis level}} \qquad [15.8]$$

The estimated indirect cost is the amount allocated to a cost center. For example, if a division has two producing departments, the total indirect cost allocated to a department is used as the numerator in Equation [15.8] to determine the department rate. Example 15.8 illustrates allocation when the cost center is a machine.

EXAMPLE 15.8

The manager of beauty products at BestWay wants to determine allocation rates for $150,000 of indirect costs for the three machines used to process beauty lotions. The following information was obtained from last year's budget for the three machines. Determine rates for each machine if the amount is equally distributed.

Cost Source	Allocation Basis	Estimated Activity Level
Machine 1	Direct labor cost	$100,000
Machine 2	Direct labor hours	2000 hours
Machine 3	Direct material cost	$250,000

Solution

Applying Equation [15.8] for each machine, the annual rates are

$$\text{Machine 1 rate} = \frac{\text{indirect budget}}{\text{direct labor cost}} = \frac{50,000}{100,000}$$

$$= \$0.50 \text{ per direct labor dollar}$$

$$\text{Machine 2 rate} = \frac{\text{indirect budget}}{\text{direct labor hours}} = \frac{50,000}{2000}$$

$$= \$25 \text{ per direct labor hour}$$

$$\text{Machine 3 rate} = \frac{\text{indirect budget}}{\text{material cost}} = \frac{50,000}{250,000}$$

$$= \$0.20 \text{ per direct material dollar}$$

Now the actual direct labor costs and hours and material costs are determined for this year, and each dollar of direct labor cost spent on machine 1 implies that $0.50 in indirect cost will be added to the cost of the product. Similarly, indirect costs are added for machines 2 and 3.

When the same allocation basis is used to distribute indirect costs to several cost centers, a **blanket rate** may be determined. For example, if direct materials are the basis for allocation to four separate processing lines, the blanket rate is

$$\text{Indirect cost rate} = \frac{\text{total indirect costs}}{\text{total direct materials cost}}$$

If $500,000 in indirect costs and $3 million in materials are estimated for the four lines, the blanket indirect rate is $500,000/3,000,000 = \$0.167$ per materials cost dollar. Blanket rates are easier to calculate and apply, but they do not account for differences in the type of activities accomplished in each cost center.

In most cases, machinery or processes add value to the end product at different rates per unit or hour of use. For example, light machinery may contribute less per hour than heavy, more expensive machinery. This is especially true when advanced technology processing, for example, an automated manufacturing cell, is used along with traditional methods, for example, nonautomated finishing equipment. The use of blanket rates in these cases is not recommended, as the indirect cost will be incorrectly allocated. The lower-value-contribution machinery will accumulate too much of the indirect cost. The approach should be the application of different bases for different machines, activities, etc., as discussed earlier and illustrated in Example 15.8. The use of different, appropriate bases is often called the **productive hour rate method** since the cost rate is determined based on the value added, not a uniform or blanket rate. Realization that more than one basis should be normally used in allocating indirect costs has led to the use of activity-based costing methods, as discussed in the next section.

Once a period of time (month, quarter, or year) has passed, the indirect cost rates are applied to determine the indirect cost *charge,* which is then added to direct costs. This results in the total cost of production, which is called the cost of goods sold, or **factory cost.** These costs are all accumulated by *cost center.*

If the total indirect cost budget is correct, the indirect costs charged to all cost centers for the period of time should equal this budget amount. However, since some error in budgeting always exists, there will be overallocation or underallocation relative to actual charges, which is termed *allocation variance.* Experience in indirect cost estimation assists in reducing the variance at the end of the accounting period.

EXAMPLE 15.9

Since the manager determined indirect cost rates for BestWay (Example 15.8), she can now compute the total factory cost for a month. Perform the computations using the data in Table 15–8. Also calculate the variance for indirect cost allocation for the month.

Solution

Start with the cost of goods sold (factory cost) relation given by Equation [B.1] in Appendix B, which is

Cost of goods sold = direct materials + direct labor + indirect costs

TABLE 15–8	Actual Monthly Data Used for Indirect Cost Allocation		
Cost Source	**Machine Number**	**Actual Cost, $**	**Actual Hours**
Material	1	3,800	
	3	19,550	
Labor	1	2,500	650
	2	3,200	750
	3	2,800	720

To determine indirect cost, the rates from Example 15.8 are applied:

$$\text{Machine 1 indirect} = (\text{labor cost})(\text{rate}) = 2500(0.50) = \$1250$$
$$\text{Machine 2 indirect} = (\text{labor hours})(\text{rate}) = 750(25.00) = \$18,750$$
$$\text{Machine 3 indirect} = (\text{material cost})(\text{rate}) = 19,550(0.20) = \$3910$$

Total charged indirect cost = \$23,910

Factory cost is the sum of actual material and labor costs from Table 15–8 of \$31,850 and the indirect cost charge for a total of \$55,760.

Based on the annual indirect cost budget of \$150,000, one month represents 1/12 of the total or

$$\text{Monthly budget} = \frac{150,000}{12}$$
$$= \$12,500$$

The allocation variance for total indirect cost is

$$\text{Variance} = 12,500 - 23,910 = \$-11,410$$

This is a large budget underallocation, since much more was actually charged than allocated. The \$12,500 budgeted for the three machines represents a 91.3% underallocation of indirect costs. This analysis for only one month of a year will most likely prompt a rapid review of the rates and the indirect cost budget.

Once estimates of indirect costs are determined, it is possible to perform an economic analysis of the present operation versus a proposed operation. Such a study is described in Example 15.10.

EXAMPLE 15.10

For several years, Cuisinart Corporation has purchased the carafe assembly of its major coffeemaker line at an annual cost of \$2.2 million. The suggestion to make the component in-house has been made. For the three departments involved, the annual indirect cost rates, estimated material, labor, and hours are found in Table 15–9. The allocated hours column is the time necessary to produce the carafes for a year.

Equipment must be purchased with the following estimates: first cost of \$2 million, salvage value of \$50,000, and life of 10 years. Perform an economic analysis for the make alternative, assuming that a market rate of 15% per year is the MARR.

Solution

For making the components in-house, the AOC is comprised of direct labor, direct material, and indirect costs. Use the data of Table 15–9 to calculate the indirect cost allocation.

$$\begin{array}{lll}
\text{Department A:} & 25,000(10) = & \$250,000 \\
\text{Department B:} & 25,000(5) \ \ = & 125,000 \\
\text{Department C:} & 10,000(15) = & \underline{150,000} \\
& & \$525,000
\end{array}$$

$$\text{AOC} = 500,000 + 300,000 + 525,000 = \$1,325,000$$

TABLE 15–9 Production Cost Estimates for Example 15.10

Department	Basis, Hours	Rate per Hour, $	Allocated Hours	Material Cost, $	Direct Labor Cost, $
			Indirect Costs		
A	Labor	10	25,000	200,000	200,000
B	Machine	5	25,000	50,000	200,000
C	Labor	15	10,000	50,000	100,000
				300,000	500,000

The make alternative annual worth is

$$AW_{make} = -P(A/P,i,n) + S(A/F,i,n) - AOC$$
$$= -2,000,000(A/P,15\%,10) + 50,000(A/F,15\%,10) - 1,325,000$$
$$= \$-1,721,037$$

Currently, the carafes are purchased with an AW of

$$AW_{buy} = \$-2,200,000$$

It is cheaper to make, because the AW of costs is less.

15.7 Activity-Based Costing (ABC) for Indirect Costs ●●○

As automation, software, and manufacturing technologies have advanced, the number of direct labor hours necessary to manufacture a product has decreased substantially. Where once as much as 35% to 45% of the final product cost was represented in labor, now the labor component is commonly 5% to 15% of total manufacturing cost. However, the indirect cost may represent as much as 35% to 45% of the total manufacturing cost. The use of bases, such as direct labor hours, to allocate indirect cost is not accurate enough for automated and technologically advanced environments. This has led to the development of methods that replace or supplement traditional cost allocations that rely upon one form or another of Equation [15.8]. Also, allocation bases different from traditional ones are commonly utilized.

A product that by traditional methods may have contributed a large portion to profit *may actually be a loser* when indirect costs are allocated more correctly. Companies that have a wide variety of products and produce some in small lots may find that traditional allocation methods have a tendency to underallocate the indirect cost to small-lot products. This may indicate that they are profitable, when in actuality they are losing money.

The best allocation method for high-overhead industries is **Activity-Based Costing (ABC).** It is designed to identify *cost centers, activities,* and *cost drivers.* Descriptions of each follow.

Cost centers: The final products or services of the corporation are called cost centers or cost pools. They *receive* the allocated indirect costs.

Activities: These are usually support departments (purchasing, quality, IT, maintenance, engineering, supervision) that *generate* the indirect costs which are then distributed to the cost centers.

Cost drivers: Commonly expressed in volumes, these *drive* the consumption of a shared resource. Examples are the number of purchase orders, cost of engineering change orders, number of machine setups, number of safety violations, and the like.

Implementing ABC involves several steps.

1. Identify each *activity* and its total cost.
2. Identify the *cost drivers* and their usage volumes.
3. Calculate the indirect cost *rate* for each activity.

$$\text{ABC indirect cost rate} = \frac{\text{total cost of activity}}{\text{total volume of cost driver}} \qquad [15.9]$$

4. Use the rate to *allocate* indirect cost to cost centers for each activity.

As an illustration, assume a company that produces two types of industrial lasers (cost centers) has three primary support departments (activities; step 1 in the procedure). The allocation for costs generated by the purchasing department, for example, is based on the number of purchase orders (step 2) to support laser production. The ABC rate (step 3) in dollars per purchase order is used to allocate indirect costs to the two laser products (step 4).

EXAMPLE 15.11

A multinational aerospace firm uses traditional methods to allocate manufacturing and management support costs for its European division. However, accounts such as business travel have historically been allocated on the basis of the number of employees at the plants in France, Italy, Germany, and Spain.

The president recently stated that some product lines are likely generating much more travel than others. The ABC system is chosen to augment the traditional method to more precisely allocate travel costs to major product lines at each plant.

(a) First, assume that allocation of total observed travel expenses of $500,000 to the plants using a traditional basis of workforce size is sufficient. If total employment of 29,100 is distributed as follows, allocate the $500,000.

Paris, France plant	12,500 employees
Florence, Italy plant	8,600 employees
Hamburg, Germany plant	4,200 employees
Barcelona, Spain plant	3,800 employees

(b) Now, assume that corporate management wants to know more about travel expenses based on product line, not merely plant location and workforce size. The ABC method will be applied to allocate travel costs to major product lines. Annual plant support budgets indicate that the following percentages are expended for travel:

Paris	5% of $2 million
Florence	15% of $500,000
Hamburg	17.5% of $1 million
Barcelona	30% of $500,000

Further, the study indicates that in 1 year a total of 500 travel vouchers were processed by the management of the major five product lines produced at the four plants. The distribution is as follows:

Paris	Product lines—1 and 2; number of vouchers—50 for line 1, 25 for 2.
Florence	Product lines—1, 3, and 5; vouchers—80 for line 1, 30 for 3, 30 for 5.
Hamburg	Product lines—1, 2 and 4; vouchers—100 for line 1, 25 for 2, 20 for 4.
Barcelona	Product line—5; vouchers—140 for line 5.

Use the ABC method to determine how the product lines drive travel costs at the plants.

Solution

(a) In this case, Equation [15.8] takes the form of a blanket rate per employee.

$$\text{Indirect cost rate} = \frac{\text{travel budget}}{\text{total workforce}}$$

$$= \frac{\$500,000}{29,100} = \$17.1821 \text{ per employee}$$

Using this traditional basis of rate times workforce size results in a plant-by-plant allocation.

Paris	$17.1821(12,500) = $214,777
Florence	$147,766
Hamburg	$72,165
Barcelona	$65,292

(b) The ABC method is more involved to apply, and the by-plant amounts will be different from those in part (a) since completely different bases are applied. Use the 4-step procedure to allocate travel costs to the five products.

> **Step 1.** The total amount to be allocated is determined from the percentages of each plant's support budget devoted to travel. The number is determined from the percent-of-budget data as follows:
>
> $$0.05(2,000,000) + \cdots + 0.30(500,000) = \$500,000$$

TABLE 15–10	ABC Allocation of Travel Cost ($ in Thousands), Example 15.11					
	Product Line					
	1	**2**	**3**	**4**	**5**	**Total**
Paris	50	25				75
Florence	80		30		30	140
Hamburg	100	25		20		145
Barcelona					140	140
Total	$230	$50	$30	$20	$170	$500

Step 2. The cost driver is the number of travel vouchers submitted by the management unit responsible for each product line at each plant. The allocation will be to the products directly, not to the plants. However, the travel allocation to the plants can be determined afterward since we know what product lines are produced at each plant.

Step 3. Equation [15.9] determines an ABC allocation rate.

$$\text{ABC allocation rate} = \frac{\text{total cost of travel}}{\text{total number of vouchers}}$$

$$= \frac{\$500,000}{500}$$

$$= \$1000 \text{ per voucher}$$

Step 4. Table 15–10 summarizes the vouchers and allocation by product line and by city. Product 1 ($230,000) and product 5 ($170,000) drive the travel costs based on the ABC analysis. Comparison of the by-plant totals in Table 15–10 (far right column) with the respective totals in part (*a*) indicates a substantial difference in the amounts allocated, especially to Paris, Hamburg, and Barcelona. This comparison verifies the president's statement that product lines, not plants, drive travel requirements.

Comment

Let's assume that product 1 has been produced in small lots at the Hamburg plant for a number of years. This analysis, when compared to the traditional cost allocation method in part (*a*), reveals a very interesting fact. In the ABC analysis, Hamburg has a total of $145,000 travel dollars allocated, $100,000 from product 1. In the traditional analysis based on workforce size, Hamburg was allocated only $72,165—about 50% of the more precise ABC analysis amount. This indicates to management the need to examine the manufacturing lot size practices at Hamburg and possibly other plants, especially when a product is currently manufactured at more than one plant.

Some proponents of the ABC method recommend discarding the traditional cost accounting methods of a company and utilizing ABC exclusively. This is not a good approach, since ABC is not a complete cost system. The ABC method provides information that assists in *cost control,* while the traditional method emphasizes cost allocation and cost estimation. The two systems work well together with the traditional methods allocating costs that have identifiable direct bases, for example, direct labor. ABC analysis is usually more expensive and time-consuming than a traditional cost allocation system, but in many cases it can assist in understanding the economic impact of management decisions and in controlling certain types of indirect costs.

15.8 Making Estimates and Maintaining Ethical Practices ●●●

Making estimates about the future of costs, revenues, cash flows, rates of return, and many other parameters is routine when one is engaged in any type of economic analysis. Public agencies, private corporations, and not-for-profit businesses all make economic decisions based on these estimates, most of which are made by employees of the organizations or by outside consultants hired to perform specific activities under contract. The opportunities for bias, poor accuracy, deception, and profit-driven or other motives are always present. The personal morals and adherence to codes of professional ethics discussed in Section 1.3 guide individuals in their work to make fair and believable estimates for the analyses and decisions that follow.

The NSPE Code of Ethics for Engineers (Appendix C) referenced previously starts with a list of six Fundamental Canons. One very relevant to estimation integrity is "Avoid deceptive acts." Acts that bias the results from experimental samples, previous cost data, or survey results for the purposes of personal gain, increased profits, or favoritism are examples of unethical behavior. Estimates of all types should be founded on practices such as the following:

Base estimates on sound information gathered over a range of situations representative of the current one.

Use accepted theory and techniques in taking statistical samples, building budget elements, and drawing conclusions that are included in proposals, applications, and recommendations.

As a consultant or contractor, keep personal and working relationships separate when making estimates and delivering the final documentation to a client or sponsor.

The second case study at the end of this chapter presents an example of some ethical challenges present when preparing estimates and proposals for contract work.

CHAPTER SUMMARY

Cost estimates are not expected to be exact, but they should be accurate enough to support a thorough economic analysis using an engineering economy approach. There are bottom-up and top-down approaches; each treats price and cost estimates differently.

Costs can be updated via a cost index, which is a ratio of costs for the same item at two separate times. The Consumer Price Index (CPI) is an often-quoted example of cost indexing. Cost estimating may also be accomplished with a variety of models called cost-estimating relationships. Two of them are

Cost-capacity equation—good for estimating costs from design variables for equipment, materials, and construction

Factor method—good for estimating total plant cost

Traditional cost allocation uses an indirect cost rate determined for a machine, department, product line, etc. Bases such as direct labor cost, direct material cost, and direct labor hours are used. With increased automation and information technology, different techniques of indirect cost allocation have been developed. The Activity-Based Costing method is an excellent technique to augment the traditional allocation method.

The ABC method allocates indirect costs on the rationale that purchase orders, inspections, machine setups, reworks, etc. *drive* the costs accumulated in departments or functions, such as quality, purchasing, accounting, and maintenance. Improved understanding of how the company or plant accumulates indirect costs is a major by-product of implementing the ABC method.

PROBLEMS

Understanding Cost Estimation

15.1 Rank the following estimate types in terms of time spent to carry out the estimate (most time to least time): partially designed, design 60% to 100% complete, order of magnitude, scoping/feasibility, detailed estimate.

15.2 Classify the following cost elements as first cost (FC) components or annual operating cost (AOC)

components for a piece of equipment on the shop floor: supplies, insurance, equipment cost, utility cost, installation, delivery charges, labor cost.

15.3 State whether actual (A) or estimated (E) costs are more likely to be used to carry out the following activities: calculate taxes, make bids, pay bonuses, determine profit or loss, predict sales, set prices, evaluate proposals, distribute resources, plan production, and set goals.

15.4 Identify the output and input variables in both the bottom-up and top-down approaches to cost estimating.

15.5 Classify the following costs as typically direct (D) or indirect (I):

Project staff	Audit and legal
Utilities	Rent
Raw materials	Equipment training
Project supplies	Labor
Administrative staff	Miscellaneous office supplies

15.6 Identify each of the following costs associated with owning an automobile as direct or indirect (assume a direct cost is one that is directly attributable to the number of miles driven): license plate, driver's license, gasoline, highway toll fee, oil change, repairs after collision, gasoline tax, monthly loan payment, annual inspection fee, and garage rental.

15.7 In the early and conceptual design stages of a project, what are the cost estimates called? Approximately how close should they be to the actual cost?

Unit Costs

15.8 Use the unit cost method to determine the preliminary cost of a guardrail for a small bridge if a total of 120 linear feet will be required at a cost of $58.19 per linear foot for material, equipment, and labor.

15.9 Sun-Metro Bus is planning to construct a 600-car parking garage on the outskirts of the city to encourage people to use the "Park 'n Ride" to work or for shopping downtown. Prepare a preliminary cost estimate for the garage if the cost per parking space is $4700.

15.10 The Office of the Undersecretary of Defense periodically releases unit cost data for use in military construction programs. If the unit cost for a satellite communications center (with shielding and generator) is $496 per square foot, what would be the estimated cost of a 6000-square-foot building?

15.11 The Department of Defense uses area cost factors (ACFs) to adjust for differences in construction costs in different parts of the country (and world). The area cost factor for Andros Island in the Bahamas is 1.70 while the ACF for Rapid City, South Dakota, is 0.93. If a cold storage processing warehouse costs $1,350,000 in Rapid City, estimate the cost for Andros Island.

15.12 Preliminary cost estimates for jails can be made using costs based on either unit area (square feet) or unit volume (cubic feet). If the unit area cost is $185 per square foot and the average height of the ceilings is 10 feet, what is the unit volume cost?

15.13 The unit area and unit volume total project costs for a library are $114 per square foot and $7.55 per cubic foot, respectively. Based on these numbers, what is the average height of library rooms?

15.14 The equipment required to place 160 cubic yards of concrete by experienced workers is two gasoline engine vibrators and one concrete pump. If the vibrators cost $76 per day and the concrete pump costs $580 per day, estimate (a) the cost of the equipment per cubic yard of concrete and (b) the equipment cost for placing 56 cubic yards of concrete.

15.15 A labor crew for placing concrete consists of 1 labor foreman at $25.85 per hour, 1 cement finisher at $28.60 per hour, 5 laborers at $23.25 per hour each, and 1 equipment operator at $31.45 per hour. Such a crew, called a C20 crew, can place 160 cubic yards of concrete per 8-hour day. Determine (a) the cost per day of labor for the C20 crew, (b) the cost of the C20 crew per cubic yard of concrete, and (c) the cost to place 250 cubic yards of concrete.

15.16 Site work activities associated with constructing a small bridge are shown in the table below. The table includes the quantity of each activity, the unit of measurement associated with each activity, and the unit cost of each activity. Use the data to determine (a) the total cost for structural excavation, (b) the total cost for the pile-driving rig, and (c) the total labor cost for the site work.

Activity	Quantity	Unit of Measurement	Labor Unit Cost, $	Equipment Unit Cost, $	Material Unit Cost, $
Excavation, unclassified	1667	cy	1.35	1.43	0
Excavation, structural	120	cy	21.31	5.00	0
Backfill, compacted	340	cy	7.78	1.72	0
Pile-driving rig	job	ls	5688	6420	300
Piling, steel, driving	2240	lf	3.13	2.93	16.57

Legend: cy = cubic yard; ls = lump sum; lf = linear foot

Cost Indexes

15.17 From historical data, you discover that the national average construction cost of middle schools is $10,500 per student. If the state index for Texas is 76.9 and for California it is 108.5, estimate the total construction cost of a middle school for 800 students in each state.

15.18 A consulting engineering firm is preparing a preliminary cost estimate for a design-construct project of a coal processing plant. The firm, which completed a similar project in 2001 with a construction cost of $30 million, wants to use the *ENR* construction cost index to update the cost. Use the values in Table 15–3 to estimate the cost of construction for a similar-size plant in mid-2010.

15.19 If the editors at *ENR* decide to redo the construction cost index so that the year 2000 has a base value of 100, determine the value for the year (*a*) 1995 and (*b*) 2009.

15.20 (*a*) Estimate the value of the *ENR* construction cost index by using the average (compounded) percentage change in its value between 1995 and 2005 to predict the value in 2009.
      (*b*) How much difference (numerical) is there between the estimated and actual 2009 values? Is this an underestimate or overestimate?

15.21 An engineer who owns a construction company that specializes in large commercial projects noticed that material costs increased at a rate of 1% per month over the past 12 months. If a material cost index were created for each month of this year with the value of the index set at 100 for the beginning of the year, what would be the value of the index at the end of the year? Express your answer to two decimal places.

15.22 Using 1913 as the base year with a value of 100, the *ENR* construction cost index (CCI) for August 2009 was 8563.35. For August 2010, the CCI value was 8837.38. (*a*) What was the inflation rate for construction for that 1-year period? (*b*) What will be the CCI value in August 2013, provided the inflation rate remains the same as it was in the 2009–2010 period?

15.23 Electropneumatic general-purpose pressure transducers convert supply pressure to regulated output pressure in direct proportion to an electrical input signal. The cost of a certain transducer was $194 in 1985 when the M&S equipment index was at 789.6. If the price increased exactly in proportion to the M&S equipment index, what would its cost be in 2010 (midyear) when the index value was 1461.3?

15.24 The *ENR* construction cost index for New York City had a value of 12,381.40 in February 2007. For Pittsburgh and Atlanta, the values were 7341.32 and 4874.06, respectively. If a general contractor in Atlanta won construction jobs totaling $54.3 million, determine their equivalent total value in New York City.

15.25 The *ENR* construction cost index (CCI) for August 2010 had a value of 8837.37 when the base year was 1913 with a value of 100. If the base year is 1967 with a value of 1.0, the CCI for August 2010 would be 8.2272. In this case, what is the CCI value for 1967 when the base year is 1913?

15.26 The *ENR* materials cost index (MCI) had a value of 2708.51 in August 2010. In the same month, the cost for cement was $96.55 per ton. If cement increased in price exactly in accordance with the MCI, what was the cost of cement per ton in 1913 when the MCI index value was 100?

15.27 A contractor purchased equipment costing $40,000 in 2010 when the M&S equipment cost index was at 1461.3. He remembers purchasing the same equipment for $21,771 many years ago, but he does not remember the year that he did so. If the M&S equipment cost index increased by 2.68% per year over that time period and the equipment increased in price exactly in proportion to the index, (*a*) in what year did he purchase the equipment and (*b*) what was the value of the index in that year?

Cost-Capacity and Factor Methods

15.28 If the materials cost index decreased by 2% over the same time period that the construction cost index increased by 2%, what does that suggest about the labor cost index, if labor represents 35% of the cost of construction?

15.29 Use the exponent values in Table 15–6 to estimate the cost for the following equipment to be placed on an offshore drilling platform.
      (*a*) The cost of a 125-hp centrifugal pump if a 200-hp pump costs $28,000.
      (*b*) The cost of a 1700-gallon stainless steel tank if a 900-gallon tank costs $4100.

15.30 A high-pressure stainless steel pump (1000 psi) with a variable-frequency drive (VFD) is installed in a seawater reverse-osmosis pilot plant that is recovering water from membrane concentrate at a rate of 4 gallons per minute (gpm). The cost of the pump was $13,000. Because of favorable results from the pilot study, the city utility wants to go with a full-scale system that will produce 500 gpm. Determine the estimated cost of the larger pump, if the exponent in the cost-capacity equation has a value of 0.37.

15.31 A 0.75 million gallon per day (MGD) induced-draft packed tower for air-stripping trihalomethanes from drinking water costs $58,890. Estimate the cost of a 2-MGD tower if the exponent in the cost-capacity equation is 0.58.

15.32 The on-site manager told you that the variable-frequency drive (VFD) for a 300-hp motor costs $20,000. Make the best cost estimate possible for the VFD for a 100-hp motor. The exponent in the cost-capacity equation is not available currently where you are located.

15.33 The cost of a 68 m^2 falling-film evaporator was 1.52 times the cost of the 30 m^2 unit. What exponent value in the cost-capacity equation yielded these results?

15.34 Reinforced concrete pipe (RCP) that is 12 inches in diameter had a cost of $12.54 per linear foot in Dallas, Texas in 2007. The cost for 24-inch RCP was $27.23 per foot. If the cross-sectional area of the pipe is considered the "capacity" in the cost-capacity equation, determine the value of the exponent in the cost-capacity equation that exactly relates the two pipe sizes.

15.35 A 100,000 barrel per day (bpd) fractionation tower cost $1.2 million in 2001 when the *Chemical Engineering* plant cost index value was 394.3. How much would a 450,000 bpd plant cost when the index value is 575.8, provided the exponent in the cost-capacity equation is 0.67?

15.36 A mini wind tunnel for calibrating vane or hot-wire anemometers cost $3750 in 2002 when the M&S equipment index value was 1104.2. If the index value is now 1620.6, estimate the cost of a tunnel twice as large. The cost-capacity equation exponent is 0.89.

15.37 In 2008, a military engineer estimated the cost for a classified laser-guided device to be $376,900. The engineer used the M&S equipment index for the years 1998 and 2008 and the cost-capacity equation with an exponent value of 0.61. If the original equipment had only one-fourth the capacity of the new equipment, what was the cost of the original equipment in 1998?

15.38 Instead of using a cost-capacity equation for relating project size and construction cost, the Department of Defense uses Size Adjustment Factors (SAFs) that are based on Size Relationship Ratios (SRRs). For example, a SRR of 2.00 has a SAF of 0.942. Determine the required exponent value in the cost-capacity equation in order for C_2 to be $0.942C_1$ when $Q_2/Q_1 = 2.00$.

15.39 The equipment cost for removing arsenic from a well that delivers 800 gallons per minute (gpm) is $1.8 million. If the overall cost factor for this type of treatment system is 2.25, what is the total plant cost expected to be?

15.40 A closed-loop filtration system for waterjet cutting industries eliminates the cost of makeup water treatments (water softeners, reverse osmosis, etc.) while maximizing orifice life and machine performance. If the equipment cost is $225,000 and the total plant cost is $1.32 million, what is the overall cost factor for the system?

15.41 The equipment cost for a laboratory that plans to specialize in analyzing for endocrine disrupters, pharmaceuticals, and personal care products is $870,000. If the direct cost factor is 1.32 and the indirect cost factor is 0.45 (applies to equipment only), determine the expected cost of the laboratory.

15.42 The equipment cost for a 10 gallon per minute farm-scale ethanol fuel production plant is $243,000. The direct cost factor for construction is 1.28 and for installation is 0.23. The indirect cost factor for licenses, insurance, etc. is 0.84 (applied to total direct cost). Determine the estimated total plant cost.

15.43 A chemical engineer at Western Refining has estimated that the total cost for a diesel fuel desulfurization system will be $2.3 million. If the direct cost factor is 1.35 and the indirect cost factor is 0.41, what is the total equipment cost? Both factors apply to delivered-equipment cost.

15.44 A mechanical engineer estimated that the equipment cost for a multitube cyclone system with a capacity of 60,000 cfm would be $400,000. If the direct cost factor is 3.1 and the indirect cost factor is 0.38, what is the estimated total plant cost? The indirect cost factor applies to the total direct cost.

15.45 Nicole is an engineer on temporary assignment at a refinery operation in Seaside. She has reviewed a cost estimate for $430,000, which covers some new processing equipment for the ethylene line. The equipment itself is estimated at $250,000 with a construction cost factor of 0.30 and an installation cost factor of 0.30. No indirect cost factor is listed, but she knows from other sites that indirect cost is a sizable amount that increases the cost of the line's equipment. (*a*) If the indirect cost factor should be 0.40, determine whether the current estimate includes a factor comparable to this value (*b*) Determine the cost estimate if the 0.40 indirect cost factor is used.

Indirect Cost Allocation

15.46 The company you work for currently allocates insurance costs on the basis of cost per direct labor hour. This indirect cost component for the year is budgeted at $36,000. If the direct labor hours for departments A, B, and C are expected to be 2000, 8000, and 5000, respectively, this year, determine the allocation to each department.

15.47 The director of public works needs to distribute the indirect cost allocation of $1.2 million to the three branches around the city. The recorded amounts for this year are as follows:

	Records for This Year	
Branch	**Miles Driven**	**Direct Labor Hours**
North	275,000	38,000
South	247,000	31,000
Midtown	395,000	55,500

The director plans to use the allocation and information from last year to determine the rates for this year. This information follows:

Branch	**Miles Driven**	**Direct Labor Hours**	**Basis**	**Indirect Cost Allocation Last Year, $**
North	350,000	40,000	Miles	300,000
South	200,000	20,000	Labor	200,000
Midtown	500,000	64,000	Labor	450,000

(a) Determine the rates for this year for each branch.

(b) Use the rate to distribute this year's total indirect cost. What percentage of this year's budget is now distributed?

15.48 A company has a processing department with 10 stations. Because of the nature and use of three of these stations, each is considered a separate cost center for indirect cost allocation. The remaining seven are grouped as one center, CC190. Machine operating hours are used as the allocation basis for all machines. A total of $250,000 is allocated to the department for next year. Use the data collected this year to determine the indirect cost rate for each center.

Cost Center	**Indirect Cost Allocated, $**	**Estimated Machine Hours**
CC100	25,000	800
CC110	50,000	200
CC120	75,000	1200
CC190	100,000	1600

15.49 All the indirect costs are allocated by accounting for a department. The manager has obtained records of allocation rates and actual charges for the prior 3 months and estimates for this month (May) and next month (see the table). The basis

of allocation is not indicated, and the company accountant has no record of the basis used. However, the accountant advises the manager to not be concerned because the allocation rates have decreased each month.

		Indirect Cost, $	
Month	**Rate**	**Allocated**	**Charged**
February	1.40	2800	2600
March	1.33	3400	3800
April	1.37	3500	3500
May	1.03	3600	
June	0.92	6000	

During the evaluation, the following additional information from departmental and accounting records is obtained.

	Direct Labor		Material	Departmental
Month	**Hours**	**Cost, $**	**Cost, $**	**Space, ft^2**
February	640	2560	5400	2000
March	640	2560	4600	2000
April	640	2560	5700	3500
May	640	2720	6300	3500
June	800	3320	6500	3500

(a) With this information determine the allocation basis used each month.

(b) Comment on the accountant's statement about decreasing allocation rates.

15.50 The mechanical components division manager asks you to recommend a make/buy decision on a major automotive subassembly that is currently purchased externally for a total of $3.9 million this year. This cost is expected to continue rising at a rate of $300,000 per year. Your manager asks that both direct and indirect costs be included when in-house manufacturing (make alternative) is evaluated. New equipment will cost $3 million and will have a salvage of $0.5 million and a life of 6 years. Estimates of materials, labor costs, and other direct costs are $1.5 million per year. Typical indirect rates, bases, and expected usage are shown below. Perform the AW evaluation at MARR = 12% per year over a 6-year study period. Show both hand and spreadsheet solutions, as directed.

Department	**Basis**	**Rate**	**Expected Usage**
X	Direct labor cost	$2.40 per $	$450,000
Y	Materials cost	$0.50 per $	$850,000
Z	Number of inspections	$20 per inspection	4,500

ABC Method

15.51 The municipal water and desalinization utility in a California city currently allocates some costs for maintenance shop workers to pumping stations based on the number of pumps at each station. At the last director's semiannual meeting, a suggestion was made to change the allocation basis to the number of trips that pump service personnel make to each station, because some stations have old pumps that require more maintenance. Information about the stations is below. The indirect cost budget is $20,000 per pump.

(a) Allocate the budget to each station based on the number of service trips.

(b) Determine the old allocation on the basis of the number of pumps, and comment on any significant differences in the amounts allocated to the stations.

Station ID	No. of Pumps	Service Trips per Year
Sylvester	5	190
Laurel	7	55
7th St	3	38
Spicewood	4	104

15.52 Factory Direct manufactures and sells manufactured homes. Traditionally, it has distributed indirect costs to its three construction plants based on materials cost. Each plant builds different models and floor plans. Advances in weight and shape of plastic and wood composite components have decreased cost and time to produce a unit. Because of these advances, the CFO plans to use build-time per unit as the new basis. However, he initially wants to determine what the allocation would have been this year had build-time been the basis prior to incorporation of the new materials. The data shown represents average costs and times. Use this data and the three bases indicated to determine the allocation rates and indirect cost distribution of $900,000 for this year.

Plant	Texas	Oklahoma	Kansas
Direct material cost, $ per unit	20,000	12,700	18,600
Previous build-time per unit, work-hours	400	415	580
New build-time per unit, work-hours	425	355	480

Problems 15.53 through 15.55 use the following information.

Jet Green Airways historically distributes the indirect costs of lost and damaged baggage to its three major hubs using a basis of annual number of flights in and out of each hub. Last year $667,500 was distributed as follows:

Hub Airport	Flights	Rate, $ per Flight	Allocation, $
DFW	55,000	6	330,000
YYZ	20,833	9	187,500
MEX	15,000	10	150,000

The airline's baggage management director suggests that an allocation on the basis of baggage traffic, not flights, will better represent the distribution, primarily based on the fact that the high fees now charged to passengers to check luggage have significantly changed the number of bags handled at the major hubs. Total number of bags handled during the year are 2,490,000 at DFW, 1,582,400 at YYZ, and 763,500 at MEX.

15.53 What are the activity and the cost driver for the suggested baggage-traffic basis?

15.54 Using the baggage-traffic basis, determine the allocation rate using last year's total of $667,500, and distribute this amount to the hubs this year.

15.55 What are the percentage changes in allocation at each hub using the two different bases?

15.56 On-line Vacation distributes advertising costs to its four resort sites in the Caribbean on the basis of the size of the resort budget. For this year, in round numbers, the budgets and allocation of $1 million advertising indirect costs are as follows:

	Site			
	A	B	C	D
Budget, $	2 million	3 million	4 million	1 million
Allocation, $	200,000	300,000	400,000	100,000

(a) Determine the allocation if the ABC method is used with a new basis. Define the activity as the advertising department at each resort. The cost driver is the number of guests during the year.

	Site			
	A	B	C	D
Guests	3500	4000	8000	1000

(b) Again use the ABC method, but now make the cost driver the total number of guest nights at each resort. The average number of lodging nights for guests at each site is as follows:

	Site			
	A	B	C	D
Length of stay, nights	3.0	2.5	1.25	4.75

(c) Comment on the distribution of advertising costs using the two methods. Identify any other cost drivers that might be considered for the ABC approach that may reflect a realistic allocation of the costs.

ADDITIONAL PROBLEMS AND FE EXAM REVIEW QUESTIONS

15.57 In the bottom-up approach to cost estimating:
- (a) Required price is an input variable.
- (b) Cost estimates are an output variable.
- (c) Required price is an output variable.
- (d) Both (a) and (b) are correct.

15.58 A ratio of the cost of something today to its cost at some time in the past is called a:
- (a) Cost-capacity index
- (b) Cost index
- (c) Buyer's guide
- (d) Bluebook index

15.59 Index values could probably be obtained from all of the following places except:
- (a) Trade organizations
- (b) *Engineering News Record* magazine
- (c) Government organizations
- (d) Home Depot or Lowe's home improvement stores

15.60 A 50-hp turbine pump was purchased for $2100. If the exponent in the cost-capacity equation has a value of 0.76, a 200-hp turbine pump could be expected to cost about:
- (a) $6020
- (b) $5320
- (c) $4890
- (d) $4260

15.61 The city built a recreation park in 1980 for $500,000. The *ENR* construction cost index had a value of 3378.17 at that time. If the city is planning to construct a similar recreation park when the index value is 5542.16, the estimated cost is closest to:
- (a) $695,800
- (b) $750,700
- (c) $820,300
- (d) $910,500

15.62 A small company bought a 250-hp compressor in 1998 for $3000 when the M&S equipment cost index had a value of 1061.9. If the exponent in the cost-capacity equation is 0.32 and the M&S index value was 1449.3 in 2008, the cost of a 500-hp compressor in 2008 was closest to:
- (a) $3744
- (b) $4094
- (c) $4627
- (d) $5110

15.63 The cost for implementing a manufacturing process that has a capacity of 6000 units per day was $550,000. If the cost for a plant with a capacity of 100,000 units per day was $3 million, the value of the exponent in the cost-capacity equation is:
- (a) 0.26
- (b) 0.39
- (c) 0.45
- (d) 0.60

15.64 The equipment for applying specialty coatings that provide a high angle of skid for the paperboard and corrugated box industries has a delivered cost of $390,000. If the overall cost factor for the complete system is 2.96, the total plant cost is approximately:
- (a) $954,400
- (b) $1,054,400
- (c) $1,154,400
- (d) $1,544,400

15.65 The delivered-equipment cost for setting up a production and assembly line for high-sensitivity, gas-damped accelerometers is $650,000. If the direct cost and indirect cost factors are 1.82 and 0.31, respectively, and both factors apply to delivered-equipment cost, the total plant cost estimate is approximately:
- (a) $2,034,500
- (b) $1,734,500
- (c) $1,384,500
- (d) $1,183,000

15.66 A police department wants to allocate the indirect cost of speed monitoring to the three toll roads around the city. An allocation basis that may not be reasonable is:
- (a) Miles of toll road monitored
- (b) Average number of cars patrolling per hour
- (c) Amount of car traffic per section of toll road
- (d) Cost to operate a patrol car

15.67 The IT department allocates indirect costs to user departments on the basis of CPU time at the rate of $2000 per second. For the first quarter, the two heaviest-use departments logged 900 and 1300 seconds, respectively. If the IT indirect budget for the year is $8.0 million, the percentage of this year's allocation consumed by these departments is closest to:
- (a) 32%
- (b) 22.5%
- (c) 55%
- (d) Not enough information to determine

15.68 If the engineering department is the activity to receive indirect cost allocation for the year, cost drivers for the ABC method that seem reasonable may be:

1. Cost of engineering changes processed
2. Size of the workforce
3. Administrative cost to process a change order

(*a*)   1
(*b*)   2
(*c*)   3
(*d*)   1 and 3

15.69   Advantages of the ABC method are:

1. It is an excellent replacement for a traditional cost accounting system.
2. It is always cheaper to operate than a traditional allocation system.
3. It can help explain the economic impact of management decisions.

(*a*)   1
(*b*)   2
(*c*)   3
(*d*)   1 and 3

CASE STUDY

INDIRECT COST ANALYSIS OF MEDICAL EQUIPMENT MANUFACTURING COSTS

Background

Three years ago Medical Dynamics, a medical equipment unit of Johnson and Sons, Inc., initiated the manufacture and sales of a portable sterilization unit (Quik-Sterz) that can be placed in the hospital room of a patient. This unit sterilizes and makes available at the bedside some of the reusable instruments that nurses and doctors usually obtain by walking to or receiving delivery from a centralized area. This new unit makes the instruments available at the point and time of use for burn and severe wound patients who are in a regular patient room.

There are two models of Quik-Sterz sold. The standard version sells for $10.75, and a premium version with customized trays and a battery backup system sells for $29.75. The product has sold well to hospitals, convalescent units, and nursing homes at the level of about 1 million units per year.

Information

Medical Dynamics has historically used an indirect cost allocation system based upon direct hours to manufacture for all its other product lines. The same was applied when Quik-Sterz was priced. However, Arnie, the person who performed the indirect cost analysis and set the sales price, is no longer at the company, and the detailed analysis is no longer available. Through e-mail and telephone conversations, Arnie said the current price was set at about 10% above the total manufacturing cost determined 2 years ago, and that some records were available in the design department files. A search of these files revealed the manufacturing and cost information in Table 15–11. It is clear from these and other records that Arnie used traditional indirect cost analysis based on direct labor hours to estimate the total manufacturing costs of $9.73 per unit for the standard model and $27.07 per unit for the premium model.

Last year management decided to place the entire plant on the ABC system of indirect cost allocation. The costs and sales figures collected for Quik-Sterz the year before were still accurate. Five activities and their cost drivers were identified for the Medical Dynamics manufacturing operations (Table 15–12). Also, the volumes for each model are summarized in this table.

TABLE 15–11   Historical Records of Direct and Indirect Cost Analyses for Quik-Sterz

	Quik-Sterz Direct Cost (DC) Evaluation			
Model	Direct Labor, $/Unit[1]	Direct Material, $/Unit	Direct Labor, Hours/Unit	Total Direct Labor Hours
Standard	5.00	2.50	0.25	187,500
Premium	10.00	3.75	0.50	125,000

	Quik-Sterz Indirect Cost (IDC) Evaluation			
Model	Direct Labor, Hours/Unit	Fraction IDC Allocated	Allocated IDC, $	Sales, Units/Year
Standard	0.25	$\frac{1}{3}$	1.67 million	750,000
Premium	0.50	$\frac{2}{3}$	3.33 million	250,000

[1]Average direct labor rate is $20 per hour.

TABLE 15–12 Quik-Sterz Activities, Cost Drivers, and Volume Levels for ABC-Based Indirect Cost Allocation

Activity	Cost Driver	Volume/Year	Actual Cost, $/Year
Quality	Inspections	20,000 inspections	800,000
Purchasing	Purchase orders	40,000 orders	1,200,000
Scheduling	Change orders	1,000 orders	800,000
Production setup	Setups	5,000 setups	1,000,000
Machine operations	Machine hours	10,000 hours	1,200,000

Cost Driver	Volume Level for the Year	
	Standard	Premium
Quality inspections	8,000	12,000
Purchase orders	30,000	10,000
Number of change orders	400	600
Production setup	1,500	3,500
Machine hours	7,000	3,000

The ABC method will be used henceforth, with the intention of determining the total cost and price based on its results. The first impression of the production manager is that the new system will show that indirect costs for Quik-Sterz are about the same as they have been for other products over the last several years when a standard model and an upgrade (premium) model were sold. Predictably, they state, the standard model will receive about 1/3 of the indirect cost, and the premium will receive the remaining 2/3. Fundamentally, there are two reasons why the production manager does not like to produce premium versions: They are less profitable for the company, and they require significantly more time and operations to manufacture.

Case Study Exercises

1. Use traditional indirect cost allocation to verify Arnie's cost and price estimates.

2. Use the ABC method to estimate the indirect cost allocation and total cost for each model.

3. If the prices and number of units sold are the same next year (750,000 standard and 250,000 premium), and all other costs remain constant, compare the profit from Quik-Sterz under the ABC method with the profit using the traditional indirect cost allocation method.

4. What prices should Medical Dynamics charge next year based on the ABC method and a 10% markup over cost? What is the total profit from Quik-Sterz predicted to be if sales hold steady?

5. Using the results above, comment on the production manager's prediction of indirect costs using ABC (1/3 standard; 2/3 premium) and the two reasons given to not produce the premium version of Quik-Sterz.

CASE STUDY

DECEPTIVE ACTS CAN GET YOU IN TROUBLE

Contributed by Dr. Paul Askenasy, Agronomist, Texas Commission on Environmental Quality

Background

Surface mining of coal is the removal of soil and sediments from underlying strata that lie above the material to be mined. The law requires that land disturbed by these types of mining activities be returned to a productive capacity that is as good as or better than its productive capacity before mining.

The productive capacity of soils is directly correlated to the textural (sand, silt, and clay content) and chemical characteristics of the soil (for example, pH). To this end, mining companies must sample the different soils found in the areas to be disturbed by mining activities. The purpose of the sampling is to establish a baseline characterizing the textural and chemical makeup of the soils prior to mining. Soils in a low pH range (pH values < 5) are indicative of low fertility. Once the natural resource, such as coal, is removed, the pit is backfilled with sediments and the terrain surface is contoured to reestablish the premine drainages. To meet the baseline for pH, the acreage of the mine soils with low pH should not exceed the acreage of the unmined soils with low pH.

Information

Yucatan Mining Company (not the actual name) planned to disturb 600 acres due to mining activities. The different soils within the 600 acres were depicted in the County Soil Survey where the mining activities were to take place. Prior to mining, the company obtained soil samples from 10 different locations within each soil type and had them analyzed for a number of parameters including pH. Assessment of the data indicated that 30% of the area (180 acres) occupied by the soils in the area to be disturbed had pH values between 4.0 and 4.9. The application for mining was approved by the State Department of Mining and Reclamation.

Six years later, 450 acres had been mined and the terrain surface had been leveled to reestablish premine slopes. Of the 450 acres leveled, 175 acres had pH values between 4.0 and 4.9. The president of Yucatan indicated that the company would submit a revised soil baseline based on new sampling in the remaining 150 acres of unmined soils because, in his opinion, the first soil baseline was biased.

The request to do more soil analyses to augment their existing soil baseline was approved. The company quickly hired a consultant to develop the new baseline, and subsequently Yucatan submitted the final report from its consultant to the State Department of Mining and Reclamation. This revised premine soil baseline indicated that 45% of the premine soils had pH values between 4.0 and 4.9. Comparative results between the old and new samples can be expressed as follows:

	Percent and Acreage of Area	
Soil Baselines	Old Soil Baseline	Revised Soil Baseline
pH: 4.0 − 4.9	30%	45%
	180 acres	270 acres

A rough statistical check between the old and revised soil baselines indicated that the results were mixed. Based on this preliminary result and the fact that there was a significant increase in the percent of area with low-productivity soil, an in-depth analysis of the revised baseline sample study was performed. Contained in the submitted new-sample package was a letter from the Yucatan consultant. It indicated to Yucatan's management that 100 separate soil samples had been obtained and analyzed and that the revised premine soil baseline had been developed using the data from the 30 samples *with the lowest pH values*.

The State Department of Mining and Reclamation staff concluded that the revised soil pH sample data had been carefully "screened" to reduce the amount of remediation work that Yucatan Mining would have to complete. Within a week, Yucatan was notified that further review of the revised soil baseline could not be pursued, because it appeared the revised soil baseline was developed using a technique that skewed results in favor of lower pH values. It was also noted that should Yucatan Mining disagree with this response, the case would be filed with the legal staff as a contested case.

Within several days, the Yucatan president responded indicating that the company was withdrawing the new application from consideration by the department.

Case Study Questions

1. Assume you are the director of the State Department of Mining and Reclamation and were informed of the findings on the new samples versus the old samples. What actions would you direct your staff to take concerning this situation?

2. Suppose there had been several cases of deceptive acts similar to this one over the last few years. What type of "audit" procedures might you want implemented to identify these possibly unethical activities?

3. Yucatan clearly would state that both the old and new samples were randomly located about the entire mining area. When the 30 lowest pH samples were used to establish the new baseline, were the samples still random, according to experimental design standards? If so, why? If not, why not?

4. You and the president of Yucatan Mining have been acquaintances for some years. You have golfed together several times, your and his children are on the same soccer team at school, and your families are members of the same community swimming pool club. What effect would this event have upon your and your family's relationships with the family of the Yucatan president? How would you handle this situation?

5. As a matter of principle and practice, do you believe there is some amount of data-altering or bias-making that is allowed before an application (such as the one described here) should be considered the result of professionally unethical acts? How would you define such a threshold limit?

CHAPTER 16

Depreciation Methods

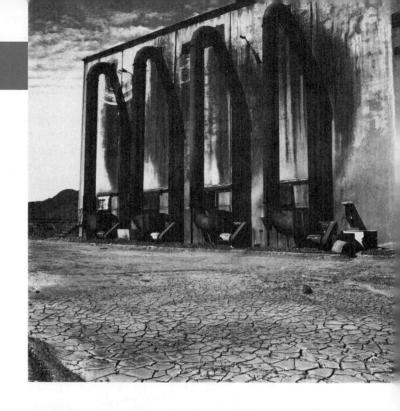

LEARNING OUTCOMES

Purpose: Use depreciation or depletion methods to reduce the book value of a capital investment in an asset and natural resource.

SECTION	TOPIC	LEARNING OUTCOME
16.1	Terminology	• Define and use the basic terms of asset depreciation.
16.2	Straight line	• Apply the straight line (SL) method of depreciation.
16.3	Declining balance	• Apply the declining balance (DB) and double declining balance (DDB) methods of depreciation.
16.4	MACRS	• Apply the modified accelerated cost recovery system for tax depreciation purposes for U.S.-based corporations.
16.5	Recovery period	• Select the asset recovery period for MACRS depreciation.
16.6	Depletion	• Explain depletion; apply cost depletion and percentage depletion methods.
Chapter 16 Appendix		
16A.1	Historical methods	• Apply the sum-of-years-digits (SYD) and unit-of-production (UOP) methods of depreciation.
16A.2	Switching	• Switch between classical depreciation methods; explain how MACRS provides for switching.
16A.3	MACRS and switching	• Calculate MACRS rates using switching between classical methods and MACRS rules.

T he capital investments of a corporation in tangible assets—equipment, computers, vehicles, buildings, and machinery—are commonly recovered on the books of the corporation through *depreciation*. Although the depreciation amount is **not an actual cash flow,** the process of depreciating an asset on the books of the corporation accounts for the decrease in an asset's value because of age, wear, and obsolescence. Even though an asset may be in excellent working condition, the fact that it is worth less through time is taken into account in after-tax economic evaluation studies. An introduction to depreciation types, terminology, and classical methods is followed by a discussion of the *Modified Accelerated Cost Recovery System (MACRS)*, which is the standard in the United States for tax purposes. Other countries commonly use the classical methods for tax computations.

Why is depreciation important to engineering economy? Depreciation is a **tax-allowed deduction** included in tax calculations in virtually all industrialized countries. Depreciation lowers income taxes via the general relation

$$\text{Taxes} = (\text{income} - \text{deductions})(\text{tax rate})$$

Income taxes are discussed further in Chapter 17.

This chapter concludes with an introduction to two methods of **depletion,** which are used to recover capital investments in deposits of natural resources such as oil, gas, minerals, ores, and timber.

The chapter appendix describes two historically useful methods of depreciation—sum-of-years-digits and unit-of-production. Additionally, the appendix includes an in-depth derivation of the MACRS depreciation rates from the straight line and declining balance rates. This is accomplished using a procedure called **switching** between classical depreciation methods.

16.1 Depreciation Terminology ● ● ○

The concept and types of depreciation are defined here. Most descriptions are applicable to corporations as well as individuals who own depreciable assets.

Depreciation is a book method (noncash) to represent the reduction in value of a tangible asset. The method used to depreciate an asset is a way to account for the decreasing value of the asset to the owner *and* to represent the diminishing value (amount) of the capital funds invested in it. The annual depreciation amount is **not an actual cash flow,** nor does it necessarily reflect the actual usage pattern of the asset during ownership.

Though the term **amortization** is sometimes used interchangeably with the term *depreciation,* they are different. Depreciation is applied to tangible assets, while amortization is used to reflect the decreasing value of intangibles, such as loans, mortgages, patents, trademarks, and goodwill. In addition, the term *capital recovery* is sometimes used to identify depreciation. This is clearly a different use of the term than what we learned in Chapter 5. The term *depreciation* is used throughout this book.

There are two different purposes for using the depreciation methods we will cover in this chapter:

Book depreciation   Used by a corporation or business for *internal* financial accounting to *track the value* of an asset or property over its life.

Tax depreciation   Used by a corporation or business to *determine taxes due* based on current tax laws of the government entity (country, state, province, etc.). Even though depreciation itself is not a cash flow, it can result in actual cash flow changes because the amount of tax depreciation is a deductible item when calculating annual income taxes for the corporation or business.

The methods applied for these two purposes may or may not utilize the same formulas. Book depreciation indicates the reduced investment in an asset based upon the usage pattern and expected useful life of the asset. There are classical, internationally accepted depreciation methods used to determine book depreciation: straight line, declining balance, and the historical sum-of-years-digits method. The amount of tax depreciation is important in an after-tax engineering economy study and will vary among nations.

> In most industrialized countries, the annual **tax depreciation is tax-deductible;** that is, it is subtracted from income when calculating the amount of taxes due each year. However, the tax depreciation amount must be calculated using a government-approved method.

Tax depreciation may be calculated and referred to differently in countries outside the United States. For example, in Canada the equivalent is CCA (capital cost allowance), which is calculated based on the undepreciated value of all corporate properties that form a particular class of assets, whereas in the United States depreciation may be determined for each asset separately.

Where allowed, tax depreciation is usually based on an **accelerated method,** whereby the depreciation for the first years of use is larger than that for later years. In the United States this method is called MACRS, as covered in later sections. In effect, accelerated methods defer some of the income tax burden to later in the asset's life; they do not reduce the total tax burden.

Common terms used in depreciation are explained here.

First cost P or **unadjusted basis B** is the delivered and installed cost of the asset including purchase price, delivery and installation fees, and other depreciable direct costs incurred to prepare the asset for use. The term *unadjusted basis,* or simply *basis,* is used when the asset is new, with the term *adjusted basis* used after some depreciation has been charged. When the first cost has no added, depreciable costs, the basis is the first cost, that is, $P = B$.

Book value BV_t represents the remaining, undepreciated capital investment on the books after the total amount of depreciation charges to date has been subtracted from the basis. The book value is determined at the end of each year t ($t = 1, 2, \ldots, n$), which is consistent with the end-of-year convention.

Recovery period n is the depreciable life of the asset in years. Often there are different n values for book and tax depreciation. Both of these values may be different from the asset's estimated productive life.

Market value MV, a term also used in replacement analysis, is the estimated amount realizable if the asset were sold on the open market. Because of the structure of depreciation laws, the book value and market value may be substantially different. For example, a commercial building tends to increase in market value, but the book value will decrease as depreciation charges are taken. However, a computer workstation may have a market value much lower than its book value due to rapidly changing technology.

Salvage value S is the estimated trade-in or market value at the end of the asset's useful life. The salvage value, expressed as an estimated dollar amount or as a percentage of the first cost, may be positive, zero, or negative due to dismantling and carry-away costs.

Depreciation rate or **recovery rate d_t** is the fraction of the first cost removed by depreciation each year t. This rate may be the same each year, which is called the straight line rate d, or different for each year of the recovery period.

Personal property, one of the two types of property for which depreciation is allowed is the income-producing, tangible possessions of a corporation used to conduct business. Included is most manufacturing and service industry property—vehicles, manufacturing equipment, materials handling devices, computers and networking equipment, communications equipment, office furniture, refining process equipment, construction assets, and much more.

Real property includes real estate and all improvements—office buildings, manufacturing structures, test facilities, warehouses, apartments, and other structures. *Land itself is considered real property, but it is not depreciable.*

Half-year convention assumes that assets are placed in service or disposed of in midyear, regardless of when these events actually occur during the year. This convention is utilized in this text and in most U.S.-approved tax depreciation methods. There are also midquarter and midmonth conventions.

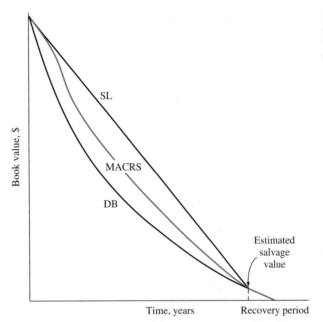

Figure 16–1
General shape of book value curves for different depreciation methods.

As mentioned before, there are several models for depreciating assets. The straight line (SL) method is used historically and internationally. Accelerated models, such as the declining balance (DB) method, decrease the book value to zero (or to the salvage value) more rapidly than the straight line method, as shown by the general book value curves in Figure 16–1.

For each of the methods—straight line, declining balance, MACRS, and sum-of-years-digits—there are spreadsheet functions available to determine annual depreciation. Each function is introduced and illustrated as the method is explained.

As expected, there are many rules and exceptions to the depreciation laws of a country. One that may be of interest to a U.S.-based *small or medium-sized business* performing an economic analysis is the *Section 179 Deduction*. This is an economic incentive that changes over the years and encourages businesses to invest capital in equipment directly used in the company. Up to a specified amount, the entire basis of an asset is treated as a business expense in the year of purchase. This tax treatment reduces federal income taxes, just as depreciation does, but it is allowed in lieu of depreciating the first cost over several years. The limit changes with time; it was $24,000 in 2002; $102,000 in 2004; $125,000 in 2007; and $250,000 in 2008–2010. The economic stimulus efforts in the United States and around the world during the latter part of the decade made many attempts to put investment capital to work in small and medium-sized businesses. Investments above these limits must be depreciated using MACRS.

In the 1980s the U.S. government standardized accelerated methods for *federal tax depreciation* purposes. In 1981, all classical methods, including straight line, declining balance, and sum-of-years-digits depreciation, were disallowed as tax deductible and replaced by the Accelerated Cost Recovery System (ACRS). In a second round of standardization, MACRS (Modified ACRS) was made the required tax depreciation method in 1986. To this date, the following is the law in the United States.

Tax depreciation must be calculated using MACRS; **book depreciation** may be calculated using any classical method or MACRS.

MACRS has the DB and SL methods, in slightly different forms, embedded in it, but these two methods cannot be used directly if the annual depreciation is to be tax deductible. Many U.S. companies still apply the classical methods for keeping their own books, because these methods are more representative of how the usage patterns of the asset reflect the remaining capital invested in it. Most other countries still recognize the classical methods of straight line and declining balance for tax or book purposes. Because of the continuing importance of the SL and DB methods, they are explained in the next two sections prior to MACRS. Appendix Section 16A.1 discusses two historical methods of depreciation.

Tax law revisions occur often, and depreciation rules are changed from time to time in the United States and other countries. For more depreciation and tax law information, consult the U.S. Department of the Treasury, Internal Revenue Service (IRS), website at www.irs.gov. Pertinent publications can be downloaded. Publication 946, *How to Depreciate Property,* is especially applicable to this chapter. MACRS and most corporate tax depreciation laws are discussed in it.

Online Presentation

16.2 Straight Line (SL) Depreciation ● ● ●

Straight line depreciation derives its name from the fact that the book value decreases **linearly with time.** The depreciation rate is the same $(1/n)$ each year of the recovery period n.

Straight line depreciation is considered the standard against which any depreciation model is compared. For *book depreciation* purposes, it offers an excellent representation of book value for any asset that is used regularly over an estimated number of years. For *tax depreciation,* as mentioned earlier, it is not used directly in the United States, but it is commonly used in most countries for tax purposes. However, the U.S. MACRS method includes a version of SL depreciation with a larger n value than that prescribed by regular MACRS (see Section 16.5).

The annual SL depreciation is determined by multiplying the first cost minus the salvage value by d_t. In equation form,

$$D_t = (B - S)d_t$$
$$= \frac{B - S}{n} \qquad [16.1]$$

where t = year $(t = 1, 2, \ldots, n)$
D_t = annual depreciation charge
B = first cost or unadjusted basis
S = estimated salvage value
n = recovery period
d_t = depreciation rate = $1/n$

Since the asset is depreciated by the same amount each year, the book value after t years of service, denoted by BV_t, will be equal to the first cost B minus the annual depreciation times t.

$$BV_t = B - tD_t \qquad [16.2]$$

Earlier we defined d_t as a depreciation rate for a specific year t. However, the SL model has the same rate for all years, that is,

$$d = d_t = \frac{1}{n} \qquad [16.3]$$

The format for the spreadsheet function to display the annual depreciation D_t in a single-cell operation is

$$= SLN(B,S,n) \qquad [16.4]$$

EXAMPLE 16.1

If an asset has a first cost of $50,000 with a $10,000 estimated salvage value after 5 years, (*a*) calculate the annual depreciation and (*b*) calculate and plot the book value of the asset after each year, using straight line depreciation.

Solution
(*a*) The depreciation each year for 5 years can be found by Equation [16.1].

$$D_t = \frac{B - S}{n} = \frac{50,000 - 10,000}{5} = \$8000$$

Enter the function = SLN(50000,10000,5) in any cell to display the D_t of $8000.

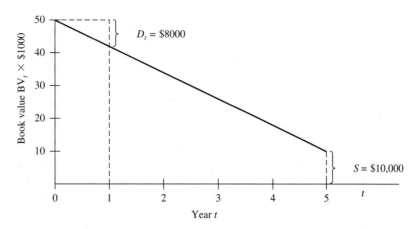

Figure 16–2
Book value of an asset using straight line depreciation, Example 16.1.

(*b*) The book value after each year *t* is computed using Equation [16.2]. The BV$_t$ values are plotted in Figure 16–2. For years 1 and 5, for example,

$$BV_1 = 50{,}000 - 1(8000) = \$42{,}000$$
$$BV_5 = 50{,}000 - 5(8000) = \$10{,}000 = S$$

16.3 Declining Balance (DB) and Double Declining Balance (DDB) Depreciation ● ● ●

The **declining balance method** is commonly applied as the book depreciation method. Like the SL method, DB is embedded in the MACRS method, but the DB method itself cannot be used to determine the annual tax-deductible depreciation in the United States. This method is used routinely in most other countries for tax and book depreciation purposes.

Declining balance is also known as the *fixed percentage* or *uniform percentage* method. DB depreciation accelerates the write-off of asset value because the annual depreciation is determined by multiplying the *book value at the beginning of a year* by a fixed (uniform) percentage *d*, expressed in decimal form. If *d* = 0.1, then 10% of the book value is removed each year. Therefore, the depreciation amount decreases each year.

The maximum annual depreciation rate for the DB method is twice the straight line rate, that is,

$$d_{max} = 2/n \qquad \text{[16.5]}$$

In this case the method is called *double declining balance (DDB)*. If *n* = 10 years, the DDB rate is 2/10 = 0.2; so 20% of the book value is removed annually. Another commonly used percentage for the DB method is 150% of the SL rate, where *d* = 1.5/*n*.

The depreciation for year *t* is the fixed rate *d* times the book value at the end of the previous year.

$$D_t = (d)BV_{t-1} \qquad \text{[16.6]}$$

The actual depreciation rate for each year *t*, relative to the basis *B*, is

$$d_t = d(1 - d)^{t-1} \qquad \text{[16.7]}$$

If BV$_{t-1}$ is not known, the depreciation in year *t* can be calculated using *B* and *d*.

$$D_t = dB(1 - d)^{t-1} \qquad \text{[16.8]}$$

Book value in year *t* is determined in one of two ways: by using the rate *d* and basis *B* or by subtracting the current depreciation charge from the previous book value. The equations are

$$BV_t = B(1 - d)^t \qquad \text{[16.9]}$$
$$BV_t = BV_{t-1} - D_t \qquad \text{[16.10]}$$

It is important to understand that the book value for the DB method never goes to zero, because the book value is always decreased by a fixed percentage. The **implied salvage value** after n years is the BV_n amount, that is,

$$\text{Implied } S = BV_n = B(1 - d)^n \tag{16.11}$$

If a salvage value is estimated for the asset, this *estimated S value is not used in the DB or DDB method* to calculate annual depreciation. However, if the implied $S <$ estimated S, it is necessary to stop charging further depreciation when the book value is at or below the estimated salvage value. In most cases, the estimated S is in the range of zero to the implied S value. (This guideline is important when the DB method can be used directly for tax depreciation purposes.)

If the fixed percentage d is not stated, it is possible to determine an implied fixed rate using the estimated S value, if $S > 0$. The range for d is $0 < d < 2/n$.

$$\text{Implied } d = 1 - \left(\frac{S}{B}\right)^{1/n} \tag{16.12}$$

The spreadsheet functions DDB and DB are used to display depreciation amounts for specific years. The function is repeated in consecutive spreadsheet cells because the depreciation amount D_t changes with t. For the double declining balance method, the format is

$$= \text{DDB}(B,S,n,t,d) \tag{16.13}$$

The entry d is the fixed rate expressed as a number between 1 and 2. If omitted, this optional entry is assumed to be 2 for DDB. An entry of $d = 1.5$ makes the DDB function display 150% declining balance method amounts. The DDB function automatically checks to determine when the book value equals the estimated S value. No further depreciation is charged when this occurs. (To allow *full* depreciation charges to be made, ensure that the S entered is between zero and the implied S from Equation [16.11].) Note that $d = 1$ is the same as the straight line rate $1/n$, but D_t *will not be the SL amount* because declining balance depreciation is determined as a fixed percentage of the previous year's book value, which is completely different from the SL calculation in Equation [16.1].

The format for the DB function is $= \text{DB}(B,S,n,t)$. Caution is needed when using this function. The fixed rate d is not entered in the DB function; d is an embedded calculation using a spreadsheet equivalent of Equation [16.12]. Also, only three significant digits are maintained for d, so the book value may go below the estimated salvage value due to round-off errors. Therefore, if the depreciation rate is known, always use the DDB function to ensure correct results. Examples 16.2 and 16.3 illustrate DB and DDB depreciation and their spreadsheet functions.

EXAMPLE 16.2

Underwater electroacoustic transducers were purchased for use in SONAR applications. The equipment will be DDB depreciated over an expected life of 12 years. There is a first cost of $25,000 and an estimated salvage of $2500. (*a*) Calculate the depreciation and book value for years 1 and 4. Write the spreadsheet functions to display depreciation for years 1 and 4. (*b*) Calculate the implied salvage value after 12 years.

Solution

(*a*) The DDB fixed depreciation rate is $d = 2/n = 2/12 = 0.1667$ per year. Use Equations [16.8] and [16.9].

Year 1: $D_1 = (0.1667)(25,000)(1 - 0.1667)^{1-1} = \$4167$
           $BV_1 = 25,000(1 - 0.1667)^1 = \$20,833$

Year 4: $D_4 = (0.1667)(25,000)(1 - 0.1667)^{4-1} = \$2411$
           $BV_4 = 25,000(1 - 0.1667)^4 = \$12,054$

The DDB functions for D_1 and D_4 are, respectively, $= \text{DDB}(25000,2500,12,1)$ and $= \text{DDB}(25000,2500,12,4)$.

(*b*) From Equation [16.11], the implied salvage value after 12 years is

$$\text{Implied } S = 25{,}000(1 - 0.1667)^{12} = \$2803$$

Since the estimated $S = \$2500$ is less than $\$2803$, the asset is not fully depreciated when its 12-year expected life is reached.

EXAMPLE 16.3

Freeport-McMoRan Copper and Gold has purchased a new ore grading unit for $80,000. The unit has an anticipated life of 10 years and a salvage value of $10,000. Use the DB and DDB methods to compare the schedule of depreciation and book values for each year. Solve by hand and by spreadsheet.

Solution by Hand

An implied DB depreciation rate is determined by Equation [16.12].

$$d = 1 - \left(\frac{10{,}000}{80{,}000}\right)^{1/10} = 0.1877$$

Note that $0.1877 < 2/n = 0.2$, so this DB model does not exceed twice the straight line rate. Table 16–1 presents the D_t values using Equation [16.6] and the BV_t values from Equation [16.10] rounded to the nearest dollar. For example, in year $t = 2$, the DB results are

$$D_2 = d(BV_1) = 0.1877(64{,}984) = \$12{,}197$$
$$BV_2 = 64{,}984 - 12{,}197 = \$52{,}787$$

Because we round off to even dollars, $2312 is calculated for depreciation in year 10, but $2318 is deducted to make $BV_{10} = S = \$10{,}000$ exactly. Similar calculations for DDB with $d = 0.2$ result in the depreciation and book value series in Table 16–1.

TABLE 16–1 D_t and BV_t Values for DB and DDB Depreciation, Example 16.3

Year *t*	Declining Balance, $		Double Declining Balance, $	
	D_t	BV_t	D_t	BV_t
0	—	80,000	—	80,000
1	15,016	64,984	16,000	64,000
2	12,197	52,787	12,800	51,200
3	9,908	42,879	10,240	40,960
4	8,048	34,831	8,192	32,768
5	6,538	28,293	6,554	26,214
6	5,311	22,982	5,243	20,972
7	4,314	18,668	4,194	16,777
8	3,504	15,164	3,355	13,422
9	2,846	12,318	2,684	10,737
10	2,318	10,000	737	10,000

Solution by Spreadsheet

The spreadsheet in Figure 16–3 displays the results for the DB and DDB methods. The chart plots book values for each year. Since the fixed rates are close—0.1877 for DB and 0.2 for DDB—the annual depreciation and book value series are approximately the same for the two methods.

The depreciation rate (cell B5) is calculated by Equation [16.12], but note in the cell tags that the DDB function is used in both columns B and D to determine annual depreciation. As mentioned earlier, the DB function automatically calculates the implied rate by Equation [16.12] and maintains it to only three significant digits. Therefore, if the DB function

Figure 16–3
Annual depreciation and book value using DB and DDB methods, Example 16.3.

were used in column B (Figure 16–3), the fixed rate applied would be 0.188. The resulting D_t and BV_t values for years 8, 9, and 10 would be as follows:

t	D_t, \$	BV_t, \$
8	3,501	15,120
9	2,842	12,277
10	2,308	9,969

Also noteworthy is the fact that the DB function uses the implied rate without a check to halt the book value at the estimated salvage value. Thus, BV_{10} will go slightly below $S = \$10,000$, as shown above. However, the DDB function uses a relation different from that of the DB function to determine annual depreciation—one that correctly stops depreciating at the estimated salvage value, as shown in Figure 16–3, cells E17–E18.

Online
Presentation

16.4 Modified Accelerated Cost Recovery System (MACRS) ● ● ○

In the 1980s, the United States introduced MACRS as the **required tax depreciation method** for all depreciable assets. Through MACRS, the 1986 Tax Reform Act defined statutory depreciation rates that take advantage of the accelerated DB and DDB methods. Corporations are free to apply any of the classical methods for book depreciation. When developed, MACRS and its predecessor ACRS were intended to create economic growth through the investment of new capital and the tax advantages that accelerated depreciation methods offer corporations and businesses.[1]

Many aspects of MACRS deal with the specific depreciation accounting aspects of tax law. This section covers only the elements that materially affect after-tax economic analysis. Additional information on how the DDB, DB, and SL methods are embedded into MACRS and how to derive the MACRS depreciation rates is presented and illustrated in the chapter appendix, Sections 16A.2 and 16A.3.

MACRS determines annual depreciation amounts using the relation

$$D_t = d_t B \qquad [16.14]$$

---

[1]R. Lundquist, "The Pedagogy of Taxes and Tax Purpose Depreciation," *Proceedings,* ASEE Annual Conference, Austin, TX, June 2009.

where the depreciation rate d_t is provided in tabulated form. As for other methods, the book value in year t is determined by subtracting the depreciation amount from the previous year's book value

$$\text{BV}_t = \text{BV}_{t-1} - D_t \qquad [16.15]$$

or by subtracting the total depreciation from the first cost.

> $\text{BV}_t = \text{first cost} - \text{sum of accumulated depreciation}$
>
> $$= B - \sum_{j=1}^{j=t} D_j \qquad [16.16]$$

MACRS has standardized and simplified many of the decisions and calculations of depreciation.

> The basis B (or first cost P) is completely depreciated; salvage is always assumed to be zero, or $S = \$0$.
>   Recovery periods are standardized to specific values:
>
> $n = 3, 5, 7, 10, 15,$ or 20 years   for personal property (e.g., equipment or vehicles)
> $n = 27.5$ or 39 years   for real property (e.g., rental property or structures)
>
> Depreciation rates provide accelerated write-off by incorporating switching between classical methods.

Section 16.5 explains how to determine an allowable MACRS recovery period. The MACRS personal property depreciation rates (d_t values) for $n = 3, 5, 7, 10, 15,$ and 20 for use in Equation [16.14] are included in Table 16–2.

   MACRS depreciation rates incorporate the DDB method ($d = 2/n$) and switch to SL depreciation during the recovery period as an inherent component for *personal property* depreciation. The MACRS rates start with the DDB rate or the 150% DB rate and switch when the SL method offers faster write-off.

TABLE 16–2   Depreciation Rates d_t Applied to the Basis B for the MACRS Method

Year	Depreciation Rate (%) for Each MACRS Recovery Period in Years					
	$n = 3$	$n = 5$	$n = 7$	$n = 10$	$n = 15$	$n = 20$
1	33.33	20.00	14.29	10.00	5.00	3.75
2	44.45	32.00	24.49	18.00	9.50	7.22
3	14.81	19.20	17.49	14.40	8.55	6.68
4	7.41	11.52	12.49	11.52	7.70	6.18
5		11.52	8.93	9.22	6.93	5.71
6		5.76	8.92	7.37	6.23	5.29
7			8.93	6.55	5.90	4.89
8			4.46	6.55	5.90	4.52
9				6.56	5.91	4.46
10				6.55	5.90	4.46
11				3.28	5.91	4.46
12					5.90	4.46
13					5.91	4.46
14					5.90	4.46
15					5.91	4.46
16					2.95	4.46
17–20						4.46
21						2.23

For *real property,* MACRS utilizes the SL method for $n = 39$ throughout the recovery period. The annual percentage depreciation rate is $d = 1/39 = 0.02564$. However, MACRS forces partial-year recovery in years 1 and 40. The MACRS real property rates in percentage amounts are

$$
\begin{array}{ll}
\text{Year 1} & 100d_1 = 1.391\% \\
\text{Years 2–39} & 100d_t = 2.564\% \\
\text{Year 40} & 100d_{40} = 1.177\%
\end{array}
$$

The real property recovery period of 27.5 years, which applies only to residential rental property, uses the SL method in a similar fashion.

Note that all MACRS depreciation rates in Table 16–2 are presented for 1 year longer than the stated recovery period. Also note that the extra-year rate is one-half of the previous year's rate. This is so because a built-in **half-year convention** is imposed by MACRS. This convention assumes that all property is placed in service at the midpoint of the tax year of installation. Therefore, only 50% of the first-year DB depreciation applies for tax purposes. This removes some of the accelerated depreciation advantage and requires that one-half year of depreciation be taken in year $n + 1$.

No specially designed spreadsheet function is present for MACRS depreciation. However, the variable declining balance (VDB) function, which is used to determine when to switch between classical methods, can be adapted to display MACRS deprecation for each year. (The VDB function is explained in detail in Section 16A.2 of this chapter and Appendix A of the text.) The MACRS depreciation format of the VDB function requires embedded MAX and MIN functions, as follows:

$$
= \text{VDB}(B, 0, n, \text{MAX}(0, t-1.5), \text{MIN}(n, t-0.5), d) \qquad [16.17]
$$

where B = first cost
  0 = salvage value of $S = 0$
  n = recovery period
  $d = \begin{cases} 2 & \text{if MACRS } n = 3, 5, 7, \text{ or } 10 \\ 1.5 & \text{if MACRS } n = 15 \text{ or } 20 \end{cases}$

The MAX and MIN functions ensure that the MACRS half-year conventions are followed; that is, only one-half of the first year's depreciation is charged in year 1, and one-half of the last year's charge is carried over to year $n + 1$.

EXAMPLE 16.4

Chevron Phillips Chemical Company in Baytown, Texas, acquired new equipment for its polyethylene processing line. This chemical is a resin used in plastic pipe, retail bags, blow molding, and injection molding. The equipment has an unadjusted basis of $B = \$400,000$, a life of only 3 years, and a salvage value of 5% of B. The chief engineer asked the finance director to provide an analysis of the difference between (1) the DDB method, which is the internal book depreciation and book value method used at the plant, and (2) the required MACRS tax depreciation and its book value. He is especially curious about the differences after 2 years of service for this short-lived, but expensive asset. Use hand and spreadsheet solutions to do the following:

(*a*) Determine which method offers the larger total depreciation after 2 years.
(*b*) Determine the book value for each method after 2 years and at the end of the recovery period.

Solution by Hand
The basis is $B = \$400,000$ and the estimated $S = 0.05(400,000) = \$20,000$. The MACRS rates for $n = 3$ are taken from Table 16–2, and the depreciation rate for DDB is $d_{max} = 2/3 = 0.6667$. Table 16–3 presents the depreciation and book values. Year 3 depreciation for DDB would be $\$44,444(0.6667) = \$29,629$, except this would make $BV_3 < \$20,000$. Only the remaining amount of $\$24,444$ is removed.

		MACRS		DDB	
		Tax	Book	Book	Book
Year	Rate	Depreciation, $	Value, $	Depreciation, $	Value, $
0			400,000		400,000
1	0.3333	133,320	266,680	266,667	133,333
2	0.4445	177,800	88,880	88,889	44,444
3	0.1481	59,240	29,640	24,444	20,000
4	0.0741	29,640	0		

TABLE 16–3 Comparing MACRS and DDB Depreciation, Example 16.4

(*a*) The 2-year accumulated depreciation values from Table 16–3 are

MACRS: $D_1 + D_2 = \$133{,}320 + 177{,}800 = \$311{,}120$

DDB: $D_1 + D_2 = \$266{,}667 + 88{,}889 = \$355{,}556$

The DDB depreciation is larger. (Remember that for tax purposes, the company does not have the choice in the United States of DDB as applied here.)

(*b*) After 2 years the book value for DDB at $44,444 is 50% of the MACRS book value of $88,880. At the end of recovery (4 years for MACRS due to the built-in half-year convention, and 3 years for DDB), the MACRS book value is $BV_4 = 0$ and for DDB, $BV_3 = \$20{,}000$. This occurs because MACRS always removes the entire first cost, regardless of the estimated salvage value. This is a tax depreciation advantage of the MACRS method (unless the asset is disposed of for more than the MACRS-depreciated book value, as discussed in Section 17.4).

Solution by Spreadsheet

Figure 16–4 presents the spreadsheet solution using the VDB function (column B) for MACRS depreciation (in lieu of the MACRS rates) and applying the DDB function in column D.

(*a*) The 2-year accumulated depreciation values are

MACRS (add cells B6 + B7): $\$133{,}333 + 177{,}778 = \$311{,}111$

DDB (add cells D6 + D7): $\$266{,}667 + 88{,}889 = \$355{,}556$

(*b*) Book values after 2 years are

MACRS (cell C7): $88,889

DDB (cell E7): $44,444

The book values are plotted in Figure 16–4. Observe that MACRS goes to zero in year 4, while DDB stops at $20,000 in year 3.

Comment

It is advisable to set up a *spreadsheet template* for use with depreciation problems in this and future chapters. The format and functions of Figure 16–4 are a good template for MACRS and DDB methods.

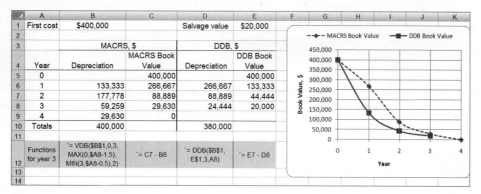

Figure 16–4
Spreadsheet screen shot of MACRS and DDB depreciation and book value, Example 16.4.

MACRS simplifies depreciation computations, but it removes much of the flexibility of method selection for a business or corporation. In general, an economic comparison that includes depreciation may be performed more rapidly and usually without altering the final decision by applying the *classical straight line method in lieu of MACRS.*

Online Presentation

16.5 Determining the MACRS Recovery Period ●●●

The expected useful life of property is estimated in years and used as the *n* value in alternative evaluation and in depreciation computations. For book depreciation the *n* value should be the expected useful life. However, when the depreciation will be claimed as tax deductible, the *n* value should be lower. There are tables that assist in determining the life and recovery period for tax purposes.

The advantage of a recovery period **shorter** than the anticipated useful life is leveraged by the accelerated depreciation methods that write off more of the basis *B* in the initial years.

The U.S. government requires that all depreciable property be classified into a **property class** which identifies its MACRS-allowed recovery period. Table 16–4, a summary of material from IRS Publication 946, gives examples of assets and the MACRS *n* values. Virtually any property considered in an economic analysis has a MACRS *n* value of 3, 5, 7, 10, 15, or 20 years.

Table 16–4 provides two MACRS *n* values for each property. The first is the **general depreciation system (GDS)** value, which we use in examples and problems. The depreciation rates in Table 16–2 correspond to the *n* values for the GDS column and provide the fastest write-off allowed. The rates utilize the DDB method or the 150% DB method with a switch to SL depreciation. Note that any asset not in a stated class is automatically assigned a 7-year recovery period under GDS.

The far right column of Table 16–4 lists the **alternative depreciation system (ADS)** recovery period range. This alternative method allows the use of *SL depreciation over a longer*

TABLE 16–4	Example MACRS Recovery Periods for Various Asset Descriptions		
		MACRS *n* Value, Years	
Asset Description (Personal and Real Property)		**GDS**	**ADS Range**
Special manufacturing and handling devices, tractors, racehorses		3	3–5
Computers and peripherals, oil and gas drilling equipment, construction assets, autos, trucks, buses, cargo containers, some manufacturing equipment		5	6–9.5
Office furniture; some manufacturing equipment; railroad cars, engines, tracks; agricultural machinery; petroleum and natural gas equipment; **all property not in another class**		7	10–15
Equipment for water transportation, petroleum refining, agriculture product processing, durable-goods manufacturing, shipbuilding		10	15–19
Land improvements, docks, roads, drainage, bridges, landscaping, pipelines, nuclear power production equipment, telephone distribution		15	20–24
Municipal sewers, farm buildings, telephone switching buildings, power production equipment (steam and hydraulic), water utilities		20	25–50
Residential rental property (house, mobile home)		27.5	40
Nonresidential real property attached to the land, but not the land itself		39	40

recovery period than the GDS. The half-year convention applies, and any salvage value is neglected, as it is in regular MACRS. The use of ADS is generally a choice left to a company, but it is required for some special asset situations. Since it takes longer to depreciate the asset, and since the SL model is required (thus removing the advantage of accelerated depreciation), ADS is usually not considered an option for the economic analysis. This electable SL option is, however, sometimes chosen by businesses that are young and do not need the tax benefit of accelerated depreciation during the first years of operation and asset ownership. If ADS is selected, tables of d_t rates are available.

16.6 Depletion Methods ● ● ◦

Online
Presentation

Previously, for all assets, facilities, and equipment that can be replaced we have applied depreciation. We now turn to irreplaceable natural resources and the equivalent of depreciation, which is called *depletion*.

> **Depletion** is a book method (noncash) to represent the decreasing value of a **natural resource** as it is recovered, removed, or felled. The two methods of depletion for book or tax purposes are used to write off the first cost, or value of the estimated quantity, of resources in mines, wells, quarries, geothermal deposits, forests, and the like.

The two methods of depletion are *cost* and *percentage depletion,* as described below. Details for U.S. taxes on depletion are found in IRS Publication 535, *Business Expenses.*

Cost depletion   Sometimes referred to as factor depletion, cost depletion is based on the level of activity or usage, not time, as in depreciation. Cost depletion may be applied to most types of natural resources and *must* be applied to timber production. The cost depletion factor for year t, denoted by CD_t, is the ratio of the first cost of the resource to the estimated number of units recoverable.

$$CD_t = \frac{\text{first cost}}{\text{resource capacity}} \qquad [16.18]$$

The annual depletion charge is CD_t times the year's usage or volume. *The total cost depletion cannot exceed the first cost of the resource.* If the capacity of the property is reestimated some year in the future, a new cost depletion factor is determined based upon the undepleted amount and the new capacity estimate.

Percentage depletion   This is a special consideration given for natural resources. A constant, stated percentage of the resource's **gross income** may be depleted each year *provided it does not exceed 50% of the company's taxable income.* The depletion amount for year t is calculated as

$$\text{Percentage depletion}_t = \text{percentage depletion rate}$$
$$\times \text{ gross income from property}$$
$$= PD \times GI_t \qquad [16.19]$$

Using percentage depletion, total depletion charges may exceed first cost with no limitation. The U.S. government does not generally allow percentage depletion to be applied to oil and gas wells (except small independent producers).

The annual percentage depletion rates for some common natural deposits are listed below per U.S. tax law.

Deposit	Percentage of Gross Income, PD
Sulfur, uranium, lead, nickel, zinc, and some other ores and minerals	22
Gold, silver, copper, iron ore, and some oil shale	15
Oil and natural gas wells (varies)	15–22
Coal, lignite, sodium chloride	10
Gravel, sand, peat, some stones	5
Most other minerals, metallic ores	14

EXAMPLE 16.5

Temple-Inland Corporation has negotiated the rights to cut timber on privately held forest acreage for $700,000. An estimated 350 million board feet of lumber is harvestable.

(a) Determine the depletion amount for the first 2 years if 15 million and 22 million board feet are removed.
(b) After 2 years the total recoverable board feet was reestimated upward to be 450 million from the time the rights were purchased. Compute the new cost depletion factor for years 3 and later.

Solution

(a) Use Equation [16.18] for CD_t in dollars per million board feet.

$$CD_t = \frac{700{,}000}{350} = \$2000 \text{ per million board feet}$$

Multiply CD_t by the annual harvest to obtain depletion of $30,000 in year 1 and $44,000 in year 2. Continue until a total of $700,000 is written off.

(b) After 2 years, a total of $74,000 has been depleted. A new CD_t value must be calculated based on the remaining $700{,}000 - 74{,}000 = \$626{,}000$ investment. Additionally, with the new estimate of 450 million board feet, a total of $450 - 15 - 22 = 413$ million board feet remains. For years $t = 3, 4, \ldots$, the cost depletion factor is

$$CD_t = \frac{626{,}000}{413} = \$1516 \text{ per million board feet}$$

EXAMPLE 16.6

A gold mine was purchased for $10 million. It has an anticipated gross income of $5.0 million per year for years 1 to 5 and $3.0 million per year after year 5. Assume that depletion charges do not exceed 50% of taxable income. Compute annual depletion amounts for the mine. How long will it take to recover the initial investment at $i = 0\%$?

Solution

The rate for gold is PD = 0.15. Depletion amounts are

Years 1 to 5:   0.15(5.0 million) = $750,000

Years thereafter: 0.15(3.0 million) = $450,000

A total of $3.75 million is written off in 5 years, and the remaining $6.25 million is written off at $450,000 per year. The total number of years is

$$5 + \frac{\$6.25 \text{ million}}{\$450{,}000} = 5 + 13.9 = 18.9$$

In 19 years, the initial investment could be fully depleted.

In many of the natural resource depletion situations, the tax law allows the larger of the two depletion amounts to be claimed each year. This is allowed provided the percentage depletion amount does not exceed 50% of taxable income. Therefore, it is wise to calculate both depletion amounts and select the larger. Use the following terminology for year t ($t = 1, 2, \ldots$).

$$CDA_t = \text{cost depletion amount}$$
$$PDA_t = \text{percentage depletion amount}$$
$$TI_t = \text{taxable income}$$

The guideline for the tax-allowed depletion amount for year t is

$$\text{Depletion} = \begin{cases} \max[CDA_t, PDA_t] & \text{if } PDA_t \leq 50\% \text{ of } TI_t \\ \max[CDA_t, 50\% \text{ of } TI_t] & \text{if } PDA_t > 50\% \text{ of } TI_t \end{cases}$$

For example, assume a medium-sized quarry owner calculates the following for 1 year.

$$\text{TI} = \$500,000 \qquad \text{CDA} = \$275,000 \qquad \text{PDA} = \$280,000$$

Since 50% of TI is $250,000, the PDA is too large and, therefore, is not allowed. For tax purposes, apply the guideline above and use the cost depletion of $275,000, since it is larger than 50% of TI.

CHAPTER SUMMARY

Depreciation may be determined for internal company records (book depreciation) or for income tax purposes (tax depreciation). In the United States, the MACRS method is the only one allowed for tax depreciation. In many other countries, straight line and declining balance methods are applied for both tax and book depreciation. Depreciation does not result in cash flow directly. It is a book method by which the capital investment in tangible property is recovered. The annual depreciation amount is tax deductible, which can result in actual cash flow changes.

Some important points about the straight line, declining balance, and MACRS methods are presented below. Common relations for each method are summarized in Table 16–5.

Straight Line (SL)
- It writes off capital investment linearly over n years.
- The estimated salvage value is always considered.
- This is the classical, nonaccelerated depreciation model.

Declining Balance (DB)
- The method accelerates depreciation compared to the straight line method.
- The book value is reduced each year by a fixed percentage.
- The most used rate is twice the SL rate, which is called double declining balance (DDB).
- It has an implied salvage that may be lower than the estimated salvage.
- It is not an approved tax depreciation method in the United States. It is frequently used for book depreciation purposes.

Modified Accelerated Cost Recovery System (MACRS)
- It is the only approved tax depreciation system in the United States.
- It automatically switches from DDB or DB to SL depreciation.
- It always depreciates to zero; that is, it assumes $S = 0$.
- Recovery periods are specified by property classes.
- Depreciation rates are tabulated.
- The actual recovery period is 1 year longer due to the imposed half-year convention.
- MACRS straight line depreciation is an option, but recovery periods are longer than those for regular MACRS.

Cost and *percentage depletion methods* recover investment in natural resources. The annual cost depletion factor is applied to the amount of resource removed. No more than the initial investment can be recovered with cost depletion. Percentage depletion, which can recover more than the initial investment, reduces the investment value by a constant percentage of gross income each year.

TABLE 16–5	Summary of Common Depreciation Method Relations		
Method	**MACRS**	**SL**	**DDB**
Fixed depreciation rate d	Not defined	$\dfrac{1}{n}$	$\dfrac{2}{n}$
Annual rate d_t	Table 16–2	$\dfrac{1}{n}$	$d(1-d)^{t-1}$
Annual depreciation D_t	$d_t B$	$\dfrac{B-S}{n}$	$d(\text{BV}_{t-1})$
Book value BV_t	$\text{BV}_{t-1} - D_t$	$B - tD_t$	$B(1-d)^t$

CHAPTER 16   APPENDIX

16A.1 Sum-of-Years-Digits (SYD) and Unit-of-Production (UOP) Depreciation ●●●

The **SYD method** is a historical accelerated depreciation technique that removes much of the basis in the first one-third of the recovery period; however, write-off is not as rapid as for DDB or MACRS. This technique may be used in an engineering economy analysis in the book depreciation of multiple-asset accounts (group and composite depreciation).

The mechanics of the method involve the sum of the year's digits from 1 through the recovery period n. The depreciation charge for any given year is obtained by multiplying the basis of the asset, less any salvage value, by the ratio of the number of years remaining in the recovery period to the sum of the year's digits SUM.

$$D_t = \frac{\text{depreciable years remaining}}{\text{sum of years digits}}(\text{basis} - \text{salvage value})$$

$$D_t = \frac{n - t + 1}{\text{SUM}}(B - S) \qquad\qquad [16A.1]$$

where SUM is the sum of the digits 1 through n.

$$\text{SUM} = \sum_{j=1}^{j=n} j = \frac{n(n + 1)}{2}$$

The book value for any year t is calculated as

$$\text{BV}_t = B - \frac{t(n - t/2 + 0.5)}{\text{SUM}}(B - S) \qquad\qquad [16A.2]$$

The rate of depreciation decreases each year and equals the multiplier in Equation [16A.1].

$$d_t = \frac{n - t + 1}{\text{SUM}} \qquad\qquad [16A.3]$$

The SYD spreadsheet function displays the depreciation for the year t. The function format is

$$= \text{SYD}(B,S,n,t)$$

EXAMPLE 16A.1

Calculate the SYD depreciation charges for year 2 for electro-optics equipment with $B = \$25,000$, $S = \$4000$, and an 8-year recovery period.

Solution
The sum of the year's digits is 36, and the depreciation amount for the second year by Equation [16A.1] is

$$D_2 = \frac{7}{36}(21,000) = \$4083$$

The SYD function is $= \text{SYD}(25000,4000,8,2)$.

Figure 16A–1 is a plot of the book values for an $80,000 asset with $S = \$10,000$ and $n = 10$ years using the four depreciation methods that we have learned. The MACRS, DDB, and SYD curves track closely except for year 1 and years 9 through 11.

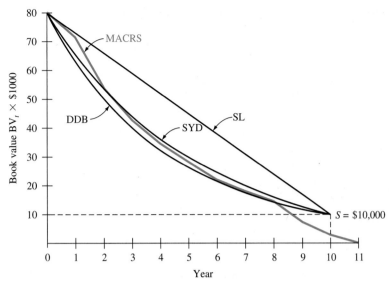

Figure 16A–1
Comparison of book values using SL, SYD, DDB, and MACRS depreciation.

A second depreciation method that is not allowed for tax purposes, but useful in some situations is the **unit-of-production (UOP) method.** When the decreasing value of equipment is **based on usage, not time,** the UOP method is quite applicable. Suppose a highway contractor has a series of state highway department contracts that will last several years and that earth moving equipment is purchased for use on all contracts. If the equipment usage goes up and down significantly over the years, the UOP method is ideal for book depreciation. For year t, UOP deprecation is calculated as

$$D_t = \frac{\text{actual usage for year } t}{\text{total lifetime usage}}(\text{basis} - \text{salvage}) \qquad \text{[16A.4]}$$

EXAMPLE 16A.2

Zachry Contractors purchased an $80,000 mixer for use during the next 10 years for contract work on IH-10 in San Antonio. The mixer will have a negligible salvage value after 10 years, and the total amount of material to process is estimated at 2 million m^3. Use the actual usage per year shown in Table 16A–1 and the unit-of-production method to determine annual depreciation.

Solution
The actual usage each year is placed in the numerator of Equation [16A.4] to determine the annual depreciation based on the estimated total lifetime amount of material, 2 million m^3 in this case. Table 16A–1 shows the annual and cumulative depreciation over the 10 years. If the mixer is continued in service after the 2 million m^3 is processed, no further depreciation is allowed.

TABLE 16A–1	Unit-of-Production Method of Depreciation, Example 16A.2		
Year t	**Actual Usage, 1000 m^3**	**Annual Depreciation D_t, $**	**Cumulative Depreciation, $**
1	400	16,000	16,000
2–8	200	8,000	72,000
9–10	100	4,000	80,000
Total	2000	80,000	

16A.2 Switching between Depreciation Methods ●●●

Switching between depreciation methods may assist in accelerated reduction of the book value. It also maximizes the present value of accumulated and total depreciation over the recovery period. Therefore, switching usually increases the tax advantage in years where the depreciation is larger. The approach below is an inherent part of MACRS.

Switching from a DB method to the SL method is the most common switch because it usually offers a real advantage, especially if the DB method is DDB. General rules of switching are summarized here.

1. Switching is recommended when the depreciation for year t by the currently used method is less than that for a new method. The selected depreciation D_t is the larger amount.
2. Only one switch can take place during the recovery period.
3. Regardless of the (classical) depreciation methods, the book value cannot go below the estimated salvage value. When switching from a DB method, the estimated salvage value, not the DB-implied salvage value, is used to compute the depreciation for the new method; we assume $S = 0$ in all cases. (This does not apply to MACRS, since it already includes switching.)
4. The undepreciated amount, that is, BV_t, is used as the new adjusted basis to select the larger D_t for the next switching decision.

In all situations, the criterion is to **maximize the present worth of the total depreciation PW_D.** The combination of depreciation methods that results in the largest present worth is the best switching strategy.

$$PW_D = \sum_{t=1}^{t=n} D_t(P/F,i,t) \qquad [16A.5]$$

This logic minimizes tax liability in the early part of an asset's recovery period.

Switching is most advantageous from a rapid write-off method such as DDB to the SL model. This switch is predictably advantageous if the implied salvage value computed by Equation [16.11] exceeds the salvage value estimated at purchase time; that is, switch if

$$BV_n = B(1 - d)^n > \text{estimated } S \qquad [16A.6]$$

Since we assume that S will be zero per rule 3 above, and since BV_n *will be greater than zero,* for a DB method a switch to SL is always advantageous. Depending upon the values of d and n, the switch may be best in the later years or last year of the recovery period, which removes the implied S inherent to the DDB model.

The procedure to switch from DDB to SL depreciation is as follows:

1. For each year t, compute the two depreciation charges.

   For DDB: $D_{DDB} = d(BV_{t-1})$ [16A.7]

   For SL: $D_{SL} = \dfrac{BV_{t-1}}{n - t + 1}$ [16A.8]

2. Select the larger depreciation value. The depreciation for each year is

   $$D_t = \max[D_{DDB}, D_{SL}] \qquad [16A.9]$$

3. If needed, determine the present worth of total depreciation, using Equation [16A.5].

It is acceptable, though not usually financially advantageous, to state that a switch will take place in a particular year, for example, a mandated switch from DDB to SL in year 7 of a 10-year recovery period. This approach is usually not taken, but the switching technique will work correctly for all depreciation methods.

To use a spreadsheet for switching, first understand the depreciation model switching rules and practice the switching procedure from declining balance to straight line. Once these are understood, the mechanics of the switching can be speeded up by applying the spreadsheet function

VDB (variable declining balance). This is a quite powerful function that determines the depreciation for 1 year or the total over several years for the DB-to-SL switch. The function format is

$$= \text{VDB}(B,S,n,\text{start\_}t,\text{end\_}t,d,\text{no\_switch}) \qquad [16A.10]$$

Appendix A explains all the fields in detail, but for simple applications, where the DDB and SL annual D_t values are needed, the following are correct entries:

start_t is the year $(t-1)$

end_t is year t

d is optional; 2 for DDB is assumed, the same as in the DDB function

no_switch is an optional logical value:

    FALSE or omitted—switch to SL occurs, if advantageous

    TRUE—DDB or DB method is applied with no switching to SL depreciation considered.

Entering TRUE for the no_switch option obviously causes the VDB function to display the same depreciation amounts as the DDB function. This is discussed in Example 16A.3d. You may notice that the VDB function is the same one used to calculate annual MACRS depreciation.

EXAMPLE 16A.3

The Outback Steakhouse main office has purchased a $100,000 online document imaging system with an estimated useful life of 8 years and a tax depreciation recovery period of 5 years. Compare the present worth of total depreciation for (a) the SL method, (b) the DDB method, and (c) DDB-to-SL switching. (d) Perform the DDB-to-SL switch using a spreadsheet and plot the book values. Use a rate of $i = 15\%$ per year.

Solution by Hand

The MACRS method is not involved in this solution.

(a) Equation [16.1] determines the annual SL depreciation.

$$D_t = \frac{100,000 - 0}{5} = \$20,000$$

Since D_t is the same for all years, the P/A factor replaces P/F to compute PW_D.

$$\text{PW}_D = 20,000(P/A,15\%,5) = 20,000(3.3522) = \$67,044$$

(b) For DDB, $d = 2/5 = 0.40$. The results are shown in Table 16A–2. The value $\text{PW}_D = \$69,915$ exceeds $67,044$ for SL depreciation. As is predictable, the accelerated depreciation of DDB increases PW_D.

(c) Use the DDB-to-SL switching procedure.

1. The DDB values for D_t in Table 16A–2 are repeated in Table 16A–3 for comparison with the D_{SL} values from Equation [16A.8]. The D_{SL} values change each year because BV_{t-1} is different. Only in year 1 is $D_{\text{SL}} = \$20,000$, the same as computed in part (a). For illustration, compute D_{SL} values for years 2 and 4. For $t = 2$, $\text{BV}_1 = \$60,000$ by the DDB method and

$$D_{\text{SL}} = \frac{60,000 - 0}{5 - 2 + 1} = \$15,000$$

For $t = 4$, $\text{BV}_3 = \$21,600$ by the DDB method and

$$D_{\text{SL}} = \frac{21,600 - 0}{5 - 4 + 1} = \$10,800$$

2. The column "Larger D_t" indicates a switch in year 4 with $D_4 = \$10,800$. The $D_{\text{SL}} = \$12,960$ in year 5 would apply *only* if the switch occurred in year 5. Total depreciation with switching is $100,000 compared to the DDB amount of $92,224$.

3. With switching, $\text{PW}_D = \$73,943$, which is an increase over both the SL and DDB methods.

TABLE 16A–2	DDB Model Depreciation and Present Worth Computations, Example 16A.3b			
Year t	D_t, $	BV_t, $	$(P/F,15\%,t)$	Present Worth of D_t, $
0		100,000		
1	40,000	60,000	0.8696	34,784
2	24,000	36,000	0.7561	18,146
3	14,400	21,600	0.6575	9,468
4	8,640	12,960	0.5718	4,940
5	5,184	7,776	0.4972	2,577
Totals	92,224			69,915

TABLE 16A–3	Depreciation and Present Worth for DDB-to-SL Switching, Example 16A.3c					
Year t	DDB Method, $ D_{DDB}	BV_t	SL Method D_{SL}, $	Larger D_t, $	P/F Factor	Present Worth of D_t, $
0	—	100,000				
1	40,000	60,000	20,000	40,000	0.8696	34,784
2	24,000	36,000	15,000	24,000	0.7561	18,146
3	14,400	21,600	12,000	14,400	0.6575	9,468
4*	8,640	12,960	10,800	10,800	0.5718	6,175
5	5,184	7,776	12,960	10,800	0.4972	5,370
Totals	92,224			100,000		73,943

*Indicates year of switch from DDB to SL depreciation.

Solution by Spreadsheet

(d) In Figure 16A–2, column D entries are the VDB functions to determine that the DDB-to-SL switch should take place in year 4. The entries "2,FALSE" at the end of the VDB function are optional (see the VDB function description). If TRUE were entered, the declining balance model would be maintained throughout the recovery period, and the annual depreciation amounts would be equal to those in column B. The plot in Figure 16A–2 indicates another difference in depreciation methods. The terminal book value in year 5 for the DDB method is $BV_5 = \$7776$, while the DDB-to-SL switch reduces the book value to zero.

The NPV function determines the PW of depreciation (row 9). The results here are the same as in parts (b) and (c) above. The DDB-to-SL switch has the larger PW_D value.

Figure 16A–2
Depreciation for DDB-to-SL switch using the VDB function, Example 16A.3.

In MACRS, recovery periods of 3, 5, 7, and 10 years apply DDB depreciation with half-year convention switching to SL. When the switch to SL takes place, which is usually in the last 1 to 3 years of the recovery period, any remaining basis is charged off in year $n + 1$ so that the book value reaches zero. Usually 50% of the applicable SL amount remains after the switch has occurred. For recovery periods of 15 and 20 years, 150% DB with the half-year convention and the switch to SL apply.

The present worth of depreciation PW_D will always indicate which method is the most advantageous. Only the MACRS rates for the GDS recovery periods (Table 16–4) utilize the DDB-to-SL switch. The MACRS rates for the alternative depreciation system (ADS) have longer recovery periods and impose the SL model for the entire recovery period.

EXAMPLE 16A.4

In Example 16A.3, parts (c) and (d), the DDB-to-SL switching method was applied to a $100,000, $n = 5$ years asset resulting in $PW_D = \$73,943$ at $i = 15\%$. Use MACRS to depreciate the same asset for a 5-year recovery period, and compare PW_D values.

Solution
Table 16A–4 summarizes the computations for depreciation (using Table 16–2 rates), book value, and present worth of depreciation. The PW_D values for all four methods are as follows:

DDB-to-SL switching	$73,943
Double declining balance	$69,916
MACRS	$69,016
Straight line	$67,044

MACRS provides a slightly less accelerated write-off. This is so, in part, because the half-year convention disallows 50% of the first-year DDB depreciation (which amounts to 20% of the basis). Also the MACRS recovery period extends to year 6, further reducing PW_D.

TABLE 16A–4 Depreciation and Book Value Using MACRS, Example 16A.4

t	d_t	D_t, $	BV_t, $
0	—	—	100,000
1	0.20	20,000	80,000
2	0.32	32,000	48,000
3	0.192	19,200	28,800
4	0.1152	11,520	17,280
5	0.1152	11,520	5,760
6	0.0576	5,760	0
	1.000	100,000	

$$PW_D = \sum_{t=1}^{t=6} D_t(P/F,15\%,t) = \$69,016$$

16A.3 Determination of MACRS Rates ● ● ●

The depreciation rates for MACRS incorporate the DB-to-SL switching for all GDS recovery periods from 3 to 20 years. In the first year, some adjustments have been made to compute the MACRS rate. The adjustments vary and are not usually considered in detail in economic analyses. The half-year convention is always imposed, and any remaining book value in year n is removed in year $n + 1$. The value $S = 0$ is assumed for all MACRS schedules.

Since different DB depreciation rates apply for different n values, the following summary may be used to determine D_t and BV_t values. The symbols D_{DB} and D_{SL} are used to identify DB and SL depreciation, respectively.

For n = 3, 5, 7, and 10   Use DDB depreciation with the half-year convention, switching to SL depreciation in year t when $D_{SL} \geq D_{DB}$. Use the switching rules of Section 16A.2, and add one-half year when computing D_{SL} to account for the half-year convention. The yearly depreciation rates are

$$d_t = \begin{cases} \dfrac{1}{n} & t = 1 \\[2ex] \dfrac{2}{n} & t = 2, 3, \ldots \end{cases} \qquad [16A.11]$$

Annual depreciation values for each year t applied to the adjusted basis, allowing for the half-year convention, are

$$D_{DB} = d_t(BV_{t-1}) \qquad [16A.12]$$

$$D_{SL} = \begin{cases} \dfrac{1}{2}\left(\dfrac{1}{n}\right)B & t = 1 \\[2ex] \dfrac{BV_{t-1}}{n - t + 1.5} & t = 2, 3, \ldots, n \end{cases} \qquad [16A.13]$$

After the switch to SL depreciation takes place—usually in the last 1 to 3 years of the recovery period—any remaining book value in year n is removed in year $n + 1$.

For n = 15 and 20   Use 150% DB with the half-year convention and the switch to SL when $D_{SL} \geq D_{DB}$. Until SL depreciation is more advantageous, the annual DB depreciation is computed using a form of Equation [16A.7]

$$D_{DB} = d_t(BV_{t-1})$$

where

$$d_t = \begin{cases} \dfrac{0.75}{n} & t = 1 \\[2ex] \dfrac{1.50}{n} & t = 2, 3, \ldots \end{cases} \qquad [16A.14]$$

EXAMPLE 16A.5

A wireless tracking system for shop floor control with a MACRS 5-year recovery period has been purchased for $10,000. (*a*) Use Equations [16A.11] through [16A.13] to obtain the annual depreciation and book value. (*b*) Determine the resulting annual depreciation rates and compare them with the MACRS rates in Table 16–2 for $n = 5$.

Solution

(*a*) With $n = 5$ and the half-year convention, use the DDB-to-SL switching procedure to obtain the results in Table 16A–5. The switch to SL depreciation, which occurs in year 4 when both depreciation values are equal, is indicated by

$$D_{DB} = 0.4(2880) = \$1152$$

$$D_{SL} = \frac{2880}{5 - 4 + 1.5} = \$1152$$

The SL depreciation of $1000 in year 1 results from applying the half-year convention included in the first relation of Equation [16A.13]. Also, the SL depreciation of $576 in year 6 is the result of the half-year convention.

(*b*) The actual rates are computed by dividing the "Larger D_t" column values by the first cost of $10,000. The rates below are the same as the Table 16–2 rates.

t	1	2	3	4	5	6
d_t	0.20	0.32	0.192	0.1152	0.1152	0.0576

TABLE 16A–5	Depreciation Amounts Used to Determine MACRS Rates for $n = 5$, Example 16A.5				
Years	DDB		SL Depreciation	Larger	
t	d_t	D_{DB}, $	D_{SL}, $	D_t, $	BV_t, $
0	—	—	—	—	10,000
1	0.2	2,000	1,000	2,000	8,000
2	0.4	3,200	1,777	3,200	4,800
3	0.4	1,920	1,371	1,920	2,880
4	0.4	1,152	1,152	1,152	1,728
5	0.4	691	1,152	1,152	576
6	—	—	576	576	0
				10,000	

It is clearly easier to use the rates in Table 16–2 or the VDB spreadsheet function than to determine each MACRS rate using the switching logic above. But the logic behind the MACRS rates is described here for those interested. The annual MACRS rates may be derived by using the applicable rate for the DB method. The subscripts DB and SL have been inserted along with the year t. For the first year $t = 1$,

$$d_{DB,1} = \frac{1}{n} \qquad \text{or} \qquad d_{SL,1} = \frac{1}{2}\left(\frac{1}{n}\right)$$

For summation purposes only, we introduce the subscript i ($i = 1, 2, \ldots, t$) on d. Then the depreciation rates for years $t = 2, 3, \ldots, n$ are

$$d_{DB,t} = d\left(1 - \sum_{i=1}^{i=t-1} d_i\right) \qquad [16A.15]$$

$$d_{SL,t} = \frac{\left(1 - \sum_{i=1}^{i=t-1} d_i\right)}{n - t + 1.5} \qquad [16A.16]$$

Also, for year $n + 1$, the MACRS rate is one-half the SL rate of the previous year n.

$$d_{SL,n+1} = 0.5(d_{SL,n}) \qquad [16A.17]$$

The DB and SL rates are compared each year to determine which is larger and when the switch to SL depreciation should occur.

EXAMPLE 16A.6

Verify the MACRS rates in Table 16–2 for a 3-year recovery period. The rates in percent are 33.33, 44.45, 14.81, and 7.41.

Solution

The fixed rate for DDB with $n = 3$ is $d = 2/3 = 0.6667$. Using the half-year convention in year 1 and Equations [16A.15] through [16A.17], the results are as follows:

d_1: $d_{DB,1} = 0.5d = 0.5(0.6667) = 0.3333$

d_2: Cumulative depreciation rate is 0.3333.

$$d_{DB,2} = 0.6667(1 - 0.3333) = 0.4445 \qquad \text{(larger value)}$$

$$d_{SL,2} = \frac{1 - 0.3333}{3 - 2 + 1.5} = 0.2267$$

d_3: Cumulative depreciation rate is $0.3333 + 0.4445 = 0.7778$.

$$d_{DB,3} = 0.6667(1 - 0.7778) = 0.1481$$

$$d_{SL,2} = \frac{1 - 0.7778}{3 - 3 + 1.5} = 0.1481$$

Both values are the same; switch to straight line depreciation.
d_4: This rate is 50% of the last SL rate.

$$d_4 = 0.5(d_{SL,3}) = 0.5(0.1481) = 0.0741$$

PROBLEMS

Fundamentals of Depreciation

16.1 How does depreciation affect a company's cash flow?

16.2 What is the difference between book value and market value?

16.3 State the difference between book depreciation and tax depreciation.

16.4 State the difference between unadjusted and adjusted basis.

16.5 Explain why the recovery period used for tax depreciation purposes may be different from the estimated n value in an engineering economy study.

16.6 Visit the U.S. Internal Revenue Service website at www.irs.gov and answer the following questions about depreciation and MACRS by consulting Publication 946, *How to Depreciate Property*.
   (a) What is the definition of depreciation according to the IRS?
   (b) What is the description of the term *salvage value*?
   (c) What are the two depreciation systems within MACRS, and what are the major differences between them?
   (d) What are the properties listed that cannot be depreciated under MACRS?
   (e) When does depreciation begin and end?
   (f) What is a Section 179 deduction?

16.7 A major energy production company has the following information regarding the acquisition of new-generation equipment.

   Purchase price = $580,000
   Transoceanic shipping and delivery cost = $4300
   Installation cost (1 technician at $1600 per day for 4 days) = $6400
   Tax recovery period = 15 years
   Book depreciation recovery period = 10 years

   Salvage value = 10% of purchase price
   Operating cost (with technician) = $185,000 per year

The manager of the department asked the newest hire to enter the appropriate data in the tax accounting program. For the MACRS method, what are the values of B, n, and S in depreciating the asset for tax purposes?

16.8 Stahmann Products paid $350,000 for a numerical controller during the last month of 2007 and had it installed at a cost of $50,000. The recovery period was 7 years with an estimated salvage value of 10% of the original purchase price. Stahmann sold the system at the end of 2011 for $45,000.
   (a) What numerical values are needed to develop a depreciation schedule at purchase time?
   (b) State the numerical values for the following: remaining life at sale time, market value in 2011, book value at sale time if 65% of the basis had been depreciated.

16.9 An asset with an unadjusted basis of $50,000 was depreciated over $n_{tax} = 10$ years for tax depreciation purposes and $n_{book} = 5$ years for book depreciation purposes. The annual depreciation was $1/n$ using the relevant life value. Use a spreadsheet to plot on one graph the annual book value for both methods of depreciation.

Straight Line Depreciation

16.10 What is the depreciation rate d_t per year for an asset that has an 8-year useful life and is straight line depreciated?

16.11 Pneumatics Engineering purchased a machine that had a first cost of $40,000, an expected useful life of 8 years, a recovery period of 10 years, and a salvage value of $10,000. The operating cost of the machine is expected to be $15,000 per year. The inflation rate is 6% per year and the company's

MARR is 11% per year. Determine (a) the depreciation charge for year 3, (b) the present worth of the third-year depreciation charge in year 0, the time of asset purchase, and (c) the book value for year 3 according to the straight line method.

16.12 An asset that is book-depreciated over a 5-year period by the straight line method has $BV_3 = \$62,000$ with a depreciation charge of $26,000 per year. Determine (a) the first cost of the asset and (b) the assumed salvage value.

16.13 Lee Company of Westbrook, Connecticut, manufactures pressure relief inserts for thermal relief and low-flow hydraulic pressure relief applications where zero leakage is required. A machine purchased 3 years ago has been book-depreciated by the straight line method using a 5-year useful life. If the book value at the end of year 3 is $30,000 and the company assumed that the machine would be worthless at the end of its 5-year useful life, (a) what is the book depreciation charge each year and (b) what was the first cost of the machine?

16.14 An asset has an unadjusted basis of $200,000, a salvage value of $10,000, and a recovery period of 7 years. Write a single-cell spreadsheet function to display the book value after 5 years of straight line depreciation. Use your function to determine the book value.

16.15 Bristol Myers Squibb purchased a tablet-forming machine in 2008 for $750,000. The company planned to use the machine for 10 years; however, due to rapid obsolescence it will be retired after only 4 years in 2012. Develop a spreadsheet for depreciation and book value amounts necessary to answer the following.
  (a) What is the amount of capital investment remaining when the asset is prematurely retired?
  (b) If the asset is sold at the end of 4 years for $175,000, what is the amount of capital investment lost based on straight line depreciation?
  (c) If the new-technology machine has an estimated cost of $300,000, how many more years should the company retain and depreciate the currently owned machine to make its book value and the first cost of the new machine equal to each other?

16.16 A special-purpose graphics workstation acquired by Busbee Consultants has $B = \$50,000$ with a 4-year recovery period. Tabulate the values for SL depreciation, accumulated depreciation, and book value for each year if (a) $S = 0$ and (b) $S = \$16,000$. (c) Use a spreadsheet to plot the book value over the 4 years on one chart for both salvage value estimates.

16.17 A company owns the same asset in a U.S. plant (New York) and in a EU plant (Paris). It has $B = \$2,000,000$ and a salvage value of 20% of B. For tax depreciation purposes, the United States allows a straight line write-off over 5 years, while the EU allows SL write-off over 8 years. The general managers of the two plants want to know the difference in (a) the depreciation amount for year 5 and (b) the book value after 5 years. Using a spreadsheet, write cell functions in only two cells to answer both questions.

Declining Balance Depreciation

16.18 When declining balance (DB) depreciation is applied, there can be three different depreciation rates involved—d, d_{max}, and d_t. Explain the differences between these rates.

16.19 Equipment for immersion cooling of electronic components has an installed value of $182,000 with an estimated trade-in value of $40,000 after 15 years. For years 2 and 10, use DDB book depreciation to determine (a) the depreciation charge and (b) the book value.

16.20 A cooling-water pumping station at the LCRA plant costs $600,000 to construct, and it is projected to have a 25-year life with an estimated salvage value of 15% of the construction cost. However, the station will be book-depreciated to zero over a recovery period of 30 years. Calculate the annual depreciation charge for years 4, 10, and 25, using (a) straight line depreciation and (b) DDB depreciation. (c) What is the implied salvage value for DDB? (d) Use a spreadsheet to build the depreciation and book value schedules for both methods to verify your answers.

16.21 A video recording system was purchased 3 years ago at a cost of $30,000. A 5-year recovery period and DDB depreciation have been used to write off the basis. The system is to be replaced this year with a trade-in value of $5000. What is the difference between the book value and the trade-in value?

16.22 An engineer with Accenture Middle East BV in Dubai was asked by her client to help him understand the difference between 150% DB and DDB depreciation. Answer these questions if $B = \$180,000$, $n = 12$ years, and $S = \$30,000$.
  (a) What are the book values after 12 years for both methods?
  (b) How do the estimated salvage and these book values compare in value after 12 years?
  (c) Which of the two methods, when calculated correctly considering $S = \$30,000$, writes off more of the first cost over 12 years?

16.23 Exactly 10 years ago, Boyditch Professional Associates purchased $100,000 in depreciable assets with an estimated salvage of $10,000. For tax depreciation, the SL method with $n = 10$ years was used, but for book depreciation, Boyditch applied the DDB method with $n = 7$ years and neglected the salvage estimate. The company sold the assets today for $12,500.

    (a)   Compare the sales price today with the book values using the SL and DDB methods.

    (b)   If the salvage of $12,500 had been estimated exactly 10 years ago, determine the depreciation for each method in year 10.

16.24 Shirley is studying depreciation in her engineering management course. The instructor asked her to graphically compare the total percent of first cost depreciated for an asset costing B dollars over a life of $n = 5$ years for DDB and 125% DB depreciation. Help her by developing the plots of percent of B depreciated versus years. Use a spreadsheet unless otherwise instructed.

MACRS Depreciation

16.25 Explain the difference between an accelerated depreciation method and one that is not accelerated. Give an example of each.

16.26 What was one of the prime reasons that MACRS depreciation was initiated in the mid-1980s?

16.27 A company just purchased an intelligent robot, which has a first cost of $80,000. Since the robot is unique in its capabilities, the company expects to be able to sell it in 4 years for $95,000.

    (a)   If the company spends $10,000 per year in maintenance and operation of the robot, what will the company's MACRS depreciation charge be in year 2? Assume the recovery period for robots is 5 years and the company's MARR is 16% per year when the inflation rate is 9% per year.

    (b)   Determine the book value of the robot at the end of year 2.

16.28 Animatics Corp. of Santa Clara, California, makes small servo systems with built-in controllers, amplifiers, and encoders so that they can control entire machines. The company purchased an asset 2 years ago that has a 5-year recovery period. The depreciation charge by the MACRS method for year 2 is $24,320.

    (a)   What was the first cost of the asset?

    (b)   How much was the depreciation charge in year 1?

    (c)   Develop the complete MACRS depreciation and book value schedule using the VDB function.

16.29 A plant manager for a large cable company knows that the remaining invested value of certain types of manufacturing equipment is more closely approximated when the equipment is depreciated linearly by the SL method compared to a rapid write-off method such as MACRS. Therefore, he keeps two sets of books, one for tax purposes (MACRS) and one for equipment management purposes (SL). For an asset that has a first cost of $80,000, a depreciable life of 5 years, and a salvage value equal to 25% of the first cost, determine the difference in the book values shown in the two sets of books at the end of year 4.

16.30 The manager of a Glidden Paint manufacturing plant is aware that MACRS and DDB are both accelerated depreciation methods; however, out of curiosity, she wants to determine which one provides the faster write-off in the first 3 years for a recently purchased mixer that has a first cost of $300,000, a 5-year recovery period, and a $60,000 salvage value. Determine which method yields the lower book value and by how much after 3 years. The annual MACRS depreciation rates are 20%, 32%, and 19.2% for years 1, 2, and 3, respectively.

16.31 Railroad cars used to transport coal from Wyoming mines to Texas power plants cost $1.2 million and have an estimated salvage value of $300,000. Develop the depreciation and book value schedules for the GDS MACRS method by using two methods on a spreadsheet—the VDB function and the MACRS rates. Are the book value series the same?

16.32 A 120-metric-ton telescoping crane that cost $320,000 is owned by Upper State Power. Salvage is estimated at $75,000. (a) Compare book values for MACRS and standard SL depreciation over a 7-year recovery period. (b) Explain how the estimated salvage is treated using MACRS.

16.33 Youngblood Shipbuilding Yard just purchased $800,000 in capital equipment for ship repairing functions on dry-docked ships. Estimated salvage is $150,000 for any year after 5 years of use. Compare the depreciation and book value for year 3 for each of the following depreciation methods.

    (a)   GDS MACRS where a recovery period of 10 years is allowed

    (b)   Double declining balance with a recovery period of 15 years

    (c)   ADS straight line as an alternative to MACRS, with a recovery period of 15 years

16.34 Basketball.com has installed $100,000 worth of depreciable software and equipment that represents the latest in Internet teaming and basket competition, intended to allow anyone to enjoy the sport on the Web or in the alley. No salvage value is estimated. The company can depreciate using MACRS for a 5-year recovery period or opt for the ADS alternate system over 10 years using the straight line method. The SL rates require the half-year convention; that is, only 50% of the regular annual rate applies for years 1 and 11.

(a) Construct the book value curves for both methods on one graph. Show hand or spreadsheet computations as instructed.

(b) After 3 years of use, what percentage of the $100,000 basis is removed for each method? Compare the two percentages.

16.35 A company has purchased special-purpose equipment for the manufacture of rubber products (asset class 30.11 in IRS Publication 946) and expects to use it predominately outside the United States. In this case, the ADS alternative to MACRS is required for tax depreciation purposes. The manager wants to understand the difference in yearly recovery rates for classical SL, MACRS, and the ADS alternative to MACRS. Using a recovery period of 3 years, except for the ADS alternative, which requires a 4-year recovery with half-year convention included, prepare a single graph showing the annual recovery rates (in percent) for the three methods.

Depletion

16.36 A coal mine purchased 3 years ago for $7 million was estimated to contain 4,000,000 tons of coal. During the past 3 years the amount of coal removed was 21,000, 18,000, and 20,000 tons, respectively. The gross income obtained in these 3 years was $257,000 for the first year, $320,000 for the second year, and $340,000 for the third year. Determine (a) the cost depletion allowance for each year and (b) the percentage of the purchase price depleted thus far.

16.37 NA Forest Resources purchased forest acreage for $500,000 from which an estimated 200 million board feet of lumber is recoverable. The company will sell the lumber for $0.10 per board foot. No lumber will be sold for the next 2 years because an environmental impact statement must be completed before harvesting can begin. In years 3 to 10, however, the company expects to remove 20 million board feet per year. The inflation rate is 8%, and the company's MARR is 10%.

(a) Determine the depletion amount in year 2 by the cost depletion method.

(b) Determine the depletion amount in year 5 by the percentage depletion method.

16.38 A sand and gravel pit purchased for $900,000 is expected to yield 50,000 tons of gravel and 80,000 tons of sand per year. The gravel will sell for $6 per ton and the sand for $9 per ton.

(a) Determine the depletion charge according to the percentage depletion method. The percentage depletion rate for sand and gravel is 5%.

(b) If taxable income is $100,000 for the year, is this depletion charge allowed? If not, how much is allowed?

16.39 Vesco Mineral Resources purchased mineral rights to land in the foothills of the Santa Cristo mountains. The cost of the purchase was $9 million. Vesco originally thought that it would be able to extract 200,000 tons of lignite from the land, but further exploration revealed that 280,000 tons could be economically removed. If the company sold 20,000 tons in year 1 and 30,000 tons in year 2, what would the depletion charges be each year according to the cost depletion method?

16.40 A company owns gold mining operations in the United States, Australia, and South Africa. The Colorado mine has the taxable income and sales results summarized below. Determine the annual percentage depletion amount for the gold mine.

Year	Taxable Income, $	Sales, Ounces	Average Sales Price, $ per Ounce
1	1,500,000	1800	1115
2	2,000,000	5500	1221
3	800,000	2300	1246

16.41 A highway construction company operates a quarry. During the last 5 years, the amount extracted each year was 60,000, 50,000, 58,000 60,000, and 65,000 tons. The mine is estimated to contain a total of 2.5 million tons of usable stones and gravel. The quarry land had an initial cost of $3.2 million. The company had a per-ton gross income of $30 for the first year, $25 for the second year, $35 for the next 2 years, and $40 for the last year.

(a) Determine the depletion charge each year, using the larger of the values for the two depletion methods. Assume all depletion amounts are less than 50% of taxable income.

(b) Compute the percent of the initial cost that has been written off in these 5 years, using the depletion charges in part (a).

(c) If the quarry operation is reevaluated after the first 3 years of operation and estimated to contain a total of 1.5 million tons remaining, rework parts (a) and (b).

ADDITIONAL PROBLEMS AND FE EXAM REVIEW QUESTIONS

16.42 All of the following types of real property are depreciable except:
 (a) Warehouses
 (b) Land
 (c) Office buildings
 (d) Test facilities

16.43 According to information on the IRS website, a taxpayer can take a depreciation deduction as long as the property meets all of the following requirements except:
 (a) The taxpayer must own the property.
 (b) The taxpayer must use the property in an income-producing activity.
 (c) The taxpayer must use the property for personal purposes.
 (d) The property must have a determinable useful life of more than 1 year.

16.44 A machine with a 5-year life has a first cost of $20,000 and a $2000 salvage value. Its annual operating cost is $8000 per year. According to the classical straight line method, the depreciation charge in year 2 is nearest to:
 (a) $2800   (b) $3600
 (c) $4500   (d) $5300

16.45 A machine with a 10-year life is MACRS-depreciated. The machine has a first cost of $40,000 with a $5000 salvage value. Its annual operating cost is $7000 per year, and d_t for years 1, 2, and 3 is 10.00%, 18.00%, and 14.40%, respectively. The depreciation charge in year 3 is nearest to:
 (a) $5800   (b) $7200
 (c) $8500   (d) $9300

16.46 A coal mine purchased for $5 million has enough coal to operate for 10 years. The annual cost is expected to be $200,000 per year. The coal is expected to sell for $150 per ton, with annual production expected to be 10,000 tons. Coal has a depletion percentage rate of 10%. The depletion charge for year 6 according to the percentage depletion method would be closest to:
 (a) $75,000
 (b) $100,000
 (c) $125,000
 (d) $150,000

16.47 The depreciation charge for a 5-year, straight line depreciated vehicle is $3000 in year 4. If the first cost was $20,000, the salvage value used in the depreciation calculation was closest to:
 (a) $0   (b) $2500
 (c) $5000   (d) $7500

16.48 The book value of an asset that was DDB-depreciated over a 10-year period was $5832 at the end of year 4. If the first cost of the asset was $80,000, the salvage value that was used in the depreciation calculation was closest to:
 (a) $0   (b) $2000
 (c) $5000   (d) $8000

16.49 For an asset that has $B = \$100,000$, $S = \$40,000$, and a 10-year depreciable life, the book value at the end of year 4 according to the MACRS method would be closest to (d_t values for years 1, 2, 3, 4, and 5 are 10.00%, 18.00%, 14.40%, 11.52%, and 9.22%, respectively):
 (a) $58,700   (b) $62,400
 (c) $53,900   (d) $46,100

16.50 An asset that was depreciated over a 5-year period by the MACRS method had a BV of $33,025 at the end of year 3. If the MACRS depreciation rates for years 1, 2, and 3, were 0.20, 0.32, and 0.192, respectively, the basis of the asset was closest to:
 (a) $158,000   (b) $172,000
 (c) $185,000   (d) $193,000

16.51 A lumber company purchased a tract of land for $70,000 that contained an estimated 25,000 usable trees. The value of the land was estimated at $20,000. In the first year of operation, the lumber company cut down 5000 trees. According to the cost depletion method, the depletion deduction for year 1 is closest to:
 (a) $2000   (b) $7000
 (c) $10,000   (d) $14,000

16.52 An asset with a first cost of $50,000 and an estimated salvage value of $10,000 is depreciated by the MACRS method. If its book value at the end of year 3 is $21,850 and its market value is $25,850, the total amount of depreciation charged against the asset up to this time is closest to:
 (a) $18,850   (b) $21,850
 (c) $25,850   (d) $28,150

16.53 Under the General Depreciation System (GDS) of asset classification, any asset that is not in a stated class is automatically assigned a recovery period of:
 (a) 5 years   (b) 7 years
 (c) 10 years   (d) 15 years

16.54 All of the following statements about the Alternative Depreciation System (ADS) are true except:
 (a) The half-year convention applies.
 (b) Salvage value is neglected.
 (c) The recovery periods are shorter than in GDS.
 (d) The straight line method is required.

APPENDIX PROBLEMS

Sum-of-Years-Digits Depreciation

16A.1 A European manufacturing company has new equipment with a first cost of 12,000 euros, an estimated salvage value of 2000 euros, and a recovery period of 8 years. Use the SYD method to tabulate annual depreciation and book value.

16A.2 Earthmoving equipment with a first cost of $150,000 is expected to have a life of 10 years. The salvage value is expected to be 10% of the first cost. Calculate (*a*) by hand and (*b*) by spreadsheet the depreciation charge and book value for years 2 and 7 using the SYD method.

16A.3 If $B = \$12,000$, $n = 6$ years, and S is estimated at 15% of B, use the SYD method to determine (*a*) the book value after 3 years and (*b*) the rate of depreciation and the depreciation amount in year 4.

Unit-of-Production Depreciation

16A.4 A robot used in simulated car crashes cost $70,000, has no salvage value, and has an expected capacity of tests not to exceed 10,000 according to the manufacturer. Volvo Motors decided to use the unit-of-production depreciation method because the number of test crashes per year in which the robot would be involved was not estimable. Determine the annual depreciation and book value for the first 3 years if the number of tests were 3810, 2720, and 5390 per year.

16A.5 A new hybrid car was purchased by Pedernales Electric Cooperative as a courier vehicle to transport items between its 12 city offices. The car cost $30,000 and was retained for 5 years. Alternatively, it could have been retained for 100,000 miles. Salvage value is nil. Five-year DDB depreciation was applied. The car pool manager stated that he prefers UOP depreciation on vehicles because it writes off the first cost faster. Use the actual annual miles driven, listed below, to plot the book values for both methods. Determine which method would have removed the $30,000 faster. Show hand or spreadsheet solution, as instructed.

Year	1	2	3	4	5
Miles Driven, 1000	15	22	16	18	24

Switching Methods

16A.6 An asset has a first cost of $45,000, a recovery period of 5 years, and a $3000 salvage value. Use the switching procedure from DDB to SL depreciation, and calculate the present worth of depreciation at $i = 18\%$ per year.

16A.7 If $B = \$45,000$, $S = \$3000$, and $n = $ 5-year recovery period, use a spreadsheet and $i = 18\%$ per year to maximize the present worth of depreciation, using the following methods: DDB-to-SL switching (this was determined in Problem 16A.6) and MACRS. Given that MACRS is the required depreciation system in the United States, comment on the results.

16A.8 Hempstead Industries has a new milling machine with $B = \$110,000$, $n = 10$ years, and $S = \$10,000$. Determine the depreciation schedule and present worth of depreciation at $i = 12\%$ per year, using the 175% DB method for the first 5 years and switching to the classical SL method for the last 5 years. Use a spreadsheet to solve this problem.

16A.9 Reliant Electric Company has erected a large portable building with a first cost of $155,000 and an anticipated salvage of $50,000 after 25 years. (*a*) Should the switch from DDB to SL depreciation be made? (*b*) For what values of the uniform depreciation rate in the DB method would it be advantageous to switch from DB to SL depreciation at some point in the life of the building?

MACRS Rates

16A.10 Verify the 5-year recovery period rates for MACRS given in Table 16–2. Start with the DDB method in year 1, and switch to SL depreciation when it offers a larger recovery rate.

16A.11 A video recording system was purchased 3 years ago at a cost of $30,000. A 5-year recovery period and MACRS depreciation have been used to write off the basis. The system is to be prematurely replaced with a trade-in value of $5000. Determine the MACRS depreciation, using the switching rules to find the difference between the book value and the trade-in value after 3 years.

16A.12 Use the computations in Equations [16A.11] through [16A.13] to determine the MACRS annual depreciation for the following asset data: $B = \$50,000$ and a recovery period of 7 years.

16A.13 The 3-year MACRS recovery rates are 33.33%, 44.45%, 14.81%, and 7.41%, respectively. (*a*) What are the corresponding rates for the alternative MACRS straight line ADS method with the half-year convention imposed? (*b*) Compare the PW_D values for these two methods if $B = \$80,000$ and $i = 15\%$ per year.

After-Tax Economic Analysis

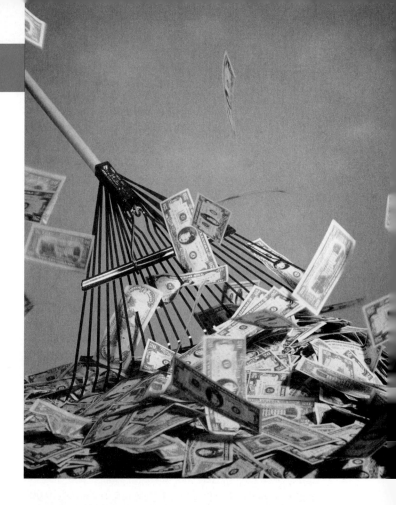

This chapter provides an overview of tax terminology, income tax rates, and tax equations pertinent to an after-tax economic analysis. The change from estimating cash flow before taxes (CFBT) to **cash flow after taxes (CFAT)** involves a consideration of significant tax effects that may alter the final decision, and estimates the magnitude of the effect on cash flow over the life of the alternative that taxes may have.

Mutually exclusive alternative comparisons using after-tax PW, AW, and ROR methods are explained with major tax implications considered. **Replacement studies** are discussed with tax effects that occur at the time that a defender is replaced. Also, the after-tax **economic value added** by an alternative is discussed in the context of annual worth analysis. All these methods use the procedures learned in earlier chapters, except now with tax effects considered.

An after-tax evaluation using any method requires more computations than in previous chapters. Templates for tabulation of cash flow after taxes by hand and by spreadsheet are developed. Additional information on U.S. federal taxes—tax law and annually updated tax rates—is available through Internal Revenue Service publications and, more readily, on the IRS website www.irs.gov. Publications 542, *Corporations*, and 544, *Sales and Other Dispositions of Assets*, are especially applicable to this chapter. Some differences in tax considerations outside the United States are summarized.

17.1 Income Tax Terminology and Basic Relations ● ● ●

Online Presentation

The perspective taken in engineering economy when performing an after-tax evaluation is that of the *project* and how relevant tax rules and allowances influence the economic decision. The perspective of a financial study is that of the *corporation* and how the tax structure and laws affect profitability. We take the engineering economy viewpoint in the sections that follow.

There are many types of taxes levied upon corporations and individuals in all countries, including the United States. Some are sales tax, valued-added tax, import tax, income tax, highway tax, gasoline tax, and property (real estate) tax. Federal governments rely on income taxes for a significant portion of their annual revenue. States, provinces, and municipal-level governments rely on sales, value-added, and property taxes to maintain services, schools, etc. for the citizenry. For a starting point on *corporate income taxes* and how they are used when performing an after-tax economic evaluation of a project or multiple alternatives, this section covers basic definitions, terms, and relations.

Income tax is the amount of the payment (taxes) on income or profit that must be delivered to a federal (or lower-level) government unit. Taxes are **real cash flows;** however, for corporations tax computation requires some noncash elements, such as depreciation. Corporate income taxes are usually submitted quarterly, and the last payment of the year is submitted with the annual tax return.

The Internal Revenue Service (IRS), a part of the U.S. Department of the Treasury, collects the taxes and enforces tax laws. The website www.irs.gov provides information on tax laws, rates, publications, etc. that are referenced in this chapter.

Though the formulas are much more complex when applied to a specific situation, two fundamental relations form the basis for income tax computations. The first involves only actual cash flows:

$$\text{Net operating income} = \text{revenue} - \text{operating expenses}$$

The second involves actual cash flows and noncash deductibles, such as depreciation.

$$\text{Taxable income} = \text{revenue} - \text{operating expenses} - \text{depreciation}$$

These terms and relations for corporations are now described. Since each term is calculated for 1 year, there can be a subscript t ($t = 1, 2, \ldots$) added; t is omitted here for simplicity.

Operating revenue R, also commonly called **gross income GI,** is the total income realized from all revenue-producing sources. These incomes are listed in the income statement. (See Appendix B on accounting reports.) Other, nonoperating revenues such as sale of assets, license fee income, and royalties are considered separately for tax purposes.

Operating expenses OE include all costs incurred in the transaction of business. These expenses are tax-deductible for corporations. For after-tax economic evaluations, the AOC (annual operating costs) and M&O (maintenance and operating) costs are applicable here. Depreciation is not included here since it is *not an operating expense*.

Net operating income NOI, often called **EBIT** (earnings before interest and income taxes), is the difference between gross income and operating expenses.

$$\text{NOI} = \text{EBIT} = \text{GI} - \text{OE} \qquad [17.1]$$

Taxable income TI is the amount of income upon which taxes are based. A corporation is allowed to **remove depreciation,** depletion and amortization, and some other deductibles from net operating income in determining the taxable income for a year. For our evaluations, we define taxable income as

$$\text{TI} = \text{gross income} - \text{operating expenses} - \text{depreciation}$$
$$= \text{GI} - \text{OE} - \text{D} \qquad [17.2]$$

Though there may be subtleties and varying interpretations over time, in essence, the differences between NOI and TI are tax-law-allowed deductibles, such as depreciation. (In keeping with the project view of engineering economics, we will primarily use the TI relation when conducting an after-tax evaluation.)

Tax rate T is a percentage, or decimal equivalent, of TI that is owed in taxes. The tax rates in many countries (including the United States) are **graduated** (or **progressive**) by level of TI; that is, higher rates apply as the TI increases. The **marginal tax rate** is the percentage paid on the *last dollar of income*. The average tax rate paid is calculated separately from the highest marginal rate used, as shown later. The general tax computation relation is

$$\text{Income taxes} = \text{applicable tax rate} \times \text{taxable income}$$
$$= (T)(\text{TI}) \qquad [17.3]$$

Net operating profit after taxes NOPAT is the amount remaining each year after taxes are subtracted from taxable income.

$$\text{NOPAT} = \text{TI} - \text{taxes} = \text{TI} - (T)(\text{TI})$$
$$= \text{TI}(1 - T) \qquad [17.4]$$

Basically, NOPAT represents the money remaining in the corporation as a result of the capital invested during the year. It is also called *net profit after taxes* (NPAT).

The graduated tax rate schedule for corporations is presented in Table 17–1 as taken from IRS Publication 542, *Corporations*. These are rates for the entire corporation, not for an individual project, though they are often applied in the after-tax analysis of a single project. The rates can change annually based upon government legislation; however, the corporate tax rate schedule has remained the same for some years. To illustrate the use of the graduated tax rate, assume a company is expected to generate a taxable income of $500,000 in 1 year. From Table 17–1, the marginal tax rate for the last dollar of TI is 34%, but the graduated rates become progressively larger as TI increases. In this case, for TI = $500,000,

$$\text{Taxes} = 113,900 + 0.34(500,000 - 335,000)$$
$$= 113,900 + 56,100$$
$$= \$170,000$$

Alternatively, the rates for each TI level can be used to calculate taxes the longer way.

$$\text{Taxes} = 0.15(50,000) + 0.25(75,000 - 50,000) + 0.34(100,000 - 75,000)$$
$$+ 0.39(335,000 - 100,000) + 0.34(500,000 - 335,000)$$
$$= 7500 + 6250 + 8500 + 91,650 + 56,100$$
$$= \$170,000$$

TABLE 17–1	U.S. Corporate Income Tax Rate Schedule (2010)		
	If Taxable Income ($) Is:		
Over	**But Not over**	**Tax Is**	**Of the Amount over**
0	50,000	15%	0
50,000	75,000	7,500 + 25%	50,000
75,000	100,000	13,750 + 34%	75,000
100,000	335,000	22,250 + 39%	100,000
335,000	10,000,000	113,900 + 34%	335,000
10,000,000	15,000,000	3,400,000 + 35%	10,000,000
15,000,000	18,333,333	5,150,000 + 38%	15,000,000
18,333,333	—	35%	0

Smaller businesses (with TI < $335,000) receive a slight tax advantage compared to large corporations. Once the TI exceeds $335,000, an effective federal tax rate of 34% applies, and when TI > $18.33 million, there is a flat tax rate of 35%.

As we move forward with after-tax analysis, it is important to keep the following in mind.

The corporate tax rates apply to a corporation as a whole, not to a specific project, unless the project *is* the company. Tax rates are usually graduated by level of taxable income. Therefore, the last dollar of TI is taxed at a marginal rate. The average federal tax rate actually paid is lower than the highest marginal rate paid because the rates, in general, graduate to higher percentages as TI increases.

Because the marginal tax rates change with TI, it is not possible to quote directly the percent of TI paid in income taxes. Alternatively, a single-value number, the *average tax rate,* is calculated as

$$\text{Average tax rate} = \frac{\text{total taxes paid}}{\text{taxable income}} = \frac{\text{taxes}}{\text{TI}} \qquad [17.5]$$

Referring to Table 17–1, for a small business with TI = $100,000, the federal income tax burden averages $22,250/100,000 = 22.25%. If TI = $15 million, the average tax rate is $5.15 million/15 million = 34.33%.

As mentioned earlier, there are federal, state, and local taxes imposed. For the sake of simplicity, the tax rate used in an economy study is often a single-figure **effective tax rate T_e,** which accounts for all taxes. Effective tax rates are in the range of 35% to 50%. One reason to use the effective tax rate is that state taxes are deductible for federal tax computation. The effective tax rate and taxes are calculated as

$$T_e = \text{state rate} + (1 - \text{state rate})(\text{federal rate}) \qquad [17.6]$$
$$\text{Taxes} = (T_e)(\text{TI}) \qquad [17.7]$$

EXAMPLE 17.1

REI (Recreational Equipment Incorporated) sells outdoor equipment and sporting goods through retail outlets, the Internet, and catalogs. Assume that for 1 year REI has the following financial results in the state of Kentucky, which has a flat tax rate of 6% on corporate taxable income.

Total revenue	$19.9 million
Operating expenses	$8.6 million
Depreciation and other allowed deductions	$1.8 million

(a) Determine the state taxes and federal taxes due using Table 17–1 rates.
(b) Find the average federal tax rate paid for the year.
(c) Determine a single-value tax rate useful in economic evaluations using the average federal tax rate determined in part (b).
(d) Estimate federal and state taxes using the single-value rate, and compare their total with the total in part (a).

Solution

(*a*) Calculate TI by Equation [17.2] and use Table 17–1 rates for federal taxes due.

$$\text{Kentucky state TI} = \text{GI} - \text{OE} - \text{D} = 19.9 \text{ million} - 8.6 \text{ million} - 1.8 \text{ million}$$
$$= \$9.5 \text{ million}$$

$$\text{Kentucky state taxes} = 0.06(\text{TI}) = 0.06(9{,}500{,}000) = \$570{,}000$$

$$\text{Federal TI} = \text{GI} - \text{OE} - \text{D} - \text{state taxes} = 9{,}500{,}000 - 570{,}000$$
$$= \$8{,}930{,}000$$

$$\text{Federal taxes} = 113{,}900 + 0.34(8{,}930{,}000 - 335{,}000) = \$3{,}036{,}200$$

$$\text{Total federal and state taxes} = 3{,}036{,}200 + 570{,}000 = \$3{,}606{,}200 \qquad [17.8]$$

(*b*) From Equation [17.5], the average tax rate paid is approximately 32% of TI.

$$\text{Average federal tax rate} = 3{,}036{,}200/9{,}500{,}000 = 0.3196$$

(*c*) By Equation [17.6], T_e is slightly over 36% per year for combined state and federal taxes.

$$T_e = 0.06 + (1 - 0.06)(0.3196) = 0.3604 \quad (36.04\%)$$

(*d*) Use the effective tax rate and TI = \$9.5 million from part (*a*) in Equation [17.7] to approximate total taxes.

$$\text{Taxes} = 0.3604(9{,}500{,}000) = \$3{,}423{,}800$$

Compared to Equation [17.8], this approximation is \$182,400 low, a 5.06% underestimate.

It is interesting to understand how corporate tax and individual tax computations differ. Gross income for an individual taxpayer is comparable if revenue is replaced by salaries and wages. However, for an individual's taxable income, most of the expenses for living and working are not tax deductible to the same degree as operating expenses are for corporations. For individual tax-payers,

$$\text{GI} = \text{salaries} + \text{wages} + \text{interest and dividends} + \text{other income}$$

$$\text{TI} = \text{GI} - \text{personal exemption} - \text{standard or itemized deductions}$$

$$\text{Taxes} = (T)(\text{TI})$$

For TI, operating expenses are replaced by individual exemptions and specific deductions. Exemptions are yourself, your spouse, your children, and your other dependents. Each exemption reduces TI by \$3500 to \$4000 per year, depending upon current exemption allowances.

In the United States, the tax rates for individuals, like those for corporations, are graduated by level of TI. In 2010, the marginal rates ranged from 10% to 35%; however, the top marginal rates are increasing for individuals with larger TI amounts. These rates and TI levels are the subject of ongoing political debates at the U.S. national level depending upon the balance of power in the congressional bodies and the economic condition of the country. Once the marginal rates are set, the TI levels are adjusted each year to account for inflation and other factors. This process is called **indexing.** Clearly, tax rates for individuals change much more frequently than the rates for corporations change. Current information is available on the IRS website www.irs.gov through Publication 17, *Your Federal Income Tax.* The current rate schedule is published in the back of this document for four filing status categories:

Unmarried individuals (single)
Married filing jointly
Married filing separately
Head of household

Online Presentation

17.2 Calculation of Cash Flow after Taxes ● ● ○

Early in the text, the term *net cash flow* (NCF) was identified as the best estimate of actual cash flow each year. The NCF is calculated as cash inflows minus cash outflows. Since then, the annual NCF amounts have been used many times to perform alternative evaluations via the PW, AW, ROR, and B/C methods. Now that the impact on cash flow of depreciation and related taxes will be considered, it is time to expand our terminology. NCF is replaced by the term **cash flow before taxes (CFBT),** and we introduce the new term **cash flow after taxes (CFAT).**

CFBT and CFAT are **actual cash flows;** that is, they represent the estimated actual flow of money into and out of the corporation that will result from the alternative. The remainder of this section explains how to transition from before-tax to after-tax cash flows for solutions by hand and by spreadsheet, using tax regulations described in the next few sections. Once the CFAT estimates are developed, the economic evaluation is performed using the same methods and selection guidelines applied previously. However, the analysis is performed on the CFAT estimates.

We learned that net operating income (NOI) does not include the purchase or sale of capital assets. However, the annual CFBT estimate *must include* the initial capital investment and salvage value for the years in which they occur. Incorporating the definitions of gross income and operating expenses from NOI, CFBT for any year is defined as

$$\text{CFBT} = \text{gross income} - \text{operating expenses} - \text{initial investment} + \text{salvage value}$$
$$= \text{GI} - \text{OE} - P + S \qquad \qquad [17.9]$$

As in previous chapters, P is the initial investment (year 0) and S is the estimated salvage value in year n. Therefore, only in year 0 will the CFBT include P, and only in year n will an S value be present. Once all taxes are estimated, the annual after-tax cash flow is simply

$$\text{CFAT} = \text{CFBT} - \text{taxes} \qquad \qquad [17.10]$$

where taxes are estimated using the relation $(T)(\text{TI})$ or $(T_e)(\text{TI})$.

We know from Equation [17.2] that depreciation D is considered when calculating TI. It is very important to understand the different roles of depreciation for income tax computations and in CFAT estimation.

Depreciation is not an operating expense and is a *non*cash flow. Depreciation is tax deductible for determining the amount of income taxes only, but it does not represent a direct, after-tax cash flow. Therefore, the after-tax engineering economy study must be based on actual cash flow estimates, that is, annual CFAT estimates that do not include depreciation as an expense (negative cash flow).

Accordingly, if the CFAT expression is determined using the TI relation, depreciation must not be included outside of the TI component. Equations [17.9] and [17.10] are now combined.

$$\text{CFAT} = \text{GI} - \text{OE} - P + S - (\text{GI} - \text{OE} - D)(T_e) \qquad \qquad [17.11]$$

Suggested table column headings for CFBT and CFAT calculations by hand or by spreadsheet are shown in Table 17–2. The equations are shown in column numbers, with the effective tax rate T_e used for income tax estimation. Expenses OE and initial investment P carry negative signs in all tables and spreadsheets.

A **negative TI value** may occur in some years due to a depreciation amount that is larger than (GI − OE). It is possible to account for this in a detailed after-tax analysis using carry-forward and carry-back rules for operating losses. It is the exception that the engineering economy study will consider this level of detail. Rather, the associated negative income tax is considered as a **tax savings for the year.** The assumption is that the negative tax will offset taxes for the same year in other income-producing areas of the corporation.

TABLE 17–2	Suggested Column Headings for Calculation of CFAT						
Year	Gross Income GI	Operating Expenses OE	Investment and Salvage P and S	CFBT	Depreciation D	Taxable Income TI	Taxes	CFAT
	(1)	(2)	(3)	(4) = (1) + (2) + (3)	(5)	(6) = (1) + (2) − (5)	(7) = T_e(6)	(8) = (4) − (7)

EXAMPLE 17.2

Wilson Security has received a contract to provide additional security for corporate and government personnel along the Arizona-Mexico border. Wilson plans to purchase listening and detection equipment for use in the 6-year contract. The equipment is expected to cost $550,000 and have a resale value of $150,000 after 6 years. Based on the incentive clause in the contract, Wilson estimates that the equipment will increase contract revenue by $200,000 per year and require an additional M&O expense of $90,000 per year. MACRS depreciation allows recovery in 5 years, and the effective corporate tax rate is 35% per year. Tabulate and plot the CFBT and CFAT series.

Solution

The spreadsheet in Figure 17–1 presents before-tax and after-tax cash flows using the format of Table 17–2. The functions for year 6 are detailed in row 11. Discussion and sample calculations follow.

CFBT: The operating expenses OE and initial investment P are shown as negative cash flows. The $150,000 salvage (resale) is a positive cash flow in year 6. CFBT is calculated by Equation [17.9]. In year 6, for example, when the equipment is sold, the function in row 11 indicates that

$$CFBT_6 = 200,000 - 90,000 + 150,000 = \$260,000$$

CFAT: Column F for MACRS depreciation, which is determined using the VDB function over the 6-year period, writes off the entire $550,000 investment. Taxable income, taxes, and CFAT are calculated, using year 4 as an example.

$$TI_4 = GI - OE - D = 200,000 - 90,000 - 63,360 = \$46,640$$

$$Taxes_4 = (0.35)(TI) = (0.35)(46,640) = \$16,324$$

$$CFAT_4 = GI - OE - taxes = 200,000 - 90,000 - 16,324 = \$93,676$$

In year 2, MACRS depreciation is large enough to cause TI to be negative ($-66,000). As mentioned above, the negative tax ($-23,100) is considered a *tax savings* in year 2, thus increasing CFAT.

Comment

MACRS depreciates to a salvage value of $S = 0$. Later we will learn about a tax implication due to "recapturing of depreciation" when an asset is sold for an amount larger than zero and MACRS was applied to fully depreciate the asset to zero.

Figure 17–1
Computation of CFBT and CFAT using MACRS depreciation and $T_e = 35\%$, Example 17.2.

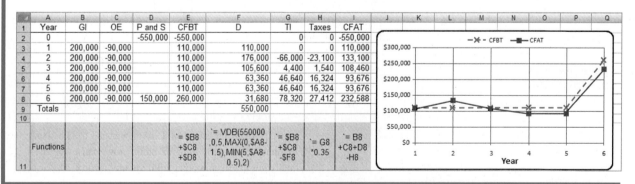

17.3 Effect on Taxes of Different Depreciation Methods and Recovery Periods ● ● ●

It is important to understand why *accelerated depreciation rates* give the corporation a tax advantage relative to the straight line method with the same recovery period. Larger rates in earlier years of the recovery period require less taxes due to the larger reductions in taxable income. The criterion of **minimizing the present worth of taxes** is used to demonstrate the tax effect. For the recovery period n, choose the depreciation rates that result in the **minimum** present worth value for taxes.

$$PW_{tax} = \sum_{t=1}^{t=n} (\text{taxes in year } t)(P/F,i,t) \qquad [17.12]$$

This is equivalent to maximizing the present worth of total depreciation PW_D.

Compare any two depreciation methods. Assume the following: (1) There is a constant single-value tax rate, (2) CFBT exceeds the annual depreciation amount, (3) both methods reduce book value to the same salvage value, and (4) the same recovery period is used. On the basis of these assumptions, the following statements are correct:

The total taxes paid are **equal** for all depreciation methods.
The present worth of taxes is **less** for accelerated depreciation methods.

As we learned in Chapter 16, MACRS is the prescribed tax depreciation method in the United States, and the only alternative is MACRS straight line depreciation with an extended recovery period. The accelerated write-off of MACRS always provides a smaller PW_{tax} compared to less accelerated methods. If the DDB method were still allowed directly, rather than embedded in MACRS, DDB would not fare as well as MACRS. This is so because DDB does not reduce the book value to zero. This is illustrated in Example 17.3.

EXAMPLE 17.3

An after-tax analysis for a new $50,000 machine proposed for a fiber optics manufacturing line is in process. The CFBT for the machine is estimated at $20,000. If a recovery period of 5 years applies, use the present worth of taxes criterion, an effective tax rate of 35%, and a return of 8% per year to compare the following: classical straight line, classical DDB, and required MACRS depreciation. Use a 6-year period for the comparison to accommodate the half-year convention imposed by MACRS.

Solution

Table 17–3 presents a summary of annual depreciation, taxable income, and taxes for each method. For classical straight line depreciation with $n = 5$, $D_t = \$10,000$ for 5 years and $D_6 = 0$ (column 3). The CFBT of $20,000 is fully taxed at 35% in year 6.

The classical DDB percentage of $d = 2/n = 0.40$ is applied for 5 years. The implied salvage value is $\$50,000 - 46,112 = \$3888$, so not all $50,000 is tax-deductible. The taxes using classical DDB will be $\$3888(0.35) = \$1361$ larger than for the classical SL method.

TABLE 17–3   Comparison of Taxes and Present Worth of Taxes for Different Depreciation Methods

| (1) | (2) | Classical Straight Line | | | Classical Double Declining Balance | | | MACRS | | |
| | | (3) | (4) | (5) = 0.35(4) | (6) | (7) | (8) = 0.35(7) | (9) | (10) | (11) = 0.35(10) |
Year t	CFBT, $	D_t, $	TI, $	Taxes, $	D_t, $	TI, $	Taxes, $	D_t, $	TI, $	Taxes, $
1	+20,000	10,000	10,000	3,500	20,000	0	0	10,000	10,000	3,500
2	+20,000	10,000	10,000	3,500	12,000	8,000	2,800	16,000	4,000	1,400
3	+20,000	10,000	10,000	3,500	7,200	12,800	4,480	9,600	10,400	3,640
4	+20,000	10,000	10,000	3,500	4,320	15,680	5,488	5,760	14,240	4,984
5	+20,000	10,000	10,000	3,500	2,592	17,408	6,093	5,760	14,240	4,984
6	+20,000	0	20,000	7,000	0	20,000	7,000	2,880	17,120	5,992
Totals		50,000		24,500	46,112		25,861*	50,000		24,500
PW_{tax}				18,386			18,549			18,162

*Larger than other values since there is an implied salvage value of $3888 not recovered.

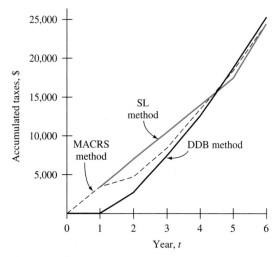

Figure 17–2
Cumulative taxes incurred by different depreciation rates for a
6-year comparison period, Example 17.3.

MACRS writes off $50,000 in 6 years using the rates of Table 16–2. Total taxes are $24,500, the same as for classical SL depreciation.

The annual taxes (columns 5, 8, and 11) are accumulated year by year in Figure 17–2. Note the pattern of the curves, especially the lower total taxes relative to the SL model after year 1 for MACRS and in years 1 through 4 for DDB. These higher tax values for SL cause PW_{tax} for SL depreciation to be larger. The PW_{tax} values at the bottom of Table 17–3 are calculated using Equation [17.12]. The MACRS PW_{tax} value is the smallest at $18,162.

To compare taxes for different *recovery periods*, change only assumption 4 at the beginning of this section to read: The same depreciation method is applied. It can be shown that a shorter recovery period will offer a tax advantage over a longer period using the criterion to minimize PW_{tax}. Comparison will indicate that

> The total taxes paid are **equal** for all n values.
> The present worth of taxes is **less** for smaller n values.

This is why corporations want to use the shortest MACRS recovery period allowed for income tax purposes. Example 17.4 demonstrates these conclusions for classical straight line depreciation, but the conclusions are correct for MACRS or any other tax depreciation method.

EXAMPLE 17.4

Grupo Grande Maquinaría, a diversified manufacturing corporation based in Mexico, maintains parallel records for depreciable assets in its European operations in Germany. This is common for multinational corporations. One set is for corporate use that reflects the estimated useful life of assets. The second set is for foreign government purposes, such as depreciation and taxes.

The company just purchased an asset for $90,000 with an estimated useful life of 9 years; however, a shorter recovery period of 5 years is allowed by German tax law. Demonstrate the tax advantage for the smaller n if net operating income (NOI) is $30,000 per year, an effective tax rate of 35% applies, invested money is returning 5% per year after taxes, and classical SL depreciation is allowed. Neglect the effect of any salvage value.

Solution
Determine the annual TI and taxes by Equations [17.2] through [17.3] and the present worth of taxes using Equation [17.12] for both n values.

Useful life $n = 9$ years:

$$D = \frac{90,000}{9} = 10,000$$

$$\text{TI} = 30,000 - 10,000 = \$20,000 \text{ per year}$$

$$\text{Taxes} = (0.35)(20,000) = \$7000 \text{ per year}$$

$$\text{PW}_{\text{tax}} = 7000(P/A,5\%,9) = \$49,755$$

$$\text{Total taxes} = (7000)(9) = \$63,000$$

Recovery period $n = 5$ years:

Use the same comparison period of 9 years, but depreciation occurs only during the first 5 years.

$$D_t = \begin{cases} \dfrac{90,000}{5} = \$18,000 & t = 1 \text{ to } 5 \\ 0 & t = 6 \text{ to } 9 \end{cases}$$

$$\text{Taxes} = \begin{cases} (0.35)(30,000 - 18,000) = \$4200 & t = 1 \text{ to } 5 \\ (0.35)(30,000) = \$10,500 & t = 6 \text{ to } 9 \end{cases}$$

$$\begin{aligned} \text{PW}_{\text{tax}} &= 4200(P/A,5\%,5) + 10,500(P/A,5\%,4)(P/F,5\%,5) \\ &= \$47,356 \end{aligned}$$

$$\text{Total taxes} = 4200(5) + 10,500(4) = \$63,000$$

A total of \$63,000 in taxes is paid in both cases. However, the more rapid write-off for $n = 5$ results in a present worth tax savings of nearly \$2400 (49,755 − 47,356).

17.4 Depreciation Recapture and Capital Gains (Losses) ●●●

Online Presentation

All the tax implications discussed here are the result of disposing of a depreciable asset before, at, or after its recovery period. In an after-tax economic analysis of large investment assets, these tax effects should be considered. The key is the size of the selling price (or salvage or market value) relative to the current book value at disposal time and relative to the first cost, which is called the *unadjusted basis B* in depreciation terminology. There are three relevant terms.

Depreciation recapture DR, also called **ordinary gain,** occurs when a depreciable asset is sold for more than the current book value BV_t. As shown in Figure 17–3,

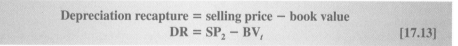

Depreciation recapture = selling price − book value
$$\text{DR} = \text{SP}_2 - \text{BV}_t \qquad\qquad [17.13]$$

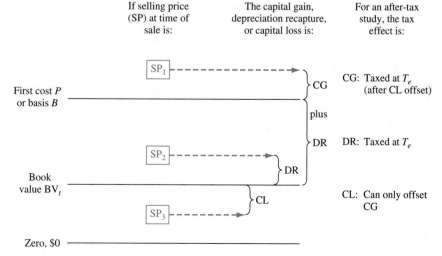

Figure 17–3
Summary of calculations and tax treatment for depreciation recapture and capital gains (losses).

Depreciation recapture is often present in the after-tax study. In the United States, an amount equal to the estimated salvage value can always be anticipated as DR when the asset is disposed of after the MACRS recovery period. This is correct simply because MACRS depreciates every asset to zero in $n + 1$ years. The amount of DR is treated as ordinary taxable income in the year of asset disposal.

Capital gain CG is an amount incurred when the selling price exceeds its (unadjusted) basis *B*. See Figure 17–3.

$$\text{Capital gain} = \text{selling price} - \text{basis}$$
$$CG = SP_1 - B \qquad [17.14]$$

Since future capital gains are difficult to predict, they are usually not detailed in an after-tax economy study. An exception is for assets that historically increase in value, such as buildings and land.

When the selling price exceeds *B*, the TI due to the sale is the capital gain *plus* the depreciation recapture, as shown in Figure 17–3. The DR is now the total amount of depreciation taken thus far, that is, $B - BV$.

Capital loss CL occurs when a depreciable asset is disposed of for less than its current book value. In Figure 17–3,

$$\text{Capital loss} = \text{book value} - \text{selling price}$$
$$CL = BV_t - SP_3 \qquad [17.15]$$

An economic analysis does not commonly account for capital loss, simply because it is not estimable for a specific alternative. However, an *after-tax replacement study* should account for any capital loss if the defender must be traded at a "sacrifice" price. For the purposes of the economic study, this provides a tax savings in the year of replacement. Use the effective tax rate to estimate the tax savings. These savings are assumed to be offset elsewhere in the corporation by other income-producing assets that generate taxes.

There are several additional points worth mentioning about capital gains and capital losses for a corporation, apart from their presence in an economic evaluation.

- U.S. tax law defines capital gains as long-term (items retained for more than 1 year) or short-term.
- Capital gains actually take place for property that is not depreciated or amortized. The term *capital gain* correctly applies at sale time to property such as investments (stocks and bonds), art, jewelry, land, and the like. When a depreciable asset's selling price is higher than the original cost (its basis), it is correctly termed an **ordinary gain.** Corporate tax treatment at this time is the same for both: taxed as ordinary income. All said, it is common to classify an expected ordinary gain on a depreciable asset as a capital gain, without altering the economic decision.
- Capital gains are taxed as ordinary taxable income at the corporation's regular tax rates.
- Capital losses do not directly reduce annual income taxes because they can only be *netted* against capital gains to the maximum extent of the capital gains for the year. The terms used then are *net* capital gains (losses).
- When capital losses exceed capital gains, the corporation can take advantage of carry-back and carry-forward tax laws for the excess, something of value to a finance or tax officer, not an engineer doing an economic analysis.
- When an asset is disposed of, the tax treatment is referred to as a *Section 1231 transaction,* which is the IRS rule section of the same number.
- IRS Publication 544, *Sales and Other Dispositions of Assets,* may be helpful if gains and losses are present in a study.
- All these rules apply to corporations. Individual tax rules and rates are different concerning asset disposition.

If the three additional income and tax elements covered here are incorporated into Equation [17.2], *taxable income* is defined as

$$\begin{aligned}
\text{TI} &= \text{gross income} - \text{operating expenses} - \text{depreciation} \\
&\quad + \text{depreciation recapture} + \text{net capital gain} - \text{net capital loss} \\
&= GI - OE - D + DR + CG - CL \qquad [17.16]
\end{aligned}$$

In keeping with our perspective of an engineering economy study rather than that of a financial study, deprecation recapture (i.e., ordinary gain) will be the primary element considered in after-tax evaluations. Only when a capital gain or loss must be included due to the nature of the problem will the calculations involve it.

EXAMPLE 17.5

Biotech, a medical imaging and modeling company, must purchase a bone cell analysis system for use by a team of bioengineers and mechanical engineers studying bone density in athletes. This particular part of a 3-year contract with the NBA will provide additional gross income of $100,000 per year. The effective tax rate is 35%. Estimates for two alternatives are summarized below.

	Analyzer 1	Analyzer 2
Basis B, $	150,000	225,000
Operating expenses, $ per year	30,000	10,000
MACRS recovery, years	5	5

Answer the following questions, solving by hand and spreadsheet.

(a) The Biotech president, who is very tax conscious, wishes to use a criterion of minimizing total taxes incurred over the 3 years of the contract. Which analyzer should be purchased?

(b) Assume that 3 years have now passed, and the company is about to sell the analyzer. Using the same total tax criterion, did either analyzer have an advantage? Assume the selling price is $130,000 for analyzer 1, or $225,000 for analyzer 2.

Solution by Hand

(a) Table 17–4 details the tax computations. First, the yearly MACRS depreciation is determined. Equation [17.2], TI = GI − OE − D, is used to calculate TI, after which the 35% tax rate is applied each year. Taxes for the 3-year period are summed, with no consideration of the time value of money.

Analyzer 1 tax total: $36,120   Analyzer 2 tax total: $38,430

The two analyzers are very close, but analyzer 1 wins with $2310 less in total taxes.

(b) When the analyzer is sold after 3 years of service, there is a depreciation recapture (DR) that is taxed at the 35% rate. This tax is in addition to the third-year tax. For each analyzer, account for the DR by Equation [17.13], SP − BV_3; then determine the TI, using Equation [17.16], TI = GI − OE − D + DR. Again, find the total taxes over 3 years, and select the analyzer with the smaller total.

TABLE 17–4   Comparison of Total Taxes for Two Alternatives, Example 17.5a

Year	Gross Income GI, $	Operating Expenses OE, $	Basis B, $	MACRS Depreciation D, $	Book Value BV, $	Taxable Income TI, $	Taxes at 0.35TI, $
				Analyzer 1			
0			150,000		150,000		
1	100,000	30,000		30,000	120,000	40,000	14,000
2	100,000	30,000		48,000	72,000	22,000	7,700
3	100,000	30,000		28,800	43,200	41,200	14,420
							36,120
				Analyzer 2			
0			225,000		225,000		
1	100,000	10,000		45,000	180,000	45,000	15,750
2	100,000	10,000		72,000	108,000	18,000	6,300
3	100,000	10,000		43,200	64,800	46,800	16,380
							38,430

Analyzer 1: $DR = 130,000 - 43,200 = \$86,800$
Year 3 $TI = 100,000 - 30,000 - 28,800 + 86,800 = \$128,000$
Year 3 taxes $= (0.35)(128,000) = \$44,800$
Total taxes $= 14,000 + 7700 + 44,800 = \$66,500$

Analyzer 2: $DR = 225,000 - 64,800 = \$160,200$
Year 3 $TI = 100,000 - 10,000 - 43,200 + 160,200 = \$207,000$
Year 3 taxes $= (0.35)(207,000) = \$72,450$
Total taxes $= 15,750 + 6300 + 72,450 = \$94,500$

Now, analyzer 1 has a considerable advantage in total taxes ($94,500 versus $66,500).

Solution by Spreadsheet

(a) Rows 5 through 9 of Figure 17–4 perform the same computations as the hand solution for analyzer 1 with total taxes of $36,120. Similar analysis in rows 14 to 18 results in total taxes of $38,430 for analyzer 2, indicating that the company should select analyzer 1, based on taxes only.

(b) Revised year 3 entries for analyzer 1 in row 10 show the sales price of $130,000, the updated TI of $128,000, and a 3-year tax total of $66,500. The new TI in year 3 has the depreciation recapture incorporated as $DR =$ selling price $-$ book value $= SP - BV_3$, which is shown in the cell tag as the last term (D10 − F10). With a similar updating for analyzer 2 (row 19), total taxes of $94,500 now show a significantly larger tax advantage for analyzer 1 over 3 years.

Comment

Note that no time value of money is considered in these analyses, as we have used in previous alternative evaluations. In Section 17.5 below we will rely upon PW, AW, and ROR analyses at an established MARR to make an after-tax decision based upon CFAT values.

	A	B	C	D	E	F	G	H	I	J	K	L	M
1													
2													
3	Year	GI	OE	B and S	D	BV	TI	Taxes					
4				**Analyzer 1**						= B6+C6−E6			
5	0			-150,000		150,000							
6	1	100,000	-30,000		30,000	120,000	40,000	14,000					
7	2	100,000	-30,000		48,000	72,000	22,000	7,700					
8	3	100,000	-30,000		28,800	43,200	41,200	14,420					
9	Total							36,120					
10	**Revised 3**	100,000	-30,000	130,000	28,800	43,200	128,000	44,800					
11	**New Total**							66,500		= B10+C10−E10+(D10−F10)			
12										Last term calculates the depreciation recapture			
13				**Analyzer 2**									
14	0			-225,000		225,000							
15	1	100,000	-10,000		45,000	180,000	45,000	15,750					
16	2	100,000	-10,000		72,000	108,000	18,000	6,300		Total tax advantage of analyzer 1 over 2 without sale and with sale in year 3			
17	3	100,000	-10,000		43,200	64,800	46,800	16,380					
18	Total							38,430					
19	**Revised 3**	100,000	-10,000	225,000	43,200	64,800	207,000	72,450					
20	**New Total**							94,500					

Figure 17–4
Impact of depreciation recapture on total taxes, Example 17.5.

17.5 After-Tax Evaluation ●●●

The required after-tax MARR is established using the market interest rate, the corporation's effective tax rate, and its weighted average cost of capital. The CFAT estimates are used to compute the PW or AW at the after-tax MARR. When *positive and negative CFAT values* are present, a PW or AW < 0 indicates the MARR is not met. For a single project or mutually exclusive alternatives, apply the same logic as in Chapters 5 and 6. The guidelines are:

One project. PW or AW $\geq$ 0, the project is financially viable because the after-tax MARR is met or exceeded.

Two or more alternatives. Select the alternative with the best (numerically largest) PW or AW value.

ME alternative selection

Equal service

If *only cost CFAT amounts* are estimated, calculate the after-tax savings generated by the operating expenses and depreciation. Assign a plus sign to each saving and apply the guidelines above.

Remember, the **equal-service** assumption requires that the PW analysis be performed over the least common multiple (LCM) of alternative lives. This requirement must be met for every analysis—before or after taxes.

Since the CFAT estimates usually vary from year to year in an after-tax evaluation, the spreadsheet offers a much speedier analysis than solution by hand.

For AW analysis: Use the PMT function with an embedded NPV function *over one life cycle.* The general format is as follows, with the NPV function in italics for the CFAT series.

$$= -\text{PMT}(MARR, n, NPV(MARR, year\_1: year\_n) + year\_0) \qquad [17.17]$$

For PW analysis: Obtain the PMT function results first, followed by the PV function taken over the LCM. (There is an LCM function in Excel.) The cell containing the PMT function result is entered as the *A* value. The general format is

$$= -\text{PV}(MARR, LCM\_years, PMT\_result\_cell) \qquad [17.18]$$

EXAMPLE 17.6

Paul is designing the interior walls of an industrial building. In some places, it is important to reduce noise transmission across the wall. Two construction options—stucco on metal lath (S) and bricks (B)—each have about the same transmission loss, approximately 33 decibels. This will reduce noise attenuation costs in adjacent office areas. Paul has estimated the first costs and after-tax savings each year for both designs. (*a*) Use the CFAT values and an after-tax MARR of 7% per year to determine which is economically better. (*b*) Use a spreadsheet to select the alternative and determine the required first cost for the plans to break even.

Plan S		Plan B	
Year	CFAT, $	Year	CFAT, $
0	−28,800	0	−50,000
1–6	5,400	1	14,200
7–10	2,040	2	13,300
10	2,792	3	12,400
		4	11,500
		5	10,600

Solution by Hand

(*a*) In this example, both AW and PW analyses are shown. Develop the AW relations using the CFAT values over each plan's life. Select the larger value.

$$\begin{aligned} AW_S &= [-28{,}800 + 5400(P/A,7\%,6) + 2040(P/A,7\%,4)(P/F,7\%,6) \\ &\quad + 2792(P/F,7\%,10)](A/P,7\%,10) \\ &= \$422 \end{aligned}$$

$$\begin{aligned} AW_B &= [-50{,}000 + 14{,}200(P/F,7\%,1) + \cdots + 10{,}600(P/F,7\%,5)](A/P,7\%,5) \\ &= \$327 \end{aligned}$$

Both plans are financially viable; select plan S because AW_S is larger.

For the PW analysis, the LCM is 10 years. Use the AW values and the *P/A* factor for the LCM of 10 years to select stucco on metal lath, plan S.

$$PW_S = AW_S(P/A,7\%,10) = 422(7.0236) = \$2964$$
$$PW_B = AW_B(P/A,7\%,10) = 327(7.0236) = \$2297$$

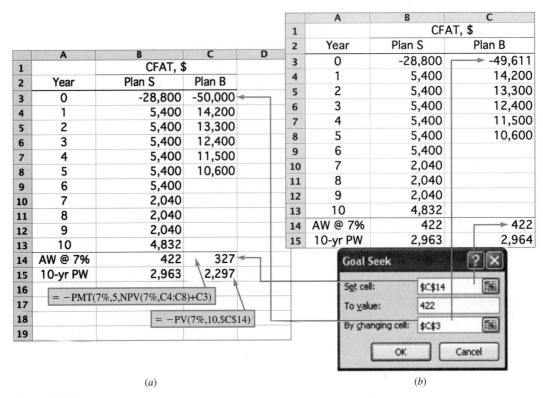

	A	B	C	D
1		CFAT, $		
2	Year	Plan S	Plan B	
3	0	-28,800	-50,000	
4	1	5,400	14,200	
5	2	5,400	13,300	
6	3	5,400	12,400	
7	4	5,400	11,500	
8	5	5,400	10,600	
9	6	5,400		
10	7	2,040		
11	8	2,040		
12	9	2,040		
13	10	4,832		
14	AW @ 7%	422	327	
15	10-yr PW	2,963	2,297	
16	= −PMT(7%,5,NPV(7%,C4:C8)+C3)			
17				
18	= −PV(7%,10,$C$14)			
19				

	A	B	C
1		CFAT, $	
2	Year	Plan S	Plan B
3	0	-28,800	-49,611
4	1	5,400	14,200
5	2	5,400	13,300
6	3	5,400	12,400
7	4	5,400	11,500
8	5	5,400	10,600
9	6	5,400	
10	7	2,040	
11	8	2,040	
12	9	2,040	
13	10	4,832	
14	AW @ 7%	422	422
15	10-yr PW	2,963	2,964

Goal Seek

Set cell: $C$14
To value: 422
By changing cell: $C$3

OK   Cancel

(a)   (b)

Figure 17–5
(a) After-tax AW and PW analysis, and (b) breakeven first cost using Goal Seek, Example 17.6.

Solution by Spreadsheet

(b) Figure 17–5a, row 14, displays the AW value calculated using the PMT function defined by Equation [17.17], and row 15 shows the 10-year PW that results from the PV function in Equation [17.18]. Plan S is chosen by a relatively small margin.

Figure 17–5b shows the Goal Seek template used to equate the AW values and determine the plan B first cost of $−49,611 that causes the plans to break even. This is a small reduction from the $−50,000 first cost initially estimated.

Comment

It is important to remember the minus signs in PMT and PV functions when utilizing them to obtain the corresponding PW and AW values. If the minus is omitted, the AW and PW values have the wrong sign and it appears that the plans are not financially viable in that they do not return at least the after-tax MARR. That would happen in this example.

To utilize the **ROR method,** apply exactly the same procedures as in Chapter 7 (single project) and Chapter 8 (two or more alternatives) to the CFAT series. A PW or AW relation is developed to estimate the rate of return i^* for a project, or Δi^* for the incremental CFAT between two alternatives. Multiple roots may exist in the CFAT series, as they can for any cash flow series. For a single project, set the PW or AW equal to zero and solve for i^*.

Project evaluation

Present worth: $0 = \sum_{t=1}^{t=n} \text{CFAT}_t (P/F,i^*,t)$ [17.19]

Annual worth: $0 = \sum_{t=1}^{t=n} \text{CFAT}_t (P/F,i^*,t)(A/P,i^*,n)$ [17.20]

If $i^* \geq$ after-tax MARR, the project is economically justified.

Spreadsheet solution for i^* is faster for most CFAT series. It is performed using the IRR function with the general format

$$= \text{IRR}(\text{year\_0\_CFAT:year\_n\_CFAT}) \qquad [17.21]$$

If the after-tax ROR is important to the analysis, but the details of an after-tax study are not of interest, the before-tax ROR (or MARR) can be adjusted with the effective tax rate T_e by using the *approximating* relation

$$\textbf{Before-tax ROR} = \frac{\textbf{after-tax ROR}}{1 - T_e} \qquad [17.22]$$

For example, assume a company has an effective tax rate of 40% and normally uses an after-tax MARR of 12% per year for economic analyses that consider taxes explicitly. To *approximate* the effect of taxes without performing the details of an after-tax study, the before-tax MARR can be estimated as

$$\text{Before-tax MARR} = \frac{0.12}{1 - 0.40} = 20\% \text{ per year}$$

If the decision concerns the economic viability of a project and the resulting PW or AW value is close to zero, the details of an after-tax analysis should be developed.

EXAMPLE 17.7

A fiber optics manufacturing company operating in Hong Kong has spent $50,000 for a 5-year-life machine that has a projected $20,000 annual NOI and annual depreciation of $10,000 for years 1 through 5. The company has a T_e of 40%. (*a*) Determine the after-tax rate of return. (*b*) Approximate the before-tax return.

Solution

(*a*) The CFAT in year 0 is $-50,000$. For years 1 through 5 there is no capital purchase or sale, so NOI = CFBT. (See Equations [17.1] and [17.9].) Determine CFAT.

$$\text{TI} = \text{NOI} - \text{D} = 20,000 - 10,000 = \$10,000$$
$$\text{Taxes} = T_e\,(\text{TI}) = 0.4(10,000) = \$4000$$
$$\text{CFAT} = \text{CFBT} - \text{taxes} = 20,000 - 4000 = \$16,000$$

Since the CFAT for years 1 through 5 has the same value, use the P/A factor in Equation [17.19].

$$0 = -50,000 + 16,000(P/A,i^*,5)$$
$$(P/A,i^*,5) = 3.125$$

Solution gives $i^* = 18.03\%$ as the after-tax rate of return.

(*b*) Use Equation [17.22] for the before-tax return estimate.

$$\text{Before-tax ROR} = \frac{0.1803}{1 - 0.40} = 0.3005 \quad (30.05\%)$$

The actual before-tax i^* using CFBT = $20,000 for 5 years is 28.65% from the relation

$$0 = -50,000 + 20,000(P/A,i^*,5)$$

The tax effect will be slightly overestimated if a MARR of 30.05% is used in a before-tax analysis.

A rate of return evaluation performed by hand on two or more alternatives utilizes a PW or AW relation to determine the incremental return Δi^* of the incremental CFAT series between two alternatives. Solution by spreadsheet is accomplished using the incremental CFAT values and the IRR function. The equations and procedures applied are the same as in Chapter 8 (Sections 8.4 through 8.6) for selection from mutually exclusive alternatives using the ROR method. You should review and understand these sections before proceeding with this section.

From this review, several important facts can be recalled:

Selection guideline: The fundamental rule of incremental ROR evaluation at a stated MARR is as follows:

> Select the one alternative that requires the largest initial investment, provided the extra investment is justified relative to another justified alternative.

Incremental ROR: Incremental analysis must be performed. Overall i^* values cannot be depended upon to select the correct alternative, unlike the PW or AW method at the MARR, which will always indicate the correct alternative.

Equal-service requirement: Incremental ROR analysis requires that the alternatives be evaluated over equal time periods. The LCM of the two alternative lives must be used to find the PW or AW of incremental cash flows. (The only exception, mentioned in Section 8.5, occurs when the AW analysis is performed on *actual cash flows, not the increments;* then one-life-cycle analysis is acceptable over the respective alternative lives.)

Revenue and cost alternatives: Revenue alternatives (positive and negative cash flows) may be treated differently from cost alternatives (cost-only cash flow estimates). For revenue alternatives, the overall i^* may be used to perform an initial screening. Alternatives with $i^* <$ MARR can be removed from further evaluation. An i^* for cost-only alternatives cannot be determined, so incremental analysis is required with all alternatives included.

Breakeven ROR

Once the CFAT series are developed, the **breakeven ROR** can be obtained using a plot of PW versus i^* by solving the PW relation for each alternative over the LCM at several interest rates. For any after-tax MARR greater than the breakeven ROR, the extra investment is not justified.

The next examples solve CFAT problems using incremental ROR analysis and the breakeven ROR plot of PW versus i.

EXAMPLE 17.8

In Example 17.6, Paul estimated the CFAT for interior wall materials to reduce sound transmission; plan S is to construct with stucco on metal lath, and plan B is to construct using brick. Figure 17–5a presented both a PW analysis over 10 years and an AW analysis over the respective lives. Plan S was selected. After reviewing this earlier solution, (*a*) perform an ROR evaluation at the after-tax MARR of 7% per year and (*b*) plot the PW versus Δi graph to determine the breakeven ROR.

Solution by Spreadsheet

(*a*) The LCM is 10 years for the incremental ROR analysis, and plan B requires the extra investment that must be justified. Apply the procedure in Section 8.6 for incremental ROR analysis. Figure 17–6 shows the estimated CFAT for each alternative and the incremental CFAT series. Since these are revenue alternatives, the overall Δi^* is calculated first to ensure that they both make at least the MARR of 7%. Row 14 indicates they do. The IRR function (cell E14) is applied to the incremental CFAT, indicating that $\Delta i^* = 6.35\%$. Since this is lower than the MARR, the extra investment in brick walls is not justified. Plan S is selected, the same as with the PW and AW methods.

(*b*) The NPV function is used to find the PW of the incremental CFAT series at various i values. The graph indicates that the breakeven Δi^* occurs at 6.35%—the same that the IRR function found. Whenever the after-tax MARR is above 6.35%, as is the case here with MARR $=$ 7%, the extra investment in plan B is not justified.

Comment

Note that the incremental CFAT series has three sign changes. The cumulative series also has three sign changes (Norstrom's criterion). Accordingly, there may be multiple Δi^* values. The application of the IRR function using the "guess" option finds no other real number roots in the normal rate of return range.

Figure 17–6
Incremental evaluation of CFAT and determination of breakeven ROR, Example 17.8.

EXAMPLE 17.9

In Example 17.5 an after-tax analysis of two bone cell analyzers was initiated due to a new 3-year NBA contract. The criterion used to select analyzer 1 was the total taxes for the 3 years. The complete solution is in Table 17–4 (hand) and Figure 17–4 (spreadsheet).

Continue the spreadsheet analysis by performing an after-tax ROR evaluation, assuming the analyzers are sold after 3 years for the amounts estimated in Example 17.5: $130,000 for analyzer 1 and $225,000 for analyzer 2. The after-tax MARR is 10% per year.

Solution

A spreadsheet solution is presented here, but a hand solution is equivalent, just slower. Figure 17–7 is an updated version of the spreadsheet in Figure 17–4 to include the sale of the analyzers in year 3. The CFAT series (column I) are determined by the relation CFAT = CFBT − taxes, with the taxable income determined using Equation [17.16], where DR is included. For example, in year 3 when analyzer 2 is sold for $S = \$225,000$, the CFAT calculation is

$$\text{CFAT}_3 = \text{CFBT} - (\text{TI})(T_e) = \text{GI} - \text{OE} - P + S - (\text{GI} - \text{OE} - D + \text{DR})(T_e)$$

The depreciation recapture DR is the amount above the year 3 book value received at sale time. Using the book value after 3 years (F14),

$$\text{DR} = \text{selling price} - \text{BV}_3 = 225,000 - 64,800 = \$160,200$$

Now the CFAT in year 3 for analyzer 2 can be determined.

$$\begin{aligned}
\text{CFAT}_3 &= 100,000 - 10,000 + 0 + 225,000 \\
&\quad -(100,000 - 10,000 - 43,200 + 160,200)(0.35) \\
&= 315,000 - 207,000(0.35) = \$242,550
\end{aligned}$$

The cell tags in row 14 of Figure 17–7 follow this same progression. The incremental CFAT is calculated in column J, ready for the after-tax incremental ROR analysis.

These are revenue alternatives, so the overall i^* values indicate that both CFAT series are acceptable. The value $\Delta i^* = 23.6\%$ (cell J17) also exceeds MARR = 10%, so **analyzer 2 is selected**. This decision applies the ROR method guideline: Select the alternative that requires the largest, *incrementally* justified investment.

Figure 17–7
Incremental ROR analysis of CFAT with depreciation recapture, Example 17.9.

Comment

In Section 8.4, we demonstrated the fallacy of selecting an alternative based solely on the overall $i*$, because of the ranking inconsistency problem of the ROR method. The incremental ROR must be used. The same fact is demonstrated in this example. If the larger $i*$ alternative is chosen, analyzer 1 is incorrectly selected. When $\Delta i*$ exceeds the MARR, the larger investment is correctly chosen—analyzer 2 in this case. For verification, the PW at 10% is calculated for each analyzer (column I). Again, analyzer 2 is the winner, based on its larger PW of $93,905.

17.6 After-Tax Replacement Study ● ● ●

When a currently installed asset (the defender) is challenged with possible replacement, the effect of taxes can have an impact upon the decision of the replacement study. The final decision may not be reversed by taxes, but the difference between before-tax AW values may be significantly different from the after-tax difference. Tax considerations in the year of the replacement are as follows:

Depreciation recapture or tax savings due to a sizable **capital loss** are possible, if it is necessary to trade the defender at a sacrifice price. Additionally, the after-tax replacement study considers tax-deductible **depreciation** and **operating expenses** not accounted for in a before-tax analysis.

The effective tax rate T_e is used to estimate the amount of annual taxes (or tax savings) from TI. The same procedure as the before-tax replacement study in Chapter 11 is applied here, but for CFAT estimates. The procedure should be thoroughly understood before proceeding. Special attention to Sections 11.3 and 11.5 is recommended.

   Example 17.10 presents a solution by hand of an after-tax replacement study using a simplifying assumption of classical SL (straight line) depreciation. Example 17.11 solves the same problem by spreadsheet, but includes the detail of MACRS depreciation. This provides an opportunity to observe the difference in the AW values between the two depreciation methods.

EXAMPLE 17.10

Midcontinent Power Authority purchased emission control equipment 3 years ago for $600,000. Management has discovered that it is technologically and legally outdated now. New equipment has been identified. If a market value of $400,000 is offered as the trade-in for the current equipment, perform a replacement study using (*a*) a before-tax MARR of 10% per year and

(b) a 7% per year after-tax MARR. Assume an effective tax rate of 34%. As a simplifying assumption, use classical straight line depreciation with $S = 0$ for both alternatives.

	Defender	Challenger
Market value, $	400,000	
First cost, $		−1,000,000
Annual cost, $/year	−100,000	−15,000
Recovery period, years	8 (originally)	5

Solution

Assume that an ESL (economic service life) analysis has determined the best life values to be 5 more years for the defender and 5 years total for the challenger.

(a) For the *before-tax replacement study,* find the AW values. The defender AW uses the market value as the first cost, $P_D = \$-400,000$.

$$AW_D = -400,000(A/P,10\%,5) - 100,000 = \$-205,520$$
$$AW_C = -1,000,000(A/P,10\%,5) - 15,000 = \$-278,800$$

Applying step 1 of the replacement study procedure (Section 11.3), we select the better AW value. The defender is retained now with a plan to keep it for the 5 remaining years. The defender has a $73,280 lower equivalent annual cost compared to the challenger. This complete solution is included in Table 17–5 (left half) for comparison with the after-tax study.

(b) For the *after-tax replacement study,* there are no tax effects other than income tax for the defender. The annual SL depreciation is $75,000, determined when the equipment was purchased 3 years ago.

$$D_t = 600,000/8 = \$75,000 \qquad t = 1 \text{ to } 8 \text{ years}$$

Table 17–5 shows the TI and taxes at 34%. The taxes are actually tax savings of $59,500 per year, as indicated by the minus sign. (Remember that for tax savings in an economic

| **TABLE 17–5** | Before-Tax and After-Tax Replacement Analyses, Example 17.10 |

		Before Taxes			After Taxes			
Defender Age	Year	Expenses OE, $	P and S, $	CFBT, $	Depreciation D, $	Taxable Income TI, $	Taxes* at 0.34TI, $	CFAT, $
Defender
3	0		−400,000	−400,000				−400,000
4	1	−100,000		−100,000	75,000	−175,000	−59,500	−40,500
5	2	−100,000		−100,000	75,000	−175,000	−59,500	−40,500
6	3	−100,000		−100,000	75,000	−175,000	−59,500	−40,500
7	4	−100,000		−100,000	75,000	−175,000	−59,500	−40,500
8	5	−100,000	0	−100,000	75,000	−175,000	−59,500	−40,500
AW at 10%				−205,520	AW at 7%			−138,056
Challenger
	0		−1,000,000	−1,000,000		+25,000†	8,500	−1,008,500
	1	−15,000		−15,000	200,000	−215,000	−73,100	+58,100
	2	−15,000		−15,000	200,000	−215,000	−73,100	+58,100
	3	−15,000		−15,000	200,000	−215,000	−73,100	+58,100
	4	−15,000		−15,000	200,000	−215,000	−73,100	+58,100
	5	−15,000	0	−15,000	200,000	−215,000‡	−73,100	+58,100
AW at 10%				−278,800	AW at 7%			−187,863

* Minus sign indicates a tax savings for the year.
† Depreciation recapture on defender trade-in.
‡ Assumes challenger's salvage actually realized is $S = 0$; no tax.

analysis it is assumed that there is positive taxable income elsewhere in the corporation to offset the saving.) Since only costs are estimated, the annual CFAT is negative, but the $59,500 tax savings has reduced it. The CFAT and AW at 7% per year are

$$CFAT = CFBT - taxes = -100,000 - (-59,500) = \$-40,500$$

$$AW_D = -400,000(A/P,7\%,5) - 40,500 = \$-138,056$$

For the challenger, depreciation recapture on the defender occurs when it is replaced because the trade-in amount of $400,000 is larger than the current book value. In year 0 for the challenger, Table 17–5 includes the following computations to arrive at a tax of $8500.

Defender book value, year 3:	$BV_3 = 600,000 - 3(75,000) = \$375,000$
Depreciation recapture:	$DR_3 = TI = 400,000 - 375,000 = \$25,000$
Taxes on trade-in, year 0:	$Taxes = 0.34(25,000) = \$8500$

The SL depreciation is $1,000,000/5 = \$200,000$ per year. This results in tax saving and CFAT as follows:

$$Taxes = (-15,000 - 200,000)(0.34) = \$-73,100$$

$$CFAT = CFBT - taxes = -15,000 - (-73,100) = \$+58,100$$

In year 5, it is assumed the challenger is sold for $0; there is no depreciation recapture. The AW for the challenger at the 7% after-tax MARR is

$$AW_C = -1,008,500(A/P,7\%,5) + 58,100 = \$-187,863$$

The defender is again selected; however, the equivalent annual advantage has decreased from $73,280 before taxes to $49,807 after taxes.

Conclusion: By either analysis, retain the defender now and plan to keep it for 5 more years. Additionally, plan to evaluate the estimates for both alternatives 1 year hence. If and when cash flow estimates change significantly, perform another replacement analysis.

Comment

If the market value (trade-in) had been less than the current defender book value of $375,000, a capital loss, rather than depreciation recapture, would occur in year 0. The resulting tax savings would decrease the CFAT (which is to reduce costs if CFAT is negative) of the challenger. For example, a trade-in amount of $350,000 would result in a TI of $350,000 - 375,000 = \$-25,000$ and a *tax savings* of $\$-8500$ in year 0. The CFAT is then $\$-1,000,000 - (-8500) = \$-991,500$.

EXAMPLE 17.11

Repeat the after-tax replacement study of Example 17.10*b* using 7-year MACRS depreciation for the defender and 5-year MACRS depreciation for the challenger. Assume either asset is sold after 5 years for exactly its book value. Determine if the answers are significantly different from those obtained when the simplifying assumption of classical SL depreciation was made.

Solution

Figure 17–8 shows the complete analysis. MACRS requires more computation than SL depreciation, but this effort is easily reduced by the use of a spreadsheet. Again the *defender is selected* for retention, but now by an advantage of $44,142 annually. This compares to the $49,807 advantage using classical SL depreciation and the $73,280 before-tax advantage of the defender. Therefore, taxes and MACRS have reduced the defender's economic advantage, but not enough to reverse the decision to retain it.

Several other differences in the results between SL and MACRS depreciation are worth noting. There is depreciation recapture in year 0 of the challenger due to trade-in of the defender at $400,000, a value larger than the book value of the 3-year-old defender. This amount,

	A	B	C	D	E	F	G	H	I	J
1	MARR =	7%								
2	Purchase =	$ 600,000		**Defender after-tax MACRS analysis**						
3	Defender		Basis B &	(Expenses)	MACRS					
4	age	Year	salvage S[(1)]	CFBT	rates	Depr	TI	Tax savings	CFAT	
5	3	0	-400,000						-400,000	
6	4	1		-100,000	0.1249	74,940	-174,940	-59,480	-40,520	
7	5	2		-100,000	0.0893	53,580	-153,580	-52,217	-47,783	
8	6	3		-100,000	0.0892	53,520	-153,520	-52,197	-47,803	
9	7	4		-100,000	0.0893	53,580	-153,580	-52,217	-47,783	
10	8	5	0	-100,000	0.0446	26,760	-126,760	-43,098	-56,902	
11	Total					262,380				
12	[(1)]Defender assumed to be sold in year 5 (year 8 of its life) for exactly BV = 0.							AW at 7%	**-$145,273**	
13	All the original B = $600,000 is depreciated over the 8 years; no tax effect.									
14								Depreciation recapture		
15	Purchase =	$1,000,000		**Challenger after-tax MACRS analysis**				= −C5 − F11		
16	Challenger		Basis B &	(Expenses)	MACRS			Taxes or		
17	age	Year	salvage S[(1)]	CFBT	rates	Depr	TI [(2)]	Tax savings	CFAT	
18	0	0	-1,000,000				137,620	46,791	-1,046,791	
19	1	1		-15,000	0.2000	200,000	-215,000	-73,100	58,100	
20	2	2		-15,000	0.3200	320,000	-335,000	-113,900	98,900	
21	3	3		-15,000	0.1920	192,000	-207,000	-70,380	55,380	
22	4	4		-15,000	0.1152	115,200	-130,200	-44,268	29,268	
23	5	5	57,600	-15,000	0.1152	115,200	-130,200	-44,268	86,868	
24	Total					942,400				
25	[(1)]Challenger assumed to be sold in year 5 for exactly							AW at 7%	**-$189,415**	
26	BV = 1,000,000-942,400 = $57,600.									
27	No tax effect, but CFAT increases in year 5.					Challenger sales price				
28	[(2)] TI of $137,620 in year 0 is depreciation recapture from					= B15 − F24				
29	trade of defender. DR = B - current BV = 400,000 - 262,380.									

Figure 17–8
After-tax replacement study with MACRS depreciation and depreciation recapture, Example 17.11.

$137,620 (cell G18), is treated as ordinary taxable income. The calculations for the DR and associated tax, by hand, are as follows:

$$BV_3 = \text{first cost} - \text{MACRS depreciation for 3 years}$$
$$= \text{total MACRS depreciation for years 4 through 8}$$
$$= \$262,380 \qquad (\text{cell F11})$$

$$DR = TI_0 = \text{trade-in} - BV_3$$
$$= 400,000 - 262,380 = \$137,620 \quad (\text{cell G18})$$

$$\text{Taxes} = (0.34)(137,620) = \$46,791 \qquad (\text{cell H18})$$

See the cell tags and table notes that duplicate this logic.

The assumption that the challenger is sold after 5 years at its book value implies a positive cash flow in year 5. The entry $57,600 (C23) reflects this assumption, since the forgone MACRS depreciation in year 6 would be $1,000,000(0.0576) = \$57,600$. The spreadsheet relation $= B15 − F24$ determines this value using the accumulated depreciation in F24. [*Note:* If the salvage $S = 0$ is anticipated after 5 years, then a capital loss of $57,600 will be incurred. This implies an additional tax saving of $57,600(0.34) = \$19,584$ in year 5. Conversely, if the salvage value exceeds the book value, a depreciation recapture and associated tax should be estimated.]

17.7 After-Tax Value-Added Analysis ●●●

When a person or company is willing to pay more for an item, it is likely that some processing has been performed on an earlier version of the item to make it more valuable now to the purchaser. This is value added.

Value added is a term used to indicate that a product or service has **added worth** from the perspective of a consumer, owner, investor, or purchaser. It is common to leverage value-adding activities on a product or service.

Value added

For an example of highly leveraged value-added activities, consider onions that are grown and sold at the farm level for cents per pound. They may be purchased by the shopper in a store at 50 cents to $1.25 per pound. But when onions are cut and coated with a special batter, they may be fried in hot oil and sold as onion rings for several dollars per pound. Thus, from the perspective of the consumer, there has been a large amount of value added by the processing from raw onions in the ground into onion rings sold at a restaurant or fast-food shop.

The value-added measure was briefly introduced in conjunction with AW analysis before taxes. When value-added analysis is performed after taxes, the approach is somewhat different from that of CFAT analysis developed previously in this chapter. However, as shown below,

> The decision about an alternative will be the same for both the value-added and CFAT methods, because the AW of economic value-added estimates is the same as the AW of CFAT estimates.

Value-added analysis starts with Equation [17.4], net operating profit after taxes (NOPAT), which includes the depreciation for *year 1* through *year n*. Depreciation D is included in that TI = GI − OE − D. This is different from CFAT, where the depreciation has been specifically removed so that only *actual* cash flow estimates are used for *year 0* through year *n*.

The term **economic value added (EVA)** indicates the monetary worth added by an alternative to the corporation's bottom line. (The term *EVA* is a trademark of Stern Stewart & Co.) The technique discussed below was first publicized in several articles[1] in the mid-1990s, and it has since become very popular as a means to evaluate the ability of a corporation to increase its economic worth, especially from the shareholders' viewpoint.

> The annual EVA is the amount of NOPAT remaining on corporate books after removing the **cost of invested capital** during the year. That is, EVA indicates the project's **contribution to the net profit** of the corporation after taxes.

The **cost of invested capital** is the after-tax rate of return (usually the MARR value) multiplied by the book value of the asset during the year. This is the interest incurred by the current level of capital invested in the asset. (If different tax and book depreciation methods are used, the *book depreciation value is used* here, because it more closely represents the remaining capital invested in the asset from the corporation's perspective.) Computationally,

$$\text{EVA} = \text{NOPAT} - \text{cost of invested capital}$$
$$= \text{NOPAT} - (\text{after-tax interest rate})(\text{book value in year } t - 1)$$
$$= \text{TI}(1 - T_e) - (i)(\text{BV}_{t-1}) \qquad [17.23]$$

Since both TI and the book value consider depreciation, EVA is a measure of worth that mingles actual cash flow with noncash flows to determine the estimated financial worth contribution to the corporation. This financial worth is the amount used in public documents of the corporation (balance sheet, income statement, stock reports, etc.). Because corporations want to present the largest value possible to the stockholders and other owners, the EVA method is often more appealing than the AW method from the financial perspective.

The result of an EVA analysis is a series of annual EVA estimates. Two or more alternatives are compared by calculating the AW of EVA estimates and selecting the alternative with the larger AW value. If only one project is evaluated, AW > 0 means the after-tax MARR is exceeded, thus making the project value-adding.

Sullivan and Needy[2] have demonstrated that the AW of EVA and the AW of CFAT are identical in amount. Thus, either method can be used to make a decision. The annual EVA estimates indicate added worth to the corporation generated by the alternative, while the annual CFAT estimates describe how cash will flow. This comparison is made in Example 17.12.

[1]A. Blair, "EVA Fever," *Management Today,* Jan. 1997, pp. 42–45; W. Freedman, "How Do You Add Up?" *Chemical Week,* Oct. 9, 1996, pp. 31–34.

[2]W. G. Sullivan and K. L. Needy, "Determination of Economic Value Added for a Proposed Investment in New Manufacturing." *The Engineering Economist,* vol. 45, no. 2 (2000), pp. 166–181.

EXAMPLE 17.12

Biotechnics Engineering has developed two mutually exclusive plans for investing in new capital equipment with the expectation of increased revenue from its medical diagnostic services to cancer patients. The estimates are summarized below. (*a*) Use classical straight line depreciation, an after-tax MARR of 12%, and an effective tax rate of 40% to perform two annual worth after-tax analyses: EVA and CFAT. (*b*) Explain the fundamental difference between the results of the two analyses.

	Plan A	Plan B
Initial investment, $	−500,000	−1,200,000
Gross income − expenses, $	170,000 per year	600,000 in year 1, decreasing by 100,000 per year thereafter
Estimated life, years	4	4
Salvage value	None	None

Solution by Spreadsheet

(*a*) Refer to the spreadsheet and function cells (row 22) in Figure 17–9.

EVA evaluation: All the necessary information for EVA estimation is determined in columns B through G. The net operating profit after taxes (NOPAT) in column H is calculated by Equation [17.4], TI − taxes. The book values (column E) are used to determine the cost of invested capital in column I, using the second term in Equation [17.23], that is, $i(BV_{t-1})$, where i is the 12% after-tax MARR. This represents the amount of interest at 12% per year, after taxes, for the currently invested capital as reflected by the book value at the beginning of the year. The EVA estimate is the sum of columns H and I for years 1 through 4. *Notice there is no EVA estimate for year 0,* since NOPAT and the cost of invested capital are estimated for years 1 through *n*. Finally, the larger AW of the EVA value is selected, which indicates that plan B is better and that plan A does not make the 12% return.

CFAT evaluation: As shown in function row 22 (plan B for year 3), CFAT estimates (column K) are calculated as (GI − OE) − P − taxes. The AW of CFAT again concludes that plan B is better and that plan A does not return the after-tax MARR of 12% (K10).

(*b*) What is the fundamental difference between the EVA and CFAT series in columns J and K? They are clearly equivalent from the time value of money perspective since the AW values are numerically the same. To answer the question, consider plan A, which has a constant CFAT estimate of $152,000 per year. To obtain the AW of EVA estimate of $−12,617 for

	A	B	C	D	E	F	G	H	I	J	K
1						PLAN A					
2									EVA analysis		CFAT analysis
3			Investment P	SL	Book value	Taxable			Cost of		
4	Year	GI - OE	(Basis B)	Depreciation	BV	income, TI	Taxes	NOPAT	inv. capital	EVA	CFAT
5	0		-500,000		500,000						-500,000
6	1	170,000		125,000	375,000	45,000	18,000	27,000	-60,000	-33,000	152,000
7	2	170,000		125,000	250,000	45,000	18,000	27,000	-45,000	-18,000	152,000
8	3	170,000		125,000	125,000	45,000	18,000	27,000	-30,000	-3,000	152,000
9	4	170,000		125,000	0	45,000	18,000	27,000	-15,000	12,000	152,000
10	AW values									-$12,617	-$12,617
11						PLAN B					
12									EVA analysis		CFAT analysis
13			Investment P	SL	Book value	Taxable			Cost of		
14	Year	GI - OE	(Basis B)	Depreciation	BV	income, TI	Taxes	NOPAT	inv. capital	EVA	CFAT
15	0		-1,200,000		1,200,000						-1,200,000
16	1	600,000		300,000	900,000	300,000	120,000	180,000	-144,000	36,000	480,000
17	2	500,000		300,000	600,000	200,000	80,000	120,000	-108,000	12,000	420,000
18	3	400,000		300,000	300,000	100,000	40,000	60,000	-72,000	-12,000	360,000
19	4	300,000		300,000	0	0	0	0	-36,000	-36,000	300,000
20	AW values									$3,388	$3,388
21											
22	Functions for Plan B, year 3			`= -$C$15/4`	`= E17-D18`	`= B18-D18`	`= F18*0.4`	`= F18-G18`	`=-0.12*E17`	`= H18+I18`	`= B18+C18-G18`

Figure 17–9
Comparison of two plans using EVA and CFAT analyses, Example 17.12.

years 1 through 4, the initial investment of $500,000 is distributed over the 4-year life using the A/P factor at 12%. That is, an equivalent amount of $500,000(A/P,12\%,4) = \$164,617$ is "charged" against the cash inflows in each of years 1 through 4. In effect, the yearly CFAT is reduced by this charge.

$$\text{CFAT} - (\text{initial investment})(A/P,12\%,4) = 152,000 - 500,000(A/P,12\%,4)$$
$$152,000 - 164,617 = \$-12,617$$
$$= \text{AW of EVA}$$

This is the AW value for both series, demonstrating that the two methods are economically equivalent. However, the EVA method indicates an alternative's yearly estimated contribution to the *value of the corporation,* whereas the CFAT method estimates the actual cash flows to the corporation. This is why the EVA method is often more popular than the cash flow method with corporate executives.

Comment

The calculation $P(A/P,i,n) = \$500,000(A/P,12\%,4)$ is exactly the same as the capital recovery in Equation [6.3], assuming an estimated salvage value of zero. Thus, the cost of invested capital for EVA is the same as the capital recovery discussed in Chapter 6. This further demonstrates why the AW method is economically equivalent to the EVA evaluation.

Online
Presentation

17.8 After-Tax Analysis for International Projects ●●●

Primary questions to be answered prior to performing a corporate-based after-tax analysis for international settings revolve around tax-deductible allowances—depreciation, business expenses, capital asset evaluation—and the effective tax rate needed for Equation [17.7], taxes = $(T_e)(\text{TI})$. As discussed in Chapter 16, most governments of the world recognize and use the straight line (SL) and declining balance (DB) methods of depreciation with some variations to determine the annual tax-deductible allowance. Expense deductions vary widely from country to country. By way of example, some of these are summarized here.

Canada

Depreciation: This is deductible and is normally based on DB calculations, although SL may be used. An equivalent of the half-year convention is applied in the first year of ownership. The annual tax-deductible allowance is termed *capital cost allowance (CCA)*. As in the U.S. system, recovery rates are standardized, so the depreciation amount does not necessarily reflect the useful life of an asset.

Class and CCA rate: Asset classes are defined and annual depreciation rates are specified by class. No specific recovery period (life) is identified, in part because assets of a particular class are grouped together and the annual CCA is determined for the entire class, not individual assets. There are some 44 classes, and CCA rates vary from 4% per year (the equivalent of a 25-year-life asset) for buildings (class 1) to 100% (1-year life) for applications software, chinaware, dies, etc. (class 12). Most rates are in the range of 10% to 30% per year.

Expenses: Business expenses are deductible in calculating TI. Expenses related to capital investments are not deductible, since they are accommodated through the CCA.

Internet: Further details are available on the Revenue Canada website at www.cra.gc.ca in the Forms and Publications section.

China (PRC)

Depreciation: Officially, SL is the primary method of tax-deductible depreciation; however, assets employed in selected industries or types of assets can utilize accelerated DB or SYD (sum-of-years-digits) depreciation, when approved by the government. The selected industries and assets can change over time; currently favored industries serve areas such as technology and oil exploration, and equipment subjected to large vibrations during normal usage is allowed accelerated depreciation.

Recovery period: Standardized recovery periods are published that vary from 3 years (electronic equipment) to 10 years (aircraft, machinery, and other production equipment) to 20 years (buildings). Shortened periods can be approved, but the minimum recovery period cannot be less than 60% for the normal period defined by current tax law.

Expenses: Business expenses are deductible with some limitations and some special incentives. Limitations are placed, for example, on advertising expense deductions (15% of sales for the year). Incentives are generous in some cases; for example, 150% of actual expenses is deductible for new technology and new product R&D activities.

Internet: Summary information for China and several other countries is available at www.worldwide-tax.com.

Mexico

Depreciation: This is a fully deductible allowance for calculating TI. The SL method is applied with an index for inflation considered each year. For some asset types, an immediate deduction of a percentage of the first cost is allowed. (This is a close equivalent to the Section 179 Deduction in the United States.)

Class and rates: Asset types are identified, though not as specifically defined as in some countries. Major classes are identified, and annual recovery rates vary from 5% for buildings (the equivalent of a 20-year life) to 100% for environmental machinery. Most rates range from 10% to 30% per year.

Profit tax: The income tax is levied on profits on income earned from carrying on business in Mexico. Most business expenses are deductible. Corporate income is taxed only once, at the federal level; no state-level taxes are imposed.

Tax on Net Assets (TNA): Under some conditions, a tax of 1.8% of the average value of assets located in Mexico is paid annually in addition to income taxes.

Internet: The best information is via websites for companies that assist international corporations located in Mexico. One example is PriceWaterhouseCoopers at www.pwcglobal.com/mx/eng.

Japan

Depreciation: This is fully deductible and based on classical SL and DB methods. To encourage capital investment for long-term economic growth, in 2007 Japan allowed assets to be depreciated using a *250% DB method;* that is, a significantly increased accelerated rate is allowed compared to historical Japanese allowances. Switching to classical SL depreciation must take place in the year that the accelerated rate amount falls below the corresponding SL amount.

Class and life: A statutory useful life ranging from 4 to 24 years, with a 50-year life for reinforced concrete buildings, is specified.

Expenses: Business expenses are deductible in calculating TI.

Internet: Further details are available on the Japanese Ministry of Finance website at www.mof.go.jp.

The effective tax rate varies considerably among countries. Some countries levy taxes only at the federal level, while others impose taxes at several levels of government (federal, state or provincial, prefecture, county, and city). A summary of international corporate average tax rates is presented in Table 17–6 for a wide range of industrialized countries. These include income taxes at all reported levels of government within each country; however, other types of taxes may be imposed by a particular government. Although these average rates of taxation will vary from year to year, especially as tax reform is enacted, it can be surmised that most corporations face effective rates of about 20% to 40% of taxable income. A close examination of international rates shows that they have decreased significantly over the last decade. In fact, the KPMG report noted in Table 17–6 indicates that the global average corporate tax rate on TI has decreased from 32.7% (1999) to 25.5% (2009). This has encouraged corporate investment and business expansion within country borders and helped soften the massive economic downturn experienced in recent years.

TABLE 17–6 Summary of International Corporate Average Tax Rates	
Tax Rate Levied on Taxable Income, %	**For These Countries**
≥ 40	United States, Japan
35 to < 40	South Africa, Pakistan
32 to < 35	Canada, France, India
28 to < 32	Australia, United Kingdom, New Zealand, Spain, Germany, Mexico, Indonesia
24 to < 28	China, Taiwan, Republic of Korea
20 to < 24	Russia, Turkey, Saudi Arabia
< 20	Singapore, Hong Kong, Chile, Ireland, Iceland, Hungary

Sources: Extracted from KPMG's *Corporate and Indirect Tax Survey 2009* (www.kpmg. com/Global/en/IssuesAndInsights/Pages/default.aspx) and from country websites on corporate taxation.

One of the prime ways that governments have been able to reduce corporate tax rates is by shifting to **indirect taxes** on goods and services for additional tax revenue. These taxes are usually in the form of a **value-added tax (VAT),** goods and services tax (GST), and taxes on products imported from outside its borders. As corporate tax rates have declined, in general the indirect tax rates have increased. This has been especially true during the first decade of the 21st century. However, the worldwide economic slump experienced in recent years has made it necessary for governments around the world to be more cautious of how they tax corporations and maintain a reasonable balance between regular tax rates (as listed in Table 17–6) and indirect tax rates. The VAT system is explained now.

Online Presentation

17.9 Value-Added Tax ● ● ●

A value-added tax (VAT) has facetiously been called a sales tax on steroids, because the VAT rates on some items in some countries that impose a VAT can be as high as 90%. It is also called a GST (goods and service tax).

A **value-added tax** is an indirect tax; that is, it is a tax on goods and services rather than on people or corporations. It differs from a sales tax in two ways: (1) when it is charged and (2) who pays (as explained below). A specific percentage, say 10%, is a charge added to the price of the item and paid by the buyer. The seller then sends this 10% VAT to the taxing entity, usually a government unit. This process of 10% VAT continues every time the item is resold—as purchased or in a modified form—thus the term *value-added.*

A value-added tax is commonly used throughout the world. In some countries, VAT is used in lieu of business or individual income taxes. There is no VAT system in the United States yet. In fact, the United States is the only major industrialized country in the world that does not have a VAT system, though other forms of indirect taxation are used liberally. There is mounting evidence, however, that a VAT/GST system will be necessary in the United States in the near future, but not without a great amount of political discord.

A **sales tax** is used by the U.S. government, by nearly all of the states, and by many local entities. A sales tax is charged on goods and services at the time the goods and services reach *the end user or consumer.* That is, businesses do not pay a sales tax on raw material, unfinished goods, or items they purchase that will ultimately be sold to an end user; only the end user pays the sales tax. Businesses *do pay* a sales tax on items for which they are the end user. Total sales tax percentages imposed by multiple government levels can range from 5% to 11%, sometimes larger on specific items. For example, when Home Depot (HD) purchases microwave ovens from Jenn-Air, Inc., Home Depot does not pay a sales tax on the microwave ovens because they will be sold to HD customers who *will* pay the sales tax. On the other hand, if Home Depot purchases a forklift from Caterpillar for loading and unloading merchandise in one of its stores, HD will pay the sales tax on the forklift, since HD is the end user. Thus, a sales tax is **paid only one time,** and that time is when the goods or services are purchased **by the end user.** The sales tax is the responsibility of the merchant to collect and remand to the taxing entity.

A **value-added tax (VAT),** on the other hand, is charged to the buyer at purchase time, whether the buyer is a business or end user. The seller sends the collected VAT to the taxing entity. If the buyer subsequently resells the goods to another buyer (as is or a modification of it), another VAT is collected by the seller. Now, this second seller will send to the taxing entity an amount that is equal to the total tax collected *minus* the amount of VAT already paid.

As an illustration, assume the U.S. government charged a 10% VAT. Here is how the VAT might work.

Northshore Mining Corporation of Babbitt, Minnesota, sells $100,000 worth of iron ore to Westfall Steel. As part of the price, Northshore collects $110,000, that is, $100,000 + 0.1 × 100,000, from Westfall Steel. Northshore remits the $10,000 VAT to the U.S. Treasury.

Westfall Steel sells all of the steel it made from the iron ore at a price of $300,000 to General Electric (GE). Westfall collects $330,000 from GE and then sends $20,000 to the U.S. Treasury, that is, $30,000 it collected in VAT from GE minus $10,000 it paid in taxes to Northshore Mining.

GE uses the steel to make refrigerators that it sells for $700,000 to retailers, such as Home Depot, Lowe's, and others. GE collects $770,000 and then remits $40,000 to the U.S. Treasury, that is, $70,000 it collected in taxes from retailers minus $30,000 it paid in VAT to Westfall Steel.

If GE purchased machines, tools, or other items to make the refrigerators during this accounting period, and it paid taxes on those items, the taxes paid would also be deducted from the VAT that GE collected before sending the money to the U.S. Treasury. For example, if GE paid $5000 in taxes on motors it purchased for the refrigerators, the amount GE would remit to the U.S. Treasury would be $35,000 (that is, $70,000 it collected from retailers minus $30,000 it paid in taxes to Westfall Steel minus $5000 it paid in taxes on the motors).

The retailers sell the refrigerators for $950,000 and collect $95,000 in taxes from end users—consumers. The retailers remit $25,000 to the U.S. Treasury (that is, $95,000 they collected minus $70,000 they paid previously).

Through this process, the U.S Treasury has received $10,000 from Northshore, $20,000 from Westfall Steel, $35,000 from GE, $5000 from the supplier of the motors, and $25,000 from the retailers, for a total of $95,000. This is 10% of the final sales price of $950,000. The VAT money was deposited into the Treasury at several different times from several different companies.

The taxes that a company *pays* for materials or items it purchases in order to produce goods or services that will subsequently be sold to another business or end user are called *input taxes*, and the company is able to recover them when it collects the VAT from the sale of its products. The taxes that a company *collects* are called *output taxes*, and these are forwarded to the taxing entity, less the amount of input taxes the company paid. Hence, the businesses incur no taxes themselves, the same as with a sales tax.

Several dimensions of a VAT distinguish it from a sales tax or corporate income taxes. Some are as follow:

- Value-added taxes are taxes on consumption, not production or taxable income.
- The end user pays all of the value-added taxes, but VATs are not as obvious as a sales tax that is added to the price of the item at the time of purchase (and displayed on the receipt). Therefore, VAT taxing entities encounter less resistance from consumers.
- Value-added taxes are generally considerably higher than sales taxes, with the average European VAT rate at 20% and the worldwide average at 15.25%.
- The VAT is essentially a "sales tax," but it is charged at each stage of the product development process instead of when the product is sold.
- With VAT, there is less evasion of taxes because it is harder for multiple entities to evade collecting and paying the taxes than it is for one entity to do so.
- VAT rates vary from country to country and from category to category. For example, in some countries, food has a 0% VAT rate, while aviation fuel is taxed at 32%.

EXAMPLE 17.13

Tata Motors is a major player in automobile manufacturing in India. It has three different manufacturing units that specialize in manufacturing different transportation-related products, such as trucks, engines and axles, commercial vehicles, utility vehicles, and passenger cars. The company buys products that fall under different sections of the Indian government VAT tax code, and therefore the products have different VAT rates. In one particular accounting period, Tata had invoices from four different suppliers (vendors A, B, C, and D) in the respective amounts of $1.5 million, $3.8 million, $1.1 million, and $900,000. The products Tata purchased were subject to VAT rates of 4%, 4%, 12.5%, and 22%, respectively.

(a) How much total VAT did Tata pay to its vendors?
(b) Assume that Tata's products have a VAT rate of 12.5%. If Tata's sales during the period were $9.2 million, how much VAT did the Indian Treasury receive from Tata?

Solution

(a) Let X equal the product price before the VAT is added. Solve for X and then subtract it from the purchase amount to determine the VAT charged by each vendor. Table 17–7 shows the VAT that Tata paid its four vendors. An example computation for vendor A is as follows:

$$X + 0.04X = 1,500,000$$
$$1.04X = 1,500,000$$
$$X = \$1,442,308$$

$$\text{VAT}_A = 1,500,000 - 1,442,308$$
$$= \$57,692$$

$$\text{Total VAT paid} = 57,692 + 146,154 + 122,222 + 162,295$$
$$= \$488,363$$

(b) Total from Tata = total VAT − VAT paid by vendors
$$= 9,200,000(0.125) - 488,363$$
$$= \$661,637$$

TABLE 17–7　VAT Computation, Example 17.13

Vendor	Purchases, $	VAT Rate, %	Price before VAT, X, $	VAT, $
A	1,500,000	4.0	1,442,308	57,692
B	3,800,000	4.0	3,653,846	146,154
C	1,100,000	12.5	977,778	122,222
D	900,000	22.0	737,705	162,295
Total				488,363

CHAPTER SUMMARY

After-tax analysis does not usually change the decision to select one alternative over another; however, it does offer a much clearer estimate of the monetary impact of taxes. After-tax PW, AW, and ROR evaluations of one or more alternatives are performed on the CFAT series using exactly the same procedures as in previous chapters.

Income tax rates for U.S. corporations and individual taxpayers are graduated or progressive— higher taxable incomes pay higher income taxes. A single-value, effective tax rate T_e is usually applied in an after-tax economic analysis. Taxes are reduced because of tax deductible items, such as depreciation and operating expenses. Because depreciation is a noncash flow, it is important to consider depreciation only in TI computations, and not directly in the CFBT and

CFAT calculations. Accordingly, key general cash flow after-tax relations for each year are as follows:

$$NOI = \text{gross income} - \text{expenses}$$
$$TI = \text{gross income} - \text{operating expenses} - \text{depreciation} + \text{depreciation recapture}$$
$$CFBT = \text{gross income} - \text{operating expenses} - \text{initial investment} + \text{salvage value}$$
$$CFAT = CFBT - \text{taxes} = CFBT - (T_e)(TI)$$

If an alternative's estimated contribution to corporate financial worth is the economic measure, the economic value added (EVA) should be determined. Unlike CFAT, the EVA includes the effect of depreciation. The equivalent annual worths of CFAT and EVA estimates are the same numerically, because they interpret the annual cost of the capital investment in different, but equivalent manners when the time value of money is taken into account.

In a replacement study, the tax impact of depreciation recapture, which may occur when the defender is traded for the challenger, is accounted for in an after-tax analysis. The replacement study procedure of Chapter 11 is applied. The tax analysis may not reverse the decision to replace or retain the defender, but the effect of taxes will likely reduce (possibly by a significant amount) the economic advantage of one alternative over the other.

International corporate tax rates have steadily decreased, but indirect taxes, such as value-added tax (VAT), have increased. The mechanism of a VAT is explained and compared to a sales tax. The United States currently has no VAT.

PROBLEMS

Basic Tax Computations

17.1   (a)   Define the following tax terms: graduated tax rates, marginal tax rate, and indexing.

      (b)   Describe a fundamental difference between each of the following terms: net operating income (NOI), taxable income (TI), and net operating profit after taxes (NOPAT).

17.2   Determine the average tax rate for a corporation that has taxable income of (a) $150,000 and (b) $12,000,000.

17.3   Determine the single-value effective tax rate for a corporation that has a federal tax rate of 35% and a state tax rate of 5%.

17.4   Identify the primary term described by each event below: gross income, depreciation, operating expense, taxable income, income tax, or net operating profit after taxes.

      (a)   A new machine had a first-year write-off of $10,500.

      (b)   A public corporation estimates that it will report a $-750,000 net profit on its annual income statement.

      (c)   An asset with a book value of $8000 was retired and sold for $8450.

      (d)   An over-the-counter software system will generate $420,000 in revenue this quarter.

      (e)   An asset with a MACRS recovery period of 7 years has been owned for 10 years. It was just sold for $2750.

      (f)   The cost of goods sold in the past year was $3,680,200.

      (g)   A convenience store collected $33,550 in lottery ticket sales last month. Based on winners holding these tickets, a rebate of $350 was sent to the manager.

      (h)   An asset with a first cost of $65,000 was utilized on a new product line to increase sales by $150,000.

      (i)   The cost to maintain equipment during the past year was $641,000.

17.5   Two companies have the following values on their annual tax returns.

	Company 1	Company 2
Sales revenue, $	1,500,000	820,000
Interest revenue, $	31,000	25,000
Expenses, $	−754,000	−591,000
Depreciation, $	48,000	18,000

      (a)   Calculate the federal income tax for the year for each company.

      (b)   Determine the percent of sales revenue each company will pay in federal income tax.

      (c)   Estimate the taxes using an effective rate of 34% of the entire TI. Determine the percentage error made relative to the exact taxes in part (a).

17.6   Last year, one division of Hagauer.com, a dot-com sports industry service firm that provides real-time analysis of mechanical stress due to athlete injury, had $250,000 in taxable income. This year, TI is estimated to be $600,000. Calculate the federal income taxes and answer the following.

(a) What was the average federal tax rate paid last year?

(b) What is the marginal federal tax rate on the additional TI this year?

(c) What will be the average federal tax rate this year?

(d) What will be the NOPAT on just the additional $350,000 in taxable income?

17.7 Yamachi and Nadler of Hawaii has a gross income of $7.5 million for the year. Depreciation and operating expenses total $4.3 million. The combined state and local tax rate is 7.2%. If an effective federal rate of 35% applies, estimate the income taxes using the effective tax rate equation.

17.8 Workman Tools reported a TI of $90,000 last year. If the state income tax rate is 7%, determine the (a) average federal tax rate, (b) overall effective tax rate, (c) total taxes to be paid based on the effective tax rate, and (d) total taxes paid to the state and paid to the federal government.

17.9 The taxable income for a motorcycle sales and repair business is estimated to be $150,000 this year. A single-value tax rate of 39% is used. A new engine diagnostics system will cost $40,000 and have an average annual depreciation of $8000. The equipment will increase gross income by an estimated $9000 and expenses by $2000 for the year. Compute the expected change in income taxes for the year if the new system is purchased.

17.10 C.F. Jordon Construction Services has operated for the last 26 years in a northern U.S. state where the state income tax on corporate revenue is 6% per year. C.F. Jordon pays an average federal tax of 23% and reports taxable income of $7 million. Because of pressing labor cost increases, the president wants to move to another state to reduce the total tax burden. The new state may have to be willing to offer tax allowances or an interest-free grant for the first couple of years in order to attract the company. You are an engineer with the company and are asked to do the following.

(a) Determine the effective tax rate for C.F. Jordon.

(b) Estimate the state tax rate that would be necessary to reduce the overall effective tax rate by 10% per year.

(c) Determine what the new state would have to do financially for C.F. Jordon to move there and to reduce its effective tax rate to 22% per year.

CFBT and CFAT

17.11 Identify which of the following items are included in the calculation of cash flow before taxes (CFBT): operating expenses, salvage value, depreciation, initial investment, gross income, tax rate.

17.12 What is the basic difference between cash flow after taxes (CFAT) and net operating profit after taxes (NOPAT)?

17.13 For a year in which there is no initial investment P or salvage value S, derive an equation for CFBT that contains only the following terms: CFAT, CFBT, D, and T_e.

17.14 Determine the cash flow before taxes for Anderson Consultants when the cash flow after taxes was $600,000, asset depreciation was $350,000 and the company's effective tax rate was 36%.

17.15 Four years ago Sierra Instruments of Monterey, California spent $200,000 for equipment to manufacture standard gas flow calibrators. The equipment was depreciated by MACRS using a 3-year recovery period. The gross income for year 4 was $100,000, with operating expenses of $50,000. Use an effective tax rate of 40% to determine the CFAT in year 4 if the asset was (a) discarded with no salvage value in year 4 and (b) sold for $20,000 at the end of year 4 (neglect any taxes that may be incurred on the sale of the equipment). The MACRS depreciation rate for year 4 is 7.41%.

17.16 Four years ago a division of Harcourt-Banks purchased an asset that was depreciated by the MACRS method using a 3-year recovery period. The total revenue for year 2 was $48 million, depreciation was $8.2 million, and operating expenses were $28 million. Use a federal tax rate of 35% and a state tax rate of 6.5% to determine (a) CFAT, (b) percentage of total revenue expended on taxes, and (c) net profit after taxes for the year.

17.17 Advanced Anatomists, Inc., researchers in medical science, is contemplating a commercial venture concentrating on proteins based on the new X-ray technology of free-electron lasers. To recover the huge investment needed, an annual $2.5 million CFAT is needed. A favored average federal tax rate of 20% is expected; however, state taxing authorities will levy an 8% tax on TI. Over a 3-year period, the deductible expenses and depreciation are estimated to total $1.3 million the first year, increasing by $500,000 per year thereafter. Of this, 50% is expenses and 50% is depreciation. What is the required gross income each year?

The following information is used in Problems 17.18 through 17.21. (Show hand and spreadsheet solutions, as instructed)

After 4 years of use, Procter and Gamble has decided to replace capital equipment used on its Zest bath soap line. The equipment was MACRS-depreciated over a 3-year recovery period. After-tax MARR is 10% per year, and T_e is 35% in the United States. The cash flow data is tabulated in $1000 units.

	Year				
	0	**1**	**2**	**3**	**4**
Purchase, $	−1900				
Gross income, $		800	950	600	300
Operating expenses, $		−100	−150	−200	−250
Salvage, $					700

17.18 Utilize the CFBT series and AW value to determine whether the equipment investment exceeded the MARR.

17.19 Calculate MACRS depreciation and estimate the CFAT series over 4 years. Neglect any tax impact caused by the $700,000 salvage received in year 4.

17.20 Utilize the CFAT series and AW value to determine whether the investment exceeded the MARR.

17.21 Compare the after-tax ROR values using both methods—CFAT series and approximation from the CFBT values using the before-tax ROR and T_e.

17.22 A Wal-Mart Distribution Center has put into service forklifts and conveyors purchased for $250,000. Use a spreadsheet to tabulate CFBT, CFAT, NOPAT, and i^* before and after taxes for 6 years of ownership. The effective tax rate is 40%, and the estimated cash flow and depreciation amounts are shown. Salvage is expected to be zero.

Year	Gross Income, $	Operating Expenses, $	MACRS Depreciation, $
1	210,000	−120,000	50,000
2	210,000	−120,000	80,000
3	160,000	−122,000	48,000
4	160,000	−124,000	28,800
5	160,000	−126,000	28,800
6	140,000	−128,000	14,400

Taxes and Depreciation

17.23 An asset purchased by Stratasys, Inc. had a first cost of $70,000 with an expected salvage value of $10,000 at the end of its 5-year life. In year 2, the revenue was $490,000 with operating expenses of $140,000. If the company's effective tax rate was 36%, determine the difference in taxes paid in year 2 if the depreciation method had been straight line instead of MACRS.

17.24 Cheryl, an electrical engineering student who is working on a business minor, is studying depreciation and finance in her engineering management course. The assignment is to demonstrate that shorter recovery periods require the same total taxes, but they offer a time value of taxes advantage for depreciable assets. Help her, using asset estimates made for a 6-year study period: $P =$ $65,000, $S =$ $5000, GI = $32,000 per year, AOC is $10,000 per year, SL depreciation, $i = 12\%$ per year, $T_e = 31\%$. Make the comparison using recovery periods of 3 and 6 years.

17.25 Complete the last four columns of the table below using an effective tax rate of 40% for an asset that has a first cost of $20,000 and a 3-year recovery period with no salvage value, using (a) straight line depreciation and (b) MACRS depreciation. All cash flows are in $1000 units.

	Estimates, $						
Year	GI	P	OE	D	TI	Taxes	CFAT
0	—	−20	—	—	—	—	−20
1	8		−2				
2	15		−4				
3	12		−3				
4	10		−5				

17.26 Use an effective tax rate of 32% to determine the CFAT and NOPAT associated with the asset shown below under two different scenarios: (a) with depreciation at $6000 per year and (b) with depreciation at $6000, $9600, $5760, and $3456 in years 1 through 4, respectively. All monetary amounts are in $1000.

	Estimates, $							
Year	GI	OE	P	D	TI	Taxes	CFAT	NOPAT
0	—	—	−30	—	—	—	−30	—
1	8	−2						
2	15	−4						
3	12	−3						
4	10	−5						

17.27 J. B. Hunt, Inc., an overland freight company, has purchased new trailers for $150,000 and expects to realize a net $80,000 in gross income over operating expenses for each of the next 3 years. The trailers have a recovery period of 3 years. Assume an effective tax rate of 35% and an interest rate of 15% per year.

(a) Show the advantage of accelerated depreciation by calculating the present worth of taxes for the MACRS method versus the classical

SL method. Since MACRS takes an additional year to fully depreciate the basis, assume no CFBT beyond year 3, but include any negative tax as a tax savings.

(b) Show that the total taxes are the same for both methods.

17.28 A bioengineer is evaluating methods used to apply adhesive onto microporous paper tape that is commonly used after surgery. The machinery costs $200,000, has no salvage value, and the CFBT estimate is $75,000 per year for up to 10 years. The $T_e = 38\%$ and $i = 8\%$ per year. The two depreciation methods to consider are: MACRS with $n = 5$ years and SL with $n = 8$ years (neglect the half-year convention effect). For a study period of 8 years, (a) determine which depreciation method offers the better tax advantage, and (b) demonstrate that the same total taxes are paid for MACRS and SL.

17.29 An asset with a first cost of $9000 is depreciated by MACRS over a 5-year recovery period. The CFBT is estimated at $10,000 for the first 4 years and $5000 thereafter as long as the asset is retained. The effective tax rate is 40%, and money is worth 10% per year. In present worth dollars, how much of the cash flow generated by the asset over its recovery period is lost to taxes?

Depreciation Recapture and Capital Gains (Losses)

17.30 Last month, a company specializing in wind power plant design and construction made a capital investment of $400,000 in physical simulation equipment that will be used for at least 5 years, after which it is expected to be sold for approximately 25% of its first cost. According to tax law, the simulation is MACRS-depreciated using a 3-year recovery period.

(a) Explain why there is a predictable tax implication when the simulator is sold.

(b) Determine by how much the sale will cause TI and taxes to change in year 5 if $T_e = 35\%$.

The following information is used in Problems 17.31 through 17.34.

Open Access, Inc. is an international provider of computer network communications gear. Different depreciation, recovery period, and tax law practices in the three countries where depreciable assets are located are summarized in the table. Also, information is provided about assets purchased 5 years ago at each location and sold this year. After-tax MARR = 9% per year and $T_e = 30\%$ can be used for all countries.

Practice or Estimate	Country 1	Country 2	Country 3
Depreciation method	SL with $n = 5$	MACRS with $n = 3$	DDB with $n = 5$
Depreciation recapture	Not taxed	Taxed as TI	Taxed as TI
First cost, $	−100,000	−100,000	−100,000
GI − OE, $ per year	25,000	25,000	25,000
Estimated salvage, $	0 in year 5	0 in year 5	20,000 in year 5
Life, years	5	5	5
Actual selling price, $	20,000 in year 5	20,000 in year 5	20,000 in year 5

17.31 For country 1, SL depreciation is $20,000 per year. Determine the (a) CFAT series and (b) PW of depreciation, taxes, and CFAT series.

17.32 For country 2, MACRS depreciation for 4 years is $33,333, $44,444, $14,815, and $7407, respectively. Determine the (a) CFAT series and (b) PW of depreciation, taxes, and CFAT series.

17.33 For country 3, DDB depreciation for 5 years is $40,000, $24,000, $14,400, $1600, and 0, respectively. Determine the (a) CFAT series and (b) PW of depreciation, taxes, and CFAT series.

17.34 If you worked Problems 17.31 through 17.33, develop a table that summarizes, for each country, the total taxes paid and the PW values of the depreciation, taxes, and CFAT series. For each criterion, select the country that provides the best PW value. Explain why the same country is not selected for all three criteria. (*Hint:* The PW should be minimized for some criteria and maximized for other criteria. Review previous material first to be sure you choose correctly.)

17.35 An asset with a first cost of $350,000 three years ago is sold for $385,000. The asset was depreciated by the MACRS method and has a book value of $100,800 at the time of sale. Determine the capital gain and depreciation recapture, if any.

17.36 Sun-Tex Truck Stop is located in the desert southwest and is 5 miles from the nearest municipal water system. In order to have fresh water at the site, the company purchased a turnkey reverse osmosis (RO) system for $355,000. The company depreciated the RO system using the MACRS method with a 10-year recovery period. Four years after the system was purchased, water lines from a local water system were extended to the truck stop, so Sun-Tex sold the RO system for $190,000. Determine which of the following apply and the amount, if any, to include in an after-tax analysis of the project: depreciation recapture, capital gain, capital loss. The MACRS depreciation rates are 10%, 18%, 14.4%, and 11.52% for years 1 through 4, respectively.

17.37 Freeman Engineering paid $28,500 for specialized equipment for use with its new GPS/GIS system. The equipment was depreciated over a 3-year recovery period using MACRS depreciation. The company sold the equipment after 2 years for $5000 when it purchased an upgraded system. (a) Determine the amount of the depreciation recapture or capital loss involved in selling the asset. (b) What tax effect will this amount have?

17.38 Determine any depreciation recapture, capital gain, or capital loss generated by each event described below. Use them to determine the amount of income tax effect, if the effective tax rate is 35%.
  (a) A MACRS-depreciated asset with a 7-year recovery period has been sold prematurely after 4 years at an amount equal to 40% of its first cost, which was $150,000.
  (b) A hi-tech machine was sold internationally for $10,000 more than its purchase price just after it was in service 1 year. The asset had $P = \$100,000$, $S = \$1000$, and $n = 3$ years and was depreciated by the MACRS method for the 1 year.
  (c) Land purchased 4 years ago for $1.8 million was sold at a 10% profit.
  (d) A 21-year-old asset was removed from service and sold for $500. When purchased, the asset was entered on the books with a basis of $P = \$180,000$, $S = \$5000$, and $n = 18$ years. Classical straight line depreciation was used for the entire recovery period.
  (e) A corporate car was depreciated using MACRS over a 3-year recovery period. It was sold in the fourth year of use for $2000.

17.39 Sunnen Products Co. of St. Louis, Missouri, makes actuator hones for gas meter tubes where light-duty metal removal is needed. The company purchased land, a building, and two depreciable assets from MPG Automation Systems Corporation, all of which have recently been disposed of. Use the information shown to determine the presence and amount of any depreciation recapture, capital gain, or capital loss.

Asset	Purchase Price, $	Recovery Period, Years	Current Book Value, $	Sales Price, $
Land	−220,000	—		295,000
Building	−900,000	27.5	300,000	275,000
Asset 1	−50,500	3	15,500	19,500
Asset 2	−20,000	3	5,000	12,500

After-Tax Economic Analysis

17.40 Compute the required before-tax return if an after-tax return of 7% per year is expected and the state and local tax rates total 4.2%. The effective federal tax rate is 34%.

17.41 Estimate the approximate after-tax rate of return (ROR) for a project that has a before-tax ROR of 24%. Assume that the company has an effective tax rate of 35% and it uses MACRS depreciation for an asset that has a $27,000 salvage value.

17.42 Estimate the approximate after-tax rate of return for a project that has a first cost of $500,000, a salvage value of 20% of the first cost after 3 years, and annual gross income less operating expenses of $(GI-OE) = \$230,000$. Assume the company has a 35% effective tax rate.

17.43 An engineer is making an annual return of 12% before taxes on the retirement investments placed through her employer. No taxes are paid on retirement earnings until they are withdrawn; however, she was told by her brother, an accountant, that this is the equivalent of an 8% per year after-tax return. What percent of taxable income is her brother assuming to be taken by income taxes?

17.44 Bart is an economic consultant to the textile industry. In both a small business and a large corporation he performed economic evaluations that have an average 18% per year before-tax return. If the stated MARR in both companies is 12% per year after taxes, determine if management at both companies should accept the projects. The before-tax return is used to approximate the after-tax return. Effective tax rates are 34% for the larger corporation and 28% for the small company.

17.45 Elias wants to perform an after-tax evaluation of equivalent methods A and B to electrostatically remove airborne particulate matter from clean rooms used to package liquid pharmaceutical products. Using the information shown, MACRS depreciation with $n = 3$ years, a 5-year study period, after-tax MARR = 7% per year, and $T_e = 34\%$ and a spreadsheet, he obtained the results $AW_A = \$-2176$ and $AW_B = \$3545$. Any tax effects when the equipment is salvaged were neglected. Thus, with MACRS depreciation, method B is the better method. Now, use classical SL depreciation with $n = 5$ years to evaluate the alternatives. Is the decision different from that reached using MACRS?

	Method A	Method B
First cost, $	−100,000	−150,000
Salvage value, $	10,000	20,000
Savings, $ per year	35,000	45,000
AOC, $ per year	−15,000	−6,000
Expected life, years	5	5

17.46 A corporation uses the following: before-tax MARR of 14% per year, after-tax MARR of 7% per year, and T_e of 50%. Two new machines have the following estimates.

	Machine A	Machine B
First cost, $	−15,000	−22,000
Salvage value, $	3,000	5,000
AOC, $ per year	−3,000	−1,500
Life, years	10	10

The machine is retained in use for a total of 10 years, then sold for the estimated salvage value. Select one machine under the following conditions:
(a) Before-tax PW analysis.
(b) After-tax PW analysis, using classical SL depreciation over the 10-year life.
(c) After-tax PW analysis, using MACRS depreciation with a 5-year recovery period.

17.47 Choose between alternatives A and B below if the after-tax MARR is 8% per year, MACRS depreciation is used, and $T_e = 40\%$. The GI−OE estimate is made for only 3 years; it is zero when each asset is sold in year 4.

	Alternative A	Alternative B
First cost, $	−8,000	−13,000
Actual salvage value, $	0	2,000
Gross income − expenses, $	3,500	5,000
Recovery period, years	3	3

17.48 Offshore platform safety equipment, designed for special jobs, will cost $2,500,000, will have no salvage value, and will be kept in service for exactly 5 years, according to company policy. Operating revenue minus expenses is estimated to be $1,500,000 in year 1 and only $300,000 each additional year. The effective tax rate for the multinational oil company is 30%. Show hand and spreadsheet solutions, as instructed, to find the after-tax ROR using (a) classical SL depreciation and (b) MACRS 5-year depreciation, neglecting any tax effects at the end of 5 years of service.

17.49 Automatic inspection equipment purchased for $78,000 by Stimson Engineering generated an average of $26,080 annually in before-tax cash flow during its 5-year estimated life. This represents a return of 20%. However, the corporate tax expert said the CFAT was only $15,000 per year. If the corporation wants to realize an after-tax return of 10% per year, for how many more years must the equipment remain in service?

After-Tax Replacement

17.50 In a replacement study between a defender and a challenger, there may be a capital gain or loss when the defender asset is sold. (a) How is the gain or loss calculated, and (b) how does it affect the AW values in the study?

17.51 In an after-tax replacement study of cost alternatives involving one challenger and one defender, how will a capital loss when selling the defender affect the AW of each alternative?

17.52 An asset that was purchased 2 years ago was expected to be kept in service for its projected life of 5 years, but a new version (the challenger) of this asset promises to be more efficient and have lower operating costs. You have been asked to figure out if it would be more economically attractive to replace the defender now or keep it for 3 more years as originally planned. The defender had a first cost of $300,000, but its market value now is only $150,000. It has operating expenses of $120,000 per year and no estimated salvage value after 3 more years. To simplify calculations for this problem only, assume that SL depreciation was charged at $60,000 per year and that it will continue at that rate for the next 3 years.

The challenger will cost $420,000, will have no salvage value after its 3-year life, will have chargeable expenses of $30,000 per year, and will be depreciated at $140,000 per year (again, using SL depreciation for simplicity in this case). Assume the company's effective tax rate is 35%, and its after-tax MARR is 15% per year.
(a) Determine the CFAT in year 0 for the challenger and defender.
(b) Determine the CFAT in years 1 through 3 for the challenger and defender.
(c) Conduct an AW evaluation to determine if the defender should be kept for 3 more years or replaced now.

17.53 The defender in a catalytic oxidizer manufacturing plant has a market value of $130,000 and expected annual operating costs of $70,000 with no salvage value after its remaining life of 3 years. The depreciation charges for the next 3 years will be $69,960, $49,960, and $35,720. Using an effective tax rate of 35%, determine the CFAT for next year only that should be used in a present worth equation to compare this defender against a challenger that also has a 3-year life. Assume the company's after-tax MARR is 12%.

17.54 Perform a PW replacement study (hand and spreadsheet solutions, if instructed) from the information shown using an after-tax MARR = 12% per year, $T_e = 35\%$, and a study period of 4 years.

(Assume that the assets will be salvaged at their original salvage estimates. Since no revenues are estimated, all taxes are negative and considered "savings" to the alternative.) All monetary values are in $1000 units.

	Defender	Challenger
First cost, $1000	−45	−24
Estimated S at purchase, $1000	5	0
Market value now, $1000	35	—
AOC, $1000 per year	−7	−8
Depreciation method	SL	MACRS
Recovery period, years	8	3
Useful life, years	8	5
Years owned	3	—

17.55 After 8 years of use, the heavy truck engine overhaul equipment at Pete's Truck Repair was evaluated for replacement. Pete's accountant used an after-tax MARR of 8% per year, $T_e = 30\%$, and a current market value of $25,000 to determine AW = $2100. The new equipment costs $75,000, uses SL depreciation over a 10-year recovery period, and has a $15,000 salvage estimate. Estimated CFBT is $15,000 per year. Pete asked his engineer son Ramon to determine if the new equipment should replace what is owned currently. From the accountant, Ramon learned the current equipment cost $20,000 when purchased and reached a zero book value several years ago. Help Ramon answer his father's question.

17.56 Apple Crisp Foods signed a contract some years ago for maintenance services on its fleet of trucks and cars. The contract is up for renewal now for a period of 1 year or 2 years only. The contract quote is $300,000 per year if taken for 1 year and $240,000 per year if taken for 2 years. The finance vice president wants to renew the contract for 2 years without further analysis, but the vice president for engineering believes it is more economical to perform the maintenance in-house. Since much of the fleet is aging and must be replaced in the near future, a fixed 3-year study period has been agreed upon. The estimates for the in-house (challenger) alternative are as follows:

First cost, $	−800,000
AOC, $ per year	−120,000
Life, years	4
Estimated salvage	Loses 25% of P annually:
	End year 1, S = $600,000
	End year 2, S = $400,000
	End year 3, S = $200,000
	End year 4, S = $0
MACRS depreciation	3-year recovery period

The effective tax rate is 35%, and the after-tax MARR is 10% per year. Perform an after-tax AW analysis, and determine which vice president has the better economic strategy over the next 3 years.

17.57 Nuclear safety devices installed several years ago have been depreciated from a first cost of $200,000 to zero using MACRS. The devices can be sold on the used equipment market for an estimated $15,000. Or they can be retained in service for 5 more years with a $9000 upgrade now and an AOC of $6000 per year. The upgrade investment will be depreciated over 3 years with no salvage value. The challenger is a replacement with newer technology at a first cost of $40,000, n = 5 years, and S = 0. The new units will have operating expenses of $7000 per year. (a) Use a 5-year study period, an effective tax rate of 40%, an after-tax MARR of 12% per year, and an assumption of classical straight line depreciation (no half-year convention) to perform an after-tax replacement study. (b) If the challenger is known to be salable after 5 years for an amount between $2000 and $4000, will the challenger AW value become more or less costly? Why?

17.58 Three years ago, Silver House Steel purchased a new quenching system for $550,000. The salvage value after 10 years at that time was estimated to be $50,000. Currently the expected remaining life is 7 years with an AOC of $27,000 per year. The new president has recommended early replacement of the system with one that costs $400,000 and has a 12-year life, a $35,000 salvage value, and an estimated AOC of $50,000 per year. The MARR for the corporation is 12% per year. The president wishes to know the replacement value that will make the recommendation to replace now economically advantageous. Use a spreadsheet and Solver to find the minimum trade-in value (a) before taxes and (b) after taxes, using an effective tax rate of 30%. For solution purposes, use classical SL depreciation for both systems. Comment on the difference in replacement value made by the consideration of taxes.

Economic Value Added

17.59 While an engineering manager may prefer to use CFAT estimates to evaluate the AW of a project, a financial manager may select AW of EVA estimates. Why are these preferences predictable?

17.60 Cardenas and Moreno Engineering is evaluating a very large flood control program for several southern U.S. cities. One component is a 4-year project for a special-purpose transport ship-crane for use in building permanent storm surge protection against hurricanes on the New Orleans coastline. The estimates are $P = \$300,000$, $S = 0$, and $n = 3$ years. MACRS depreciation with a 3-year recovery is indicated. Gross income and operating expenses are estimated at $200,000 and $80,000, respectively, for each of 4 years. The CFAT is shown below. Calculate the AW values of the CFAT and EVA series. They should have the same value. The after-tax MARR is 9.75% and $T_e = 35\%$.

Year	GI, $	OE, $	P and S, $	D, $	TI, $	Taxes, $	CFAT, $
0			−300,000				−300,000
1	200,000	−80,000		99,990	20,010	7,003	112,997
2	200,000	−80,000		133,350	−13,350	−4,673	124,673
3	200,000	−80,000	0	44,430	75,570	26,450	93,551
4	200,000	−80,000		22,230	97,770	34,220	85,781

17.61 Triple Play Innovators Corporation (TPIC) plans to offer IPTV (Internet Protocol TV) service to North American customers starting soon. Perform an AW analysis of the EVA series for the two alternative suppliers available for the hardware and software. Let $T_e = 30\%$ and after-tax MARR = 8%; use SL depreciation (neglect half-year convention and MACRS, for simplicity) and a study period of 8 years.

	Vendor	
	Hong Kong	**Japan**
First cost, $	4.2 million	3.6 million
Recovery period, years	8	5
Salvage value, $	0	0
GI – OE, $ per year	1,500,000 in year 1; increasing by 300,000 per year up to 8 years	

17.62 Sun Microsystems has developed partnerships with several large manufacturing corporations to use Java software in their consumer and industrial products. A new corporation will be formed to manage these applications. One major project involves using Java in commercial and industrial appliances that store and cook food. The gross income and operating expenses are expected to follow the relations shown for the estimated life of 6 years. For $t = 1$ to 6 years,

Annual gross income, GI = $2,800,000 − 100,000t$
Annual operating expenses, OE = $950,000 + 50,000t$

The effective tax rate is 30%, the interest rate is 12% per year, and the depreciation method chosen for the $3,000,000 capital investment is a 5-year MACRS ADS alternative that allows straight line write-off with the half-year convention in years 1 and 6. Using a spreadsheet, estimate (a) the annual economic contribution of the project to the new corporation and (b) the equivalent annual worth of these contributions.

Value-Added Tax

17.63 What is the primary difference between a sales tax and a value-added tax?

17.64 In Denmark, VAT is applied at a rate of 25%, with few exceptions. Vendor A sells raw materials to vendor B for $60,000 plus VAT, vendor B sells a product to vendor C for $130,000 plus VAT, and vendor C sells an improved, value-added product to an end user for $250,000 plus VAT. Determine the following.
(a) The amount of VAT collected by vendor B.
(b) The amount of tax vendor B sends to Denmark's Treasury.
(c) The total amount of tax collected by the Treasury department.

The following information is used in Problems 17.65 through 17.70

Ajinkya Electronic Systems, a company in India that manufactures many different electronic products, has to purchase goods and services from a variety of suppliers (wire, diodes, LED displays, plastic components, etc.). The table below shows several suppliers and the VAT rates associated with each. It also shows the purchases in $1000 units that Ajinkya made (before taxes) from each supplier in the previous accounting period. Ajinkya's sales to end users was $9.2 million, and Ajinkya's products carry a VAT of 12.5%.

Supplier	VAT Rate, %	Purchases by Ajinkya, $1000
A	4.0	350
B	12.5	870
C	12.5	620
D	21.3	90
E	32.6	50

17.65 How much VAT did supplier C collect?

17.66 How much tax did Ajinkya keep from the tax it collected based on the purchases it made from supplier A?

17.67 What was the total amount of taxes paid by Ajinkya to the suppliers?

17.68 What was the average VAT rate paid by Ajinkya in purchasing goods and services?

17.69 What was the amount of taxes Ajinkya sent to the Treasury of India?

17.70 What was the total amount of taxes collected by the Treasury of India from Ajinkya and Ajinkya's suppliers?

ADDITIONAL PROBLEMS AND FE EXAM REVIEW QUESTIONS

17.71 A graduated income tax system means:
  (a) Only taxable incomes above a certain level pay any taxes.
  (b) A higher flat rate goes with all of the taxable income.
  (c) Higher tax rates go with higher taxable incomes.
  (d) Rates are indexed each year to keep up with inflation.

17.72 All of the following are characteristics of a value-added tax system except:
  (a) Value-added taxes are taxes on consumption.
  (b) The end user pays value-added taxes.
  (c) Value-added taxes are charged at each stage of product development.
  (d) Value-added taxes are charged only on the raw materials for product development.

17.73 A small company has a taxable income that places it in the 35% tax bracket. The amount of taxes that a depreciation charge of $16,000 would save is closest to:
  (a) $0
  (b) $3200
  (c) $5600
  (d) $10,400

17.74 A company that has a 50% effective tax rate had income of $200 million in each of the last 2 years. In one of those years, the company had deductions of $100 million. In the other year, the company had deductions of only $80 million. The difference in income taxes paid by the company in those 2 years was closest to:
  (a) $10 million
  (b) $20 million
  (c) $50 million
  (d) $60 million

17.75 Taxable income (TI) is defined as:
  (a) TI = revenue + operating expenses − depreciation
  (b) TI = revenue − operating expenses + depreciation

  (c) TI = revenue − operating expenses − depreciation + amortization
  (d) TI = revenue − operating expenses − depreciation

17.76 The marginal tax rate is defined as:
  (a) The percentage paid on the last dollar of income
  (b) The tax rate that applies to a questionable investment
  (c) The tax rate that includes federal, state, and local taxes
  (d) The percentage paid on the first dollar of income

17.77 A subcontractor with an effective tax rate of 25% has gross income of $55,000, other income of $4000, operating expenses of $13,000, and other deductions and exemptions of $11,000. The income tax due is closest to:
  (a) $11,750
  (b) $8,750
  (c) $10,750
  (d) $13,750

17.78 When a depreciable asset is disposed of for less than its current book value, the transaction is known as:
  (a) An after-tax expense
  (b) Capital loss
  (c) Capital gain
  (d) Depreciation recapture

17.79 The after-tax analysis for a $60,000 investment with associated gross income minus expenses (GI−OE) is shown below for the first 2 years only. If the effective tax rate is 40%, the values for depreciation (D), taxable income (TI), and taxes for year 1 are closest to:

Year	Investment, $	GI − OE, $	D, $	TI, $	Taxes, $	CFAT, $
0	−60,000					−60,000
1		30,000				26,000
2		35.000	15,000	6000		29,000

(a) $D = \$5,000$, TI $= \$25,000$, taxes $= \$10,000$
(b) $D = \$30,000$, TI $= \$30,000$, taxes $= \$4,000$
(c) $D = \$20,000$, TI $= \$50,000$, taxes $= \$20,000$
(d) $D = \$20,000$, TI $= \$10,000$, taxes $= \$4,000$

17.80 If the after-tax rate of return for a cash flow series is 11.9% and the corporate effective tax rate is 34%, the approximated before-tax rate of return is closest to:
(a) 6.8%
(b) 5.4%
(c) 18.0%
(d) 28.7%

17.81 An asset purchased for $100,000 with $S = \$20,000$ after 5 years was depreciated using the 5-year MACRS rates. Expenses averaged $18,000 per year, and the effective tax rate is 30%. The asset was actually sold after 5 years of service for $22,000. MACRS rates in years 5 and 6 are 11.53% and 5.76%, respectively. The after-tax cash flow from the sale is closest to:

(a) $27,760
(b) $17,130
(c) $26,870
(d) $20,585

17.82 When accelerated depreciation methods or shortened recovery periods are applied, there are impacts on the income taxes due. Of the following, the statements that are commonly *incorrect* are:
1. Total taxes paid are the same for all depreciation methods.
2. Present worth of taxes is lower for shorter recovery periods.
3. Accelerated depreciation imposes more taxes in the later years of the recovery period.
4. Present worth of taxes is higher for shorter recovery periods.
(a) 1, 2, and 3
(b) 1 and 4
(c) 2
(d) 4

CASE STUDY

AFTER-TAX ANALYSIS FOR BUSINESS EXPANSION

Background

Charles was always a hands-on type of person. Within a couple of years of graduating from college, he started his own business. After some 20 years, it has grown significantly. He owns and operates Pro-Fence, Inc. in the Metroplex, specializing in custom-made metal and stone fencing for commercial and residential sites. For some time, Charles has thought he should expand into a new geographic region, with the target area being another large metropolitan area about 500 miles north, called Victoria.

Pro-Fence is privately owned by Charles; therefore, the question of how to finance such an expansion has been, and still is, the major challenge. Debt financing would not be a problem in that the Victoria Bank has already offered a loan of up to $2 million. Taking capital from the retained earnings of Pro-Fence is a second possibility, but taking too much will jeopardize the current business, especially if the expansion were not an economic success and Pro-Fence were stuck with a large loan to repay.

This is where you come in as a long-time friend of Charles. He knows you are quite economically oriented and that you understand the rudiments of debt and equity financing and economic analysis. He wants you to advise him on the balance between using Pro-Fence funds and borrowed funds. You have agreed to help him, as much as you can.

Information

Charles has collected some information that he shares with you. Between his accountant and a small market survey of the

business opportunities in Victoria, the following generalized estimates seem reasonable.

Initial capital investment $= \$1.5$ million

Annual gross income $= \$700,000$

Annual operating expenses $= \$100,000$

Effective income tax rate for Pro-Fence $= 35\%$

Five-year MACRS depreciation for all $1.5 million investment

The terms of the Victoria Bank loan would be 6% per year simple interest based on the initial loan principal. Repayment would be in 5 equal payments of interest and principal. Charles comments that this is not the best loan arrangement he hopes to get, but it is a good worst-case scenario upon which to base the debt portion of the analysis. A range of D-E mixes should be analyzed. Between Charles and yourself, you have developed the following viable options.

Debt		Equity	
Percentage	Loan Amount, $	Percentage	Investment Amount, $
0		100	1,500,000
50	750,000	50	750,000
70	1,050,000	30	450,000
90	1,350,000	10	150,000

Case Study Exercises

1. For each funding option, perform a spreadsheet analysis that shows the total CFAT and its present worth over a 6-year period, the time it will take to realize the full advantage of MACRS depreciation. An after-tax return of 10% is expected. Which funding option is best for Pro-Fence? (*Hint:* For the spreadsheet, sample column headings are: Year, GI – OE, Loan interest, Loan principal, Equity investment, Depreciation rate, Depreciation, Book value, TI, Taxes, and CFAT.)

2. Observe the changes in the total 6-year CFAT as the D-E percentages change. If the time value of money is neglected, what is the constant amount by which this sum changes for every 10% increase in equity funding?

3. Charles noticed that the CFAT total and PW values go in opposite directions as the equity percentage increases. He wants to know why this phenomenon occurs. How should you explain this to Charles?

4. After deciding on the 50-50 split of debt and equity financing, Charles wants to know what additional bottom-line contributions to the economic worth of the company may be added by the new Victoria site. What are the best estimates at this time?

CHAPTER 18

Sensitivity Analysis and Staged Decisions

LEARNING OUTCOMES

Purpose: Perform a sensitivity analysis of parameters; use expected values to evaluate staged funding options.

SECTION	TOPIC	LEARNING OUTCOME
18.1	Sensitivity to variation	• Use a measure of worth to explain sensitivity to variation in one or more parameters.
18.2	Three-estimate variation	• Choose the better alternative using three estimates of variation in selected parameters.
18.3	$E(X)$	• Calculate the expected value of a variable.
18.4	$E(X)$ of cash flows	• Evaluate a project or alternatives using expected values of cash flows.
18.5	Decision trees	• Use a decision tree to evaluate alternatives stage by stage.
18.6	Real options	• Explain a real option in engineering economics and evaluate a staged decision using real options analysis.

T his chapter includes several related topics about alternative evaluation. Initially, we expand our capability to perform a **sensitivity analysis** of one or more parameters and of an entire alternative. Then the determination and use of the **expected value** of a cash flow series are treated. The techniques of **decision trees** help make a series of economic decisions for alternatives that have different, but closely connected stages.

Economic decisions that involve **staged funding** are very common in professional and everyday life. The last topic of **real options analysis** introduces a method useful in these circumstances.

18.1 Determining Sensitivity to Parameter Variation ● ● ●

The term *parameter* is used in this chapter to represent any variable or factor for which an estimate or stated value is necessary. Example parameters are first cost, salvage value, AOC, estimated life, production rate, and materials costs. Estimates such as the loan interest rate and the inflation rate can also be parameters of the analysis.

Economic analysis uses estimates of a parameter's future value to assist decision makers. Since future estimates are always incorrect to some degree, inaccuracy is present in the economic projections. The effect of variation may be determined by using sensitivity analysis.

> **Sensitivity analysis** determines how a measure of worth—PW, AW, FW, ROR, B/C, or CER—is altered when one or more parameters vary over a **selected range of values.** Usually one parameter at a time is varied, and independence with other parameters is assumed. Though this approach is an oversimplification in real-world situations, since the dependencies are difficult to accurately model, the end results are usually correct.

Sensitivity analysis

In reality, we have applied this approach (informally) throughout previous chapters to determine the response to variation in a variety of parameters. Variation in a parameter such as MARR will not alter the decision to select an alternative when all compared alternatives return considerably more than the MARR; thus, the decision is relatively insensitive to the MARR. However, variation in the n or AOC value may indicate that the alternative's measure of worth is very sensitive to the estimated life or annual operating costs.

Usually the variations in life, annual costs, and revenues result from variations in selling price, operation at different levels of capacity, inflation, etc. For example, if an operating level of 90% of airline seating capacity for a domestic route is compared with 70% for a proposed international route, the operating cost and revenue per passenger-mile will increase, but anticipated aircraft life will probably decrease only slightly. Usually several important parameters are studied to learn how the uncertainty of estimates affects the economic analysis.

Sensitivity analysis routinely concentrates on the variation expected in estimates of P, AOC, S, n, unit costs, unit revenues, and similar parameters. These parameters are often the result of design questions and their answers, as discussed in Chapter 15. Parameters that are interest rate–based are not treated in the same manner.

> Parameters such as MARR and other interest rates (loan rates, inflation rate) are more stable from project to project. If performed, sensitivity analysis on them is for specific values or over a **narrow range of values.** This point is important to remember if simulation is used for decision making under risk (Chapter 19).

Plotting the sensitivity of PW, AW, or ROR versus the parameter(s) studied is very helpful. Two alternatives can be compared with respect to a given parameter and the breakeven point. This is the value at which the two alternatives are economically equivalent. However, the breakeven chart commonly represents only one parameter per chart. Thus, several charts are constructed, and independence of each parameter is assumed. In previous uses of breakeven analysis, we often computed the measure of worth at only two values of a parameter and connected the points with a straight line. However, if the results are sensitive to the parameter value, several intermediate points should be used to better evaluate the sensitivity, especially if the relationships are not linear.

When several parameters are studied, sensitivity analysis can become quite complex. It may be performed one parameter at a time using a spreadsheet or computations by hand or calculator.

The computer facilitates comparison of multiple parameters and multiple measures of worth, and the software can rapidly plot the results.

Here is a general procedure to follow when conducting a thorough sensitivity analysis.

1. Determine which parameter(s) of interest might vary from the most likely estimated value.
2. Select the probable range and an increment of variation for each parameter.
3. Select the measure of worth.
4. Compute the results for each parameter, using the measure of worth as a basis.
5. To better interpret the sensitivity, graphically display the parameter versus the measure of worth.

This sensitivity analysis procedure should indicate the parameters that warrant closer study or require additional information. When there are two or more alternatives, it is better to use the PW or AW measure of worth in step 3. If ROR is used, it requires the extra efforts of incremental analysis between alternatives. Example 18.1 illustrates sensitivity analysis for one project.

EXAMPLE 18.1

Wild Rice, Inc. expects to purchase a new asset for automated rice handling. Most likely estimates are a first cost of $80,000, zero salvage value, and a cash flow before taxes (CFBT) per year t that follows the relation $27,000 - 2000t$. The MARR for the company varies over a wide range from 10% to 25% per year for different types of investments. The economic life of similar machinery varies from 8 to 12 years. Evaluate the sensitivity of PW by varying (a) MARR, while assuming a constant n value of 10 years, and (b) n, while MARR is constant at 15% per year. Perform the analysis by hand and by spreadsheet.

Solution by Hand

(a) Follow the procedure above to understand the sensitivity of PW to MARR variation.
1. MARR is the parameter of interest.
2. Select 5% increments to evaluate sensitivity to MARR; the range is 10% to 25%.
3. The measure of worth is PW.
4. Set up the PW relation for 10 years. When MARR = 10%,

$$PW = -80,000 + 25,000(P/A,10\%,10) - 2000(P/G,10\%,10)$$
$$= \$27,830$$

The PW for all MARR values at 5% intervals is as follows:

MARR, %	PW, $
10	27,830
15	11,512
20	−962
25	−10,711

5. A plot of MARR versus PW is shown in Figure 18–1. The steep negative slope indicates that the decision to accept the proposal based on PW is quite sensitive to variations in the MARR. If the MARR is established at the upper end of the range, the investment is not attractive.

(b) 1. Asset life n is the parameter.
2. Select 2-year increments to evaluate PW sensitivity over the range 8 to 12 years.
3. The measure of worth is PW.
4. Set up the same PW relation as in part (a) at $i = 15\%$. The PW results are

n	PW, $
8	7,221
10	11,511
12	13,145

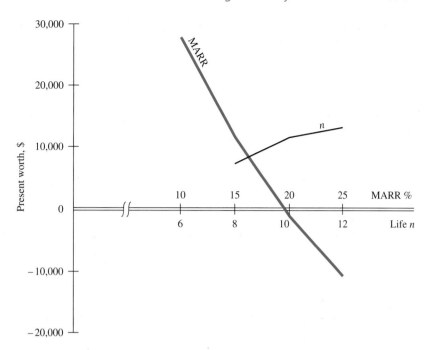

Figure 18–1
Plot of PW versus MARR
and *n* for sensitivity
analysis, Example 18.1.

5. Figure 18–1 presents the plot of PW versus *n*. Since the PW measure is positive for all values of *n*, the decision to invest is not materially affected by the estimated life. The PW curve levels out above *n* = 10. This insensitivity to changes in cash flow in the distant future is a predictable observation, because the P/F factor gets smaller as *n* increases.

Solution by Spreadsheet

Figure 18–2 presents two spreadsheets and accompanying plots of PW versus MARR (fixed *n*) and PW versus *n* (fixed MARR). The NPV function calculates PW for *i* values from 10% to 25% and *n* values from 8 to 12 years. As the solution by hand indicated, so do the charts; PW is sensitive to changes in MARR values, but not very sensitive to variations in *n*.

(*a*)

Figure 18–2
Sensitivity analysis of
PW to variation in
(*a*) MARR values and
(*b*) life estimates,
Example 18.1.

(*b*)

Figure 18–3
Sensitivity analysis graph of percent variation from the most likely estimate.

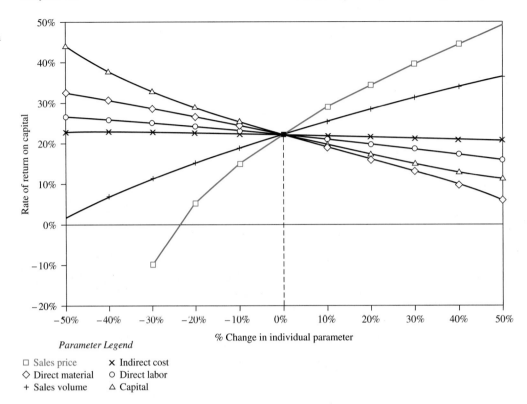

Parameter Legend

□ Sales price   × Indirect cost
◇ Direct material ○ Direct labor
+ Sales volume   △ Capital

When the sensitivity of *several parameters* is considered for *one alternative* using a *single measure of worth*, it is helpful to **graph percentage change** for each parameter versus the measure of worth. This is sometimes called a **spider graph.** Figure 18–3 illustrates ROR versus six different parameters for one alternative. The variation in each parameter is indicated as a percentage deviation from the most likely estimate on the horizontal axis. If the ROR response curve is flat and approaches horizontal over the range of total variation graphed for a parameter, there is little sensitivity of ROR to changes in the parameter's value. This is the conclusion for indirect cost in Figure 18–3. On the other hand, ROR is very sensitive to sales price. A reduction of 30% from the expected sales price reduces the ROR from approximately 20% to −10%, whereas a 10% increase in price raises the ROR to about 30%.

If *two alternatives* are compared and the sensitivity to *one parameter* is sought, the graph may show quite nonlinear results. Observe the general shape of the sample sensitivity graphs in Figure 18–4. The plots are shown as linear segments between specific computation points. The graph indicates that the PW of each plan is a nonlinear function of hours of operation. Plan A is

Figure 18–4
Sample PW sensitivity to hours of operation for two alternatives.

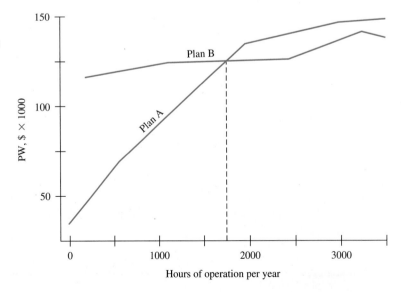

very sensitive in the range of 0 to 2000 hours, but it is comparatively insensitive above 2000 hours. Plan B is more attractive due to its relative insensitivity. The breakeven point is at about 1750 hours per year. It may be necessary to plot the measure of worth at intermediate points to better understand the nature of the sensitivity.

EXAMPLE 18.2

Columbus, Ohio needs to resurface a 3-kilometer stretch of highway. Knobel Construction has proposed two methods of resurfacing. The first method is a concrete surface for a cost of $1.5 million and an annual maintenance cost of $10,000. The second method is an asphalt covering with a first cost of $1 million and a yearly maintenance of $50,000. However, Knobel requests that every third year the asphalt highway be touched up at a cost of $75,000.

    The city uses the interest rate on bonds, 6% on its last bond issue, as the discount rate.

(a) Determine the breakeven number of years of the two methods. If the city expects an interstate to replace this stretch of highway in 10 years, which method should be selected?
(b) If the touch-up cost increases by $5000 per kilometer every 3 years, is the decision sensitive to this increase?

Solution

(a) Use PW analysis to determine the breakeven n value.

$$\text{PW of concrete} = \text{PW of asphalt}$$

$$-1{,}500{,}000 - 10{,}000(P/A,6\%,n) = -1{,}000{,}000 - 50{,}000(P/A,6\%,n)$$

$$-75{,}000\left[\sum_j (P/F,6\%,j)\right]$$

where $j = 3, 6, 9, \ldots, n$. The relation can be rewritten to reflect the incremental cash flows.

$$-500{,}000 + 40{,}000(P/A,6\%,n) + 75{,}000\left[\sum_j (P/F,6\%,j)\right] = 0 \qquad [18.1]$$

The breakeven n value can be determined by hand solution by increasing n until Equation [18.1] switches from negative to positive PW values. Alternatively, a spreadsheet solution using the NPV function can find the breakeven n value (Figure 18–5). The NPV functions in column C are the same each year, except that the cash flows are extended 1 year for each

	A	B	C	D	E
1		Part (a)		Part (b)	
2		Incremental	PW for	Incremental	PW for
3	Year, n	cash flow, $	n years, $	cash flow, $	n years, $
4	0	-500,000		-500,000	
5	1	40,000	-462,264	40,000	-462,264
6	2	40,000	-426,664	40,000	-426,664
7	3	115,000	-330,108	115,000	-330,108
8	4	40,000	-298,424	40,000	-298,424
9	5	40,000	-268,534	40,000	-268,534
10	6	115,000	-187,464	130,000	-176,889
11	7	40,000	-160,861	40,000	-150,287
12	8	40,000	-135,765	40,000	-125,190
13	9	115,000	-67,696	145,000	-39,365
14	10	40,000	-45,361	40,000	-17,029
15	11	40,000	-24,289	40,000	4,042
16	12	115,000	32,862	160,000	83,557
17	13	40,000	51,616	40,000	102,311
18	14	40,000	69,308	40,000	120,003
19	15	115,000	117,293	175,000	193,024
20	16	40,000	133,039	40,000	208,770
21					
22				**PW for 12 years**	
23		**PW for 11 years**		= NPV(6%,$B$5:$B16)+$B$4	
24		= NPV(6%,$B$5:$B15)+$B$4			
25					
26			Year counter advances by 1		
27					

Figure 18–5
Sensitivity of the breakeven life between two alternatives, Example 18.2.

present worth calculation. At approximately $n = 11.4$ years, concrete and asphalt resurfacing break even economically. Since the road is needed for 10 more years, the extra cost of concrete is not justified; select the asphalt alternative.

(b) The total touch-up cost will increase by $15,000 every 3 years. Equation [18.1] is now

$$-500{,}000 + 40{,}000(P/A,6\%,n) + \left[75{,}000 + 15{,}000\left(\frac{j-3}{3}\right)\right]\left[\sum_j (P/F,6\%,j)\right] = 0$$

Now the breakeven n value is between 10 and 11 years—10.8 years using linear interpolation (Figure 18–5, column E). The decision has become marginal for asphalt, since the interstate is planned for 10 years hence.

Noneconomic considerations may be used to determine if asphalt is still the better alternative. One conclusion is that the asphalt decision becomes more questionable as the asphalt alternative maintenance costs increase; that is, the PW value is sensitive to increasing touch-up costs.

18.2 Sensitivity Analysis Using Three Estimates ● ● ●

We can thoroughly examine the economic advantages and disadvantages among two or more alternatives by borrowing from the field of project scheduling the concept of making three estimates for each parameter: **a pessimistic, a most likely, and an optimistic estimate.** Depending upon the nature of a parameter, the pessimistic estimate may be the lowest value (alternative life is an example) or the largest value (such as asset first cost).

This approach allows us to study measure of worth and alternative selection sensitivity within a predicted range of variation for each parameter. Usually the most likely estimate is used for all other parameters when the measure of worth is calculated for one particular parameter or one alternative.

EXAMPLE 18.3

An engineer is evaluating three alternatives for new equipment at Emerson Electronics. She has made three estimates for the salvage value, annual operating cost, and life. The estimates are presented on an alternative-by-alternative basis in Table 18–1. For example, alternative B has pessimistic estimates of $S = \$500$, AOC $= \$-4000$, and $n = 2$ years. The first costs are known, so they have the same value. Perform a sensitivity analysis and determine the most economical alternative, using AW analysis at a MARR of 12% per year.

TABLE 18–1	Competing Alternatives with Three Estimates Made for Salvage Value, AOC, and Life Parameters			
Strategy	First Cost, $	Salvage Value S, $	AOC, $ per Year	Life n, Years
Alternative A				
Estimates P	−20,000	0	−11,000	3
Estimates ML	−20,000	0	−9,000	5
Estimates O	−20,000	0	−5,000	8
Alternative B				
Estimates P	−15,000	500	−4,000	2
Estimates ML	−15,000	1,000	−3,500	4
Estimates O	−15,000	2,000	−2,000	7
Alternative C				
Estimates P	−30,000	3,000	−8,000	3
Estimates ML	−30,000	3,000	−7,000	7
Estimates O	−30,000	3,000	−3,500	9

P = pessimistic; ML = most likely; O = optimistic.

	Alternative AW Values, $		
TABLE 18–2 Annual Worth Values, Example 18.3			
Estimates	**A**	**B**	**C**
P	−19,327	−12,640	−19,601
ML	−14,548	−8,229	−13,276
O	−9,026	−5,089	−8,927

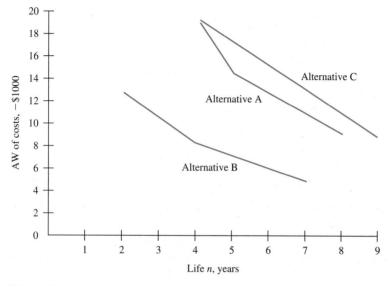

Figure 18–6
Plot of AW of costs for different-life estimates, Example 18.3.

Solution

For each alternative in Table 18–1, calculate the AW value of costs. For example, the AW relation for alternative A, pessimistic estimates, is

$$AW = -20,000(A/P,12\%,3) - 11,000 = \$-19,327$$

Table 18–2 presents all AW values. Figure 18–6 is a plot of AW versus the three estimates of *life* for each alternative. Since the AW calculated using the ML estimates for alternative B ($−8229) is economically better than even the optimistic AW value for alternatives A and C, alternative B is clearly favored.

Comment

While the alternative that should be selected here is quite obvious, this is not normally the case. For example, in Table 18–2, if the pessimistic alternative B equivalent AW were much higher, say, $−21,000 per year (rather than $−12,640), and the optimistic AW values for alternatives A and C were less than that for B ($−5089), the choice of B would not be apparent or correct. In this case, it would be necessary to select one set of estimates (P, ML, or O) upon which to base the decision. Alternatively, the different estimates can be used in an expected value analysis, which is introduced next.

18.3 Estimate Variability and the Expected Value ●●●

Engineers and economic analysts usually deal with estimate variation and risk about an uncertain future by placing appropriate reliance on past data, if any exist. This means that *probability* and *samples* are used. Actually the use of probabilistic analysis is not as common as might be expected. The reason is not that the computations are difficult to perform or understand, but that realistic probabilities associated with cash flow estimates are difficult to assign. Experience and judgment can often be used in conjunction with probabilities and expected values to evaluate the desirability of an alternative.

> The **expected value** can be interpreted as a long-run average observable if the project is repeated many times. Since a particular alternative is evaluated or implemented only once, the expected value results in a **point estimate.** However, even for a single occurrence, the expected value is a meaningful number.

The expected value $E(X)$ is computed using the relation

$$E(X) = \sum_{i=1}^{i=m} X_i P(X_i) \qquad [18.2]$$

where

X_i = value of the variable X for i from 1 to m different values
$P(X_i)$ = probability that a specific value of X will occur

Probabilities are always correctly stated in decimal form, but they are routinely spoken of in percentages and often referred to as **chance,** such as *the chances are about 10%*. When placing the probability value in Equation [18.2] or any other relation, use the decimal equivalent of 10%, that is, 0.1. In all probability statements the $P(X_i)$ values for a variable X must total to 1.0.

$$\sum_{i=1}^{i=m} P(X_i) = 1.0$$

We may frequently omit the subscript i on X for simplicity.

If X represents the estimated cash flows, some will be positive and others will be negative. If a cash flow sequence includes revenues and costs, and the measure of worth is present worth calculated at the MARR, the result is the expected value of the discounted cash flows $E(PW)$. If the expected value is negative, the overall outcome is expected to be a cash outflow. For example, if $E(PW) = \$-1500$, this indicates that the proposal is not *expected* to return the MARR.

EXAMPLE 18.4

ANA airlines plans to offer several new electronic services on flights between Tokyo and selected European destinations. The marketing director estimates that for a typical 24-hour period there is a 50% chance of having a net cash flow of $5000 and a 35% chance of $10,000. He also estimates there is a small 5% chance of no cash flow and a 10% chance of a loss of $1000, which is the estimated extra personnel and utility costs to offer the services. Determine the expected net cash flow.

Solution

Let NCF be the net cash flow in dollars, and let $P(NCF)$ represent the associated probabilities. Using Equation [18.2],

$$E(NCF) = 5000(0.5) + 10,000(0.35) + 0(0.05) - 1000(0.1) = \$5900$$

Although the "no cash flow" possibility does not increase or decrease $E(NCF)$, it is included because it makes the probability values sum to 1.0 and it makes the computation complete.

Online Presentation

18.4 Expected Value Computations for Alternatives ● ● ◉

The expected value computation $E(X)$ is utilized in a variety of ways. Two prime ways are to:

- Prepare information for use in an economic analysis.
- Evaluate the expected viability of a fully formulated alternative.

Example 18.5 illustrates the first situation, and Example 18.6 determines the expected PW when the entire cash flow series and its probabilities are estimated.

EXAMPLE 18.5

There are many government incentives to become more energy-efficient. Installing solar panels on homes, business buildings, and multiple-family dwellings is one of them. The owner pays a portion of the total installation costs, and the government agency pays the rest. Nichole works for the Department of Energy and is responsible for approving solar panel incentive payouts. She has exceeded the annual budgeted amount of $50 million per year in each of the previous 2 years. Disappointed with this situation, Nichole and her boss decided to collect data to determine what size increase in annual budget the incentive program needs in the future. Over the last 36 months, the amount of average monthly payout and number of months are shown in Table 18–3. She categorized by level the monthly averages according to her experience with the program. Provided the same pattern continues, what is the expected value of the dollar increase in annual budget that is needed to meet the requests?

TABLE 18–3 Solar Panel Incentive Payouts, Example 18.5

Level	Average Payout, $ Million per Month	Months over Past 3 Years
Very high	6.5	15
High	4.7	10
Moderate	3.2	7
Low	2.9	4

Solution

Use the 36 months of payouts PO_j (j = low, . . . ,very high) to estimate the probability $P(PO_j)$ for each level, and make sure the total is 1.0.

Level, j	Probability of Payout Level, $P(PO_j)$
Very high	$P(PO_1) = 15/36 = 0.417$
High	$P(PO_2) = 10/36 = 0.278$
Moderate	$P(PO_3) = 7/36 = 0.194$
Low	$P(PO_4) = 4/36 = \underline{0.111}$
	1.000

The expected monthly payout is calculated using Equation [18.2]. In $ million units,

$$E[PO] = 6.5(0.417) + 4.7(0.278) + 3.2(0.194) + 2.9(0.111)$$
$$= 2.711 + 1.307 + 0.621 + 0.322$$
$$= \$4.961 \quad (\$4,961,000)$$

The annual expected budget need is 12×4.961 million = $59.532 million. The current budget of $50 million should be increased by an average of $9.532 million per year.

EXAMPLE 18.6

Lite-Weight Wheelchair Company has a substantial investment in tubular steel bending equipment. A new piece of equipment costs $5000 and has a life of 3 years. Estimated cash flows (Table 18–4) depend on economic conditions classified as receding, stable, or expanding. A probability is estimated that each of the economic conditions will prevail during the 3-year period. Apply expected value and PW analysis to determine if the equipment should be purchased. Use a MARR of 15% per year.

TABLE 18–4	Equipment Cash Flow and Probabilities, Example 18.6		
	Economic Condition		
	Receding (Prob. = 0.4)	**Stable** (Prob. = 0.4)	**Expanding** (Prob. = 0.2)
Year	**Annual Cash Flow Estimates, $**		
0	−5000	−5000	−5000
1	+2500	+2500	+2000
2	+2000	+2500	+3000
3	+1000	+2500	+3500

Solution

First determine the PW of the cash flows in Table 18–4 for each economic condition, and then calculate $E(PW)$ using Equation [18.2]. Define subscripts R for receding economy, S for stable, and E for expanding. The PW values for the three scenarios are

$$PW_R = -5000 + 2500(P/F,15\%,1) + 2000(P/F,15\%,2) + 1000(P/F,15\%,3)$$
$$= -5000 + 4344 = \$-656$$
$$PW_S = -5000 + 5708 = \$+708$$
$$PW_E = -5000 + 6309 = \$+1309$$

Only in a receding economy will the cash flows not return the 15% to justify the investment. The expected present worth is

$$E(PW) = \sum_{j=R,S,E} PW_j[P(j)]$$
$$= -656(0.4) + 708(0.4) + 1309(0.2)$$
$$= \$283$$

At 15%, $E(PW) > 0$; the equipment is justified, using an expected value analysis.

Comment

It is also correct to calculate the E(cash flow) for each year and then determine PW of the E(cash flow) series, because the PW computation is a linear function of cash flows. Computing E(cash flow) first may be easier in that it reduces the number of PW computations. In this example, calculate $E(CF_t)$ for each year, then determine $E(PW)$.

$$E(CF_0) = \$-5000$$
$$E(CF_1) = 2500(0.4) + 2500(0.4) + 2000(0.2) = \$2400$$
$$E(CF_2) = \$2400$$
$$E(CF_3) = \$2100$$
$$E(PW) = -5000 + 2400(P/F,15\%,1) + 2400(P/F,15\%,2) + 2100(P/F,15\%,3)$$
$$= \$283$$

Online Presentation

18.5 Staged Evaluation of Alternatives Using a Decision Tree ● ● ●

Alternative evaluation may require a series of decisions in which the outcome from one stage is important to the next stage of decision making. When each alternative is clearly defined and probability estimates can be made to **account for risk,** it is helpful to perform the evaluation using a **decision tree.**

A decision tree includes:

- More than one stage of alternative selection.
- Selection of an alternative at one stage that leads to another stage.
- Expected results from a decision at each stage.
- Probability estimates for each outcome.
- Estimates of economic value (cost or revenue) for each outcome.
- Measure of worth as the selection criterion, such as $E(PW)$.

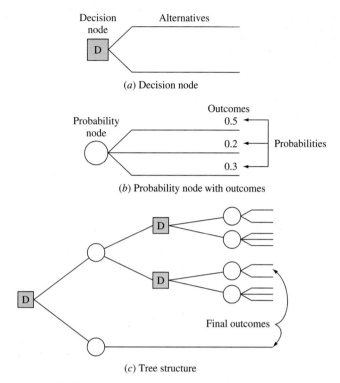

Figure 18–7
Decision and probability nodes used to construct a decision tree.

The decision tree is constructed left to right and includes each possible decision and outcome.

- A square represents a **decision node** with the possible alternatives indicated on the **branches** from the decision node (Figure 18–7a).
- A circle represents a **probability node** with the possible **outcomes** and estimated **probabilities** on the branches (Figure 18–7b).
- The treelike structure in Figure 18–7c results, with outcomes following a decision.

Usually each branch of a decision tree has some estimated economic value (often referred to as *payoff*) in cost, revenue, saving, or benefit. These cash flows are expressed in terms of PW, AW, or FW values and are shown to the right of each final outcome branch. The cash flow and probability estimates on each outcome branch are used in calculating the expected economic value of each decision branch. This process, called **solving the tree** or **rollback,** is explained after Example 18.7, which illustrates the construction of a decision tree.

EXAMPLE 18.7

Jerry Hill is president and CEO of a U.S.-based food processing company, Hill Products and Services. He was recently approached by an international supermarket chain that wants to market in-country its own brand of frozen microwaveable dinners. The offer made to Jerry by the supermarket corporation requires that a series of two decisions be made, now and 2 years hence. The current decision involves two alternatives: (1) *Lease* a facility in the United Arab Emirates (UAE) from the supermarket chain, which has agreed to convert a current processing facility for immediate use by Jerry's company; or (2) *build and own* a processing and packaging facility in the UAE. Possible outcomes of this first decision stage are good market or poor market depending upon the public's response.

The decision choices 2 years hence are dependent upon the lease-or-own decision made now. If Hill *decides to lease,* good market response means that the future decision alternatives are to produce at twice, equal to, or one-half of the original volume. This will be a mutual decision between the supermarket chain and Jerry's company. A poor market response will indicate a one-half level of production, or complete removal from the UAE market. Outcomes for the future decisions are, again, good and poor market responses.

As agreed by the supermarket company, the current decision for Jerry *to own* the facility will allow him to set the production level 2 years hence. If market response is good, the decision alternatives are four or two times original levels. The reaction to poor market response will be production at the same level or no production at all.

Construct the tree of decisions and outcomes for Hill Products and Services.

Solution

This is a two-stage decision tree that has alternatives now and 2 years hence. Identify the decision nodes and branches, and then develop the tree using the branches and the outcomes of good and poor market for each decision. Figure 18–8 details the decision stages and outcome branches.

Stage 1 (decision now):

Label it D1.

Alternatives: lease (L) and own (O).

Outcomes: good and poor markets.

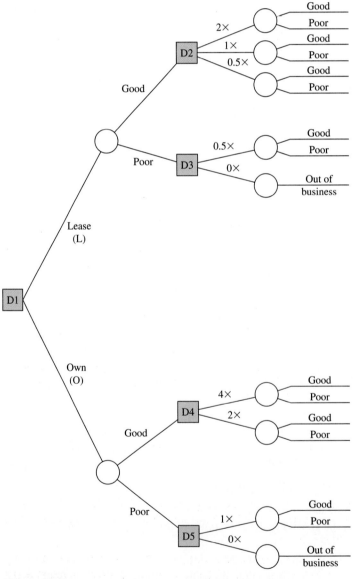

Figure 18–8
A two-stage decision tree identifying alternatives and possible outcomes.

Stage 2 (Decisions 2 years hence):
  Label them D2 through D5.
  Outcomes: good market, poor market, and out of business.

  Choice of production levels for D2 through D5:
  Quadruple production (4×); double production (2×); level production
  (1×); one-half production (0.5×); stop production (0×)

The alternatives for future production levels (D2 through D5) are added to the tree and followed by the market responses of good and poor. If the stop-production (0×) decision is made at D3 or D5, the only outcome is out of business.

To utilize the decision tree for alternative evaluation and selection, the following additional information is necessary for each branch:

- The **estimated probability** that each outcome may occur. These probabilities must sum to 1.0 for each set of outcomes (branches) that result from a decision.
- **Economic information** for each decision alternative and possible outcome, such as initial investment and estimated cash flows.

Decisions are made using the probability estimate and economic value estimate for each outcome branch. Commonly the present worth at the MARR is used in an expected value computation of the type in Equation [18.2]. This is the general procedure to solve the tree using PW analysis:

1. Start at the top right of the tree. Determine the PW value for each outcome branch considering the time value of money.
2. Calculate the expected value for each decision alternative.

$$E(\text{decision}) = \Sigma(\text{outcome estimate})P(\text{outcome}) \qquad [18.3]$$

where the summation is taken over all possible outcomes for each decision alternative.

3. At each decision node, select the best E(decision) value—minimum cost or maximum value (if both costs and revenues are estimated).

4. Continue moving to the left of the tree to the root decision in order to select the best alternative.

5. Trace the best decision path through the tree.

EXAMPLE 18.8

A decision is needed to either market or sell a new invention. If the product is marketed, the next decision is to take it international or national. Assume the details of the outcome branches result in the decision tree of Figure 18–9. The probabilities for each outcome and PW of CFBT (cash flow before taxes) are indicated. These payoffs are in millions of dollars. Determine the best decision at the decision node D1.

Solution
Use the procedure above to determine that the D1 decision alternative to sell the invention should maximize E(PW of CFBT).

1. Present worth of CFBT is supplied.
2. Calculate the expected PW for alternatives from nodes D2 and D3, using Equation [18.3]. In Figure 18–9, to the right of decision node D2, the expected values of 14 and 0.2 in ovals are determined as

$$E(\text{international decision}) = 12(0.5) + 16(0.5) = 14$$
$$E(\text{national decision}) = 4(0.4) - 3(0.4) - 1(0.2) = 0.2$$

The expected PW values of 4.2 and 2 for D3 are calculated in a similar fashion.

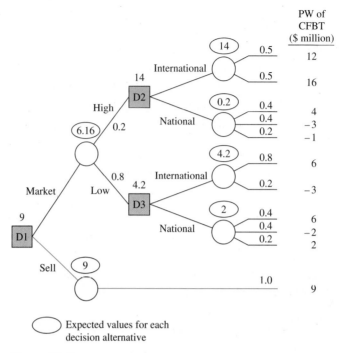

Figure 18–9
Solution of a decision tree with present worth of estimated CFBT values,
Example 18.8.

3. Select the larger expected value at each decision node. These are 14 (international) at D2
   and 4.2 (international) at D3.
4. Calculate the expected PW for the two D1 branches.

$$E(\text{market decision}) = 14(0.2) + 4.2(0.8) = 6.16$$
$$E(\text{sell decision}) = 9(1.0) = 9$$

The expected value for the sell decision is simple since the one outcome has a payoff of 9.
The sell decision yields the larger expected PW of 9.

5. The largest expected PW of CFBT path is to select the sell branch at D1 for a guaranteed
   $9,000,000.

18.6 Real Options in Engineering Economics ● ● ●

As we learned with decision trees, many of the problems in engineering economy can be viewed
as staged decisions. When the decision to invest more or less can be delayed into the future, the
problem is called **staged funding.** As an illustration, assume a large company expects to sell an
energy-saving, window-mounted residential air conditioning unit at the rate of 100,000 per
month by the end of 2 years on the market. Decision makers may opt to (1) build the capacity to
supply 100,000 per month to market immediately or (2) build capacity to supply 25,000 per
month now and test the market's receptivity. If positive, they can stage the increase by 25,000
additional units each 6 months to meet current demand. Of course, if an aggressive competitor
enters the scene, or the economy falters, the staged funding decision will change as warranted.
These alternatives provide time-based options to the company. Before we go further, some definitions are needed.

An **option** is a purchase or investment that contractually provides the privilege to take a
stated action by some stated time in the future, or the right to not accept the offer and forfeit
the option.

A **real option,** in engineering economy terms, is the investment (cost) in a project, process, or system. The options usually involve **physical (real)** assets, buildings, equipment, materials, and the like, thus, the word *real*. Options may also be leases, subcontracts, or franchises. The investment alternatives present varying amounts of **risk,** which is estimated by probabilities of occurrence for predictable future events.

Real options analysis is the application of techniques to determine the economic consequences of **delaying** the funding decisions as allowed by the option. The estimated cash flows and other consequences of these delays are analyzed with risk taken into account to the degree possible. A measure of worth, e.g., PW or AW, is the criterion used to make the staged funding decisions. A decision may be to expand, continue as is, contract, abandon, or replicate the alternative at the time the option must be exercised.

An inherent part of real options analysis is the uncertainty of future estimates, as it is for most economic analyses. After some illustrations of real options, we will discuss the probabilistic dimensions. Samples from industry and everyday personal life that can be formulated as real options follow.

Industrial Setting

New markets—Purchase equipment and staff to enter an expanding international market over the next 5 years.

New planes—Purchase commercial airplanes now with an option to buy an additional 5 planes over the next 3 years at the same price as that paid for the current order.

Removing car models—Ford Motor Company can decide to maintain production on an established car model with dwindling sales for the next 3 years or can opt to discontinue the model in stages over a 1- or 2-year period.

Drilling lease—Buy a drilling option contract from landowners to drill for oil and gas at some time in the next 10 years. The drilling may not be justified at this time, but the contract offers the option to drill were it to become economically advantageous based on events such as increased oil prices or improved recovery technology.

Personal Decision Making

Extended car warranty—When purchasing a new car, the option to buy an extended-coverage warranty beyond the manufacturer's warranty is always an option. The price of the option is the cost of the extended warranty. The uncertainties and risks are the future unknown costs for repairs and failed components.

House insurance—When a homeowner has no mortgage to pay, maintaining house insurance is an option. Deductibles are high enough, e.g., 1% to 5% of the fully appraised value, that insurance primarily covers only catastrophic damage to the structure. Self-insurance, where money is set aside for potential damages while accepting the risk that a major event will take place, is an option for the homeowner.

Some of the primary characteristics (with an example) of a real options analysis performed within the context of engineering economics are as follows:

- **Cost** to obtain the option to delay a decision (PW of initial investment, lease cost, or future investment amount).
- Anticipated future **options** and **cash flow** estimates (double production with annual net cash flows estimated).
- **Time period** for follow-on decisions (staged decision time, such as 1 year or a 3-year test period).
- Market and risk-free **interest rates** (expected market MARR of 12% per year and inflation rate estimate of 4% per year).
- Estimates of risk and future **uncertainty** for each option (probability that an estimated cash flow series will actually occur, if a specific option is selected).
- **Economic criterion** used to make a decision (PW, ROR, or other measure of worth).

It is common to use a decision tree to record and understand the options prior to performing a real options analysis with risk included. Example 18.9 demonstrates the use of a decision tree and PW analysis.

EXAMPLE 18.9

A start-up company in the solar energy production business, SolarScale Energy, Inc., has developed and field-tested a modularized, scalar solar thermal electric (STE) generation system that is relatively inexpensive to purchase and has an efficiency considerably better than traditional photovoltaic (PV) panels. The technology is promising enough that Capital Investor Funds (CIF) has provided $10 million for manufacturing. Additionally, a contract with a consortium of sunbelt states has been offered, but not accepted thus far, for a total of $1.5 million per year for a 2-year test period. By contract, the units will be marketed through the state energy departments with all revenue going to the state treasuries. The lead engineer at SolarScale, the manager of CIF, and a conservation representative for the state consortium have developed the following staged-funding options, based on the delayed decision to increase manufacturing production level until preliminary results of the 2-year contract are in hand.

Condition	Option	CIF Funding	Consortium Contract
Sales are excellent (> 5000 units/year)	2 × production level	Additional $10 million in year 2	Additional 8 years; $4 million in years 3–10
Sales are excellent (3000–5000 units/year)	1 × production level	Nothing; no salvage after 10 years	Additional 8 years; $1.5 million in years 3–10
Sales are poor (2000–3000 units/year)	Reduce to ½ × production	Nothing; sell for $2.5 million after 5 years	Additional 3 years; $1.5 million in years 3–5
Sales are poor (< 2000 units/year)	Stop after 2 years	Nothing; sell for $5 million after 2 years	Nothing

(a) Develop the two-stage decision tree for the options described.
(b) The base case is the 1 × production level with the 8-year follow-on contract from the consortium. If the estimates for this option are considered the most likely (expected value) estimates, determine the present worth at a MARR of 10% per year.
(c) Determine the PW values for each possible final outcome at 10% per year, and identify the best economic option when the stage 2 funding decision must be made.

Solution
(a) Figure 18–10 details the options with the year shown at the bottom. There are 2 outcome branches initially (accept option; decline option) and four final branches for the accept decision at D1, based upon sales level. The decline option has a $0 outcome. (SolarScale has other ways to pursue revenue that are not represented in this abbreviated example.)
(b) Perform a PW evaluation as we have in all previous chapters, assuming the estimates are point estimates over the 10-year life of the project. The resulting $PW_{1\times} < 0$ as shown below indicates the contract is not justified economically. In $ millions,

$$PW_{1\times} = -10 + 1.5(P/A,10\%,10)$$
$$= \$-0.78 \quad (\$-780,000)$$

(c) Figure 18–11 is a spreadsheet screen shot that calculates the PW for each option using the NPV function. The i^* values are also shown using the IRR function. Note that the sale of production assets after 5 years for $2.5 million (½ × level) or after 2 years for $5 million (stop) is included. Also, the extra $10 million investment in year 2 for the 2 × level option is included.

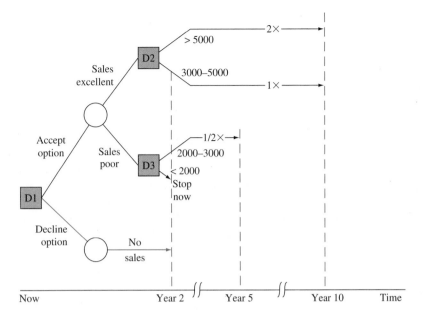

Figure 18–10
Decision tree showing real options over 10-year period, Example 18.9.

	A	B	C	D	E	F
1		Net cash flow estimates, $ million per year				
2	Year	2X level	1X level	1/2X level	Stop	
3	0	-10	-10	-10	-10	
4	1	1.5	1.5	1.5	1.5	
5	2	-8.5	1.5	1.5	6.5	
6	3	4.0	1.5	1.5		
7	4	4.0	1.5	1.5		
8	5	4.0	1.5	4.0		
9	6	4.0	1.5			
10	7	4.0	1.5		Sale of assets in this year	
11	8	4.0	1.5			
12	9	4.0	1.5			
13	10	4.0	1.5			
14	i*	12.6%	8.1%	0.0%	-11.5%	
15	PW @ 10%	1.97	-0.78	-2.76	-3.26	

Figure 18–11
PW analysis of real options without risk considered, Example 18.9.

Only the 2 × production-level option is justified at MARR = 10% per year. If SolarScale and CIF, the financial backers, are not convinced that the sales level will exceed 5000 units per year, the contract option should be declined. Some marketing survey information and risk analysis may be very helpful before this important decision is made.

Of the characteristics listed above for real option situations, the primary one absent in Example 18.9 is that of estimate variation and some measure of **risk.** Decision making under risk is covered more extensively in the next chapter; however, we can use the following definition for this discussion of real options analysis.

When a parameter can take on more than one value and there is any estimate of **chance** or **probability** about the opportunity that each value may be observed, risk is present.

Risk

A coin has two sides. If it is a perfectly balanced coin, a flip of the coin should result in heads 50% of the time and tails 50% of the time. If the coin is intentionally biased in weight, such that 58% of the time it lands heads up, then the long-run probability of heads is $P(\text{heads}) = 0.58$. Since the sum of probabilities across all possible values must add to 1, the biased coin has $P(\text{tails}) = 0.42$.

There are a couple of points worth mentioning about risk and the calculated PW values for real options analysis.

- When the risk is higher and the stakes are larger, the real options analysis is often more valuable. (We shall see this in Example 18.10.)
- Since one of the objectives of real options analysis is to evaluate the economic consequences of delaying a decision, a PW value of the base case that is *moderately positive* means that the project is justified and should be accepted immediately, without the decision delay. On the other hand, if the PW is *largely negative,* the delay is likely not worthwhile, as it would take a very large positive PW to result in $E(\text{PW}) > 0$. Thus, the project should be rejected now, not delayed for a future decision.

We return to the previous example with some risk assessment added to determine if the contract option should be declined, as indicated by the base case.

EXAMPLE 18.10

Before the stage 1 decision is made in Example 18.9 about SolarScale's state-consortium contract offer, some expected sales information was collected. Once the results were reviewed by the three individuals in attendance at the final decision-making meeting—one representative each for SolarScale, CIF, and the consortium—each person recorded her or his estimated probability as a measure of risk that the sales would be excellent or poor, which represents the two outcome possibilities were the option accepted. The results are as follows:

	Probability of Outcome	
	Excellent	Poor
SolarScale	0.5	0.5
CIF	0.8	0.2
Consortium	0.6	0.4

Use these probability estimates to determine the expected PW value, provided equal weighting is given to each representative's input.

Solution

For each outcome (excellent and poor), select the best PW value from Figure 18–11, and then find $E(\text{PW})$ for each representative.

Excellent: From 2 × level and 1 × level, select 2 × with PW = $1.97 million.

Poor: From ½ × level and stop now, select ½ × with PW = $-2.76 million.

In $ million, $E(\text{PW})$ for each organization is

$$E(\text{PW for SolarScale}) = 1.97(0.5) - 2.76(0.5) = \$-0.40$$
$$E(\text{PW for CIF}) = 1.97(0.8) - 2.76(0.2) = \$1.02$$
$$E(\text{PW for consortium}) = 1.97(0.6) - 2.76(0.4) = \$0.08$$

With a 1/3 chance assigned to each representative, the overall $E(\text{PW of stage 2 decision})$ is

$$E(\text{PW of stage 2 decision}) = 0.33(-0.40 + 1.02 + 0.08)$$
$$= \$0.23 \quad (\$230{,}000)$$

The base case of 1 × level production in part (*b*) of Example 18.9 resulted in $E(\text{PW}) = \$-780{,}000$. When compared with the positive $E(\text{PW})$ result here, we see that with consideration of the different options of production level and probabilities for sales level, the expected PW has increased to a positive value. All other things being equal, the state consortium offer should be accepted; that is, accept the real option of the contract.

There are many other examples and dimensions of real options analysis in engineering economics and in the area of financial analysis, where options analysis got its start some years ago. If you are interested in this new and interesting area of analysis, consult more advanced texts and journal articles on the topic of real options.

CHAPTER SUMMARY

In this chapter the emphasis is on sensitivity to variation in one or more parameters using a specific measure of worth. When two alternatives are compared, compute and graph the measure of worth for different values of the parameter to determine when each alternative is better.

When several parameters are expected to vary over a predictable range, the measure of worth is plotted and calculated using three estimates for a parameter—most likely, pessimistic, and optimistic. This approach can help determine which alternative is best among several. Independence between parameters is assumed in all these analyses.

The combination of parameter and probability estimates results in the expected value relation

$$E(X) = \Sigma XP(X)$$

This expression is also used to calculate E(revenue), E(cost), E(cash flow), and E(PW) for the entire cash flow sequence of an alternative.

Decision trees are used to make a series of alternative selections. This is a way to explicitly take risk into account. It is necessary to make several types of estimates for a decision tree: outcomes for each possible decision, cash flows, and probabilities. Expected value computations are coupled with those for the measure of worth to solve the tree and find the best alternatives stage by stage.

Staged funding over time can be approached using the evolving area of real options. Delaying an investment decision and considering the risks of the future can improve the overall E(PW) of a project, process, or system.

PROBLEMS

Sensitivity to Parameter Variation

18.1 Kahn Instruments is considering an investment of $500,000 in a new product line. The company will make the investment only if it will result in a rate of return of 15% per year or higher. If the revenue is expected to be between $135,000 and $165,000 per year for 5 years, determine if the decision to invest is sensitive to the projected range of income using a present worth analysis.

18.2 A young couple planning ahead for their retirement has decided that $2,600,000 is the amount they will need to retire comfortably 20 years from now. For the past 5 years they have been able to invest one of their salaries ($50,000 per year, which includes employer contributions) while living off the other one. They plan to start a family sometime in the next 10 years, and when they have their first child, one of the parents will quit working, causing the savings to decrease to $15,000 per year thereafter. If they have gotten a rate of return of 10% per year on their investments and expect to continue at this ROR, is reaching their goal of $2.6 million 20 years from now sensitive to when they have their first child (i.e., between now and 10 years from now)? Use an FW analysis.

18.3 A company that manufactures high-speed submersible rotary indexing spindles is considering upgrading the production equipment to reduce costs over a 6-year planning horizon. The company can invest $80,000 now, 1 year from now, or 2 years from now. Depending on when the investment is made, the savings will vary. That is, the savings will be $25,000, $26,000, or $29,000 per year if the investment is made now (year 0), in 1 year, or in 2 years, respectively. Will the timing of the investment affect the request to make at least a 20% per year return? Use future worth analysis.

18.4 Different membrane systems are under consideration for treating 3 million gallons per day (MGD) of cooling tower blowdown water to reduce its volume. Option 1 is a low-pressure seawater reverse osmosis (SWRO) system that will operate at 500 psi with a fixed cost of $465 per day and an operating cost of $0.67 per 1000 gallons. A second option is a higher-pressure SWRO system offered by vendor X that operates at 800 psi and will have a lower fixed cost of $328 per day (because of fewer membranes); however, its operating cost will be $1.35 per 1000 gallons. A third option is also a high-pressure SWRO system from vendor Y, who claims that its system will have a lower operating cost of $1.28 per 1000 gallons and the same fixed cost as that of vendor X. Determine if the selection of a low- or high-pressure system is dependent on the lower operating cost offered by vendor Y.

18.5   A machine that is currently used in manufacturing circuit board card locks has AW = $–63,000 per year. A possible replacement is under consideration with a first cost of $64,000 and an operating cost of $38,000 per year for the next 3 years. Three different engineers have given their opinion, about what the salvage value of the new machine will be 3 years from now: $10,000, $13,000, and $18,000. Is the decision to replace the machine sensitive to the salvage value estimates at the company's MARR of 15% per year?

18.6   An equipment alternative is being economically evaluated separately by three engineers at Raytheon. The first cost will be $77,000, and the life is estimated at 6 years with a salvage value of $10,000. The engineers disagree, however, on the estimated revenue the equipment will generate. Joe has made an estimate of $10,000 per year. Jane states that this is too low and estimates $14,000, while Carlos estimates $18,000 per year. If the before-tax MARR is 8% per year, use PW to determine if these different estimates will change the decision to purchase the equipment.

18.7   The owner of a small construction company is planning to purchase specialized equipment to complete a contract he just received. The first cost of the equipment is $250,000, and it will likely have a salvage value of $90,000 in 3 years, at which time he will not need the equipment anymore. The operating cost is expected to be $75,000 per year. Alternatively, the owner can subcontract the work for $175,000 per year. Because the equipment is specialized, the owner is not sure about the salvage value. He thinks it might be worth as little as $10,000 in 3 years (a scrap value). If his minimum attractive rate of return is 15% per year, determine if the decision to buy the equipment is sensitive to the salvage value.

18.8   A company planning to borrow $10.5 million for a plant expansion is not sure what the interest rate will be when it applies for the loan. The rate could be as low as 10% per year or as high as 12% per year for a 5-year loan. The company will only move forward with the project if the annual worth of the expansion is below $5.7 million. The M&O cost is fixed at $3.1 million per year. The salvage could be $2 million if the interest rate is 10% or $2.5 million if it is 12% per year. Is the decision to move forward with the project sensitive to the interest rate and salvage value estimates?

18.9   A company that manufactures clear PVC pipe is investigating the production options of batch and continuous processes. Estimated cash flows are as follows:

	Batch	Continuous
First cost, $	−80,000	−130,000
Annual cost, $ per year	−55,000	−30,000
Salvage value for any year, $	10,000	40,000
Life, years	3–10	5

The chief operating officer (COO) has asked you to determine if the batch option would ever have a lower annual worth than the continuous flow system, using interest rates over a range of 5% to 15% for the batch option but only 15% for the continuous flow system. (*Note*: The continuous flow process was previously determined to have its lowest cost over a 5-year life cycle; the batch process can be used from 3 to 10 years.)

18.10 An engineer collected average cost and revenue data for Arenson's FC1 handheld financial calculator.

   Fixed cost = $300,000 per year

   Cost per unit = $40

   Revenue per unit = $70

   (*a*) What is the range in breakeven quantity if there is possible variation in the fixed cost from $200,000 to $400,000 per year? (Use $50,000 increments.)

   (*b*) What is the incremental change in the breakeven quantity for each $50,000 change in fixed cost?

The following information is used for Problems 18.11 through 18.14.

A new online patient diagnostics system for surgeons will cost $200,000 to install, cost $5000 annually to maintain and will have an expected life of 5 years. The added revenue is estimated to be $60,000 per year, and the MARR is 10% per year. Examine the sensitivity of present worth to variation in selected parameter estimates, while others remain constant.

18.11 Sensitivity to first cost variation: $150,000 to $250,000 (−25% to +25%).

18.12 Sensitivity to revenue variation: $45,000 to $75,000 (−25% to +25%).

18.13 Sensitivity to life variation: 4 years to 7 years (−20% to +40%).

18.14 Plot the results on a graph similar to Figure 18–3 and comment on the relative sensitivity of each parameter.

18.15 Charlene plans to place an annual savings amount of $A = \$27,185$ into a retirement program at the end of each year for 20 years starting next year. She expects to retire and start to draw a total of $R = \$60,000$ per year 1 year after the 20th deposit. Assume an effective earning rate of $i = 6\%$ per year on the retirement investments and an infinite life. Determine and comment on the sensitivity of the size of the annual withdrawal R for variations in A and i. Show hand and spreadsheet solutions.
  (a) Variation of ±5% in the annual deposit A.
  (b) Variation of ±1% in the effective earning rate i, that is, ranging from 5% to 7% per year.

18.16 Ned Thompson Labs performs tests on super alloys, titanium, aluminum, and most metals. Tests on metal composites that rely upon scanning electron microscope results can be subcontracted, or the labs can purchase new equipment. Evaluate the sensitivity of the economic decision to purchase the equipment over a range of ±20% (in 10% increments) of the estimates for P, AOC, R, n, and MARR (range on MARR is 12% to 16%). Use the AW method and plot the results on a sensitivity graph (like Figure 18–3). For which parameter(s) is the AW most sensitive? Least sensitive?

  First cost $P = \$-220,000$

  Salvage $S = \$20,000$

  Life $n = 10$ years

  Annual operating cost   AOC = $\$-30,000$/year

  Annual revenue $R = \$70,000$ per year

  MARR $i = 15\%$ per year

18.17 Titan manufactures and sells gas-powered electricity generators. It can purchase a new line of fuel injectors from either of two companies. Cost and savings estimates are made, but the savings estimate is unreliable at this time. Use an AW analysis at 10% per year to determine if the selection between company A and company B changes when the savings per year may vary as much as 40% from the best estimates made thus far.

	Company A	Company B
First cost, $	−50,000	−37,500
AOC, $ per year	−7,500	−8,000
Savings best estimate, $ per year	15,000	13,000
Salvage, $	5,000	3,700
Life, years	5	5

18.18 (a) Graph the sensitivity of what a person should be willing to pay now for a 9%, $10,000 bond due in 10 years if there is a 30% change in (1) face value, (2) dividend rate, or (3) required nominal rate of return, which is expected to be 8% per year, compounded semiannually. The bond pays dividends semiannually.
  (b) If the investor did purchase the $10,000 face value bond at a premium of 5% (i.e., 5% above face value) and all your other estimates were correct, that is, 0% change, did he pay too much or too little? How much?

Three Estimates

18.19 DVH Technologies purchases several parts for the instruments it makes via a fixed-price contract of $190,000 per year from a local supplier. The company is considering making the parts in-house through the purchase of equipment that will have a first cost of $240,000 with an estimated salvage value of $30,000 after 5 years. The operating cost is difficult to estimate, but company engineers have made optimistic, most likely, and pessimistic estimates of $60,000, $85,000, and $120,000 per year, respectively. Determine if the company should purchase the equipment under any of the operating cost scenarios. The MARR is 20% per year.

18.20 Astor Engineering recently merged with another firm and could lease additional office space or purchase its own building. The $30,000 per year lease agreement will be a net, net, net lease, which means that the lessee (Astor) will pay the real estate taxes on the leased space, the building insurance on the leased space, and the common area maintenance. Since these costs are about the same if Astor owned the building, they do not need to be considered in the analysis. A new building will cost $880,000 to purchase, but there is considerable uncertainty about what it will be worth in 20 years, which is the planning period selected. The individuals involved in the discussion made optimistic, most likely, and pessimistic estimates of $2,400,000, $1,400,000, and $900,000, respectively. Determine if Astor should purchase the building under any of the estimated resale values at $i = 10\%$ per year.

18.21 Holly Farms is considering two environmental chambers to accomplish detailed laboratory confirmations of online bacteria tests in chicken meat for the presence of *E. coli* 0157:H7 and *Listeria monocytogenes*. There is some uncertainty about how long the D103 chamber will be useful. A realistic estimate is 3 years, but pessimistic and optimistic estimates of 2 years and 6 years, respectively, are also possible. The estimated salvage value will remain the same. Using an interest rate of 10% per year, determine if any of the D103 estimates would result in a lower cost than that of the 490G chamber for a 6-year planning period.

	Chamber D103	Chamber 490G
Installed cost, $	−400,000	−250,000
AOC, $ per year	−4,000	−3,000
Salvage value at 10% of *P*, $	40,000	25,000
Life, years	2, 3, or 6	2

18.22 When the country's economy is expanding, AB Investment Company is optimistic and expects a MARR of 15% for new investments. However, in a receding economy the expected return is 8%. Normally a 10% return is required. An expanding economy causes the estimates of asset life to go down about 20%, and a receding economy makes the *n* values increase about 10%. Calculate and observe or plot the sensitivity of PW values versus (*a*) the MARR and (*b*) the life values for the two plans detailed below, using the most likely estimates for the other factors. (*c*) Considering all the analyses, under which scenario, if any, should plan M or Q be rejected?

	Plan M	Plan Q
Initial investment, $	−100,000	−110,000
Cash flow, $ per year	+15,000	+19,000
Life, years	20	20

Expected Value

18.23 Determine the expected net operating income (NOI) from sales of micro turbine components. The probabilities are 20%, 50%, and 30% for revenues of $800,000, $1,000,000, and $1,100,000 per year, respectively, and operating expenses are constant at $200,000 per year.

18.24 A company that manufactures amplified pressure transducers is trying to decide between a dual-speed and a variable-speed machine. The engineers are not sure about the salvage value of the variable-speed machine, so they have asked several different used-equipment dealers for

estimates. The results can be summarized as follows: there is a 32% chance of getting $20,000, a 45% chance of getting $28,000, and a 13% chance of getting $34,000. Also, there is a 10% chance that the company may have to pay $5000 to dispose of the equipment. Calculate the expected salvage value.

18.25 The average success probability for a wildcat oil well drilled in the Wind River basin 7 miles from the nearest existing production well is estimated to be 13%. If the value of the oil has equal chances of being $1.5 million, $1.9 million, and $2.4 million, what is the expected income from the well?

18.26 Nationwide income from monthly sales data (rounded to the nearest $100,000) of Stay Flat vacuum hold-down tables for last year is shown below. Determine the expected value of the monthly income, if economic conditions remain the same.

Income, $ per Month	Number of Months
500,000	4
600,000	2
700,000	1
800,000	2
900,000	3

18.27 Determine the expected maximum rainfall intensity in El Paso, Texas for the month of July using the estimated probabilities shown.

Rainfall Rate, inches per hour	3	4	5	6
Probability	0.4	0.3	0.2	0.1

18.28 There are four estimates made for the anticipated cycle time to produce a subcomponent. The estimates, in seconds, are 10, 20, 30, and 50. (*a*) If equal weight is placed on each estimate, what is the expected cycle time? (*b*) If the largest time is disregarded, what is the percent reduction in the expected time?

18.29 The PW value for an alternative is expected to be one of two values based on bids from two vendors. Your office partner told you that the low bid is $3200 per year. If she indicates a chance of 70% of accepting the high bid and that her expected PW is $5875, what is the PW of the high bid?

18.30 A total of 40 different proposals were evaluated by the IRAD (Industrial Research and Development) committee during the past year. Twenty were funded. Their rate of return estimates are summarized with the *i** values rounded to the nearest integer. For the accepted proposals, calculate the expected rate of return $E(i)$.

Proposal ROR, $i*\%$	Number of Proposals
−8	1
−5	1
0	5
5	5
8	2
10	3
15	3
	20

18.31 Beckman Electronics has performed an economic analysis of proposed service in a new region of the country. The three-estimate approach to sensitivity analysis has been applied. The optimistic and pessimistic values each have an estimated 20% chance of occurring. Use the FW values shown to determine the expected FW.

	Optimistic	Most Likely	Pessimistic
FW value, $	300,000	50,000	−25,000

18.32 A very successful health and recreation club wants to construct a mock mountain for climbing and exercise outside for its customers' use. Because of its location, there is a 30% chance of a 120-day season of good outdoor weather, a 50% chance of a 150-day season, and a 20% chance of a 165-day season. The mountain will be used by an estimated 350 persons each day of the 4-month (120-day) season, but by only 100 per day for each extra day the season lasts. The feature will cost $375,000 to construct and require a $25,000 rework each 4 years; and the annual maintenance and insurance costs will be $56,000. The climbing fee will be $5 per person. If a life of 10 years is anticipated and a 12% per year return is expected, determine if the addition is economically justified.

18.33 The owner of Ace Roofing may invest $200,000 in new equipment. A life of 6 years and a salvage value of 12% of first cost are anticipated. The annual extra revenue will depend upon the state of the housing and construction industry. The extra revenue is expected to be only $20,000 per year if the current slump in the industry continues. Real estate economists estimate a 50% chance of the slump lasting 3 years and they give it a 20% chance of continuing for 3 additional years. However, if the depressed market does improve, during either the first or second 3-year period, the revenue of the investment is expected to increase by a total of $35,000 per year. Can the company expect to make a return of 8% per year on its investment? Use present worth analysis.

18.34 A flagship hotel in Cedar Falls must construct a retaining wall next to its parking lot due to the widening of the city's main thoroughfare located in front of the hotel. The amount of rainfall experienced in a short time may cause damage in varying amounts, and the wall increases in cost in order to protect against larger and faster rainfalls. The probabilities of a specific amount of rainfall in a 30-minute period and wall cost estimates are as follows:

Rainfall, Inches per 30 Minutes	Probability of Greater Rainfall	First Cost of Wall, $
2.0	0.3	200,000
2.25	0.1	225,000
2.5	0.05	300,000
3.0	0.01	400,000
3.25	0.005	450,000

The wall will be financed through a 6% per year loan. The principal and interest will be repaid over a 10-year period. Records indicate an average damage of $50,000 has occurred with heavy rains, due to the relatively poor cohesive properties of the soil along the thoroughfare. A discount rate of 6% per year is applicable. Find the amount of rainfall to protect against by choosing the retaining wall with the smallest AW value over the 10-year period.

Decision Trees

18.35 For the decision tree branch shown, determine the expected values of the two outcomes if decision D3 is already selected and the maximum outcome value is sought. (This decision branch is part of a larger tree.)

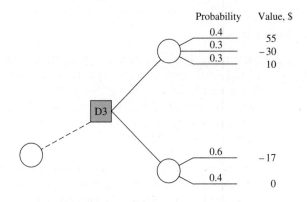

18.36 A large decision tree has an outcome branch detailed (next page). If decisions D1, D2, and D3 are all options in a 1-year period, find the decision path that maximizes the outcome value. There are specific investments necessary for decision nodes D1, D2, and D3, as indicated on each branch.

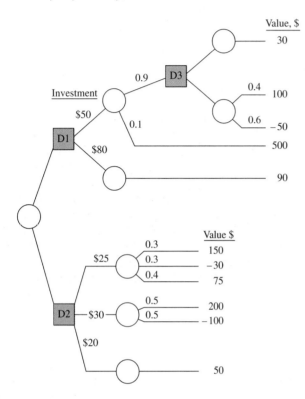

18.37 Decision D4, which has three possible alternatives—x, y, or z—must be made in year 3 of a 6-year study period in order to maximize the expected value of present worth. Using a rate of return of 15% per year, the investment required in year 3 and the estimated cash flows for years 4 through 6, determine which decision should be made in year 3.

	Investment Required, $	Cash flow, $1000				
		Years				Outcome
		3	4	5	6	probability
High	−200,000	50	50	50		0.7
Low		40	30	20		0.3
High	−75,000	30	40	50		0.45
Low		30	30	30		0.55
High	−350,000	190	170	150		0.7
Low		−30	−30	−30		0.3

18.38 A total of 5000 mechanical subassemblies are needed annually on a final assembly line. The subassemblies can be obtained in one of three ways: (1) *Make them* in one of three plants owned by the company; (2) *buy them off the shelf* from the one and only manufacturer; or (3) *contract to have them made* to specifications by a vendor. The estimated

annual equivalent cost for each alternative is dependent upon specific circumstances of the plant, producer, or contractor. The information shown details the circumstance, a probability of occurrence, and the estimated annual cost. Construct and solve a decision tree to determine the least-cost alternative to provide the subassemblies.

Decision Alternative	Outcomes		Probability	Annual Cost for 5000 Units, $ per Year
1. Make	Plant:			
		A	0.3	−250,000
		B	0.5	−400,000
		C	0.2	−350,000
2. Buy off the shelf	Quantity:			
		<5000, pay premium	0.2	−550,000
		5000 available	0.7	−250,000
		>5000, forced to buy	0.1	−290,000
3. Contract	Delivery:			
		Timely delivery	0.5	−175,000
		Late delivery, then buy some off shelf	0.5	450,000

18.39 The president of ChemTech is trying to decide whether to start a new product line or purchase a small company. It is not financially possible to do both. To make the product for a 3-year period will require an initial investment of $250,000. The expected annual cash flows with probabilities in parentheses are: $75,000 (0.5), $90,000 (0.4), and $150,000 (0.1). To purchase the small company will cost $450,000 now. Market surveys indicate a 55% chance of increased sales for the company and a 45% chance of severe decreases with an annual cash flow of $25,000. If decreases are experienced in the first year, the company will be sold immediately (during year 1) at a price of $200,000. Increased sales could be $100,000 the first 2 years. If this occurs, a decision to expand after 2 years at an additional investment of $100,000 will be considered. This expansion could generate cash flows with indicated probabilities as follows: $120,000 (0.3), $140,000 (0.3), and $175,000 (0.4). If expansion is not chosen, the current size will be maintained with anticipated sales to continue. Assume there are no salvage values on any investments. Use the description given and a 15% per year return to do the following.

(a) Construct a decision tree with all values and probabilities shown.

(b) Determine the expected PW values at the "expansion/no expansion" decision node after 2 years, provided sales are up.

(c) Determine what decision should be made now to offer the greatest return possible for ChemTech.

(d) Explain in words what would happen to the expected values at each decision node if the planning horizon were extended beyond 3 years and all cash flow values continued as forecasted in the description.

Real Options

18.40 A privately held company that makes chips that are essential for high-volume data storage is valued at $3 billion. A computer company that wants to get into cloud computing is considering purchasing the company, but because of the uncertain economy, it would prefer to purchase an option that will allow it to buy the company for up to 1 year from now at a cost of $3.1 billion. What is the maximum amount the company should be willing to pay for the option, if its MARR is 12% per year?

18.41 A company that is considering adding a new product line has determined that the first cost would be $80 million. The company is not sure about how the product will be received, so it has projected revenues using optimistic, most likely, and pessimistic estimates of $35 million, $25 million, and $10 million, respectively, with equal probability for each. Instead of expanding now, the company could implement a test program for 1 year in a limited area that will cost $4 million. (The full-scale project will still cost $80 million if implemented after the test program is over.) This will provide the company with the option to move forward or cancel the project. The criterion identified to move ahead with full-scale implementation is that revenues must exceed $900,000. In this case, the pessimistic estimate will be eliminated, and equal probability will be placed on the remaining revenue projections. If the company uses a 5-year planning horizon and a MARR of 12% per year, should the company go ahead with the full-scale project now or take the option to implement the test program for 1 year?

18.42 Dow Chemical is considering licensing a low liquid discharge (LLD) water treatment system from a small company that developed the process. Dow can purchase a 1-year option for $150,000 that will

give it time to pilot-test the LLD process, or Dow can acquire the license now at a cost of $1.8 million plus 25% of sales. If Dow waits 1 year, the cost will increase to $1.9 million plus 30% of sales. If Dow projects the sales to be $1,000,000 per year over the 5-year license period, should the company license the process now or purchase the option to license it after the 1-year test period? Assume the MARR is 15% per year.

18.43 Abby has just negotiated a $15,000 price on a 2-year-old car and is with the salesman closing the deal. There is a 1-year sales warranty with the purchase; however, an extended warranty is available for $2500 that will cover the same repairs and component failures as the 1-year warranty for 3 additional years. Abby understands this to be a real options situation with the price of the option ($2500) paid to avoid future, unknown costs. To help with her decision, the salesman provided three typical sets of historical data on estimated repair costs for used cars. The first-year costs are shown as zero because they will be covered by the sales warranty.

Repair cost, $ per year:	Year			
	1	2	3	4
A	0	−500	−1200	−850
B	0	−1000	−1400	−400
C	0	0	−500	−2000

The salesman said case C is the base case, since it shows that the extended warranty is not needed because the cost of repairs equals the warranty cost. Abby immediately recognized this to be the case only when $i = 0\%$.

(a) If Abby assumes that each repair cost scenario has equal probability of occurring with her car, and money is worth 5% per year to her, how much should she be willing to pay for the extended warranty that is offered at $2500?

(b) If the base case actually occurs for her car and she does not purchase the warranty, what is the PW value of the expected future costs at $i = 5\%$ per year?

ADDITIONAL PROBLEMS AND FE EXAM REVIEW QUESTIONS

18.44 In conducting a sensitivity analysis, all of the following could be used as a measure of worth except:
(a) Present worth
(b) Cost-capacity equations
(c) Annual worth
(d) Benefit-cost ratio

18.45 When the measure of worth is plotted versus percent change for several parameters, the parameter that is the most sensitive in the economic analysis is the one:
(a) That has the steepest curve
(b) That has the flattest curve
(c) With the largest present worth
(d) With the shortest life

18.46 In conducting a formalized sensitivity analysis using three estimates, the three estimates should be:
    (a) Strategic, pessimistic, real likely:
    (b) Deterministic, realistic, optimistic
    (c) Optimistic, pessimistic, most likely
    (d) Authentic, realistic, probabilistic

18.47 For annual worth values of $30,000, $40,000, and $50,000 with chances of 20%, 20%, and 60%, respectively, the expected AW is closest to:
    (a) $34,000
    (b) $40,000
    (c) $44,000
    (d) $48,000

18.48 The AW of a 3-year-old machine that is used in the manufacture of modular magnetic encoders, which provide data on the speed and positioning of rotating motor shafts, is $-48,000. A challenger will have a first cost of $90,000, an operating cost of $29,000, and a salvage value after 5 years that may vary considerably. For an interest rate of 10% per year and optimistic, most likely, and pessimistic salvage values of $15,000, $10,000, and $2000, respectively, the salvage value(s) for which the challenger AW will be lower than that of the defender are:
    (a) All of them
    (b) Only the optimistic one

    (c) Only the optimistic and pessimistic ones
    (d) None of them

18.49 A decision tree includes all of the following except:
    (a) Probability estimates for each outcome
    (b) Measure of worth as the selection criterion
    (c) Expected results from a decision at each stage
    (d) The MARR

18.50 A real options analysis is most valuable when:
    (a) The risk is low and stakes are high
    (b) The stakes are low and risk is high
    (c) The stakes are high and risk is high
    (d) The stakes are low and risk is low

18.51 A small manufacturing company needs to purchase a machine that will have a first cost of $70,000. The company wants to buy an option that will allow it to purchase the machine for the same price of $70,000 for up to 1 year from now. If the company's MARR is 10% per year, the maximum amount the company should pay for the option is closest to:
    (a) $5850
    (b) $6365
    (c) $6845
    (d) $7295

CASE STUDY

SENSITIVITY TO THE ECONOMIC ENVIRONMENT

Background and Information

Berkshire Controllers usually finances its engineering projects with a combination of debt and equity capital. The resulting MARR ranges from a low of 4% per year, if business is slow, to a high of 10% per year. Normally, a 7% per year return is expected. Also the life estimates for assets tend to go down about 20% from normal in a vigorous business environment and up about 10% in a receding economy. The following estimates are the most likely values for two expansion plans currently being evaluated. Plan A will be executed at one location; Plan B will require two locations. All monetary estimates are in $1000 units.

Case Study Questions

At the weekly meeting, you were asked to examine the following questions from Berkshire's president.
1. Are the PW values for plans A and B sensitive to changes in the MARR?
2. Are the PW values sensitive to varying life estimates?
3. Is the breakeven point for the first cost of plan A sensitive to the changes in MARR as business goes from vigorous to receding?

| | Plan A | Plan B | |
		Location 1	Location 2
First cost, $	−10,000	−30,000	−5,000
AOC, $ per year	−500	−100	−200
Salvage value, $	1,000	5,000	−200
Estimated life, years	40	40	20

CASE STUDY

SENSITIVITY ANALYSIS OF PUBLIC SECTOR PROJECTS—WATER SUPPLY PLANS

Background

One of the most basic services provided by municipal governments is the delivery of a safe, reliable water supply. As cities grow and extend their boundaries to outlying areas, they often inherit water systems that were not constructed according to city codes. The upgrading of these systems is sometimes more expensive than installing one correctly in the first place. To avoid these problems, city officials sometimes install water systems beyond the existing city limits in anticipation of future growth. This case study was extracted from such a countywide water and wastewater management plan and is limited to only some of the water supply alternatives.

From about a dozen suggested plans, five methods were developed by an executive committee as alternative ways of providing water to the study area. These methods were then subjected to a preliminary evaluation to identify the most promising alternatives. Six attributes or factors were used in the initial rating: ability to serve the area, relative cost, engineering feasibility, institutional issues, environmental considerations, and lead time requirement. Each factor carried the same weighting and had values ranging from 1 to 5, with 5 being best. After the top three alternatives were identified, each was subjected to a detailed economic evaluation for selection of the best alternative. These detailed evaluations included an estimate of the capital investment of each alternative amortized over 20 years at 8% per year interest and the annual maintenance and operation (M&O) costs. The annual cost (an AW value) was then divided by the population served to arrive at a monthly cost per household.

Information

Table 18–5 presents the results of the screening using the six factors rated on a scale of 1 to 5. Alternatives 1A, 3, and 4 were determined to be the three best and were chosen for further evaluation.

Detailed Cost Estimates

All amounts are cost estimates.

Alternative 1A

Capital cost	
Land with water rights: 1720 hectares @ $5000 per hectare	$8,600,000
Primary treatment plant	2,560,000
Booster station at plant	221,425
Reservoir at booster station	50,325
Site cost	40,260
Transmission line from river	3,020,000
Transmission line right-of-way	23,350
Percolation beds	2,093,500
Percolation bed piping	60,400
Production wells	510,000
Well field gathering system	77,000
Distribution system	1,450,000
Additional distribution system	3,784,800
Reservoirs	250,000
Reservoir site, land, and development	17,000
Subtotal	22,758,060
Engineering and contingencies	5,641,940
Total capital investment	$28,400,000

TABLE 18–5   Results of Rating Six Factors for Each Alternative, Case Study

		Factors						
Alternative	Description	Ability to Supply Area	Relative Cost	Engineering Feasibility	Institutional Issues	Environmental Considerations	Lead Time Requirement	Total
1A	Receive city water and recharge wells	5	4	3	4	5	3	24
3	Joint city and county plant	5	4	4	3	4	3	23
4	County treatment plant	4	4	3	3	4	3	21
8	Desalt groundwater	1	2	1	1	3	4	12
12	Develop military water	5	5	4	1	3	1	19

Maintenance and operation costs (annual)

Pumping 9,812,610 kWh per year

@ $0.08 per kWh	$ 785,009
Fixed operating cost	180,520
Variable operating cost	46,730
Taxes for water rights	48,160
Total annual M&O cost	$1,060,419

Total annual cost = equivalent capital
investment + M&O cost

$$= 28,400,000(A/P,8\%,20)$$
$$+ 1,060,419$$
$$= 2,892,540 + 1,060,419$$
$$= \$3,952,959$$

Average monthly household cost to serve 95% of 4980 households is

$$\text{Household cost} = (3,952,959)\left(\frac{1}{12}\right)\left(\frac{1}{4980}\right)\left(\frac{1}{0.95}\right)$$
$$= \$69.63 \text{ per month}$$

Alternative 3

Total capital investment	= $29,600,000
Total annual M&O cost	= $867,119
Total annual cost	= 29,600,000(A/P,8%,20)
	+ 867,119
	= 3,014,760 + 867,119
	= $3,881,879
Household cost	= $68.38 per month

Alternative 4

Total capital investment	= $29,000,000
Total annual M&O cost	= $1,063,449
Total annual cost	= 29,000,000(A/P,8%,20)
	+ 1,063,449
	= 2,953,650 + 1,063,449
	= $4,017,099
Household cost	= $70.76 per month

On the basis of the lowest monthly household cost, alternative 3 (joint city and county plant) is the most economically attractive.

Case Study Exercises

1. If the environmental considerations factor is to have a weighting of twice as much as any of the other five factors, what is its percentage weighting?

2. If the ability to supply area and relative cost factors were each weighted 20% and the other four factors 15% each, which alternatives would be ranked in the top three?

3. By how much would the capital investment of alternative 4 have to decrease to make it more attractive than alternative 3?

4. If alternative 1A served 100% of the households instead of 95%, by how much would the monthly household cost decrease?

5. (a) Perform a sensitivity analysis on the two parameters of M&O costs and number of households to determine if alternative 3 remains the best economic choice. Three estimates are made for each parameter in Table 18–6. M&O costs may vary up

TABLE 18–6	Pessimistic, Most Likely, and Optimistic Estimates for Two Parameters	
	Annual M&O Costs	**Number of Households**
Alternative 1A		
Pessimistic	+1%	4980
Most likely	$1,060,419	+2%
Optimistic	−1%	+5%
Alternative 3		
Pessimistic	+5%	4980
Most likely	$867,119	+2%
Optimistic	0%	+5%
Alternative 4		
Pessimistic	+2%	4980
Most likely	$1,063,449	+2%
Optimistic	−10%	+5%

(pessimistic) or down (optimistic) from the most likely estimates presented in the case statement. The estimated number of households (4980) is determined to be the pessimistic estimate. Growth of 2% up to 5% (optimistic) will tend to lower the monthly cost per household.

(b) Consider the monthly cost per household for alternative 4, the optimistic estimate. The number of households is 5% above 4980, or 5230. What is the number of households that would have to be available in order for this option to have exactly the same monthly household cost as that for alternative 3 at the optimistic estimate of 5230 households?

CHAPTER 19

More on Variation and Decision Making under Risk

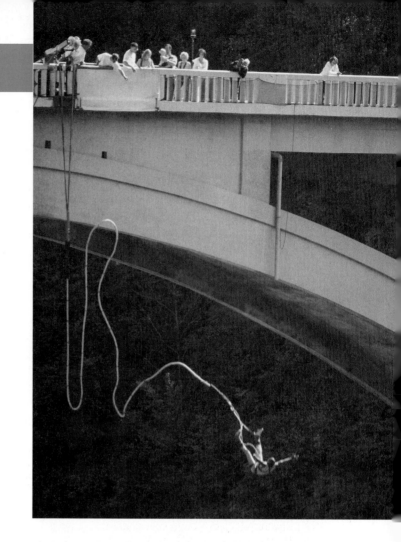

LEARNING OUTCOMES

Purpose: Incorporate decision making under risk into an engineering economy evaluation using probability, sampling, and simulation.

SECTION	TOPIC	LEARNING OUTCOME
19.1	Risk versus certainty	• Understand the approaches to decision making under risk and certainty.
19.2	Probability and distributions	• Construct a probability distribution and cumulative distribution for one variable.
19.3	Random sample	• Obtain a random sample from a cumulative distribution using a random number table.
19.4	μ, σ, and σ^2	• Estimate the population expected value, standard deviation, and variance from a random sample.
19.5	Simulation	• Use Monte Carlo sampling and spreadsheet-based simulation for alternative evaluation.

This chapter further expands our ability to analyze variation in estimates, to consider probability, and to **make decisions under risk.** Fundamentals discussed include variables; probability distributions, especially their graphs and properties of expected value and dispersion; random sampling; and the use of simulation to account for estimate variation in engineering economy studies.

Through coverage of variation and probability, this chapter complements topics in the first sections of Chapter 1: the role of engineering economy in decision making and economic analysis in the problem-solving process. These techniques are more time-consuming than using estimates made with certainty, so they should be used primarily for critical parameters.

19.1 Interpretation of Certainty, Risk, and Uncertainty ● ● ●

All things in the world vary—one from another, over time, and with different environments. We are guaranteed that variation will occur in engineering economy due to its emphasis on decision making for the future. Except for the use of breakeven analysis, sensitivity analysis, and a very brief introduction to expected values, virtually all our estimates have been **certain; that is, no variation** in the amount has entered into the computations of PW, AW, ROR, or any relations used. For example, the estimate that cash flow next year will be $+4500 is one of certainty. Decision making under certainty is, of course, not present in the real world now and surely not in the future. We can observe outcomes with a high degree of certainty, but even this depends upon the accuracy and precision of the scale or measuring instrument.

To allow a parameter of an engineering economy study to vary implies that *risk*, and possibly *uncertainty*, is introduced.

> When there may be two or more observable values for a parameter *and* it is possible to estimate the chance that each value may occur, **risk** is present. Virtually all decision making is performed *under risk*.

Risk

As an illustration, decision making under risk is introduced when an annual cash flow estimate has a 50-50 chance of being either $−1000 or $+500.

> Decision making under **uncertainty** means there are two or more values observable, but the chances of their occurring cannot be estimated or no one is willing to assign the chances. The observable values in uncertainty analysis are often referred to as *states of nature*.

For example, consider the states of nature to be the rate of national inflation in a particular country during the next 2 to 4 years: remain low, increase 2% to 6% annually, or increase 6% to 8% annually. If there is absolutely no indication that the three values are equally likely, or that one is more likely than the others, this is a statement that indicates decision making under uncertainty.

Example 19.1 explains how a parameter can be described and graphed to prepare for decision making under risk.

EXAMPLE 19.1

CMS in Fairfield, Virginia received three bids each from vendors for two different pieces of large equipment, A and B. One of each piece of equipment must be purchased. Tom, an engineer at CMS, performed an evaluation of each bid and assigned it a rating between 0 and 100, with 100 points being the best of the three. The total for each piece of equipment is 100%. The bid amounts and ratings are shown at the top of Figure 19–1.

(*a*) Consider the ratings as the chance out of 100 that the bid will be chosen, and plot cost versus chance for each vendor.

(*b*) Since one each of A and B must be purchased, the total cost will vary somewhere between the sum of the lowest bids ($11 million) and the sum of the highest bids ($25 million). Plot this range with an equal chance of 1 in 14 that any amount in between these limits is possible.

(*c*) Discuss the significant difference between the values of the cost (*x* axis values) in the graphs in (*a*) and (*b*) above and how the chances are stated (*y* axis values).

Figure 19–1

Plot of cost estimates versus chance for (*a*) each piece of equipment and (*b*) total cost range, Example 19.1.

Equipment A		Equipment B	
Bid, $1000	**Rating, %**	**Bid, $1000**	**Rating, %**
3,000	65	8,000	33.3
5,000	25	10,000	33.3
10,000	10	15,000	33.3

(*a*) Specific values

(*b*) Continuous range

Solution

(*a*) Figure 19–1*a* plots the specific bids for equipment A and B. The chances (ratings) for A and for B add to 100%. No values between the specific bids have any chance of occurring, according to the single-estimate bids from the three vendors.

(*b*) The range of total cost is between $11 million and $25 million, as shown in Figure 19–1*b*. Tom decided to make his estimate of total cost continuous between these two extremes. This means that the discrete sums of bids ($11 million, $15 million, and $25 million) are no longer used. Rather the entire range from $11 million to $25 million with a chance for every total cost in between is included. Every value has a chance of 1 in 14 of being observed. Now, the sum is a continuous value.

(*c*) In the graph for bid values (Figure 19–1*a*), only specific or discrete estimates are included on the *x* axis. In the graph for the sum of the cost for equipment A and B (Figure 19–1*b*), the *y* axis values are continuous over a specific range.

In the next section, the term *variable* is defined and two types of variables are explained—*discrete* and *continuous*—as illustrated here in an elementary form.

Before initiating an engineering economy study, it is important to decide if the analysis will be conducted with certainty for all parameters or if risk will be introduced. A summary of the meaning and use for each type of analysis follows.

Decision Making under Certainty This is what we have done in most analyses thus far. Deterministic estimates are made and entered into measure of worth relations—PW, AW, FW, ROR, B/C—and decision making is based on the results. The values estimated can be considered the most likely to occur with all chance placed on the single-value estimate. A typical example is an asset's first cost estimate made with certainty, say, $P = \$50,000$. A plot of P versus chance has the general form of Figure 19–1a with one vertical bar at \$50,000 and 100% chance placed on it. The term *deterministic,* in lieu of *certainty,* is often used when **single-value** or **single-point estimates** are used exclusively.

In fact, sensitivity analysis using different values of an estimate is simply another form of analysis with certainty, except that the analysis is repeated with different values, *each estimated with certainty.* The resulting measure of worth values are calculated and graphically portrayed to determine the decision's sensitivity to different estimates for one or more parameters.

Decision Making under Risk Now the element of chance is formally taken into account. However, it is more difficult to make a clear decision because the analysis attempts to accommodate **variation.** One or more parameters in an alternative will be allowed to vary. The estimates will be expressed as in Example 19.1 or in slightly more complex forms. Fundamentally, there are two ways to consider risk in an analysis:

Expected value analysis. Use the chance and parameter estimates to calculate expected values $E(\text{parameter})$ via formulas such as Equation [18.2]. Analysis results in $E(\text{cash flow})$, $E(\text{AOC})$, and the like; and the final result is the expected value for a measure of worth, such as $E(\text{PW})$, $E(\text{AW})$, $E(\text{ROR})$, $E(\text{B/C})$. To select the alternative, choose the most favorable expected value of the measure of worth. In an elementary form, this is what we learned about expected values in Chapter 18. The computations may become more elaborate, but the principle is fundamentally the same.

Simulation analysis. Use the chance and parameter estimates to generate repeated computations of the measure of worth relation by randomly sampling from a plot for each varying parameter similar to those in Figure 19–1. When a representative and random sample is complete, an alternative is selected utilizing a table or plot of the results. Usually, graphics are an important part of decision making via simulation analysis. Basically, this is the approach discussed in the rest of this chapter.

Decision Making under Uncertainty When chances are not known for the identified states of nature (or values) of the uncertain parameters, the use of expected value–based decision making under risk as outlined above is *not an option.* In fact, it is difficult to determine what criterion to use to even make the decision. If it is possible to agree that each state is equally likely, then all states have the same chance, and the situation reduces to one of decision making under risk, because expected values can be determined. Because of the relatively inconclusive approaches necessary to incorporate decision making under uncertainty into an engineering economy study, the techniques can be quite useful but are beyond the intended scope of this text.

In an engineering economy study, observed parameter values will vary from the value estimated at the time of the study. However, when performing the analysis, not all parameters should be considered as probabilistic (or at risk). Those that are estimable with a relatively high degree of certainty should be fixed for the study. Accordingly, the methods of sampling, simulation, and statistical data analysis are selectively used on parameters deemed important to the decision-making process. Parameters such as P, AOC, material and unit costs, sales price, revenues, etc., are the targets of decision making under risk. Anticipated variation in interest rates is more commonly addressed by sensitivity analysis.

The remainder of this chapter concentrates on decision making under risk as applied in an engineering economy study. Sections 19.2 to 19.4 provide foundation material necessary to design and correctly conduct a simulation analysis (Section 19.5).

Figure 19–2
(*a*) Discrete and continuous variable scales and (*b*) scales for a variable versus its probability.

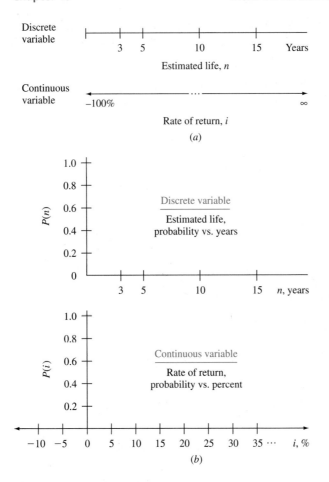

19.2 Elements Important to Decision Making under Risk ●●●

Some basics of probability and statistics are essential to correctly perform decision making under risk via expected value or simulation analysis. They are the *random variable, probability, probability distribution*, and *cumulative distribution*, as defined here. (If you are already familiar with them, this section will provide a review.)

A **random variable** or **variable** is a characteristic or parameter that can take on any one of several values. Variables are classified as *discrete* or *continuous*. Discrete variables have several specific, isolated values, while continuous variables can assume any value between two stated limits, called the *range* of the variable.

The estimated life of an asset is a discrete variable. For example, n may be expected to have values of $n = 3$, 5, 10, or 15 years, and no others. The rate of return is an example of a continuous variable; i can vary from -100% to ∞, that is, $-100\% \leq i < \infty$. The ranges of possible values for n (discrete) and i (continuous) are shown as the x axes in Figure 19–2a. (In probability texts, capital letters symbolize a variable, say X, and small letters x identify a specific value of the variable. Though correct, this level of rigor in terminology is not applied in this chapter.)

Probability is a number between 0 and 1.0 that expresses the chance in decimal form that a random variable (discrete or continuous) will take on any value from those identified for it. Probability is simply the amount of chance, divided by 100.

Probabilities are commonly identified by $P(X_i)$ or $P(X = X_i)$, which is read as the probability that the variable X takes on the value X_i. (Actually, for a continuous variable, the probability at a single value is zero, as shown in a later example.) The sum of all $P(X_i)$ for a variable must be 1.0, a requirement already discussed. The probability scale, like the percentage scale for chance in Figure 19–1, is indicated on the ordinate (y axis) of a graph. Figure 19–2b shows the 0 to 1.0 range of probability for the variables n and i.

A **probability distribution** describes how probability is distributed over the different values of a variable. Discrete variable distributions look significantly different from continuous variable distributions, as indicated by the inset at the right.

The individual probability values are stated as

$$P(X_i) = \text{probability that } X \text{ equals } X_i \qquad [19.1]$$

The distribution may be developed in one of two ways: by listing each probability value for each possible variable value (see Example 19.2) or by a mathematical description or expression that states probability in terms of the possible variable values (Example 19.3).

Cumulative distribution, also called the **cumulative probability distribution,** is the accumulation of probability over all values of a variable up to and including a specified value.

Identified by $F(X_i)$, each cumulative value is calculated as

$$\begin{aligned} F(X_i) &= \text{sum of all probabilities through the value } X_i \\ &= P(X \le X_i) \end{aligned} \qquad [19.2]$$

As with a probability distribution, cumulative distributions appear differently for discrete (stair-stepped) and continuous variables (smooth curve). Examples 19.2 and 19.3 illustrate cumulative distributions that correspond to specific probability distributions. These fundamentals about $F(X_i)$ are applied in the next section to develop a random sample.

EXAMPLE 19.2

Alvin is a medical doctor and biomedical engineering graduate who practices at Medical Center Hospital. He is planning to start prescribing an antibiotic that may reduce infection in patients with flesh wounds. Tests indicate the drug has been applied up to 6 times per day without harmful side effects. If no drug is used, there is always a positive probability that the infection will be reduced by a person's own immune system.

Published drug test results provide good probability estimates of positive reaction (i.e., reduction in the infection count) within 48 hours for increased treatments per day. Use the probabilities listed below to construct a probability distribution and a cumulative distribution for the total number of treatments per day.

Number of Added Treatments per Day	Probability of Infection Reduction for Each Added Treatment
0	0.07
1	0.08
2	0.10
3	0.12
4	0.13
5	0.25
6	0.25

Solution

Define the random variable T as the number of added treatments per day. Since T can take on only seven different values, it is a **discrete variable.** The probability of infection reduction is listed for each value in column 2 of Table 19–1. The cumulative probability $F(T_i)$ is determined using Equation [19.2] by adding all $P(T_i)$ values through T_i, as indicated in column 3.

Figure 19–3a and b shows plots of the probability distribution and cumulative distribution, respectively. The summing of probabilities to obtain $F(T_i)$ gives the cumulative distribution the stair-stepped appearance, and in all cases the final $F(T_i) = 1.0$, since the total of all $P(T_i)$ values must equal 1.0.

TABLE 19–1	Probability Distribution and Cumulative Distribution for Example 19.2	
(1)	(2)	(3)
Number per Day T_i	Probability $P(T_i)$	Cumulative Probability $F(T_i)$
0	0.07	0.07
1	0.08	0.15
2	0.10	0.25
3	0.12	0.37
4	0.13	0.50
5	0.25	0.75
6	0.25	1.00

Figure 19–3
(*a*) Probability distribution $P(T_i)$ and (*b*) cumulative distribution $F(T_i)$ for Example 19.2.

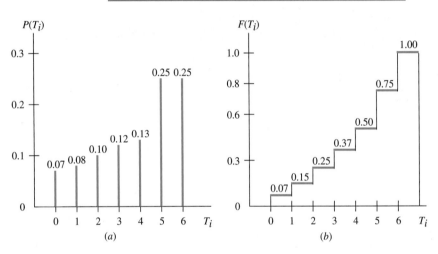

Comment

Rather than use a tabular form as in Table 19–1 to state $P(T_i)$ and $F(T_i)$ values, it is possible to express them for each value of the variable.

$$P(T_i) = \begin{cases} 0.07 & T_1 = 0 \\ 0.08 & T_2 = 1 \\ 0.10 & T_3 = 2 \\ 0.12 & T_4 = 3 \\ 0.13 & T_5 = 4 \\ 0.25 & T_6 = 5 \\ 0.25 & T_7 = 6 \end{cases} \qquad F(T_i) = \begin{cases} 0.07 & T_1 = 0 \\ 0.15 & T_2 = 1 \\ 0.25 & T_3 = 2 \\ 0.37 & T_4 = 3 \\ 0.50 & T_5 = 4 \\ 0.75 & T_6 = 5 \\ 1.00 & T_7 = 6 \end{cases}$$

In basic engineering economy situations, the probability distribution for a **continuous variable** is commonly expressed as a mathematical function, such as a *uniform distribution,* a *triangular distribution* (both discussed in Example 19.3 in terms of cash flow), or the more complex, but commonly used, *normal distribution.* For continuous variable distributions, the symbol $f(X)$ is routinely used instead of $P(X_i)$, and $F(X)$ is used instead of $F(X_i)$, simply because the point probability for a continuous variable is zero. Thus, $f(X)$ and $F(X)$ are continuous lines and curves.

EXAMPLE 19.3

As president of a manufacturing systems consultancy, Sallie has observed the monthly cash flows that have occurred over the last 3 years into company accounts from two longstanding clients. Sallie has concluded the following about the distribution of these monthly cash flows:

Client 1

Estimated low cash flow: $10,000
Estimated high cash flow: $15,000
Most likely cash flow: same for all values
Distribution of probability: uniform

Client 2

Estimated low cash flow: $20,000
Estimated high cash flow: $30,000
Most likely cash flow: $28,000
Distribution of probability: mode at $28,000

The *mode* is the most frequently observed value for a variable. Sallie assumes cash flow to be a continuous variable referred to as C. (*a*) Write and graph the two probability distributions and cumulative distributions for monthly cash flow, and (*b*) determine the probability that monthly cash flow is no more than $12,000 for client 1 and at least $25,000 for client 2.

Solution

All cash flow values are expressed in $1000 units.

Client 1: monthly cash flow distribution

(*a*) The distribution of cash flows for client 1, identified by the variable C_1, follows the *uniform distribution*. Probability and cumulative probability take the following general forms.

$$f(C_1) = \frac{1}{\text{high} - \text{low}} \qquad \text{low value} \leq C_1 \leq \text{high value}$$

$$f(C_1) = \frac{1}{H - L} \qquad L \leq C_1 \leq H \qquad\qquad [19.3]$$

$$F(C_1) = \frac{\text{value} - \text{low}}{\text{high} - \text{low}} \qquad \text{low value} \leq C_1 \leq \text{high value}$$

$$F(C_1) = \frac{C_1 - L}{H - L} \qquad L \leq C_1 \leq H \qquad\qquad [19.4]$$

For client 1, monthly cash flow is uniformly distributed with $L = \$10$, $H = \$15$, and $\$10 \leq C_1 \leq \$15$. Figure 19–4 is a plot of $f(C_1)$ and $F(C_1)$ from Equations [19.3] and [19.4].

$$f(C_1) = \frac{1}{5} = 0.2 \qquad \$10 \leq C_1 \leq \$15$$

$$F(C_1) = \frac{C_1 - 10}{5} \qquad \$10 \leq C_1 \leq \$15$$

(*b*) The probability that client 1 has a monthly cash flow of no more than $12 is easily determined from the $F(C_1)$ plot as 0.4, or a 40% chance. If the $F(C_1)$ relation is used directly, the computation is

$$F(\$12) = P(C_1 \leq \$12) = \frac{12 - 10}{5} = 0.4$$

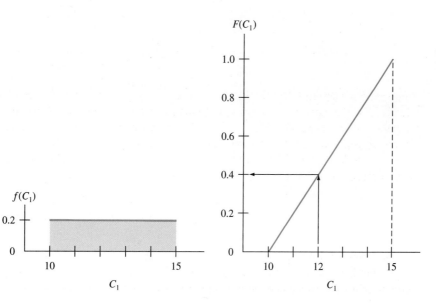

Figure 19–4
Uniform distribution for client 1 monthly cash flow, Example 19.3.

Client 2: monthly cash flow distribution

(a) The distribution of cash flows for client 2, identified by the variable C_2, follows the *triangular distribution*. This probability distribution has the shape of an upward-pointing triangle with the peak at the mode M, and downward-sloping lines joining the x axis on either side at the low (L) and high (H) values. The mode of the triangular distribution has the maximum probability value.

$$f(\text{mode}) = f(M) = \frac{2}{H - L} \qquad [19.5]$$

The cumulative distribution is comprised of two curved line segments from 0 to 1 with a break point at the mode, where

$$F(\text{mode}) = F(M) = \frac{M - L}{H - L} \qquad [19.6]$$

For C_2, the low value is $L = \$20$, the high is $H = \$30$, and the most likely cash flow is the mode $M = \$28$. The probability at M from Equation [19.5] is

$$f(28) = \frac{2}{30 - 20} = \frac{2}{10} = 0.2$$

The break point in the cumulative distribution occurs at $C_2 = 28$. Using Equation [19.6],

$$F(28) = \frac{28 - 20}{30 - 20} = 0.8$$

Figure 19–5 presents the plots for $f(C_2)$ and $F(C_2)$. Note that $f(C_2)$ is skewed, since the mode is not at the midpoint of the range $H - L$, and $F(C_2)$ is a smooth S-shaped curve with an inflection point at the mode.

(b) From the cumulative distribution in Figure 19–5, there is an estimated 31.25% chance that cash flow is $25 or less. Therefore,

$$F(\$30) - F(\$25) = P(C_2 \geq \$25) = 1 - 0.3125 = 0.6875$$

Comment

The general relations $f(C_2)$ and $F(C_2)$ are not developed here. The variable C_2 is *not* a uniform distribution; it is triangular. Therefore, it requires the use of an integral to find cumulative probability values from the probability distribution $f(C_2)$.

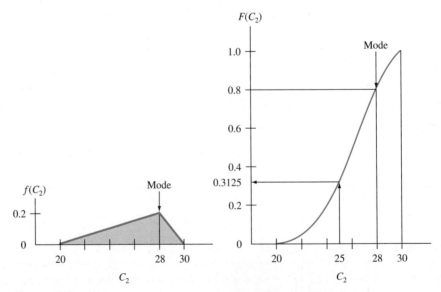

Figure 19–5
Triangular distribution for client 2 monthly cash flow, Example 19.3.

19.3 Random Samples ● ● ◦

Estimating a parameter with a single value in previous chapters is the equivalent of taking a *random sample of size 1 from an entire population* of possible values. As an illustration, assume that estimates of first cost, annual operating cost, interest rate, and other parameters are used to compute one PW value in order to accept or reject an alternative. Each estimate is a sample of size 1 from an entire population of possible values for each parameter. Now, if a second estimate is made for each parameter and a second PW value is determined, a sample of size 2 has been taken. If all values in the population were known, the probability distribution and cumulative distribution would be known. Then a sample would not be necessary.

When we perform an engineering economy study and utilize decision making under certainty, we use one estimate for each parameter to calculate a measure of worth (i.e., a sample of size 1 for each parameter). The estimate is the most likely value, that is, one estimate of the expected value. We know that all parameters will vary somewhat; yet some are important enough, or will vary enough, that a probability distribution should be determined or assumed for it and the parameter treated as a random variable. This is using risk, and a sample from the parameter's probability distribution—$P(X)$ for discrete or $f(X)$ for continuous—helps formulate probability statements about the estimates. This approach complicates the analysis somewhat; however, it also provides a sense of confidence (or possibly a lack of confidence in some cases) about the decision made concerning the economic viability of the alternative based on the varying parameter. (We will further discuss this aspect later, after we learn how to correctly take a random sample from any probability distribution.)

> A **random sample** of size n is the selection in a random fashion of n values from a population with an assumed or known probability distribution, such that the values of the variable have the **same chance of occurring** in the sample as they are expected to occur in the population.

Suppose Yvon is an engineer with 20 years of experience working for the Aircraft Safety Commission. For a two-crew aircraft, there are three parachutes on board. The safety standard states that 99% of the time, all three chutes must be "fully ready for emergency deployment." Yvon is relatively sure that nationwide the probability distribution of N, the specific number of chutes fully ready, may be described by the probability distribution

$$P(N = N_i) = \begin{cases} 0.005 & N = 0 \text{ chutes ready} \\ 0.015 & N = 1 \text{ chute ready} \\ 0.060 & N = 2 \text{ chutes ready} \\ 0.920 & N = 3 \text{ chutes ready} \end{cases}$$

This means that the safety standard is clearly not met nationwide. Yvon is in the process of sampling 200 (randomly selected) corporate and private aircraft across the nation to determine how many chutes are classified as fully ready. If the sample is truly random and Yvon's probability distribution is a correct representation of actual parachute readiness, the observed N values in the 200 aircraft will approximate the same proportions as the population probabilities, that is, 1 aircraft with 0 chutes ready, etc. Since this is a sample, it is likely that the results won't track the population exactly. However, if the results are relatively close, the study indicates that the sample results may be useful in predicting parachute safety across the nation.

To develop a random sample, use **random numbers (RN)** generated from a uniform probability distribution for the discrete numbers 0 through 9, that is,

$$P(X_i) = 0.1 \qquad \text{for } X_i = 0, 1, 2, \ldots, 9$$

In tabular form, the random digits so generated are commonly clustered in groups of two digits, three digits, or more. Table 19–2 is a sample of 264 random digits clustered into two-digit numbers. This format is very useful because the numbers 00 to 99 conveniently relate to the cumulative distribution values 0.01 to 1.00. This makes it easy to select a two-digit RN and enter $F(X)$ to determine a value of the variable with the same proportions as it occurs in the probability distribution. To apply this logic manually and develop a random sample of size n from a known

TABLE 19–2	Random Digits Clustered into Two-Digit Numbers

51	82	88	18	19	81	03	88	91	46	39	19	28	94	70	76	33	15	64	20	14	52
73	48	28	59	78	38	54	54	93	32	70	60	78	64	92	40	72	71	77	56	39	27
10	42	18	31	23	80	80	26	74	71	03	90	55	61	61	28	41	49	00	79	96	78
45	44	79	29	81	58	66	70	24	82	91	94	42	10	61	60	79	30	01	26	31	42
68	65	26	71	44	37	93	94	93	72	84	39	77	01	97	74	17	19	46	61	49	67
75	52	14	99	67	74	06	50	97	46	27	88	10	10	70	66	22	56	18	32	06	24

discrete probability distribution $P(X)$ or a continuous variable distribution $f(X)$, the following procedure may be used.

1. Develop the cumulative distribution $F(X)$ from the probability distribution. Plot $F(X)$.
2. Assign the RN values from 00 to 99 to the $F(X)$ scale (the y axis) in the same proportion as the probabilities. For the parachute safety example, the probabilities from 0.0 to 0.15 are represented by the random numbers 00 to 14. Indicate the RNs on the graph.
3. To use a table of random numbers, determine the scheme or sequence of selecting RN values—down, up, across, diagonally. Any direction and pattern is acceptable, but the scheme should be used consistently for one entire sample.
4. Select the first number from the RN table, enter the $F(X)$ scale, and observe and record the corresponding variable value. Repeat this step until there are n values of the variable that constitute the random sample.
5. Use the n sample values for analysis and decision making under risk. These may include

   - Plotting the sample probability distribution.
   - Developing probability statements about the parameter.
   - Comparing sample results with the assumed population distribution.
   - Determining sample statistics (Section 19.4).
   - Performing a simulation analysis (Section 19.5).

EXAMPLE 19.4

Develop a random sample of size 10 for the variable N, number of months, as described by the probability distribution

$$P(N = N_i) = \begin{cases} 0.20 & N = 24 \\ 0.50 & N = 30 \\ 0.30 & N = 36 \end{cases} \qquad [19.7]$$

Solution

Apply the procedure above, using the $P(N = N_i)$ values in Equation [19.7].

1. The cumulative distribution, Figure 19–6, is for the discrete variable N, which can assume three different values.
2. Assign 20 numbers (00 through 19) to $N_1 = 24$ months, where $P(N = 24) = 0.2$; 50 numbers to $N_2 = 30$; and 30 numbers to $N_3 = 36$.
3. Initially select any position in Table 19–2, and go across the row to the right and onto the row below toward the left. (Any routine can be developed, and a different sequence for each random sample may be used.)
4. Select the initial number 45 (4th row, 1st column), and enter Figure 19–6 in the RN range of 20 to 69 to obtain $N = 30$ months.
5. Select and record the remaining nine values from Table 19–2 as shown below.

RN	45	44	79	29	81	58	66	70	24	82
N	30	30	36	30	36	30	30	36	30	36

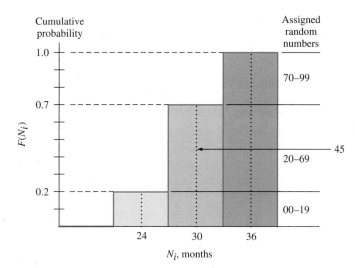

Figure 19–6
Cumulative distribution with random number values assigned in proportion to probabilities, Example 19.4.

Now, using the 10 values, develop the sample probabilities.

Months N	Times in Sample	Sample Probability	Equation [19.7] Probability
24	0	0.00	0.2
30	6	0.60	0.5
36	4	0.40	0.3

With only 10 values, we can expect the sample probability estimates to be different from the values in Equation [19.7]. Only the value $N = 24$ months is significantly different, since no RN of 19 or less occurred. A larger sample will definitely make the probabilities closer to the original data.

To take a *random sample of size n for a continuous variable,* the procedure above is applied, except the random number values are assigned to the cumulative distribution on a continuous scale of 00 to 99 corresponding to the $F(X)$ values. As an illustration, consider Figure 19–4, where C_1 is the *uniformly distributed* cash flow variable for client 1 in Example 19.3. Here $L = \$10$, $H = \$15$, and $f(C_1) = 0.2$ for all values between L and H (all values are divided by $\$1000$). The $F(C_1)$ is repeated as Figure 19–7 with the assigned random number values shown on the right scale. If the two-digit RN of 45 is chosen, the corresponding C_1 is graphically estimated to be $\$12.25$. It can also be linearly interpolated as $\$12.25 = 10 + (45/100)(15 - 10)$.

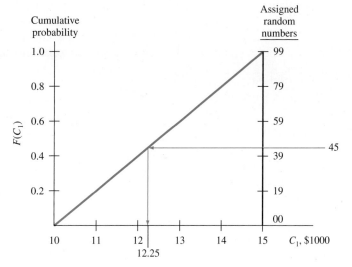

Figure 19–7
Random numbers assigned to the continuous variable of client 1 cash flows in Example 19.3.

For greater accuracy when developing a random sample, especially for a continuous variable, it is possible to use 3-, 4-, or 5-digit RNs. These can be developed from Table 19–2 simply by combining digits in the columns and rows or obtained from tables with RNs printed in larger clusters of digits. In computer-based sampling, most simulation software packages have an RN generator built in that will generate values in the range of 0 to 1 from a continuous variable uniform distribution, usually identified by the symbol $U(0, 1)$. The RN values, usually between 0.00000 and 0.99999, are used to sample directly from the cumulative distribution employing essentially the same procedure explained here. The Excel functions RAND and RANDBETWEEN are described in Appendix A, Section A.3.

An initial question in random sampling usually concerns the **minimum size of n** required to ensure confidence in the results. Without detailing the mathematical logic, sampling theory, which is based upon the law of large numbers and the central limit theorem (check a basic statistics book to learn about these), indicates that an n of 30 is sufficient. However, since reality does not follow theory exactly, and since engineering economy often deals with sketchy estimates, samples in the *range of 100 to 200* are the common practice. But samples as small as 10 to 25 provide a much better foundation for decision making under risk than the single-point estimate for a parameter that is known to vary widely.

19.4 Expected Value and Standard Deviation ● ● ●

Two very important measures or properties of a random variable are the expected value and standard deviation. If the entire population for a variable were known, these properties would be calculated directly. Since they are usually not known, random samples are commonly used to estimate them via the sample mean and the sample standard deviation, respectively. The following is a brief introduction to the interpretation and calculation of these properties using a random sample of size n from the population.

The usual symbols are Greek letters for the true population measures and English letters for the sample estimates.

	True Population Measure		Sample Estimate	
	Symbol	Name	Symbol	Name
Expected value	μ or $E(X)$	Mu or true mean	$\overline{X}$	Sample mean
Standard deviation	σ or $\sqrt{\mathrm{Var}(X)}$ or $\sqrt{\sigma^2}$	Sigma or true standard deviation	s or $\sqrt{s^2}$	Sample standard deviation

The **expected value** *E(X)* is the long-run expected average if the variable is sampled many times.

The population expected value is not known exactly, since the population itself is not known completely, so μ is estimated either by $E(X)$ from a distribution or by $\overline{X}$, the sample mean. Equation [18.2], repeated here as Equation [19.8], is used to compute the $E(X)$ of a probability distribution, and Equation [19.9] is the sample mean, also called the *sample average*.

Population:	μ	
Probability distribution:	$E(X) = \Sigma X_i P(X_i)$	[19.8]
Sample:	$\overline{X} = \dfrac{\text{sum of sample values}}{\text{sample size}}$	
	$= \dfrac{\Sigma X_i}{n} = \dfrac{\Sigma f_i X_i}{n}$	[19.9]

The f_i in the second form of Equation [19.9] is the frequency of X_i, that is, the number of times each value occurs in the sample. The resulting $\overline{X}$ is not necessarily an observed value of the variable; it is the long-run average value and can take on any value within the range of the variable. (We omit the subscript i on X and f when there is no confusion introduced.)

EXAMPLE 19.5

Kayeu, an engineer with Pacific NW Utilities, is planning to test several hypotheses about residential electricity bills in North American and Asian countries. The variable of interest is X, the monthly residential bill in U.S. dollars (rounded to the nearest dollar). Two small samples have been collected from different countries of North America and Asia. Estimate the population expected value. Do the samples (from a nonstatistical viewpoint) appear to be drawn from one population of electricity bills or from two different populations?

North American, Sample 1, $	40	66	75	92	107	159	275
Asian, Sample 2, $	84	90	104	187	190		

Solution

Use Equation [19.9] for the sample mean.

Sample 1: $n = 7$ $\Sigma X_i = 814$ $\overline{X} = \$116.29$
Sample 2: $n = 5$ $\Sigma X_i = 655$ $\overline{X} = \$131.00$

Based solely on the small sample averages, the approximate $15 difference, which is only 11% of the larger average bill, does not seem sufficiently large to conclude that the two populations are different. There are several statistical tests available to determine if samples come from the same or different populations. (Check a basic statistics text to learn about them.)

Comment

There are three commonly used measures of central tendency for data. The sample average is the most popular, but the *mode* and the *median* are also good measures. The mode, which is the most frequently observed value, was utilized in Example 19.3 for a triangular distribution. There is no specific mode in Kayeu's two samples, since all values are different. The *median is the middle value* of the sample. It is not biased by extreme sample values, as is the mean. The two medians in the samples are $92 and $104. Based solely on the medians, the conclusion is still that the samples do not necessarily come from two different populations of electricity bills.

The **standard deviation** s or $s(X)$ is the dispersion or spread of values **about the expected value** $E(X)$ or sample average $\overline{X}$.

The sample standard deviation s estimates the property σ, which is the population measure of dispersion about the expected value of the variable. A probability distribution for data with strong central tendency is more closely clustered about the center of the data, and has a smaller s, than a wider, more dispersed distribution. In Figure 19–8, the samples with larger s values—s_1 and s_4—have a flatter, wider probability distribution.

Figure 19–8
Sketches of distributions with different separate lines standard deviation values.

Actually, the variance s^2 is often quoted as the measure of dispersion. The standard deviation is simply the square root of the variance, so either measure can be used. The s value is what we use routinely in making computations about risk and probability. Mathematically, the formulas and symbols for variance and standard deviation of a discrete variable and a random sample of size n are as follows:

$$\text{Population:} \quad \sigma^2 = \text{Var}(X) \quad \text{and} \quad \sigma = \sqrt{\sigma^2} = \sqrt{\text{Var}(X)}$$

$$\text{Probability distribution:} \quad \text{Var}(X) = \Sigma[X_i - E(X)]^2 P(X_i) \qquad [19.10]$$

$$\text{Sample:} \quad s^2 = \frac{\text{sum of (sample value} - \text{sample average)}^2}{\text{sample size} - 1}$$

$$= \frac{\Sigma(X_i - \overline{X})^2}{n - 1} \qquad [19.11]$$

$$s = \sqrt{s^2}$$

Equation [19.11] for sample variance is usually applied in a more computationally convenient form.

$$s^2 = \frac{\Sigma X_i^2}{n - 1} - \frac{n}{n - 1}\overline{X}^2 = \frac{\Sigma f_i X_i^2}{n - 1} - \frac{n}{n - 1}\overline{X}^2 \qquad [19.12]$$

The standard deviation uses the sample average as a basis about which to measure the spread or dispersion of data via the calculation $(X - \overline{X})$, which can have a minus or plus sign. To accurately measure the dispersion in both directions from the average, the quantity $(X - \overline{X})$ is squared. To return to the dimension of the variable itself, the square root of Equation [19.11] is extracted. The term $(X - \overline{X})^2$ is called the *mean-squared deviation,* and s has historically also been referred to as the *root-mean-square deviation.* The f_i in the second form of Equation [19.12] uses the frequency of each X_i value to calculate s^2.

One simple way to combine the average and standard deviation is to determine the percentage or fraction of the sample that is within ± 1, ± 2, or ± 3 standard deviations of the average, that is,

$$\overline{X} \pm ts \qquad \text{for } t = 1, 2, \text{ or } 3 \qquad [19.13]$$

In probability terms, this is stated as

$$P(\overline{X} - ts \leq X \leq \overline{X} + ts) \qquad [19.14]$$

Virtually all the sample values will always be within the $\pm 3s$ range of $\overline{X}$, but the percent within $\pm 1s$ will vary depending on how the data points are distributed about $\overline{X}$. Example 19.6 illustrates the calculation of s to estimate σ and incorporates s with the sample average using $\overline{X} \pm ts$.

EXAMPLE 19.6

(a) Use the two samples of Example 19.5 to estimate population variance and standard deviation for electricity bills. (b) Determine the percentages of each sample that are inside the ranges of 1 and 2 standard deviations from the mean.

Solution

(a) For illustration purposes only, apply the two different relations to calculate s for the two samples. For sample 1 (North American) with $n = 7$, use X to identify the values. Table 19–3 presents the computation of $\Sigma(X - \overline{X})^2$ for Equation [19.11], with $\overline{X} = \$116.29$. The resulting s^2 and s values are

$$s^2 = \frac{37,743.40}{6} = 6290.57$$

$$s = \$79.31$$

TABLE 19–3	Computation of Standard Deviation Using Equation [19.11] with $\overline{X}$ = $116.29, Example 19.6	
X, $	**$X - \overline{X}$**	**$(X - \overline{X})^2$**
40	−76.29	5,820.16
66	−50.29	2,529.08
75	−41.29	1,704.86
92	−24.29	590.00
107	−9.29	86.30
159	+42.71	1,824.14
275	+158.71	25,188.86
814		37,743.40

TABLE 19–4	Computation of Standard Deviation Using Equation [19.12] with $\overline{Y}$ = $131, Example 19.6
Y, $	**Y^2**
84	7,056
90	8,100
104	10,816
187	34,969
190	36,100
655	97,041

For sample 2 (Asian), use Y to identify the values. With $n = 5$ and $\overline{Y} = 131$, Table 19–4 shows ΣY^2 for Equation [19.12]. Then

$$s^2 = \frac{97,041}{4} - \frac{5}{4}(131)^2 = 42,260.25 - 1.25(17,161) = 2809$$

$$s = \$53$$

The dispersion is smaller for the Asian sample ($53) than for the North American sample ($79.31).

(b) Equation [19.13] determines the ranges of $\overline{X} \pm 1s$ and $\overline{X} \pm 2s$. Count the number of sample data points between the limits, and calculate the corresponding percentage. See Figure 19–9 for a plot of the data and the standard deviation ranges.

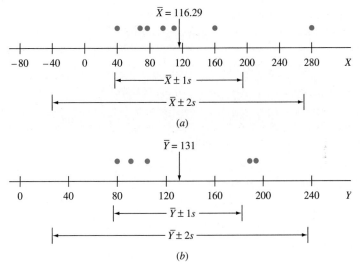

(a)

(b)

Figure 19–9
Values, averages, and standard deviation ranges for (a) North American and (b) Asian samples, Example 19.6.

North American sample

$$\overline{X} \pm 1s = 116.29 \pm 79.31 \qquad \text{for a range of \$36.98 to \$195.60}$$

Six out of seven values are within this range, so the percentage is 85.7%.

$$\overline{X} \pm 2s = 116.29 \pm 158.62 \qquad \text{for a range of \$-42.33 to \$274.91}$$

There are still six of the seven values within the $\overline{X} \pm 2s$ range. The limit \$-42.33 is meaningful only from the probabilistic perspective; from the practical viewpoint, use zero, that is, no amount billed.

Asian sample

$$\overline{Y} \pm 1s = 131 \pm 53 \qquad \text{for a range of \$78 to \$184}$$

There are three of five values, or 60%, within the range.

$$\overline{Y} \pm 2s = 131 \pm 106 \qquad \text{for a range of \$25 to \$237}$$

All five of the values are within the $\overline{Y} \pm 2s$ range.

Comment

A second common measure of dispersion is the *range,* which is simply the largest minus the smallest sample values. In the two samples here, the range estimates are \$235 and \$106.

Only the hand computations for $E(X)$, s, and s^2 have been demonstrated here. Calculators and spreadsheets all have functions to determine these values by simply entering the data.

Before we perform simulation analysis in engineering economy, it may be of use to summarize the expected value and standard deviation relations for a continuous variable, since Equations [19.8] through [19.12] address only discrete variables. The primary differences are that the summation symbol is replaced by the integral over the defined range of the variable, which we identify as R, and that $P(X)$ is replaced by the differential element $f(X)\,dX$. For a stated continuous probability distribution $f(X)$, the formulas are

Expected value: $\qquad\qquad E(X) = \int_R Xf(X)\,dX \qquad\qquad\qquad$ [19.15]

Variance: $\qquad\qquad \text{Var}(X) = \int_R X^2 f(X)\,dX - [E(X)]^2 \qquad$ [19.16]

For a numerical example, again use the uniform distribution in Example 19.3 (Figure 19–4) over the range R from \$10 to \$15. If we identify the variable as X, rather than C_1, the following are correct.

$$f(X) = \frac{1}{5} = 0.2 \qquad \$10 \le X \le \$15$$

$$E(X) = \int_R X(0.2)\,dX = 0.1X^2 \Big|_{10}^{15} = 0.1(225 - 100) = \$12.5$$

$$\text{Var}(X) = \int_R X^2(0.2)\,dX - (12.5)^2 = \frac{0.2}{3}X^3 \Big|_{10}^{15} - (12.5)^2$$

$$= 0.06667(3375 - 1000) - 156.25 = 2.08$$

$$\sigma = \sqrt{2.08} = \$1.44$$

Therefore, the uniform distribution between $L = \$10$ and $H = \$15$ has an expected value of \$12.5 (the midpoint of the range, as expected) and a standard deviation of \$1.44.

EXAMPLE 19.7

Christy is the regional safety engineer for a chain of franchise-based gasoline and food stores. The home office has had many complaints and several legal actions from employees and customers about slips and falls due to liquids (water, oil, gas, soda, etc.) on concrete surfaces.

Corporate management has authorized each regional engineer to contract locally to apply to all exterior concrete surfaces a newly marketed product that absorbs up to 100 times its own weight in liquid and to charge a home office account for the installation. The authorizing letter to Christy states that, based upon their simulation and random samples that assume a normal population, the cost of the locally arranged installation should be about $10,000 and almost always is within the range of $8000 to $12,000.

You have been asked to write a brief but thorough summary about the normal distribution, explain the $8000 to $12,000 range statement, and explain the phrase "random samples that assume a normal population."

Solution

The following summaries about the normal distribution and sampling will help explain the authorization letter.

Normal distribution, probabilities, and random samples

The normal distribution is also referred to as the *bell-shaped curve,* the *Gaussian distribution,* or the *error distribution.* It is, by far, the most commonly used probability distribution in all applications. It places exactly one-half of the probability on either side of the mean or expected value. It is used for continuous variables over the entire range of numbers. The normal distribution is found to accurately predict many types of outcomes, such as IQ values; manufacturing errors about a specified size, volume, weight, etc.; and the distribution of sales revenues, costs, and many other business parameters around a specified mean, which is why it may apply in this situation.

The normal distribution, identified by the symbol $N(\mu, \sigma^2)$, where μ is the expected value or mean and σ^2 is the variance, or measure of spread, can be described as follows:

- The mean μ locates the probability distribution (Figure 19–10a), and the spread of the distribution varies with variance (Figure 19–10b), growing wider and flatter for larger variance values.
- When a sample is taken, the estimates are identified as sample mean $\overline{X}$ for μ and sample standard deviation s for σ.
- The normal probability distribution $f(X)$ for a variable X is quite complicated, because its formula is

$$f(X) = \frac{1}{\sigma\sqrt{2\pi}} \exp\left\{-\left[\frac{(X - \mu)^2}{2\sigma^2}\right]\right\}$$

where exp represents the number $e = 2.71828$.

Since $f(X)$ is so unwieldy, random samples and probability statements are developed using a transformation, called the *standard normal distribution (SND),* which uses μ and σ (population) or $\overline{X}$ and s (sample) to compute values of the variable Z.

Population : $Z = \dfrac{\text{deviation from mean}}{\text{standard deviation}} = \dfrac{X - \mu}{\sigma}$ [19.17]

Sample: $Z = \dfrac{X - \overline{X}}{s}$ [19.18]

The SND for Z (Figure 19–10c) is the same as for X, except that it always has a mean of 0 and a standard deviation of 1, and it is identified by the symbol $N(0, 1)$. Therefore, the probability values under the SND curve can be stated exactly. It is always possible to transfer back to the original values from sample data by solving Equation [19.17] for X:

$$X = Z\sigma + \mu \qquad\qquad [19.19]$$

Figure 19–10
Normal distribution showing (*a*) different mean values μ, (*b*) different standard deviation values σ, and (*c*) relation of normal *X* to standard normal *Z*.

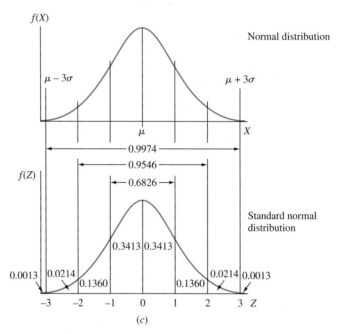

Several probability statements for *Z* and *X* are summarized in the following table and are shown on the distribution curve for *Z* in Figure 19–10*c*.

Variable *X* Range	Probability	Variable *Z* Range
$\mu + 1\sigma$	0.3413	0 to +1
$\mu \pm 1\sigma$	0.6826	−1 to +1
$\mu + 2\sigma$	0.4773	0 to +2
$\mu \pm 2\sigma$	0.9546	−2 to +2
$\mu + 3\sigma$	0.4987	0 to +3
$\mu \pm 3\sigma$	0.9974	−3 to +3

As an illustration, probability statements from this tabulation and Figure 19–10*c* for *X* and *Z* are as follows:

The probability that *X* is within 2σ of its mean is 0.9546.

The probability that *Z* is within 2σ of its mean, which is the same as between the values −2 and +2, is also 0.9546.

In order to take a random sample from a normal $N(\mu, \sigma^2)$ population, a specially prepared table of SND random numbers is used. (Tables of SND values are available in many statistics books.) The numbers are actually values from the Z or $N(0, 1)$ distribution and have values such as -2.10, $+1.24$, etc. Translation from the Z value back to the sample values for X is via Equation [19.19].

Interpretation of the home office memo

The statement that virtually all the local contract amounts should be between \$8000 and \$12,000 may be interpreted as follows: A normal distribution is assumed with a mean of $\mu = \$10,000$ and a standard deviation for $\sigma = \$667$, or a variance of $\sigma^2 = (\$667)^2$; that is, an $N[\$10,000, (\$667)^2]$ distribution is assumed. The value $\sigma = \$667$ is calculated using the fact that virtually all the probability (99.74%) is within 3σ of the mean, as stated above. Therefore,

$$3\sigma = \$2000 \quad \text{and} \quad \sigma = \$667 \quad \text{(rounded off)}$$

As an illustration, if six SND random numbers are selected and used to take a sample of size 6 from the normal distribution $N[\$10,000, (\$667)^2]$, the results are as follows:

SND Random Number Z	X Using Equation [19.19] $X = Z\sigma + \mu$
-2.10	$X = (-2.10)(667) + 10,000 = \$8599$
$+3.12$	$X = (+3.12)(667) + 10,000 = \$12,081$
-0.23	$X = (-0.23)(667) + 10,000 = \$9847$
$+1.24$	$X = (+1.24)(667) + 10,000 = \$10,827$
-2.61	$X = (-2.61)(667) + 10,000 = \$8259$
-0.99	$X = (-0.99)(667) + 10,000 = \$9340$

In this sample of six typical concrete surfacing contract amounts for sites in our region, the average is \$9825 and five of six values are within the range of \$8000 to \$12,000, with the sixth being only \$81 above the upper limit.

19.5 Monte Carlo Sampling and Simulation Analysis ●●○

Up to this point, all alternative selections have been made using estimates with certainty, possibly followed by some testing of the decision via sensitivity analysis or expected values. In this section, we will use a simulation approach that incorporates the material of the previous sections to facilitate the engineering economy decision about one alternative or between two or more alternatives.

The random sampling technique discussed in Section 19.3 is called **Monte Carlo sampling.** The general procedure outlined below uses Monte Carlo sampling to obtain samples of size n for selected parameters of formulated alternatives. These parameters, expected to vary according to a stated probability distribution, warrant decision making under risk. All other parameters in an alternative are considered certain; that is, they are known, or they can be estimated with enough precision to consider them certain. An important assumption is made, usually without realizing it.

All parameters are **independent;** that is, one variable's distribution does not affect the value of any other variable of the alternative. This is referred to as the property of **independent random variables.**

The simulation approach to engineering economy analysis is summarized in the following basic steps.

Step 1. Formulate alternative(s).   Set up each alternative in the form to be considered using engineering economic analysis, and select the measure of worth upon which to base the decision. Determine the form of the relation(s) to calculate the measure of worth.

Step 2. Parameters with variation.   Select the parameters in each alternative to be treated as random variables. Estimate values for all other (certain) parameters for the analysis.

Step 3. **Determine probability distributions.**   Determine whether each variable is discrete or continuous, and describe a probability distribution for each variable in each alternative. Use standard distributions, where possible, to simplify the sampling process and to prepare for computer-based simulation.

Step 4. **Random sampling.**   Incorporate the random sampling procedure of Section 19.3 (the first four steps) into this procedure. This results in the cumulative distribution, assignment of RNs, selection of the RNs, and a sample of size *n* for each variable.

Step 5. **Measure of worth calculation.**   Compute *n* values of the selected measure of worth from the relation(s) determined in step 1. Use the estimates made with certainty and the *n* sample values for the varying parameters. (This is when the property of independent random variables is actually applied.)

Step 6. **Measure of worth description.**   Construct the probability distribution of the measure of worth, using between 10 and 20 cells of data, and calculate measures such as $\overline{X}, s, \overline{X} \pm ts$, and relevant probabilities.

Step 7. **Conclusions.**   Draw conclusions about each alternative, and decide which is to be selected. If the alternative(s) has (have) been previously evaluated under the assumption of certainty for all parameters, comparison of results may help with the final decision.

Example 19.8 illustrates this procedure using an abbreviated manual simulation analysis, and Example 19.9 utilizes spreadsheet simulation for the same estimates.

EXAMPLE 19.8

Yvonne Ramos is the CEO of a chain of 50 fitness centers in the United States and Canada. An equipment salesperson has offered Yvonne two long-term opportunities on new aerobic exercise systems, for which the usage is charged to customers on a per-use basis on top of the monthly fees paid by customers. As an enticement, the offer includes a guarantee of annual revenue for one of the systems for the first 5 years.

Since this is an entirely new and risky concept of revenue generation, Yvonne wants to do a careful analysis of each alternative. Details for the two systems follow:

System 1. First cost is $P = \$12,000$ for a set period of $n = 7$ years with no salvage value. No guarantee for annual net revenue is offered.

System 2. First cost is $P = \$8000$, there is no salvage value, and there is a guaranteed annual net revenue of $1000 for each of the first 5 years, but after this period, there is no guarantee. The equipment with updates may be useful up to 15 years, but the exact number is not known. Cancellation anytime after the initial 5 years is allowed, with no penalty.

For either system, new versions of the equipment will be installed with no added costs. If the MARR is 15% per year, use PW analysis to determine if neither, one, or both of the systems should be installed.

Solution by Hand

Estimates that Yvonne makes to use the simulation analysis procedure are included in the following steps.

Step 1. **Formulate alternatives.**   Using PW analysis, the relations for system 1 and system 2 are developed. The symbol NCF identifies the net cash flows (revenues), and NCF_G is the guaranteed NCF of $1000 for system 2.

$$PW_1 = -P_1 + NCF_1(P/A,15\%,n_1) \qquad [19.20]$$

$$PW_2 = -P_2 + NCF_G(P/A,15\%,5)$$
$$+ NCF_2(P/A,15\%,n_2-5)(P/F,15\%,5) \qquad [19.21]$$

Step 2. **Parameters with variation.**   Yvonne summarizes the parameters estimated with certainty and makes distribution assumptions about three parameters treated as random variables.

System 1

Certainty. $P_1 = \$12,000$; $n_1 = 7$ years.

Variable. NCF_1 is a continuous variable, uniformly distributed between $L = \$-4000$ and $H = \$6000$ per year, because this is considered a high-risk venture.

System 2

Certainty. $P_2 = \$8000$; $NCF_G = \$1000$ for first 5 years.

Variable. NCF_2 is a discrete variable, uniformly distributed over the values $L = \$1000$ to $H = \$6000$ only in $\$1000$ increments, that is, $\$1000$, $\$2000$, etc.

Variable. n_2 is a continuous variable that is uniformly distributed between $L = 6$ and $H = 15$ years.

Now, rewrite Equations [19.20] and [19.21] to reflect the estimates made with certainty.

$$\begin{aligned} PW_1 &= -12{,}000 + NCF_1(P/A,15\%,7) \\ &= -12{,}000 + NCF_1(4.1604) \end{aligned} \qquad [19.22]$$

$$\begin{aligned} PW_2 &= -8000 + 1000(P/A,15\%,5) \\ &\quad + NCF_2(P/A,15\%,n_2-5)(P/F,15\%,5) \\ &= -4648 + NCF_2(P/A,15\%,n_2-5)(0.4972) \end{aligned} \qquad [19.23]$$

Step 3. **Determine probability distributions.** Figure 19–11 (left side) shows the assumed probability distributions for NCF_1, NCF_2, and n_2.

Step 4. **Random sampling.** Yvonne decides on a sample of size 30 and applies the first four of the random sample steps in Section 19.3. Figure 19–11 (right side) shows the cumulative distributions (step 1) and assigns RNs to each variable (step 2). The RNs for NCF_2 identify the x axis values so that all net cash flows will be in even $\$1000$ amounts. For the continuous variable n_2, three-digit RN values are used to make the numbers come out evenly, and they are shown in cells only as "indexers" for easy reference when a RN is used to find a variable value. However, we round the number to the next higher value of n_2 because it is likely the contract may be canceled on an anniversary date. Also, now the tabulated compound interest factors for $(n_2 - 5)$ years can be used directly (see Table 19–5).

Once the first RN is selected randomly from Table 19–2, the sequence (step 3) used will be to proceed down the RN table column and then up the column to the left. Table 19–5 shows only the first five RN values selected for each sample and the corresponding variable values taken from the cumulative distributions in Figure 19–11 (step 4).

Step 5. **Measure of worth calculation.** With the five sample values in Table 19–5, calculate the PW values using Equations [19.22] and [19.23].

1. $PW_1 = -12{,}000 + (-2200)(4.1604)$ $= \$-21{,}153$
2. $PW_1 = -12{,}000 + 2000(4.1604)$ $= \$-3679$
3. $PW_1 = -12{,}000 + (-1100)(4.1604)$ $= \$-16{,}576$
4. $PW_1 = -12{,}000 + (-900)(4.1604)$ $= \$-15{,}744$
5. $PW_1 = -12{,}000 + 3100(4.1604)$ $= \$+897$

1. $PW_2 = -4648 + 1000(P/A,15\%,7)(0.4972)$ $= \$-2579$
2. $PW_2 = -4648 + 1000(P/A,15\%,5)(0.4972)$ $= \$-2981$
3. $PW_2 = -4648 + 5000(P/A,15\%,8)(0.4972)$ $= \$+6507$
4. $PW_2 = -4648 + 3000(P/A,15\%,10)(0.4972)$ $= \$+2838$
5. $PW_2 = -4648 + 4000(P/A,15\%,3)(0.4972)$ $= \$-107$

Now, 25 more RNs are selected for each variable from Table 19–2, and the PW values are calculated.

Step 6. **Measure of worth description.** Figure 19–12a and b presents the PW_1 and PW_2 probability distributions for the 30 samples with 14 and 15 cells, respectively, as well as the range of individual PW values and the $\overline{X}$ and s values.

Figure 19-11
Distributions used for random samples, Example 19.8.

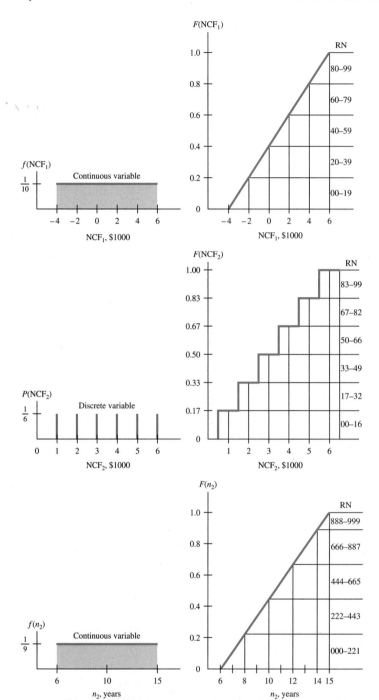

TABLE 19–5 Random Numbers and Variable Values for NCF_1, NCF_2, and n_2, Example 19.8

NCF₁		NCF₂		n₂		
RN*	Value, $	RN†	Value, $	RN‡	Value, Year	Rounded§
18	−2200	10	1000	586	11.3	12
59	+2000	10	1000	379	9.4	10
31	−1100	77	5000	740	12.7	13
29	−900	42	3000	967	14.4	15
71	+3100	55	4000	144	7.3	8

*Randomly start with row 1, column 4 in Table 19–2.
†Start with row 6, column 14.
‡Start with row 4, column 6.
§The n_2 value is rounded up.

Figure 19–12
Probability distributions of simulated PW values for a sample of size 30, Example 19.8.

PW_1. Sample values range from $\$-24,481$ to $\$+12,962$. The calculated measures of the 30 values are

$$\bar{X}_1 = \$-7729$$
$$s_1 = \$10,190$$

PW_2. Sample values range from $\$-3031$ to $\$+10,324$. The sample measures are

$$\bar{X}_2 = \$2724$$
$$s_2 = \$4336$$

Step 7. Conclusions. Additional sample values will surely make the central tendency of the PW distributions more evident and may reduce the s values, which are quite large. Of course, many conclusions are possible once the PW distributions are known, but the following seem clear at this point.

System 1. Based on this small sample of 30 observations, *do not accept* this alternative. The likelihood of making the MARR = 15% is relatively small, since the sample indicates a probability of 0.27 (8 out of 30 values) that the PW will be positive, and $\bar{X}_1$ is a large negative. Though appearing large, the standard deviation may be used to determine that about 20 of the 30 sample PW values (two-thirds) are within the limits $\bar{X} \pm 1s$, which are $\$-17,919$ and $\$2461$. A larger sample may alter this analysis somewhat.

System 2. If Yvonne is willing to accept the longer-term commitment that may increase the NCF some years out, the sample of 30 observations indicates to *accept* this alternative. At a MARR of 15%, the simulation approximates the chance for a positive PW as 67% (20 of the 30 PW values in Figure 19–12b are positive). However, the probability of observing PW within the $\bar{X} \pm 1s$ limits ($\$-1612$ and $\$7060$) is 0.53 (16 of 30 sample values).

Conclusion at this point. Reject system 1; accept system 2; and carefully watch net cash flow, especially after the initial 5-year period.

Comment

The estimates in Example 13.5 are very similar to those here, except all estimates were made with certainty ($NCF_1 = \$3000$, $NCF_2 = \$3000$, and $n_2 = 14$ years). The alternatives were evaluated by the payback period method at MARR = 15%, and the first alternative was selected. However, the subsequent PW analysis in Example 13.5 selected alternative 2 based, in part, upon the anticipated larger cash flow in the later years.

EXAMPLE 19.9

Help Yvonne Ramos set up a spreadsheet simulation for the three random variables and PW analysis in Example 19.8. Does the PW distribution vary appreciably from that developed using manual simulation? Do the decisions to reject the system 1 proposal and accept the system 2 proposal still seem reasonable?

Solution by Spreadsheet

Figures 19–13 and 19–14 are spreadsheet screen shots that accomplish the simulation portion of the analysis described above in steps 3 (determine probability distribution) through 6 (measure of worth description). Most spreadsheet systems are limited in the variety of distributions they can accept for sampling, but common ones such as uniform and normal are available.

Figure 19–13 shows the results of a small sample of 30 values from the three distributions using the RAND and IF functions. (See Section A.3 in Appendix A.)

NCF_1: Continuous uniform from -4000 to 6000. The spreadsheet relation in column B translates RN1 values (column A) into NCF1 amounts.

NCF_2: Discrete uniform in $1000 increments from $1000 to $6000. Column D cells display NCF2 in the $1000 increments using the logical IF function to translate from the RN2 values.

n_2: Continuous uniform from 6 to 15 years. The results in column F are integer values obtained using the INT function operating on the RN3 values.

	A	B	C	D	E	F
1			Sample of size 30	simulated values		
2	RN1	NCF1, $	RN2	NCF2, $	RN3	N2, years
3	12.5625	-2,800	83.6176	6,000	556.2768	12
4	25.0262	-1,500	99.5425	6,000	8.7883	7
5	9.3856	-3,100	26.4693	2,000	507.3598	11
6	38.0199	-200	36.8475	3,000	681.5397	13
7	71.5088	3,100	83.4610	6,000	369.0917	10
8	66.7820	2,600	77.8699	5,000	91.3044	7
9	48.3324	800	8.4308	1,000	457.7487	11
10	39.3886	-100	52.8630	4,000	914.5432	15
11	21.5429	-1,900	57.4819	4,000	698.7624	13
12	44.4996	400	1.9322	1,000	744.2622	13
13	32.9911	-800	70.6307	5,000	190.8139	8
14	96.0249	5,600	61.0023	4,000	714.6685	13
15	99.6675	5,900	55.7741	4,000	648.2268	12
16	13.9560	-2,700	98.9107	6,000	199.9491	8
17	99.8535	5,900	10.7429	1,000	716.5830	13
18	63.2953	2,300	4.6540	1,000	133.4986	8
19	93.0860	5,300	56.7425	4,000	553.2489	11
20	52.6539	1,200	17.1873	2,000	809.5778	14
21	34.1609	-600	46.3758	3,000	810.4792	14
22	86.0288	4,600	99.6569	6,000	950.9657	15
23	46.9626	600	24.9754	2,000	0.9088	7
24	77.4690	3,700	52.2862	4,000	339.4470	10
25	28.5200	-1,200	52.3113	4,000	514.9377	11
26	82.9615	4,200	99.0275	6,000	912.6720	15
27	17.8793	-2,300	50.9493	4,000	800.4352	14
28	89.2411	4,900	6.2988	1,000	118.2531	8
29	9.0495	-3,100	54.6552	4,000	56.4377	7
30	1.4597	-3,900	24.6463	2,000	716.2222	13
31	66.2177	2,600	6.2064	1,000	91.4505	7
32	98.4803	5,800	94.8061	6,000	120.0419	8
33	`= RAND()*100`	`= INT((100*A32 -4000)/100)*100`	`= RAND()*100`	`= IF(C32<=16,1000, IF(C32<=32,2000, IF(C32<=49,3000, IF(C32<=66,4000, IF(C32<=82,5000, IF(C32<=100,6000,6000))))))`	`= RAND()*1000`	`= INT(0.009* E32+1)+6`

⊢ ⊣ ▸ ▸‖ Fig 19-13 Random Numbers Fig 19-14 PW values Sheet6 Sheet8 Sheet9 Sheet10 Sheet11 Sheet ‖‖‖

Figure 19–13
Random sample of 30 values generated for spreadsheet simulation, Example 19.9.

	A	B	C	D	E	F	G	H	I
1				Information about Alternatives					
2	System 1, P1	$12,000		System 2, P2		$8,000			
3	n	7	years		NCF	$1,000	5	years	
4	MARR	15%			MARR	15%			
5									
6				Results of Analysis					
7	# PW ≥ MARR		10					19	
8	# PW < MARR			20					11
9	PW Average, $	-7,105				1,649			
10	PW Std Dev, $	13,199				3,871			
11									
12				Present Worth Computations					
13	PW1	-23,649	0	1	PW2	7,763	1	0	
14	PW1	-18,241	0	1	PW2	202	1	0	
15	PW1	-24,897	0	1	PW2	-885	0	1	
16	PW1	-12,832	0	1	PW2	2,045	1	0	
17	PW1	897	1	0	PW2	5,352	1	0	
18	PW1	-1,183	0	1	PW2	-607	0	1	
19	PW1	-8,672	0	1	PW2	-2,766	0	1	
20	PW1	-12,416	0	1	PW2	5,333	1	0	
21	PW1	-19,905	0	1	PW2	4,276	1	0	
22	PW1	-10,336	0	1	PW2	-2,417	0	1	
23	PW1	-15,328	0	1	PW2	1,028	1	0	
24	PW1	11,298	1	0	PW2	4,276	1	0	
25	PW1	12,546	1	0	PW2	3,626	1	0	
26	PW1	-23,233	0	1	PW2	2,163	1	0	
27	PW1	12,546	1	0	PW2	-2,417	0	1	
28	PW1	-2,431	0	1	PW2	-3,513	0	1	
29	PW1	10,050	1	0	PW2	2,878	1	0	
30	PW1	-7,007	0	1	PW2	97	1	0	
31	PW1	-14,496	0	1	PW2	2,469	1	0	
32	PW1	7,138	1	0	PW2	10,323	1	0	
33	PW1	-9,504	0	1	PW2	-3,031	0	1	
34	PW1	3,394	1	0	PW2	2,019	1	0	
35	PW1	-16,993	0	1	PW2	2,878	1	0	
36	PW1	5,474	1	0	PW2	10,323	1	0	
37	PW1	-21,569	0	1	PW2	4,841	1	0	
38	PW1	8,386	1	0	PW2	-3,513	0	1	
39	PW1	-24,897	0	1	PW2	-1,415	0	1	
40	PW1	-28,226	0	1	PW2	-186	0	1	
41	PW1	-1,183	0	1	PW2	-3,840	0	1	
42	PW1	12,130	1	0	PW2	2,163	1	0	
43	Functions for last row	`= (PV($B$4,$B$3, -'Random Numbers'!B32))-$B$2`	`= IF(B42 >= 0,1,0)`	`= IF(B42 < 0,1,0)`	`= -$F$2-PV($F$4,$G$3,$F$3) +PV($F$4,$G$3,,(PV($F$4, 'Random Numbers'!F32-$G$3, 'Random Numbers'!D32)))`		`= IF(F42 >= 0,1,0)`	`= IF(F42 < 0,1,0)`	

Annotations on spreadsheet:
- `=SUM(G13:G42)` (points to G7)
- `=AVERAGE(F13:F42)` (points to F9)
- `=STDEV(F13:F42)` (points to F10)

Tabs: Fig 19-13 Random Numbers / **Fig 19-14 PW values** / Sheet6 / Sheet8 / Sheet9 / Sheet10 / Sheet11 / Sheet

Figure 19–14
Simulation results for 30 PW values, Example 19.9.

Figure 19–14 presents the two alternatives' estimates in the top section. The PW1 and PW2 computations for the 30 repetitions of NCF1, NCF2, and N2 are the spreadsheet equivalent of Equations [19.22] and [19.23]. The tabular approach used here tallies the number of PW values below zero ($0) and equal to or exceeding zero using the IF operator. For example, cell C17 contains a 1, indicating PW1 > 0 when NCF1 = $3100 (in cell B7 of Figure 19–13), which was used to calculate PW1 = $897 by Equation [19.22]. Cells in rows 7 and 8 show the number of times in the 30 samples that system 1 and system 2 may return at least the MARR = 15% because the corresponding PW ≥ 0. Sample averages and standard deviations are also indicated.

Comparison between the hand and spreadsheet simulations is presented below.

	System 1 PW			System 2 PW		
	$\overline{X}$, $	s, $	No. of PW ≥ 0	$\overline{X}$, $	s, $	No. of PW ≥ 0
Hand	−7,729	10,190	8	2,724	4,336	20
Spreadsheet	−7,105	13,199	10	1,649	3,871	19

For the spreadsheet simulation, 10 (33%) of the PW1 values exceed zero, while the manual simulation included 8 (27%) positive values. These comparative results will change every time this spreadsheet is activated since the RAND function is set up (in this case) to produce a new RN each time. (It is possible to define RAND to keep the same RN values. See the Excel User's Guide.)

The conclusion to reject the system 1 proposal and accept system 2 is still appropriate for the spreadsheet simulation as it was for the hand solution, since there are comparable chances that PW ≥ 0.

CHAPTER SUMMARY

To perform decision making under risk implies that some parameters of an engineering alternative are treated as random variables. Assumptions about the shape of the variable's probability distribution are used to explain how the estimates of parameter values may vary. Additionally, measures such as the expected value and standard deviation describe the characteristic shape of the distribution. In this chapter, we learned several of the simple, but useful, discrete and continuous population distributions used in engineering economy—uniform and triangular—as well as specifying our own distribution or assuming the normal distribution.

Since the population's probability distribution for a parameter is not fully known, a random sample of size n is usually taken, and its sample average and standard deviation are determined. The results are used to make probability statements about the parameter, which help make the final decision with risk considered.

The Monte Carlo sampling method is combined with engineering economy relations for a measure of worth such as PW to implement a simulation approach to risk analysis. The results of such an analysis can then be compared with decisions when parameter estimates are made with certainty.

PROBLEMS

Certainty, Risk, and Uncertainty

19.1 Identify the following variables as either discrete or continuous.
  (a) The interest rates available in the marketplace for jumbo certificates of deposit
  (b) Optimistic, most likely, and pessimistic estimates of salvage value
  (c) The number of cars that are red in the first 100 that pass through a certain intersection
  (d) The weight of the purse or wallet carried when a person leaves her or his residence
  (e) The gallons of water that evaporate from Lake Erie in a given day

19.2 For each situation below, determine (1) if the variable is discrete or continuous and (2) if the information involves certainty, risk, and/or uncertainty.
  (a) The first cost of a new front-end loader is $34,000 or $38,000 depending on the size purchased.
  (b) The raises for engineers and technical staff employees will be 3%, or 5%, with one-half getting 3% and one-half getting 5%.
  (c) Revenue from a new product line is expected to be between $350,000 and $475,000 per year.
  (d) The salvage value for an old machine will be $500 (i.e., its asking price) or $0 (it will be thrown away).
  (e) Profits are equally likely to be up anywhere from 25% to 60% this year.

19.3 An engineer learned that production output is between 1000 and 2000 units per week 90% of the time, and it may fall below 1000 or go above 2000. He wants to use E(output) in the decision-making process. Identify at least two additional pieces of information that must be obtained or assumed to finalize the output information for this use.

Probability and Distributions

19.4 Royalties received by an investor in an oil well vary according to the price of oil. Data collected from stripper wells in an established oil field were used to develop the probability-royalty relationship shown below.

Royalties, $ per Year	6200	8500	9600	10,300	12,600	15,500
Probability	0.10	0.21	0.32	0.24	0.09	0.04

  (a) Calculate the expected value of royalty income (RI) per year.
  (b) Determine the probability that the royalty income will be at least $12,600 per year.

19.5 Daily revenue from vending machines placed in various buildings of a major university is as follows:

20, 75, 43, 62, 51, 52, 78, 33, 28, 39, 61, 56, 43, 49, 48, 49, 71, 53, 57, 46, 42, 41, 63, 36, 51, 59, 40, 32, 37, 29, 26

  (a) Construct a frequency distribution table with a cell size of 12 starting with 19.5 (i.e., first cell is 19.5–31.5, next is 31.5–43.5, etc.).
  (b) Determine the probability distribution.
  (c) What is the probability of revenue from a machine being less than $44?
  (d) What is the probability that revenue from a machine will equal or exceed $44?

19.6 A survey of households included a question about the number of operating automobiles N currently owned by people living at the residence and the interest rate i on the lowest-rate loan for the cars. The results for 100 households are shown.

Number of Cars N	Households
0	12
1	56
2	26
3	3
≥ 4	3

Loan Rate i, %	Households
0.0−2	13
2.01−4	14
4.01−6	19
6.01−8	38
8.01−10	12
10.01−12	4

(a) State whether each variable is discrete or continuous.

(b) Plot the probability distributions and cumulative distributions for N and i.

(c) From the data collected, what is the probability that a household has 1 or 2 cars? Three or more cars?

(d) Use the data for i to estimate the chances that the interest rate is between 7% and 11% per year.

19.7 An officer of the state lottery commission sampled lottery ticket purchasers over a 1-week period at one location. The amounts distributed back to the purchasers and the associated probabilities for 5000 tickets are as follows:

Distribution, $	0	2	5	10	100
Probability	0.91	0.045	0.025	0.013	0.007

(a) Plot the cumulative distribution of winnings.

(b) Calculate the expected value of the distribution of dollars per ticket.

(c) If tickets cost $2, what is the expected long-term income to the state per ticket, based upon this sample?

19.8 Bob is working on two separate probability-related projects. The first involves a variable N, which is the number of consecutively manufactured parts that weigh in above the weight specification limit. The variable N is described by the formula $(0.5)^N$ because each unit has a 50-50 chance of being below or above the limit. The second involves a battery life L that varies between 2 and 5 months. The probability distribution is triangular with the mode at

5 months, which is the design life. Some batteries fail early, but 2 months is the smallest life experienced thus far. (a) Write out and plot the probability distributions and cumulative distributions for Bob. (b) Determine the probability of N being 1, 2, or 3 consecutive units above the weight limit.

19.9 An alternative to buy and an alternative to lease hydraulic lifting equipment have been formulated. Use the parameter estimates and assumed distribution data shown to plot the probability distributions on one graph for each corresponding parameter. Label the parameters carefully.

	Purchase Alternative		
	Estimated Value		**Assumed**
Parameter	**High**	**Low**	**Distribution**
First cost, $	25,000	20,000	Uniform; continuous
Salvage value, $	−3,000	2,000	Triangular; mode at $2500
Life, years	8	4	Triangular; mode at 6
AOC, $ per year	9,000	5,000	Uniform; continuous

	Lease Alternative		
	Estimated Value		**Assumed**
Parameter	**High**	**Low**	**Distribution**
Lease first cost, $	2000	1800	Uniform; continuous
AOC, $ per year	9000	5000	Triangular; mode at $7000
Lease term, years	3	3	Certainty

19.10 Carla is a statistician with a bank. She has collected debt-to-equity mix data on mature (M) and young (Y) companies. The debt percentages vary from 20% to 80% in her sample. Carla has defined D_M as a variable for the mature companies from 0 to 1, with $D_M = 0$ interpreted as the low of 20% debt and $D_M = 1.0$ as the high of 80% debt. The variable for young corporation debt percentages D_Y is similarly defined. The probability distributions used to describe D_M and D_Y are

$$f(D_M) = 3(1 - D_M)^2 \qquad 0 \leq D_M \leq 1$$
$$f(D_Y) = 2D_Y \qquad 0 \leq D_Y \leq 1$$

(a) Use different values of the debt percentage between 20% and 80% to calculate values for the probability distributions and then plot them. (b) What can you comment about the probability that a mature company or a young company will have a low debt percentage? A high debt percentage?

19.11 A discrete variable X can take on integer values of 1 to 10. A sample of size 50 results in the following probability estimates:

X_i	1	2	3	6	9	10
$P(X_i)$	0.2	0.2	0.2	0.1	0.2	0.1

(a) Write out and graph the cumulative distribution.
(b) Calculate the following probabilities using the cumulative distribution: X is between 6 and 10, and X has the values 4, 5, or 6.
(c) Use the cumulative distribution to show that $P(X = 7 \text{ or } 8) = 0.0$. Even though this probability is zero, the statement about X is that it can take on integer values of 1 to 10. How do you explain the apparent contradiction in these two statements?

Random Samples

19.12 A discrete variable X can take on integer values of 1 to 5. A sample of size 100 results in the following probability estimates.

X_i	1	2	3	4	5
$P(X_i)$	0.2	0.3	0.1	03	0.1

(a) Use the following random numbers to estimate the probabilities for each value of X.
(b) Determine the sample probabilities for $X = 1$ and $X = 5$. Compare the sample results with the probabilities in the problem statement.

RN: 10, 42, 18, 31, 23, 80, 80, 26, 74, 71, 03, 90, 55, 61, 61, 28, 41, 49, 00, 79, 96, 78, 42, 31, 26

19.13 The percent price increase p on a variety of retail food prices over a 1-year period varied from 5% to 10% in all cases. Because of the distribution of p values, the assumed probability distribution for the next year is

$$f(X) = 2X \qquad 0 \le X \le 1$$

where

$$X = \begin{cases} 0 & \text{when } p = 5\% \\ 1 & \text{when } p = 10\% \end{cases}$$

For a continuous variable the cumulative distribution $F(X)$ is the integral of $f(X)$ over the same range of the variable. In this case

$$F(X) = X^2 \qquad 0 \le X \le 1$$

(a) Graphically assign RNs to the cumulative distribution, and take a sample of size 30 for the variable. Transform the X values into interest rates.
(b) Calculate the average p value for the sample.

19.14 Develop a discrete probability distribution of your own for the variable G, the expected grade in this course, where $G = $ A, B, C, D, F, or I (incomplete). Assign random numbers to $F(G)$, and take a sample from it. Now plot the probability values from the sample for each G value.

19.15 Use the RAND function in Excel to generate 100 values from a $U(0,1)$ distribution.
(a) Calculate the average and compare it to 0.5, the expected value for a random sample between 0 and 1.
(b) For the RAND function sample, cluster the results into cells of 0.1 width, that is 0.0–0.1, 0.1–0.2, etc., where the upper-limit value is excluded from each cell. Determine the probability for each grouping from the results. Does your sample come close to having approximately 10% in each cell?

Sample Estimates

19.16 An engineer was asked to determine whether the average air quality in a vehicle assembly plant was within OSHA guidelines. The following air quality readings were collected:

81, 86, 80, 91, 83, 83, 96, 85, 89

(a) Determine the sample mean.
(b) Calculate the standard deviation.
(c) Determine the number of values and percent of values that fall within ± 1 standard deviation of the mean.

19.17 Carol sampled the monthly maintenance costs for automated soldering machines a total of 100 times during 1 year. She clustered the costs into $200 cells, for example, $500 to $700, with cell midpoints of $600, $800, $1000, etc. She indicated the number of times (frequency) each cell value was observed. The costs and frequency data are as follows.

Cell Midpoint, $	Frequency
600	6
800	10
1000	7
1200	15
1400	28
1600	15
1800	9
2000	10

(a) Estimate the expected value and standard deviation of the maintenance costs the company should anticipate based on Carol's sample.
(b) What is the best estimate of the percentage of costs that will fall within 2 standard deviations of the mean?

(c)   Develop a probability distribution of the monthly maintenance costs from Carol's sample, and indicate the answers to the previous two questions on it.

(d)   Use a spreadsheet to display the sample mean.

19.18 (a) Determine the values of sample average and standard deviation of the data in Problem 19.11. (b) Determine the values 1 and 2 standard deviations from the mean. Of the 50 sample points, how many fall within these two ranges?

19.19 (a) Use the relations in Section 19.4 for continuous variables to determine the expected value and standard deviation for the distribution of $f(D_Y)$ in Problem 19.10. (b) It is possible to calculate the probability of a continuous variable X between two points (a, b) using the following integral:

$$P(a \le X \le b) = \int_a^b f(X)\, dx$$

Determine the probability that D_Y is within 2 standard deviations of the expected value.

19.20 (a) Use the relations in Section 19.4 for continuous variables to determine the expected value and variance for the distribution of D_M in Problem 19.10.

$$f(D_M) = 3(1 - D_M)^2 \qquad 0 \le D_M \le 1$$

(b) Determine the probability that D_M is within 2 standard deviations of the expected value. Use the relation in Problem 19.19.

19.21 Calculate the expected value for the variable N in Problem 19.8.

19.22 A newsstand manager is tracking Y, the number of weekly magazines left on the shelf when the new edition is delivered. Data collected over a 30-week period are summarized by the following probability distribution. Plot the distribution and the estimates for expected value and 1 standard deviation on either side of $E(Y)$ on the plot.

Y copies	3	7	10	12
P(Y)	1/3	1/4	1/3	1/12

Simulation

19.23 Carl, an engineering colleague, estimated net cash flow after taxes (CFAT) for the project he is working on. The additional CFAT of $2800 in year 10 is the salvage value of capital assets.

Year	CFAT, $
0	−28,800
1−6	5,400
7−10	2,040
10	2,800

The PW value at the current MARR of 7% per year is

$$\begin{aligned} PW = {} & -28{,}800 + 5400(P/A,7\%,6) \\ & + 2040(P/A,7\%,4)(P/F,7\%,6) \\ & + 2800(P/F,7\%,10) \\ = {} & \$2966 \end{aligned}$$

Carl believes the MARR will vary over a relatively narrow range, as will the CFAT, especially during the out years of 7 through 10. He is willing to accept the other estimates as certain. Use the following probability distribution assumptions for MARR and CFAT to perform a simulation—hand- or spreadsheet-based.

MARR. Uniform distribution over the range 6% to 10%.

CFAT, years 7 through 10. Uniform distribution over the range $1600 to $2400 for each year.

Plot the resulting PW distribution. Should the plan be accepted using decision making under certainty? Under risk?

19.24 Repeat Problem 19.23, except use the normal distribution for the CFAT in years 7 through 10 with an expected value of $2000 and a standard deviation of $500.

ADDITIONAL PROBLEMS AND FE EXAM REVIEW QUESTIONS

19.25 When there are at least two values for a parameter and it is possible to estimate the chance that each may occur, this situation is known as:
(a)   Uncertainty
(b)   Risk
(c)   Standard deviation
(d)   Cost estimating

19.26 A *deterministic* economic analysis is one wherein:
(a)   Single-value estimates are used exclusively
(b)   Risk is taken into account
(c)   A range of values for each parameter is included in the analysis
(d)   Taxes and inflation are not considered in the cash flow estimates

19.27 Decision making under risk includes all of the following except:
  (a) Expected value analysis
  (b) Simulation
  (c) Using only single-value estimates
  (d) Probability

19.28 All of the following are elements in decision making under risk except:
  (a) Random variable
  (b) Cost indexes
  (c) Probability
  (d) Cumulative distribution

19.29 For the income and probability values shown in the table, the probability that the income in any year will be less than $9600 is closest to:
  (a) 0.15
  (b) 0.23
  (c) 0.32
  (d) 0.38

Income, $ per Year	6200	8500	9600	10,300	12,600	15,500
Probability	0.15	0.23	0.32	0.24	0.09	0.04

19.30 The symbol that represents the true population mean is:
  (a) σ
  (b) s
  (c) μ
  (d) $\bar{X}$

19.31 The revenue from an oil dispersant product has averaged $15,000 per month for the past 12 months. If the value of $\Sigma(X_i - \bar{X})^2$ is $1,600,000, the standard deviation is closest to:
  (a) $381
  (b) $652
  (c) $958
  (d) $1265

19.32 A survey of the types of cars parked at an NFL football stadium revealed that there were equal probabilities of finding cars identified as type A, B, C, and D. Car types were assigned random numbers as follows.

Car Type	Assigned RN
A	0 through 24
B	25 through 49
C	50 through 74
D	75 through 99

The sample probability of a type B car from the 12 random numbers shown is closest to:

RN: 75, 52, 14, 99, 67, 74, 06, 50, 97, 46, 27, 88

  (a) 0.17
  (b) 0.25
  (c) 0.33
  (d) 0.42

CASE STUDY

USING SIMULATION AND THREE-ESTIMATE SENSITIVITY ANALYSIS

Background

The Knox Brewing company makes specialty-named sodas and flavored drinks for retail grocery chains throughout North and Central America. During the past year, it has become obvious that a new bottle-capping machine is needed to replace the current 10-year-old system. Dr. Knox, the owner and president, knows the business quite well. You just handed him the first cost bids from three vendors for the machine. He looked carefully at the numbers and asked you to sit down. You were quite surprised, as this was the first time you had been in his office, and most other engineers at Knox have a great fear of "the Old Man."

Information

As he examined the three bids on first cost of the machine, he started to write some numbers, which, he explained, were his estimates of the annual operating cost, useful life, and possible salvage value for each of the machines sold by the three vendors.

After a few minutes, he told you to take these numbers and use some of that "new engineering knowledge" you acquired in college to determine which, if any, of these three bids made the best economic sense. He also told you to be innovative and use a computer and some probability to come up with a robust recommendation by tomorrow at 2 P.M.

You have used the estimates from the president to develop Table 19–6 of pessimistic (P), most likely (ML), and optimistic (O) estimates for each vendor's machine. In addition, you developed some possible distributions for the parameters that Dr. Knox estimated, namely, AOC, life, and salvage value. These are summarized in Table 19–7. You plan to use a simple Monte Carlo simulation to help formulate your recommendation for tomorrow.

Case Study Exercises

First, learn to use the RNG (random number generator) in Excel, if you have not already done so. It is necessary to

TABLE 19–6	Parameter Estimates for Bottle-Capping Machine			
	First Cost, $	AOC, $ per Year	Salvage, $	Life, Years
Vendor 1				
P	−200,000	−11,000	0	3
ML	−200,000	−10,000	0	5
O	−200,000	−6,000	0	8
Vendor 2				
P	−150,000	−5,000	0	2
ML	−150,000	−3,500	5,000	4
O	−150,000	−2,000	8,000	7
Vendor 3				
P	−300,000	−8,000	5,000	5
ML	−300,000	−6,000	5,000	7
O	−300,000	−4,500	8,000	9

TABLE 19–7	Distribution Assumptions about AOC, Life, and Salvage		
Parameter	Vendor 1	Vendor 2	Vendor 3
AOC, $ per year	Normal Mean: 10,000 Std. dev.: 500	Normal Mean: 3500 Std. dev.: 500	Normal Mean: 6000 Std. dev.: 500
Salvage, $	Uniform 0 to 1000	Uniform 0 to 8000	Uniform 5000 to 8000
Life, years	Discrete uniform 3 to 8, equal probability	Discrete uniform 2 to 7, equal probability	Discrete uniform 5 to 9, equal probability

sample from the normal distributions that you have specified in Table 19–7. RNG is part of the Analysis Tool-Pak accessed through the Office Button, Excel Options, Add-Ins path.

1. Prepare the simulation using a spreadsheet; determine which of the vendors offers the best machine from an economic perspective, and take into account the estimates made by Dr. Knox. Use a sample size of at least 50, and base your conclusions on the AW measure of worth.

2. Prepare a short presentation for Dr. Knox (and class) using your analysis.

A company is looking to invest $480,000 in a new machine with a 10-year life and a salvage value of $50000. A) If the machine saves $100,000 per year, what is the IRR? (Show the IRR to 2 decimal places. It should be between 16% and 18%). B) What is the undiscounted payback period? (Show payback periods in years, e.g., "between 2 and 3 years", or "between 10 and 11 years.")

A) *NPV18% = 100,000(P/A, 18%, 10) + 50,000(P/F, 18%, 10) – 480,000 ➜ -21,035*
   NPV16% = 100,000(P/A, 16%, 10) + 50,000(P/F, 16%, 10) – 480,000 ➜ 14,655
   (x – 0.16) / (0.18 – 0.16) = (0 – 14,655) / (-21,035 – 14,655) ➜ X = 16.82%

B) *480,000 = X(100,000) + 50,000*
   430,000 = x(100,000)
   X = 4.3 years ➜ Payback is between years 4 & 5

Breakeven

A	B
FC = 10,000/yr	FC = 21,500/yr
VC = 50/test	VC = 10/test

$-10,000 - 50x = -21,500 - 10x$

$40x = 11,500$

$\boxed{x = 287.5}$

Incremental ROR of 2 products

	x	y
FC	-270	-245
AOC	-135	-134
SV	75	35
n	6	6

Take x-y, then interpolate for i

$0 = -25\left(\frac{P}{P}, i, 6\right) + 4 + 40\left(\frac{P}{F}, i, 6\right)$

calculate MARR & compare with zero

Break-even analysis

Profit = Revenue - cost = Revenue - (FC + VC)

$Q_{BE} = \frac{FC}{r-V} = \left(\frac{2.90\$/gal}{18 \; mi/gal}\right) \times miles = 1.6 \frac{mi}{gal} (20 gal)$

IRR interpolation

5%
$PW = -5000 + 250\left(\frac{P}{A}, 5\%, 6\right) + 4600\left(\frac{P}{F}, 5\%, 2\right) = -270$

4%

4% ↔ 44

i	PW
4	44
x	0
5	-270

$B\begin{bmatrix} A\begin{bmatrix} 4 \\ x \\ 5 \end{bmatrix} & \begin{bmatrix} 44 \\ 0 \\ -270 \end{bmatrix}C \end{bmatrix}D$

$\frac{A}{B} = \frac{x-4}{5-4} = \frac{0-44}{-270-44} = \frac{C}{D}$

$x = 4.14\%$

Capitalized C

if it is only a one time cost use $\left(\frac{P}{F}, \%, 11\right)$ if it will be redone every — yrs put in AW & then use $P = \frac{A}{i}$

Payback Analysis

$i = 3\%$ month

$n_p = \frac{28000 \; investment}{3500 \; NCF}$ = 8 months if MARR = 0%

$n_p = \frac{P}{NCF} (p.349)$ $0 = -28,000 + 3500\left(\frac{P}{A}, 3\%, n_p\right)$

NCF = income
= 5000 - 1
= 3500/

PW ch.5 p(n33) $CC = \frac{A}{i}$

USING SPREADSHEETS AND MICROSOFT EXCEL©

This appendix explains the layout of a spreadsheet and the use of Microsoft Excel (hereafter called Excel) functions in engineering economy. Refer to the Excel help system for your particular computer and version of Excel. Some specific commands and entries refer to Excel 2007 and may differ slightly from your version.

A.1 Introduction to Using Excel ● ● ◉

Enter a Formula or Use an Excel Function

The **= sign is required** to perform any formula or function computation in a cell. The formulas and functions on the worksheet can be displayed by simultaneously pressing Ctrl and `. The symbol ` is usually in the upper left of the keyboard with the ~ (tilde) symbol. Pressing Ctrl+` a second time hides the formulas and functions.

1. Run Excel.
2. Move to cell C3. (Move the pointer to C3 and left-click.)
3. Type = PV(5%,12,10) and <Enter>. This function will calculate the present value of 12 payments of $10 at a 5% per year interest rate.

Another example: To calculate the future value of 12 payments of $10 at 6% per year interest, do the following:

1. Move to cell B3, and type INTEREST.
2. Move to cell C3, and type 6% or = 6/100.
3. Move to cell B4, and type PAYMENT.
4. Move to cell C4, and type 10 (to represent the size of each payment).
5. Move to cell B5, and type NUMBER OF PAYMENTS.
6. Move to cell C5, and type 12 (to represent the number of payments).
7. Move to cell B7, and type FUTURE VALUE.
8. Move to cell C7, and type = FV(C3,C5,C4) and hit <Enter>. The answer will appear in cell C7.

To edit the values in cells

1. Move to cell C3 and type =5/100 (the previous value will be replaced).
2. The value in cell C7 will update.

Cell References in Formulas and Functions

If a cell reference is used in lieu of a specific number, it is possible to change the number once and perform sensitivity analysis on any variable that is referenced by the cell number, such as C5. This approach defines the referenced cell as a **global variable** for the worksheet. There are two types of cell references—relative and absolute.

Relative References   If a cell reference is entered, for example, A1, into a formula or function that is copied or dragged into another cell, the reference is changed relative to the movement of the original cell. If the formula in C5 is = A1 and it is copied into cell C6, the formula is changed to = A2. This feature is used when dragging a function through several cells, and the source entries must change with the column or row.

Absolute References   If adjusting cell references is not desired, place a **$ sign** in front of the part of the cell reference that is not to be adjusted—the column, row, or both. For example,

= \$A\$1 will retain the formula when it is moved anywhere on the worksheet. Similarly, = \$A1 will retain the column A, but the relative reference on 1 will adjust the row number upon movement around the worksheet.

Absolute references are used in engineering economy for sensitivity analysis of parameters such as MARR, first cost, and annual cash flows. In these cases, a change in the absolute-reference cell entry can help determine the sensitivity of a result, such as PW or AW.

Print the Spreadsheet

First define the portion (or all) of the spreadsheet to be printed.

1. Move the pointer to the top left cell of your spreadsheet.
2. Hold down the left-click button. (Do not release the left-click button.)
3. Drag the mouse to the lower right corner of your spreadsheet or to wherever you want to stop printing.
4. Release the left-click button. (It is ready to print.)
5. Left-click the Office button (see Figure A–1).
6. Move the pointer down to select Print and left-click.
7. In the dialog box, left-click the Print option (or similar command).

Depending on your computer environment, you may have to select a network printer and queue your printout through a server.

Save the Spreadsheet

You can save your spreadsheet at any time during or after completing your work. It is recommended that you save your work regularly.

1. Left-click the Office button.
2. To save the spreadsheet the first time, left-click the Save As . . . option.
3. Type the file name, e.g., Prob 7.9, and left-click the Save button.

To save the spreadsheet after it has been saved the first time, i.e., a file name has been assigned to it, left-click the Office button, move the pointer down, and left-click on Save.

Create an *xy* (Scatter) Chart

This chart is one of the most commonly used in scientific analysis, including engineering economy. It plots pairs of data and can place multiple series of entries on the *Y* axis. The *xy* scatter chart is especially useful for results such as the PW versus *i* graph, where *i* is the *X* axis and the *Y* axis displays the results of the NPV function for several alternatives.

1. Run Excel.
2. Enter the following numbers in columns A, B, and C, respectively.
   Column A, cell A1 through A6: Rate *i*%, 4, 6, 8, 9, 10
   Column B, cell B1 through B6: \$ for A, 40, 55, 60, 45, 10
   Column C, cell C1 through C6: \$ for B, 100, 70, 65, 50, 30.
3. Move the mouse to A1, left-click, and hold while dragging to cell C6. All cells will be highlighted, including the title cell for each column.
4. If not all the columns for the chart are adjacent to one another, first press and hold the Control key on the keyboard during the entirety of step 3. After dragging over one column of data, momentarily release the left click, then move to the top of the next (nonadjacent) column of the chart. Do not release the Control key until all columns to be plotted have been highlighted.
5. Left-click on the Insert button on the toolbar.
6. Select the Scatter option and choose a subtype of scatter chart. The graph appears with a legend (Figure A–1).

Now a large number of styling effects can be introduced for axis titles, legend, data series, etc. Note that only the bottom row of the title can be highlighted. If titles are not highlighted, the data sets are generically identified as series 1, series 2, etc. on the legend.

Figure A–1
Scatter chart for data entries and location of commonly used buttons.

Obtain Help While Using Excel

1. To get general help information, left-click on the "?" (upper right).
2. Enter the topic or phrase. For example, if you want to know more about how to save a file, type the word Save.
3. Select the appropriate matching words. You can browse through the options by left-clicking on any item.

A.2 Organization (Layout) of the Spreadsheet ●●●

A spreadsheet can be used in several ways to obtain answers to numerical questions. The first is as a rapid solution tool, often with the entry of only a few numbers or one predefined function. For example, to find the future worth in a single-cell operation, move the pointer to any cell and enter = FV(8%,5,-2500). The display of $14,666.50 is the 8% future worth at the end of year 5 of five equal payments of $2500 each.

A second use is more formal; it presents data, solutions, graphs, and tables developed on the spreadsheet and ready for presentation to others. Some fundamental guidelines in spreadsheet organization are presented here. A sample layout is presented in Figure A–2. As solutions become more complex, organization of the spreadsheet becomes increasingly important, especially for presentation to an audience via PowerPoint or similar software.

Cluster the data and the answers. It is advisable to organize the given or estimated data in the top left of the spreadsheet. A very brief label should be used to identify the data, for example, MARR = in cell A1 and the value, 12%, in cell B1. Then B1 can be the referenced cell for all entries requiring the MARR. Additionally, it may be worthwhile to cluster the answers into one area and frame it. Often, the answers are best placed at the bottom or top of the column of entries used in the formula or predefined function.

Enter titles for columns and rows. Each column or row should be labeled so its entries are clear to the reader. It is very easy to select from the wrong column or row when no brief title is present at the head of the data.

Enter income and cost cash flows separately. When there are both income and cost cash flows involved, it is strongly recommended that the cash flow estimates for revenue (usually positive)

Figure A–2
Spreadsheet layout with cash flow estimates, results of functions, function formula detailed, and a scatter chart.

and first cost, salvage value, and annual costs (usually negative, with salvage a positive number) be entered into two adjacent columns. Then a formula combining them in a third column displays the net cash flow. There are two immediate advantages to this practice: fewer errors are made when performing the summation and subtraction mentally, and changes for sensitivity analysis are more easily made.

Use cell references. The use of absolute and relative cell references is a must when any changes in entries are expected. For example, suppose the MARR is entered in cell B1 and three separate references are made to the MARR in functions on the spreadsheet. The absolute cell reference entry $B$1 in the three functions allows the MARR to be changed one time, not three.

Obtain a final answer through summing and embedding. When the formulas and functions are kept relatively simple, the final answer can be obtained using the SUM function. For example, if the present worth values (PW) of two columns of cash flows are determined separately, then the total PW is the SUM of the subtotals. This practice is especially useful when the cash flow series are complex.

Prepare for a chart. If a chart (graph) will be developed, plan ahead by leaving sufficient room on the right of the data and answers. Charts can be placed on the same worksheet or on a separate worksheet. Placement on the same worksheet is recommended, especially when the results of sensitivity analysis are plotted.

A.3 Excel Functions Important to Engineering Economy (alphabetical order) ● ● ●

DB (Declining Balance)

Calculates the depreciation amount for an asset for a specified period n using the declining balance method. The depreciation rate d used in the computation is determined from asset values S (salvage value) and B (basis or first cost) as $d = 1 - (S/B)^{1/n}$. This is Equation [16.12]. Three-decimal-place accuracy is used for d.

$$= DB(\text{cost, salvage, life, period, month})$$

cost	First cost or basis of the asset.
salvage	Salvage value.
life	Depreciation life (recovery period).
period	The period, year, for which the depreciation is to be calculated.
month	(optional entry) If this entry is omitted, a full year is assumed for the first year.

Example   A new machine costs $100,000 and is expected to last 10 years. At the end of 10 years, the salvage value of the machine is $50,000. What is the depreciation of the machine in the first year and the fifth year?

Depreciation for the first year: = DB(100000,50000,10,1)

Depreciation for the fifth year: = DB(100000,50000,10,5)

Because of the manner in which the DB function determines the fixed percentage d and the accuracy of the computations, it is **recommended that the DDB function (below) be used** for all declining balance depreciation rates. Simply use the optional factor entry for rates other than $d = 2/n$.

DDB (Double Declining Balance)

Calculates the depreciation of an asset for a specified period n using the double declining balance method. A factor can also be entered for some other declining balance depreciation method by specifying a factor in the function.

$$= \text{DDB(cost, salvage, life, period, factor)}$$

cost	First cost or basis of the asset.
salvage	Salvage value of the asset.
life	Depreciation life.
period	The period, a year, for which the depreciation is to be calculated.
factor	(optional entry) If this entry is omitted, the function will use a double declining method with 2 times the straight line rate. If, for example, the entry is 1.5, the 150% declining balance method will be used.

Example   A new machine costs $200,000 and is expected to last 10 years. The salvage value is $10,000. Calculate the depreciation of the machine for the first and the eighth years. Finally, calculate the depreciation for the fifth year using the 175% declining balance method.

Depreciation for the first year: = DDB(200000,10000,10,1)

Depreciation for the eighth year: = DDB(200000,10000,10,8)

Depreciation for the fifth year using 175% DB: = DDB(200000,10000,10,5,1.75)

EFFECT (Effective Interest Rate)

Calculates the effective annual interest rate for a stated nominal annual rate and a given number of compounding periods per year. Excel uses Equation [4.7] to calculate the effective rate.

$$= \text{EFFECT(nominal, npery)}$$

nominal	Nominal interest rate for the year.
npery	Number of times interest is compounded per year.

Example   Claude has applied for a $10,000 loan. The bank officer told him that the interest rate is 8% per year and that interest is compounded monthly to conveniently match his monthly payments. What effective annual rate will Claude pay?

Effective annual rate: = EFFECT(8%,12)

EFFECT can also be used to find **effective rates other than annually.** Enter the nominal rate for the time period of the required effective rate; npery is the number of times compounding occurs during the time period of the effective rate.

Example   Interest is stated as 3.5% per quarter with quarterly compounding. Find the effective semiannual rate.

The 6-month nominal rate is 7%, and compounding is 2 times per 6 months.

Effective semiannual rate: = EFFECT(7%,2)

FV (Future Value)

Calculates the future value (worth) based on periodic payments at a specific interest rate.

$$= FV(rate, nper, pmt, pv, type)$$

rate	Interest rate per compounding period.
nper	Number of compounding periods.
pmt	Constant payment amount.
pv	The present value amount. If pv is not specified, the function will assume it to be 0.
type	(optional entry) Either 0 or 1. A 0 represents payments made at the end of the period, and 1 represents payments at the beginning of the period. If omitted, 0 is assumed.

Example   Jack wants to start a savings account that can be increased as desired. He will deposit $12,000 to start the account and plans to add $500 to the account at the beginning of each month for the next 24 months. The bank pays 0.25% per month. How much will be in Jack's account at the end of 24 months?

Future value in 24 months: $= FV(0.25\%,24,500,12000,1)$

IF (IF Logical Function)

Determines which of two entries is entered into a cell based on the outcome of a logical check on the outcome of another cell. The logical test can be a function or a simple value check, but it must use an equality or inequality sense. If the response is a text string, place it between quote marks (" "). The responses can themselves be IF functions. Up to seven IF functions can be nested for very complex logical tests.

$$= IF(logical\_test, value\_if\_true, value\_if\_false)$$

logical_test	Any worksheet function can be used here, including a mathematical operation.
value_if_true	Result if the logical_test argument is true.
value_if_false	Result if the logical_test argument is false.

Example   The entry in cell B4 should be "selected" if the PW value in cell B3 is greater than or equal to zero and "rejected" if PW < 0.

Entry in cell B4: $= IF(B3 >=0,\text{"selected"},\text{"rejected"})$

Example   The entry in cell C5 should be "selected" if the PW value in cell C4 is greater than or equal to zero, "rejected" if PW < 0, and "fantastic" if PW ≥ 200.

Entry in cell C5: $= IF(C4<0,\text{"rejected"},IF(C4>=200,\text{"fantastic"},\text{"selected"}))$

IPMT (Interest Payment)

Calculates the interest accrued for a given period n based on constant periodic payments and interest rate.

$$= IPMT(rate, per, nper, pv, fv, type)$$

rate	Interest rate per compounding period.
per	Period for which interest is to be calculated.
nper	Number of compounding periods.
pv	Present value. If pv is not specified, the function will assume it to be 0.
fv	Future value. If fv is omitted, the function will assume it to be 0. The fv can also be considered a cash balance after the last payment is made.
type	(optional entry) Either 0 or 1. A 0 represents payments made at the end of the period, and 1 represents payments made at the beginning of the period. If omitted, 0 is assumed.

Example   Calculate the interest due in the 10th month for a 48-month, $20,000 loan. The interest rate is 0.25% per month.

Interest due: = IPMT(0.25%,10,48,20000)

IRR (Internal Rate of Return)

Calculates the internal rate of return between -100% and infinity for a series of cash flows at regular periods.

$$= \text{IRR(values, guess)}$$

values   A set of numbers in a spreadsheet column (or row) for which the rate of return will be calculated. The set of numbers must consist of at least *one* positive and *one* negative number. Negative numbers denote a payment made or cash outflow, and positive numbers denote income or cash inflow.

guess   (optional entry) To reduce the number of iterations, a *guessed rate of return* can be entered. In most cases, a guess is not required, and a 10% rate of return is initially assumed. If the #NUM! error appears, try using different values for guess. Inputting different guess values makes it possible to determine the multiple roots for the rate of return equation of a nonconventional cash flow series.

Example   John wants to start a printing business. He will need $25,000 in capital and anticipates that the business will generate the following incomes during the first 5 years. Calculate his rate of return after 3 years and after 5 years.

Year 1 $5,000
Year 2 $7,500
Year 3 $8,000
Year 4 $10,000
Year 5 $15,000

Set up an array in the spreadsheet.

In cell A1, type -25000 (negative for payment).
In cell A2, type 5000 (positive for income).
In cell A3, type 7500.
In cell A4, type 8000.
In cell A5, type 10000.
In cell A6, type 15000.

Therefore, cells A1 through A6 contain the array of cash flows for the first 5 years, including the capital outlay. *Note that any years with a zero cash flow must have a zero entered* to ensure that the year value is correctly maintained for computation purposes.

To calculate the internal rate of return after 3 years, move to cell A7, and type = IRR(A1:A4).

To calculate the internal rate of return after 5 years and specify a guess value of 5%, move to cell A8, and type = IRR(A1:A6,5%).

MIRR (Modified Internal Rate of Return)

Calculates the modified internal rate of return for a series of cash flows and reinvestment of income and interest at a stated rate.

$$= \text{MIRR(values, finance\_rate, reinvest\_rate)}$$

values   Refers to an array of cells in the spreadsheet. Negative numbers represent payments, and positive numbers represent income. The series of payments and income must occur at regular periods and must contain at least *one* positive number and *one* negative number.

finance_rate   Interest rate on funds borrowed from external sources (i_b in Equation [7.9]).

reinvest_rate   Interest rate for reinvestment on positive cash flows (i_i in Equation [7.9]). (This is not the same reinvestment rate on the net investments when the cash flow series is nonconventional. See Section 7.5 for comments.)

Example   Jane opened a hobby store 4 years ago. When she started the business, Jane borrowed $50,000 from a bank at 12% per year. Since then, the business has yielded $10,000 the first year, $15,000 the second year, $18,000 the third year, and $21,000 the fourth year. Jane reinvests her profits, earning 8% per year. What is the modified rate of return after 3 years and after 4 years?

In cell A1, type -50000.

In cell A2, type 10000.

In cell A3, type 15000.

In cell A4, type 18000.

In cell A5, type 21000.

To calculate the modified rate of return after 3 years, move to cell A6, and type = MIRR(A1:A4,12%,8%).

To calculate the modified rate of return after 4 years, move to cell A7, and type = MIRR(A1:A5,12%,8%).

NOMINAL (Nominal Interest Rate)

Calculates the nominal **annual** interest rate for a stated effective **annual** rate and a given number of compounding periods per year. *This function is designed to display only nominal annual rates.*

$$= NOMINAL(effective, npery)$$

effective   Effective interest rate for the year.

npery   Number of times that interest is compounded per year.

Example   Last year, a corporate stock earned an effective return of 12.55% per year. Calculate the nominal annual rate, if interest is compounded quarterly and compounded continuously.

Nominal annual rate, quarterly compounding: = NOMINAL(12.55%,4)

Nominal annual rate, continuous compounding: = NOMINAL(12.55%,100000)

NPER (Number of Periods)

Calculates the number of periods for the present worth of an investment to equal the future value specified, based on uniform regular payments and a stated interest rate.

$$= NPER(rate, pmt, pv, fv, type)$$

rate   Interest rate per compounding period.

pmt   Amount paid during each compounding period.

pv   Present value (lump-sum amount).

fv   (optional entry) Future value or cash balance after the last payment. If fv is omitted, the function will assume a value of 0.

type   (optional entry) Enter 0 if payments are due at the end of the compounding period and 1 if payments are due at the beginning of the period. If omitted, 0 is assumed.

Example   Sally plans to open a savings account that pays 0.25% per month. Her initial deposit is $3000, and she plans to deposit $250 at the beginning of every month. How many payments does she have to make to accumulate $25,000 to buy a new car?

Number of payments: $= \text{NPER}(0.25\%, -250, -3000, 25000, 1)$

NPV (Net Present Value)

Calculates the net present value of a series of future cash flows at a stated interest rate.

$$= \text{NPV(rate, series)}$$

    rate   Interest rate per compounding period.
    series   Series of costs and incomes set up in a range of cells in the spreadsheet.

Example   Mark is considering buying a sports store for $100,000 and expects to receive the following income during the next 6 years of business: $25,000, $40,000, $42,000, $44,000, $48,000, $50,000. The interest rate is 8% per year.
  In cells A1 through A7, enter $-100,000$, followed by the six annual incomes.

Present value: $= \text{NPV}(8\%, \text{A2:A7}) + \text{A1}$

The cell A1 value is already a present value. *Any year with a zero cash flow must have a 0 entered* to ensure a correct result.

PMT (Payments)

Calculates equivalent periodic amounts based on present value and/or future value at a constant interest rate.

$$= \text{PMT(rate, nper, pv, fv, type)}$$

    rate   Interest rate per compounding period.
    nper   Total number of periods.
    pv   Present value.
    fv   Future value.
    type   (optional entry) Enter 0 for payments due at the end of the compounding period and 1 if payment is due at the start of the compounding period. If omitted, 0 is assumed.

Example   Jim plans to take a $15,000 loan to buy a new car. The interest rate is 7% per year. He wants to pay the loan off in 5 years (60 months). What are his monthly payments?

Monthly payments: $= \text{PMT}(7\%/12, 60, 15000)$

PPMT (Principal Payment)

Calculates the payment on the principal based on uniform payments at a specified interest rate.

$$= \text{PPMT(rate, per, nper, pv, fv, type)}$$

    rate   Interest rate per compounding period.
    per   Period for which the payment on the principal is required.
    nper   Total number of periods.
    pv   Present value.
    fv   Future value.
    type   (optional entry) Enter 0 for payments that are due at the end of the compounding period and 1 if payments are due at the start of the compounding period. If omitted, 0 is assumed.

Example   Jovita is planning to invest $10,000 in equipment which is expected to last 10 years with no salvage value. The interest rate is 5%. What is the principal payment at the end of year 4 and year 8?

At the end of year 4: = PPMT(5%,4,10,−10000)

At the end of year 8: = PPMT(5%,8,10,−10000)

PV (Present Value)

Calculates the present value of a future series of equal cash flows and a single lump sum in the last period at a constant interest rate.

$$= PV(\text{rate, nper, pmt, fv, type})$$

rate	Interest rate per compounding period.
nper	Total number of periods.
pmt	Cash flow at regular intervals. Negative numbers represent payments (cash outflows), and positive numbers represent income.
fv	Future value or cash balance at the end of the last period.
type	(optional entry) Enter 0 if payments are due at the end of the compounding period and 1 if payments are due at the start of each compounding period. If omitted, 0 is assumed.

There are two primary differences between the PV function and the NPV function: PV allows for end or beginning of period cash flows, and PV requires that all amounts have the same value, whereas they may vary for the NPV function.

Example   Jose is considering leasing a car for $300 a month for 3 years (36 months). After the 36-month lease, he can purchase the car for $12,000. Using an interest rate of 8% per year, find the present value of this option.

Present value: = PV(8%/12,36,−300,−12000)

Note the minus signs on the pmt and fv amounts.

RAND (Random Number)

Returns an evenly distributed number that is (1) ≥ 0 and < 1; (2) ≥ 0 and < 100; or (3) between two specified numbers.

= RAND()	for range 0 to 1
= RAND()*100	for range 0 to 100
= RAND()*(b−a)+a	for range a to b

a = minimum integer to be generated
b = maximum integer to be generated

The Excel function RANDBETWEEN(a,b) may also be used to obtain a random number between two values.

Example   Grace needs random numbers between 5 and 10 with 3 digits after the decimal. What is the Excel function? Here a = 5 and b = 10.

Random number: = RAND()*5 + 5

Example   Randi wants to generate random numbers between the limits of −10 and 25. What is the Excel function? The minimum and maximum values are $a = -10$ and $b = 25$, so $b - a = 25 - (-10) = 35$.

Random number: = RAND()*35 − 10

RATE (Interest Rate)

Calculates the interest rate per compounding period for a series of payments or incomes.

$$= \text{RATE(nper, pmt, pv, fv, type, guess)}$$

nper	Total number of periods.
pmt	Payment amount made each compounding period.
pv	Present value.
fv	Future value (not including the pmt amount).
type	(optional entry) Enter 0 for payments due at the end of the compounding period and 1 if payments are due at the start of each compounding period. If omitted, 0 is assumed.
guess	(optional entry) To minimize computing time, include a guessed interest rate. If a value of guess is not specified, the function will assume a rate of 10%. This function usually converges to a solution if the rate is between 0% and 100%.

Example   Alysha wants to start a savings account at a bank. She will make an initial deposit of $1000 to open the account and plans to deposit $100 at the beginning of each month. She plans to do this for the next 3 years (36 months). At the end of 3 years, she wants to have at least $5000. What is the minimum interest required to achieve this result?

Interest rate: $= \text{RATE}(36, -100, -1000, 5000, 1)$

SLN (Straight Line Depreciation)

Calculates the straight line depreciation of an asset for a given year.

$$= \text{SLN(cost, salvage, life)}$$

cost	First cost or basis of the asset.
salvage	Salvage value.
life	Depreciation life.

Example   Maria purchased a printing machine for $100,000. The machine has an allowed depreciation life of 8 years and an estimated salvage value of $15,000. What is the depreciation each year?

Depreciation: $= \text{SLN}(100000, 15000, 8)$

SYD (Sum-of-Years-Digits Depreciation)

Calculates the sum-of-years-digits depreciation of an asset for a given year.

$$= \text{SYD(cost, salvage, life, period)}$$

cost	First cost or basis of the asset.
salvage	Salvage value.
life	Depreciation life.
period	The year for which the depreciation is sought.

Example   Jack bought equipment for $100,000 that has a depreciation life of 10 years. The salvage value is $10,000. What is the depreciation for year 1 and year 9?

Depreciation for year 1: $= \text{SYD}(100000, 10000, 10, 1)$

Depreciation for year 9: $= \text{SYD}(100000, 10000, 10, 9)$

VDB (Variable Declining Balance)

Calculates the depreciation using the declining balance method with a switch to straight line depreciation in the year in which straight line has a larger depreciation amount. This function automatically implements the switch from DB to SL depreciation, unless specifically instructed to not switch.

$$= \text{VDB (cost, salvage, life, start\_period, end\_period, factor, no\_switch)}$$

cost	First cost of the asset.
salvage	Salvage value.
life	Depreciation life.
start_period	First period for depreciation to be calculated.
end_period	Last period for depreciation to be calculated.
factor	(optional entry) If omitted, the function will use the double declining rate of $2/n$, or twice the straight line rate. Other entries define the declining balance method, for example, 1.5 for 150% declining balance.
no_switch	(optional entry) If omitted or entered as FALSE, the function will switch from declining balance to straight line depreciation when the latter is greater than DB depreciation. If entered as TRUE, the function will not switch to SL depreciation at any time during the depreciation life.

Example   Newly purchased equipment with a first cost of $300,000 has a depreciable life of 10 years with no salvage value. Calculate the 175% declining balance depreciation for the first year and the ninth year if switching to SL depreciation is acceptable and if switching is not permitted.

Depreciation for first year, with switching: $= \text{VDB}(300000,0,10,0,1,1.75)$

Depreciation for ninth year, with switching: $= \text{VDB}(300000,0,10,8,9,1.75)$

Depreciation for first year, no switching: $= \text{VDB}(300000,0,10,0,1,1.75,\text{TRUE})$

Depreciation for ninth year, no switching: $= \text{VDB}(300000,0,10,8,9,1.75,\text{TRUE})$

VDB (for MACRS Depreciation)

The VDB function can be adapted to generate the MACRS annual depreciation amount, when the start_period and end_period are replaced with the MAX and MIN functions, respectively. As above, the factor option should be entered if other than DDB rates start the MACRS depreciation. The VDB format is

$$= \text{VDB}(\text{cost},0,\text{life},\text{MAX}(0,t-1.5),\text{MIN}(\text{life},t-0.5),\text{factor})$$

Example   Determine the MACRS depreciation for year 4 for a $350,000 asset that has a 20% salvage value and a MACRS recovery period of 3 years. $D_4 = \$25,926$ is the display.

Depreciation for year 4: $= \text{VDB}(350000,0,3,\text{MAX}(0,4-1.5),\text{MIN}(3,4-0.5),2)$

Example   Find the MACRS depreciation in year 16 for a $350,000-asset with a recovery period of $n = 15$ years. The optional factor 1.5 is required here, since MACRS starts with 150% DB for $n = 15$-year and 20-year recovery periods. $D_{16} = \$10,334$.

Depreciation for year 16: $= \text{VDB}(350000,0,15,\text{MAX}(0,16-1.5),\text{MIN}(15,16-0.5),1.5)$

Other Functions

There are numerous additional financial functions available on Excel, as well as engineering, mathematics, trigonometry, statistics, data and time, logical, and information functions. These can be viewed by clicking the Formulas tab on the Excel toolbar.

A.4 Goal Seek—A Tool for Breakeven and Sensitivity Analysis ● ● ◌

Goal Seek is found on the Excel toolbar labeled Data, followed by What-if Analysis. This tool changes the value in a specific cell based on a numerical value in another (changing) cell as input by the user. It is a good tool for **sensitivity analysis, breakeven analysis,** and **"what if?" questions**

Figure A–3
Goal Seek template used
to specify a cell, a value,
and the changing cell.

Figure A–4
Use of Goal Seek to determine an annual cash flow to increase the rate of return.

when no constraint relations or inequalities are needed. The initial Goal Seek template is pictured in Figure A–3. One of the cells (set or changing cell) must contain an equation or spreadsheet function that uses the other cell to determine a numeric value. Only a single cell can be identified as the changing cell; however, this limitation can be avoided by using equations rather than specific numerical inputs in any additional cells also to be changed. This is demonstrated below.

Example   A new asset will cost $25,000, generate an annual cash flow of $6000 over its 5-year life, and have an estimated $500 salvage value. The rate of return using the IRR function is 6.94%. Determine the annual cash flow necessary to raise the return to 10% per year.

Figure A–4 (top left) shows the cash flows and return displayed using the function = IRR(B4:B9) prior to the use of Goal Seek. Note that the initial $6000 is input in cell B5, but other years' cash flows are input as equations that refer to B5. The $500 salvage is added for the last year. This format allows Goal Seek to change only cell B5 while making the other cash flows have the same value. The tool finds the required cash flow of $6506 to approximate the 10% per year return. The Goal Seek Status inset indicates that a solution is found. Clicking OK saves all changed cells; clicking Cancel returns to the original values.

A.5 Solver—An Optimizing Tool for Capital Budgeting, Breakeven, and Sensitivity Analysis ●●●

Solver is a powerful spreadsheet tool to change the value in multiple (one or more) cells based on the value in a specific (target) cell. It is excellent when solving a **capital budgeting problem** to select from independent projects where budget constraints are present. (Section 12.4 details this application.) The initial Solver template is shown in Figure A–5.

Figure A–5
Solver template used to specify optimization in a target cell, multiple changing cells, and constraint relations.

Set Target Cell box. Enter a cell reference or name. The target cell itself must contain a formula or function. The value in the cell can be maximized (Max), minimized (Min), or restricted to a specified value (Value of).

By Changing Cells box. Enter the cell reference for each cell to be adjusted, using commas between nonadjacent cells. Each cell must be directly or indirectly related to the target cell. Solver proposes a value for the changing cell based on input provided about the target cell. The Guess button will list all possible changing cells related to the target cell.

Subject to the Constraints box. Enter any constraints that may apply, for example, $C$1 < $50,000. Integer and binary variables are determined in this box.

Options box. Choices here allow the user to specify various parameters of the solution: maximum time and number of iterations allowed, the precision and tolerance of the values determined, and the convergence requirements as the final solution is determined. Also, linear and nonlinear model assumptions can be set here. *If integer or binary variables are involved, the tolerance option must be set to a small number, say, 0.0001.* This is especially important for the binary variables when selecting from independent projects (Chapter 12). If tolerance remains at the default value of 5%, a project may be incorrectly included in the solution set at a very low level.

Solver Results box. This appears after Solve is clicked and a solution appears. It is possible, of course, that no solution can be found for the scenario described. It is possible to update the spreadsheet by clicking Keep Solver Solution, or return to the original entries using Restore Original Values.

A.6 Error Messages ● ● ◉

If Excel is unable to complete a formula or function computation, an error message is displayed. Some of the common messages are:

#DIV/0!	Requires division by zero.
#N/A	Refers to a value that is not available.
#NAME?	Uses a name that Excel doesn't recognize.
#NULL!	Specifies an invalid intersection of two areas.
#NUM!	Uses a number incorrectly.
#REF!	Refers to a cell that is not valid.
#VALUE!	Uses an invalid argument or operand.
#####	Produces a result, or includes a constant numeric value, that is too long to fit in the cell. (Widen the column.)

BASICS OF ACCOUNTING REPORTS AND BUSINESS RATIOS

This appendix provides a fundamental description of financial statements. The documents discussed here will assist in reviewing or understanding basic financial statements and in gathering information useful in an engineering economy study.

B.1 The Balance Sheet ● ● ○

The fiscal year and the tax year are defined identically for a corporation or an individual—12 months in length. The fiscal year (FY) is commonly not the calendar year (CY) for a corporation. The U.S. government uses October through September as its FY. For example, October 2011 through September 2012 is FY2012. The fiscal or tax year is always the calendar year for an individual citizen.

At the end of each fiscal year, a company publishes a **balance sheet.** A sample balance sheet for JAGBA Corporation is presented in Table B–1. This is a yearly presentation of the state of the firm at a particular time, for example, May 31, 2012; however, a balance sheet is also usually prepared quarterly and monthly. Three main categories are used.

Assets. This section is a summary of all resources owned by or owed to the company. There are two main classes of assets. *Current assets* represent shorter-lived working capital (cash, accounts receivable, etc.), which is more easily converted to cash, usually within 1 year. Longer-lived assets are referred to as *fixed assets* (land, equipment, etc.). Conversion of these holdings to cash in a short time would require a major corporate reorientation.

Liabilities. This section is a summary of all *financial obligations* (debts, mortgages, loans, etc.) of a corporation. Bond indebtedness is included here.

Net worth. Also called *owner's equity,* this section provides a summary of the financial value of ownership, including stocks issued and earnings retained by the corporation.

TABLE B–1 Sample Balance Sheet

JAGBA CORPORATION
Balance Sheet
May 31, 2012

Assets		Liabilities	
Current			
Cash	$10,500	Accounts payable	$19,700
Accounts receivable	18,700	Dividends payable	7,000
Interest accrued receivable	500	Long-term notes payable	16,000
Inventories	52,000	Bonds payable	20,000
Total current assets	$81,700	Total liabilities	$62,700
		Net Worth	
Fixed			
Land	$25,000	Common stock	$275,000
Building and equipment	438,000	Preferred stock	100,000
Less: Depreciation		Retained earnings	25,000
allowance $82,000	356,000		
Total fixed assets	381,000	Total net worth	400,000
Total assets	$462,700	Total liabilities and net worth	$462,700

The balance sheet is constructed using the relation

> Assets = liabilities + net worth

In Table B–1 each major category is further divided into standard subcategories. For example, current assets is comprised of cash, accounts receivable, etc. Each subdivision has a specific interpretation, such as accounts receivable, which represents all money owed to the company by its customers.

B.2 Income Statement and Cost of Goods Sold Statement ● ● ●

A second important financial statement is the **income statement** (Table B–2). The income statement summarizes the profits or losses of the corporation for a stated period of time. Income statements always accompany balance sheets. The major categories of an income statement are

Revenues. This includes all *sales and interest revenue* that the company has received in the past accounting period.

Expenses. This is a summary of *all expenses* (operating and others, including taxes) for the period. Some expense amounts are itemized in other statements, for example, cost of goods sold.

The final result of an income statement is the net profit after taxes (NPAT), or NOPAT (O for operating), the amount used in Chapter 17, Sections 17.1 and 17.7. The income statement, published at the same time as the balance sheet, uses the basic equation

> Revenues − expenses = profit (or loss)

The **cost of goods sold** is an important accounting term. It represents the net cost of producing the product marketed by the firm. Cost of goods sold may also be called *factory cost*. A statement of the cost of goods sold, such as that shown in Table B–3, is useful in determining exactly how much it costs to make a particular product over a stated time period, usually a year. The total of the cost of goods sold statement is entered as an expense item on the income statement. This total is determined using the relations

$$\text{Cost of goods sold} = \text{prime cost} + \text{indirect cost}$$
$$\text{Prime cost} = \text{direct materials} + \text{direct labor} \qquad \text{[B.1]}$$

TABLE B–2	Sample Income Statement

JAGBA CORPORATION
Income Statement
Year Ended May 31, 2012

Revenues		
Sales	$505,000	
Interest revenue	3,500	
Total revenues		$508,500
Expenses		
Cost of goods sold (from Table B–3)	$290,000	
Selling	28,000	
Administrative	35,000	
Other	12,000	
Total expenses		365,000
Income before taxes		143,500
Taxes for year		64,575
Net profit after taxes (NPAT)		$ 78,925

TABLE B–3	Sample Cost of Goods Sold Statement

JAGBA CORPORATION
Statement of Cost of Goods Sold
Year Ended May 31, 2012

Materials		
Inventory June 1, 2011	$ 54,000	
Purchases during year	174,500	
Total	$228,500	
Less: Inventory May 31, 2012	50,000	
Cost of materials		$178,500
Direct labor		110,000
Prime cost		288,500
Indirect costs		7,000
Factory cost		295,500
Less: Increase in finished goods inventory during year		5,500
Cost of goods sold (into Table B–2)		$290,000

Indirect costs include all indirect and overhead charges made to a product, process, or cost center. Indirect cost allocation methods are discussed in Chapter 15.

B.3 Business Ratios ●●○

Accountants, financial analysts, and engineering economists frequently utilize business ratio analysis to evaluate the financial health (status) of a company over time and in relation to industry norms. Because the engineering economist must continually communicate with others, she or he should have a basic understanding of several ratios. For comparison purposes, it is necessary to compute the ratios for several companies in the same industry. Industrywide median ratio values are published annually by firms such as Dun and Bradstreet in *Industry Norms and Key Business Ratios.* The ratios are classified according to their role in measuring the corporation.

Solvency ratios. Assess ability to meet short-term and long-term financial obligations.

Efficiency ratios. Measure management's ability to use and control assets.

Profitability ratios. Evaluate the ability to earn a return for the owners of the corporation.

Numerical data for several important ratios are discussed here and are extracted from the JAGBA balance sheet and income statement, Tables B–1 and B–2.

Current Ratio   This ratio is utilized to analyze the company's working capital condition. It is defined as

$$\text{Current ratio} = \frac{\text{current assets}}{\text{current liabilities}}$$

Current liabilities include all short-term debts, such as accounts and dividends payable. Note that only balance sheet data are utilized in the current ratio; that is, no association with revenues or expenses is made. For the balance sheet of Table B–1, current liabilities amount to $19,700 + $7000 = $26,700 and

$$\text{Current ratio} = \frac{81,700}{26,700} = 3.06$$

Since current liabilities are those debts payable in the next year, the current ratio value of 3.06 means that the current assets would cover short-term debts approximately 3 times. Current ratio values of 2 to 3 are common.

The current ratio assumes that the working capital invested in inventory can be converted to cash quite rapidly. Often, however, a better idea of a company's *immediate* financial position can be obtained by using the acid test ratio.

Acid Test Ratio (Quick Ratio)   This ratio is

$$\text{Acid-test ratio} = \frac{\text{quick assets}}{\text{current liabilities}}$$

$$= \frac{\text{current assets} - \text{inventories}}{\text{current liabilities}}$$

It is meaningful for the emergency situation when the firm must cover short-term debts using its readily convertible assets. For JAGBA Corporation,

$$\text{Acid test ratio} = \frac{81,700 - 52,000}{26,700} = 1.11$$

Comparison of this and the current ratio shows that approximately 2 times the current debt of the company is invested in inventories. However, an acid test ratio of approximately 1.0 is generally regarded as a strong current position, regardless of the amount of assets in inventories.

Debt Ratio   This ratio is a measure of financial strength since it is defined as

$$\text{Debt ratio} = \frac{\text{total liabilities}}{\text{total assets}}$$

For JAGBA Corporation,

$$\text{Debt ratio} = \frac{62,700}{462,700} = 0.136$$

JAGBA is 13.6% creditor-owned and 86.4% stockholder-owned. A debt ratio in the range of 20% or less usually indicates a sound financial condition, with little fear of forced reorganization because of unpaid liabilities. However, a company with virtually no debts, that is, one with a very low debt ratio, may not have a promising future, because of its inexperience in dealing with short-term and long-term debt financing. The debt-equity (D-E) mix is another measure of financial strength.

Return on Sales Ratio   This often quoted ratio indicates the profit margin for the company. It is defined as

$$\text{Return on sales} = \frac{\text{net profit}}{\text{net sales}} (100\%)$$

Net profit is the after-tax value from the income statement. This ratio measures profit earned per sales dollar and indicates how well the corporation can sustain adverse conditions over time, such as falling prices, rising costs, and declining sales. For JAGBA Corporation,

$$\text{Return on sales} = \frac{78,925}{505,000} (100\%) = 15.6\%$$

Corporations may point to small return on sales ratios, say, 2.5% to 4.0%, as indications of sagging economic conditions. In truth, for a relatively large-volume, high-turnover business, an income ratio of 3% is quite healthy. Of course, a steadily decreasing ratio indicates rising company expenses, which absorb net profit after taxes.

Return on Assets Ratio   This is the key indicator of profitability since it evaluates the ability of the corporation to transfer assets into operating profit. The definition and value for JAGBA are

$$\text{Return on assets} = \frac{\text{net profit}}{\text{total assets}} (100\%)$$

$$= \frac{78,925}{462,700} (100\%) = 17.1\%$$

Efficient use of assets indicates that the company should earn a high return, while low returns usually accompany lower values of this ratio compared to the industry group ratios.

Inventory Turnover Ratio   Two different ratios are used here. They both indicate the number of times the average inventory value passes through the operations of the company. If turnover of inventory to *net sales* is desired, the formula is

$$\text{Net sales to inventory} = \frac{\text{net sales}}{\text{average inventory}}$$

where average inventory is the figure recorded in the balance sheet. For JAGBA Corporation this ratio is

$$\text{Net sales to inventory} = \frac{505{,}000}{52{,}000} = 9.71$$

This means that the average value of the inventory has been sold 9.71 times during the year. Values of this ratio vary greatly from one industry to another.

If inventory turnover is related to *cost of goods sold,* the ratio to use is

$$\text{Cost of goods sold to inventory} = \frac{\text{cost of goods sold}}{\text{average inventory}}$$

Now, average inventory is computed as the average of the beginning and ending inventory values in the statement of cost of goods sold. This ratio is commonly used as a measure of the inventory turnover rate in manufacturing companies. It varies with industries, but management likes to see it remain relatively constant as business increases. For JAGBA, using the values in Table B–3,

$$\text{Cost of goods sold to inventory} = \frac{290{,}000}{\frac{1}{2}(54{,}000 + 50{,}000)} = 5.58$$

There are, of course, many other ratios to use in various circumstances; however, the ones presented here are commonly used by both accountants and economic analysts.

EXAMPLE B.1

Sample values for financial ratios or percentages of four industry sectors are presented below. Compare the corresponding JAGBA Corporation values with these norms, and comment on differences and similarities.

Ratio or Percentage	Motor Vehicles and Parts Manufacturing 336105*	Air Transportation (Medium-Sized) 481000*	Industrial Machinery Manufacturing 333200*	Home Furnishings 442000*
Current ratio	2.4	0.4	2.2	2.6
Quick ratio	1.6	0.3	1.5	1.2
Debt ratio	59.3%	96.8%	49.1%	52.4%
Return on assets	40.9%	8.1%	8.0%	5.1%

*North American Industry Classification System (NAICS) code for this industry sector.
SOURCE: L. Troy, *Almanac of Business and Industrial Financial Ratios,* CCH, Wolters Kluwer, USA.

Solution

It is not correct to compare ratios for one company with indexes in different industries, that is, with indexes for different NAICS codes. So the comparison below is for illustration purposes only. The corresponding values for JAGBA are

$$\text{Current ratio} = 3.06$$
$$\text{Quick ratio} = 1.11$$
$$\text{Debt ratio} = 13.5\%$$
$$\text{Return on assets} = 17.1\%$$

JAGBA has a current ratio larger than all four of these industries, since 3.06 indicates it can cover current liabilities 3 times compared with 2.6 and much less in the case of the "average" air transportation corporation. JAGBA has a significantly lower debt ratio than that of any of the sample industries, so it is likely more financially sound. Return on assets, which is a measure of ability to turn assets into profitability, is not as high at JAGBA as motor vehicles, but JAGBA competes well with the other industry sectors.

To make a fair comparison of JAGBA ratios with other values, it is necessary to have norm values *for its industry type* as well as ratio values for other corporations *in the same NAICS category* and about the same size in total assets. Corporate assets are classified in categories by $100,000 units, such as 100 to 250, 1001 to 5000, over 250,000, etc.

CODE OF ETHICS FOR ENGINEERS

Source: National Society of Professional Engineers (www.nspe.org).

**National Society of
Professional Engineers®**

Code of Ethics for Engineers

Preamble

Engineering is an important and learned profession. As members of this profession, engineers are expected to exhibit the highest standards of honesty and integrity. Engineering has a direct and vital impact on the quality of life for all people. Accordingly, the services provided by engineers require honesty, impartiality, fairness, and equity, and must be dedicated to the protection of the public health, safety, and welfare. Engineers must perform under a standard of professional behavior that requires adherence to the highest principles of ethical conduct.

I. Fundamental Canons

Engineers, in the fulfillment of their professional duties, shall:
1. Hold paramount the safety, health, and welfare of the public.
2. Perform services only in areas of their competence.
3. Issue public statements only in an objective and truthful manner.
4. Act for each employer or client as faithful agents or trustees.
5. Avoid deceptive acts.
6. Conduct themselves honorably, responsibly, ethically, and lawfully so as to enhance the honor, reputation, and usefulness of the profession.

II. Rules of Practice

1. Engineers shall hold paramount the safety, health, and welfare of the public.
   a. If engineers' judgment is overruled under circumstances that endanger life or property, they shall notify their employer or client and such other authority as may be appropriate.
   b. Engineers shall approve only those engineering documents that are in conformity with applicable standards.
   c. Engineers shall not reveal facts, data, or information without the prior consent of the client or employer except as authorized or required by law or this Code.
   d. Engineers shall not permit the use of their name or associate in business ventures with any person or firm that they believe is engaged in fraudulent or dishonest enterprise.
   e. Engineers shall not aid or abet the unlawful practice of engineering by a person or firm.
   f. Engineers having knowledge of any alleged violation of this Code shall report thereon to appropriate professional bodies and, when relevant, also to public authorities, and cooperate with the proper authorities in furnishing such information or assistance as may be required.
2. Engineers shall perform services only in the areas of their competence.
   a. Engineers shall undertake assignments only when qualified by education or experience in the specific technical fields involved.
   b. Engineers shall not affix their signatures to any plans or documents dealing with subject matter in which they lack competence, nor to any plan or document not prepared under their direction and control.
   c. Engineers may accept assignments and assume responsibility for coordination of an entire project and sign and seal the engineering documents for the entire project, provided that each technical segment is signed and sealed only by the qualified engineers who prepared the segment.
3. Engineers shall issue public statements only in an objective and truthful manner.
   a. Engineers shall be objective and truthful in professional reports, statements, or testimony. They shall include all relevant and pertinent information in such reports, statements, or testimony, which should bear the date indicating when it was current.
   b. Engineers may express publicly technical opinions that are founded upon knowledge of the facts and competence in the subject matter.
   c. Engineers shall issue no statements, criticisms, or arguments on technical matters that are inspired or paid for by interested parties, unless they have prefaced their comments by explicitly identifying the interested parties on whose behalf they are speaking, and by revealing the existence of any interest the engineers may have in the matters.

4. Engineers shall act for each employer or client as faithful agents or trustees.
   a. Engineers shall disclose all known or potential conflicts of interest that could influence or appear to influence their judgment or the quality of their services.
   b. Engineers shall not accept compensation, financial or otherwise, from more than one party for services on the same project, or for services pertaining to the same project, unless the circumstances are fully disclosed and agreed to by all interested parties.
   c. Engineers shall not solicit or accept financial or other valuable consideration, directly or indirectly, from outside agents in connection with the work for which they are responsible.
   d. Engineers in public service as members, advisors, or employees of a governmental or quasi-governmental body or department shall not participate in decisions with respect to services solicited or provided by them or their organizations in private or public engineering practice.
   e. Engineers shall not solicit or accept a contract from a governmental body on which a principal or officer of their organization serves as a member.
5. Engineers shall avoid deceptive acts.
   a. Engineers shall not falsify their qualifications or permit misrepresentation of their or their associates' qualifications. They shall not misrepresent or exaggerate their responsibility in or for the subject matter of prior assignments. Brochures or other presentations incident to the solicitation of employment shall not misrepresent pertinent facts concerning employers, employees, associates, joint venturers, or past accomplishments.
   b. Engineers shall not offer, give, solicit, or receive, either directly or indirectly, any contribution to influence the award of a contract by public authority, or which may be reasonably construed by the public as having the effect or intent of influencing the awarding of a contract. They shall not offer any gift or other valuable consideration in order to secure work. They shall not pay a commission, percentage, or brokerage fee in order to secure work, except to a bona fide employee or bona fide established commercial or marketing agencies retained by them.

III. Professional Obligations

1. Engineers shall be guided in all their relations by the highest standards of honesty and integrity.
   a. Engineers shall acknowledge their errors and shall not distort or alter the facts.
   b. Engineers shall advise their clients or employers when they believe a project will not be successful.
   c. Engineers shall not accept outside employment to the detriment of their regular work or interest. Before accepting any outside engineering employment, they will notify their employers.
   d. Engineers shall not attempt to attract an engineer from another employer by false or misleading pretenses.
   e. Engineers shall not promote their own interest at the expense of the dignity and integrity of the profession.
2. Engineers shall at all times strive to serve the public interest.
   a. Engineers are encouraged to participate in civic affairs; career guidance for youths; and work for the advancement of the safety, health, and well-being of their community.
   b. Engineers shall not complete, sign, or seal plans and/or specifications that are not in conformity with applicable engineering standards. If the client or employer insists on such unprofessional conduct, they shall notify the proper authorities and withdraw from further service on the project.
   c. Engineers are encouraged to extend public knowledge and appreciation of engineering and its achievements.
   d. Engineers are encouraged to adhere to the principles of sustainable development[1] in order to protect the environment for future generations.

3. Engineers shall avoid all conduct or practice that deceives the public.
   a. Engineers shall avoid the use of statements containing a material misrepresentation of fact or omitting a material fact.
   b. Consistent with the foregoing, engineers may advertise for recruitment of personnel.
   c. Consistent with the foregoing, engineers may prepare articles for the lay or technical press, but such articles shall not imply credit to the author for work performed by others.
4. Engineers shall not disclose, without consent, confidential information concerning the business affairs or technical processes of any present or former client or employer, or public body on which they serve.
   a. Engineers shall not, without the consent of all interested parties, promote or arrange for new employment or practice in connection with a specific project for which the engineer has gained particular and specialized knowledge.
   b. Engineers shall not, without the consent of all interested parties, participate in or represent an adversary interest in connection with a specific project or proceeding in which the engineer has gained particular specialized knowledge on behalf of a former client or employer.
5. Engineers shall not be influenced in their professional duties by conflicting interests.
   a. Engineers shall not accept financial or other considerations, including free engineering designs, from material or equipment suppliers for specifying their product.
   b. Engineers shall not accept commissions or allowances, directly or indirectly, from contractors or other parties dealing with clients or employers of the engineer in connection with work for which the engineer is responsible.
6. Engineers shall not attempt to obtain employment or advancement or professional engagements by untruthfully criticizing other engineers, or by other improper or questionable methods.
   a. Engineers shall not request, propose, or accept a commission on a contingent basis under circumstances in which their judgment may be compromised.
   b. Engineers in salaried positions shall accept part-time engineering work only to the extent consistent with policies of the employer and in accordance with ethical considerations.
   c. Engineers shall not, without consent, use equipment, supplies, laboratory, or office facilities of an employer to carry on outside private practice.
7. Engineers shall not attempt to injure, maliciously or falsely, directly or indirectly, the professional reputation, prospects, practice, or employment of other engineers. Engineers who believe others are guilty of unethical or illegal practice shall present such information to the proper authority for action.
   a. Engineers in private practice shall not review the work of another engineer for the same client, except with the knowledge of such engineer, or unless the connection of such engineer with the work has been terminated.
   b. Engineers in governmental, industrial, or educational employ are entitled to review and evaluate the work of other engineers when so required by their employment duties.
   c. Engineers in sales or industrial employ are entitled to make engineering comparisons of represented products with products of other suppliers.
8. Engineers shall accept personal responsibility for their professional activities, provided, however, that engineers may seek indemnification for services arising out of their practice for other than gross negligence, where the engineer's interests cannot otherwise be protected.
   a. Engineers shall conform with state registration laws in the practice of engineering.
   b. Engineers shall not use association with a nonengineer, a corporation, or partnership as a "cloak" for unethical acts.

9. Engineers shall give credit for engineering work to those to whom credit is due, and will recognize the proprietary interests of others.
   a. Engineers shall, whenever possible, name the person or persons who may be individually responsible for designs, inventions, writings, or other accomplishments.
   b. Engineers using designs supplied by a client recognize that the designs remain the property of the client and may not be duplicated by the engineer for others without express permission.
   c. Engineers, before undertaking work for others in connection with which the engineer may make improvements, plans, designs, inventions, or other records that may justify copyrights or patents, should enter into a positive agreement regarding ownership.
   d. Engineers' designs, data, records, and notes referring exclusively to an employer's work are the employer's property. The employer should indemnify the engineer for use of the information for any purpose other than the original purpose.
   e. Engineers shall continue their professional development throughout their careers and should keep current in their specialty fields by engaging in professional practice, participating in continuing education courses, reading in the technical literature, and attending professional meetings and seminars.

Footnote 1 "Sustainable development" is the challenge of meeting human needs for natural resources, industrial products, energy, food, transportation, shelter, and effective waste management while conserving and protecting environmental quality and the natural resource base essential for future development.

As Revised July 2007

"By order of the United States District Court for the District of Columbia, former Section 11(c) of the NSPE Code of Ethics prohibiting competitive bidding, and all policy statements, opinions, rulings or other guidelines interpreting its scope, have been rescinded as unlawfully interfering with the legal right of engineers, protected under the antitrust laws, to provide price information to prospective clients; accordingly, nothing contained in the NSPE Code of Ethics, policy statements, opinions, rulings or other guidelines prohibits the submission of price quotations or competitive bids for engineering services at any time or in any amount."

Statement by NSPE Executive Committee

In order to correct misunderstandings which have been indicated in some instances since the issuance of the Supreme Court decision and the entry of the Final Judgment, it is noted that in its decision of April 25, 1978, the Supreme Court of the United States declared: "The Sherman Act does not require competitive bidding."

It is further noted that as made clear in the Supreme Court decision:
1. Engineers and firms may individually refuse to bid for engineering services.
2. Clients are not required to seek bids for engineering services.
3. Federal, state, and local laws governing procedures to procure engineering services are not affected, and remain in full force and effect.
4. State societies and local chapters are free to actively and aggressively seek legislation for professional selection and negotiation procedures by public agencies.
5. State registration board rules of professional conduct, including rules prohibiting competitive bidding for engineering services, are not affected and remain in full force and effect. State registration boards with authority to adopt rules of professional conduct may adopt rules governing procedures to obtain engineering services.
6. As noted by the Supreme Court, "nothing in the judgment prevents NSPE and its members from attempting to influence governmental action . . ."

Note: In regard to the question of application of the Code to corporations vis-a-vis real persons, business form or type should not negate nor influence conformance of individuals to the Code. The Code deals with professional services, which services must be performed by real persons. Real persons in turn establish and implement policies within business structures. The Code is clearly written to apply to the Engineer, and it is incumbent on members of NSPE to endeavor to live up to its provisions. This applies to all pertinent sections of the Code.

National Society of Professional Engineers®

1420 King Street
Alexandria, Virginia 22314-2794
703/684-2800 • Fax:703/836-4875
www.nspe.org
Publication date as revised: July 2007 • Publication #1102

ALTERNATE METHODS FOR EQUIVALENCE CALCULATIONS

Throughout the text, engineering economy factor formulas, tabulated factor values, or built-in spreadsheet functions have been used to obtain a value of P, F, A, i, or n. Because of advances in programmable and scientific calculators, many of the equivalence computations can be performed without the use of tables or spreadsheets, but rather with a handheld calculator. An overview of the possibilities is presented here.

Alternatively, the recognition that all equivalence calculations involve geometric series can likewise remove the need for tabulated values or spreadsheet functions. From the summation of the series, it is possible to perform calculator-based computations to obtain P, F, or A values. A brief introduction to this technique is presented in Section D.2.

D.1 Using Programmable Calculators ● ● ●

A basic way to calculate one parameter, given the other four, is to use a calculator that allows a present worth relation to be encoded. The software can then solve for any one of the parameters, when the remaining four are entered. For example, consider a PW relation in which all five parameters are included.

$$A(P/A,i,n) + F(P/F,i,n) + P = 0$$

The A, P, and F values can be positive (cash inflow) or negative (cash outflow) or zero, as long as there is at least one value with each sign. The interest rate i can be coded for entry as a percent or decimal. When the unknown variable is identified, the calculator's software can solve the equation for zero, thus providing the answer.

This is the approach taken by relatively simple scientific calculators, such as the Hewlett-Packard (HP) scientific series, for example, HP 33s. By substituting the formulas for the factors, the actual relation entered into the calculator is

$$A\left[\frac{1 - (1 + i/100)^{-n}}{i/100}\right] + F[1 + (i/100)]^{-n} + P = 0$$

The HP calculator uses a slightly different symbol set than we have used thus far. The initial investment is called B rather than P, and the equal uniform amount is termed P rather than A. Once entered, the relation can be solved for any one variable, given values for the other four.

Another example that offers freedom from the spreadsheet and tables is an engineering calculator that has the same functions built in as those on a spreadsheet to determine P, F, A, i, or n. An example of this higher level is Texas Instrument's TI-Nspire series. The functions are basically the same as those on a spreadsheet. They are clustered under the heading of tvm (time value of money) functions. For example, the tvmPV function format is

$$\text{tvmPV}(n,i,\text{Pmt},\text{FV},\text{PpY},\text{CpY},\text{PmtAt})$$

where n = number of periods
          i = annual interest rate as a percent
      Pmt = equal uniform periodic amount A
        FV = future amount F
      PpY = payments per year (optional; default is 1)
      CpY = compounding periods per year (optional; default is 1)
   PmtAt = beginning- or end-of-period payments (optional; default is 0 = end)

It is easy to understand why it is possible to do a lot on a calculator with relatively well-behaved cash flow series. As the series become more complex, it is necessary to move to a spreadsheet for speed and versatility. However, the use of tables or factor formulas is not necessary.

D.2 Using the Summation of a Geometric Series ● ● ●

A geometric progression is a series of n terms with a common ratio or base r. If c is a constant for each term, the series is written in the form

$$cr^a + cr^{a+1} + \cdots + cr^n = c\sum_{j=a}^{j=n} r^j$$

The sum S of a geometric series adds the terms using the closed-end form

$$S = \frac{r^{n+1} - r^a}{r - 1} \qquad \text{[D.1]}$$

Ristroph[1] and others have explained how the recognition that equivalence computations are simple applications of geometric series can be used to determine F, P, and A using Equation [D.1] and a simple handheld calculator with exponentiation capability.

Before explaining how to apply this approach, we define the base r as follows when a future worth F or present worth P is sought.

To find F: $r = 1 + i$

To find P: $r = (1 + i)^{-1}$

A very familiar application of geometric series is the determination of the equivalent future worth F in year n for a single present worth amount P in year 0. This is the same as using a geometric series of only one term. As shown in Figure D–1, if $P = \$100$, $n = 10$ years, and $i = 10\%$ per year, when $r = 1 + i$,

$$F = P(1 + i)^n = P(r)^n = 100(1.1)^{10}$$

$$= 100(2.5937)$$

$$= \$259.37$$

This is identical to using the tabulated value (or formula) for the F/P factor.

$$F = P(F/P,i,n) = 100(F/P,10\%,10) = 100(2.5937)$$

$$= \$259.37$$

Figure D–2 shows a uniform annual series $A = \$100$ for 10 years. To calculate F, we can move each A value forward to year 10. Place a subscript j on each A value to indicate the year of occurrence, and determine F for each A_j value.

$$F = A_1(1 + i)^9 + A_2(1 + i)^8 + \cdots + A_9(1 + i)^1 + A_{10}(1 + i)^0$$

Figure D–1
Future worth of a single amount in year 0.

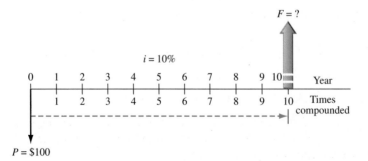

[1]J. H. Ristroph, "Engineering Economics: Time for New Directions?" *Proceedings*, ASEE Annual Conference, Austin, TX, June 2009.

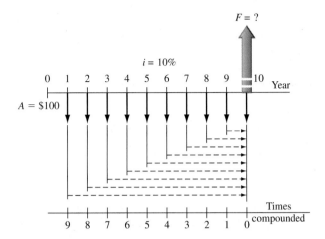

Figure D–2
Future worth of an A series from year 1 through year 10.

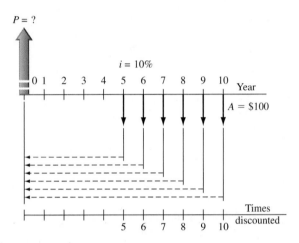

Figure D–3
Present worth of a shifted uniform series.

Note that the first value A_1 is compounded 9 times, not 10. With $r = 1 + i$, the geometric series and its sum S (from Equation [D.1]) can be developed. Removing the subscript on A,

$$F = A[r^9 + r^8 + \cdots + r^1 + r^0] = A\sum_{j=0}^{j=9} r^j$$

$$S = \frac{r^{10} - r^0}{r - 1} = \frac{(1.1)^{10} - 1}{0.1} = 15.9374$$

The F value is the same whether using the geometric series sum or the tabulated F/A factor.

$$F = 100(S) = 100(15.9374) = \$1593.74$$
$$F = 100(F/A,10\%,10) = 100(15.9374) = \$1593.74$$

As a final demonstration of the geometric series approach to equivalence computations, consider the shifted cash flow series in Figure D–3. The A series is present from years 5 through 10, and the P value is sought. If the tables are used, the solution is

$$P = A(P/A,10\%,6)(P/F,10\%,4) = 100(4.3553)(0.6830)$$
$$= \$297.47$$

As shown in the figure, this is a geometric series with the first A value discounted 5 years, therefore, $a = 5$. The last A term is discounted 10 years, making $n = 10$. Since P is sought, the geometric series base is $r = (1 + i)^{-1}$. Again using Equation [D.1] for the summation, we have

$$P = A\left[\sum_{j=5}^{j=10} r^j\right] = 100\left[\frac{r^{11} - r^5}{r - 1}\right] = 100\left[\frac{(1/1.1)^{11} - (1/1.1)^5}{(1/1.1) - 1}\right]$$

$$= 100\left[\frac{(0.9091)^{11} - (0.9091)^5}{0.9091 - 1}\right] = 100(2.9747)$$

$$= \$297.47$$

It is possible to develop similar relations to handle arithmetic and geometric series, conversions from P to A, and vice versa. Proponents of this approach point out the use of standard mathematical notation; the removal of a need to derive any factors; no need to remember the placement of P, F, and A values based on factor formula development; and the easy use of a calculator to determine the equivalence relations. As with the use of programmable calculators, the technique is excellent for well-behaved and reasonably complex series. When the series become quite involved, or when sensitivity analysis is required to reach an economic decision, it may be beneficial to use a spreadsheet. This often helps in performing sidebar and ancillary calculations that assist in the understanding of the problem, not just the math computations necessary to obtain an answer. However, once again, the use of tables and factors is not necessary.

APPENDIX E

GLOSSARY OF CONCEPTS AND TERMS

E.1 Important Concepts and Guidelines ● ● ●

Title

The following elements of engineering economy are identified throughout the text in the margin by this checkmark and a title below it. The numbers in parentheses indicate chapters where the concept or guideline is introduced or essential to obtaining a correct solution.

Time Value of Money   It is a fact that money *makes* money. This concept explains the change in the amount of money *over time* for both owned and borrowed funds. (1)

Economic Equivalence   A combination of time value of money and interest rate that makes different sums of money at different times have *equal economic value*. (1)

Cash Flow   The flow of money into and out of a company, project, or activity. *Revenues are cash inflows* and carry a positive (+) sign; *expenses are outflows* and carry a negative (−) sign. If only costs are involved, the − sign may be omitted, e.g., benefit/cost (B/C) analysis. (1, 9)

End-of-Period Convention   To simplify calculations, cash flows (revenues and costs) are assumed to occur at the *end of a time period*. An interest period or fiscal period is commonly 1 *year*. A half-year convention is often used in depreciation calculations. (1)

Cost of Capital   The interest rate incurred to obtain capital investment funds. COC is usually a *weighted average* that involves the cost of debt capital (loans, bonds, and mortgages) and equity capital (stocks and retained earnings). (1, 10)

Minimum Attractive Rate of Return (MARR)   A reasonable rate of return established for the evaluation of an economic alternative. Also called the *hurdle rate,* MARR is based on cost of capital, market trend, risk, etc. The inequality ROR ≥ MARR > COC is correct for an economically viable project. (1, 10)

Opportunity Cost   A forgone opportunity caused by the inability to pursue a project. Numerically, it is the *largest rate of return* of all the projects not funded due to the lack of capital funds. Stated differently, it is the ROR of the first project rejected because of unavailability of funds. (1, 10)

Nominal or Effective Interest Rate (r or i)   A nominal interest rate *does not include any compounding;* for example, 1% per month is the same as nominal 12% per year. Effective interest rate is the actual rate over a period of time because *compounding is imputed;* for example, 1% per month, compounded monthly, is an effective 12.683% per year. Inflation or deflation is not considered. (4)

Placement of Present Worth (P; PW)   In applying the (P/A,i%,n) factor, P or PW is always located *one interest period (year) prior to the first A amount*. The A or AW is a series of equal, end-of-period cash flows for n consecutive periods, expressed as money per time (say, $/year; €/year). (2, 3)

Placement of Future Worth (F; FW)   In applying the (F/A,i%,n) factor, F or FW is always located at the *end of the last interest period (year) of the A series.* (2, 3)

Placement of Gradient Present Worth (P_G; P_g)   The $(P/G,i\%,n)$ factor for an *arithmetic gradient* finds the P_G of only the gradient series *2 years prior* to the first appearance of the constant gradient G. The base amount A is treated separately from the gradient series.

The $(P/A,g,i,n)$ factor for a *geometric gradient* determines P_g for the gradient and initial amount A_1 *two years prior* to the appearance of the first gradient amount. The initial amount A_1 *is* included in the value of P_g. (2, 3)

Equal-Service Requirement   Identical capacity of all alternatives operating over the *same amount of time* is mandated by the equal-service requirement. Estimated costs and revenues for equal service must be evaluated. PW analysis requires evaluation over the same number of years (periods) using the LCM (least common multiple) of lives; AW analysis is performed over one life cycle. Further, equal service assumes that all costs and revenues rise and fall in accordance with the overall rate of inflation or deflation over the total time period of the evaluation. (5, 6, 8)

LCM or Study Period   To select from mutually exclusive alternatives under the equal-service requirement for PW computations, use the *LCM of lives with repurchase(s)* as necessary. For a stated study period (planning horizon), evaluate cash flows *only over this period*, neglecting any beyond this time; estimated market values at termination of the study period are the salvage values. (5, 6, 11)

Salvage/Market Value   Expected trade-in, market, or scrap value at the *end of the estimated life* or the *study period*. In a replacement study, the defender's estimated market value at the end of a year is considered its "first cost" at the beginning of the next year. MACRS depreciation always reduces the book value to a salvage of zero. (6, 11)

Do Nothing   The DN alternative is always an option, unless one of the defined alternatives *must* be selected. DN is status quo; it generates *no new costs, revenues, or savings*. (5)

Revenue or Cost Alternative   Revenue alternatives have *costs and revenues* estimated; savings are considered negative costs and carry a + sign. Incremental evaluation requires comparison with DN for revenue alternatives. Cost (or service) alternatives have *only costs* estimated; revenues and savings are assumed equal between alternatives. (5)

Rate of Return   An interest rate that equates a PW or AW relation to *zero*. Also defined as the rate on the unpaid balance of borrowed money, or rate earned on the unrecovered balance of an investment such that the *last cash flow brings the balance exactly to zero*. (7, 8)

Project Evaluation   *For a specified* MARR, determine a measure of worth for net cash flow series over the life or study period. Guidelines for a *single project* to be economically justified at the MARR (or discount rate) follow. (5, 6, 7, 9, 17)

Present worth: If PW $\geq$ 0   **Annual worth:** If AW $\geq$ 0

Future worth: If FW $\geq$ 0   **Rate of return:** If $i^* \geq$ MARR

Benefit/cost: If B/C $\geq$ 1.0   **Profitability index:** If PI $\geq$ 1.0

ME Alternative Selection   For mutually exclusive (select only one) alternatives, compare *two alternatives* at a time by determining a measure of worth for the incremental (Δ) cash flow series over the life or study period, adhering to the equal-service requirement. (5, 6, 8, 9, 10, 17)

Present worth or annual worth: Find PW or AW values at MARR; *select numerically largest* (least negative or most positive).

Rate of return: Order by *initial cost,* perform pairwise Δi^* comparison; if $\Delta i^* \geq$ MARR, select *larger cost* alternative; continue until one remains.

Benefit/cost: Order by *total equivalent cost,* perform pairwise ΔB/C comparison; if ΔB/C $\geq$ 1.0, select *larger cost* alternative; continue until one remains.

Cost-effectiveness ratio: For service sector alternatives; order by *effectiveness measure;* perform pairwise ΔC/E comparison using *dominance;* select from nondominated alternatives without exceeding budget.

Independent Project Selection   No comparison between projects; only against DN. Calculate a measure of worth and select using the guidelines below. (5, 6, 8, 9, 12)

Present worth or annual worth: Find PW or AW at MARR; select all projects with PW or AW ≥ 0.

Rate of return: No incremental comparison; select all projects with overall $i^* \geq$ MARR.

Benefit/cost: No incremental comparison; select all projects with overall B/C ≥ 1.0.

Cost-effectiveness ratio: For service sector projects; no incremental comparison; order by CER and select projects to not exceed budget.

When a capital budget limit is defined, independent projects are selected using the *capital budgeting process* based on PW values. The Solver spreadsheet tool is useful here.

Capital Recovery   CR is the equivalent annual amount an asset or system must earn to *recover the initial investment plus a stated rate of return.* Numerically, it is the AW value of the initial investment at a stated rate of return. The salvage value is considered in CR calculations. (6)

Economic Service Life   The ESL is the number of years n at which the *total AW of costs,* including salvage and AOC, is at its *minimum,* considering all the years the asset may provide service. (11)

Sunk Cost   Capital (money) that is lost and cannot be recovered. Sunk costs are not included when making decisions about the future. They should be handled using tax laws and write-off allowances, not the economic study. (11)

Inflation   Expressed as a percentage per time (% per year), it is an *increase* in the amount of money required to purchase the *same amount* of goods or services *over time.* Inflation occurs when the value of a currency decreases. Economic evaluations are performed using either a market (inflation-adjusted) interest rate or an inflation-free rate (constant-value terms). (1, 14)

Breakeven   For a single project, the value of a parameter that makes *two elements equal,* e.g., sales necessary to equate revenues and costs. For two alternatives, breakeven is the value of a common variable at which the two are equally acceptable. Breakeven analysis is fundamental to make-buy decisions, replacement studies, payback analysis, sensitivity analysis, breakeven ROR analysis, and many others. The Goal Seek spreadsheet tool is useful in breakeven analysis. (8, 13)

Payback Period   Amount of time n before *recovery of the initial capital investment* is expected. Payback with $i > 0$ or simple payback at $i = 0$ is useful for preliminary or screening analysis to determine if a full PW, AW, or ROR analysis is needed. (13)

Direct/Indirect Costs   Direct costs are primarily human labor, machines, and materials associated with a product, process, system, or service. Indirect costs, which include support functions, utilities, management, legal, taxes, and the like, are more difficult to associate with a specific product or process. (15)

Value Added   Activities have added worth to a product or service from the perspective of a consumer, owner, or investor who is willing to pay more for an enhanced value. (17)

Sensitivity Analysis   Determination of how a measure of worth is affected by changes in estimated values of a parameter over a stated range. Parameters may be any cost factor, revenue, life, salvage value, inflation rate, etc. (18)

Risk   Variation from an expected, desirable, or predicted value that may be detrimental to the product, process, or system. Risk represents an *absence of or deviation from certainty*. Probability estimates of variation (values) help evaluate risk and uncertainty using statistics and simulation. (10, 18, 19, 20)

E.2 Symbols and Terms

This section identifies and defines the common terms and their symbols used throughout the text. The numbers in parentheses indicate sections where the term is introduced and used in various applications.

Term	Symbol	Description
Annual amount or worth	A or AW	Equivalent uniform annual worth of all cash inflows and outflows over estimated life (1.5, 6.1).
Annual operating cost	AOC	Estimated annual costs to maintain and support an alternative (1.3).
Benefit/cost ratio	B/C	Ratio of a project's benefits to costs expressed in PW, AW, or FW terms (9.2).
Book value	BV	Remaining capital investment in an asset after depreciation is accounted for (16.1).
Breakeven point	Q_{BE}	Quantity at which revenues and costs are equal, or two alternatives are equivalent (13.1).
Capital budget	b	Amount of money available for capital investment projects (12.1).
Capital recovery	CR or A	Equivalent annual cost of owning an asset plus the required return on the initial investment (6.2).
Capitalized cost	CC or P	Present worth of an alternative that will last forever (or a long time) (5.5).
Cash flow	CF	Actual cash amounts that are receipts (inflow) and disbursements (outflow) (1.6).
Cash flow before or after taxes	CFBT or CFAT	Cash flow amount before relevant taxes or after taxes are applied (17.2).
Compounding frequency	m	Number of times interest is compounded per period (year) (4.1).
Cost-effectiveness ratio	CER	Ratio of equivalent cost to effectiveness measure to evaluate service sector projects (9.5).
Cost estimating relationships	C_2 or C_T	Relations that use design variables and changing costs over time to estimate current and future costs (15.3–4).
Cost of capital	WACC	Interest rate paid for the use of capital funds; includes both debt and equity funds. For debt and equity considered, it is weighted average cost of capital (1.9, 10.2).
Debt-equity mix	D-E	Percentages of debt and equity investment capital used by a corporation (10.2).
Depreciation	D	Reduction in the value of assets using specific models and rules; there are book and tax depreciation methods (16.1).
Depreciation rate	d_t	Annual rate for reducing the value of assets using different depreciation methods (16.1).
Economic service life	ESL or n	Number of years at which the AW of costs is a minimum (11.2).
Effectiveness measure	E	A nonmonetary measure used in the cost-effectiveness ratio for service sector projects (9.5).

Term	Symbol	Description
Expected value (average)	$\bar{X}$, μ, or $E(X)$	Long-run expected average if a random variable is sampled many times (18.3, 19.4).
Expenses, operating	OE	All corporate costs incurred in transacting business (17.1).
First cost	P	Total initial cost—purchase, construction, setup, etc. (1.3, 16.1).
Future amount or worth	F or FW	Amount at some future date considering time value of money (1.5, 5.4).
Gradient, arithmetic	G	Uniform change ($+$ or $-$) in cash flow each time period (2.5).
Gradient, geometric	g	Constant rate of change ($+$ or $-$) each time period (2.6).
Gross income	GI	Income from all sources for corporations or individuals (17.1).
Inflation rate	f	Rate that reflects changes in the value of a currency over time (14.1).
Interest	I	Amount earned or paid over time based on an initial amount and interest rate (1.4).
Interest rate	i or r	Interest expressed as a percentage of the original amount per time period; nominal (r) and effective (i) rates (1.4, 4.1).
Interest rate, inflation-adjusted	i_f	Interest rate adjusted to take inflation into account (14.1).
Life (estimated)	n	Number of years or periods over which an alternative or asset will be used; the evaluation time (1.5).
Life-cycle cost	LCC	Evaluation of costs for a system over all stages: feasibility to design to phaseout (6.5).
Measure of worth	Varies	Value, such as PW, AW, i^*, used to judge economic viability (1.1).
Minimum attractive rate of return	MARR	Minimum value of the rate of return for an alternative to be financially viable (1.9, 10.1).
Modified ROR	i' or MIRR	Unique ROR when a reinvestment rate i_i and external borrowing rate i_b are applied to multiple-rate cash flows (7.5).
Net cash flow	NCF	Resulting, actual amount of cash that flows in or out during a time period (1.6).
Net operating income	NOI	Difference between gross income and operating expenses (17.1).
Net operating profit after taxes	NOPAT or NPAT	Amount remaining after taxes are removed from taxable income (17.1).
Net present value	NPV	Another name for the present worth, PW.
Payback period	n_p	Number of years to recover the initial investment and a stated rate of return (13.3).
Present amount or worth	P or PW	Amount of money at the current time or a time denoted as *present* (1.5, 5.2).
Probability distribution	$P(X)$	Distribution of probability over different values of a variable (19.2).
Profitability index	PI	Ratio of PW of net cash flows to initial investment used for revenue projects; rewritten modified B/C ratio (9.2, 12.5).
Random variable	X	Parameter or characteristic that can take on any one of several values; discrete and continuous (19.2).
Rate of return	i^* or ROR	Compound interest rate on unpaid or unrecovered balances such that the final amount results in a zero balance (7.1).

Term	Symbol	Description
Recovery period	n	Number of years to completely depreciate an asset (16.1).
Return on invested capital	i'' or ROIC	Unique ROR when a reinvestment rate i_i is applied to multiple-rate cash flows (7.5).
Salvage/market value	S or MV	Expected trade-in or market value when an asset is traded or disposed of (6.2, 11.1, 16.1).
Standard deviation	s or σ	Measure of dispersion or spread about the expected value or average (19.4).
Study period	n	Specified number of years over which an evaluation takes place (5.3, 11.5).
Taxable income	TI	Amount upon which income taxes are based (17.1).
Tax rate	T	Decimal rate, usually graduated, used to calculate corporate or individual taxes (17.1).
Tax rate, effective	T_e	Single-figure tax rate incorporating several rates and bases (17.1).
Time	t	Indicator for a time period (1.7).
Unadjusted basis	B	Depreciable amount of first cost, delivery, and installation costs of an asset (18.1).
Value added	EVA	Economic value added reflects net profit after taxes (NPAT) after removing cost of invested capital during the year (17.7).
Value-added tax	VAT	An indirect consumption tax collected at each stage of production/distribution process; different from a sales tax paid by end user at purchase time (17.9).

REFERENCE MATERIALS

Textbooks on Related Topics ● ● ●

Blank, L. T., and A. Tarquin: *Basics of Engineering Economy,* McGraw-Hill, New York, 2008.

Bowman, M. S.: *Applied Economic Analysis for Technologists, Engineers, and Managers,* 2d ed., Pearson Prentice Hall, Upper Saddle River, NJ, 2003.

Bussey, L. E., and T. G. Eschenbach: *The Economic Analysis of Industrial Projects,* 2d ed., Pearson Prentice Hall, Upper Saddle River, NJ, 1992.

Canada, J. R., W. G. Sullivan, D. J. Kulonda, and J. A. White: *Capital Investment Analysis for Engineering and Management,* 3d ed., Pearson Prentice Hall, Upper Saddle River, NJ, 2005.

Collier, C. A., and C. R. Glagola: *Engineering Economic and Cost Analysis,* 3d ed., Pearson Prentice Hall, Upper Saddle River, NJ, 1999.

Cushman, R. F., and M. Loulakis: *Design-Build Contracting Handbook,* 2d ed., Aspen Publishers, New York, 2001.

Eschenbach, T. G.: *Engineering Economy: Applying Theory to Practice,* 3d ed., Oxford University Press, New York, 2010.

Fabrycky, W. J., G. J. Thuesen, and D. Verma: *Economic Decision Analysis,* 3d ed., Pearson Prentice Hall, Upper Saddle River, NJ, 1998.

Fraser, N. M., E. M. Jewkes, I. Bernhardt, and M. Tajima: *Engineering Economics in Canada,* Pearson Prentice Hall, Upper Saddle River, NJ, 2006.

Hartman, J. C.: *Engineering Economy and the Decision Making Process,* Pearson Prentice Hall, Upper Saddle River, NJ, 2007.

Levy, S. M.: *Build, Operate, Transfer: Paving the Way for Tomorrow's Infrastructure,* John Wiley & Sons, New York, 1996.

Newnan, D. G., J. P. Lavelle, and T. G. Eschenbach: *Engineering Economic Analysis,* 10th ed., Oxford University Press, New York, 2009.

Ostwald, P. F.: *Construction Cost Analysis and Estimating,* Pearson Prentice Hall, Upper Saddle River, NJ, 2001.

Ostwald, P. F., and T. S. McLaren: *Cost Analysis and Estimating for Engineering and Management,* Pearson Prentice Hall, Upper Saddle River, NJ, 2004.

Park, C. S.: *Contemporary Engineering Economics,* 5th ed., Pearson Prentice Hall, Upper Saddle River, NJ, 2011.

Park, C. S.: *Fundamentals of Engineering Economics,* 2d ed., Pearson Prentice Hall, Upper Saddle River, NJ, 2008.

Peurifoy, R. L., and G. D. Oberlender: *Estimating Construction Costs,* 5th ed., McGraw-Hill, New York, 2002.

Riggs, J. L., D. D. Bedworth, and S. U. Randhawa: *Engineering Economics,* 4th ed., McGraw-Hill, New York, 1996.

Stewart, R. D., R. M. Wyskida, and J. D. Johannes: *Cost Estimator's Reference Manual,* 2d ed., John Wiley & Sons, New York, 1995.

Sullivan, W. G., E. M. Wicks, and C. P. Koelling: *Engineering Economy,* 15th ed., Pearson Prentice Hall, Upper Saddle River, NJ, 2011.

Thuesen, G. J., and W. J. Fabrycky: *Engineering Economy,* 9th ed., Pearson Prentice Hall, Upper Saddle River, NJ, 2001.

White, J. A., K. E. Case, D. B. Pratt, and M. H. Agee: *Principles of Engineering Economic Analysis,* 5th ed., John Wiley & Sons, New York, 2010.

Using Excel 2007 ● ● ●

Gottfried, B. S.: *Spreadsheet Tools for Engineers Using Excel® 2007,* McGraw-Hill, New York, 2010.

Materials on Engineering Ethics ● ● ●

Harris, C. E., M. S. Pritchard, and M. J. Rabins: *Engineering Ethics: Concepts and Cases,* 4th ed., Wadsworth Cengage Learning, Belmont, CA, 2009.

Martin, M. W., and R. Schinzinger: *Introduction to Engineering Ethics*, 2d ed., McGraw-Hill, New York, 2010.

Websites ● ● ●

Construction cost estimation index: www.construction.com
The Economist: www.economist.com
For this textbook: www.mhhe.com/blank
Plant cost estimation index: www.che.com/pci
Revenue Canada: www.cra.gc.ca
U.S. Internal Revenue Service: www.irs.gov
Wall Street Journal: www.online.wsj.com

U.S. Government Publications (available at www.irs.gov) ● ● ●

Corporations, Publication 544, Internal Revenue Service, GPO, Washington, DC, annually.
Sales and Other Dispositions of Assets, Publication 542, Internal Revenue Service, GPO, Washington, DC, annually.
Your Federal Income Tax, Publication 17, Internal Revenue Service, GPO, Washington, DC, annually.

Selected Journals and Other Publications ● ● ●

The Engineering Economist, joint publication of ASEE and IIE, published by Taylor and Francis, Philadelphia, quarterly.
Harvard Business Review, Harvard University Press, Boston, bimonthly.
Journal of Finance, American Finance Association, published by John Wiley & Sons, New York, bimonthly.

0.25%			TABLE 1	Discrete Cash Flow: Compound Interest Factors				0.25%
	Single Payments		Uniform Series Payments				Arithmetic Gradients	
	Compound Amount F/P	Present Worth P/F	Sinking Fund A/F	Compound Amount F/A	Capital Recovery A/P	Present Worth P/A	Gradient Present Worth P/G	Gradient Uniform Series A/G
n								
1	1.0025	0.9975	1.00000	1.0000	1.00250	0.9975		
2	1.0050	0.9950	0.49938	2.0025	0.50188	1.9925	0.9950	0.4994
3	1.0075	0.9925	0.33250	3.0075	0.33500	2.9851	2.9801	0.9983
4	1.0100	0.9901	0.24906	4.0150	0.25156	3.9751	5.9503	1.4969
5	1.0126	0.9876	0.19900	5.0251	0.20150	4.9627	9.9007	1.9950
6	1.0151	0.9851	0.16563	6.0376	0.16813	5.9478	14.8263	2.4927
7	1.0176	0.9827	0.14179	7.0527	0.14429	6.9305	20.7223	2.9900
8	1.0202	0.9802	0.12391	8.0704	0.12641	7.9107	27.5839	3.4869
9	1.0227	0.9778	0.11000	9.0905	0.11250	8.8885	35.4061	3.9834
10	1.0253	0.9753	0.09888	10.1133	0.10138	9.8639	44.1842	4.4794
11	1.0278	0.9729	0.08978	11.1385	0.09228	10.8368	53.9133	4.9750
12	1.0304	0.9705	0.08219	12.1664	0.08469	11.8073	64.5886	5.4702
13	1.0330	0.9681	0.07578	13.1968	0.07828	12.7753	76.2053	5.9650
14	1.0356	0.9656	0.07028	14.2298	0.07278	13.7410	88.7587	6.4594
15	1.0382	0.9632	0.06551	15.2654	0.06801	14.7042	102.2441	6.9534
16	1.0408	0.9608	0.06134	16.3035	0.06384	15.6650	116.6567	7.4469
17	1.0434	0.9584	0.05766	17.3443	0.06016	16.6235	131.9917	7.9401
18	1.0460	0.9561	0.05438	18.3876	0.05688	17.5795	148.2446	8.4328
19	1.0486	0.9537	0.05146	19.4336	0.05396	18.5332	165.4106	8.9251
20	1.0512	0.9513	0.04882	20.4822	0.05132	19.4845	183.4851	9.4170
21	1.0538	0.9489	0.04644	21.5334	0.04894	20.4334	202.4634	9.9085
22	1.0565	0.9466	0.04427	22.5872	0.04677	21.3800	222.3410	10.3995
23	1.0591	0.9442	0.04229	23.6437	0.04479	22.3241	243.1131	10.8901
24	1.0618	0.9418	0.04048	24.7028	0.04298	23.2660	264.7753	11.3804
25	1.0644	0.9395	0.03881	25.7646	0.04131	24.2055	287.3230	11.8702
26	1.0671	0.9371	0.03727	26.8290	0.03977	25.1426	310.7516	12.3596
27	1.0697	0.9348	0.03585	27.8961	0.03835	26.0774	335.0566	12.8485
28	1.0724	0.9325	0.03452	28.9658	0.03702	27.0099	360.2334	13.3371
29	1.0751	0.9301	0.03329	30.0382	0.03579	27.9400	386.2776	13.8252
30	1.0778	0.9278	0.03214	31.1133	0.03464	28.8679	413.1847	14.3130
36	1.0941	0.9140	0.02658	37.6206	0.02908	34.3865	592.4988	17.2306
40	1.1050	0.9050	0.02380	42.0132	0.02630	38.0199	728.7399	19.1673
48	1.1273	0.8871	0.01963	50.9312	0.02213	45.1787	1040.06	23.0209
50	1.1330	0.8826	0.01880	53.1887	0.02130	46.9462	1125.78	23.9802
52	1.1386	0.8782	0.01803	55.4575	0.02053	48.7048	1214.59	24.9377
55	1.1472	0.8717	0.01698	58.8819	0.01948	51.3264	1353.53	26.3710
60	1.1616	0.8609	0.01547	64.6467	0.01797	55.6524	1600.08	28.7514
72	1.1969	0.8355	0.01269	78.7794	0.01519	65.8169	2265.56	34.4221
75	1.2059	0.8292	0.01214	82.3792	0.01464	68.3108	2447.61	35.8305
84	1.2334	0.8108	0.01071	93.3419	0.01321	75.6813	3029.76	40.0331
90	1.2520	0.7987	0.00992	100.7885	0.01242	80.5038	3446.87	42.8162
96	1.2709	0.7869	0.00923	108.3474	0.01173	85.2546	3886.28	45.5844
100	1.2836	0.7790	0.00881	113.4500	0.01131	88.3825	4191.24	47.4216
108	1.3095	0.7636	0.00808	123.8093	0.01058	94.5453	4829.01	51.0762
120	1.3494	0.7411	0.00716	139.7414	0.00966	103.5618	5852.11	56.5084
132	1.3904	0.7192	0.00640	156.1582	0.00890	112.3121	6950.01	61.8813
144	1.4327	0.6980	0.00578	173.0743	0.00828	120.8041	8117.41	67.1949
240	1.8208	0.5492	0.00305	328.3020	0.00555	180.3109	19399	107.5863
360	2.4568	0.4070	0.00172	582.7369	0.00422	237.1894	36264	152.8902
480	3.3151	0.3016	0.00108	926.0595	0.00358	279.3418	53821	192.6699

0.5%   **TABLE 2**   Discrete Cash Flow: Compound Interest Factors   **0.5%**

	Single Payments		Uniform Series Payments				Arithmetic Gradients	
	Compound Amount	Present Worth	Sinking Fund	Compound Amount	Capital Recovery	Present Worth	Gradient Present Worth	Gradient Uniform Series
n	F/P	P/F	A/F	F/A	A/P	P/A	P/G	A/G
1	1.0050	0.9950	1.00000	1.0000	1.00500	0.9950		
2	1.0100	0.9901	0.49875	2.0050	0.50375	1.9851	0.9901	0.4988
3	1.0151	0.9851	0.33167	3.0150	0.33667	2.9702	2.9604	0.9967
4	1.0202	0.9802	0.24813	4.0301	0.25313	3.9505	5.9011	1.4938
5	1.0253	0.9754	0.19801	5.0503	0.20301	4.9259	9.8026	1.9900
6	1.0304	0.9705	0.16460	6.0755	0.16960	5.8964	14.6552	2.4855
7	1.0355	0.9657	0.14073	7.1059	0.14573	6.8621	20.4493	2.9801
8	1.0407	0.9609	0.12283	8.1414	0.12783	7.8230	27.1755	3.4738
9	1.0459	0.9561	0.10891	9.1821	0.11391	8.7791	34.8244	3.9668
10	1.0511	0.9513	0.09777	10.2280	0.10277	9.7304	43.3865	4.4589
11	1.0564	0.9466	0.08866	11.2792	0.09366	10.6770	52.8526	4.9501
12	1.0617	0.9419	0.08107	12.3356	0.08607	11.6189	63.2136	5.4406
13	1.0670	0.9372	0.07464	13.3972	0.07964	12.5562	74.4602	5.9302
14	1.0723	0.9326	0.06914	14.4642	0.07414	13.4887	86.5835	6.4190
15	1.0777	0.9279	0.06436	15.5365	0.06936	14.4166	99.5743	6.9069
16	1.0831	0.9233	0.06019	16.6142	0.06519	15.3399	113.4238	7.3940
17	1.0885	0.9187	0.05651	17.6973	0.06151	16.2586	128.1231	7.8803
18	1.0939	0.9141	0.05323	18.7858	0.05823	17.1728	143.6634	8.3658
19	1.0994	0.9096	0.05030	19.8797	0.05530	18.0824	160.0360	8.8504
20	1.1049	0.9051	0.04767	20.9791	0.05267	18.9874	177.2322	9.3342
21	1.1104	0.9006	0.04528	22.0840	0.05028	19.8880	195.2434	9.8172
22	1.1160	0.8961	0.04311	23.1944	0.04811	20.7841	214.0611	10.2993
23	1.1216	0.8916	0.04113	24.3104	0.04613	21.6757	233.6768	10.7806
24	1.1272	0.8872	0.03932	25.4320	0.04432	22.5629	254.0820	11.2611
25	1.1328	0.8828	0.03765	26.5591	0.04265	23.4456	275.2686	11.7407
26	1.1385	0.8784	0.03611	27.6919	0.04111	24.3240	297.2281	12.2195
27	1.1442	0.8740	0.03469	28.8304	0.03969	25.1980	319.9523	12.6975
28	1.1499	0.8697	0.03336	29.9745	0.03836	26.0677	343.4332	13.1747
29	1.1556	0.8653	0.03213	31.1244	0.03713	26.9330	367.6625	13.6510
30	1.1614	0.8610	0.03098	32.2800	0.03598	27.7941	392.6324	14.1265
36	1.1967	0.8356	0.02542	39.3361	0.03042	32.8710	557.5598	16.9621
40	1.2208	0.8191	0.02265	44.1588	0.02765	36.1722	681.3347	18.8359
48	1.2705	0.7871	0.01849	54.0978	0.02349	42.5803	959.9188	22.5437
50	1.2832	0.7793	0.01765	56.6452	0.02265	44.1428	1035.70	23.4624
52	1.2961	0.7716	0.01689	59.2180	0.02189	45.6897	1113.82	24.3778
55	1.3156	0.7601	0.01584	63.1258	0.02084	47.9814	1235.27	25.7447
60	1.3489	0.7414	0.01433	69.7700	0.01933	51.7256	1448.65	28.0064
72	1.4320	0.6983	0.01157	86.4089	0.01657	60.3395	2012.35	33.3504
75	1.4536	0.6879	0.01102	90.7265	0.01602	62.4136	2163.75	34.6679
84	1.5204	0.6577	0.00961	104.0739	0.01461	68.4530	2640.66	38.5763
90	1.5666	0.6383	0.00883	113.3109	0.01383	72.3313	2976.08	41.1451
96	1.6141	0.6195	0.00814	122.8285	0.01314	76.0952	3324.18	43.6845
100	1.6467	0.6073	0.00773	129.3337	0.01273	78.5426	3562.79	45.3613
108	1.7137	0.5835	0.00701	142.7399	0.01201	83.2934	4054.37	48.6758
120	1.8194	0.5496	0.00610	163.8793	0.01110	90.0735	4823.51	53.5508
132	1.9316	0.5177	0.00537	186.3226	0.01037	96.4596	5624.59	58.3103
144	2.0508	0.4876	0.00476	210.1502	0.00976	102.4747	6451.31	62.9551
240	3.3102	0.3021	0.00216	462.0409	0.00716	139.5808	13416	96.1131
360	6.0226	0.1660	0.00100	1004.52	0.00600	166.7916	21403	128.3236
480	10.9575	0.0913	0.00050	1991.49	0.00550	181.7476	27588	151.7949

0.75%		TABLE 3	Discrete Cash Flow: Compound Interest Factors				0.75%
	Single Payments		Uniform Series Payments			Arithmetic Gradients	
	Compound Amount F/P	Present Worth P/F	Sinking Fund A/F	Compound Amount F/A	Capital Recovery A/P	Present Worth P/A	Gradient Present Worth P/G	Gradient Uniform Series A/G
n								
1	1.0075	0.9926	1.00000	1.0000	1.00750	0.9926		
2	1.0151	0.9852	0.49813	2.0075	0.50563	1.9777	0.9852	0.4981
3	1.0227	0.9778	0.33085	3.0226	0.33835	2.9556	2.9408	0.9950
4	1.0303	0.9706	0.24721	4.0452	0.25471	3.9261	5.8525	1.4907
5	1.0381	0.9633	0.19702	5.0756	0.20452	4.8894	9.7058	1.9851
6	1.0459	0.9562	0.16357	6.1136	0.17107	5.8456	14.4866	2.4782
7	1.0537	0.9490	0.13967	7.1595	0.14717	6.7946	20.1808	2.9701
8	1.0616	0.9420	0.12176	8.2132	0.12926	7.7366	26.7747	3.4608
9	1.0696	0.9350	0.10782	9.2748	0.11532	8.6716	34.2544	3.9502
10	1.0776	0.9280	0.09667	10.3443	0.10417	9.5996	42.6064	4.4384
11	1.0857	0.9211	0.08755	11.4219	0.09505	10.5207	51.8174	4.9253
12	1.0938	0.9142	0.07995	12.5076	0.08745	11.4349	61.8740	5.4110
13	1.1020	0.9074	0.07352	13.6014	0.08102	12.3423	72.7632	5.8954
14	1.1103	0.9007	0.06801	14.7034	0.07551	13.2430	84.4720	6.3786
15	1.1186	0.8940	0.06324	15.8137	0.07074	14.1370	96.9876	6.8606
16	1.1270	0.8873	0.05906	16.9323	0.06656	15.0243	110.2973	7.3413
17	1.1354	0.8807	0.05537	18.0593	0.06287	15.9050	124.3887	7.8207
18	1.1440	0.8742	0.05210	19.1947	0.05960	16.7792	139.2494	8.2989
19	1.1525	0.8676	0.04917	20.3387	0.05667	17.6468	154.8671	8.7759
20	1.1612	0.8612	0.04653	21.4912	0.05403	18.5080	171.2297	9.2516
21	1.1699	0.8548	0.04415	22.6524	0.05165	19.3628	188.3253	9.7261
22	1.1787	0.8484	0.04198	23.8223	0.04948	20.2112	206.1420	10.1994
23	1.1875	0.8421	0.04000	25.0010	0.04750	21.0533	224.6682	10.6714
24	1.1964	0.8358	0.03818	26.1885	0.04568	21.8891	243.8923	11.1422
25	1.2054	0.8296	0.03652	27.3849	0.04402	22.7188	263.8029	11.6117
26	1.2144	0.8234	0.03498	28.5903	0.04248	23.5422	284.3888	12.0800
27	1.2235	0.8173	0.03355	29.8047	0.04105	24.3595	305.6387	12.5470
28	1.2327	0.8112	0.03223	31.0282	0.03973	25.1707	327.5416	13.0128
29	1.2420	0.8052	0.03100	32.2609	0.03850	25.9759	350.0867	13.4774
30	1.2513	0.7992	0.02985	33.5029	0.03735	26.7751	373.2631	13.9407
36	1.3086	0.7641	0.02430	41.1527	0.03180	31.4468	524.9924	16.6946
40	1.3483	0.7416	0.02153	46.4465	0.02903	34.4469	637.4693	18.5058
48	1.4314	0.6986	0.01739	57.5207	0.02489	40.1848	886.8404	22.0691
50	1.4530	0.6883	0.01656	60.3943	0.02406	41.5664	953.8486	22.9476
52	1.4748	0.6780	0.01580	63.3111	0.02330	42.9276	1022.59	23.8211
55	1.5083	0.6630	0.01476	67.7688	0.02226	44.9316	1128.79	25.1223
60	1.5657	0.6387	0.01326	75.4241	0.02076	48.1734	1313.52	27.2665
72	1.7126	0.5839	0.01053	95.0070	0.01803	55.4768	1791.25	32.2882
75	1.7514	0.5710	0.00998	100.1833	0.01748	57.2027	1917.22	33.5163
84	1.8732	0.5338	0.00859	116.4269	0.01609	62.1540	2308.13	37.1357
90	1.9591	0.5104	0.00782	127.8790	0.01532	65.2746	2578.00	39.4946
96	2.0489	0.4881	0.00715	139.8562	0.01465	68.2584	2853.94	41.8107
100	2.1111	0.4737	0.00675	148.1445	0.01425	70.1746	3040.75	43.3311
108	2.2411	0.4462	0.00604	165.4832	0.01354	73.8394	3419.90	46.3154
120	2.4514	0.4079	0.00517	193.5143	0.01267	78.9417	3998.56	50.6521
132	2.6813	0.3730	0.00446	224.1748	0.01196	83.6064	4583.57	54.8232
144	2.9328	0.3410	0.00388	257.7116	0.01138	87.8711	5169.58	58.8314
240	6.0092	0.1664	0.00150	667.8869	0.00900	111.1450	9494.12	85.4210
360	14.7306	0.0679	0.00055	1830.74	0.00805	124.2819	13312	107.1145
480	36.1099	0.0277	0.00021	4681.32	0.00771	129.6409	15513	119.6620

1%				TABLE 4	Discrete Cash Flow: Compound Interest Factors			1%

	Single Payments		Uniform Series Payments				Arithmetic Gradients	
	Compound Amount	Present Worth	Sinking Fund	Compound Amount	Capital Recovery	Present Worth	Gradient Present Worth	Gradient Uniform Series
n	F/P	P/F	A/F	F/A	A/P	P/A	P/G	A/G
1	1.0100	0.9901	1.00000	1.0000	1.01000	0.9901		
2	1.0201	0.9803	0.49751	2.0100	0.50751	1.9704	0.9803	0.4975
3	1.0303	0.9706	0.33002	3.0301	0.34002	2.9410	2.9215	0.9934
4	1.0406	0.9610	0.24628	4.0604	0.25628	3.9020	5.8044	1.4876
5	1.0510	0.9515	0.19604	5.1010	0.20604	4.8534	9.6103	1.9801
6	1.0615	0.9420	0.16255	6.1520	0.17255	5.7955	14.3205	2.4710
7	1.0721	0.9327	0.13863	7.2135	0.14863	6.7282	19.9168	2.9602
8	1.0829	0.9235	0.12069	8.2857	0.13069	7.6517	26.3812	3.4478
9	1.0937	0.9143	0.10674	9.3685	0.11674	8.5660	33.6959	3.9337
10	1.1046	0.9053	0.09558	10.4622	0.10558	9.4713	41.8435	4.4179
11	1.1157	0.8963	0.08645	11.5668	0.09645	10.3676	50.8067	4.9005
12	1.1268	0.8874	0.07885	12.6825	0.08885	11.2551	60.5687	5.3815
13	1.1381	0.8787	0.07241	13.8093	0.08241	12.1337	71.1126	5.8607
14	1.1495	0.8700	0.06690	14.9474	0.07690	13.0037	82.4221	6.3384
15	1.1610	0.8613	0.06212	16.0969	0.07212	13.8651	94.4810	6.8143
16	1.1726	0.8528	0.05794	17.2579	0.06794	14.7179	107.2734	7.2886
17	1.1843	0.8444	0.05426	18.4304	0.06426	15.5623	120.7834	7.7613
18	1.1961	0.8360	0.05098	19.6147	0.06098	16.3983	134.9957	8.2323
19	1.2081	0.8277	0.04805	20.8109	0.05805	17.2260	149.8950	8.7017
20	1.2202	0.8195	0.04542	22.0190	0.05542	18.0456	165.4664	9.1694
21	1.2324	0.8114	0.04303	23.2392	0.05303	18.8570	181.6950	9.6354
22	1.2447	0.8034	0.04086	24.4716	0.05086	19.6604	198.5663	10.0998
23	1.2572	0.7954	0.03889	25.7163	0.04889	20.4558	216.0660	10.5626
24	1.2697	0.7876	0.03707	26.9735	0.04707	21.2434	234.1800	11.0237
25	1.2824	0.7798	0.03541	28.2432	0.04541	22.0232	252.8945	11.4831
26	1.2953	0.7720	0.03387	29.5256	0.04387	22.7952	272.1957	11.9409
27	1.3082	0.7644	0.03245	30.8209	0.04245	23.5596	292.0702	12.3971
28	1.3213	0.7568	0.03112	32.1291	0.04112	24.3164	312.5047	12.8516
29	1.3345	0.7493	0.02990	33.4504	0.03990	25.0658	333.4863	13.3044
30	1.3478	0.7419	0.02875	34.7849	0.03875	25.8077	355.0021	13.7557
36	1.4308	0.6989	0.02321	43.0769	0.03321	30.1075	494.6207	16.4285
40	1.4889	0.6717	0.02046	48.8864	0.03046	32.8347	596.8561	18.1776
48	1.6122	0.6203	0.01633	61.2226	0.02633	37.9740	820.1460	21.5976
50	1.6446	0.6080	0.01551	64.4632	0.02551	39.1961	879.4176	22.4363
52	1.6777	0.5961	0.01476	67.7689	0.02476	40.3942	939.9175	23.2686
55	1.7285	0.5785	0.01373	72.8525	0.02373	42.1472	1032.81	24.5049
60	1.8167	0.5504	0.01224	81.6697	0.02224	44.9550	1192.81	26.5333
72	2.0471	0.4885	0.00955	104.7099	0.01955	51.1504	1597.87	31.2386
75	2.1091	0.4741	0.00902	110.9128	0.01902	52.5871	1702.73	32.3793
84	2.3067	0.4335	0.00765	130.6723	0.01765	56.6485	2023.32	35.7170
90	2.4486	0.4084	0.00690	144.8633	0.01690	59.1609	2240.57	37.8724
96	2.5993	0.3847	0.00625	159.9273	0.01625	61.5277	2459.43	39.9727
100	2.7048	0.3697	0.00587	170.4814	0.01587	63.0289	2605.78	41.3426
108	2.9289	0.3414	0.00518	192.8926	0.01518	65.8578	2898.42	44.0103
120	3.3004	0.3030	0.00435	230.0387	0.01435	69.7005	3334.11	47.8349
132	3.7190	0.2689	0.00368	271.8959	0.01368	73.1108	3761.69	51.4520
144	4.1906	0.2386	0.00313	319.0616	0.01313	76.1372	4177.47	54.8676
240	10.8926	0.0918	0.00101	989.2554	0.01101	90.8194	6878.60	75.7393
360	35.9496	0.0278	0.00029	3494.96	0.01029	97.2183	8720.43	89.6995
480	118.6477	0.0084	0.00008	11765	0.01008	99.1572	9511.16	95.9200

1.25%		TABLE 5	Discrete Cash Flow: Compound Interest Factors				1.25%
	Single Payments		Uniform Series Payments			Arithmetic Gradients	
	Compound Amount F/P	Present Worth P/F	Sinking Fund A/F	Compound Amount F/A	Capital Recovery A/P	Present Worth P/A	Gradient Present Worth P/G	Gradient Uniform Series A/G
n								
1	1.0125	0.9877	1.00000	1.0000	1.01250	0.9877		
2	1.0252	0.9755	0.49680	2.0125	0.50939	1.9631	0.9755	0.4969
3	1.0380	0.9634	0.32920	3.0377	0.34170	2.9265	2.9023	0.9917
4	1.0509	0.9515	0.24536	4.0756	0.25786	3.8781	5.7569	1.4845
5	1.0641	0.9398	0.19506	5.1266	0.20756	4.8178	9.5160	1.9752
6	1.0774	0.9282	0.16153	6.1907	0.17403	5.7460	14.1569	2.4638
7	1.0909	0.9167	0.13759	7.2680	0.15009	6.6627	19.6571	2.9503
8	1.1045	0.9054	0.11963	8.3589	0.13213	7.5681	25.9949	3.4348
9	1.1183	0.8942	0.10567	9.4634	0.11817	8.4623	33.1487	3.9172
10	1.1323	0.8832	0.09450	10.5817	0.10700	9.3455	41.0973	4.3975
11	1.1464	0.8723	0.08537	11.7139	0.09787	10.2178	49.8201	4.8758
12	1.1608	0.8615	0.07776	12.8604	0.09026	11.0793	59.2967	5.3520
13	1.1753	0.8509	0.07132	14.0211	0.08382	11.9302	69.5072	5.8262
14	1.1900	0.8404	0.06581	15.1964	0.07831	12.7706	80.4320	6.2982
15	1.2048	0.8300	0.06103	16.3863	0.07353	13.6005	92.0519	6.7682
16	1.2199	0.8197	0.05685	17.5912	0.06935	14.4203	104.3481	7.2362
17	1.2351	0.8096	0.05316	18.8111	0.06566	15.2299	117.3021	7.7021
18	1.2506	0.7996	0.04988	20.0462	0.06238	16.0295	130.8958	8.1659
19	1.2662	0.7898	0.04696	21.2968	0.05946	16.8193	145.1115	8.6277
20	1.2820	0.7800	0.04432	22.5630	0.05682	17.5993	159.9316	9.0874
21	1.2981	0.7704	0.04194	23.8450	0.05444	18.3697	175.3392	9.5450
22	1.3143	0.7609	0.03977	25.1431	0.05227	19.1306	191.3174	10.0006
23	1.3307	0.7515	0.03780	26.4574	0.05030	19.8820	207.8499	10.4542
24	1.3474	0.7422	0.03599	27.7881	0.04849	20.6242	224.9204	10.9056
25	1.3642	0.7330	0.03432	29.1354	0.04682	21.3573	242.5132	11.3551
26	1.3812	0.7240	0.03279	30.4996	0.04529	22.0813	260.6128	11.8024
27	1.3985	0.7150	0.03137	31.8809	0.04387	22.7963	279.2040	12.2478
28	1.4160	0.7062	0.03005	33.2794	0.04255	23.5025	298.2719	12.6911
29	1.4337	0.6975	0.02882	34.6954	0.04132	24.2000	317.8019	13.1323
30	1.4516	0.6889	0.02768	36.1291	0.04018	24.8889	337.7797	13.5715
36	1.5639	0.6394	0.02217	45.1155	0.03467	28.8473	466.2830	16.1639
40	1.6436	0.6084	0.01942	51.4896	0.03192	31.3269	559.2320	17.8515
48	1.8154	0.5509	0.01533	65.2284	0.02783	35.9315	759.2296	21.1299
50	1.8610	0.5373	0.01452	68.8818	0.02702	37.0129	811.6738	21.9295
52	1.9078	0.5242	0.01377	72.6271	0.02627	38.0677	864.9409	22.7211
55	1.9803	0.5050	0.01275	78.4225	0.02525	39.6017	946.2277	23.8936
60	2.1072	0.4746	0.01129	88.5745	0.02379	42.0346	1084.84	25.8083
72	2.4459	0.4088	0.00865	115.6736	0.02115	47.2925	1428.46	30.2047
75	2.5388	0.3939	0.00812	123.1035	0.02062	48.4890	1515.79	31.2605
84	2.8391	0.3522	0.00680	147.1290	0.01930	51.8222	1778.84	34.3258
90	3.0588	0.3269	0.00607	164.7050	0.01857	53.8461	1953.83	36.2855
96	3.2955	0.3034	0.00545	183.6411	0.01795	55.7246	2127.52	38.1793
100	3.4634	0.2887	0.00507	197.0723	0.01757	56.9013	2242.24	39.4058
108	3.8253	0.2614	0.00442	226.0226	0.01692	59.0865	2468.26	41.7737
120	4.4402	0.2252	0.00363	275.2171	0.01613	61.9828	2796.57	45.1184
132	5.1540	0.1940	0.00301	332.3198	0.01551	64.4781	3109.35	48.2234
144	5.9825	0.1672	0.00251	398.6021	0.01501	66.6277	3404.61	51.0990
240	19.7155	0.0507	0.00067	1497.24	0.01317	75.9423	5101.53	67.1764
360	87.5410	0.0114	0.00014	6923.28	0.01264	79.0861	5997.90	75.8401
480	388.7007	0.0026	0.00003	31016	0.01253	79.7942	6284.74	78.7619

1.5%			TABLE 6	Discrete Cash Flow: Compound Interest Factors				1.5%
	Single Payments		Uniform Series Payments				Arithmetic Gradients	
	Compound Amount F/P	Present Worth P/F	Sinking Fund A/F	Compound Amount F/A	Capital Recovery A/P	Present Worth P/A	Gradient Present Worth P/G	Gradient Uniform Series A/G
n								
1	1.0150	0.9852	1.00000	1.0000	1.01500	0.9852		
2	1.0302	0.9707	0.49628	2.0150	0.51128	1.9559	0.9707	0.4963
3	1.0457	0.9563	0.32838	3.0452	0.34338	2.9122	2.8833	0.9901
4	1.0614	0.9422	0.24444	4.0909	0.25944	3.8544	5.7098	1.4814
5	1.0773	0.9283	0.19409	5.1523	0.20909	4.7826	9.4229	1.9702
6	1.0934	0.9145	0.16053	6.2296	0.17553	5.6972	13.9956	2.4566
7	1.1098	0.9010	0.13656	7.3230	0.15156	6.5982	19.4018	2.9405
8	1.1265	0.8877	0.11858	8.4328	0.13358	7.4859	25.6157	3.4219
9	1.1434	0.8746	0.10461	9.5593	0.11961	8.3605	32.6125	3.9008
10	1.1605	0.8617	0.09343	10.7027	0.10843	9.2222	40.3675	4.3772
11	1.1779	0.8489	0.08429	11.8633	0.09929	10.0711	48.8568	4.8512
12	1.1956	0.8364	0.07668	13.0412	0.09168	10.9075	58.0571	5.3227
13	1.2136	0.8240	0.07024	14.2368	0.08524	11.7315	67.9454	5.7917
14	1.2318	0.8118	0.06472	15.4504	0.07972	12.5434	78.4994	6.2582
15	1.2502	0.7999	0.05994	16.6821	0.07494	13.3432	89.6974	6.7223
16	1.2690	0.7880	0.05577	17.9324	0.07077	14.1313	101.5178	7.1839
17	1.2880	0.7764	0.05208	19.2014	0.06708	14.9076	113.9400	7.6431
18	1.3073	0.7649	0.04881	20.4894	0.06381	15.6726	126.9435	8.0997
19	1.3270	0.7536	0.04588	21.7967	0.06088	16.4262	140.5084	8.5539
20	1.3469	0.7425	0.04325	23.1237	0.05825	17.1686	154.6154	9.0057
21	1.3671	0.7315	0.04087	24.4705	0.05587	17.9001	169.2453	9.4550
22	1.3876	0.7207	0.03870	25.8376	0.05370	18.6208	184.3798	9.9018
23	1.4084	0.7100	0.03673	27.2251	0.05173	19.3309	200.0006	10.3462
24	1.4295	0.6995	0.03492	28.6335	0.04992	20.0304	216.0901	10.7881
25	1.4509	0.6892	0.03326	30.0630	0.04826	20.7196	232.6310	11.2276
26	1.4727	0.6790	0.03173	31.5140	0.04673	21.3986	249.6065	11.6646
27	1.4948	0.6690	0.03032	32.9867	0.04532	22.0676	267.0002	12.0992
28	1.5172	0.6591	0.02900	34.4815	0.04400	22.7267	284.7958	12.5313
29	1.5400	0.6494	0.02778	35.9987	0.04278	23.3761	302.9779	12.9610
30	1.5631	0.6398	0.02664	37.5387	0.04164	24.0158	321.5310	13.3883
36	1.7091	0.5851	0.02115	47.2760	0.03615	27.6607	439.8303	15.9009
40	1.8140	0.5513	0.01843	54.2679	0.03343	29.9158	524.3568	17.5277
48	2.0435	0.4894	0.01437	69.5652	0.02937	34.0426	703.5462	20.6667
50	2.1052	0.4750	0.01357	73.6828	0.02857	34.9997	749.9636	21.4277
52	2.1689	0.4611	0.01283	77.9249	0.02783	35.9287	796.8774	22.1794
55	2.2679	0.4409	0.01183	84.5296	0.02683	37.2715	868.0285	23.2894
60	2.4432	0.4093	0.01039	96.2147	0.02539	39.3803	988.1674	25.0930
72	2.9212	0.3423	0.00781	128.0772	0.02281	43.8447	1279.79	29.1893
75	3.0546	0.3274	0.00730	136.9728	0.02230	44.8416	1352.56	30.1631
84	3.4926	0.2863	0.00602	166.1726	0.02102	47.5786	1568.51	32.9668
90	3.8189	0.2619	0.00532	187.9299	0.02032	49.2099	1709.54	34.7399
96	4.1758	0.2395	0.00472	211.7202	0.01972	50.7017	1847.47	36.4381
100	4.4320	0.2256	0.00437	228.8030	0.01937	51.6247	1937.45	37.5295
108	4.9927	0.2003	0.00376	266.1778	0.01876	53.3137	2112.13	39.6171
120	5.9693	0.1675	0.00302	331.2882	0.01802	55.4985	2359.71	42.5185
132	7.1370	0.1401	0.00244	409.1354	0.01744	57.3257	2588.71	45.1579
144	8.5332	0.1172	0.00199	502.2109	0.01699	58.8540	2798.58	47.5512
240	35.6328	0.0281	0.00043	2308.85	0.01543	64.7957	3870.69	59.7368
360	212.7038	0.0047	0.00007	14114	0.01507	66.3532	4310.72	64.9662
480	1269.70	0.0008	0.00001	84580	0.01501	66.6142	4415.74	66.2883

2%				TABLE 7	Discrete Cash Flow: Compound Interest Factors			2%
	Single Payments		Uniform Series Payments				Arithmetic Gradients	
	Compound Amount F/P	Present Worth P/F	Sinking Fund A/F	Compound Amount F/A	Capital Recovery A/P	Present Worth P/A	Gradient Present Worth P/G	Gradient Uniform Series A/G
n								
1	1.0200	0.9804	1.00000	1.0000	1.02000	0.9804		
2	1.0404	0.9612	0.49505	2.0200	0.51505	1.9416	0.9612	0.4950
3	1.0612	0.9423	0.32675	3.0604	0.34675	2.8839	2.8458	0.9868
4	1.0824	0.9238	0.24262	4.1216	0.26262	3.8077	5.6173	1.4752
5	1.1041	0.9057	0.19216	5.2040	0.21216	4.7135	9.2403	1.9604
6	1.1262	0.8880	0.15853	6.3081	0.17853	5.6014	13.6801	2.4423
7	1.1487	0.8706	0.13451	7.4343	0.15451	6.4720	18.9035	2.9208
8	1.1717	0.8535	0.11651	8.5830	0.13651	7.3255	24.8779	3.3961
9	1.1951	0.8368	0.10252	9.7546	0.12252	8.1622	31.5720	3.8681
10	1.2190	0.8203	0.09133	10.9497	0.11133	8.9826	38.9551	4.3367
11	1.2434	0.8043	0.08218	12.1687	0.10218	9.7868	46.9977	4.8021
12	1.2682	0.7885	0.07456	13.4121	0.09456	10.5753	55.6712	5.2642
13	1.2936	0.7730	0.06812	14.6803	0.08812	11.3484	64.9475	5.7231
14	1.3195	0.7579	0.06260	15.9739	0.08260	12.1062	74.7999	6.1786
15	1.3459	0.7430	0.05783	17.2934	0.07783	12.8493	85.2021	6.6309
16	1.3728	0.7284	0.05365	18.6393	0.07365	13.5777	96.1288	7.0799
17	1.4002	0.7142	0.04997	20.0121	0.06997	14.2919	107.5554	7.5256
18	1.4282	0.7002	0.04670	21.4123	0.06670	14.9920	119.4581	7.9681
19	1.4568	0.6864	0.04378	22.8406	0.06378	15.6785	131.8139	8.4073
20	1.4859	0.6730	0.04116	24.2974	0.06116	16.3514	144.6003	8.8433
21	1.5157	0.6598	0.03878	25.7833	0.05878	17.0112	157.7959	9.2760
22	1.5460	0.6468	0.03663	27.2990	0.05663	17.6580	171.3795	9.7055
23	1.5769	0.6342	0.03467	28.8450	0.05467	18.2922	185.3309	10.1317
24	1.6084	0.6217	0.03287	30.4219	0.05287	18.9139	199.6305	10.5547
25	1.6406	0.6095	0.03122	32.0303	0.05122	19.5235	214.2592	10.9745
26	1.6734	0.5976	0.02970	33.6709	0.04970	20.1210	229.1987	11.3910
27	1.7069	0.5859	0.02829	35.3443	0.04829	20.7069	244.4311	11.8043
28	1.7410	0.5744	0.02699	37.0512	0.04699	21.2813	259.9392	12.2145
29	1.7758	0.5631	0.02578	38.7922	0.04578	21.8444	275.7064	12.6214
30	1.8114	0.5521	0.02465	40.5681	0.04465	22.3965	291.7164	13.0251
36	2.0399	0.4902	0.01923	51.9944	0.03923	25.4888	392.0405	15.3809
40	2.2080	0.4529	0.01656	60.4020	0.03656	27.3555	461.9931	16.8885
48	2.5871	0.3865	0.01260	79.3535	0.03260	30.6731	605.9657	19.7556
50	2.6916	0.3715	0.01182	84.5794	0.03182	31.4236	642.3606	20.4420
52	2.8003	0.3571	0.01111	90.0164	0.03111	32.1449	678.7849	21.1164
55	2.9717	0.3365	0.01014	98.5865	0.03014	33.1747	733.3527	22.1057
60	3.2810	0.3048	0.00877	114.0515	0.02877	34.7609	823.6975	23.6961
72	4.1611	0.2403	0.00633	158.0570	0.02633	37.9841	1034.06	27.2234
75	4.4158	0.2265	0.00586	170.7918	0.02586	38.6771	1084.64	28.0434
84	5.2773	0.1895	0.00468	213.8666	0.02468	40.5255	1230.42	30.3616
90	5.9431	0.1683	0.00405	247.1567	0.02405	41.5869	1322.17	31.7929
96	6.6929	0.1494	0.00351	284.6467	0.02351	42.5294	1409.30	33.1370
100	7.2446	0.1380	0.00320	312.2323	0.02320	43.0984	1464.75	33.9863
108	8.4883	0.1178	0.00267	374.4129	0.02267	44.1095	1569.30	35.5774
120	10.7652	0.0929	0.00205	488.2582	0.02205	45.3554	1710.42	37.7114
132	13.6528	0.0732	0.00158	632.6415	0.02158	46.3378	1833.47	39.5676
144	17.3151	0.0578	0.00123	815.7545	0.02123	47.1123	1939.79	41.1738
240	115.8887	0.0086	0.00017	5744.44	0.02017	49.5686	2374.88	47.9110
360	1247.56	0.0008	0.00002	62328	0.02002	49.9599	2482.57	49.7112
480	13430	0.0001			0.02000	49.9963	2498.03	49.9643

3%								3%
			TABLE 8	Discrete Cash Flow: Compound Interest Factors				

	Single Payments		Uniform Series Payments				Arithmetic Gradients	
	Compound Amount	Present Worth	Sinking Fund	Compound Amount	Capital Recovery	Present Worth	Gradient Present Worth	Gradient Uniform Series
n	F/P	P/F	A/F	F/A	A/P	P/A	P/G	A/G
1	1.0300	0.9709	1.00000	1.0000	1.03000	0.9709		
2	1.0609	0.9426	0.49261	2.0300	0.52261	1.9135	0.9426	0.4926
3	1.0927	0.9151	0.32353	3.0909	0.35353	2.8286	2.7729	0.9803
4	1.1255	0.8885	0.23903	4.1836	0.26903	3.7171	5.4383	1.4631
5	1.1593	0.8626	0.18835	5.3091	0.21835	4.5797	8.8888	1.9409
6	1.1941	0.8375	0.15460	6.4684	0.18460	5.4172	13.0762	2.4138
7	1.2299	0.8131	0.13051	7.6625	0.16051	6.2303	17.9547	2.8819
8	1.2668	0.7894	0.11246	8.8923	0.14246	7.0197	23.4806	3.3450
9	1.3048	0.7664	0.09843	10.1591	0.12843	7.7861	29.6119	3.8032
10	1.3439	0.7441	0.08723	11.4639	0.11723	8.5302	36.3088	4.2565
11	1.3842	0.7224	0.07808	12.8078	0.10808	9.2526	43.5330	4.7049
12	1.4258	0.7014	0.07046	14.1920	0.10046	9.9540	51.2482	5.1485
13	1.4685	0.6810	0.06403	15.6178	0.09403	10.6350	59.4196	5.5872
14	1.5126	0.6611	0.05853	17.0863	0.08853	11.2961	68.0141	6.0210
15	1.5580	0.6419	0.05377	18.5989	0.08377	11.9379	77.0002	6.4500
16	1.6047	0.6232	0.04961	20.1569	0.07961	12.5611	86.3477	6.8742
17	1.6528	0.6050	0.04595	21.7616	0.07595	13.1661	96.0280	7.2936
18	1.7024	0.5874	0.04271	23.4144	0.07271	13.7535	106.0137	7.7081
19	1.7535	0.5703	0.03981	25.1169	0.06981	14.3238	116.2788	8.1179
20	1.8061	0.5537	0.03722	26.8704	0.06722	14.8775	126.7987	8.5229
21	1.8603	0.5375	0.03487	28.6765	0.06487	15.4150	137.5496	8.9231
22	1.9161	0.5219	0.03275	30.5368	0.06275	15.9369	148.5094	9.3186
23	1.9736	0.5067	0.03081	32.4529	0.06081	16.4436	159.6566	9.7093
24	2.0328	0.4919	0.02905	34.4265	0.05905	16.9355	170.9711	10.0954
25	2.0938	0.4776	0.02743	36.4593	0.05743	17.4131	182.4336	10.4768
26	2.1566	0.4637	0.02594	38.5530	0.05594	17.8768	194.0260	10.8535
27	2.2213	0.4502	0.02456	40.7096	0.05456	18.3270	205.7309	11.2255
28	2.2879	0.4371	0.02329	42.9309	0.05329	18.7641	217.5320	11.5930
29	2.3566	0.4243	0.02211	45.2189	0.05211	19.1885	229.4137	11.9558
30	2.4273	0.4120	0.02102	47.5754	0.05102	19.6004	241.3613	12.3141
31	2.5001	0.4000	0.02000	50.0027	0.05000	20.0004	253.3609	12.6678
32	2.5751	0.3883	0.01905	52.5028	0.04905	20.3888	265.3993	13.0169
33	2.6523	0.3770	0.01816	55.0778	0.04816	20.7658	277.4642	13.3616
34	2.7319	0.3660	0.01732	57.7302	0.04732	21.1318	289.5437	13.7018
35	2.8139	0.3554	0.01654	60.4621	0.04654	21.4872	301.6267	14.0375
40	3.2620	0.3066	0.01326	75.4013	0.04326	23.1148	361.7499	15.6502
45	3.7816	0.2644	0.01079	92.7199	0.04079	24.5187	420.6325	17.1556
50	4.3839	0.2281	0.00887	112.7969	0.03887	25.7298	477.4803	18.5575
55	5.0821	0.1968	0.00735	136.0716	0.03735	26.7744	531.7411	19.8600
60	5.8916	0.1697	0.00613	163.0534	0.03613	27.6756	583.0526	21.0674
65	6.8300	0.1464	0.00515	194.3328	0.03515	28.4529	631.2010	22.1841
70	7.9178	0.1263	0.00434	230.5941	0.03434	29.1234	676.0869	23.2145
75	9.1789	0.1089	0.00367	272.6309	0.03367	29.7018	717.6978	24.1634
80	10.6409	0.0940	0.00311	321.3630	0.03311	30.2008	756.0865	25.0353
84	11.9764	0.0835	0.00273	365.8805	0.03273	30.5501	784.5434	25.6806
85	12.3357	0.0811	0.00265	377.8570	0.03265	30.6312	791.3529	25.8349
90	14.3005	0.0699	0.00226	443.3489	0.03226	31.0024	823.6302	26.5667
96	17.0755	0.0586	0.00187	535.8502	0.03187	31.3812	858.6377	27.3615
108	24.3456	0.0411	0.00129	778.1863	0.03129	31.9642	917.6013	28.7072
120	34.7110	0.0288	0.00089	1123.70	0.03089	32.3730	963.8635	29.7737

4%			TABLE 9	Discrete Cash Flow: Compound Interest Factors				4%
	Single Payments		Uniform Series Payments				Arithmetic Gradients	
	Compound Amount F/P	Present Worth P/F	Sinking Fund A/F	Compound Amount F/A	Capital Recovery A/P	Present Worth P/A	Gradient Present Worth P/G	Gradient Uniform Series A/G
n								
1	1.0400	0.9615	1.00000	1.0000	1.04000	0.9615		
2	1.0816	0.9246	0.49020	2.0400	0.53020	1.8861	0.9246	0.4902
3	1.1249	0.8890	0.32035	3.1216	0.36035	2.7751	2.7025	0.9739
4	1.1699	0.8548	0.23549	4.2465	0.27549	3.6299	5.2670	1.4510
5	1.2167	0.8219	0.18463	5.4163	0.22463	4.4518	8.5547	1.9216
6	1.2653	0.7903	0.15076	6.6330	0.19076	5.2421	12.5062	2.3857
7	1.3159	0.7599	0.12661	7.8983	0.16661	6.0021	17.0657	2.8433
8	1.3686	0.7307	0.10853	9.2142	0.14853	6.7327	22.1806	3.2944
9	1.4233	0.7026	0.09449	10.5828	0.13449	7.4353	27.8013	3.7391
10	1.4802	0.6756	0.08329	12.0061	0.12329	8.1109	33.8814	4.1773
11	1.5395	0.6496	0.07415	13.4864	0.11415	8.7605	40.3772	4.6090
12	1.6010	0.6246	0.06655	15.0258	0.10655	9.3851	47.2477	5.0343
13	1.6651	0.6006	0.06014	16.6268	0.10014	9.9856	54.4546	5.4533
14	1.7317	0.5775	0.05467	18.2919	0.09467	10.5631	61.9618	5.8659
15	1.8009	0.5553	0.04994	20.0236	0.08994	11.1184	69.7355	6.2721
16	1.8730	0.5339	0.04582	21.8245	0.08582	11.6523	77.7441	6.6720
17	1.9479	0.5134	0.04220	23.6975	0.08220	12.1657	85.9581	7.0656
18	2.0258	0.4936	0.03899	25.6454	0.07899	12.6593	94.3498	7.4530
19	2.1068	0.4746	0.03614	27.6712	0.07614	13.1339	102.8933	7.8342
20	2.1911	0.4564	0.03358	29.7781	0.07358	13.5903	111.5647	8.2091
21	2.2788	0.4388	0.03128	31.9692	0.07128	14.0292	120.3414	8.5779
22	2.3699	0.4220	0.02920	34.2480	0.06920	14.4511	129.2024	8.9407
23	2.4647	0.4057	0.02731	36.6179	0.06731	14.8568	138.1284	9.2973
24	2.5633	0.3901	0.02559	39.0826	0.06559	15.2470	147.1012	9.6479
25	2.6658	0.3751	0.02401	41.6459	0.06401	15.6221	156.1040	9.9925
26	2.7725	0.3607	0.02257	44.3117	0.06257	15.9828	165.1212	10.3312
27	2.8834	0.3468	0.02124	47.0842	0.06124	16.3296	174.1385	10.6640
28	2.9987	0.3335	0.02001	49.9676	0.06001	16.6631	183.1424	10.9909
29	3.1187	0.3207	0.01888	52.9663	0.05888	16.9837	192.1206	11.3120
30	3.2434	0.3083	0.01783	56.0849	0.05783	17.2920	201.0618	11.6274
31	3.3731	0.2965	0.01686	59.3283	0.05686	17.5885	209.9556	11.9371
32	3.5081	0.2851	0.01595	62.7015	0.05595	17.8736	218.7924	12.2411
33	3.6484	0.2741	0.01510	66.2095	0.05510	18.1476	227.5634	12.5396
34	3.7943	0.2636	0.01431	69.8579	0.05431	18.4112	236.2607	12.8324
35	3.9461	0.2534	0.01358	73.6522	0.05358	18.6646	244.8768	13.1198
40	4.8010	0.2083	0.01052	95.0255	0.05052	19.7928	286.5303	14.4765
45	5.8412	0.1712	0.00826	121.0294	0.04826	20.7200	325.4028	15.7047
50	7.1067	0.1407	0.00655	152.6671	0.04655	21.4822	361.1638	16.8122
55	8.6464	0.1157	0.00523	191.1592	0.04523	22.1086	393.6890	17.8070
60	10.5196	0.0951	0.00420	237.9907	0.04420	22.6235	422.9966	18.6972
65	12.7987	0.0781	0.00339	294.9684	0.04339	23.0467	449.2014	19.4909
70	15.5716	0.0642	0.00275	364.2905	0.04275	23.3945	472.4789	20.1961
75	18.9453	0.0528	0.00223	448.6314	0.04223	23.6804	493.0408	20.8206
80	23.0498	0.0434	0.00181	551.2450	0.04181	23.9154	511.1161	21.3718
85	28.0436	0.0357	0.00148	676.0901	0.04148	24.1085	526.9384	21.8569
90	34.1193	0.0293	0.00121	827.9833	0.04121	24.2673	540.7369	22.2826
96	43.1718	0.0232	0.00095	1054.30	0.04095	24.4209	554.9312	22.7236
108	69.1195	0.0145	0.00059	1702.99	0.04059	24.6383	576.8949	23.4146
120	110.6626	0.0090	0.00036	2741.56	0.04036	24.7741	592.2428	23.9057
144	283.6618	0.0035	0.00014	7066.55	0.04014	24.9119	610.1055	24.4906

5%					TABLE 10	Discrete Cash Flow: Compound Interest Factors		5%
	Single Payments		Uniform Series Payments				Arithmetic Gradients	
	Compound Amount F/P	Present Worth P/F	Sinking Fund A/F	Compound Amount F/A	Capital Recovery A/P	Present Worth P/A	Gradient Present Worth P/G	Gradient Uniform Series A/G
n								
1	1.0500	0.9524	1.00000	1.0000	1.05000	0.9524		
2	1.1025	0.9070	0.48780	2.0500	0.53780	1.8594	0.9070	0.4878
3	1.1576	0.8638	0.31721	3.1525	0.36721	2.7232	2.6347	0.9675
4	1.2155	0.8227	0.23201	4.3101	0.28201	3.5460	5.1028	1.4391
5	1.2763	0.7835	0.18097	5.5256	0.23097	4.3295	8.2369	1.9025
6	1.3401	0.7462	0.14702	6.8019	0.19702	5.0757	11.9680	2.3579
7	1.4071	0.7107	0.12282	8.1420	0.17282	5.7864	16.2321	2.8052
8	1.4775	0.6768	0.10472	9.5491	0.15472	6.4632	20.9700	3.2445
9	1.5513	0.6446	0.09069	11.0266	0.14069	7.1078	26.1268	3.6758
10	1.6289	0.6139	0.07950	12.5779	0.12950	7.7217	31.6520	4.0991
11	1.7103	0.5847	0.07039	14.2068	0.12039	8.3064	37.4988	4.5144
12	1.7959	0.5568	0.06283	15.9171	0.11283	8.8633	43.6241	4.9219
13	1.8856	0.5303	0.05646	17.7130	0.10646	9.3936	49.9879	5.3215
14	1.9799	0.5051	0.05102	19.5986	0.10102	9.8986	56.5538	5.7133
15	2.0789	0.4810	0.04634	21.5786	0.09634	10.3797	63.2880	6.0973
16	2.1829	0.4581	0.04227	23.6575	0.09227	10.8378	70.1597	6.4736
17	2.2920	0.4363	0.03870	25.8404	0.08870	11.2741	77.1405	6.8423
18	2.4066	0.4155	0.03555	28.1324	0.08555	11.6896	84.2043	7.2034
19	2.5270	0.3957	0.03275	30.5390	0.08275	12.0853	91.3275	7.5569
20	2.6533	0.3769	0.03024	33.0660	0.08024	12.4622	98.4884	7.9030
21	2.7860	0.3589	0.02800	35.7193	0.07800	12.8212	105.6673	8.2416
22	2.9253	0.3418	0.02597	38.5052	0.07597	13.1630	112.8461	8.5730
23	3.0715	0.3256	0.02414	41.4305	0.07414	13.4886	120.0087	8.8971
24	3.2251	0.3101	0.02247	44.5020	0.07247	13.7986	127.1402	9.2140
25	3.3864	0.2953	0.02095	47.7271	0.07095	14.0939	134.2275	9.5238
26	3.5557	0.2812	0.01956	51.1135	0.06956	14.3752	141.2585	9.8266
27	3.7335	0.2678	0.01829	54.6691	0.06829	14.6430	148.2226	10.1224
28	3.9201	0.2551	0.01712	58.4026	0.06712	14.8981	155.1101	10.4114
29	4.1161	0.2429	0.01605	62.3227	0.06605	15.1411	161.9126	10.6936
30	4.3219	0.2314	0.01505	66.4388	0.06505	15.3725	168.6226	10.9691
31	4.5380	0.2204	0.01413	70.7608	0.06413	15.5928	175.2333	11.2381
32	4.7649	0.2099	0.01328	75.2988	0.06328	15.8027	181.7392	11.5005
33	5.0032	0.1999	0.01249	80.0638	0.06249	16.0025	188.1351	11.7566
34	5.2533	0.1904	0.01176	85.0670	0.06176	16.1929	194.4168	12.0063
35	5.5160	0.1813	0.01107	90.3203	0.06107	16.3742	200.5807	12.2498
40	7.0400	0.1420	0.00828	120.7998	0.05828	17.1591	229.5452	13.3775
45	8.9850	0.1113	0.00626	159.7002	0.05626	17.7741	255.3145	14.3644
50	11.4674	0.0872	0.00478	209.3480	0.05478	18.2559	277.9148	15.2233
55	14.6356	0.0683	0.00367	272.7126	0.05367	18.6335	297.5104	15.9664
60	18.6792	0.0535	0.00283	353.5837	0.05283	18.9293	314.3432	16.6062
65	23.8399	0.0419	0.00219	456.7980	0.05219	19.1611	328.6910	17.1541
70	30.4264	0.0329	0.00170	588.5285	0.05170	19.3427	340.8409	17.6212
75	38.8327	0.0258	0.00132	756.6537	0.05132	19.4850	351.0721	18.0176
80	49.5614	0.0202	0.00103	971.2288	0.05103	19.5965	359.6460	18.3526
85	63.2544	0.0158	0.00080	1245.09	0.05080	19.6838	366.8007	18.6346
90	80.7304	0.0124	0.00063	1594.61	0.05063	19.7523	372.7488	18.8712
95	103.0347	0.0097	0.00049	2040.69	0.05049	19.8059	377.6774	19.0689
96	108.1864	0.0092	0.00047	2143.73	0.05047	19.8151	378.5555	19.1044
98	119.2755	0.0084	0.00042	2365.51	0.05042	19.8323	380.2139	19.1714
100	131.5013	0.0076	0.00038	2610.03	0.05038	19.8479	381.7492	19.2337

| 6% | | TABLE 11 | Discrete Cash Flow: Compound Interest Factors | | | | 6% |

	Single Payments		Uniform Series Payments				Arithmetic Gradients	
	Compound Amount	Present Worth	Sinking Fund	Compound Amount	Capital Recovery	Present Worth	Gradient Present Worth	Gradient Uniform Series
n	F/P	P/F	A/F	F/A	A/P	P/A	P/G	A/G
1	1.0600	0.9434	1.00000	1.0000	1.06000	0.9434		
2	1.1236	0.8900	0.48544	2.0600	0.54544	1.8334	0.8900	0.4854
3	1.1910	0.8396	0.31411	3.1836	0.37411	2.6730	2.5692	0.9612
4	1.2625	0.7921	0.22859	4.3746	0.28859	3.4651	4.9455	1.4272
5	1.3382	0.7473	0.17740	5.6371	0.23740	4.2124	7.9345	1.8836
6	1.4185	0.7050	0.14336	6.9753	0.20336	4.9173	11.4594	2.3304
7	1.5036	0.6651	0.11914	8.3938	0.17914	5.5824	15.4497	2.7676
8	1.5938	0.6274	0.10104	9.8975	0.16104	6.2098	19.8416	3.1952
9	1.6895	0.5919	0.08702	11.4913	0.14702	6.8017	24.5768	3.6133
10	1.7908	0.5584	0.07587	13.1808	0.13587	7.3601	29.6023	4.0220
11	1.8983	0.5268	0.06679	14.9716	0.12679	7.8869	34.8702	4.4213
12	2.0122	0.4970	0.05928	16.8699	0.11928	8.3838	40.3369	4.8113
13	2.1329	0.4688	0.05296	18.8821	0.11296	8.8527	45.9629	5.1920
14	2.2609	0.4423	0.04758	21.0151	0.10758	9.2950	51.7128	5.5635
15	2.3966	0.4173	0.04296	23.2760	0.10296	9.7122	57.5546	5.9260
16	2.5404	0.3936	0.03895	25.6725	0.09895	10.1059	63.4592	6.2794
17	2.6928	0.3714	0.03544	28.2129	0.09544	10.4773	69.4011	6.6240
18	2.8543	0.3503	0.03236	30.9057	0.09236	10.8276	75.3569	6.9597
19	3.0256	0.3305	0.02962	33.7600	0.08962	11.1581	81.3062	7.2867
20	3.2071	0.3118	0.02718	36.7856	0.08718	11.4699	87.2304	7.6051
21	3.3996	0.2942	0.02500	39.9927	0.08500	11.7641	93.1136	7.9151
22	3.6035	0.2775	0.02305	43.3923	0.08305	12.0416	98.9412	8.2166
23	3.8197	0.2618	0.02128	46.9958	0.08128	12.3034	104.7007	8.5099
24	4.0489	0.2470	0.01968	50.8156	0.07968	12.5504	110.3812	8.7951
25	4.2919	0.2330	0.01823	54.8645	0.07823	12.7834	115.9732	9.0722
26	4.5494	0.2198	0.01690	59.1564	0.07690	13.0032	121.4684	9.3414
27	4.8223	0.2074	0.01570	63.7058	0.07570	13.2105	126.8600	9.6029
28	5.1117	0.1956	0.01459	68.5281	0.07459	13.4062	132.1420	9.8568
29	5.4184	0.1846	0.01358	73.6398	0.07358	13.5907	137.3096	10.1032
30	5.7435	0.1741	0.01265	79.0582	0.07265	13.7648	142.3588	10.3422
31	6.0881	0.1643	0.01179	84.8017	0.07179	13.9291	147.2864	10.5740
32	6.4534	0.1550	0.01100	90.8898	0.07100	14.0840	152.0901	10.7988
33	6.8406	0.1462	0.01027	97.3432	0.07027	14.2302	156.7681	11.0166
34	7.2510	0.1379	0.00960	104.1838	0.06960	14.3681	161.3192	11.2276
35	7.6861	0.1301	0.00897	111.4348	0.06897	14.4982	165.7427	11.4319
40	10.2857	0.0972	0.00646	154.7620	0.06646	15.0463	185.9568	12.3590
45	13.7646	0.0727	0.00470	212.7435	0.06470	15.4558	203.1096	13.1413
50	18.4202	0.0543	0.00344	290.3359	0.06344	15.7619	217.4574	13.7964
55	24.6503	0.0406	0.00254	394.1720	0.06254	15.9905	229.3222	14.3411
60	32.9877	0.0303	0.00188	533.1282	0.06188	16.1614	239.0428	14.7909
65	44.1450	0.0227	0.00139	719.0829	0.06139	16.2891	246.9450	15.1601
70	59.0759	0.0169	0.00103	967.9322	0.06103	16.3845	253.3271	15.4613
75	79.0569	0.0126	0.00077	1300.95	0.06077	16.4558	258.4527	15.7058
80	105.7960	0.0095	0.00057	1746.60	0.06057	16.5091	262.5493	15.9033
85	141.5789	0.0071	0.00043	2342.98	0.06043	16.5489	265.8096	16.0620
90	189.4645	0.0053	0.00032	3141.08	0.06032	16.5787	268.3946	16.1891
95	253.5463	0.0039	0.00024	4209.10	0.06024	16.6009	270.4375	16.2905
96	268.7590	0.0037	0.00022	4462.65	0.06022	16.6047	270.7909	16.3081
98	301.9776	0.0033	0.00020	5016.29	0.06020	16.6115	271.4491	16.3411
100	339.3021	0.0029	0.00018	5638.37	0.06018	16.6175	272.0471	16.3711

7%				TABLE 12	Discrete Cash Flow: Compound Interest Factors			7%	

	Single Payments		Uniform Series Payments				Arithmetic Gradients		
	Compound Amount F/P	Present Worth P/F	Sinking Fund A/F	Compound Amount F/A	Capital Recovery A/P	Present Worth P/A	Gradient Present Worth P/G	Gradient Uniform Series A/G
n								
1	1.0700	0.9346	1.00000	1.0000	1.07000	0.9346		
2	1.1449	0.8734	0.48309	2.0700	0.55309	1.8080	0.8734	0.4831
3	1.2250	0.8163	0.31105	3.2149	0.38105	2.6243	2.5060	0.9549
4	1.3108	0.7629	0.22523	4.4399	0.29523	3.3872	4.7947	1.4155
5	1.4026	0.7130	0.17389	5.7507	0.24389	4.1002	7.6467	1.8650
6	1.5007	0.6663	0.13980	7.1533	0.20980	4.7665	10.9784	2.3032
7	1.6058	0.6227	0.11555	8.6540	0.18555	5.3893	14.7149	2.7304
8	1.7182	0.5820	0.09747	10.2598	0.16747	5.9713	18.7889	3.1465
9	1.8385	0.5439	0.08349	11.9780	0.15349	6.5152	23.1404	3.5517
10	1.9672	0.5083	0.07238	13.8164	0.14238	7.0236	27.7156	3.9461
11	2.1049	0.4751	0.06336	15.7836	0.13336	7.4987	32.4665	4.3296
12	2.2522	0.4440	0.05590	17.8885	0.12590	7.9427	37.3506	4.7025
13	2.4098	0.4150	0.04965	20.1406	0.11965	8.3577	42.3302	5.0648
14	2.5785	0.3878	0.04434	22.5505	0.11434	8.7455	47.3718	5.4167
15	2.7590	0.3624	0.03979	25.1290	0.10979	9.1079	52.4461	5.7583
16	2.9522	0.3387	0.03586	27.8881	0.10586	9.4466	57.5271	6.0897
17	3.1588	0.3166	0.03243	30.8402	0.10243	9.7632	62.5923	6.4110
18	3.3799	0.2959	0.02941	33.9990	0.09941	10.0591	67.6219	6.7225
19	3.6165	0.2765	0.02675	37.3790	0.09675	10.3356	72.5991	7.0242
20	3.8697	0.2584	0.02439	40.9955	0.09439	10.5940	77.5091	7.3163
21	4.1406	0.2415	0.02229	44.8652	0.09229	10.8355	82.3393	7.5990
22	4.4304	0.2257	0.02041	49.0057	0.09041	11.0612	87.0793	7.8725
23	4.7405	0.2109	0.01871	53.4361	0.08871	11.2722	91.7201	8.1369
24	5.0724	0.1971	0.01719	58.1767	0.08719	11.4693	96.2545	8.3923
25	5.4274	0.1842	0.01581	63.2490	0.08581	11.6536	100.6765	8.6391
26	5.8074	0.1722	0.01456	68.6765	0.08456	11.8258	104.9814	8.8773
27	6.2139	0.1609	0.01343	74.4838	0.08343	11.9867	109.1656	9.1072
28	6.6488	0.1504	0.01239	80.6977	0.08239	12.1371	113.2264	9.3289
29	7.1143	0.1406	0.01145	87.3465	0.08145	12.2777	117.1622	9.5427
30	7.6123	0.1314	0.01059	94.4608	0.08059	12.4090	120.9718	9.7487
31	8.1451	0.1228	0.00980	102.0730	0.07980	12.5318	124.6550	9.9471
32	8.7153	0.1147	0.00907	110.2182	0.07907	12.6466	128.2120	10.1381
33	9.3253	0.1072	0.00841	118.9334	0.07841	12.7538	131.6435	10.3219
34	9.9781	0.1002	0.00780	128.2588	0.07780	12.8540	134.9507	10.4987
35	10.6766	0.0937	0.00723	138.2369	0.07723	12.9477	138.1353	10.6687
40	14.9745	0.0668	0.00501	199.6351	0.07501	13.3317	152.2928	11.4233
45	21.0025	0.0476	0.00350	285.7493	0.07350	13.6055	163.7559	12.0360
50	29.4570	0.0339	0.00246	406.5289	0.07246	13.8007	172.9051	12.5287
55	41.3150	0.0242	0.00174	575.9286	0.07174	13.9399	180.1243	12.9215
60	57.9464	0.0173	0.00123	813.5204	0.07123	14.0392	185.7677	13.2321
65	81.2729	0.0123	0.00087	1146.76	0.07087	14.1099	190.1452	13.4760
70	113.9894	0.0088	0.00062	1614.13	0.07062	14.1604	193.5185	13.6662
75	159.8760	0.0063	0.00044	2269.66	0.07044	14.1964	196.1035	13.8136
80	224.2344	0.0045	0.00031	3189.06	0.07031	14.2220	198.0748	13.9273
85	314.5003	0.0032	0.00022	4478.58	0.07022	14.2403	199.5717	14.0146
90	441.1030	0.0023	0.00016	6287.19	0.07016	14.2533	200.7042	14.0812
95	618.6697	0.0016	0.00011	8823.85	0.07011	14.2626	201.5581	14.1319
96	661.9766	0.0015	0.00011	9442.52	0.07011	14.2641	201.7016	14.1405
98	757.8970	0.0013	0.00009	10813	0.07009	14.2669	201.9651	14.1562
100	867.7163	0.0012	0.00008	12382	0.07008	14.2693	202.2001	14.1703

8%			TABLE 13	Discrete Cash Flow: Compound Interest Factors				8%
	Single Payments			Uniform Series Payments			Arithmetic Gradients	
	Compound Amount	Present Worth	Sinking Fund	Compound Amount	Capital Recovery	Present Worth	Gradient Present Worth	Gradient Uniform Series
n	F/P	P/F	A/F	F/A	A/P	P/A	P/G	A/G
1	1.0800	0.9259	1.00000	1.0000	1.08000	0.9259		
2	1.1664	0.8573	0.48077	2.0800	0.56077	1.7833	0.8573	0.4808
3	1.2597	0.7938	0.30803	3.2464	0.38803	2.5771	2.4450	0.9487
4	1.3605	0.7350	0.22192	4.5061	0.30192	3.3121	4.6501	1.4040
5	1.4693	0.6806	0.17046	5.8666	0.25046	3.9927	7.3724	1.8465
6	1.5869	0.6302	0.13632	7.3359	0.21632	4.6229	10.5233	2.2763
7	1.7138	0.5835	0.11207	8.9228	0.19207	5.2064	14.0242	2.6937
8	1.8509	0.5403	0.09401	10.6366	0.17401	5.7466	17.8061	3.0985
9	1.9990	0.5002	0.08008	12.4876	0.16008	6.2469	21.8081	3.4910
10	2.1589	0.4632	0.06903	14.4866	0.14903	6.7101	25.9768	3.8713
11	2.3316	0.4289	0.06008	16.6455	0.14008	7.1390	30.2657	4.2395
12	2.5182	0.3971	0.05270	18.9771	0.13270	7.5361	34.6339	4.5957
13	2.7196	0.3677	0.04652	21.4953	0.12652	7.9038	39.0463	4.9402
14	2.9372	0.3405	0.04130	24.2149	0.12130	8.2442	43.4723	5.2731
15	3.1722	0.3152	0.03683	27.1521	0.11683	8.5595	47.8857	5.5945
16	3.4259	0.2919	0.03298	30.3243	0.11298	8.8514	52.2640	5.9046
17	3.7000	0.2703	0.02963	33.7502	0.10963	9.1216	56.5883	6.2037
18	3.9960	0.2502	0.02670	37.4502	0.10670	9.3719	60.8426	6.4920
19	4.3157	0.2317	0.02413	41.4463	0.10413	9.6036	65.0134	6.7697
20	4.6610	0.2145	0.02185	45.7620	0.10185	9.8181	69.0898	7.0369
21	5.0338	0.1987	0.01983	50.4229	0.09983	10.0168	73.0629	7.2940
22	5.4365	0.1839	0.01803	55.4568	0.09803	10.2007	76.9257	7.5412
23	5.8715	0.1703	0.01642	60.8933	0.09642	10.3711	80.6726	7.7786
24	6.3412	0.1577	0.01498	66.7648	0.09498	10.5288	84.2997	8.0066
25	6.8485	0.1460	0.01368	73.1059	0.09368	10.6748	87.8041	8.2254
26	7.3964	0.1352	0.01251	79.9544	0.09251	10.8100	91.1842	8.4352
27	7.9881	0.1252	0.01145	87.3508	0.09145	10.9352	94.4390	8.6363
28	8.6271	0.1159	0.01049	95.3388	0.09049	11.0511	97.5687	8.8289
29	9.3173	0.1073	0.00962	103.9659	0.08962	11.1584	100.5738	9.0133
30	10.0627	0.0994	0.00883	113.2832	0.08883	11.2578	103.4558	9.1897
31	10.8677	0.0920	0.00811	123.3459	0.08811	11.3498	106.2163	9.3584
32	11.7371	0.0852	0.00745	134.2135	0.08745	11.4350	108.8575	9.5197
33	12.6760	0.0789	0.00685	145.9506	0.08685	11.5139	111.3819	9.6737
34	13.6901	0.0730	0.00630	158.6267	0.08630	11.5869	113.7924	9.8208
35	14.7853	0.0676	0.00580	172.3168	0.08580	11.6546	116.0920	9.9611
40	21.7245	0.0460	0.00386	259.0565	0.08386	11.9246	126.0422	10.5699
45	31.9204	0.0313	0.00259	386.5056	0.08259	12.1084	133.7331	11.0447
50	46.9016	0.0213	0.00174	573.7702	0.08174	12.2335	139.5928	11.4107
55	68.9139	0.0145	0.00118	848.9232	0.08118	12.3186	144.0065	11.6902
60	101.2571	0.0099	0.00080	1253.21	0.08080	12.3766	147.3000	11.9015
65	148.7798	0.0067	0.00054	1847.25	0.08054	12.4160	149.7387	12.0602
70	218.6064	0.0046	0.00037	2720.08	0.08037	12.4428	151.5326	12.1783
75	321.2045	0.0031	0.00025	4002.56	0.08025	12.4611	152.8448	12.2658
80	471.9548	0.0021	0.00017	5886.94	0.08017	12.4735	153.8001	12.3301
85	693.4565	0.0014	0.00012	8655.71	0.08012	12.4820	154.4925	12.3772
90	1018.92	0.0010	0.00008	12724	0.08008	12.4877	154.9925	12.4116
95	1497.12	0.0007	0.00005	18702	0.08005	12.4917	155.3524	12.4365
96	1616.89	0.0006	0.00005	20199	0.08005	12.4923	155.4112	12.4406
98	1885.94	0.0005	0.00004	23562	0.08004	12.4934	155.5176	12.4480
100	2199.76	0.0005	0.00004	27485	0.08004	12.4943	155.6107	12.4545

9%			TABLE 14	Discrete Cash Flow: Compound Interest Factors				9%
	Single Payments		Uniform Series Payments				Arithmetic Gradients	
n	Compound Amount F/P	Present Worth P/F	Sinking Fund A/F	Compound Amount F/A	Capital Recovery A/P	Present Worth P/A	Gradient Present Worth P/G	Gradient Uniform Series A/G
1	1.0900	0.9174	1.00000	1.0000	1.09000	0.9174		
2	1.1881	0.8417	0.47847	2.0900	0.56847	1.7591	0.8417	0.4785
3	1.2950	0.7722	0.30505	3.2781	0.39505	2.5313	2.3860	0.9426
4	1.4116	0.7084	0.21867	4.5731	0.30867	3.2397	4.5113	1.3925
5	1.5386	0.6499	0.16709	5.9847	0.25709	3.8897	7.1110	1.8282
6	1.6771	0.5963	0.13292	7.5233	0.22292	4.4859	10.0924	2.2498
7	1.8280	0.5470	0.10869	9.2004	0.19869	5.0330	13.3746	2.6574
8	1.9926	0.5019	0.09067	11.0285	0.18067	5.5348	16.8877	3.0512
9	2.1719	0.4604	0.07680	13.0210	0.16680	5.9952	20.5711	3.4312
10	2.3674	0.4224	0.06582	15.1929	0.15582	6.4177	24.3728	3.7978
11	2.5804	0.3875	0.05695	17.5603	0.14695	6.8052	28.2481	4.1510
12	2.8127	0.3555	0.04965	20.1407	0.13965	7.1607	32.1590	4.4910
13	3.0658	0.3262	0.04357	22.9534	0.13357	7.4869	36.0731	4.8182
14	3.3417	0.2992	0.03843	26.0192	0.12843	7.7862	39.9633	5.1326
15	3.6425	0.2745	0.03406	29.3609	0.12406	8.0607	43.8069	5.4346
16	3.9703	0.2519	0.03030	33.0034	0.12030	8.3126	47.5849	5.7245
17	4.3276	0.2311	0.02705	36.9737	0.11705	8.5436	51.2821	6.0024
18	4.7171	0.2120	0.02421	41.3013	0.11421	8.7556	54.8860	6.2687
19	5.1417	0.1945	0.02173	46.0185	0.11173	8.9501	58.3868	6.5236
20	5.6044	0.1784	0.01955	51.1601	0.10955	9.1285	61.7770	6.7674
21	6.1088	0.1637	0.01762	56.7645	0.10762	9.2922	65.0509	7.0006
22	6.6586	0.1502	0.01590	62.8733	0.10590	9.4424	68.2048	7.2232
23	7.2579	0.1378	0.01438	69.5319	0.10438	9.5802	71.2359	7.4357
24	7.9111	0.1264	0.01302	76.7898	0.10302	9.7066	74.1433	7.6384
25	8.6231	0.1160	0.01181	84.7009	0.10181	9.8226	76.9265	7.8316
26	9.3992	0.1064	0.01072	93.3240	0.10072	9.9290	79.5863	8.0156
27	10.2451	0.0976	0.00973	102.7231	0.09973	10.0266	82.1241	8.1906
28	11.1671	0.0895	0.00885	112.9682	0.09885	10.1161	84.5419	8.3571
29	12.1722	0.0822	0.00806	124.1354	0.09806	10.1983	86.8422	8.5154
30	13.2677	0.0754	0.00734	136.3075	0.09734	10.2737	89.0280	8.6657
31	14.4618	0.0691	0.00669	149.5752	0.09669	10.3428	91.1024	8.8083
32	15.7633	0.0634	0.00610	164.0370	0.09610	10.4062	93.0690	8.9436
33	17.1820	0.0582	0.00556	179.8003	0.09556	10.4644	94.9314	9.0718
34	18.7284	0.0534	0.00508	196.9823	0.09508	10.5178	96.6935	9.1933
35	20.4140	0.0490	0.00464	215.7108	0.09464	10.5668	98.3590	9.3083
40	31.4094	0.0318	0.00296	337.8824	0.09296	10.7574	105.3762	9.7957
45	48.3273	0.0207	0.00190	525.8587	0.09190	10.8812	110.5561	10.1603
50	74.3575	0.0134	0.00123	815.0836	0.09123	10.9617	114.3251	10.4295
55	114.4083	0.0087	0.00079	1260.09	0.09079	11.0140	117.0362	10.6261
60	176.0313	0.0057	0.00051	1944.79	0.09051	11.0480	118.9683	10.7683
65	270.8460	0.0037	0.00033	2998.29	0.09033	11.0701	120.3344	10.8702
70	416.7301	0.0024	0.00022	4619.22	0.09022	11.0844	121.2942	10.9427
75	641.1909	0.0016	0.00014	7113.23	0.09014	11.0938	121.9646	10.9940
80	986.5517	0.0010	0.00009	10951	0.09009	11.0998	122.4306	11.0299
85	1517.93	0.0007	0.00006	16855	0.09006	11.1038	122.7533	11.0551
90	2335.53	0.0004	0.00004	25939	0.09004	11.1064	122.9758	11.0726
95	3593.50	0.0003	0.00003	39917	0.09003	11.1080	123.1287	11.0847
96	3916.91	0.0003	0.00002	43510	0.09002	11.1083	123.1529	11.0866
98	4653.68	0.0002	0.00002	51696	0.09002	11.1087	123.1963	11.0900
100	5529.04	0.0002	0.00002	61423	0.09002	11.1091	123.2335	11.0930

	Single Payments		Uniform Series Payments				Arithmetic Gradients	
	Compound Amount	Present Worth	Sinking Fund	Compound Amount	Capital Recovery	Present Worth	Gradient Present Worth	Gradient Uniform Series
n	F/P	P/F	A/F	F/A	A/P	P/A	P/G	A/G
1	1.1000	0.9091	1.00000	1.0000	1.10000	0.9091		
2	1.2100	0.8264	0.47619	2.1000	0.57619	1.7355	0.8264	0.4762
3	1.3310	0.7513	0.30211	3.3100	0.40211	2.4869	2.3291	0.9366
4	1.4641	0.6830	0.21547	4.6410	0.31547	3.1699	4.3781	1.3812
5	1.6105	0.6209	0.16380	6.1051	0.26380	3.7908	6.8618	1.8101
6	1.7716	0.5645	0.12961	7.7156	0.22961	4.3553	9.6842	2.2236
7	1.9487	0.5132	0.10541	9.4872	0.20541	4.8684	12.7631	2.6216
8	2.1436	0.4665	0.08744	11.4359	0.18744	5.3349	16.0287	3.0045
9	2.3579	0.4241	0.07364	13.5795	0.17364	5.7590	19.4215	3.3724
10	2.5937	0.3855	0.06275	15.9374	0.16275	6.1446	22.8913	3.7255
11	2.8531	0.3505	0.05396	18.5312	0.15396	6.4951	26.3963	4.0641
12	3.1384	0.3186	0.04676	21.3843	0.14676	6.8137	29.9012	4.3884
13	3.4523	0.2897	0.04078	24.5227	0.14078	7.1034	33.3772	4.6988
14	3.7975	0.2633	0.03575	27.9750	0.13575	7.3667	36.8005	4.9955
15	4.1772	0.2394	0.03147	31.7725	0.13147	7.6061	40.1520	5.2789
16	4.5950	0.2176	0.02782	35.9497	0.12782	7.8237	43.4164	5.5493
17	5.0545	0.1978	0.02466	40.5447	0.12466	8.0216	46.5819	5.8071
18	5.5599	0.1799	0.02193	45.5992	0.12193	8.2014	49.6395	6.0526
19	6.1159	0.1635	0.01955	51.1591	0.11955	8.3649	52.5827	6.2861
20	6.7275	0.1486	0.01746	57.2750	0.11746	8.5136	55.4069	6.5081
21	7.4002	0.1351	0.01562	64.0025	0.11562	8.6487	58.1095	6.7189
22	8.1403	0.1228	0.01401	71.4027	0.11401	8.7715	60.6893	6.9189
23	8.9543	0.1117	0.01257	79.5430	0.11257	8.8832	63.1462	7.1085
24	9.8497	0.1015	0.01130	88.4973	0.11130	8.9847	65.4813	7.2881
25	10.8347	0.0923	0.01017	98.3471	0.11017	9.0770	67.6964	7.4580
26	11.9182	0.0839	0.00916	109.1818	0.10916	9.1609	69.7940	7.6186
27	13.1100	0.0763	0.00826	121.0999	0.10826	9.2372	71.7773	7.7704
28	14.4210	0.0693	0.00745	134.2099	0.10745	9.3066	73.6495	7.9137
29	15.8631	0.0630	0.00673	148.6309	0.10673	9.3696	75.4146	8.0489
30	17.4494	0.0573	0.00608	164.4940	0.10608	9.4269	77.0766	8.1762
31	19.1943	0.0521	0.00550	181.9434	0.10550	9.4790	78.6395	8.2962
32	21.1138	0.0474	0.00497	201.1378	0.10497	9.5264	80.1078	8.4091
33	23.2252	0.0431	0.00450	222.2515	0.10450	9.5694	81.4856	8.5152
34	25.5477	0.0391	0.00407	245.4767	0.10407	9.6086	82.7773	8.6149
35	28.1024	0.0356	0.00369	271.0244	0.10369	9.6442	83.9872	8.7086
40	45.2593	0.0221	0.00226	442.5926	0.10226	9.7791	88.9525	9.0962
45	72.8905	0.0137	0.00139	718.9048	0.10139	9.8628	92.4544	9.3740
50	117.3909	0.0085	0.00086	1163.91	0.10086	9.9148	94.8889	9.5704
55	189.0591	0.0053	0.00053	1880.59	0.10053	9.9471	96.5619	9.7075
60	304.4816	0.0033	0.00033	3034.82	0.10033	9.9672	97.7010	9.8023
65	490.3707	0.0020	0.00020	4893.71	0.10020	9.9796	98.4705	9.8672
70	789.7470	0.0013	0.00013	7887.47	0.10013	9.9873	98.9870	9.9113
75	1271.90	0.0008	0.00008	12709	0.10008	9.9921	99.3317	9.9410
80	2048.40	0.0005	0.00005	20474	0.10005	9.9951	99.5606	9.9609
85	3298.97	0.0003	0.00003	32980	0.10003	9.9970	99.7120	9.9742
90	5313.02	0.0002	0.00002	53120	0.10002	9.9981	99.8118	9.9831
95	8556.68	0.0001	0.00001	85557	0.10001	9.9988	99.8773	9.9889
96	9412.34	0.0001	0.00001	94113	0.10001	9.9989	99.8874	9.9898
98	11389	0.0001	0.00001		0.10001	9.9991	99.9052	9.9914
100	13781	0.0001	0.00001		0.10001	9.9993	99.9202	9.9927

10%   **TABLE 15**   Discrete Cash Flow: Compound Interest Factors   **10%**

11%				TABLE 16	Discrete Cash Flow: Compound Interest Factors			11%
	Single Payments		Uniform Series Payments				Arithmetic Gradients	
	Compound Amount F/P	Present Worth P/F	Sinking Fund A/F	Compound Amount F/A	Capital Recovery A/P	Present Worth P/A	Gradient Present Worth P/G	Gradient Uniform Series A/G
n								
1	1.1100	0.9009	1.00000	1.0000	1.11000	0.9009		
2	1.2321	0.8116	0.47393	2.1100	0.58393	1.7125	0.8116	0.4739
3	1.3676	0.7312	0.29921	3.3421	0.40921	2.4437	2.2740	0.9306
4	1.5181	0.6587	0.21233	4.7097	0.32233	3.1024	4.2502	1.3700
5	1.6851	0.5935	0.16057	6.2278	0.27057	3.6959	6.6240	1.7923
6	1.8704	0.5346	0.12638	7.9129	0.23638	4.2305	9.2972	2.1976
7	2.0762	0.4817	0.10222	9.7833	0.21222	4.7122	12.1872	2.5863
8	2.3045	0.4339	0.08432	11.8594	0.19432	5.1461	15.2246	2.9585
9	2.5580	0.3909	0.07060	14.1640	0.18060	5.5370	18.3520	3.3144
10	2.8394	0.3522	0.05980	16.7220	0.16980	5.8892	21.5217	3.6544
11	3.1518	0.3173	0.05112	19.5614	0.16112	6.2065	24.6945	3.9788
12	3.4985	0.2858	0.04403	22.7132	0.15403	6.4924	27.8388	4.2879
13	3.8833	0.2575	0.03815	26.2116	0.14815	6.7499	30.9290	4.5822
14	4.3104	0.2320	0.03323	30.0949	0.14323	6.9819	33.9449	4.8619
15	4.7846	0.2090	0.02907	34.4054	0.13907	7.1909	36.8709	5.1275
16	5.3109	0.1883	0.02552	39.1899	0.13552	7.3792	39.6953	5.3794
17	5.8951	0.1696	0.02247	44.5008	0.13247	7.5488	42.4095	5.6180
18	6.5436	0.1528	0.01984	50.3959	0.12984	7.7016	45.0074	5.8439
19	7.2633	0.1377	0.01756	56.9395	0.12756	7.8393	47.4856	6.0574
20	8.0623	0.1240	0.01558	64.2028	0.12558	7.9633	49.8423	6.2590
21	8.9492	0.1117	0.01384	72.2651	0.12384	8.0751	52.0771	6.4491
22	9.9336	0.1007	0.01231	81.2143	0.12231	8.1757	54.1912	6.6283
23	11.0263	0.0907	0.01097	91.1479	0.12097	8.2664	56.1864	6.7969
24	12.2392	0.0817	0.00979	102.1742	0.11979	8.3481	58.0656	6.9555
25	13.5855	0.0736	0.00874	114.4133	0.11874	8.4217	59.8322	7.1045
26	15.0799	0.0663	0.00781	127.9988	0.11781	8.4881	61.4900	7.2443
27	16.7386	0.0597	0.00699	143.0786	0.11699	8.5478	63.0433	7.3754
28	18.5799	0.0538	0.00626	159.8173	0.11626	8.6016	64.4965	7.4982
29	20.6237	0.0485	0.00561	178.3972	0.11561	8.6501	65.8542	7.6131
30	22.8923	0.0437	0.00502	199.0209	0.11502	8.6938	67.1210	7.7206
31	25.4104	0.0394	0.00451	221.9132	0.11451	8.7331	68.3016	7.8210
32	28.2056	0.0355	0.00404	247.3236	0.11404	8.7686	69.4007	7.9147
33	31.3082	0.0319	0.00363	275.5292	0.11363	8.8005	70.4228	8.0021
34	34.7521	0.0288	0.00326	306.8374	0.11326	8.8293	71.3724	8.0836
35	38.5749	0.0259	0.00293	341.5896	0.11293	8.8552	72.2538	8.1594
40	65.0009	0.0154	0.00172	581.8261	0.11172	8.9511	75.7789	8.4659
45	109.5302	0.0091	0.00101	986.6386	0.11101	9.0079	78.1551	8.6763
50	184.5648	0.0054	0.00060	1668.77	0.11060	9.0417	79.7341	8.8185
55	311.0025	0.0032	0.00035	2818.20	0.11035	9.0617	80.7712	8.9135
60	524.0572	0.0019	0.00021	4755.07	0.11021	9.0736	81.4461	8.9762
65	883.0669	0.0011	0.00012	8018.79	0.11012	9.0806	81.8819	9.0172
70	1488.02	0.0007	0.00007	13518	0.11007	9.0848	82.1614	9.0438
75	2507.40	0.0004	0.00004	22785	0.11004	9.0873	82.3397	9.0610
80	4225.11	0.0002	0.00003	38401	0.11003	9.0888	82.4529	9.0720
85	7119.56	0.0001	0.00002	64714	0.11002	9.0896	82.5245	9.0790

12%			TABLE 17	Discrete Cash Flow: Compound Interest Factors				12%
	Single Payments		Uniform Series Payments				Arithmetic Gradients	
	Compound Amount F/P	Present Worth P/F	Sinking Fund A/F	Compound Amount F/A	Capital Recovery A/P	Present Worth P/A	Gradient Present Worth P/G	Gradient Uniform Series A/G
n								
1	1.1200	0.8929	1.00000	1.0000	1.12000	0.8929		
2	1.2544	0.7972	0.47170	2.1200	0.59170	1.6901	0.7972	0.4717
3	1.4049	0.7118	0.29635	3.3744	0.41635	2.4018	2.2208	0.9246
4	1.5735	0.6355	0.20923	4.7793	0.32923	3.0373	4.1273	1.3589
5	1.7623	0.5674	0.15741	6.3528	0.27741	3.6048	6.3970	1.7746
6	1.9738	0.5066	0.12323	8.1152	0.24323	4.1114	8.9302	2.1720
7	2.2107	0.4523	0.09912	10.0890	0.21912	4.5638	11.6443	2.5512
8	2.4760	0.4039	0.08130	12.2997	0.20130	4.9676	14.4714	2.9131
9	2.7731	0.3606	0.06768	14.7757	0.18768	5.3282	17.3563	3.2574
10	3.1058	0.3220	0.05698	17.5487	0.17698	5.6502	20.2541	3.5847
11	3.4785	0.2875	0.04842	20.6546	0.16842	5.9377	23.1288	3.8953
12	3.8960	0.2567	0.04144	24.1331	0.16144	6.1944	25.9523	4.1897
13	4.3635	0.2292	0.03568	28.0291	0.15568	6.4235	28.7024	4.4683
14	4.8871	0.2046	0.03087	32.3926	0.15087	6.6282	31.3624	4.7317
15	5.4736	0.1827	0.02682	37.2797	0.14682	6.8109	33.9202	4.9803
16	6.1304	0.1631	0.02339	42.7533	0.14339	6.9740	36.3670	5.2147
17	6.8660	0.1456	0.02046	48.8837	0.14046	7.1196	38.6973	5.4353
18	7.6900	0.1300	0.01794	55.7497	0.13794	7.2497	40.9080	5.6427
19	8.6128	0.1161	0.01576	63.4397	0.13576	7.3658	42.9979	5.8375
20	9.6463	0.1037	0.01388	72.0524	0.13388	7.4694	44.9676	6.0202
21	10.8038	0.0926	0.01224	81.6987	0.13224	7.5620	46.8188	6.1913
22	12.1003	0.0826	0.01081	92.5026	0.13081	7.6446	48.5543	6.3514
23	13.5523	0.0738	0.00956	104.6029	0.12956	7.7184	50.1776	6.5010
24	15.1786	0.0659	0.00846	118.1552	0.12846	7.7843	51.6929	6.6406
25	17.0001	0.0588	0.00750	133.3339	0.12750	7.8431	53.1046	6.7708
26	19.0401	0.0525	0.00665	150.3339	0.12665	7.8957	54.4177	6.8921
27	21.3249	0.0469	0.00590	169.3740	0.12590	7.9426	55.6369	7.0049
28	23.8839	0.0419	0.00524	190.6989	0.12524	7.9844	56.7674	7.1098
29	26.7499	0.0374	0.00466	214.5828	0.12466	8.0218	57.8141	7.2071
30	29.9599	0.0334	0.00414	241.3327	0.12414	8.0552	58.7821	7.2974
31	33.5551	0.0298	0.00369	271.2926	0.12369	8.0850	59.6761	7.3811
32	37.5817	0.0266	0.00328	304.8477	0.12328	8.1116	60.5010	7.4586
33	42.0915	0.0238	0.00292	342.4294	0.12292	8.1354	61.2612	7.5302
34	47.1425	0.0212	0.00260	384.5210	0.12260	8.1566	61.9612	7.5965
35	52.7996	0.0189	0.00232	431.6635	0.12232	8.1755	62.6052	7.6577
40	93.0510	0.0107	0.00130	767.0914	0.12130	8.2438	65.1159	7.8988
45	163.9876	0.0061	0.0074	1358.23	0.12074	8.2825	66.7342	8.0572
50	289.0022	0.0035	0.00042	2400.02	0.12042	8.3045	67.7624	8.1597
55	509.3206	0.0020	0.00024	4236.01	0.12024	8.3170	68.4082	8.2251
60	897.5969	0.0011	0.00013	7471.64	0.12013	8.3240	68.8100	8.2664
65	1581.87	0.0006	0.00008	13174	0.12008	8.3281	69.0581	8.2922
70	2787.80	0.0004	0.00004	23223	0.12004	8.3303	69.2103	8.3082
75	4913.06	0.0002	0.00002	40934	0.12002	8.3316	69.3031	8.3181
80	8658.48	0.0001	0.00001	72146	0.12001	8.3324	69.3594	8.3241
85	15259	0.0001	0.00001		0.12001	8.3328	69.3935	8.3278

14%			TABLE 18	Discrete Cash Flow: Compound Interest Factors				14%
	Single Payments		Uniform Series Payments				Arithmetic Gradients	
	Compound Amount F/P	Present Worth P/F	Sinking Fund A/F	Compound Amount F/A	Capital Recovery A/P	Present Worth P/A	Gradient Present Worth P/G	Gradient Uniform Series A/G
n								
1	1.1400	0.8772	1.00000	1.0000	1.14000	0.8772		
2	1.2996	0.7695	0.46729	2.1400	0.60729	1.6467	0.7695	0.4673
3	1.4815	0.6750	0.29073	3.4396	0.43073	2.3216	2.1194	0.9129
4	1.6890	0.5921	0.20320	4.9211	0.34320	2.9137	3.8957	1.3370
5	1.9254	0.5194	0.15128	6.6101	0.29128	3.4331	5.9731	1.7399
6	2.1950	0.4556	0.11716	8.5355	0.25716	3.8887	8.2511	2.1218
7	2.5023	0.3996	0.09319	10.7305	0.23319	4.2883	10.6489	2.4832
8	2.8526	0.3506	0.07557	13.2328	0.21557	4.6389	13.1028	2.8246
9	3.2519	0.3075	0.06217	16.0853	0.20217	4.9464	15.5629	3.1463
10	3.7072	0.2697	0.05171	19.3373	0.19171	5.2161	17.9906	3.4490
11	4.2262	0.2366	0.04339	23.0445	0.18339	5.4527	20.3567	3.7333
12	4.8179	0.2076	0.03667	27.2707	0.17667	5.6603	22.6399	3.9998
13	5.4924	0.1821	0.03116	32.0887	0.17116	5.8424	24.8247	4.2491
14	6.2613	0.1597	0.02661	37.5811	0.16661	6.0021	26.9009	4.4819
15	7.1379	0.1401	0.02281	43.8424	0.16281	6.1422	28.8623	4.6990
16	8.1372	0.1229	0.01962	50.9804	0.15962	6.2651	30.7057	4.9011
17	9.2765	0.1078	0.01692	59.1176	0.15692	6.3729	32.4305	5.0888
18	10.5752	0.0946	0.01462	68.3941	0.15462	6.4674	34.0380	5.2630
19	12.0557	0.0829	0.01266	78.9692	0.15266	6.5504	35.5311	5.4243
20	13.7435	0.0728	0.01099	91.0249	0.15099	6.6231	36.9135	5.5734
21	15.6676	0.0638	0.00954	104.7684	0.14954	6.6870	38.1901	5.7111
22	17.8610	0.0560	0.00830	120.4360	0.14830	6.7429	39.3658	5.8381
23	20.3616	0.0491	0.00723	138.2970	0.14723	6.7921	40.4463	5.9549
24	23.2122	0.0431	0.00630	158.6586	0.14630	6.8351	41.4371	6.0624
25	26.4619	0.0378	0.00550	181.8708	0.14550	6.8729	42.3441	6.1610
26	30.1666	0.0331	0.00480	208.3327	0.14480	6.9061	43.1728	6.2514
27	34.3899	0.0291	0.00419	238.4993	0.14419	6.9352	43.9289	6.3342
28	39.2045	0.0255	0.00366	272.8892	0.14366	6.9607	44.6176	6.4100
29	44.6931	0.0224	0.00320	312.0937	0.14320	6.9830	45.2441	6.4791
30	50.9502	0.0196	0.00280	356.7868	0.14280	7.0027	45.8132	6.5423
31	58.0832	0.0172	0.00245	407.7370	0.14245	7.0199	46.3297	6.5998
32	66.2148	0.0151	0.00215	465.8202	0.14215	7.0350	46.7979	6.6522
33	75.4849	0.0132	0.00188	532.0350	0.14188	7.0482	47.2218	6.6998
34	86.0528	0.0116	0.00165	607.5199	0.14165	7.0599	47.6053	6.7431
35	98.1002	0.0102	0.00144	693.5727	0.14144	7.0700	47.9519	6.7824
40	188.8835	0.0053	0.00075	1342.03	0.14075	7.1050	49.2376	6.9300
45	363.6791	0.0027	0.00039	2590.56	0.14039	7.1232	49.9963	7.0188
50	700.2330	0.0014	0.00020	4994.52	0.14020	7.1327	50.4375	7.0714
55	1348.24	0.0007	0.00010	9623.13	0.14010	7.1376	50.6912	7.1020
60	2595.92	0.0004	0.00005	18535	0.14005	7.1401	50.8357	7.1197
65	4998.22	0.0002	0.00003	35694	0.14003	7.1414	50.9173	7.1298
70	9623.64	0.0001	0.00001	68733	0.14001	7.1421	50.9632	7.1356
75	18530	0.0001	0.00001		0.14001	7.1425	50.9887	7.1388
80	35677				0.14000	7.1427	51.0030	7.1406
85	68693				0.14000	7.1428	51.0108	7.1416

15%			TABLE 19	Discrete Cash Flow: Compound Interest Factors				15%
	Single Payments		Uniform Series Payments				Arithmetic Gradients	
	Compound Amount F/P	Present Worth P/F	Sinking Fund A/F	Compound Amount F/A	Capital Recovery A/P	Present Worth P/A	Gradient Present Worth P/G	Gradient Uniform Series A/G
n								
1	1.1500	0.8696	1.00000	1.0000	1.15000	0.8696		
2	1.3225	0.7561	0.46512	2.1500	0.61512	1.6257	0.7561	0.4651
3	1.5209	0.6575	0.28798	3.4725	0.43798	2.2832	2.0712	0.9071
4	1.7490	0.5718	0.20027	4.9934	0.35027	2.8550	3.7864	1.3263
5	2.0114	0.4972	0.14832	6.7424	0.29832	3.3522	5.7751	1.7228
6	2.3131	0.4323	0.11424	8.7537	0.26424	3.7845	7.9368	2.0972
7	2.6600	0.3759	0.09036	11.0668	0.24036	4.1604	10.1924	2.4498
8	3.0590	0.3269	0.07285	13.7268	0.22285	4.4873	12.4807	2.7813
9	3.5179	0.2843	0.05957	16.7858	0.20957	4.7716	14.7548	3.0922
10	4.0456	0.2472	0.04925	20.3037	0.19925	5.0188	16.9795	3.3832
11	4.6524	0.2149	0.04107	24.3493	0.19107	5.2337	19.1289	3.6549
12	5.3503	0.1869	0.03448	29.0017	0.18448	5.4206	21.1849	3.9082
13	6.1528	0.1625	0.02911	34.3519	0.17911	5.5831	23.1352	4.1438
14	7.0757	0.1413	0.02469	40.5047	0.17469	5.7245	24.9725	4.3624
15	8.1371	0.1229	0.02102	47.5804	0.17102	5.8474	26.6930	4.5650
16	9.3576	0.1069	0.01795	55.7175	0.16795	5.9542	28.2960	4.7522
17	10.7613	0.0929	0.01537	65.0751	0.16537	6.0472	29.7828	4.9251
18	12.3755	0.0808	0.01319	75.8364	0.16319	6.1280	31.1565	5.0843
19	14.2318	0.0703	0.01134	88.2118	0.16134	6.1982	32.4213	5.2307
20	16.3665	0.0611	0.00976	102.4436	0.15976	6.2593	33.5822	5.3651
21	18.8215	0.0531	0.00842	118.8101	0.15842	6.3125	34.6448	5.4883
22	21.6447	0.0462	0.00727	137.6316	0.15727	6.3587	35.6150	5.6010
23	24.8915	0.0402	0.00628	159.2764	0.15628	6.3988	36.4988	5.7040
24	28.6252	0.0349	0.00543	184.1678	0.15543	6.4338	37.3023	5.7979
25	32.9190	0.0304	0.00470	212.7930	0.15470	6.4641	38.0314	5.8834
26	37.8568	0.0264	0.00407	245.7120	0.15407	6.4906	38.6918	5.9612
27	43.5353	0.0230	0.00353	283.5688	0.15353	6.5135	39.2890	6.0319
28	50.0656	0.0200	0.00306	327.1041	0.15306	6.5335	39.8283	6.0960
29	57.5755	0.0174	0.00265	377.1697	0.15265	6.5509	40.3146	6.1541
30	66.2118	0.0151	0.00230	434.7451	0.15230	6.5660	40.7526	6.2066
31	76.1435	0.0131	0.00200	500.9569	0.15200	6.5791	41.1466	6.2541
32	87.5651	0.0114	0.00173	577.1005	0.15173	6.5905	41.5006	6.2970
33	100.6998	0.0099	0.00150	664.6655	0.15150	6.6005	41.8184	6.3357
34	115.8048	0.0086	0.00131	765.3654	0.15131	6.6091	42.1033	6.3705
35	133.1755	0.0075	0.00113	881.1702	0.15113	6.6166	42.3586	6.4019
40	267.8635	0.0037	0.00056	1779.09	0.15056	6.6418	43.2830	6.5168
45	538.7693	0.0019	0.00028	3585.13	0.15028	6.6543	43.8051	6.5830
50	1083.66	0.0009	0.00014	7217.72	0.15014	6.6605	44.0958	6.6205
55	2179.62	0.0005	0.00007	14524	0.15007	6.6636	44.2558	6.6414
60	4384.00	0.0002	0.00003	29220	0.15003	6.6651	44.3431	6.6530
65	8817.79	0.0001	0.00002	58779	0.15002	6.6659	44.3903	6.6593
70	17736	0.0001	0.00001		0.15001	6.6663	44.4156	6.6627
75	35673				0.15000	6.6665	44.4292	6.6646
80	71751				0.15000	6.6666	44.4364	6.6656
85					0.15000	6.6666	44.4402	6.6661

16%			TABLE 20	Discrete Cash Flow: Compound Interest Factors				16%
	Single Payments		Uniform Series Payments				Arithmetic Gradients	
	Compound Amount F/P	Present Worth P/F	Sinking Fund A/F	Compound Amount F/A	Capital Recovery A/P	Present Worth P/A	Gradient Present Worth P/G	Gradient Uniform Series A/G
n								
1	1.1600	0.8621	1.00000	1.0000	1.16000	0.8621		
2	1.3456	0.7432	0.46296	2.1600	0.62296	1.6052	0.7432	0.4630
3	1.5609	0.6407	0.28526	3.5056	0.44526	2.2459	2.0245	0.9014
4	1.8106	0.5523	0.19738	5.0665	0.35738	2.7982	3.6814	1.3156
5	2.1003	0.4761	0.14541	6.8771	0.30541	3.2743	5.5858	1.7060
6	2.4364	0.4104	0.11139	8.9775	0.27139	3.6847	7.6380	2.0729
7	2.8262	0.3538	0.08761	11.4139	0.24761	4.0386	9.7610	2.4169
8	3.2784	0.3050	0.07022	14.2401	0.23022	4.3436	11.8962	2.7388
9	3.8030	0.2630	0.05708	17.5185	0.21708	4.6065	13.9998	3.0391
10	4.4114	0.2267	0.04690	21.3215	0.20690	4.8332	16.0399	3.3187
11	5.1173	0.1954	0.03886	25.7329	0.19886	5.0286	17.9941	3.5783
12	5.9360	0.1685	0.03241	30.8502	0.19241	5.1971	19.8472	3.8189
13	6.8858	0.1452	0.02718	36.7862	0.18718	5.3423	21.5899	4.0413
14	7.9875	0.1252	0.02290	43.6720	0.18290	5.4675	23.2175	4.2464
15	9.2655	0.1079	0.01936	51.6595	0.17936	5.5755	24.7284	4.4352
16	10.7480	0.0930	0.01641	60.9250	0.17641	5.6685	26.1241	4.6086
17	12.4677	0.0802	0.01395	71.6730	0.17395	5.7487	27.4074	4.7676
18	14.4625	0.0691	0.01188	84.1407	0.17188	5.8178	28.5828	4.9130
19	16.7765	0.0596	0.01014	98.6032	0.17014	5.8775	29.6557	5.0457
20	19.4608	0.0514	0.00867	115.3797	0.16867	5.9288	30.6321	5.1666
22	26.1864	0.0382	0.00635	157.4150	0.16635	6.0113	32.3200	5.3765
24	35.2364	0.0284	0.00467	213.9776	0.16467	6.0726	33.6970	5.5490
26	47.4141	0.0211	0.00345	290.0883	0.16345	6.1182	34.8114	5.6898
28	63.8004	0.0157	0.00255	392.5028	0.16255	6.1520	35.7073	5.8041
30	85.8499	0.0116	0.00189	530.3117	0.16189	6.1772	36.4234	5.8964
32	115.5196	0.0087	0.00140	715.7475	0.16140	6.1959	36.9930	5.9706
34	155.4432	0.0064	0.00104	965.2698	0.16104	6.2098	37.4441	6.0299
35	180.3141	0.0055	0.00089	1120.71	0.16089	6.2153	37.6327	6.0548
36	209.1643	0.0048	0.00077	1301.03	0.16077	6.2201	37.8000	6.0771
38	281.4515	0.0036	0.00057	1752.82	0.16057	6.2278	38.0799	6.1145
40	378.7212	0.0026	0.00042	2360.76	0.16042	6.2335	38.2992	6.1441
45	795.4438	0.0013	0.00020	4965.27	0.16020	6.2421	38.6598	6.1934
50	1670.70	0.0006	0.00010	10436	0.16010	6.2463	38.8521	6.2201
55	3509.05	0.0003	0.00005	21925	0.16005	6.2482	38.9534	6.2343
60	7370.20	0.0001	0.00002	46058	0.16002	6.2492	39.0063	6.2419

18%	TABLE 21		Discrete Cash Flow: Compound Interest Factors					18%
	Single Payments		**Uniform Series Payments**				**Arithmetic Gradients**	
	Compound Amount	Present Worth	Sinking Fund	Compound Amount	Capital Recovery	Present Worth	Gradient Present Worth	Gradient Uniform Series
n	F/P	P/F	A/F	F/A	A/P	P/A	P/G	A/G
1	1.1800	0.8475	1.00000	1.0000	1.18000	0.8475		
2	1.3924	0.7182	0.45872	2.1800	0.63872	1.5656	0.7182	0.4587
3	1.6430	0.6086	0.27992	3.5724	0.45992	2.1743	1.9354	0.8902
4	1.9388	0.5158	0.19174	5.2154	0.37174	2.6901	3.4828	1.2947
5	2.2878	0.4371	0.13978	7.1542	0.31978	3.1272	5.2312	1.6728
6	2.6996	0.3704	0.10591	9.4420	0.28591	3.4976	7.0834	2.0252
7	3.1855	0.3139	0.08236	12.1415	0.26236	3.8115	8.9670	2.3526
8	3.7589	0.2660	0.06524	15.3270	0.24524	4.0776	10.8292	2.6558
9	4.4355	0.2255	0.05239	19.0859	0.23239	4.3030	12.6329	2.9358
10	5.2338	0.1911	0.04251	23.5213	0.22251	4.4941	14.3525	3.1936
11	6.1759	0.1619	0.03478	28.7551	0.21478	4.6560	15.9716	3.4303
12	7.2876	0.1372	0.02863	34.9311	0.20863	4.7932	17.4811	3.6470
13	8.5994	0.1163	0.02369	42.2187	0.20369	4.9095	18.8765	3.8449
14	10.1472	0.0985	0.01968	50.8180	0.19968	5.0081	20.1576	4.0250
15	11.9737	0.0835	0.01640	60.9653	0.19640	5.0916	21.3269	4.1887
16	14.1290	0.0708	0.01371	72.9390	0.19371	5.1624	22.3885	4.3369
17	16.6722	0.0600	0.01149	87.0680	0.19149	5.2223	23.3482	4.4708
18	19.6733	0.0508	0.00964	103.7403	0.18964	5.2732	24.2123	4.5916
19	23.2144	0.0431	0.00810	123.4135	0.18810	5.3162	24.9877	4.7003
20	27.3930	0.0365	0.00682	146.6280	0.18682	5.3527	25.6813	4.7978
22	38.1421	0.0262	0.00485	206.3448	0.18485	5.4099	26.8506	4.9632
24	53.1090	0.0188	0.00345	289.4945	0.18345	5.4509	27.7725	5.0950
26	73.9490	0.0135	0.00247	405.2721	0.18247	5.4804	28.4935	5.1991
28	102.9666	0.0097	0.00177	566.4809	0.18177	5.5016	29.0537	5.2810
30	143.3706	0.0070	0.00126	790.9480	0.18126	5.5168	29.4864	5.3448
32	199.6293	0.0050	0.00091	1103.50	0.18091	5.5277	29.8191	5.3945
34	277.9638	0.0036	0.00065	1538.69	0.18065	5.5356	30.0736	5.4328
35	327.9973	0.0030	0.00055	1816.65	0.18055	5.5386	30.1773	5.4485
36	387.0368	0.0026	0.00047	2144.65	0.18047	5.5412	30.2677	5.4623
38	538.9100	0.0019	0.00033	2988.39	0.18033	5.5452	30.4152	5.4849
40	750.3783	0.0013	0.00024	4163.21	0.18024	5.5482	30.5269	5.5022
45	1716.68	0.0006	0.00010	9531.58	0.18010	5.5523	30.7006	5.5293
50	3927.36	0.0003	0.00005	21813	0.18005	5.5541	30.7856	5.5428
55	8984.84	0.0001	0.00002	49910	0.18002	5.5549	30.8268	5.5494
60	20555			114190	0.18001	5.5553	30.8465	5.5526

20%			TABLE 22	Discrete Cash Flow: Compound Interest Factors				20%
	Single Payments		Uniform Series Payments				Arithmetic Gradients	
n	Compound Amount F/P	Present Worth P/F	Sinking Fund A/F	Compound Amount F/A	Capital Recovery A/P	Present Worth P/A	Gradient Present Worth P/G	Gradient Uniform Series A/G
1	1.2000	0.8333	1.00000	1.0000	1.20000	0.8333		
2	1.4400	0.6944	0.45455	2.2000	0.65455	1.5278	0.6944	0.4545
3	1.7280	0.5787	0.27473	3.6400	0.47473	2.1065	1.8519	0.8791
4	2.0736	0.4823	0.18629	5.3680	0.38629	2.5887	3.2986	1.2742
5	2.4883	0.4019	0.13438	7.4416	0.33438	2.9906	4.9061	1.6405
6	2.9860	0.3349	0.10071	9.9299	0.30071	3.3255	6.5806	1.9788
7	3.5832	0.2791	0.07742	12.9159	0.27742	3.6046	8.2551	2.2902
8	4.2998	0.2326	0.06061	16.4991	0.26061	3.8372	9.8831	2.5756
9	5.1598	0.1938	0.04808	20.7989	0.24808	4.0310	11.4335	2.8364
10	6.1917	0.1615	0.03852	25.9587	0.23852	4.1925	12.8871	3.0739
11	7.4301	0.1346	0.03110	32.1504	0.23110	4.3271	14.2330	3.2893
12	8.9161	0.1122	0.02526	39.5805	0.22526	4.4392	15.4667	3.4841
13	10.6993	0.0935	0.02062	48.4966	0.22062	4.5327	16.5883	3.6597
14	12.8392	0.0779	0.01689	59.1959	0.21689	4.6106	17.6008	3.8175
15	15.4070	0.0649	0.01388	72.0351	0.21388	4.6755	18.5095	3.9588
16	18.4884	0.0541	0.01144	87.4421	0.21144	4.7296	19.3208	4.0851
17	22.1861	0.0451	0.00944	105.9306	0.20944	4.7746	20.0419	4.1976
18	26.6233	0.0376	0.00781	128.1167	0.20781	4.8122	20.6805	4.2975
19	31.9480	0.0313	0.00646	154.7400	0.20646	4.8435	21.2439	4.3861
20	38.3376	0.0261	0.00536	186.6880	0.20536	4.8696	21.7395	4.4643
22	55.2061	0.0181	0.00369	271.0307	0.20369	4.9094	22.5546	4.5941
24	79.4968	0.0126	0.00255	392.4842	0.20255	4.9371	23.1760	4.6943
26	114.4755	0.0087	0.00176	567.3773	0.20176	4.9563	23.6460	4.7709
28	164.8447	0.0061	0.00122	819.2233	0.20122	4.9697	23.9991	4.8291
30	237.3763	0.0042	0.00085	1181.88	0.20085	4.9789	24.2628	4.8731
32	341.8219	0.0029	0.00059	1704.11	0.20059	4.9854	24.4588	4.9061
34	492.2235	0.0020	0.00041	2456.12	0.20041	4.9898	24.6038	4.9308
35	590.6682	0.0017	0.00034	2948.34	0.20034	4.9915	24.6614	4.9406
36	708.8019	0.0014	0.00028	3539.01	0.20028	4.9929	24.7108	4.9491
38	1020.67	0.0010	0.00020	5098.37	0.20020	4.9951	24.7894	4.9627
40	1469.77	0.0007	0.00014	7343.86	0.20014	4.9966	24.8469	4.9728
45	3657.26	0.0003	0.00005	18281	0.20005	4.9986	24.9316	4.9877
50	9100.44	0.0001	0.00002	45497	0.20002	4.9995	24.9698	4.9945
55	22645		0.00001		0.20001	4.9998	24.9868	4.9976

22%	TABLE 23		Discrete Cash Flow: Compound Interest Factors					22%
	Single Payments		Uniform Series Payments				Arithmetic Gradients	
	Compound Amount F/P	Present Worth P/F	Sinking Fund A/F	Compound Amount F/A	Capital Recovery A/P	Present Worth P/A	Gradient Present Worth P/G	Gradient Uniform Series A/G
n								
1	1.2200	0.8197	1.00000	1.0000	1.22000	0.8197		
2	1.4884	0.6719	0.45045	2.2200	0.67045	1.4915	0.6719	0.4505
3	1.8158	0.5507	0.26966	3.7084	0.48966	2.0422	1.7733	0.8683
4	2.2153	0.4514	0.18102	5.5242	0.40102	2.4936	3.1275	1.2542
5	2.7027	0.3700	0.12921	7.7396	0.34921	2.8636	4.6075	1.6090
6	3.2973	0.3033	0.09576	10.4423	0.31576	3.1669	6.1239	1.9337
7	4.0227	0.2486	0.07278	13.7396	0.29278	3.4155	7.6154	2.2297
8	4.9077	0.2038	0.05630	17.7623	0.27630	3.6193	9.0417	2.4982
9	5.9874	0.1670	0.04411	22.6700	0.26411	3.7863	10.3779	2.7409
10	7.3046	0.1369	0.03489	28.6574	0.25489	3.9232	11.6100	2.9593
11	8.9117	0.1122	0.02781	35.9620	0.24781	4.0354	12.7321	3.1551
12	10.8722	0.0920	0.02228	44.8737	0.24228	4.1274	13.7438	3.3299
13	13.2641	0.0754	0.01794	55.7459	0.23794	4.2028	14.6485	3.4855
14	16.1822	0.0618	0.01449	69.0100	0.23449	4.2646	15.4519	3.6233
15	19.7423	0.0507	0.01174	85.1922	0.23174	4.3152	16.1610	3.7451
16	24.0856	0.0415	0.00953	104.9345	0.22953	4.3567	16.7838	3.8524
17	29.3844	0.0340	0.00775	129.0201	0.22775	4.3908	17.3283	3.9465
18	35.8490	0.0279	0.00631	158.4045	0.22631	4.4187	17.8025	4.0289
19	43.7358	0.0229	0.00515	194.2535	0.22515	4.4415	18.2141	4.1009
20	53.3576	0.0187	0.00420	237.9893	0.22420	4.4603	18.5702	4.1635
22	79.4175	0.0126	0.00281	356.4432	0.22281	4.4882	19.1418	4.2649
24	118.2050	0.0085	0.00188	532.7501	0.22188	4.5070	19.5635	4.3407
26	175.9364	0.0057	0.00126	795.1653	0.22126	4.5196	19.8720	4.3968
28	261.8637	0.0038	0.00084	1185.74	0.22084	4.5281	20.0962	4.4381
30	389.7579	0.0026	0.00057	1767.08	0.22057	4.5338	20.2583	4.4683
32	580.1156	0.0017	0.00038	2632.34	0.22038	4.5376	20.3748	4.4902
34	863.4441	0.0012	0.00026	3920.20	0.22026	4.5402	20.4582	4.5060
35	1053.40	0.0009	0.00021	4783.64	0.22021	4.5411	20.4905	4.5122
36	1285.15	0.0008	0.00017	5837.05	0.22017	4.5419	20.5178	4.5174
38	1912.82	0.0005	0.00012	8690.08	0.22012	4.5431	20.5601	4.5256
40	2847.04	0.0004	0.00008	12937	0.22008	4.5439	20.5900	4.5314
45	7694.71	0.0001	0.00003	34971	0.22003	4.5449	20.6319	4.5396
50	20797		0.00001	94525	0.22001	4.5452	20.6492	4.5431
55	56207				0.22000	4.5454	20.6563	4.5445

24%				TABLE 24	Discrete Cash Flow: Compound Interest Factors			24%
	Single Payments		Uniform Series Payments				Arithmetic Gradients	
	Compound Amount F/P	Present Worth P/F	Sinking Fund A/F	Compound Amount F/A	Capital Recovery A/P	Present Worth P/A	Gradient Present Worth P/G	Gradient Uniform Series A/G
n								
1	1.2400	0.8065	1.00000	1.0000	1.24000	0.8065		
2	1.5376	0.6504	0.44643	2.2400	0.68643	1.4568	0.6504	0.4464
3	1.9066	0.5245	0.26472	3.7776	0.50472	1.9813	1.6993	0.8577
4	2.3642	0.4230	0.17593	5.6842	0.41593	2.4043	2.9683	1.2346
5	2.9316	0.3411	0.12425	8.0484	0.36425	2.7454	4.3327	1.5782
6	3.6352	0.2751	0.09107	10.9801	0.33107	3.0205	5.7081	1.8898
7	4.5077	0.2218	0.06842	14.6153	0.30842	3.2423	7.0392	2.1710
8	5.5895	0.1789	0.05229	19.1229	0.29229	3.4212	8.2915	2.4236
9	6.9310	0.1443	0.04047	24.7125	0.28047	3.5655	9.4458	2.6492
10	8.5944	0.1164	0.03160	31.6434	0.27160	3.6819	10.4930	2.8499
11	10.6571	0.0938	0.02485	40.2379	0.26485	3.7757	11.4313	3.0276
12	13.2148	0.0757	0.01965	50.8950	0.25965	3.8514	12.2637	3.1843
13	16.3863	0.0610	0.01560	64.1097	0.25560	3.9124	12.9960	3.3218
14	20.3191	0.0492	0.01242	80.4961	0.25242	3.9616	13.6358	3.4420
15	25.1956	0.0397	0.00992	100.8151	0.24992	4.0013	14.1915	3.5467
16	31.2426	0.0320	0.00794	126.0108	0.24794	4.0333	14.6716	3.6376
17	38.7408	0.0258	0.00636	157.2534	0.24636	4.0591	15.0846	3.7162
18	48.0386	0.0208	0.00510	195.9942	0.24510	4.0799	15.4385	3.7840
19	59.5679	0.0168	0.00410	244.0328	0.24410	4.0967	15.7406	3.8423
20	73.8641	0.0135	0.00329	303.6006	0.24329	4.1103	15.9979	3.8922
22	113.5735	0.0088	0.00213	469.0563	0.24213	4.1300	16.4011	3.9712
24	174.6306	0.0057	0.00138	723.4610	0.24138	4.1428	16.6891	4.0284
26	268.5121	0.0037	0.00090	1114.63	0.24090	4.1511	16.8930	4.0695
28	412.8642	0.0024	0.00058	1716.10	0.24058	4.1566	17.0365	4.0987
30	634.8199	0.0016	0.00038	2640.92	0.24038	4.1601	17.1369	4.1193
32	976.0991	0.0010	0.00025	4062.91	0.24025	4.1624	17.2067	4.1338
34	1500.85	0.0007	0.00016	6249.38	0.24016	4.1639	17.2552	4.1440
35	1861.05	0.0005	0.00013	7750.23	0.24013	4.1664	17.2734	4.1479
36	2307.71	0.0004	0.00010	9611.28	0.24010	4.1649	17.2886	4.1511
38	3548.33	0.0003	0.00007	14781	0.24007	4.1655	17.3116	4.1560
40	5455.91	0.0002	0.00004	22729	0.24004	4.1659	17.3274	4.1593
45	15995	0.0001	0.00002	66640	0.24002	4.1664	17.3483	4.1639
50	46890		0.00001		0.24001	4.1666	17.3563	4.1653
55					0.24000	4.1666	17.3593	4.1663

25%			TABLE 25	Discrete Cash Flow: Compound Interest Factors				25%
	Single Payments		**Uniform Series Payments**				**Arithmetic Gradients**	
	Compound Amount F/P	Present Worth P/F	Sinking Fund A/F	Compound Amount F/A	Capital Recovery A/P	Present Worth P/A	Gradient Present Worth P/G	Gradient Uniform Series A/G
n								
1	1.2500	0.8000	1.00000	1.0000	1.25000	0.8000		
2	1.5625	0.6400	0.44444	2.2500	0.69444	1.4400	0.6400	0.4444
3	1.9531	0.5120	0.26230	3.8125	0.51230	1.9520	1.6640	0.8525
4	2.4414	0.4096	0.17344	5.7656	0.42344	2.3616	2.8928	1.2249
5	3.0518	0.3277	0.12185	8.2070	0.37185	2.6893	4.2035	1.5631
6	3.8147	0.2621	0.08882	11.2588	0.33882	2.9514	5.5142	1.8683
7	4.7684	0.2097	0.06634	15.0735	0.31634	3.1611	6.7725	2.1424
8	5.9605	0.1678	0.05040	19.8419	0.30040	3.3289	7.9469	2.3872
9	7.4506	0.1342	0.03876	25.8023	0.28876	3.4631	9.0207	2.6048
10	9.3132	0.1074	0.03007	33.2529	0.28007	3.5705	9.9870	2.7971
11	11.6415	0.0859	0.02349	42.5661	0.27349	3.6564	10.8460	2.9663
12	14.5519	0.0687	0.01845	54.2077	0.26845	3.7251	11.6020	3.1145
13	18.1899	0.0550	0.01454	68.7596	0.26454	3.7801	12.2617	3.2437
14	22.7374	0.0440	0.01150	86.9495	0.26150	3.8241	12.8334	3.3559
15	28.4217	0.0352	0.00912	109.6868	0.25912	3.8593	13.3260	3.4530
16	35.5271	0.0281	0.00724	138.1085	0.25724	3.8874	13.7482	3.5366
17	44.4089	0.0225	0.00576	173.6357	0.25576	3.9099	14.1085	3.6084
18	55.5112	0.0180	0.00459	218.0446	0.25459	3.9279	14.4147	3.6698
19	69.3889	0.0144	0.00366	273.5558	0.25366	3.9424	14.6741	3.7222
20	86.7362	0.0115	0.00292	342.9447	0.25292	3.9539	14.8932	3.7667
22	135.5253	0.0074	0.00186	538.1011	0.25186	3.9705	15.2326	3.8365
24	211.7582	0.0047	0.00119	843.0329	0.25119	3.9811	15.4711	3.8861
26	330.8722	0.0030	0.00076	1319.49	0.25076	3.9879	15.6373	3.9212
28	516.9879	0.0019	0.00048	2063.95	0.25048	3.9923	15.7524	3.9457
30	807.7936	0.0012	0.00031	3227.17	0.25031	3.9950	15.8316	3.9628
32	1262.18	0.0008	0.00020	5044.71	0.25020	3.9968	15.8859	3.9746
34	1972.15	0.0005	0.00013	7884.61	0.25013	3.9980	15.9229	3.9828
35	2465.19	0.0004	0.00010	9856.76	.025010	3.9984	15.9367	3.9858
36	3081.49	0.0003	0.00008	12322	0.25008	3.9987	15.9481	3.9883
38	4814.82	0.0002	0.00005	19255	0.25005	3.9992	15.9651	3.9921
40	7523.16	0.0001	0.00003	30089	0.25003	3.9995	15.9766	3.9947
45	22959		0.00001	91831	0.25001	3.9998	15.9915	3.9980
50	70065				0.25000	3.9999	15.9969	3.9993
55					0.25000	4.0000	15.9989	3.9997

30%				TABLE 26	Discrete Cash Flow: Compound Interest Factors			30%	
	Single Payments			Uniform Series Payments				Arithmetic Gradients	
n	Compound Amount F/P	Present Worth P/F	Sinking Fund A/F	Compound Amount F/A	Capital Recovery A/P	Present Worth P/A	Gradient Present Worth P/G	Gradient Uniform Series A/G
1	1.3000	0.7692	1.00000	1.0000	1.30000	0.7692		
2	1.6900	0.5917	0.43478	2.3000	0.73478	1.3609	0.5917	0.4348
3	2.1970	0.4552	0.25063	3.9900	0.55063	1.8161	1.5020	0.8271
4	2.8561	0.3501	0.16163	6.1870	0.46163	2.1662	2.5524	1.1783
5	3.7129	0.2693	0.11058	9.0431	0.41058	2.4356	3.6297	1.4903
6	4.8268	0.2072	0.07839	12.7560	0.37839	2.6427	4.6656	1.7654
7	6.2749	0.1594	0.05687	17.5828	0.35687	2.8021	5.6218	2.0063
8	8.1573	0.1226	0.04192	23.8577	0.34192	2.9247	6.4800	2.2156
9	10.6045	0.0943	0.03124	32.0150	0.33124	3.0190	7.2343	2.3963
10	13.7858	0.0725	0.02346	42.6195	0.32346	3.0915	7.8872	2.5512
11	17.9216	0.0558	0.01773	56.4053	0.31773	3.1473	8.4452	2.6833
12	23.2981	0.0429	0.01345	74.3270	0.31345	3.1903	8.9173	2.7952
13	30.2875	0.0330	0.01024	97.6250	0.31024	3.2233	9.3135	2.8895
14	39.3738	0.0254	0.00782	127.9125	0.30782	3.2487	9.6437	2.9685
15	51.1859	0.0195	0.00598	167.2863	0.30598	3.2682	9.9172	3.0344
16	66.5417	0.0150	0.00458	218.4722	0.30458	3.2832	10.1426	3.0892
17	86.5042	0.0116	0.00351	285.0139	0.30351	3.2948	10.3276	3.1345
18	112.4554	0.0089	0.00269	371.5180	0.30269	3.3037	10.4788	3.1718
19	146.1920	0.0068	0.00207	483.9734	0.30207	3.3105	10.6019	3.2025
20	190.0496	0.0053	0.00159	630.1655	0.30159	3.3158	10.7019	3.2275
22	321.1839	0.0031	0.00094	1067.28	0.30094	3.3230	10.8482	3.2646
24	542.8008	0.0018	0.00055	1806.00	0.30055	3.3272	10.9433	3.2890
25	705.6410	0.0014	0.00043	2348.80	0.30043	3.3286	10.9773	3.2979
26	917.3333	0.0011	0.00033	3054.44	0.30033	3.3297	11.0045	3.3050
28	1550.29	0.0006	0.00019	5164.31	0.30019	3.3312	11.0437	3.3153
30	2620.00	0.0004	0.00011	8729.99	0.30011	3.3321	11.0687	3.3219
32	4427.79	0.0002	0.00007	14756	0.30007	3.3326	11.0845	3.3261
34	7482.97	0.0001	0.00004	24940	0.30004	3.3329	11.0945	3.3288
35	9727.86	0.0001	0.00003	32423	0.30003	3.3330	11.0980	3.3297

35%			TABLE 27		Discrete Cash Flow: Compound Interest Factors			35%
	Single Payments		Uniform Series Payments				Arithmetic Gradients	
	Compound Amount F/P	Present Worth P/F	Sinking Fund A/F	Compound Amount F/A	Capital Recovery A/P	Present Worth P/A	Gradient Present Worth P/G	Gradient Uniform Series A/G
n								
1	1.3500	0.7407	1.00000	1.0000	1.35000	0.7407		
2	1.8225	0.5487	0.42553	2.3500	0.77553	1.2894	0.5487	0.4255
3	2.4604	0.4064	0.23966	4.1725	0.58966	1.6959	1.3616	0.8029
4	3.3215	0.3011	0.15076	6.6329	0.50076	1.9969	2.2648	1.1341
5	4.4840	0.2230	0.10046	9.9544	0.45046	2.2200	3.1568	1.4220
6	6.0534	0.1652	0.06926	14.4384	0.41926	2.3852	3.9828	1.6698
7	8.1722	0.1224	0.04880	20.4919	0.39880	2.5075	4.7170	1.8811
8	11.0324	0.0906	0.03489	28.6640	0.38489	2.5982	5.3515	2.0597
9	14.8937	0.0671	0.02519	39.6964	0.37519	2.6653	5.8886	2.2094
10	20.1066	0.0497	0.01832	54.5902	0.36832	2.7150	6.3363	2.3338
11	27.1439	0.0368	0.01339	74.6967	0.36339	2.7519	6.7047	2.4364
12	36.6442	0.0273	0.00982	101.8406	0.35982	2.7792	7.0049	2.5205
13	49.4697	0.0202	0.00722	138.4848	0.35722	2.7994	7.2474	2.5889
14	66.7841	0.0150	0.00532	187.9544	0.35532	2.8144	7.4421	2.6443
15	90.1585	0.0111	0.00393	254.7385	0.35393	2.8255	7.5974	2.6889
16	121.7139	0.0082	0.00290	344.8970	0.35290	2.8337	7.7206	2.7246
17	164.3138	0.0061	0.00214	466.6109	0.35214	2.8398	7.8180	2.7530
18	221.8236	0.0045	0.00158	630.9247	0.35158	2.8443	7.8946	2.7756
19	299.4619	0.0033	0.00117	852.7483	0.35117	2.8476	7.9547	2.7935
20	404.2736	0.0025	0.00087	1152.21	0.35087	2.8501	8.0017	2.8075
22	736.7886	0.0014	0.00048	2102.25	0.35048	2.8533	8.0669	2.8272
24	1342.80	0.0007	0.00026	3833.71	0.35026	2.8550	8.1061	2.8393
25	1812.78	0.0006	0.00019	5176.50	0.35019	2.8556	8.1194	2.8433
26	2447.25	0.0004	0.00014	6989.28	0.35014	2.8560	8.1296	2.8465
28	4460.11	0.0002	0.00008	12740	0.35008	2.8565	8.1435	2.8509
30	8128.55	0.0001	0.00004	23222	0.35004	2.8568	8.1517	2.8535
32	14814	0.0001	0.00002	42324	0.35002	2.8569	8.1565	2.8550
34	26999		0.00001	77137	0.35001	2.8570	8.1594	2.8559
35	36449		0.00001		0.35001	2.8571	8.1603	2.8562

40%			TABLE 28	Discrete Cash Flow: Compound Interest Factors				40%
	Single Payments		Uniform Series Payments				Arithmetic Gradients	
	Compound Amount F/P	Present Worth P/F	Sinking Fund A/F	Compound Amount F/A	Capital Recovery A/P	Present Worth P/A	Gradient Present Worth P/G	Gradient Uniform Series A/G
n								
1	1.4000	0.7143	1.00000	1.0000	1.40000	0.7143		
2	1.9600	0.5102	0.41667	2.4000	0.81667	1.2245	0.5102	0.4167
3	2.7440	0.3644	0.22936	4.3600	0.62936	1.5889	1.2391	0.7798
4	3.8416	0.2603	0.14077	7.1040	0.54077	1.8492	2.0200	1.0923
5	5.3782	0.1859	0.09136	10.9456	0.49136	2.0352	2.7637	1.3580
6	7.5295	0.1328	0.06126	16.3238	0.46126	2.1680	3.4278	1.5811
7	10.5414	0.0949	0.04192	23.8534	0.44192	2.2628	3.9970	1.7664
8	14.7579	0.0678	0.02907	34.3947	0.42907	2.3306	4.4713	1.9185
9	20.6610	0.0484	0.02034	49.1526	0.42034	2.3790	4.8585	2.0422
10	28.9255	0.0346	0.01432	69.8137	0.41432	2.4136	5.1696	2.1419
11	40.4957	0.0247	0.01013	98.7391	0.41013	2.4383	5.4166	2.2215
12	56.6939	0.0176	0.00718	139.2348	0.40718	2.4559	5.6106	2.2845
13	79.3715	0.0126	0.00510	195.9287	0.40510	2.4685	5.7618	2.3341
14	111.1201	0.0090	0.00363	275.3002	0.40363	2.4775	5.8788	2.3729
15	155.5681	0.0064	0.00259	386.4202	0.40259	2.4839	5.9688	2.4030
16	217.7953	0.0046	0.00185	541.9883	0.40185	2.4885	6.0376	2.4262
17	304.9135	0.0033	0.00132	759.7837	0.40132	2.4918	6.0901	2.4441
18	426.8789	0.0023	0.00094	1064.70	0.40094	2.4941	6.1299	2.4577
19	597.6304	0.0017	0.00067	1491.58	0.40067	2.4958	6.1601	2.4682
20	836.6826	0.0012	0.00048	2089.21	0.40048	2.4970	6.1828	2.4761
22	1639.90	0.0006	0.00024	4097.24	0.40024	2.4985	6.2127	2.4866
24	3214.20	0.0003	0.00012	8033.00	0.40012	2.4992	6.2294	2.4925
25	4499.88	0.0002	0.00009	11247	0.40009	2.4994	6.2347	2.4944
26	6299.83	0.0002	0.00006	15747	0.40006	2.4996	6.2387	2.4959
28	12348	0.0001	0.00003	30867	0.40003	2.4998	6.2438	2.4977
30	24201		0.00002	60501	0.40002	2.4999	6.2466	2.4988
32	47435		0.00001		0.40001	2.4999	6.2482	2.4993
34	92972				0.40000	2.5000	6.2490	2.4996
35					0.40000	2.5000	6.2493	2.4997

50%			TABLE 29	Discrete Cash Flow: Compound Interest Factors				50%
	Single Payments		Uniform Series Payments				Arithmetic Gradients	
	Compound Amount	Present Worth	Sinking Fund	Compound Amount	Capital Recovery	Present Worth	Gradient Present Worth	Gradient Uniform Series
n	F/P	P/F	A/F	F/A	A/P	P/A	P/G	A/G
1	1.5000	0.6667	1.00000	1.0000	1.50000	0.6667		
2	2.2500	0.4444	0.40000	2.5000	0.90000	1.1111	0.4444	0.4000
3	3.3750	0.2963	0.21053	4.7500	0.71053	1.4074	1.0370	0.7368
4	5.0625	0.1975	0.12308	8.1250	0.62308	1.6049	1.6296	1.0154
5	7.5938	0.1317	0.07583	13.1875	0.57583	1.7366	2.1564	1.2417
6	11.3906	0.0878	0.04812	20.7813	0.54812	1.8244	2.5953	1.4226
7	17.0859	0.0585	0.03108	32.1719	0.53108	1.8829	2.9465	1.5648
8	25.6289	0.0390	0.02030	49.2578	0.52030	1.9220	3.2196	1.6752
9	38.4434	0.0260	0.01335	74.8867	0.51335	1.9480	3.4277	1.7596
10	57.6650	0.0173	0.00882	113.3301	0.50882	1.9653	3.5838	1.8235
11	86.4976	0.0116	0.00585	170.9951	0.50585	1.9769	3.6994	1.8713
12	129.7463	0.0077	0.00388	257.4927	0.50388	1.9846	3.7842	1.9068
13	194.6195	0.0051	0.00258	387.2390	0.50258	1.9897	3.8459	1.9329
14	291.9293	0.0034	0.00172	581.8585	0.50172	1.9931	3.8904	1.9519
15	437.8939	0.0023	0.00114	873.7878	0.50114	1.9954	3.9224	1.9657
16	656.8408	0.0015	0.00076	1311.68	0.50076	1.9970	3.9452	1.9756
17	985.2613	0.0010	0.00051	1968.52	0.50051	1.9980	3.9614	1.9827
18	1477.89	0.0007	0.00034	2953.78	0.50034	1.9986	3.9729	1.9878
19	2216.84	0.0005	0.00023	4431.68	0.50023	1.9991	3.9811	1.9914
20	3325.26	0.0003	0.00015	6648.51	0.50015	1.9994	3.9868	1.9940
22	7481.83	0.0001	0.00007	14962	0.50007	1.9997	3.9936	1.9971
24	16834	0.0001	0.00003	33666	0.50003	1.9999	3.9969	1.9986
25	25251		0.00002	50500	0.50002	1.9999	3.9979	1.9990
26	37877		0.00001	75752	0.50001	1.9999	3.9985	1.9993
28	85223		0.00001		0.50001	2.0000	3.9993	1.9997
30					0.50000	2.0000	3.9997	1.9998
32					0.50000	2.0000	3.9998	1.9999
34					0.50000	2.0000	3.9999	2.0000
35					0.50000	2.0000	3.9999	2.0000

PHOTO CREDITS

Chapter 1
Opener: Royalty-Free/CORBIS.

Chapter 2
Opener: Royalty-Free/CORBIS.

Chapter 3
Opener: Getty Images.

Chapter 4
Opener: Chad Baker/Getty Images.

Chapter 5
Opener: © Digital Vision/PunchStock.

Chapter 6
Opener: Stockdisc.

Chapter 7
Opener: Getty Images.

Chapter 8
Opener: PhotoLink/Getty Images.

Chapter 9
Opener: Royalty-Free/CORBIS.

Chapter 10
Opener: © Digital Vision/PunchStock.

Chapter 11
Opener: (a) Royalty-Free/CORBIS; (b) Royalty-Free/CORBIS.

Chapter 12
Opener: Ariel Skelley/Blend Images/Getty Images.

Chapter 13
Opener: Randy Allbritton/Getty Images.

Chapter 14
Opener: Photodisc/Getty Images.

Chapter 15
Opener: Ryan McVay/Getty Images.

Chapter 16
Opener: Arthur S. Aubry/Getty Images.

Chapter 17
Opener: Photodisc/Getty Images.

Chapter 18
Opener: Rubberball/SuperStock.

Chapter 19
Opener: © Brand X/JupiterImages.

INDEX